Mosquitoes of India

This is an up-to-date and comprehensive handbook that presents a wealth of information on the different aspects of one of the largest dipterous family, Culicidae (Mosquitoes). India shares more than ten percent of the global mosquito fauna and this book provides an extensive inventory of extant taxa, along with a detailed description of the key identification features of medically important mosquitoes. This book illustrates updated information on insecticide-based mechanisms of resistance development in vectors and showcases varied host–parasite interactions. This authoritative account is a crucial reference source for mosquito-borne disease control and prevention. This book is meant for researchers, university students, medical entomologists, parasitologists, and public health professionals.

Mosquitoes of India

Biology, Phylotaxonomy, Host-Parasite Interaction and Resistance Development

Edited by

B.K. Tyagi

CRC Press
Taylor & Francis Group
Boca Raton London New York

CRC Press is an imprint of the
Taylor & Francis Group, an **informa** business

Designed cover image: Shutterstock

by CRC Press
2385 NW Executive Center Drive, Suite 320, Boca Raton FL 33431

and by CRC Press
4 Park Square, Milton Park, Abingdon, Oxon, OX14 4RN

CRC Press is an imprint of Taylor & Francis Group, LLC

ISBN: 978-1-032-35224-4 (hbk)
ISBN: 978-1-032-35458-3 (pbk)
ISBN: 978-1-003-32699-1 (ebk)

DOI: 10.1201/9781003326991

Typeset in Times
by SPi Technologies India Pvt Ltd (Straive)

Dedicated to

All mosquito researchers who spent their lives in understanding the strengths and weaknesses of the man's deadliest foe on our planet, Earth!

and

All those mosquito combaters who lived to see the mosquito relinquish her past and don a new role in bettering human lives!

Contents

SECTION IV BIOLOGY

SECTION V ECOLOGY AND BEHAVIOR

SECTION VI MOSQUITO GENOMES, GENOMICS, AND PROTEOMICS

SECTION VII VECTOR DYNAMICS UNDER CLIMATE CHANGE

SECTION VIII RESISTANCE DEVELOPMENT IN VECTORS

Foreword

Anastasia Mosquito Control District of St. Johns County
120 EOC Drive, St. Augustine
Web: www.amcdsjc.org

BOARD OF COMMISSIONERS
Jeanne Moeller, Chairperson
Trish Becker, Vice Chairperson
Gayle Gardner, Secretary/Treasurer
Catherina Brandhorst, Commissioner
Gina LeBlanc, Commissioner

DISTRICT DIRECTOR
Dr. Rui-De Xue

The history of mosquito research is very rich in India – the country where Dr. Ronald Ross, a physician in the British Indian Medical Service, demonstrated for the first time an inextricable link between malaria and mosquitoes, thereby laying the foundation for successful research on this disease and methods of combating it. It is noteworthy here that for his epochal research on malaria in the Indian city of Secunderabad, Dr Ross was awarded the Nobel Prize in 1902. Ever since, India has been making great progress in malaria control and is presently heading for malaria elimination by 2030, which I believe will be achieved based on Sri Lanka and China's experiences. Integrated vector management (IVM) of mosquitoes has played an integral role in reducing the incidence of 6.45 million cases in 1976 to almost half a million cases in 2022. Alongside malaria elimination, India is remarkably also on the verge of eliminating another serious mosquito-borne disease, lymphatic filariasis, although diseases like dengue, chikungunya, zika, and Japanese encephalitis, also transmitted to man by mosquitoes, continue to pose a serious public health threat. It is not surprising, therefore, that owing to the mosquito-driven humongous losses to the global economy, health, and intelligentsia, the mosquito is regarded as the man's deadliest foe on the Earth and, as a priority, a thorough knowledge about all aspects of mosquito biology, in the broadest sense, is absolutely essentially inevitably necessitated to effectively deal with their control and the infections they transmit.

During the 20th century, some highly interesting publications, based on taxonomy and identification, but also with occasional references to biology, ecology, distribution, and biodiversity, appeared either on Anophelinae (e.g., Puri, 1931; Christophers, 1933, Wattal and Kalra, 1961, Das et al., 1990) or Culicinae (Barraud, 1934), as assorted and abridged editions. Thus, a comprehensive single treatment including various different aspects of the biology of the mosquito was completely wanting, despite recent inventories by Bhattacharyya et al. (2014) and Tyagi et al. (2015, 2020). Currently, not only have the quantum of mosquito species enhanced to over 400 taxa but also nearly three dozen vector species are known to be involved in transmitting diseases. Control of these deadly mosquitoes has been a serious constraint due to the development of resistance to various insecticides, on the one hand, and changes in their behavior, on the other. These failures to control vector mosquitoes effectively have been a driving force to update knowledge on all aspects of life as well as the strengths and weaknesses of this deadly enemy of human beings.

I am very pleased to see that Prof. B.K. Tyagi has sincerely attempted to fill up a long-standing vacuum by way of updating knowledge on Indian mosquitoes through consilience (e.g., biology, ecology, taxonomy, identification, distribution, host–parasite relationship, the transmission of disease, mosquito-borne diseases, control and regulatory provisions, etc.) and concerted efforts of about one hundred experts in mosquito science from both India and abroad. In my well-researched opinion, Prof. Tyagi's book *Mosquitoes of India*, in two separate volumes, is not only the first comprehensive compilation of available information on Indian mosquitoes to date but is also a novel treatment ever. The book, a sure *vade mecum*, will prove to be a permanent reference work for every serious mosquito researcher in India, in particular, and elsewhere, in general. I am more than confident that this book will be appreciated by both amateurs (undergraduates and post-graduates) and professionals alike in universities and research organizations. I wish the book a great success.

January 25th, 2023

Dr. Rui-De Xue
Anastasia Mosquito Control District
120 EOC Drive, St. Augustine, FL 32092
USA

Preface

Mosquitoes form one of the largest dipterous families, Culicidae, and are among the most important creatures to man on Earth. In fact, no other animal whatsoever has ever directly or indirectly affected his own survival to an extent where the mosquito is regarded as the deadliest foe of man on the planet! There are over 3,600 species of mosquitoes worldwide. They are present in all regions and counties of the world, save for the Island. They transmit some of the deadliest vector-borne diseases and are reasons for 17% of all communicable diseases globally. Mosquito-borne infections affect almost 60% of the world's population, mostly in the tropics and subtropics; malaria alone infecting 229 mill. and killing over 400,000 people annually. Their dangers are on the increase day by day due to increased global transcontinental travel and trade, besides a new-found range of distribution of the dominant vectors. Invasion of malaria vector *Anopheles stephensi* in Africa and Sri Lanka (Tyagi, 2023), and of dengue/chikungunya/zika vector *Aedes albopictus* in Europe and the Indian Ocean region is threatening enough to raise alert signals for future epidemics. The annual losses brought about by mosquitoes to humans are astronomical; malaria alone levied a phenomenal estimate of the disease burden of 464,00,000 Disability Adjusted Life Years (DALYs) in 2019.

India shares more than 10% (420 species) of the global fauna. Some of these mosquito species are responsible for transmitting several deadly and/or debilitating diseases such as malaria, filariasis, dengue, chikungunya, Japanese Encephalitis, West Nile, zika, etc., exacting death and sickness to millions of people, besides sapping off the country's crucial national resources of economy, health and intelligentsia. Unfortunately, there has never been a single comprehensive book in the past nearly 100 years that would offer updated information on various facets of the life of a mosquito such as, for example, taxonomy, phylogeny, biogeography, biology, ecology, physiology, and genetics as well as mosquito-borne diseases and their treatment and control – so important for strategizing their effective control in future for a healthy and prosperous nation! Over the years many vector mosquitoes have rapidly changed their behavior both under the continuous and injudicious use of insecticides for their control, on the one hand, and the unavoidable impact of climate change and anthropization, on the other. Such behavioral changes in vector mosquitoes have rendered them resistant to the insecticides of choice, thereby resulting in an unhindered increase in their population. Many other vectors have shifted their resting and feeding behavior from indoors to outdoors, and have again questioned the validity of the Indoor Residual Spraying (IRS). Moreover, we have no information on the exotic species sneaking into India, on the one hand, and the status of endangered species (read "genome"), on the other! Needless to say, with the advancement of scientific and technological tools in the recent past, the phylogeny and classification of mosquitoes have been thoroughly churned out with an altogether new interpretation. Thus, mosquitoes of India, or for that matter of the world, have altogether a new face to present today, and it is inevitably indispensable to know these highly important animals in their current form of knowledge. Therefore, it is now considered commensurate to be *au fait* about the mosquitoes of India in totality, and the present book *Mosquitoes of India*, in two parts, has been designed and reinforced with updated knowledge on all-important subjects that will make it an important reference work for not only Indian researchers but also for their contemporaries in other parts of the world.

Jodhpur, January 1, 2025

Prof. Dr. B.K. Tyagi
Professor of Practice

Acknowledgment

A book on *Mosquitoes of India*, in the wake of as glittering masterpieces as those produced by Christophers (1933), Barraud (1934), and Rao (1984), just to mention a few examples, on one hand, and a nearly complete vacuum of a comprehensive treatment embodying within its folds an updated description of all the important disciplines of Indian Culicidae, on the other, was truly a Himalayan responsibility full of endless challenges, but still hugely gratifying when I thought of it becoming useful to a large number of students and researchers, both in India and abroad! This uphill task was, however, made largely easy by over a hundred dedicated authors engaged in completing as many as 40 chapters for the two volumes describing nearly all-important facets of life of mosquitoes, especially with reference to India. To all of them I am grateful from the bottom of my heart.

For permission to use figures and/or tables from their publications my heartfelt gratitude goes to a large number friends and colleagues, namely, Dr. Dale E. Greenwalt (Departments of Paleobiology, National Museum of Natural History, Washington, DC 20013, USA), Dr. Derek E.G. Briggs (G. Evelyn Hutchinson Professor of Earth and Planetary Sciences, Curator, Yale Peabody Museum, Mailing address (U.S. Postal Service), Department of Earth and Planetary Sciences, Yale University, New Haven, CT 06520-8109, USA), Dr. Leonardo Puerta (Institute for Immunological Research, University of Cartagena, Cartagena, Colombia), Dr. R. Harbach (British Museum Natural History, London), Dr. Marc J. Klowden (Editor, *Journal of Vector Ecology*, Gainesville, USA), and a host of other generous permitters who are all duly acknowledged on appropriate places in the book. Further, I am also deeply indebted to the universal policy of organizations like the World Health Organization, Geneva, Switzerland, Centers for Disease Control, USA, etc. to use their material with due acknowledgment, which I am respectfully doing herewith. They are all duly cited on appropriate legend of the figure(s), table(s) or other places in the book.

Above all, my cordial remembrances with heartfelt thanks go to the world famous malariologist and friend, the late Professor Dr Felix P. Amerasinghe of International Water Management Institute, Colombo, Sri Lanka, who appreciated my malaria research in the Thar Desert, remarking on 7th April, 2004, "*Tyagi is best known internationally among malaria and medical entomology circles for his long years of research into mosquitoes, malaria and irrigated agriculture in the Thar Desert of Rajasthan, India.*" His appreciation was my motivation for writing this book on 'Mosquitoes of India.'

I am extremely grateful to Dr. Rui-De Xue (Anastasia Mosquito Control District of St. Johns County, St. Augustine, FL 32092, USA) for writing the Foreword and giving encouragement to me throughout the course of completing this book.

As always, here too I wish to thank my wife, Ajita, who gave me full freedom of time and movement so that I could put my whole attention to the accomplishment of this long and meandering journey into the *Mosquitoes of India: Biology, Phylotaxonomy, Host-Parasite Interaction and Resistance Development*.

Last, but not least, I duly thank the publisher Taylor & Francis, USA, and their highly competent and able editorial team for their patient guidance throughout the course of processing this book for publication.

January 1, 2025 Prof. Dr. B.K. Tyagi

About the Editor

Dr. Brij Kishore Tyagi currently serves as the Professor of Practice, Department of Biosciences, University Institute of Biotechnology, Chandigarh University, Mohali (Punjab), India as well as the Advisor, Vellore Institute of Technology, India, and the Consultant, Reckitt Benckiser, Gurugram (Haryana), India. He completed his MSc (Zoology) in 1974 and Ph.D. (Entomology/Zoology) in 1978. He joined the Malaria Research Center (now National Institute of Malaria Research) in 1981 on a WHO-TDR research project on tribal malaria in south Gujarat, and subsequently the ICMR mainstream as Senior Research Officer at Vector Control Research Center, Pondicherry in 1984, and as Assistant Director and later Deputy Director at Desert Medicine Research Center, Jodhpur in 1988. He finally joined as a Scientist 'F' (Joint Director) (2000) and subsequently Scientist 'G' (Director) and Director-in-Charge (2005/07) at the Center for Research in Medical Entomology, Madurai (Tamil Nadu State) through 35 years' long and highly distinguished career in diverse medico-entomological research, particularly the various vector-borne diseases (VBDs), in most of States and UTs in the country. He emerited from the ICMR Service in 2013, and was rewarded in continuity with two years' extension to 2015 with all administrative powers. Post-emeritation, he was appointed first as the Visiting Fellow followed by Visiting Professor at Bharathidasan Univ., Trichy (TN) (2015–2016) and subsequently as Visiting Professor at Punjabi Univ., Patiala (Punjab) (2017–2019). He has authored more than 700 scientific titles, 43 books, including a training manual, *Biosafety for Human Health and the Environment in the Context of the Potential Use of Genetically Modified Mosquitoes (GMMs): A tool for biosafety training based on courses in Africa, Asia and Latin America, 2008–2011*, published by the World Health Organization (2015).

List of Contributors

Nazni bt. Hj. Wasi Ahmad
Medical Entomology Unit
Institute of Medical Research
Kuala Lumpur, Malaysia

Vivek Anand
Department of Community Medicine
Armed Forces Medical College
Pune, India

Syed Afrin Azmi
Ex-DH & FWS, CMOH Office
Purba Bardhaman, India

Minu Bharati
Department of Zoology
S.B.S. Govt. College
Dakshin Dinajpur, India

Sajal Bhattacharya
Department of Zoology
Asutosh College
Kolkata, India

Prerana Bhujel
Insect Biochemistry and Molecular Biology Laboratory
Department of Zoology
University of North Bengal, Siliguri
Darjeeling, India

Chandrima Bose
Department of Zoology
Asutosh College (Affiliated to Calcutta University)
Kolkata, India

Kailash Chandra
Ex-Zoological Survey of India
Prani Vigyan Bhawan, M-Block New Alipore
Kolkata, India

Neil Roy Chaudhary
Division of Cell Biology and Immunology
Biomedical Parasitology and Translational-immunology Lab
CSIR-Institute of Microbial Technology
Chandigarh, India

Sambashiva Daravath
Department of Biotechnology
Nizam College (OU)
Hyderabad, India

Prapti Das
Insect Biochemistry and Molecular Biology Laboratory
Department of Zoology
University of North Bengal, Siliguri
Darjeeling, India

Subhajit Das
Insect Biochemistry and Molecular Biology Laboratory
Department of Zoology
University of North Bengal, Siliguri
Darjeeling, India

Vas Dev
Ex-ICMR-National Institute of Malaria Research
Field Unit
Guwahati, India

Ramesh C. Dhiman
Ex-ICMR-National Institute of Malaria Research
Sector 8, Dwarka
New Delhi, India

Diwakar Singh Dinesh
Ex-ICMR-Rajendra Memorial Research Institute of Medical Science
Patna, India

Rajnikant Dixit
ICMR-National Institute of Malaria Research
Sector-8, Dwarka
Delhi, India

Tanuka Ghosh
Department of Zoology
Asutosh College (Affiliated to Calcutta University)
Kolkata, India

M. Govindaraju
Environmental Biotechnology
School of Environmental Science
Bharathidasan University
Tiruchirappalli, India

Lalita Gupta
Department of Zoology
Chaudhary Bansi Lal University
Bhiwani, India

S.L. Hoti
ICMR-Vector Control Research Center
Puducherry, India

S.S. Areeb Hussain
ICMR-National Institute of Malaria Research
Sector 8, Dwarka
New Delhi, India

C. Kandasamy
Ex-Consultant – Agrochemicals
Chennai, India

S.A. Khan
Medical Entomology, Arbovirology and Rickettsial Diseases
ICMR-Regional Medical Research Center
Dibrugarh, India

Jagbir Singh Kirti
Ex-Department of Zoology & Environment Science
Punjabi University
Patiala, India

Sanjeev Kumar
Department of Biotechnology
Chaudhary Bansi Lal University
Bhiwani, India

Vikas Kumar
Zoological Survey of India
Prani Vigyan Bhawan
Kolkata, India

Seena Kumari
ICMR-National Institute of Malaria Research
Sector-8, Dwarka
Delhi, India

Nilu Limboo
Insect Biochemistry and Molecular Biology Laboratory
Department of Zoology
University of North Bengal, Siliguri
Darjeeling, India

Mihir Makwana
Institute of Science
Nirma University
Ahmedabad, India

Abhijit Mazumdar
Entomology Research Unit
Department of Zoology
University of Burdwan
Burdwan, India

Manas Pratim Modak
Insect Biochemistry and Molecular Biology Laboratory
Department of Zoology
University of North Bengal, Siliguri
Darjeeling, India

Ritwik Mondal
Ex-Department of Zoology
North Bengal University
Darjeeling, India

B. Reddya Naik
Zoology Department
Osmania University
Hyderabad, India

Anil Prakash
Ex-ICMR-National Institute for Research in Environmental Health
Bhopal, India

E. Pushpalatha
Biopesticides and Toxicology Division
Department of Zoology
University of Calicut
Malappuram, India

Debayan Raha
Insect Biochemistry and Molecular Biology Laboratory
Department of Zoology
University of North Bengal, Siliguri
Darjeeling, India

Priyanka Rai
Insect Biochemistry and Molecular Biology Laboratory
Department of Zoology
University of North Bengal, Siliguri
Darjeeling, India

Pooja Rohilla
ICMR-National Institute of Malaria Research
Sector-8, Dwarka
Delhi, India

Abhirup Saha
Insect Biochemistry and Molecular Biology Laboratory
Department of Zoology
University of North Bengal, Siliguri
Darjeeling, West Bengal, India

Dhiraj Saha
Insect Biochemistry and Molecular Biology Laboratory
Department of Zoology
University of North Bengal, Siliguri
Darjeeling, West Bengal, India

Vaishali Saini
ICMR-National Institute of Malaria Research
Sector-8, Dwarka
Delhi, India

Manas Sarkar
Reckitt Benckiser Ltd.
Gurugram, India

Nirmala Sankhala
ICMR-National Institute of Malaria Research
Delhi, India

R.S. Sharma
Ex-National Center for Disease Control
Faridabad, Haryana, India

M. Siddaiah
Department of Biotechnology
Nizam College (OU)
Hyderabad, India

Devinder Singh
Pro Vice Chancellor
Chandigarh University
Gharuan, India

Poonam Singh
Ex-ICMR-National Institute of Malaria Research
Sector 8, Dwarka
New Delhi, India

Tanvi Singh
ICMR-National Institute of Malaria Research
Sector-8, Dwarka
New Delhi, India

Shakya Sinha
Department of Zoology
Asutosh College (Affiliated to Calcutta University)
Kolkata, West Bengal, India
&
Biophysical and Structural Genomics Division
Saha Institute of Nuclear Physics (DAE)
Kolkata, India

Abhisekh Subba
Insect Biochemistry and Molecular Biology Laboratory
Department of Zoology
University of North Bengal, Siliguri
Darjeeling, India

P.K. Sumodan
Mohanam
Hi Rises Homes, Cheruvatta
Calicut, India

Nikunj Tandel
Institute of Science
Nirma University
Ahmedabad, India

Sanjay Tevatiya
ICMR-National Institute of Malaria Research
Sector-8, Dwarka
Delhi, India

Rina Tilak
Ex-Department of Community Medicine
Armed Forces Medical College
Pune, India

B.K. Tyagi
Professor of Practice
Department of Biosciences, University Institute of Biotechnology
Chandigarh University
Mohali, Punjab, India

Rajeev K. Tyagi
Biomedical Parasitology and Translational-immunology Lab
CSIR-Institute of Microbial Technology
Chandigarh, India

Varun Tyagi
Entomology-Household & Public Health Pesticides Testing Services
Eurofins Agroscience Services Pvt. Ltd
Tiruppur, India

Tejaswini Vaidya
Institute of Science
Nirma University
Ahmedabad, India

Vasuki Venkatesan
Ex-ICMR-Vector Control Research Center
Puducherry, India

Indra Vythilingam
Parasitology Department
Universiti Malaya
Kuala Lumpur, Malaysia

Pooja Yadav
ICMR-National Institute of Malaria Research
Sector-8, Dwarka
New Delhi, India

Section I

Origin, Evolution, and Distribution

1 History of Mosquito Research

B.K. Tyagi
University Institute of Biotechnology, Chandigarh University, Mohali, India

Indra Vythilingam
Universiti Malaya, Kuala Lumpur, Malaysia

1.1 INTRODUCTION

Mosquitoes are undeniably one of the most studied groups of insects due to their great impact on human health as the agents that transmit the pathogens that cause malaria, filariasis, and numerous viral diseases, such as yellow fever and dengue fever (Howard et al., 1917; Howard, 2019; Harbach, 2018). The dominance of mosquitoes among the world's hematophagous animals is a fundamental scientific insight which means that in chorogeographic and bio-ecologic attributes of biodiversity beyond our comprehension these creatures not only surpass to function as a self-sustaining biological system but permeate our lives inextricably so much that the human history can be very likely traced back in the impact of this little creature – the mosquito (Harrison, 1978; Winegard, 2019). They are, in fact, an integral and complex part of the terrestrial, aerial, and freshwater ecosystems with which the future of human beings is inseparably interlinked. They are indeed an enormous source of a great deal of scientific information to benefit, contrary to the conventional belief, human life, if viewed holistically (Tyagi, 2023a; Tyagi et al., 2024).

There are lots of reasons why people should be interested in mosquitoes (Howard, 2019). Mosquitoes are some of the deadliest predators on the planet. They associate humans not only with their irritating and painful bites, bite-associated allergies, vexation of various different kinds like uneasiness and loss of sleep, as well as mosquito phobia involving ill temperament and psychological disturbance but, in the process at least some of them, also mediate transmission of a large number of life-threatening and/or debilitating disease pathogens (Becker et al., 2020; Tyagi, 2020). They are branded objects of deep abhorrence and dislike for all they offer to humankind – the annoyance, the pain, the nuisance, and the deadly infections as well as responsible for wasting precious national resources by impoverishing human health, reducing economy, and emaciating intelligentsia, especially in less resourceful countries (Sinton, 1935; Spielman and D'Antonio, 2001; Tyagi 2004). A mosquito bite *per se* may not bring instant havoc, but it is capable of causing disasters when it (only the female mosquito needs blood primarily to sustain maturing ova) carries disease pathogens of any of the deadly and/or incapacitating diseases such as, for example, malaria, lymphatic filariasis, yellow fever, dengue, chikungunya, Zika, West Nile, Japanese encephalitis, and other types of encephalitides. India has been experiencing for centuries the colossal burden of infections like malaria and lymphatic filariasis, while others (e.g., Zika) have either emerged *de novo* or re-emerged (e.g., chikungunya) with the vengeance of a severe epidemic during the past few decades. Still, infections like dengue, with its two fatal syndromes (dengue hemorrhagic syndrome and dengue shock syndrome) and Japanese encephalitis have been continuously spreading into even those regions which were hitherto a *terra incognito* for both the pathogen and the vector. Interestingly all these infections involve multiple vector species with different potentials to transmit a pathogen (Tyagi, 2008).

DOI: 10.1201/9781003326991-2

No animal on Earth has affected human lives so directly and profusely as is brought about by the mosquito. In fact, according to Winegard (2019), mosquitoes are to blame for killing nearly half of the entire world's cumulative population, which amounted to approximately 52 billion people. He also noted that there are more than 100 trillion individual mosquitoes living in the world today. With the average adult mosquito measuring less than one-half of an inch long and weighing between 2 and 2.5 milligrams, mosquitoes are an especially small pest. But the impact they have on humans, as well as their evolutionary history, is anything but small. A mosquito is surely small but is giant of a killer. It is indeed no small enemy of humans to whom they carry viruses, bacteria, and parasites, which they, along with their other arthropod allies, transmit through bites, infecting close to 700 million people and killing more than one million every year (WHO, 2014). A mosquito that hardly measures the size and weight of a grape seed, is ubiquitously abhorred for the nuisance, pain, and death it hails every year on millions of humans. It is amazing to know that this tiny and feeble-looking organism has in fact assumed the reputation of the most formidable envy of man in modern times. Rao (1984), a celebrated Indian medical entomologist, cautioned us not to overwhelmingly indulge in self-aggrandization in our understanding about the biology of mosquito – her strengths and weaknesses – and warned us not to underestimate the miniatured enemy who has so far outwitted humans in their fight against various mosquito-borne and life-threatening diseases. In past, they were responsible, among others, for (i) felling the kingdoms of great leaders, for example, Alexander "The Great" was suspected to have succumbed to the infection of malaria parasite *Plasmodium falciparum* while retreating after a decisive battle against the Indian king Porus in the mosquito-ridden marshy lands of the Sindhu river valley of Punjab, in 323 BC (vide SOVE Newsletter), and similarly the much dreaded and ruthless ruler, Sultan Muhammed bin Tughluk contracted malaria in 1351 AD while on a military campaign against rebels, and within a short time succumbed to the disease, (ii) decimating armies such as, for example, the deadly yellow fever was stated to have taken its toll on thousands of U.S. Army recruits during the brief Spanish-American War, and that death from disease among American troops was much more common than death from bullets, and (iii) deciding the fates of nations such as, for example the slim Panama, which still belonged to Colombia until late nineteenth century, raged a war for independence, following which a bitter civil war ensued between 1900 and 1903, when it finally won its independence (Spielman and D'Antonio, 2001). Furthermore, the creation of the sovereign country of Bangladesh, from out of the erstwhile East Pakistan with the help of India, will remain in the annals of human history a spectacle when about 1,00,000 Pakistani soldiers were strategically engaged in a bitter war, surrounded and forced to lay down their arms by surrendering to the Indian forces on 16th December 1971, but few realize the fact that mosquitoes also made a hell of living for the Pakistan armed forces having been deployed there through seaways from the "West Pakistan." The exit of the British regime from India in 1947 was finally caused by several factors including world's first and unique "non-violence movement" led by Mahatma Gandhi (Mohandas Karamchand Gandhi), but one major reason was the deadly mosquito-supported hostile climate and environment of the vast Indian land. Throughout their regime in India, the Britishers were troubled by mosquitoes. Therefore, as soon as the British government took over from the East India Company, it started pondering on this perennially vexing issue – the mosquito menace and mosquito-borne diseases. The government then instructed experts to study mosquitoes thoroughly so as to be able to create a good environment for its regime, and that is how the great works of Theobald (1901a, 1901b, 1901c, 1903, 1905, 1907, 1910), Christophers (1933), and Barraud (1934), just to cite a few standalone examples, took to their birth.

Mosquitoes are also blamed for procrastinating several beneficial hydro-electrical and hydro-agricultural development programs among which Indira Gandhi Nehar Pariyojana (IGNP), initiated in 1958 to be the first of its kind in the world in a xeric ecosystem, which could never be completed in all the initially proposed stages (only two of three stages have so far been completed) with a network of 10,000 km of distributaries across the water and agriculture-deficient The Great Indian Thar Desert in western Rajasthan State, India (Anonymous, 1994).

1.2 DEFINITION OF MOSQUITO

Mosquitoes are unisexual animals, that is, male and female reproductive systems are present in different individuals (Figure 1.1). They are bilaterally symmetrical. The adult mosquito's body, covered with an exoskeleton, is organized into three distinct body regions: head, thorax, and abdomen. The head is ovoid in shape, with two large compound eyes and five appendages such as two antennae, two maxillary palpi, and the proboscis. The thorax has three segments: the prothorax, mesothorax, and metathorax. Each segment has a pair of jointed legs, while the mesothorax has a pair of functional wings and the metathorax, a pair of knobbled structures called halters. The abdomen has 10 segments, of which the three terminal segments are specialized for reproduction and excretion (Figure 1.2).

There are some 3,724 valid mosquito species (as of February 6th, 2024; see R.E. Harbach's *Mosquito Taxonomic Inventory*), with a worldwide estimated population of more than 100–200 trillion. They belong to the Family Culicidae, comprising two subfamilies, viz., Anophelinae and Culicinae, which embrace all "true" mosquitoes, in Class Insecta. It is noteworthy here that Reinert (2009) has proposed submergence of the classical third and erstwhile subfamily Toxorhynchitinae with the subfamily Culicinae (https://mosquito-taxonomic-inventory.info/).

Mosquitoes are ubiquitous in distribution and are found in all regions of the world; currently only Iceland is the only place on Earth where mosquitoes do not occur. Most of them exist in tropical climates, but there are mosquitoes in arctic and desert regions (Tyagi, 2023b), ranging in distribution from sea level to altitudes as high as 10,800 feet (3,600 meters). In India mosquitoes are found in all States and Union Territories.

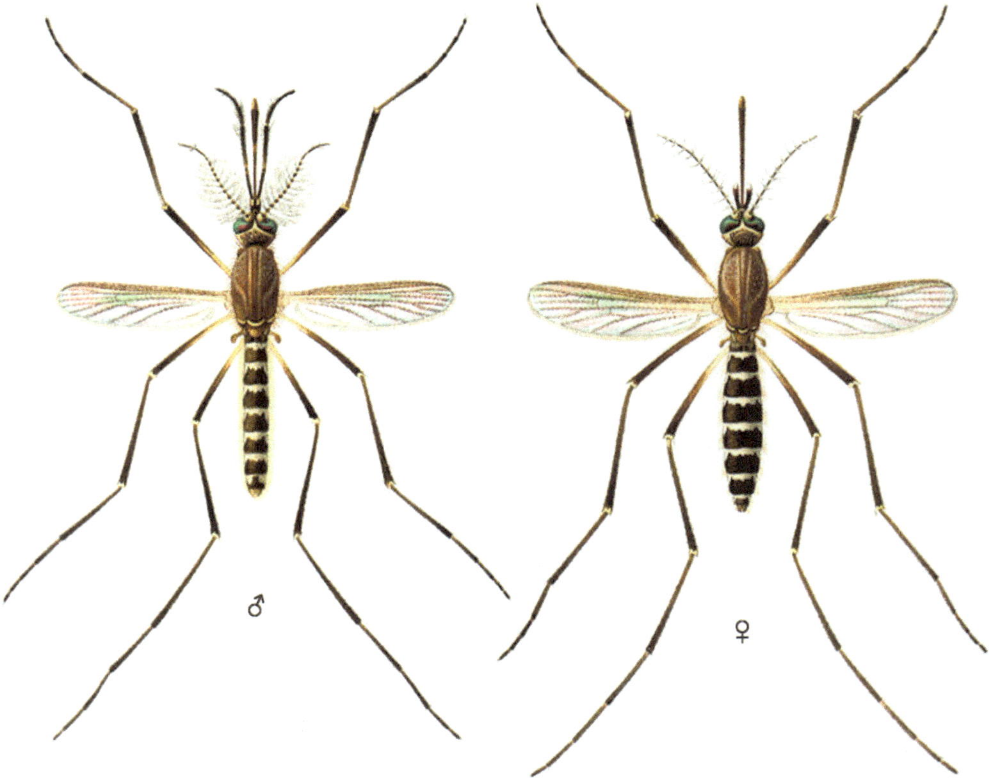

FIGURE 1.1 Male and female sexes of the universal black mosquito, *Culex quinquefasciatus*, a global pest and vector of human lymphatic filariasis.

Source: **Free Wikipedia.**

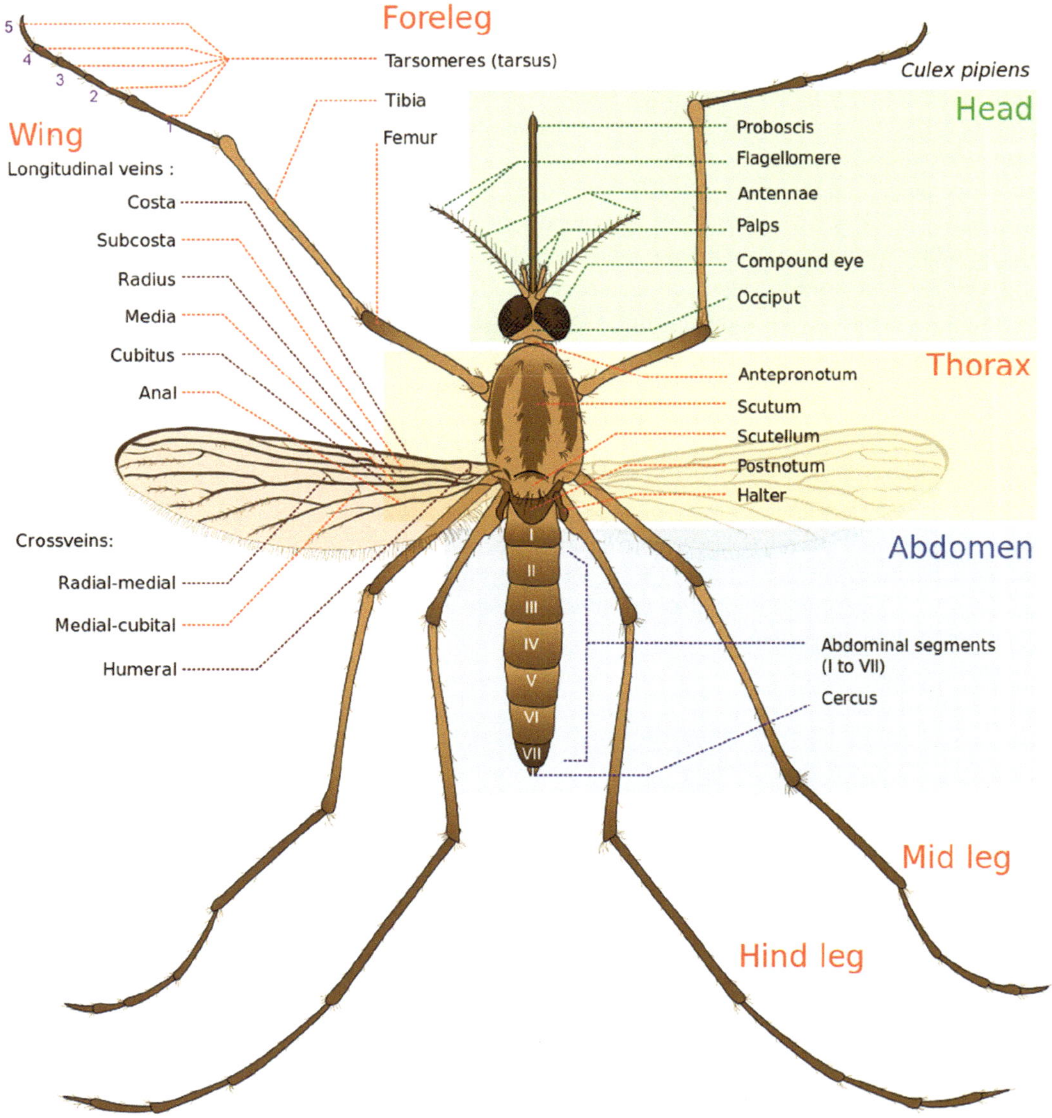

FIGURE 1.2 Mosquitoes have a symmetric body, e.g., *Culex quinquefasciatus.*

Source: Free Wikipedia.

1.3 ORIGIN OF MOSQUITOES

Knight and Stone (1977a,b) were among the first to have enlisted a total of 16 culicid species in some kind of fossil form, under genera *Anopheles* (1), *Culex* (8), *Mansonia* (3), *Aedes* (1), *Toxorhynchtes* (1), and *Culiseta* (2). They were mostly of European and Western World origin. Later, Briggs (2013) enlisted 25 fossil species from around the world. The earliest fossil mosquito was discovered in the Mid-Cretaceous Burmese amber (Borkent and Grimaldi, 2004). So far, no fossil mosquito has been discovered in India (Rao, 1984).

Very recently, paleontological evidence from the fossil mosquitoes, two from Myanmar and one from Canada, point toward a possible birth of Culicidae during the Mesozoic Era, 252–266 million years ago (Edwards, 1923). The two mosquitoes recovered from the amber in Myanmar were: (i) *Burmaculex antiquus* – Borkent and Grimaldi, 2004, described it as both a new genus and a new species from a single partially preserved adult female in Burmese amber which dated as back as 100–90 million years ago approximately between Upper Albian to Turonian sediments. The presence of palpi extending beyond the apex of a rather shorter proboscis indicates the fossil mosquito might be an *Anopheles* species. Morphology of mouthparts and antennae allude toward the femalized fossil to be a blood feeder on vertebrates, and (ii) *Priscoculex burmanicus* (Poinar, Zavortink, & Brown, 2019), a new genus and species of mosquito described from a female from mid-Cretaceous Myanmar amber, with proboscis subequal in length to the maxillary palpi. The various body parts' morphological attributes pointedly assign Priscoculex as an early lineage of the Anophelinae that existed about 100–110 Mya. Of the two fossils that emerged from Myanmar, at least *Priscoculex burmanicus* can be definitively assigned to an *Anopheles* species (Figure 1.3). Recently, Zavortink and Poinar (2000) have described a new species, *Anopheles (Nyssorhynchus) dominicanus*, from Dominican amber.

The third and the last fossil mosquito recovered was a male of *Paleoculicis minutus* (Poinar et al., 2000), found in Canadian amber from Alberta, Canada, dating as far back to the Campanian stage of the Late Cretaceous, around 79 million years ago (Eberth and Hamblin, 1993). It is a definite identity of Culicidae, being closer to Culicinae than either Anophelinae or Toxorhynchitinae, and is therefore easily distinguishable from the extant taxa on the basis of a number of morphological features on various body parts. Although no concrete inference is drawn about its hematophagous behavior which is, though, the nearest possibility bearing in mind the structures of a mouthpart. If this was so, then it is likely that *P. minutus* was engaged in blood-sucking on a variety of terrestrial hosts that would include multituberculates, monotremes, marsupials, lizards, snakes, turtles, crocodilians, eosuchans, and the first placentals, as well as representatives of the Dinosauria. Extant mosquitoes transmit a wide range of micro-organisms, including viruses, protozoans, and nematodes. Such vector relationships might have been established by the Mesozoic. This hypothesis gives rise to the assumption that a split between the two subfamilies might have occurred 197.5 million years ago, during the Early Jurassic period, but that major diversification could possibly not take place until the Cretaceous (Lorenz et al., 2021).

Greenwalt et al. (2013), for the first time, identified a fossil of a female mosquito, a *Culiseta* sp., with traces of blood in its engorged abdomen (Figure 1.4). Earlier than this find, only fossils of suspected blood-sucking insects were discovered and the creatures' feeding habits had mostly been

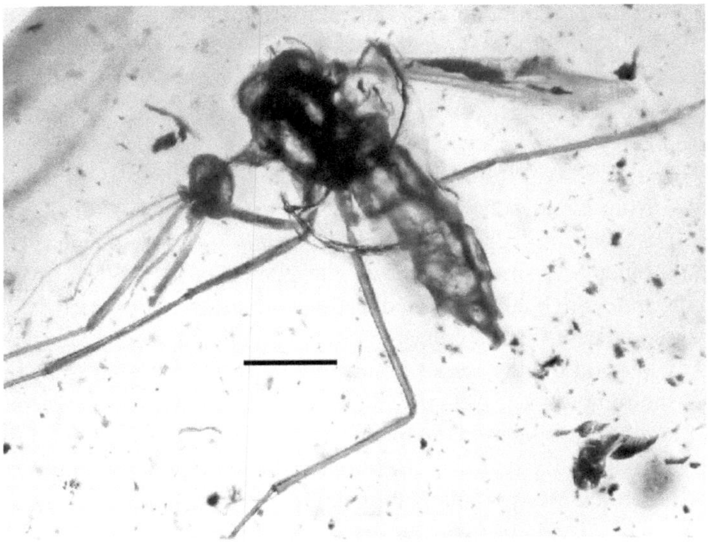

FIGURE 1.3 *Priscoculex burmanicus* (Poinar, Zavortink and Brown, 2019), a recently discovered fossil mosquito from the Cretaceous period.

***Source*: Poinar et al., 2019; permission obtained.**

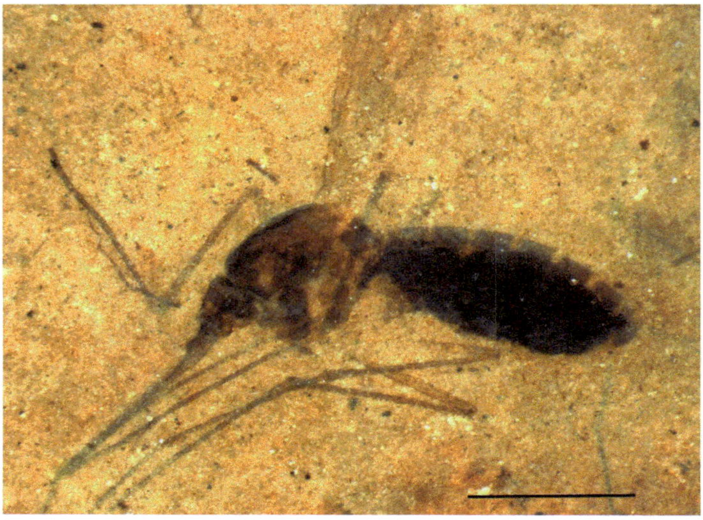

FIGURE 1.4 *Culiseta* sp. (Diptera: Culicidae), a blood-engorged female from the Middle Eocene Kishenehn Formation of northwestern Montana.

Source: **Greenwalt et al., 2013; Permission obtained.**

inferred from their anatomy or the presence of blood-borne parasites in their guts. But Greenwalt's fossilized mosquito contains molecules that provide strong evidence of blood-feeding (hematophagy) among ancient insects back to 46 million years ago (Yong, 2013).

1.4 TAPHONOMIC INTERPRETATION OF THE FOSSILIZED MOSQUITO INFORMATION

Mosquitoes are not only small and soft-bodied delicate creatures, but their fossilization and the taphonomy involved in the process are also bereft with so many challenges. Therefore, any likelihood of their postprandial preservation are vanishingly small (Briggs, 2013) (Figure 1.5).

Recently, Reidenbach et al. (2009) have thrown light on rare fossil occurrences based on molecular phylogeny and postulated that mosquitoes radiated in the early Cretaceous, or even much earlier. In spite of a long geological history, about two dozen mosquito fossil species are recognized from different kinds of preserved media, such as amber (the oldest fossil, a single specimen, from mid-Cretaceous ca. 93.9–100.5 Mya) from Burma and a male fossil (the second one, younger, and solitary specimen also from Cretaceous 72.1–83.6 Mya) from Alberta, Canada, which are the sole occurrences from the Mesozoic age. These findings clearly evidence about mosquitoes coexisting with dinosaurs and other ectoparasites, such as the giant flea-like insects from the Jurassic and Cretaceous periods, and more likely blood-fed on them, although not yet analyzed for hematophagy. Present-day mosquitoes seem to have appeared rather recently, after the Cretaceous, in the Eocene.

Chronic scarcity of good-quality fossils of mosquitoes is a serious handicap in interpreting their natural history in the remote past. Actually, their preservation depended not only on exceptional conditions or processes of fossilization or taphonomy but also the sustenance of their tissues' chemical components. Unlike hard and cuticularized insects whose body is composed largely of longer chain hydrocarbons formed by a polymerization process which, though likely to be altered, remain deep into the Palaeozoic fossils in traces of chitin and protein, mosquitoes' DNA is likely to be much more decay-prone and, thus, hardly sustainable over long geological ages. During hematophagy

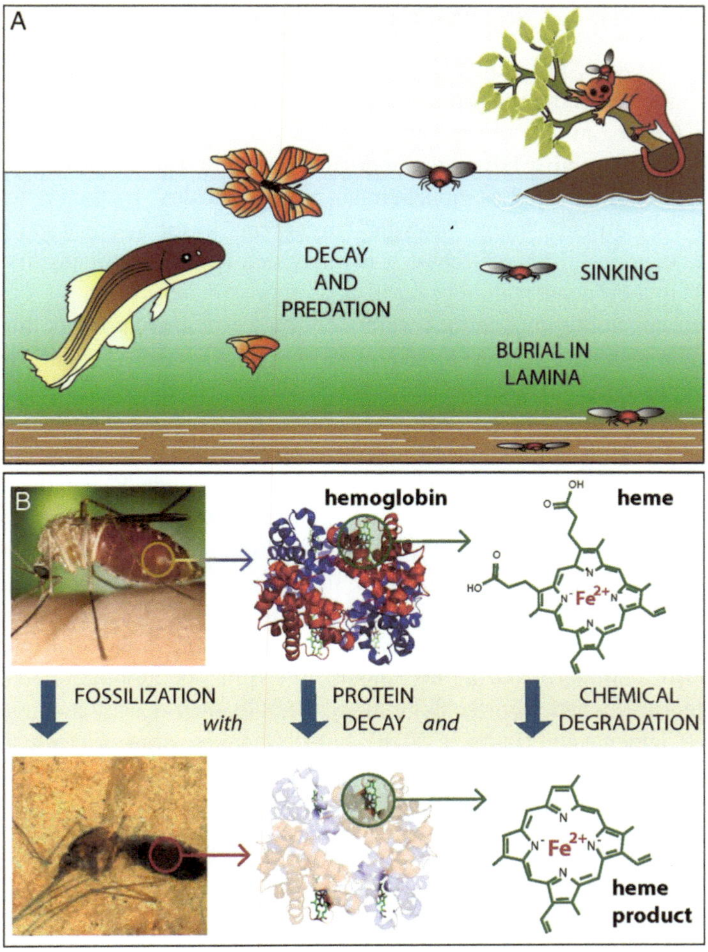

FIGURE 1.5 Taphonomy (process of fossilization) of mosquitoes during which they sink and are buried in the seasonal varves. The mosquito is flattened in the oil shale while the protein component of the blood hemoglobin decays, leaving the heme which is modified slightly during fossilization.

***Source*: Briggs 2013; Permission obtained.**

on the late Cretaceous dinosaurs whose bone comprised heme, confirmed through nuclear magnetic resonance (NMR), Raman spectroscopy, and immunological methods, mosquitoes engorged with blood, and the red blood cells consisted mainly of hemoglobin combining a protein with heme, the porphyrin ring structure that accommodates the iron involved in oxygen exchange. Although the protein component of the hemoglobin in the mosquito's last meal lost out through decay, nevertheless the heme had survived the stresses of vanishing processes of fossilization (Briggs, 2013). Much higher concentrations of iron are present in the female mosquito abdomen than elsewhere, as would be expected if heme were present. Though rare, preservation of heme is likely, as evidenced by the more advanced, time-of-flight secondary ion MS (ToF SIMS) technique. New discoveries, like the well-fed mosquito from the Eocene of Montana, have prompted scientists to have a relook on their postulations and hypotheses about the limits of fossilization. Taphonomy based on future excavation and collecting can generate a huge source of information on conditions of preservation and fossilization.

1.5 ORIGIN OF THE WORD "MOSQUITO"

The modern English word "mosquito" (plural "mosquitos" in Spanish or "mosquitoes" in English, but both are in vogue and correct) is a distortion from the Spanish or Portuguese "musketas," meaning little fly. In the beginning, several hundred years ago, the terminology was not uniformly accepted worldwide; the native Hispanic Americans called them "zancudos," which means long-legged, "gnats" by the English, "Les moucherons" or "Les cousins" by the French, "Stechmucken" or "Schnacke" by the Germans, "myg," "myyga," "myggorne" by the Swedish and other Scandinavian inhabitants, and "konopus" by the Greeks. It is likely that North Americans first began to use the word "mosquito" in their conversation as far back as approximately 1583.

In India where apparently mosquitoes were well-known in ancient times throughout the Vedic era, they carry multiple names such as, for example, Mashâ, Mashakâh (in Sanskrit), Mashâ, Mashak (in Assamese, Bengali, Oriya), Dās (in Gujarati), Machchar (in Hindi), Solle (Kannada), Moh (in Kashmiri), Kothuku (in Malayalam), Dās (in Marathi), Machhad (in Punjabi), Koshu, Koschu (in Tamil), Doma (in Telugu) and Machchad (in Urdu). "Sushruta Samhita," an ancient Sanskrit text on medicine and surgery of great authority and significance written during the 6th century BCE, described mosquitoes as "*Their bite is painful as that of serpents, and causes disease resulting from the three humours together: the bite as if burnt with caustic or fire, is red, yellow, white, and pink color accompanied by fever, pain of limbs, hair standing on end, pain, vomiting, diarrhea, thirst, heat, giddiness, yawning, shivering, hiccup, burning sensation, intense cold, vesicles or pustules increasing, swelling, knots under the skin, circles etc.*" Hasselquist, one of Linnaeus' favorite disciples, also described similarly, during his sojourn in Cyprus, being bitten fiercely by mosquitoes (Swedish, "myggorne") when near rice fields along the Nile, their density making them "intolerable and invincible." (Clements and Harbach, 2018). In fact, there are numerous allusions to mosquitoes in Indian writings from the Vedic era (Yadav et al., 2022). The "*Atharvaveda*" makes special reference to mosquito shape, habitat, many varieties, behavior, and even their presence throughout the year.

कण्डूमान्मशकैरीषच्छवयथुमान्दवेदनाः।
असाध्यंकीटसद्दशमसाध्यमशकितम्।।

(अ.स.उ.43/17)

The 67 different forms of *Keet* (insects) and their genesis were documented by Acharya Susruta in "*Kalpasthana,*" while "*Keet Visha*" describes five different types of "Mashak" in which "*Parvatiya Mashak*" is the most hazardous to people. In Atharvaveda, the world renowned medical classic, there is a detailed discussion about the habitat of the mosquito, that is rain-fed groundwater, as mentioned in the following Mantra.

ओको अस्य मूज्वन्त ओको अस्य महावषृ:।
यावज्जातस्तक्मंस्तावानभसबतल्हकेषुन्योचि: ।।

(AV 5.22.5)

There is also the "*Mashak*" description in "*Charaka Samhita*" of chapter "*Visha Chikitsa.*" The mosquitoes have been known for a long time also in the West as well as in China (Russell, 1955).

1.6 EARLIEST MOSQUITO KNOWLEDGE IN THE WORLD

Medical literature from many countries across the world have given references to mosquitoes and the likely human maladies associated with them (Winegard, 2019; Tyagi, 2004, 2020; Yadav et al., 2022). One of the oldest accounts about different types of mosquitoes is found in India's world famous ayurvedic compendium *Sushruta Samhita* or *Suśrutasaṃhitā* (सुश्रुतसंहिता, literally "Suśruta's

Compendium"), a Sanskrit work of yore, believed to have been composed 600 BC (Meulenbeld, 2000). This ancient medical book refers to the mosquito by a variety of names – all in masculine gender: *Sãmudrah* (marine or coastal mosquito); *Parimandalah* ("Reginal" or "Circular" mosquito); *Hastimashakah* (Elephant mosquito, perhaps because of large size); *Krishnah* (Black or dark mosquito, *Pãrwatiya* (Hill mosquito). Outside India Aristotle (384–322 BCE; Before Common or Current Era), popularly known as the "Father of Biology and Zoology," and Ulisse Aldrovandi (1522–1605), who had worked well over a millennium and more than one century, respectively, before the great Swedish botanist Carolus Linnaeus, contributed significantly to the knowledge of animals, particularly arthropods including insects, during the 18th century. Aristotle's works "The History of the Animals" and "On the Parts of Animals," the latter written in 350 BCE, provided detailed descriptions of its life cycle, including development, growth, metamorphosis, and inter-relationships of animals of various different kinds. The group he named *εητομον* (in Latin "entomon"), meaning notch or insect, included not only insects (incorporating locusts, cicadas, bees, and flies) but also myriapods and arachnids. Aristotle (citing Gessner, 1551–1558) referred to mosquitoes as "empis" in his *Historia Animalium*, iterated 300 BCE, where he had also described their biology.

1.6.1 Mosquitoes in Mideaval and Modern Literature

During the sixteenth century, Ulisse Aldrovandi (1522–1605), an Italian zoologist, in his *magnum opus* entitled "*De animalibvs insectis libri septem, cvm cingvlorvm iconibvs ad viuum expressis*," classified in a greater reformative manner animals including insects. Among insects he particularly described dipterans as the two-winged animals or flies; the latter including the current dipteran names *Culex* (*Culex communis* and *Culex maximus*) and *Tabanus*. This implied that binomials such as *Culex communis* and *Lumbricus terrestris* indicated his recognition of species as constituents of genera. Interestingly, centuries later, the name *Culex* provided the generic name used on the first identification and naming of the yellow fever mosquito, i.e., *Culex mosquito* Robineau-Desvoidy (1827) (Knab, 1916). So, one thing which is now absolutely, certainly, inevitably documented is the inexorable fact that it was the lesser known Italian Aldrovandi (1602, 1638), rather than the Swedish taxonomic giant Linnaeus, who had apparently introduced to the scientific world the generic name of the mosquito for the first time.

Linnaeus, on the other hand, has himself extensively made and also received from other countries all over the world mosquito collections largely through his correspondents, associates, and students such as, for example, Frederic Hasselquist (1722–1752) who actually innovatively described in detail the anatomical characters of mosquitoes some of which were later found to be accessioned in Linnaeus' collections. Linnaeus' collection of mosquitoes contains two mosquito specimens, one of which is a male of the *Anopheles maculipennis* complex, and the other, a female and damaged unidentifiably to the species, belongs to *Aedes* (*Ochlerotatus*) (Harbach et al., 1985). Linnaeus (1758) applied binomial nomenclature to several flies in his collections then considered as mosquitoes, of which two were true mosquitoes (Culicidae) with valid names, namely, *Culex pipiens* Linn., 1758 and *Culex aegypti* Linn., 1758 – the latter species today being known as *Aedes aegypti* (Linn, 1762) (a-e′dēz a-jip′tē; *Aedes* in Greek means "unpleasant," while *aegypti* suggests an origin of its description from "Egypt")), after numerous nomenclatural oscillations; the "year 1762" was ascribed to Hasselquist rather than Linnaeus (Figure 1.6). It is because of the fact that Hasselquist (1762) already taxonomically described in Latin two species of Culex in *Iter Palæstinum* and named the unnamed mosquito, endemic to Cyprus, as *Culex aegypti*. Linnaeus (1707–1778), on the other hand, never wrote about this species in any of his compositions including his magisterial *Systema Natura* – 10th edition, the most significant *vade mecum* of all the 12 volumes of the classic, or any of the later editions or works completed by others after his death, albeit at least two valid mosquito species, viz., *Culex pipiens* and *Culex argenteus* Poiret, 1787 being present in his collections (Harbach et al., 1985; Gardiner and Morris 2007).

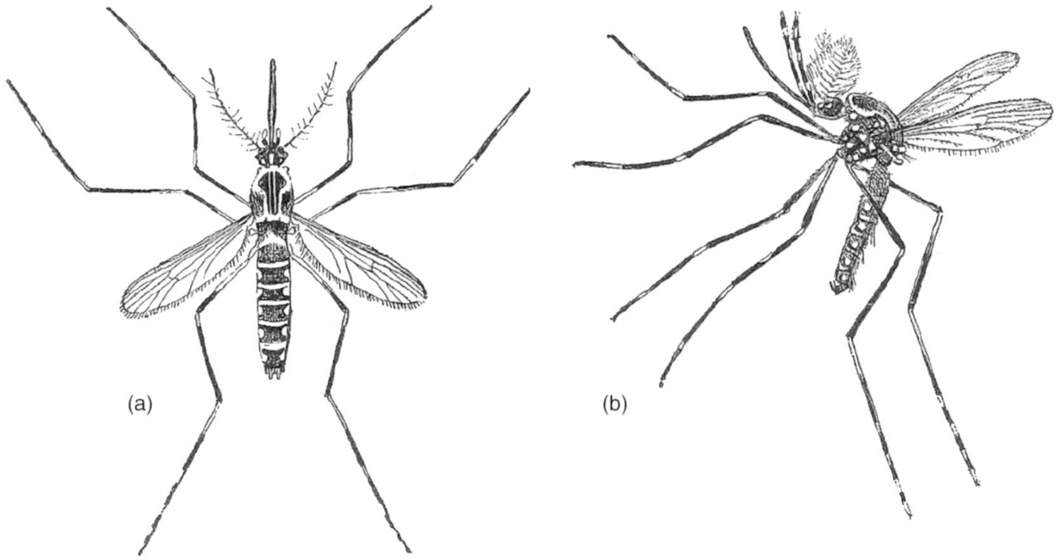

FIGURE 1.6 The yellow fever mosquito, which should be *Stegomyia aegypti* (Hasselquist), originally named *Culex fasciatus* Fabricius, 1805 = *Stegomyia fasciata* of Theobald (1901b). The drawings and the following captions, cited verbatim, are from Howard (1901). (a) *Stegomyia fasciata*, adult female, (b) Adult Male of *Stegomyia fasciata* from side. The inset shows the ungues (claws) from the tip of a leg, probably a foreleg. Characteristics that distinguish the yellow fever mosquito from *Culex aegypti* Linnaeus, as it was described by Hasselquist (1757), are the distinct white scale patterns over a black background on the scutum, notably the two lyre-shaped marks, and the distribution of white rings on the hindlegs.

***Source*: Clements and Harbach, 2018.**

According to Clements and Harbach (2018), at least two sets of evidence point out that Hasselquist, not Linnaeus, was almost certainly the author of the name *Culex aegypti*.

(i) In *Iter Palæstinum*, Hasselquist taxonomically described *Culex aegypti* in great detail compared to Linnaeus' much briefer descriptions.
(ii) Linnaeus, in his Introduction to *Iter Palæstinum* (Hasselquist, 1757), conceded to have adopted the names of animals and plants appearing there for his own *Systema Naturae* 10th edition (1758), albeit the fact that *Culex aegypti*, which did not appear in 9th edition (1756), also failed to figure out in any of the subsequent editions of *Systema Naturae*, or in his another work *Animalium Specierum* (Linnaeus, 1759b).

The question then arises how the name *Aedes aegypti* Linnaeus, 1762 is validated (Figure 1.6; Table 1.1). The factual situation is that after the republication of *Iter Palæstinum* under the title *Reise nach Palästina* in 1762, both in Swedish and German, the name *Culex aegypti* was nearly lost from the literature until it was listed as "*Culex aegypti* Linnaeus in Hasselquist, 1762" in Sherborn's *Index Animalium* for the period 1758–1800.

It is to be remembered here that Theobald (1901a) introduced the genus *Stegomyia*, with a brief description of *Stegomyia fasciata* (Fabricius), although it was not immediately accepted and the name *Culex fasciata* continued to appear in subsequent writings. More than 60 years later, this was formally recognized by the International Commission on Zoological Nomenclature (1964), "The generic name *Stegomyia* Theobald, 1901 (gender: feminine), type-species, by designation by Neveu-Lemaire, 1902, *Culex fasciatus* Fabricius, 1805, is hereby placed on the Official List of Generic Names in Zoology." Since the species *Stegomyia fasciata* (Fabricius) was a logotype, i.e., determined

TABLE 1.1
Standard Taxonomic Hierarchic and Nomenclature for the Yellow Fever Mosquito, *Aedes aegypti* (Linnaeus, 1762)

Taxon	Nomenclature (Authority)
Kingdom	Animalia (Margulis & Schwartz, 1998)
Subkingdom	Bilateria (Hatschek, 1888)
Infrakingdom	Protostomia (Grobben, 1908)
Superphylum	Ecdysozoa (Aguinaldo et al., 1997)
Phylum	Arthropoda (Latreille, 1829)
Subphylum	Hexapoda (Latreille, 1825)
Class	Insecta (Linnaeus, 1758)
Subclass	Pterygota (Lang, 1888)
Infraclass	Neoptera (Martynov, 1923)
Superorder	Endopterygota (Sharp, 1898)
Order	Diptera (Linnaeus, 1758)
Suborder	Nematocera (Berthold, 1827)
Infraorder	Culicomorpha (Wood & Borkent, 1989)
Family	Culicidae (Stephens, 1829)
Subfamily	Culicinae (Meigen, 1818)
Tribe	Aedini (Neveu-Lemaire, 1902)
Genus	*Aedes* (Meigen, 1818)
Subgenus	Stegomyia (Theobald, 1901)
Species	*Aedes aegypti* (Linnaeus, 1762)

Source: OECD, 2018.

from a written description in the absence of both a specimen and an illustration, therefore, the ICZN further clarified, "It has been suggested that those unfamiliar with the nomenclatural procedure may form the impression that the above declaration prejudices the use of the name *aegypti* in such combinations as *Aedes aegypti* (Linnaeus) or *Aedes (Stegomyia) aegypti* (Linnaeus) (Table 1.1). This is not the case. It remains perfectly proper to employ the name in these combinations or any others that further taxonomic study may render desirable." Thus, the International Commission on Zoological Nomenclature (1964) validated the specific name *aegypti* Linnaeus. It is noteworthy here that the birth of ICZN occurred on the same date as the publication of Linnaeus' 10th edition of *Systema Naturae* – that is, January 1, 1758 – and that ICZN accepted binomial names of species published only after this date.

1.7 MOSQUITO – THE TARGET

1.7.1 Mosquito Under Suspicion

Mosquitoes were hideously suspected for long to have a connection with certain illnesses, particularly intermittent fevers. Malaria was already known widely and adequately characterized clinically among people in various parts of the world, particularly in China, India, and South America. It indeed occurred, in the mind of natives, as a special malignity of the environment. Europeans had found out through expeditions to South America from the 17th century onwards that cinchona bark extraction was able to cure intermittent fevers but they had classified all fevers as one phenomenon. Europeans, whose dominance around the world then was phenomenal, associated malarial sickness with wetlands, hence the name "*paludisme*" [which in French means from swamp; *palus* (Latin) = swamp].

Rome was particularly a malarial inferno owing to Pontine swamps in vicinity, hence the local name "Roman fever." Clearly malarial fevers related to swamps, as unhealthy and notorious fever breeders, along with disturbed lands, forests under denudation, and lands under plowing, particularly the last one, hence the emergence and acceptance of the "telluric" and "miasmic" theories for centuries. Yet, though traditional epidemiologists stuck to their unhealthy soil (telluric) and malicious air (miasmas) theories, some noted 18th century Italian physicians, like Lancisi, tried to find out etiology of the disease by explaining the fevers were caused due to inhalation of a living contagium or animalcules, called "serafici" invisible to naked eyes, which multiplied in human body under unmeasurable conditions. A hypothesis subsequently advanced by Rasori noted that different kinds of fevers were caused by different types of species – an extraordinary thought so close to the true interpretation of malaria which would still take several decades until discovery of advanced microscopes good enough to spot microorganisms. The relationship between swamp-emanated inhaled bad air, comprising whatever toxin of Lancisi and others, with the fever was so obvious that the Italians, possibly in the 18th century, naturally saw the cause of fever as mal' aria (bad air), which was later refined as malaria in Great Britain. Subsequently, in the late 19th century when Laveran and others discovered causative parasites, *Plasmodium* species, the word "malaria" was readily accepted unchanged for the disease itself. Both the telluric and miasmic explanations circulated for a long period were added by another explanation of mixing the toxin or poison from the soil into water which resulted in fever after being drunk. This hypothesis changed the course of thinking about malaria, which was now classified along with other important gut diseases as cholera and typhoid. This meant that, as typhoid and cholera – which were associated with unhealthy aquatic, land, and living environments – had to be taken care of through improved sanitary measures, malaria too – considered as a disease of rural decay, spawn of an ill-kempt, savage, and uncultivated land, yielding to good husbandry and civilization – fit the pattern and invited similar countermeasures. This thinking of control of disease through good sanitation is still largely practiced. So, water was understood beyond doubt as the key factor in malaria prevalence. Italians prepared a detailed map of the occurrence of malaria, and it was discovered inadvertently that malaria distribution virtually overlapped the distribution of anopheline mosquitoes. But all these were conjectures. There were too many mosquitoes, and too few experts on anopheline taxonomy.

Why the miasmas were associated with these mosquitoes is because most of the malaria-carrying species bite following the night is set in, a pointer to the fears surrounding night miasmas. Another interpretation is the feeble-bodied adult females, in particular, face the danger of desiccation if exposed to the heat of the day, hence they venture out and about more often when the air is damp than when it is dry, which is very much likely during the daytime. While a school of thought of the day prudently tried to build the connection between intermittent fevers and mosquito prevalence, another school silently advocated the presence of the pathogenic microorganisms. Unfortunately, the fewer doctors who ventured out to bridge observed associations between mosquitoes and intermittent fevers were scientifically greatly impoverished due to a lack of demonstrable facts. Some approached the solution in a different way; Dr. John Crawford, an American physician, while working on "yellow fever mosquito theory" opined in 1807 that these are the eggs "insinuated into our bodies" which caused the malarial fever. He was faced with all-round criticism, but little did the world know that he was more nearly right than anyone else would be for 90 years in the future.

In 1854, Louis Beauperthuy, a French physician, better known as the prototypical author of the mosquito theory, boldly connected mosquitoes with intermittent fever – in warm weather but not in the cold. He was however firm in narrating that mosquitoes injected toxins, picked up from the putrefied vegetation in swamps, into humans. Rasori, an Italian physician, thought mosquitoes might carry contagium or animalcules. All these early thinkers had no idea or a scrap of evidence of what exactly that imaginative contagium was or how and where it came from, albeit the basic notion that parasites might shuttle between different animal species having been already proposed in 1790 by Peter Christian Abilgaard!

1.7.2 Mosquitoes Under Scrutiny

Mosquitoes have been conceived to cast evils on man for centuries, and by 1870–1880, some concrete evidence also began to emerge manifesting the mosquito's role in transmitting certain diseases. The first important work on the role of insects was done by Dr. Patrick Manson in China during the 1870s when he showed that mosquitoes might be vectors of the disease colloquially known as elephantiasis from the gross swellings, particularly of the legs and scrotum, that it often produces (Manson, 1978; Harrison, 1978). The hypothesis that he forwarded was absolutely irrefutable: he found out the young microfilariae which, unlike their procreator, moved freely in the bloodstream and were always enclosed in a delicate membrane, could neither feed nor move of their own volition and therefore must complete their life cycles somewhere else – the mosquito? In early 1877, Manson succeeded in demonstrating that the mosquito *Culex pipiens* was responsible to effect man-to-man transfer of a highly incapacitating infection in humans – lymphatic filariasis (Manson, 1878). This discovery was by all standards a great victory in furthering human knowledge about mosquito's placement in the man-mosquito-disease triangle. However, despite the partly true nature of his discovery, Dr. Manson, a surgeon-physician, had made two major mistakes while proclaiming the mosquito's intermediate role in disease transmission:

(i) First, he thought that these worms completed their development in mosquitoes by five days after feeding – which we today know is more than 10 days – which implied that worms had still not matured, and
(ii) Secondly, believing that the mosquito fed only once in her lifetime, he postulated that the parasites infected people only when they consumed drinking water containing the disintegrating body of an infected mosquito.

Manson's faulty deduction bore no adverse significance, considering Manson's phenomenal recognition as the tropical medicine genius and his firm demonstration that mosquitoes performed an essential role in generating human diseases, particularly his guidance to Dr. Ronald Ross about malaria and mosquito – a subject that would soon reveal the mosquito's role in producing explosive outbreaks of the disease, both in terms of human maladies of morbidity and mortality exacted by the *Plasmodium-Anopheles* nexus, as well as the unprecedented scientific rivalry between the geniuses of the day. The filariasis mosquito bond's discovery by Manson went largely unopposed.

Taking a cue from his "filariasis mosquito connection" theory, Manson began to speculate on malaria. Like Laveran, he too was baffled with the encounter of "exflagellation" phenomenon of the parasite. Following Laveran's assertion that these flagellae were the mature stage of the parasite, as well as the speculation that mosquitoes picked them up in that stage for transfer to a new host, Manson, in 1894, began to navigate his thinking the filariae way, though quite faultily. Already, for the benefit of Manson, the emergence of flagella in only half a minute, or more, old drawn blood was highlighted by the Russian Sakharov in 1889 and confirmed by Grassi in 1892. By this time, too, Manson – a warm, patient, and avuncular doctor of 50 years of mature yet youthful age – had emerged without question an outstanding expert in tropical diseases in England, and it had become widely known in medical circles that he, following his remarkable "mosquito-filariasis theory," was seriously embarking to explain the malaria tangle by way of mosquito's role in transmission. However, Manson's newly found filaria-related mosquito theory weighed heavily on him and he drew the conclusion that the red blood cells which enclosed the malaria parasite were comparable to the filarial sheaths. Consequently, he landed with a totally misplaced explanation. Even so, Manson's mosquito theory (first presented preliminarily in 1894 and later in the fully developed form in lectures of 1896) was a brilliant imaginative construction to pave the way for future research. Ross picked up the evidence where Manson had left!

Almost around the same period, precisely in 1880 in Havana, Dr. Carlos Finlay aired his finding of *Aedes aegypti* (then known as *Culex fasciatus* or *Stegomyia fasciatus*) transmitting the yellow fever disease. He arrived at this adrenaline-charged conclusion by pursuing a scientific methodology, otherwise next to impossible at present due to ethical reasons. He made yellow fever patients to be bitten by mosquitoes, only to be let loose to interact in course of time with the uninfected healthy persons. He did succeed partially in effecting yellow fever in some healthy subjects, but failed to realize the minimum time period the infected mosquitoes needed to be fully infective (actually, Finlay only conjectured about the disease based on its jaundice-implicated nature impaling the patient, without any knowledge about the pathogen, a virus, which was not yet known to the science). But, unlike Manson in case of filariasis mosquito, Finlay rightly discovered that mosquitoes bite more than once – a trait that truly pointed toward the mosquito's role as a biological transmitter of the disease. However, as it would normally go with every exciting new discovery, Finlay a Cuban, was not only widely ignored, he was ferociously repudiated for this observation, particularly by the then dominating English and German workers but also his contemporaries in Latin America.

Finlay's yellow fever mosquito theory meandered a long journey through all thick and thin to be finally accepted, almost in the name of another person, a truly famous mosquito-man himself though, the great Dr. Walter Reed, when in 1990 a universal acceptance was accorded to this great piece of scientific inquiry. Such was the deep impact of worldwide recognition of the scientific genius of Dr. Reed, a U.S. Army officer, that the "discovery of yellow fever mosquito theory" was without any further innuendoes accepted globally. However, unlike malaria where the pathogen, a protozoan parasite belonging to genus *Plasmodium*, was discovered much earlier than its mosquito vector, the world had to wait for some 30 years before the yellow fever virus could be identified, along with another au fait discovery regarding feral primate populations in the American and African forests harboring a reservoir of the disease. Luckily to Finlay, his notion on mosquito theory for yellow fever was propped up by the ultimate discovery of mosquito's prowess as a vector of malaria by Ross and the Italians, among which Grassi was most prominent, in 1897.

1.7.3 The International Competition

The three major mosquito-related researches in connection with demystifying transmission cycles of some of the most deadly and debilitating tropical diseases included well-placed and rather globally important "mosquito-yellow fever theory," "mosquito-filariasis theory" and "mosquito-malaria theory" – all during the fourth quarter of the 19th century. Of these three milestones, the yellow fever and filariasis connections to mosquitoes went rather peaceful or unchallenged, though low-voiced disagreements or oppositions could always be audible. Both these mosquito-borne revelations were made much earlier than the one for malaria or, more importantly than anything else, the establishment of the highest prize money in the name of the Nobel Prize in 1896, to be commenced for awarding from the year 1901. Malaria thus asserted some special claim, as already was paralleled – with still greater attention due to its prowess to stay on year after year – with the population decimating deadly plague, a reputation no other tropical disease was bestowed with! Enemies not in clear vision then included *Plasmodium* parasites and the *Anopheles* vector mosquito species, and, true to the demand of Nature, they were to be found out by none other than the two military men, Doctors Alphonse Laveran of the French Army and Ronald Ross of the British Indian Medical Service, who made basic discoveries that unlocked the secrets of malaria. Both scientists had characteristic opposite basic human qualities (Harrison, 1978):

(i) Dr. Ronald Ross, a complicated and chronically maladjusted man, was nevertheless filled with ceaseless energy, endurance, positive thinking and, above all, the desire to do something extraordinary – something that nobody had earlier done and which would single him out among the world's greatest men of honor. He took life as a struggle and a challenge, and himself as a soldier eternally pitched against the people and forces who sought, from stupidity or maliciousness, to frustrate him and hold back the course of human progress. Ross was exaggeratingly ambitious for progress and would not accept anyone as a hindrance in his way, and

(ii) Dr. Alphonse Laveran, on the other hand, was by nature a sober and noble human with a special penchant for scientific research and professional progress.

Initially, it all started as a pure international competition, with the noted German pathologists Dr. Heinrich Meckel, in 1847 and Dr. T. Frerichs, in 1858, along with Dr. Rudolf Virchow, having described and characterized the black particles in the blood smears of malaria patients of being formed of degraded hemoglobin from the affected red blood corpuscles, respectively, and many a French doctor like F.C. Mallot, Leon Colin, P.J. Keiner, and Acheille Kelsch, particularly the last one, who had enormous experience in malaria pathology, showing presence of innumerable tiny black particles, especially conglomerated in spleen, liver and, sometimes, brain to give the affected organ a tinge of slate gray.

Many explanations were floored as to the origin of these malaria-specific black particles, and the science of understanding moved ahead without many hassles. Dr. Kelsch, who worked in Algeria in the early 1870s, apparently missed the discovery of the first malarial parasite viewed in the human, which was some six years later done by Laveran in Algeria only. Professor Kelsch, true to the nature of his profession as a pathologist, looked at everything malarial with a classical pathologist's eye only, even having probably routinely killed the contents of the freshly drawn blood from the malaria patients – in which there existed every likelihood of finding what Laveran had subsequently discovered, the "crescents." Though, Kelsch had passed by the brink of discovery, he nevertheless had demonstrated that the black pigments emerged with the onset of fever. He obviously did not follow the question of uniqueness of black pigment in malaria, and focused on other important lines of malaria research pertaining to the disease's impacts on red and white blood cells. Thus, the central theme was left unattended and unsolved, for the man who had a keen eye for observation and a determination to find answers to most baffling queries – Charles Louis Alphonse Laveran.

Laveran, sent to Algeria in 1878 as an official necessity and certainly not on any mission to research on malaria, was nonetheless forced to go into studying malaria as military bosses were desperately keen to cease the outbreaks (of malaria) that constantly wrecked their soldiers there and exacted astronomical costs in replacing dead and the sick as well as in boosting up their soldiers' morale. He was familiar with the work of Kelsch and others and soon understood the value of studying alive the contents of the freshly drawn blood to which he had access in plenty. He met with success rather early on November 5th, 1880, when he clearly saw "spheres" (the exflagellating male gametes). For the next four years he worked to know more about the malaria parasite, which he described in extenso in a book titled *La Nature parasitaire des accidents de l'impaludisme*, published in 1881, followed by two more elaborative books in 1884 and 1898, respectively. When he reported his findings to the Academy of Medicine in Paris, he did not forget to underline his claim to have been the first to discover the malaria parasite. It was indeed the first glimpse of one form of the malarial parasite, *Plasmodium falciparum* (later it had been confirmed that he actually found stages of development of both *P. falciparum* and *P. malariae*), a protozoan – still not known by that name. ironically, even such a spectacular discovery had to go unattended in appreciation. There were contemporary scientists who stooped low to belittle the significance of Laveran's grand discovery by giving one or another bizarre explanation. Others, more vocal, including the celebrated hematologist Dr. William Osler, Baltimore (USA), totally refused to believe his observations and an event such as the "exflagellation" in blood.

Disagreement with Laveran's single-celled protozoan (which he referred to as the germ) led to divert the attention of several scientists of the day to consider an alternative possibility – the bacteria. The bacillus theory of *Bacillus malariae* prevalent in Italy's infamous malarious swamps, raised by two Italians, was as soon side-lined as it arose despite the undercurrent of support from the newly emerging highly promising discipline of bacteriology, directly under the guidance of immensely famous bacteriologists such as Louis Pasteur and Robert Koch. However, this dethronement of the malarial bacillus hypothesis did not quite help ready acceptance of Laveran's germ theory but, on the contrary, Laveran had to take upon himself the responsibility to convince in person by practical demonstration of "exflagellation" to his fellow scientists both in Algeria (Dr. Richard) and in Italy

(Drs Ettore Marchiafava and Angelo Celli), but largely in vain, since both the Italian experts had their opinion reserved in favor of *Bacillus malariae*, thought as the most likely agent of malaria. Not only this, Marchiafava and Celli mischievously hid from him their own research outcome in the next two years conforming to Laveran's findings, which they published without any reference to Laveran's remarkable original work with a viewpoint of making their claim as first for the discovery. Laveran was immensely exasperated by this fraudulent act of Italians who, in turn, believed what they experienced in fresh blood preparations was not seen by Laveran and was certainly different. To put their stake for an original claim of the discovery, they named their form *Plasmodium malariae*. Dr. Camillo Golgi, whose primary interest was the nervous system, confirmed their findings, with some startling advancement in knowledge. With this, the initial coolness of trust in comradeship and scientific fraternity seemed to have volatilized, and the first seeds of an embittered future world war of priorities and primacy had been sown minus healthy international competition, even when new human malaria parasite species continued to be discovered with many excellent scientific inputs by Golgi and others.

Laveran, as said earlier, was a never-say-die type of scientist and his consistently good relationship with scientists worldwide paid dividends before long. Starting in 1885, Laveran's discovery soon spread to America, Great Britain, and India, and by the end of 1890 it was said that acceptance had been universally granted. The technique, along with a lot of malaria-related knowledge, would in time pass on to Dr. Patrick Manson in Great Britain who, in return, descended the wealth of facts to his protege, Dr. Ronald Ross. Both Grassi and Ross would later take their own time to accept or reject these wonderful discoveries, but never to sit quite even a bit moment when it came to their own research findings. The stage for the gruesome "First Mosquito World War" had just been set in.

1.7.4 Mosquito Pinned Down in a Battle for Priority and Primacy

Dr. Ronald Ross, who had been ushered into the medical profession against his intention to become a writer or a musician, did not even imagine ever pursuing research in his early life in British Indian Medical Service, leaving aside the exemplary Nobel-Prize-winning mosquito-malaria research, with which of course he did beat everybody else with his unmatched discipline and focused attention. The transformation was not subtle. In fact, Ross seemed to have never come to terms with the extreme tropical climate of India, rife with blood-sucking mosquitoes everywhere almost year-round. It was after all his father's plan, not his, to serve in the British Medical Services in India, where the senior Ross was lucky to have risen himself to the rank of a full General. Ronald Ross, a good son, obeyed his father to this extent, but felt eternally distant from medicine and research, and quite closer to matters deeply imprinted on his mind – the music, poem, and novel writing (he did succeed in publishing a romance novel, *The Spirit in the Storm*, in London in 1896) and, of course, mathematics. In his initial days Ross had felt tormented by both the dusty heat of Indian plains (he once even got a brief posting to Nasirabad close to the eastern edge of the Thar Desert in the erstwhile Rajaputana Estate) and the callous behavior of his seniors in sending him quite frequently on postings to less cozy and recreational, and monetarily least attractive places still mostly as a forced bachelor having had to leave the family behind either in the native Great Britain or the previous posting.

Consequently, he silently confronted his military bosses on one or another pretext, and, out of agony and exasperation, even threatened to quit the military job and return to England. Frustration enveloped Ross – *hook, line, and sinker*! He had plenty of time to golf away and compose music, but the impact of continued humiliation in service kept constantly blasting off his ever-alert mind to do something extraordinary that nobody in the world had earlier done and doing what he could rise to the top honors of the world. He was eternally restless but did not know where to go or navigate his life to reach his objective. Greatly lost in the labyrinthine medical disciplines and research, he decided to coolly think meticulously within the precincts of his medical assignments and carefully narrowed down to malaria research which indeed was the hottest topic of the day. Ross immediately jumped with a burning desire in his eyes. He thought he had got the objective, but still did not know how to strike the bull's eye! The problem was, however, that Dr. Ross, with almost total ignorance of the literature, knew very little about malaria, had never seen a parasite, did not know the necessary techniques to examine malaria parasites, despite the fact that Laveran and, subsequently, the Italians and Americans have demonstrated microscopically the presence of various stages of *Plasmodium* species throughout the second half of the 1880s, and could not at all distinguish one mosquito from another. He did overcome most of these one by one as the inexorable laws of time and destiny had something in store for him that humanity would be later proud of. As if providentially designed to reach his objective, he first met the great Dr. Patrick Manson under whose careful tutelage Ross rose to enormous heights of success, winning finally the race to pinpoint *Anopheles* mosquito's role in malaria transmission during 1897 and, more rewardingly for Ross, the Nobel Prize in 1902 (awarded in 1904). His discovery of the role of *Anopheles* mosquito (which he ignorantly referred to as "dapple wing") in malaria transmission and later a complete life cycle of the bird malaria parasite in an avian using a natural culicine vector mosquito species are a classical saga of exemplary determination, perseverance, courage, patience, and ambitiously focused application.

At a time when Ronald Ross was groping in the dark and ceaselessly endeavoring to grapple with the necessary microbiological and microscopic techniques, the Italians, particularly the genius of Giovanni Battista Grassi and his team of as celebrated scientists as Drs Marchiafava, Celli, and Golgi, were already at the problem of solving the malaria mystery through implication of mosquito. It appears that both Ross and Grassi knew about each other's moves in demystifying the hidden secrets of the malaria cycle, but none really showed any signs of this awareness. Ross, true to his military training was meticulous in keeping his day-to-day account of mosquito-malaria experiments as well as bringing the findings to the notice of all through Manson in England and quick dispatches to the scientific journals of international fame. Thus, not only his phenomenal express reporting of the pigmented cells in two mosquitoes fed on malaria patients (Ross, 1897), but even the publication of his proof of mosquito's involvement in an avian model did not take long. Within a few months, he ensured its publication following the revelation of pigmented bodies in *Anopheles* mosquito (Ross, 1898a,b). He carefully planned with every possible, logical means in the international race for staking the claim for the unique discovery. It coincided knowingly or unknowingly with the news of the institution of the Nobel Prizes, including one in Physiology and Medicine, following the death of their winner, the dynamite king Alfred Nobel, in 1896, which, if won, meant unimaginably colossal prospects ahead both monetarily and from the honors point of view.

Around the same time, and possibly a bit earlier to Ross's revelation of August 20th, 1897, Professor Grassi, who had been working relentlessly on the mosquito-malaria connection for several months, then immediately unaware of Ross's discovery, true to his reputation and genius patiently waited for the last iota of truth to emerge to fool-proof his own "first" discovery of the same matter as Ross's. Grassi's first publication from his observations taken since mid-July, 1898 through summer 1899 came to light in October, 1899 and then in December, 1899, wherein he announced, referring to his associate colleagues (Celli, Bignami, Dionisi and Bastianelli), having arrived at a definitive solution of the question of mosquito-malaria riddle and that the victory was completely theirs. The rush to publish the results and air the news was unusual and inclined to beat Robert Koch, who too had been advocating his own mosquito-malaria theory since as early as 1883. Ironically, it was Ross, and not Koch, who was hit. Ross felt treacherously ignored by the Italians for his "first and original" discovery. Italians' superior techniques and work infrastructure were always highly revered in scientific circles. Even Ross was full of praise for both Grassi's and Bignami's ingenuity in medical research and in identifying the vector and confirming the mosquito cycle of the *Plasmodium*. In return, Grassi returned courtesies full of applause to Ross for his efforts.

These pleasantries between Grassi and Ross were only ephemeral and all the sense of mutual respectability effervesced with the initial news aired about Ross's having brought first his claim for the primacy of the discovery of *Anopheles* mosquito being invariably responsible for transmitting malaria. The competition readily reached its lowest and ugliest point with the war for credit of priority and primacy of the discovery. Ross charged Grassi's work of several years as childish which he could have accomplished in a few weeks. Grassi, who was rightly considered by many to be the preeminent medical zoologist of his time, shot back calling Ross as an amateur and an interloper. Ross was stung to an instant wrath. He was a person who, once angered, was never to pull back. Dr. Ronald Ross and Professor Giovanni Battista Grassi envied each other for the rest of their lives, broaching over priority for various mosquito malaria-related researches. Ross, feeling eternally insecure about his place on various discoveries, was more vociferous and hostile, having spared no opportunity to attack Grassi and inflating his own achievements, Ross engaged in a lifelong polemic to quash all Italian pretensions and reassert with ever-diminishing qualifications that he and only he had demonstrated the transmission of malaria by mosquitoes.

Grassi, who in the course of the continuing controversy on primacy lost the association of two of his best colleagues, Bignami and Bastianelli, the chief architects of Grassi's mosquito-malaria work due mostly to his growing arrogant and egoistical contentions, battled Ross ceaselessly with matching intensity. Ross's obsession with garnering every possible laurel was so intense that he did not even hesitate to hurl insinuations on his mentor, Dr. Patrick Manson, his science and scientific integrity, just because Manson, truly a generous, impartial and honest grand man of tropical disease sciences, indicated to credit Grassi for his outstanding contributions in malaria research. Ross, in spite of having lost many friends in the course of the never-ending battle with Grassi, did have many others to support his cause and value. One such well-wisher was Dr. Robert Koch. Haunted by Grassi's ignominious treatment of his Italian visit in the late 1890s and later surreptitious correspondence, Koch had turned into an eternally implacable foe of Grassi. Koch's persistently vehement attacks on Grassi's scientific integrity, climaxing into his violent opposition to the Italian genius against consideration for awarding the 1902 Nobel Prize jointly to both Grassi and Ross, in all likelihood cost the former the fortune. Ross won the Nobel Prize, besides the prestigious knighthood a few years later. Tyagi et al. (2020) have claimed that Dr. Ronald Ross was the first Indian to win the Nobel Prize.

Whatever happened between these two geniuses, humanity greatly benefited from their tireless inquiries into the mystique mosquito-malaria relationship. What benefits we reap today in saving precious lives to the savage attacks of the insidious malarial disease is due entirely to their, and other similarly devoted scientists', tireless and focused hard work.

1.8 SOME EXEMPLARY GLOBAL AND INDIAN MOSQUITO CONTROL CAMPAIGNS

1.8.1 When Mosquito Was Defeated by Man

Mosquitoes have been despised for their blood-sucking habits for times immemorial, and besides being a great source of nuisance and disturbance all over the world, their involvement in transmitting several deadly and/or debilitating infections has branded them to be man's deadliest foe. Many significant attempts have been executed in the Americas, Africa, and Asia to wipe out the mosquito which, barring a fewer examples, remained invincible (Ross, 1903; Spielman and D'Antonio, 2001; Tyagi, 2004, 2020). From humankind's point of view, "he" overpowered "her" in some encounters, but not without tasting a pinch of defeat or frustration in several others. Notwithstanding victory or defeat each of these ventures was a saga of man's enormous courage and determination to wipe off the villainous mosquito to inherit a life without any bite or infection.

Even though there are a large number of instances where man had clearly beaten both the mosquito and diseases transmitted by it such as dengue fever: (i) the vertically structured paramilitary hemispheric eradication campaign by the Pan American Sanitary Board (1946–70), (ii) rigorous top-down military like vector (*Aedes aegypti*) control operation in Cuba that was based on intensified insecticidal treatment followed by reduction of available larval habitats (1981), and lastly the Singapore model of vector (*Aedes aegypti* and *Ae. albopictus*) and dengue control, for reasons of comprehension, only certain famous events are accounted for in the following pages.

1.8.1.1 Yellow Fever at Panama Canal

This was the daunting task that the Yellow Fever Commission scientists (Reed, Caroll, Agramonte, and Lazear) faced in Cuba more than 100 years ago. How could these four men possibly stop this dreaded killer? They had spent their first full month, July 1900, in unsuccessful attempts to prove a relationship between yellow fever and the bacteria, *Bacillus*

icteriodes. Because there was little hope of identifying the etiologic agent, the commission members decided to focus their efforts on determining how yellow fever was spread. They found mounting evidence against the commonly held fomite theory of transmission. Among the several other theories, Dr. Jesse Lazear was most intrigued by the possibility of a mosquito vector. The mosquito theory of transmission of yellow fever has been around for many years. Possibly the first to discuss it was Dr. Josiah Nott of Mobile, Ala, in 1848. It had been proposed again by Dr. Carlos Finlay of Havana in 1881. But try as he might, Finlay had been unable to prove his theory in almost 100 experiments on human subjects. Many in the medical field belittled him because of his passion for his mosquito theory, but the discovery in 1897 by Dr. Ronald Ross of the British Indian Medical Service that malaria was transmitted by the mosquito gave added weight to this possibility.

The tale of the Panama Canal, or "the big ditch," as it was then locally called, is a passionate one, embodying within its innumerable folds dramas of every kind – the joy, the sorrow, the disappointment, rewards, and honors, and even comeuppance and death. It is the story of an impassioned and persistent struggle against the forces of nature, marked with the blood, sweat, and tears of tens of thousands of laborers. Engineering challenges, political turmoil, and bureaucratic headaches each added its own seasoning to an already spicy and exciting chapter in history exemplifying man's indefatigable ventures to command victory over the seemingly invincible looking land mass, and, of course, the real villain – the yellow fever disease and its vampish carrier, *Aedes aegypti*.

Cutting a nearly 50-mile swath through mangrove swamp and mountains, the Panama Canal was considered the most daring and ambitious engineering feat of its time. But before completion of "the big ditch" in 1914, tens of thousands would die, international tempers would flare, and the biggest villain had to be held at bay. There were many starts and stops before scientists, sanitary engineers, and canal builders would come to agree that the biggest enemy was actually so small as to pass unnoticed, yet it was so powerful that it very nearly destroyed the centuries-old dream of a passage connecting the world's two greatest oceans – the Atlantic and the Pacific. Before the canal was finally built, more than 30,000 people would succumb to this tiny twin-enemy, *Anopheles* spp. and its distant ally *Aedes aegypti*, the mosquitoes that carry malaria and yellow fever, respectively. The dream of opening a waterway across the isthmus of Panama dates back to 1524, when Emperor Charles V commissioned a study for constructing a canal to link the Chagres River, which flowed into the Atlantic Ocean, with the Pacific Ocean. Charles' successor, Philip II, quashed the project, justifying that the act would be against God's wish. Interest in the idea waned until the mid-nineteenth century, when gold was discovered in California, igniting man's rapacious desire to rush to the new-found golden land. This penchant for gold inevitably also engendered a rejuvenated interest in finding a quicker east-to-west route than the tedious trip around Cape Hom at the tip of South America. France, Great Britain, and the United States – the three commerce giants at the time – drew up different plans for cutting the distance. One resulted in a 50-mile railway built across the Panamanian isthmus in the 1850s. Here, railroad workers had their first run-in with what canal laborers would later face: malaria and yellow fever waged war on all comers to their tropical soil. The non-immune laborer population was particularly sickened. The French were the first to tackle canal-building in Panama.

Ferdinand de Lesseps, a great engineer who had just successfully completed the Suez Canal, linking the Mediterranean and the Red Sea, was basking in the international glory of bringing the Far East nearer to the West. He was naturally the first choice to undertake the job and, looking at the enormous financial prospects, he was only too eager to repeat the feat without having done good homework. But he soon realized that Panama was not Egypt. He was met with thick jungle vegetation, torrential rains, and floods, and oppressive heat and humidity, compounded by vastly different soil conditions than those of the arid desert. The Culebra and the Chagres, the mountain and river that barred his path, could not be overcome in any other way. Disease and death were taking a toll on those brought from France to work at the Panama Canal with such a frenzy that the site infamously earned the ghost name of a pest house. Once the digging began, the mosquitoes were driven out of their habitats, bringing them in direct contact with canal workers. Yellow fever and malaria killed an estimated 3,500 people a year during 1881–1889, forcing de Lesseps to terminate the attempt precociously, at huge costs of character assassination, charges of financial bungling, and amassing public wealth, and of course international repudiation about his engineering skills. The malaria and dengue mosquitoes were obviously the main villains in driving out the genius of de Lessips, who, following felonious charges of fund-swindling and failure to complete the task of building the canal, was abominably tried out, with comeuppance settled – heavy fines and imprisonment – by the French Government. De Lesseps died on December 7th, 1894, in his ninetieth year. Mosquito-triggered malaria and yellow fever had beaten man in their first major encounter!

At the time, widespread disagreement existed about what caused these epidemics. One popular belief was the "marsh miasma" theory, which held that the invisible mists and vapors given off by swamps and decaying organic matter produced yellow fever and malaria. Even men of science became locked in fierce disagreement. Debate raged between those who believed that these pestilences were communicable diseases caused by some infectious agent and spread by materials contaminated by patients, and those who clung passionately to the idea that they were air- or waterborne. These differences in opinion gave rise to wildly different approaches to prevention and treatment. For instance, during a yellow fever epidemic in Philadelphia in 1793, Dr. Benjamin Rush, a well-known professor and physician, attempted to cure his patients by bleeding them to get rid of the environmental contaminants that he believed had invaded their bodies. Around the same time, in Washington, D.C., Dr. Albert King hypothesized that mosquitoes were the real culprits, but he became the object of much ridicule when he proposed that a gigantic mesh screen be erected around the U.S. capital to keep the insects out. Another hundred years would pass before scientists realized that King was on the right track.

In 1880, as de Lesseps was *en route* to Panama to begin work on the canal, Dr. Charles Alphonse Laveran, a French physician working in the then-French territory of Algeria, discovered *Plasmodium*, the organism that causes malaria, in the blood of a patient. The news came too late for de Lesseps. On the other hand, it might not have mattered because de Lesseps did not believe in the mosquito theory. The next year, a Cuban physician, Dr. Carlos Finlay, made a discovery that would revolutionize disease control and ultimately make possible the construction of the Panama Canal. He identified the *Aedes aegypti* mosquito as the carrier of yellow fever. Finlay's breakthrough had glitches, however. He had correctly surmised that the mosquito was spreading the disease by biting people previously infected with the disease and then transmitting the disease agent to other people through its bite. But when he attempted to prove his theory by having infected mosquitoes inoculate healthy volunteers with the disease, he failed.

What Finlay did not know was that the virus needed to incubate for approximately 12 days inside the insect to develop. He was attempting to inoculate his volunteers with the virus using mosquitoes that had only been infected a few hours before. More years would pass before these missing pieces were added to decipher the enigma. His discovery got a boost when Sir Ronald Ross, a British physician in India, identified a malaria parasite in the stomach of an *Anopheles* mosquito. Finlay formulated a detailed plan for combating both yellow fever and malaria.

In 1898, the United States declared war on Spain and colonized Cuba. During the hostilities, it soon became apparent to the U.S. military that yellow fever was claiming more lives than the actual war itself. True to its name, the diseased person turned yellow and, in most instances, readied to follow a nearly certain death. It had taken its toll on thousands of U.S. Army recruits during the brief Spanish-American War, and it was thought that death due to yellow fever among American troops was much more common than that from bullets. Yellow fever had been the scourge of the Caribbean and the coastal and southern United States for centuries. Despite years of efforts by the best medical experts, no one knew how it was spread, why some were stricken, and other family, friends, or neighbors simply passed over. It appeared that no defense, no barrier, and no precaution could stop its deadly progression. In response, the U.S. Government dispatched a medical commission, whose members comprised Walter Reed, James Carroll, A. Agramonte and Lazear, to find a way of halting the epidemic. The commission mounted a vigorous campaign to rout the disease, but since Reed did not put much stock in Finlay's mosquito theory, efforts focused initially on ridding the Cuban capital of the filth and contamination presumed to cause the disease. But when the yellow fever death toll rose despite a massive clean-up effort, Reed decided to test Finlay's theory. Reed constructed two small buildings through which air could circulate freely (to first show that the disease was not airborne). In one building he had volunteers sleep for 20 nights in the foul bedsheets and garments of yellow fever patients. In the other building, volunteers slept next to infected mosquitoes but were separated by screens. In this experiment, Reed did not use recently infected mosquitoes like Finlay had, but rather those mosquitoes that had been infected for some time. Then one volunteer was exposed to infected mosquitoes for three successive nights. This volunteer was the only one to develop yellow fever. Proof at last was conclusive: *Aedes aegypti* was the culprit. Reed had yielded the knowledge that would free Havana from yellow fever. Despite this proof, many in the medical community, including General Gorgas, remained unconvinced of Finlay's theory. But he set about eliminating the mosquito and its breeding places in Cuba where he succeeded in virtually eradicating yellow fever within one year, forcing him to agree with Finlay's mosquito theory. Gorgas then proceeded to destroy *Anopheles* species' habitats and scored similarly dramatic success against malaria. The Havana anti-mosquito campaigns, the first large-scale systematic public health effort of its kind anywhere in the world by that time, not only provided a formula for controlling epidemics elsewhere, but they also served as an early model for effective health promotion and community involvement.

This victory over yellow fever laid the foundation for yet another significant and mammoth task of cleaning up the area of the future Panama Canal. However, there were many serious political constraints. At the time, Panama still belonged to Columbia. A bitter civil war ensued from 1900 until 1903 when, with U.S. support, it finally won its independence. Meanwhile, a second French company had bought the canal franchise in 1894, largely to keep the project alive. But they sold their holdings to the U.S. for $40 million in 1904, where after plans to resume construction began. Taking stock of the huge numbers of workers the French had lost to malaria, yellow fever, and other ills related to Panama's swampy environment, the newly formed Panama Canal Commission decided that the first step was to tackle the sanitary situation in the Canal Zone. President Theodore Roosevelt sent a group of experts that included Gorgas, Reed, and others who had taken part in the Havana triumph. Since they knew now who the enemy was, they made plans to wage an all-out mosquito war, not without being shown a contretemps by the bureaucrats who dismissed the mosquito theory – *hook, line, and sinker!*

In the meantime, a veritable health crisis was brewing. As hundreds of non-immune workers reached Panama, the number of yellow fever cases and deaths began to grow steadily, raising the specter of an uncontrolled epidemic. By June 1905, fear of the disease had grown so intense that three-quarters of the canal workforce, including the chief engineer, had fled Panama. It appeared that, once again, *Aedes aegypti* would foil attempts to build a canal. Faced with the prospect of failure, President Roosevelt fired all the commissioners, appointed new ones, and named John F. Stevens the new chief engineer. Stevens believed in Gorgas and gave him free rein.

Without the bureaucracy, Gorgas was able to get the isthmus rid of the mosquitoes and yellow fever. His group began by eliminating *Aedes aegypti* breeding places. At the same time, they carried out a sweeping fumigation campaign. In Havana, which had 200,000 inhabitants, only the houses of yellow fever patients and neighboring houses had been fumigated. Panama City, which had only 20,000 inhabitants, and Colon, which had a much smaller population, were fumigated house by house. For the first time in the history of the isthmus, yellow fever was stamped out and the victory was truly fantastic. Yet, it was eclipsed by an even bigger threat that remained – Panama's most serious problem was malaria, which most believed to have killed even more people than yellow fever.

The two diseases differed in several ways. Yellow fever confers immunity, but malaria does not. Repeated attacks are frequent, and people who have been infected remain capable for years of transmitting the disease to others through the bite of an anopheline mosquito. Gorgas estimated that most of the 12,000 people living along the Canal Zone carried the infection. The mosquito carriers also had very different habits. Whereas *Aedes aegypti* laid its eggs in fresh, still water, *Anopheles* mosquitoes bred in swamps which were prodigious in Panama. So, Gorgas and his men researched and, noting that *Anopheles* mosquitoes could fly only short distances, ordered the drainage or destruction of all potential breeding places located close to human dwellings which resulted in their elimination. Obligatorily doors and windows were screened to keep the adult mosquitoes out, and underbrush that might shelter them was cut. In addition, the antimalarial drug quinine was dispensed to all inhabitants as a preventive measure. The results were simply spectacular: the number of malaria deaths fell to zero and the number of cases among canal workers also dropped dramatically. The isthmus had been freed from the grip of yellow fever and malaria. At last, the canal could be completed after a long peregrination of time spanning over 34 years, on August 15, 1914!

Thus, the making of the Panama Canal is also a saga of the fierce battle between humans and mosquitoes, in which the former finally edged out the latter by freeing the isthmus from both malaria and yellow fever, largely due to the invaluable public health contributions, directly or indirectly, of indefatigable men like Laveran, Ross, Finlay, Gorgas, and Reed, but never to forget in the least men of great courage like Carroll, Agramonte, and Lazear, the last of whom even sacrificed his life for a noble cause of deciphering a centuries-old puzzle – the cause of transmission of yellow fever. The malaria and yellow fever campaigns carried out in Panama provided a working blueprint for the control of these diseases that was subsequently applied around the world.

1.8.1.2 Mosquito/Malaria Control in Kheda District (Gujarat State), India

At the beginning of the first half of the 20th century, there were an estimated 100,000,000 cases of malaria each year in India – roughly 25% of the population. India used to bear a huge disease burden due to malaria (Sinton, 1935). The insecticide DDT, introduced in health programs in 1950s, was dramatically successful in controlling malaria vectors and the incidence plummeted to 100,000 cases per year in the mid-1960s. The momentum of the then National Malaria Eradication Program (NMEP), which was first rechristened as the National Anti-Malaria Program in the late 1990s and more recently subsumed along with other similar national programs into a newly erected integrated scheme called the "National Vector Borne Disease Control Program" (NVBDCP) by a government decree dated December 2nd, 2003, could not be, however, sustained due to multiple reasons. These included the vector resistance to insecticides, decreasing community acceptance of household spraying and the high cost of alternative insecticides, compounded by various financial and administrative constraints. At the same time, malaria incidence rose rapidly to around two million per year, with the *Plasmodium falciparum* proportion rising steadily, notwithstanding.

Recognizing plausible drawbacks in controlling malaria vectors with insecticide spraying, the Indian Government through the Indian Council of Medical Research (ICMR/Malaria Research Center (now National Institute of Malaria Research) launched, under the able stewardship of Dr. V.P. Sharma, Director of Malaria Research Center in Delhi, a malaria control program with bio-environmental approaches to vector control (including engineering modifications, environmental manipulation involving local communities and the use of larvivorous fish and other predators to reduce mosquito abundance) at a dozen sites throughout India in a variety of urban and rural locations, of which Kheda district was most significant.

Kheda is basically a rural site, typical of much of the high malaria risk areas, located close to the industrial capital of Gujarat State in western India. The project was started in mid-1983 and soon various details were gathered about mapping, surveys, malaria stratification, site selection, fish production, requirement, and procurement of equipment. The project carried out work in seven villages with a population of 25,000 but expanded yearly to cover a population of 700,000 by 1987. The project undertook several field-level activities, like (a) environmental management to reduce mosquito breeding sites by minor engineering works, drainage, soak pits and application of polystyrene beads to cover water sources, (b) biological control of mosquito larvae through the introduction of larvivorous fish, (c) education of the community concerning malaria, its transmission and control, (d) surveillance and chemotherapy to reduce transmission of malaria, and (e) other environmental improvements and income generating activities (e.g., social forestry, fish breeding, the introduction of smokeless stoves). In addition, the project also undertook a considerable amount of research, management, supervision, and training activities.

A substantial case reduction was achieved in environmental control area villages than those that were conventionally insecticide-sprayed by the NMEP, although the annual parasite incidence (API) remained similar in both types of localities (*cf.* Tyagi, 2004). It was effectively demonstrated that environmental control measures have an edge over the insecticide-spraying methodology. The main vector of malaria in the Kheda district is *Anopheles culicifacies* which has developed resistance against DDT, dieldrin/HCH, and malathion. Therefore, entire emphasis was laid on reducing populations of this vector through various environmental measures.

There are obviously a variety of associated benefits from the environmental intervention. A few examples are worth quoting here; (i) the introduction of fish to reduce larvae and mosquito densities has provided a lucrative additional benefit in the form of fish for consumption. The gross income from the ponds during 1986–1987 was Rs. 78,000 and was retained within the villages for their development fund, (ii) tree plantations in wasteland and marshy areas to reduce malaria breeding sites have encouraged the development of village nurseries which are expected to generate income either during their lifetime or on felling (in one village alone the tree yield is estimated to be worth Rs. 200,000). The net cost per head for environmental control is thus clearly less than Rs. 4.5. This project was obviously a great success, igniting a fresh hope to abate both mosquito population and the malarial disease they transmit, in a true "Gandhian" way – by practicing safe environmental measures. Dr. V.P. Sharma, the dynamic crusader of this unique experiment, was a clear victor! Later, during the late 1980s/early 1990s, Dr. V.P. Sharma had done a similar wonder in Chennai City where *An. stephensi* was the main vector for urban malaria.

1.8.1.3 Mosquito/Bancroftian Filariasis Control in Pondicherry, India

Most parts of the Coromandel coastal region along the southeast peninsular India are endemic for lymphatic filariasis caused by *Wuchereria bancrofti*, a nematode belonging to family Filariidae, and transmitted to man by the infective bites of hematophagous mosquitoes. *Culex quinquefasciatus* is the major vector responsible for the transmission of bancroftian filariasis in India. As such, the areas in and around Pondicherry Union Territory have also been affected badly by both the disease and mosquito infestation, and in order to develop a model for their containment, the ICMR through its institute known as Vector Control Research Center, Pondicherry had undertaken a filariasis control demonstration project, better known as the Pondicherry Project, through an integrated disease vector control approach, under the pioneering leadership of the Center's Director, Dr. P.K. Rajagopalan, ably assisted by Dr. P.K. Das. Project's Assistant Director (and later the Director, VCRC) and a team of dedicated scientists and technicians. Dr. B.K. Tyagi was associated with this project between 1984 and 1986, and made many a field-oriented exploration including high-level meetings with Chief Secretary and Health Secretary, Pondicherry.

Pondicherry (now Puducherry), a former French dominion with a total area of 59.38 sq.km and population of 272,250 (at 1981 census level), lies between 11.45° to 12.15° latitude north and 79.35° to 80.00° longitude east on the Coromandel coast, about 165 km south of Chennai (old name Madras). Its climate is dissymmetric type, with high humidity, rainy days, and a relatively cool season. Despite general hygiene in the urban agglomeration of Pondicherry being of a relatively high order, it nevertheless bore enormous mosquitogenic conditions largely due to unplanned urban growth, choking or obliterating in the process of the existing storm-water drainage canal system of the township. The major contributory breeding sites, of course were cesspits, cesspools, and choked drains, etc. Over the years Pondicherry presented a grave situation of the disease prevalence. In 1957, of about 12% people surveyed, out of a total population of 59,835 resident in the town, for various epidemiologic parameters the *mf* rate was found to be as high as 10.3%, with the disease rate, infection rate, and infectivity rate as 27%, 3.1%, and 1.4%, respectively. In 1980, according to the National Filariasis Control Project, the *mf* rate, disease rate, infection rate, and infectivity rate obviously declined to 4.89%, 0.62%, 1.20%, and 0.24%. With both *mf* (aver. 8.59%) and disease rate (45.81%) being highest in the core urban area, the vector mosquito, *Cx. quinquefasciatus*, also was found to be the highest per man-hour density (3224). The control of disease was brought about mainly by reducing the mosquito population through an integrated vector management strategy comprising (i) inter-sectoral cooperation, (ii) community participation, and (iii) integrated vector control by source reduction and allied methods.

A detailed plan of activities was drawn which was subjected to evaluation and re-evaluation periodically to assess the vector preponderance in each classified area. Amidst a control program utilizing extensively only environmental and biological methods, insecticides were also deployed where inevitably required. At the conclusion of the project, in 1985, the *mf* rate dropped in Pondicherry from 8.41% in 1981 to 6.34% in 1985, and to 6.35 in 1986 (showing a reduction by 24.51%) from 8.93 in 1981 in the VCRC control demonstration area, which was comparable to the National Filariasis Control Program (NFCP) zone figures of 6.33 in 1986 and 8.98 in 1981.

Of great interest was the observation that *mf* rate in the age group 0–5 yrs, was 2.22% in 1981 and 1.76% in 1986 in the NFCP area, but dropped to 0.35% in 1986 in the VCRC control demonstration area. The disease rate in the VCRC area was found to be 2.26% in females and 13.67% in males, with older age group people manifesting more disease symptoms. Both the diseases and *mf* rate had clearly been shown to decline. The ingenious leadership of Dr. P.K. Rajagopalan, ably assisted by his team, especially Dr. P.K. Das, emphatically demonstrated that the filariasis vector mosquito, *Cx. quinquefasciatus*, could be substantially controlled through known, simple, and yet strictly implemented environmental measures and with minimum use of insecticides. Dr. B.K. Tyagi, being a part of the project team, witnessed the major part of the development of this project between 1984 and 1986. Victory in the end was surely theirs – Dr. Rajagopalan and his team.

1.8.1.4 Control of Lymphatic Filariasis in Tamil Nadu, South India

India has been suffering from the scourge of lymphatic filariasis since time immemorial and until the 1990s was one of the leading countries in reporting the highest incidence of the infection (*Wuchereria bancrofti*) throughout the world (Tyagi, 2018). The World Health Assembly in 1997 had targeted the elimination of lymphatic filariasis (LF) by 2020 (which has now once more shifted to 2030). The goal for India was originally set for 2015, to be achieved by annual single-dose mass drug administration (MDA). The country needed a model which could be replicated all over the affected parts of the country. The Center for Research in Medical Entomology, where I was Director in-Charge between 2005 and 2015, came out with a very successful model (Tyagi, 2014). Since the control/elimination of LF was, under policy decision, based on an anti-filariasis drug, diethylcarbamazine citrate (DEC), and the community acceptance of the drug as such was not encouraging, it was necessitated to demonstrate if the community acceptance and use of this drug, with or without another anti-helminthic, albendazole, could be escalated or not. The Center carried out a long-duration experiment in six villages of Trikoilur and proved the point that an annual single dose of DEC+-albendazole could reduce the microfilaria (*mf*) rate to less than 1% and the corresponding antigenaemia to less than 0.1% (Rajendran et al., 2004) (Figure 1.7).

Sunish et al. (2006) offered a new insight that vector control could keep the incidence under check in case the MDA could not be implemented on time, since vector control was able to reduce the Transmission Intensity Index (TII) in vectors to zero. This (VC) was also additionally able to impact the antigenemia rate substantially which indeed was a good sign. The Trikoilur Project (Tyagi, 2014) was a highly successful and path-breaking masterpiece and a model in the field of LF elimination through vector control.

Still 100% control/elimination was not being achieved merely with the anti-filarial drug (+albendazole), therefore, due emphasis was laid on the vector control to demonstrate for the first time in the country if the vector control (VC) could be used as an adjunct to DEC+albendazole in our fight against LF (Table 1.2).

FIGURE 1.7 Map showing six filaria endemic study villages in Tirukoilur, south India.

Source: Sunish et al., 2006.

TABLE 1.2
Entomological Parameters Estimated from the Indoor Resting *Culex quinquefasciatus* in Tirukoilur (India)

Treatment in Village	Period	Mosquitoes Caught Per Worker Hour	Number Dissected	Number Infected (%) with 95% CI[a]	Number Infected (%) with 95% CI	Mean Number of Third-Stage Larvae (L^3)	Transmission Intensity Index[b]
Mass drug administration (MDA) alone	July 1999–March 2002	23.70	2108	44 (2.09)[c] [1.6–2.8]	9 (0.43)[d] [0.02–0.8]	3.44	0.3486
	April 2002–August 2005	39.15	3073	43 (1.40)[c] [1.0–1.9]	6 (0.20)[d] [0.1–1.9]	4.33	0.3312
MDA with vector control	July 1999–March 2002	1.49	373	6 (1.61)[c] [0.7–3.5]	2 (0.54)[d] [0.1–1.9]	1.00	0.0080
	April 2002–August 2005	0.83	354	3 (0.85)[c] [0.3–2.5]	0 (0)[d] [1.1]	0	0

Source: Sunish et al. (2006).

[a] 95% binomial confidence interval.
[b] Transmission intensity index = mosquitoes caught per man-hour × proportion infective × mean L3.
[c] No significant difference among % infected (based on c2 or Fisher's exact test).
[d] No significant difference among % infective (based on c2 or Fisher's exact test).

1.8.1.5 The Klang and Port Swettenham Project

Almost parallel to the infamous Mian Mir experiment of mosquito and malaria control in India at the beginning of the early 20th century, malaria control activities were also initiated in Malaya in 1901, aided by the new information on the role of mosquitoes in the transmission of malaria (*cf.* Konradsen et al., 2004). The town of Klang, harboring 3,500 people, was located in hilly terrain on the Klang River in one of the most important districts of rubber and coffee cultivation. The town experienced an enormous malaria problem afflicting both civilian and service personnel as well as laborers who were largely incapacitated and unable to work. Anophelines were found breeding copiously in the swamps in and around the town, as well as a large number of open wells and pools along the road. The foothill lands, being rich in high water table, often created flooded conditions following rains.

Control measures aiming at bringing down the mosquito population and reducing malaria burden included (Reid and Ahmad, 1958):

(i) Dense forest vegetation and undergrowth from inside and outside the town was removed.
(ii) As many as possible swamps present in the township were filled or drained.
(iii) Large-scale drainage systems were put into use, based on the construction of concrete turf-lined inverts. This type of drain was found to be relatively inexpensive and the turf sides facilitated water inflow from subsoil.
(iv) Tidal gates were designed to prevent seawater from entering the drains.
(v) Open drains that followed the contour of the hill at its lowest point were constructed along the foothills. The idea behind this was to capture the water from the springs and surface run-off before it could accumulate on the plains outside the town.

Within a few years the entire town perimeter was drained, with the cost primarily covered by the state and, to a lesser extent, by private landowners and enterprises. After some years experiments with subsoil drainage systems were introduced with a good result in terms of improved drainage and reduced mosquito populations. Mosquito menace and malaria were also rampant further down the Klang River at Port Swettenham surrounded by mangrove swamps. Though not very old, the port immediately began to report malaria cases, with virtually 100% labor force falling sick. The infection later spread even to crews from the laborers returning to work. To abate vector mosquito breeding, embankments were raised at the edge of the port township to prevent inflow of seawater, aided by a network of surface drains to channel any tidal water. An iron pipe was attached to each drain in such a way that the intruded water would be fluxed out during low tide, but the pipe was closed during high tide. However, the mosquito menace could be partially checked because of a railway line on a raised bund, which prevented the clearing of deposited water across the bund albeit frequent culvert constructions. Mosquito breeding was prevented, therefore, by both chemical and environmental methods, and the cases were treated with quinine. All in all, it was a joint effort of several organizations working in coordination with a meticulously drawn plan. The result was spectacular, both in the Klang township and Port Swettenham, as both malaria cases and the mosquito menace were reduced substantially. This experiment proved to be highly successful and replicated in many parts of Malaysia, Singapore, and elsewhere in Asia, as well as in the United States of America.

1.8.1.6 Control of Urban Malaria and Mosquitoes in Bangalore City

The general impression one carries about Bangalore is that the salubrious city was free from malaria during a time when it earned the reputation as, "A retired army officer's first choice to settle for the rest of life" during the early part of the 20th century. The malaria situation was alarming, with a spleen rate of 23.2%, and parasite rates ranging from 6.5 to 21.2% (Sweet and Rao, 1934). *Anopheles stephensi*, breeding in several hundreds of wells, was regarded as the vector for disease. Malaria in the city was controlled by the use of fishes like *Gambusia affinis* in wells, reducing the spleen rate below 1% and parasite rates to about 5%. It is noteworthy that after a brief and benign quiescence, malaria built up gradually through the 1970s and 1980s until the infection virtually engulfed the entire and fast-developing metropolis, when not only the original vector *An. stephensi* but also the newly discovered vector *An. culicifacies* enabled transmission there (Ghosh et al., 2012).

1.8.1.7 Control of Urban Malaria in Pune City

Pune City has been notorious for malaria for a long time (Barber and Rice, 1938). The main vector was *An. culicifacies* and the secondary vector was *An. stephensi* var. *mysorensis*. Following extensive surveys and larval control in 1944–1945, both the spleen and parasite rates fell by 33% and 20% of what they were before. Later, during the 1950s DDT was brought into use to further the impact of overall control measures.

1.8.1.8 Control of Urban Malaria in Bombay City

Bombay (now Mumbai) was nearly always a malariated town, especially at the beginning of the 20th century (Bentley, 1911). For example, in 1903, about 74% were infected with a malaria parasite. *Anopheles stephensi* was the vector, breeding in all kinds of stored and potable water sources, particularly wells. Thorough laws under Municipal Act were deployed and all the wells were managed to reduce vector breeding completely, resulting in the absolute elimination of malaria from Bombay (Covell, 1928).

1.8.1.9 Control of Urban Malaria and Mosquitoes in Delhi

Delhi city was notorious for the prevalence of malaria, so much so that at the outset of the 20th century the army garrison had to be shifted to the Ridge about 3 km to north and west during the worst months (Covell, 1934). Close at the heels of the inevitable Second World War, a planned anti-malaria operation was carried out and *An. culicifacies* was confirmed as the main vector, assisted by *An. stephensi* in the disease transmission. Malaria and mosquito control was affected by the use of fishes (*Gambusia affinis*), engineering works with drains, and canals of all kinds, including straightening and lining of all nullahs, and filling of wells and ditches. After DDT became available, it was used in peripheral semi-urban areas, along with the antilarval works. The malaria problem of Delhi City could never be overcome completely but it had surely dipped a little after the advent of DDT and better vector management.

1.8.1.10 Control of Irrigation Malaria in the Sarda Canal Project

The paradox of irrigation and malaria has been well known and, in India, the Sarda Canal Project, built on the Sarda River in Banbasa near the Indo-Nepal border, is a glaring example (Clyde, 1931). The canal construction, aiming mainly to provide water for irrigation in Uttar Pradesh with a network of over a command area of 7 million acres, was accomplished between 1920 and 1929. Using environmental and insecticidal methods malaria was significantly controlled even though there was an outbreak again in 1928 under exceptional weather conditions.

1.8.1.11 Control of Irrigation Malaria in Irwin Canal Project (Now Visveswarya Canals)

In this highly noteworthy irrigation project in Mandya District, Karnataka, the main vectors were *An. culicifacies* and *An. fluviatilis*, which increased transmission of disease in irrigated lands. Control measures exercised included management of irrigation waters, rotation of crops, and use of Paris Green in addition to therapeutic intervention (Rao, 1945). It appeared that the disease cases were greatly reduced particularly with the deployment of antilarval methods. This was further improved subsequently by the pyrethrum spraying during the 1940s and the DDT in 1947–1948. One of the most encouraging examples of controlling *An. culicifacies*-induced malaria comes from the irrigated area in Pattukkottai taluk, Tanjore District of the Madras Presidency, once the taluk was free from malaria before the introduction of irrigation in 1933 (Knipe and Russell, 1942). A malaria outbreak occurred in the project area in the mid-1930s. An exemplary four-year-long work in the control of *An. culicifacies*-driven malaria was directed against the breeding of the vector in a group of villages with a population of 3,390 in an area of 7 sq.m. In this area there were 3.3 miles of main canal, 4.6 miles of branch canal, and many miles of poorly designed and ill-kept field channels. There were 27 tanks varying in size from a few hundred to several thousand square feet. There were 1,300 wells used for irrigation of groundnuts between January and June; the canal water was used for the irrigation of rice from June to January. There were 1,900 acres of irrigated land, including 175 acres of swamp and 25 acres of seepage. Countless borrow pits produced *An. culicifacies*. The malaria control measures employed in undertaking a formidable task included filling and draining, tidying of channels, the use of *Gambusia*, organic pollution of breeding places, the use of Paris Green by dustless and automatic methods, and intermittent irrigation. Some 55,000 yards of filling were used. The measures were applied step by step and with the additional aim of raising the standard of living by securing better agriculture, better roads, and better washing and bathing facilities. Noteworthy results were achieved. In three villages, the spleen rates in the malaria season of 1937 were 48, 54, and 56; in the malaria season of 1941, the spleen rates of these villages were 4, 19, and 11. In a comparable untreated village, the corresponding spleen rates were 53 and 57. The fall in the parasite rates during the same period was from 42, 50, and 49 to 0, 10, and 3; in the untreated village, it rose from 43 to 48. The total cost of the four-year program was Rs. 4.60, which is equivalent to an annual cost of Rs. 1.15 per capita. This cost was considered low.

1.8.1.12 Control of Rural Malaria in North Kanara District, Karnataka

It's a fable of malaria control that *An. fluviatilis* was eliminated in the difficult terrain of Western Ghat extending to the North Kanara District (Viswanathan, 1946, 1950). The antilarval measures, along with pyrethrum spraying, were continued for two years and moderately good results were obtained. Spleen rate reduced to 32% from 62% and parasite rates to 14% from 22% in Haliyal town within one year of operation. A good example of malaria control hails from the erstwhile Bengal-Nagpur Railway (now South East Railways), where *An. fluviatilis*, along with secondary vectors like *An. varuna*, and *An. annularis* were controlled by the use of antilarval measures such as filling of borrow pits, canalization, and use of Paris Green as well as oil (Venkat Rao, 1949). *Anopheles sundaicus* was also an important vector in certain areas, such as coastal Orissa, but its control via antilarval measures was not satisfactory. It was DDT which eventually helped eliminate malaria from the area and complete the project.

1.8.1.13 Control of Malaria in Coal Fields

Malaria in coal fields offered a very challenging situation and needed special attention in areas such as the Raniganj Coal Field (now in west Bengal) in the year 1931. Though not highly effective early intervention measures included antilarval initiatives, consisting of the application of Paris Green, oiling, de-weeding, canalization, and minor engineering works. Later these were reinforced with pyrethrum spraying to achieve moderate success. However, with the advent of DDT for adult killing, it was introduced against vectors from 1948 onwards and the rest is mere history – a great saga of victory over malaria in coal fields (Sen and Azeez, 1963).

1.8.1.14 Bangalore Mosquito Control – Master Plan

Bangalore city, once famous as a retired military officer's preferred abode with its everlasting salubrious environment and verdant fields, numerous lakes and ponds, was a virtual paradise until the mid-1950s when it slowly began to turn into a mosquito haven. This was thanks to the senseless and unplanned development of urbanization, pollution of water bodies, blocked drains, and population boom (from less than 10 lakhs in 1950 to more than 25 lakhs by 1981). Though the major nuisance was caused by *Culex quinquefasciatus* (97%), nevertheless other mosquitoes such as for example *Cx. gelidus*, *Cx vishnui* group mosquitoes, *Cx (Lutzia) fuscanus*, *Mansonia annulifera*, *Ae. aegypti*, *Ae. albopictus*, *Armegeres subalbatus*, *An. subpictus* and *An. vagus* escalated the loss of tranquility of the inhabitants. Incidentally, two of these mosquitoes, *Ae. aegypti* and Ae. albopictus, unleashed a spate of outbreaks of dengue and chikungunaya fevers in Karnataka State, including Bangalore (now Bangaluru) city. The problem of nuisance due to mosquitoes became so grave in Bangalore that Mr. Ramakrishna Hegde, Chief Minister of Karnataka State had to call a high-level meeting on August 23rd 1983, with the ICMR-Vector Control Research Center to study the problem of mosquito nuisance and prepare a Master Plan for Control of Mosquitoes in the Bangalore City. A two-year-long strenuous study in the 151.16 sq. km Bangalore City Corporation area plus an additional 496.64 sq.km area under the Urban Malaria Scheme was detailed in the document "Bangalore Mosquito Control: Master Plan" (Rajagopalan et al., 1987), the first-ever such guidance book in the history of Indian Metropolises. This document was a corroboration of Dr. Ross's well-judged dogma of public health responsibility, commitment, and execution with the absolute resolution:

> A genuine campaign does not consist merely in the formation of expert committees, the passing of ordinances ...; and the issue of wide advice to the public... it must always be permanent concern of the state, requiring careful measurement of the amount of the sickness present, a nice appreciation of measures most suitable for the locality, exact estimates of their cost compared with the cost of the disease, a well-considered organization, and, above all, a fixed determination to succeed.

Bangalore is today largely free from the mosquito menace.

1.8.2 When Mosquito Defeated Man

In comparison to the successful campaigns against mosquito, failures are rather prodigious. Yet, instead of disappointment and dejection, they had largely taught and equipped scientists for future face-off on several occasions. Any failure, if taken in the right spirit, teaches a good lesson, and make you well prepared to make the onslaught!

1.8.2.1 Freetown, Accara, and Lagos in West Africa

Freetown, Sierra Leone, the was first of Britain's West African colonies. The Freetown malaria control program had failed because Dr. Ronald Ross, who was posted there to fight malaria, underestimated the number of breeding pools and the sheer number of vectors that he was trying to control. Ross had very limited funding and the best available technology was to pour oil on the numerous breeding sites around Freetown. The problem was not only that Ross severely misjudged the severity of the number of breeding sites, but he didn't appreciate that as soon as the oil treatments stopped, breeding would begin again. Ross redoubled his efforts with increased funding from private sources and ensured the removal of all potential breeding sites, including rubbish, broken bottles, and other potential water containers. Despite these concerted efforts, the program failed and was remembered more for its impact on the Freetown's rubbish than with malaria control. Ironically, even though Stephens and Christophers had seen the program fail in Freetown, however, they both believed that malaria could be dramatically reduced at malaria-ridden Mian Mir way back in India.

1.8.2.2 Mosquito Control in Mian Mir

One of the earliest and highly challenging mosquito control campaigns was undertaken in India, or for that matter in the entire Asia, at the beginning of the 20th century. Then Mian Mir was a part of the undivided Punjab in pre-independent India and is now a suburb in Lahore in Pakistan. Malaria was particularly severe in Mian Mir and was therefore taken up as an experimental site and model for intervention programs elsewhere with equally high malaria endemicity. Mian Mir comprised a huge and sprawling army garrison, which continued to be severely hit by malaria despite extensive sanitation activities, and traditional farming areas punctuated with scattered residential localities.

The physiography of Mian Mir, though alluding toward an arid environment, included very flat terrain allowing limited natural drainage (*cf.* Konradsen et al., 2005). The soil of the region was impervious to water after 1–2 hr of rains, which would result in the formation of indiscriminate puddles, encouraging mosquito breeding if it continued. Since the land gradient was very poor, even the storm-water drains did not keep constant flow, leading to the formation of a perennial breeding ground for certain mosquitoes. Another significant feature of Mian Mir was its extensive irrigation network of shallow channels from which peasants drew water by means of Persian wheels to irrigate field crops and home gardens. Christophers (1904) has described in detail the baseline investigations carried out before initiating VC activities, which included assessment of malaria endemicity in the area, seasonal transmission dynamics of malaria, and stratification of malariogenic conditions – particularly the vector breeding sites. Besides malaria being rampant among the civilians, the infection also came down heavily on the army personnel. Thus, a special focus of the investigation was the burden of disease among the service men and the vector breeding in the vicinity of army barracks. Soon, it was revealed that *An. culicifacies*, which bred profusely in the smaller irrigation channels growing some kind of vegetation along the edges, was the vector of endemic malaria in Mian Mir, signifying the importance of special attention on types of irrigation network in the region.

Sometimes, these water courses would overflow, or even get breached, creating breeding opportunities along the channels. The poorly designed and maintained small earthen drainage canals were other important breeding sites. The practice of providing excess irrigation water on a daily basis to small gardens immediately surrounding homesteads created breeding opportunities, too. The army quarters being very close to the irrigation canals, numbering ten, were the worst affected. The Mian Mir mosquito-malaria menace was conceived as a problem of environmental degradation, therefore, most of the methods adopted to reduce vector breeding comprised repairing the banks of irrigation channels, and clearing silt from the irrigation channels to maintain good flow without allowing overflow;

(i) Reclamation of low-lying areas otherwise tending to collect water with the smallest of rains;
(ii) Removal of detritus and/or debris from the irrigation canals near the army barracks to reduce chances of pool formation;
(iii) Drying up of seasonal water pools, as much as possible;
(iv) Emptying the domestic water containers undesired for any good domestic purpose and which unnecessarily collected rainwater or otherwise for unduly long periods;
(v) Creating superficial small drainage lines from large temporary ground pools to dry them up by thinning them under the sun; and
(vi) During the most troublesome rainy period of June and July, sustained efforts were made to reduce the breeding potential of large pools formed in drainage canals either by using buckets to empty the pools or by applying kerosene. oil. Simultaneously, malaria cases were treated with thorough mass distribution of quinine.

The Mian Mir case was one of the first documented cases, resulting in a heated debate over the success achieved in reducing vector population and malaria incidence through mainly environment-based methods. Since, however, very little reduction in the adult vector population was achieved, along with a nearly indiscernible reduction in malaria cases particularly in the army residential areas, the prestigious Mian Mir campaign was branded a failure. Ross and others were critical from the outset about the improper design of the operation. Ross (1910) claimed that not many efforts were actually put into environmentally sanitate the region, besides insignificant investment into the campaign. If only the costs on sick soldiers at Mian Mir as made by the government were to be considered, then the overall expenditure for malaria prevention made up only a 1/400th proportion of this cost. The Mian Mir failure offered opportunities for many to raise their eyebrows and question the *modus operandi* as well as the expertise of the authorities in reference. It was argued that the area was never sufficiently drained and proper outlets should have been constructed from the drainage canals, possibly with pumping devices, to assist the drying of canals. The Mian Mir case was also one of the first cases resulting in a debate over the role of irrigation in the spread of malaria and how to combine the necessity of food production and livelihood with public health. It also gave rise to some of the fundamental questions for effective malaria VC, *i.e.*, how large an area should he covered to influence the protection of human settlements and how much reduction in adult population will have to be achieved to measure the impact of malaria on human beings? The subject of cost-effectiveness in environment-based control methodology became the central point of subsequent discussions and inquiry – a damnifying and marooning experience for both Dr. Christophers and Dr. Stephens, the most experienced and energetic malariologists of their time!

1.9 ROLE AND IMPACT OF INDIAN SCIENTISTS' INNOVATIVE RESEARCH ON THE GLOBAL LEVEL

Discoveries, innovations, and inventions are the core constituents of science, and mosquito science is no exception. Many of these and their protagonists faced stern opposition, but eventually braced the tape. Winston Churchill, the famed British Prime Minister during the Second World War, often used to say, "*You have an enemy? Good. That means you have stood up for something, sometime in your life*." In fact, the intensity of criticism is a direct proof of its importance; that is, the more fierce the criticism, the bigger in importance the discovery is. While the early guessers of the "mosquito-malaria connection" like Lancisi, Beauperthuy, and Rasori seemed to have had a low-keyed repudiation for their offbeat ideas during the 18th century, scientists like Laveran and Koch faced stern opposition and criticism pertaining to their initial discoveries along the same line. Crawford was unfortunately being maltreated and ostracized, while Finlay's "mosquito-yellow fever theory" was almost hijacked by another genius in the field. Manson was a rather exception, though not without a trace of fault-finding, if not direct criticism, in both his "mosquito-filariasis theory" and the undemonstrated "mosquito-malaria theory." The Grassi-Ross classical bickering, full of mud-slinging and professional assassination for each other's contributions in deciphering man-mosquito-malaria mystery, is well known worldwide. A hundred years later, the scenario has not changed a bit, and the originator of a discovery is senselessly made to face oft-repeated worthless criticism even today.

1.9.1 Dr. Ronald Ross's Discovery of Oocysts in the Midgut of a Mosquito

The most famous discovery in mosquito science ever has crowned Dr. Ronald Ross for his demonstration of the inextricable link between mosquito and malaria, paving the way for a rather easier and direct way of eradication of the disease through VC. This revelation was, however, not so simple a task as it appears, and Ross had had to endure all kinds of hardships in achieving this goal. His courage, displayed itself in silent effort and endeavor, was more truly heroic than the achievements of physical valor, which are rewarded by honors, titles, and laurels. Dr. Ross' combating spirit can be put forward than that of his life.

Dr. Ross, a medical officer in the British Indian Army since 1881, had routinely joined inter alia his regiment 19th Madras Infantry at Begumpet (now Begumpet) Lines, situated near the outflow of the Husein Saugar (now Hussain Sagar) Lake on June 18th, 1897, after arranging accommodation for his wife and children at Ootacamund (now Ooty or Udhagamandalam), where, to his shattering dismay, the latter contracted contorting whooping cough after a while. Hyderabad was exceptionally dry. He initially whiled away his spare time playing cricket and bicycling, and otherwise inditing long letters to his mentor and good guide Dr. Manson: "Company of wise men always offers good knowledge." He followed Manson's advice: "Follow the flagellum" and was almost convinced that he had seen the spectacular entity in his numerous malaria-infected human blood slides' preparations. The message about his malariological obsession had gone all around in the unit. For the next few weeks or so, nothing very notable happened in his monotonous life except that, on one particular occasion when he was challenged by one of his colleagues to give evidence for the presence of malaria parasites in human blood, he miserably failed to demonstrate the prevalence of the malaria parasite in the known malaria patient's blood under the microscope. He was crestfallen and ridiculed in silence. The humiliation penetrated deep in his heart but he valiantly fought to douse the anger within him. Over the next few days he, in a silent attempt to overcome depression, spent the time doing almost nothing in malariology and found some solace in writing poetry. Nonetheless. the continuous feeling of drifting away from his objective forced him to find his track to tread, lest the Italians, French, and Americans might take away all the hard-earned discovery. He became restless and pondered pensively, "One more effort and the thing will be done."

Ross religiously followed Dr. Manson's thesis to "follow the flagellum" and demonstrated as early as 1895 that the sexual bodies found in the human blood had unsuspectingly transformed into sphericals – distinctly different form, shape, and size – and in many the flagella were formed. Beyond that point he could not build a correct pathway of the flagellum in the stomach of the mosquito. Ironically, this latter part of Dr. Manson's thesis that the parasite somehow gets into the groundwater which, on drinking, makes its way into the human system, was never demonstrated in any of Dr. Ross's experiments since 1895. So, he thought to review afresh the whole issue of parasite biology in mosquitoes. He also understood that to prove his viewpoint he needed to have, in addition to his "A" (gray) and "B" (brindled)-type mosquitoes the third, "C" type of mosquito (the "dappled wing") which he had earlier encountered in Sigur Ghat. A major shortcoming in his seeking mosquito-related wisdom was that despite all efforts, he could neither study any book or literature nor was there any person he knew he could contact to solve his doubts. He had had to trust his own confidence, and by mid-July–July 16th, to be precise, when the monsoon had commenced – he decided to go forward, with the assistance of three natives. Before he could have encountered the "dappled wings" mosquitoes once again, he continued working with the mosquitoes in his hot office-cum-laboratory; he did not use "*punkha*," fearing his dissected specimens of mosquitoes might be blown away! The result was invariably failure after failure, and he became dizzy and even hallucinated due to overwork, leaving for his quarters almost exasperated. At the dining table as well as in bed, he continuously pondered over the issue of whether he was using the right mosquito group for the malaria result? His microscope, too, was showing signs of "fatigue" since its screws had rusted and the eyepiece was cracked. But Dr. Ross refused to stop and rather pushed forward with a reinvigorated charge to nail the answer once more. On August 15th, 1897, one of his three naive native attendants brought him a bottle containing water full of wriggling mosquito larvae, some of which were quite different from those of other two types, i.e., gray and brindled types. Again, on August 16th, his hospital assistant, an Indian native, showed him a "dappled winged mosquito" (*An. stephensi*?) resting on the wall. He bottled it carefully and, after quickly incapacitating with tobacco smoke, dissected the mosquito hoping to see parasites under the microscope. Alas! He could see nothing of that sort. Failure, once again. The hospital assistant looked at his face which painted a dark shade of dismay and dejection. "*Where there is a will, there is a way*" goes the saying as one of his attendants almost yelled with pleasure, "Sir, I know a place where many such critters can be seen." They were the same as those that emerged from the bottle, Ross quickly found out. Leaving everything as it was in the office he dashed to the place he was guided to by his hospital assistant. He was amazed and elated to see about a dozen "dappled winged mosquitoes," his "C" type (*An. stephensi*?), trying hopelessly to pass through a gauze. He collected them; his first face-off with the "dappled winged" mosquitoes ever in Secunderabad. Hussein Khan, his hospital patient with copious crescents in the blood, was immediately brought to a secluded room and made to lie down on a cot under the mosquito net. Before putting Husein on the cot, he was body-bandaged to restrict mobility. At 12.20 p.m. he squeezed from the inside the bed net 10 fully blood-fed mosquitoes, and put each one in a separate test tube. Between 12.40 and 12.50 pm, he first subjected two to systematic operation or surgery to see under the microscope any crescent-sphere flagella metamorphosis in them. He succeeded little and for a disappointingly ephemeral time. His disappointment doubled the next morning when he found two of his gorged mosquitoes had perished. For a moment he panicked and hurriedly dissected two of the remaining mosquitoes, but in vain! In the raze of excitement he dissected, or rather wasted, all the balance sans any sign of the crescent-sphere flagella metamorphosis in them. The man was so obsessed with dissection that he left no stone unturned in the process and even began dissecting fleas caught from a rat for trypanosomes. He was becoming infinitely fearless and his mind was growing stronger. Summoning his energies, he sent for his hospital assistant and the three native attendants and asked them to take him to the place where one of the latter had originally found the "dappled wing" mosquito. To his grave misfortune and sheer disappointment, they miserably failed to spot the site again! Ross sank in his chair, deeply pensive. The next moment he jumped out of his comfort zone and began to jot down details of his work to Dr. Manson.

Providentially, on the morning of August 19th, 1897, Dr. Ross once again had in his hands some adults of his "dappled wings" mosquitoes. They were the outcome the mosquito larvae earlier fetched to him in a bottle by one of his native attendants (they were his "dappled wings" type; *An. stephensi*?). He quickly set them all to feed on the feverish Hussein Khan's blood. On August 18th, although the blood was still not digested, he ventured to select one of them for dissection to see crescent-sphere flagella metamorphosis in them. Failure, once more. He became stressed as time was running out and he desperately wanted results. From his already-shrinking reserve, he picked up a third mosquito on the third day (August 19th) of blood-feeding (so that the blood in the stomach was by that time completely digested), and dissected it with utmost care. His dissection was not reasonably neat, but he found there some "peculiar vacuolated cells in stomach about 10 μmm in diameter." It was certainly a "eureka" moment for him, but he rather feared to immediately announce it lest he would pre-empt the occasion! Dr. Ross was still subscribing to Dr. Manson's thesis that mosquito feed only once in their lifetime, and then die after laying eggs on water. He hoped for better results with next dissections on the remaining samples the following day. Only three mosquitoes were left from that batch and he could hardly sleep that night.

The next day, August 20th, Dr. Ross was informed of the worst news of his life: all three ready-for-dissection mosquitoes had died the previous night. More so also because his native Indian attendants could not succeed in finding a place to collect more larvae of "dappled wings" mosquitoes. For a moment, Ross felt like he'd been evanesced – but he was determined to see the parasites in two mosquito cadavers, but failed as expected. He then took the 7th and the last one left behind from the test of his August 16th blood-fed ones (i.e., on the 4th day after blood-feeding)! Already, due to poor sleep the previous night, his eyes were sore and he yawned continuously. Nevertheless, with a phlebotomist's precision he was perfectly flawless this time; on cursory examination, he did not find anything unusual and felt betrayed by his belief that the "dappled winged mosquitoes" were the key channel to regulate parasite biology before re-entering a fresh human being. Lunchtime had passed and he continued on an empty stomach. Then just as he was once again dipping in distress, he casually opened up wide the body wall covering the stomach. "Eureka!" he cried spontaneously, at the top of his voice, since this time lay before him in no uncertain way the oval-shaped spheres with black pigment on the stomach wall, which represented one advanced stage in the development of the malarial parasite. He, as per army training, immediately jotted down and drew in his diary what he, in his moment of triumph, had just seen – the oocysts!

Of the "Oocyst," there were a total of 12, with each measuring about 12 μmm in diameter, embedded in the stomach wall and scrupulously identifiable due to the black pigment similar to the ones in the parasite crescents in the blood. He questioned himself as to how parasite flagella could move to the stomach wall, retaining the black pigment (hemoglobin) numbering about 12 to 16? He was as sure as death about those cysts on the stomach wall representing the parasite, but there was a remotely sounding minuscule doubt, too. To answer this irrefutably. he needed to demonstrate growth in these cysts with the passing time in his last specimen – the eighth mosquito. Still, as far as his concept of malaria transmission was concerned, his "Great Problem" had been thus solved, i.e., the mosquito-malaria connection was established. He did not stop at this and went ahead to sacrifice on August 21st, on the 5th day after feeding, the 8th and last survivor of the mosquito's batch fed on August 16th and witnessed the presence of more pronounced 'cells' (oocysts; 12–14 μmm; the largest being 20 μmm) numbering 21 in all each of which bearing the characteristic black pigments. The "cells" (oocysts) were therefore parasites, and as they contained the characteristic black pigment, they were almost certainly the malaria parasites growing in the mosquito tissue – the stomach wall. The coordination of his mind's eyes with hands scripted an immortal verse:

This day designing God
Hath put into my hand
A wondrous thing. And God
Be praised. At his command,
I have found thy secret deeds
Oh million-murdering Death.
I know that this little thing
A million men will save –
Oh death where is thy sting?
Thy Victory, oh grave?

Dr. Ross prepared a manuscript on his discovery, titled, "On some peculiar, pigmented cells found in two mosquitoes fed on malaria blood" which was published in *The British Medical Journal* on December 18th, 1897. Later he confirmed the finding in another type of "dappled winged mosquito" (*An. culicifacies*?).

1.9.2 Mosquito Pupae Separator

Dr. V.P. Sharma developed a new technique for the sex separation of mosquitoes at the pupal stage, for which he was awarded the Young Scientist Award.

1.9.3 Mosquito Sampler

Sampling is among the most important research activities warranted to determine a lot of various bio-ecological characteristics of mosquitoes, especially ones acting as vector species for communicable disease pathogens. In mosquito science, adult sampling, represented by per human-hour density, has long been in use, and in India, it is still practiced using a mouth-operated aspirator, though outdated, unethical, and not practiced in other countries. This aspirator device, comprising a plastic or glass tube connected to one end with a rubber tube, and separated from the latter by a thin mosquito plastic or metallic netting at the joint (so that mosquitoes sucked are held there at the end of the tube), was in the earlier days recommended by even the World Health Organization for all countries combating malaria scourge. Even though this device has in course undergone enormous modification and, in some developed countries, like the USA and UK, has also been successfully replaced by certain motorized sampling devices, however, in India the same mouth-operated device had more or less been exclusively employed in the exercise (except in some institutes where, too, the motorized samplers were obtained by purchase from abroad!) – totally unaware of the fact that the aspirator was a health hazard, besides of course not fool-proof to basic entomological vagaries generally encountered during adult sampling in field. This mouth-operated aspirator, though simple in appearance, could be operated up to nearly two feet above the erected hand (meaning a sampling person could collect adult mosquitoes everywhere but the roof, which in most urban set up is about 9–10 ft. above the ground) but with the sampler held aloft he could barely access 8 ft., considering the average human height of 5′6″. Consequently only a part of the adults actually present indoors could be sampled during a specified period of time (meaning insufficient and misleading sampling), and, above all, the risk of contracting throat infection by the sampling human (the inhaled mosquito carries a lot of dust fouled with fungi, bacteria, and or similar unwanted things). The entire sampling procedure involving a mouth-operated aspirator seemed a farce in the name of scientific investigation.

Dr. B.K. Tyagi invented a "Mechanical Mosquito Sampler" (or "Tyagi Sampler") in 1992 (Tyagi, 1993) that was highly effective in sampling all kinds of resting mosquitoes, economical, and transportable. He also obtained patent for the same (India Patent # 191635) in 2005.

1.9.4 Transfer of Technologies Based on Mosquitoes

Dr. B.K. Tyagi, during his tenure as Scientist "G" (Director) and Director in-Charge, Center for Research in Medical Entomology, Madurai, had executed a Transfer of Technology (ToT), namely, "*The JE detection in wild-caught mosquitoes using TOXO-IFA System for JEV surveillance.*" The true promoter and exponent of this methodology, accomplished in two stages, is Dr. V. Thenmozhi (*cf.* Thenmozhi et al., 2016), even though Dr. A. Gajanana could be credited to have invented it (Gajanana et al., 1997). Under this technique, wild-caught mosquitoes are separated species-wise, and while unfed mosquitoes were pooled (1–50) on the same day of collection, engorged female mosquitoes were held for 48 h for digestion of blood meals before pooling. Mosquito (only females) abundance was calculated as density (number collected per man-hour). Mosquitoes were pooled (minimum 5 specimens per pool) and stored at –80°C until processed for virus detection and isolation. Two systems were used:

(i) Antigen capture ELISA: Monoclonal antibody 6B4A-10 (reactive against all viruses in JE/WN/SLE/MVE complex) was used as capture antibody and monoclonal antibody peroxidase conjugate SLE MAB 6B6C-1 (reactive against all flaviviruses) as detector antibodies (supplied by Dr. T.F. Tsai, Centers for Disease Control and Prevention, Fort Collins Co., USA). A mosquito pool was considered ELISA positive if its optical density value was ` mean + 4 standard deviation of the six normal pools.
(ii) Insect bioassay (Toxo-IFA): Toxorhynchites splendens mosquito larvae were inoculated with ELISA positive pools intracerebrally and incubated for 7–10 days at 29°C and then tested by the indirect immunofluorescence assay (IFA) on head squeeze preparations (Toxo-IFA). Smears were tested with JEV-specific monoclonal antibody, MAB 112 (supplied by Dr. Kimura Kuroda, Tokyo Metropolitan Institute of Neurosciences, Japan) and detected by Fluorescein isothiocyanate (FITC) conjugated anti-mouse immunoglobulin (Dakoppats, Denmark).

The virus infection rate in mosquitoes was expressed as the minimum infection rate (MIR, i.e., Number of positive pools/Total number of mosquitoes tested × 1000).

The Technology Transfer (TT) was made to Dr. K. Durairaj, Addl. Director, representing the Health Directorate, Tamil Nadu government on July 25, 2010, in the presence of Dr. V.M. Katoch, Director General and Mr. Sanjeev, Dutta. Senior Financial Advisor, ICMR. This was for the first time the ICMR was transferring a technology directly to a state health directorate – for their and people's direct benefit (Figure 1.8).

1.9.5 N,N-diethylphenylacetamide (DEPA)

Mosquito-borne diseases, including malaria, filariasis, dengue, and encephalitis viruses continue to cause morbidity and mortality in humans, domestic animals, and wildlife. Arthropod "eradication" measures envisaged five decades ago failed to make any significant impact. A change to the philosophy of insect "control" or "management" has continued to evolve. The use of insect repellents seems to be one of the best available means for personal protection against annoyance and diseases associated with hematophagous organisms, especially because repellents protect subjects without having a major adverse impact on the environment. Amides, imides, esters, ethers, and poly-functional compounds are known to be good repellents. Selected petroleum oil fractions and essential oils repel *Ae. aegypti* (L.). Dimethylphthalate (DMP) was the first synthetic chemical that had considerable repellent efficacy and has been used widely as a mosquito repellent. N,N-Diethyl m-toluamide (DEET) emerged in 1954 as a superior arthropod repellent. Cost considerations arising from the unavailability of m-toluic acid for the production of DEET in India prompted an investigation into an array of alternative compounds bearing high repellence, safety, and cost-effectiveness.

FIGURE 1.8 (a) Event of Transfer of Technology by Dr. V.M. Katoch, Director General, ICMR & Secretary Department of Health Research, MOH&FW, GOI to Dr. K. Durairaj, Addl. Director, the Representative from the Health Directorate, Tamil Nadu, in the presence of (from left to right) Dr. D.V.P. Raja, Dr. V.M. Katoch, Dr. K. Durairaj, Dr. B.K. Tyagi, Dr. V. Thenmozhi, Mr. Sanjeev Dutta, and Dr. R. Paramasivan.

Source: Dr. B.K. Tyagi, pers. archive.

Consequently, this resulted in the development of a new insect repellent, N,N-diethylphenylacetamide (DEPA) (Kalyanasundaram, 1982). N,N-diethylphenylacetamide, popularly called as DEPA, has repellent activity against hematophagous insects including mosquitoes, black flies, horse flies, muscoid flies, rat fleas, and ticks, as well as land leeches and cockroaches. The efficacy of DEPA is comparable to that of N,N-diethyl-3-methylbenzamide (DEET) in both laboratory and field tests (Kandasamy, 2022). Toxicological studies indicate that DEPA is safe for human use. Different formulations as well as fabrics impregnated with DEPA are effective for personal protection from insect bites. Dr. Vijay Veer, former Director, DRDO, during his deputation in the DRDO's Defense Research Development Establishment (DRDE) at Gwalior, sweated hard to swing the use of this great repellent in favor of Indian armed personnel serving in difficult pest and vector-infested front areas, saving the nation in the process crores of rupees otherwise going out in the purchase of DEET, etc.

1.9.6 "MosTyag" – an *Aedes albopictus* Breeding Preventer for Rubber-Harvesting Cups

Tyagi et al. (2006) found out that in the Kerala State, where dengue had erupted in a severe epidemic during 2003, the vector mosquito *Aedes albopictus* bred predominantly in the latex collecting cups in the rubber plantation fields grown over nearly 1000,000 ha in the mountainous and sylvatic terrains. The mosquito biology in the sylvan environment of the rubber plantation posed several challenges in controlling it. Tyagi (2020) invented a trap called "*AlboTyag*," which maneuvered to completely check the breeding of the vector in the rubber-collecting cups even during the intermittent rains (Figure 1.9).

FIGURE 1.9 "*AlboTyag*," a simple and manually operated trap developed to prevent breeding by *Aedes albopictus* in the rubber-collecting cups tethered with the main plank to harvest rubber.

***Source*: Tyagi, 2020.**

1.10 FINAL REMARKS

Mosquitoes are ubiquitous in distribution, occurring both in tropical and temperate regions of the world except Antarctica. They transmit several different types of infections to both humans and animals. Mosquito-borne diseases in humans persist predominantly in tropical countries, although the industrialized temperate countries are also not free from many of these. While these are the deadly and incapacitating mosquito-transmitted diseases that are the focus of control or elimination in the tropical world, yet, human discomfort from mosquito bites is often the chief concern in developed nations such as the United States of America, where hundreds of millions of dollars are spent annually to get rid of them for this reason alone. Additionally, large populations of mosquitoes can cause intense irritation and extensive blood loss to livestock and wildlife, resulting in reduced productivity and even death (Foster and Walker, 2002).

BIBLIOGRAPHY

Aldrovandi, U. (1602). *De animalibvs insectis libri septem, cvm cingvlorvm iconibvs ad viuum expressis*. Apud I.B. Bellagambam, Bononiae.

Aldrovandi, U. (1638). De animalibvs insectis libri septem cvm singvlorvm iconibvs ad viuum expressis. Apud C. Ferronium, Bononiae.

Anonymous (1994). *Resource Atlas of Rajasthan*. Department of Science & Technology, Government of Rajasthan, Jaipur, 231 pp.

Barber, M.A. and Rice J.B. (1938). Malaria in Poona and its vicinity. *Journal of Malaria Institute of India*, 1, 37–55.

Barraud, P.J. (1934). *The fauna of British India, including Ceylon and Burma. Diptera: Family Culicidae; Tribes Megarhinini and Culicini*. Vol. 5, Taylor & Francis, London, 463 pp.

Becker, Nn., Petrić, D., Zgomba, M., Boase, C., Madon, M.B., Dahl, C., Kaiser, A. (2020). *Mosquitoes: Identification, Ecology and Control*. 3rd Edition. Springer Nature, 570 pp.

Bentley, C.A. (1911). *Report on malaria in Bombay*. Government Press, Bombay.

Borkent, A. and Grimaldi, D.A. (2004). The earliest fossil mosquito (Diptera: Culicidae), in Mid-Cretaceous Burmese amber. *Annals of the Entomological Society of America*, 97 (5), 882–888. https://doi.org/10.1603/0013-8746(2004)097[0882:TEFMDC]2.0.CO;2

Briggs, D.E. (2013). A mosquito's last supper reminds us not to underestimate the fossil record. *Proceedings of National Academy of Sciences USA* 110(46), 18353–4. https://doi.org/10.1073/pnas.1319306110

Christophers, S.R. (1904). Second report of the anti malarial operations at Mian Mir, 1901–1903. ScientiHc Memoirs by the Office of the Medical and Sanitary Departments of the Government of India, Calcutta, pp. 1–37.

Christophers, S.R. (1933). *The Fauna of British India, including Ceylon and Burma. Diptera: Family Culicidae; Tribe Anophelini*. Vol. 4, Taylor & Francis, London, 271 pp.

Clements A.N. and Harbach, R.E. (2018). Controversies over the scientific name of the principal mosquito vector of yellow fever virus – expediency versus validity. *Journal of Vector Ecology* 43(1): 1–14.

Clyde, D. (1931). Reports on the control of malaria during Sarda Canal construction (1920–1929). *Records of Malaria Survey of India*, 2, 49–110.

Covell, G. (1928). *Malaria in Bombay*. Government Centre Press, Bombay, 113 pp.

Covell, G. (1934). Notes on the control of mosquitoes and malaria in Delhi. *Records of Malaria Survey of India*, 4, 273–289.

Eberth, D.A. and Hamblin, A.P. (1993). Tectonic, stratigraphic, and sedimentological significance of a regional discontinuity in the Judith River Group (Belly River Wedge) of southern Alberta, Saskatchewan and northern Montana. *Canadian Journal of Earth Sciences*, 30, 174–200.

Edwards, F.W. (1921). A revision of the mosquitos [sic] of the Palaearctic Region. *Bulletin of Entomological Research*, 12, 263–351.

Edwards, F.W. (1923). Oligocene mosquitoes in the British Museum; with a summary of our present knowledge concerning fossil Culicidae. *The Quarterly journal of the Geological Society of London*, 79, 139–155.

Foster, W.A. and E.D. Walker (2002). "Mosquitoes (Culicidae)", in G. Mullen and L. Durden (eds.), *Medical and Veterinary Entomology*, Academic Press, San Diego, pp. 203–262.

Gajanana, A., Rajendran, R., Samuel, P.P., Thenmozhi, V., Tsai, T.F., Kimura-Kuroda, J., et al. (1997). Japanese encephalitis in South Arcot district, Tamil Nadu, India: a three-year longitudinal study of vector abundance and infection frequency. *Journal of Medical Entomology*, 34, 651–9.

Gardiner, B. and M. Morris (2007). *The Linnean Collections*. The Linnean Special Issue, No. 7. The Linnean Society of London, London.

Ghosh, S.K., Tiwari, S. and Ojha, V.P. (2012). A renewed way of malaria control in Karnataka, South India. *Frontiers in Physiology*, 3, 1–3. https://doi.org/10.3389/fphys.2012.00194

Greenwalt, D. E., Goreva, Y.S., Siljeström S.M. and Harbach R.E. (2013). Hemoglobin-derived porphyrins preserved in a Middle Eocene blood-engorged mosquito. *Proceedings of National Academy of Sciences USA*, 110 (46), 18496–18500. http://doi.org/10.1073/pnas.1310885110

Harbach, R.E., Dahl, C., and White, G.B. (1985). *Culex (Culex) pipiens* Linnaeus (Diptera: Culicidae): Concepts, type designations, and descriptions. *Proceedings of the Entomological Society of Washington*, 87(1), 1–24.

Harbach, R.E. (2018). *Culicipedia: species-group, genus-group and family-group names in Culicidae (Diptera)*. CABI, 378 pp.

Harrison, G. (1978). *Mosquitoes, Malaria, and Man: A History of the Hostilities since 1880*. E.P. Dutton, New York. 314 pp.

Hasselquist, F. (1757). Friedrich Hasselquists … Iter Palæstinum, eller Resa til Heliga Landet förrättad ifrån År 1749 Til 1752, med beskrifningar, rön, anmärkningar, öfver de märkvärdigaste Naturalier, på Hennes Kongl. Maj:ts befallning, utgifven af Carl Linnæus. Trykt på Lars Salvii, Stockholm.

Hasselquist, F. (1762). D. Friedrich Hasselquists … Reise nach Palästina in den Jahren von 1749 bis 1752. Auf Befehl Ihro Majestät der Königinn von Schweden herausgegeben von Carl Linnæus. Aus dem Schwedischen. [This is the German translation by T. H. Gadebusch of Hasselquist's 1757 Iter Palæstinum]. Johann Christian Koppe, Rostock. [Culex aegypti at p. 470].

Howard L.O. (2019) *Mosquitoes: How They Live; How They Carry Disease; How They Are Classified; How They May Be Destroyed*. (Reprint). Forgotten Books. 263 pp.

Howard, L.O., Dyar H.G. and Knab F. (1917). The mosquitoes of North and Central America and the West Indies. Vol. iv. Systematic description (in two parts). Part II. Publication No. 159. Carnegie Institution of Washington, Washington, DC.s

Kandasamy, C. (2022). DEPA (N, N-Diethyl Phenylacetamide): Commercial Journey of India's First Indigenous Mosquito Repellent with Novel Properties. *In*: *Genetically Modified and other Innovative Vector Control Technologies*. B.K. Tyagi (Ed.), Springer-Nature, Singapore, pp. 369–378.

Knab, F. (1916). The earliest name of the yellow fever mosquito. *Insec. Insc. Menstr.*, 4, 59–60.

Knight, K.L. and Stone, A. (1977a). *A Catalog of the Mosquitoes of the World (Diptera: Culicidae)*. Entomological Society of America, Maryland, 621 pp.

Knight, K.L. and Stone, A. (1977b). A catalog of the mosquitoes of the world (Diptera: Culicidae). 2nd edition. *The Thomas Say Foundation. Volume VI*. Entomological Society of America, College Park, Maryland.

Knipe, F.W. and Russell, P.F. (1942). A Demonstration Project in the Control of Rural Irrigation Malaria by Antilarval Measures. *Journal of the Malaria Institute of India*, 4(4), 615–631.

Konradsen, F., van der Hoek, W., Amerasinghe, F.P., Mutero, C., and Boelee, E. (2004). Engineering and malaria control: learning from the past 100 years. *Acta Tropica*, 89, 99–108.

Konradsen, F., van der Hoek, W., Amerasinghe, F.P., Mutero, C., and Boelee, E. (2005). Engineering and malaria control: learning from the past 100 years. *Papers and Abstracts for the SIMA Special Seminar at the ICID 18th International Congress on Irrigation and Drainage*, Montreal, 23 July 2005. Malaria in Irrigated Agriculture. Working Paper 47. SIMA Document 2. p.5.

Linnaeus, C. [as Caroli a Linné]. 1766–1772. Systema naturæ per regna tria naturæ, secundum classes, ordines, genera, species, cum characteribus, differentiis, synonymis, locis. Editio Duodecima, Reformata. Impensis Direct, Laurentii Salvii, Holmiæ. [12th edition, in three volumes, Culex in Volume 1, Part 2, pp. 1002–1003.]

Linnaeus, C. [as Caroli a Linné]. 1788–1793. Systema naturae per regna tria naturae, secundum classes, ordines, genera, species cum characteribus, differentiis, synonymis, locis. Editio Decima Tertia, Aucta, Reformata. Cura Jo. Frid. Gmelin. Impensis Georg. Emanuel Beer, Lipsiae. [13th edition, in three volumes; Culex in Volume 1, Part 5, pp. 2886–2888.]

Linnaeus, C. [as Caroli Linnæi]. (1735). *Caroli Linnæi, Sveci, Doctoris Medicinæ. Systema naturæ, sive, regna tria naturæ systematice proposita per classes, ordines, genera, & species*. Apud Theodorum Haak. Ex Typographia Joannis Wilhelmi de Groot, Lugduni Batavorum [Leiden, Netherlands]. [1st edition].

Linnaeus, C. [as Caroli Linnæi]. (1758). *Systema naturæ per regna tria naturæ, secundum classes, ordines, genera, species, cum characteribus, differentiis, synonymis, locis*. Tomus I. Editio Decima, Reformata. Impensis Direct. Laurentii Salvii, Holmiæ [Stockholm]. [10th edition; *Culex* at pp. 602–603].

Lorenz, C., Alves, J.M.P., Foster, P.G., Suesdek, L. and Sallum, M.A.M. (2021). Phylogeny and temporal diversification of mosquitoes (Diptera: Culicidae) with an emphasis on the Neotropical fauna. *Systematic Entomology*, 46(4), 798–811. https://doi.org/10.1111/syen.12489

Manson, P. (1878) On the development of *Filaria sanguis hominis* and on the mosquito considered as a nurse. *Journal of Linnean Society (Zoology)*, 14, 304–311.

Meulenbeld, G. J. (2000). *A history of Indian medical literature*. Vol. IA. Groningen. pp. 342–344.

Neveu-Lemaire, M. (1902). Classification de la famille des Culicidae. *Memoirs of Society of Zoology, France*, 15, 195–227.

OECD (2018). Consensus document on the biology of mosquito *Aedes aegypti*. Series on Harmonisation of Regulatory Oversight in Biotechnology, No. 65. Environment directorate joint meeting of the chemicals committee and the working party on chemicals, pesticides and biotechnology. Environment Directorate Organisation for Economic Co-operation and Development Paris. ENV/JM/MONO(2018)23.

Poinar, G., Zavortink, T.J. and Brown, A. (2019). *Priscoculex burmanicus* n. gen. et sp. (Diptera: Culicidae: Anophelinae) from mid-Cretaceous Myanmar amber. *Historical Biology*, 32(9), 1157–1162. https://doi.org/10.1080/08912963.2019.1570185

Poinar, G., Zavortinik, T.J. and Pike, T. and Johnston, P. (2000). *Paleoculicis minututs* (Diptera: Culicidae) n. gen., n. sp., from Cretaceous Canadian amber, with a summary of described fossil mosquitoes. *Acta Geológica Hispánica*, 35(1–2), 119–128. pp. 1–56.

Rajagopalan, P.K., Das, P.K., Kalyanasundaram, M., Tyagi, B.K., Arunachalam, N., Somachary, N., Reddy, C.B.S. and Reddy C.M.R., 1987. *Bangalore mosquito control project: Master plan*. Vector Control Research Centre, Pondicherry, pp. 309.

Rajendran, R., Sunish, I.P., Mani, T.R., Munirathinam, A., Abdullah, S. Md., Arunachalam, N. and Satyanarayana, K. (2004). Impact of two annual single-dose mass drug administrations with diethylcarbamazine alone or in combination with albendazole on Wuchereria bancrofti microfilaraemia and antigenaemia in south India. *Transactions of Royal Society of Tropical Medicine & Hygiene*, 98(3), 174–81. https://doi.org/10.1016/s0035-9203(03)00042-7

Rao, B.A. (1945). Malaria in the Irwin Canal area in Mysore State. Part I. *Journal of Malaria Institute of India*, 6, 101–108.

Rao, T.R. (1984). *The Anopheline of India* (Revised Edition.), Malaria Research Centre, Indian Council of Medical Research, New Delhi, 518 pp.

Reid, J.A. and Ahmad, M.D. (1958). Malaria control in the Federation of Malaya. *Medical Journal of Malaya*, 12 (4), 569–584.

Reidenbach, K.R., Cook, S., Bertone M.A., Harbach, R.E., Wiegmann, B.M. and Besansky, N.J. (2009). Phylogenetic analysis and temporal diversification of mosquitoes (Diptera: Culicidae) based on nuclear genes and morphology. *BMC Evolutionary Bioliology*, 9, 298. https://doi.org/10.1186/1471-2148-9-298

Robineau-Desvoidy, J.-B. (1827). Essai sur la tribu des culicides. *Mémoires de la Société d'Histoire Naturelle de Paris*, 3, 390–413.

Ross, R. (1897). On some peculiar pigmented cells found in two mosquitoes fed on malaria blood. *British Medical Journal*, 2, 1786–1788.

Ross, R. (1898a). Further observations on the transformation of crescents. *Indian Medical Gazette*, 33, 14–16.

Ross, R. (1898b). Preliminary report on the infection of birds with *Proteosoma* by the bites of mosquitoes. Government Press Calcutta (11 Oct, 1898); *Indian Medical Gazette* (Jan. 1899); *British Medical Journal* (18 Feb., 1899).

Ross, R., 1903. *Report on malaria at Ismailia and Suez*. Memoir. Liverpool School of Tropical Medicine.

Ross, R., 1910. Report on the measures taken against malaria in the Lahore (Mian Mir) cantonment. *The Lancet*, 2, 1371–3.

Russell, P.F. (1955). *Man's mastery of malaria*. Oxford University Press, London, pp. 308.

Sen, R.N. and Azeez, S.A. 1963. History of malaria and its control in different coal fields of India. A review of 15 years in retrospect. *Indian Journal of Malariology*, 17, 157–178.

Sinton, J. A. (1935). *What Malaria Costs India, Nationally, Socially and Economically*. Thacker's Press & Directories, 77 pp.

Spielman, A. and D'Antonio, M. (2001). *Mosquito: the story of man's deadliest foe*. London: Faber & Faber, vi+247 pp.

Sunish, I.P., Rajendran, R., Mani, T.R., Munirathinam, A., Dash, A.P. and Tyagi, B.K. (2006). Vector control complements mass drug administration against bancroftian filariasis in Tirukoilur, India. *Bulletin of the World Health Organization*, 85(2), 138–45. https://doi.org/10.2471/blt.06.029389

Thenmozhi, V., Balaji, T., Venkatasubramani, K., Dhananjeyan, K.J., Selvam, A., Rajamannar, V. and Tyagi, B.K. (2016). Role of *Anopheles subpictus* Grassi in Japanese encephalitis virus transmission in Tirunelveli, South India. *Indian Journal of Medical Research*, 144(3), 477–481. https://doi.org/10.4103/0971-5916.198684

Theobald, F.V. (1901a). The classification of mosquitoes. *Journal of Tropical Medicine*, 4, 229–235.

Theobald, F.V. (1901b). *A monograph of the Culicidae or mosquitoes*. Volume 1. British Museum (Natural History), London.

Theobald, F.V. (1901c). *A monograph of the Culicidae or mosquitoes*. Volume 2. British Museum (Natural History), London.

Theobald, F.V. (1903). *A monograph of the Culicidae or mosquitoes*. Volume 3. British Museum (Natural History), London.

Theobald, F.V. (1905). Genera Insectorum. Diptera, Fam. Culicidae. Fascicle 26. Belgium.

Theobald, F.V. (1907). *A monograph of the Culicidae or mosquitoes*. Volume 4. British Museum (Natural History), London.

Theobald, F.V. (1910). *A monograph of the Culicidae or mosquitoes*. Volume 5. British Museum (Natural History), London.

Tyagi, B.K., 1993. An indigenously developed mechanical device for sampling the resting adults of mosquitoes. *Journal of Applied Zoological Researches*, 4, 180–184.

Tyagi, B.K. (2004). *The Invincible Deadly Mosquitoes*. Scientific Publishers (India), Jodhpur, pp. 265.

Tyagi, B.K., J. Hiriyan, P. Philip Samuel, S.C. Tewari and R. Paramasivan. 2006. Dengue in Kerala: A critical review. *ICMR Bulletin*, 36(4–5), 13–22.

Tyagi, B.K. (2008). *Vector-borne diseases: Epidemiology and control*. Scientific Publishers (India), 298 pp.

Tyagi, B.K. (2014). *The Tirukoilur Project*. Centre for Research in Medical Entomology, Madurai, pp. 291.

Tyagi, B.K. (2018). *Lymphatic filariasis: epidemiology, treatment and prevention – the Indian perspective*. Springer, pp 314.

Tyagi, B.K. (2020). *Mosquito hunters: A history of hostilities against man's deadliest foe – the mosquito – since 1881*. Scientific Publishers (India), pp. 474.

Tyagi, B.K. (2023a). *Male Mosquito: A zaddy and subaltern sex in the disease epidemiology and control*. Scientific Publishers (India), 232 pp.

Tyagi, B.K. (2023b). *Desert Malaria: An Emerging Malaria Paradigm and Its Global Impact on Disease Elimination*. Springer-Nature, Singapore, 416 pp.

Tyagi, B.K., Naik B. Reddya, Bhattacharya Sajal and Pushpalatha, E. (2024). *Mosquito: Ugly, Bold and Beautiful*. Scientific Publishers (India), Jodhpur (in press).

Venkat Rao, V., 1949. A critical review of malaria control measures in India. *Indian Journal of Malariology*, 3, 313–326.

Viswanathan, D.K. (1946). The seasonal prevalence of malaria transmitted by Anopheles fluviatlis James, 1902, in different regions of Bombay Presidency. *India Journal of Malaria Institute of India*, 6, 253–257.

Viswanathan, D.K. (1950). *Malaria and its Control in Bombay State*. Aryabhushan Press, Pune.

WHO (2014). *A global brief on vector-borne diseases*. World Health Organization, Geneva, pp. 1–56.

Winegard, T.C. (2019). *The Mosquito: A Human History of Our Deadliest Predator*. Dutton, 496 pp.

Yadav, A., Tiwari, R.C., Dikshit, M., Mittal, B. and Sharma, V.B. (2022). Contemporary and Ancient Review of Mosquitoes - A Review. *Journal of Ayurveda and Integrated Medical Sciences*, 7(8), 81–85. https://jaims.in/jaims/article/view/1939

Yong, E. (2013). Blood-filled mosquito is a fossil first. *Nature*. https://doi.org/10.1038/nature.2013.13946

Zavortink, T.J. and Poinar Jr. G.O. (2000). *Anopheles (Nyssorhynchus) dominicanus* sp. n. (Diptera: Culicidae) from Dominican amber. *Annals of the Entomological Society of America*, 93, 1230–1235.

2 Phylogeny, Evolution, and Classification

B.K. Tyagi
University Institute of Biotechnology, Chandigarh University, Mohali, India

2.1 INTRODUCTION

The phylogeny of mosquitoes (Culicidae) is the fulcrum that underlies various macroevolutionary phenomena – such as the evolution of two-winged flight, mating, co-speciation with plants, morphological specializations for diverse modes of feeding, differential habitat selection for resting and breeding, to name but a few attributes – the knowledge about which, as far as Indian mosquitoes are concerned, comes comprehensively only from the two major works, one each on the tribe "*Anophelini*" by Christophers (1933) and the tribes "*Megarhinini and Culicini*" by Barraud (1934), albeit far from being satisfactory in view of the modern analyses and interpretations. Therefore, a careful scrutiny of phylogeny of Culicidae on a worldwide basis informs on evolution of various different mosquito groups (Anophelinae and Culicinae), their inter-relationship and temporal and spatial diversification as well as diverse adaptability to numerous pathogens, thus reflecting on public health implications as this family includes numerous species of medical and veterinary importance (Lorenz et al., 2021). However, because of immense influence of *Anopheles* species on human health by way of transmitting several deadly and/or debilitating infections, it is not surprising, therefore, that most of the phylogenetic studies thus far have dealt with *in extenso* species Series/Groups, Subgroups, and Complexes that include vectors of human malarial protozoa. It is obvious that the evolutionary relationships of malaria vectors and their closest allies have received more attention than other groups. However, none of these studies can be regarded as complete in terms of taxonomic coverage of any group, and, therefore, the field of disease vector systematics presents many opportunities for further research. Phylogenetic patterns are used to interpret bionomic features such as differences in the nature of blood-feeding by adult females, feeding behavior, and the occurrence of immature stages in aquatic habitats.

Mosquitoes are dipterous insects – the "true" mosquitoes, that is, the ones with two wings, and are classified under the family Culicidae, order Diptera and Class Insecta (which is very recently raised to the level of Super Class). The term Culicidae has been derived from the Latin word "*culex*," meaning a "gnat." The family Culicidae is a member of one of the main stocks of Nematocera, the infraorder Culicomorpha. It consists of two superfamilies that include all of the piercing/sucking nematocerans, both predators and blood-feeding biters. The superfamily Chironomoidea comprises the families Chironomidae and Thaumaleidae, which have nonpiercing mouthparts, and Simuliidae and Ceratopogonidae, which blood-feed on either vertebrates or invertebrates. The superfamily Culicoidea comprises the families Dixidae, Chaoboridae, Corethrellidae and Culicidae, the last two of which are characterized for feeding on vertebrate blood. Although, several of these families are superficially similar with reference to appearance of adults and occurrence of immature stages in aquatic habitats, nevertheless, among all the culicomorphs, the long proboscis of mosquitoes is distinctively characteristic of the group only (Wood and Borkent, 1989).

 DOI: 10.1201/9781003326991-3

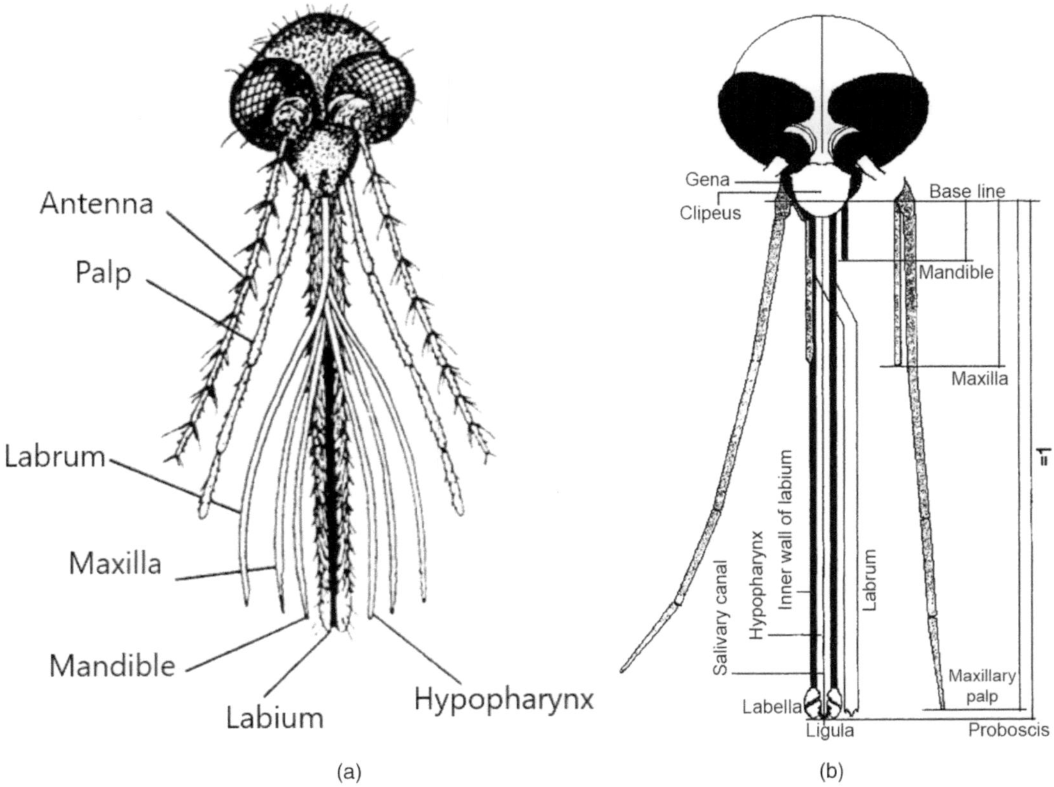

FIGURE 2.1 Head and mouthparts of a mosquitoes; (a) piercing and sucking type of a female *Anopheles* mosquito, and (b) a male *Aedes* mosquito.

Further, in as much as immature stages' taxonomy is concerned, it is also important to acknowledge here that there are very few works, even outside India, providing identification keys based on pupal taxonomy (Harrison and Scanlon, 1975). Pupal characteristics are considered more stable and distinctive compared to those found in larvae.

The proboscis is considered the most specialized of biting mouthparts among Nematocera and its specialized morphology, anatomy and physiology indicate a long and close association of mosquitoes with vertebrate animals. In essence, mosquitoes, are distinguished from such similar looking dipterous insects by the presence of scales on the wing veins and wing margins and by their long proboscis that is adapted for piercing and sucking vertebrate blood – by females only (Figure 2.1a). Female mouthparts are different from that of the male which feeds on plant nectar etc. (Figure 2.1b).

In contrast to an adult, the larva (Figure 2.2) is mainly composed of soft, membranous tissues in thorax and abdomen and hardened, sclerotized plates in the head (Rueda and Debboun, 2020).

Despite the fact that mosquitoes are the only creatures which have directly or indirectly influenced existence of man in his entire history so impressively as to earn the sobriquet "*man's deadliest foe on Earth ever*," there is a lack of information about the evolutionary history and phylogenetic positioning of the majority of mosquito species. Mosquitoes compose a large group of insects from the family Culicidae (Meigen, 1818).

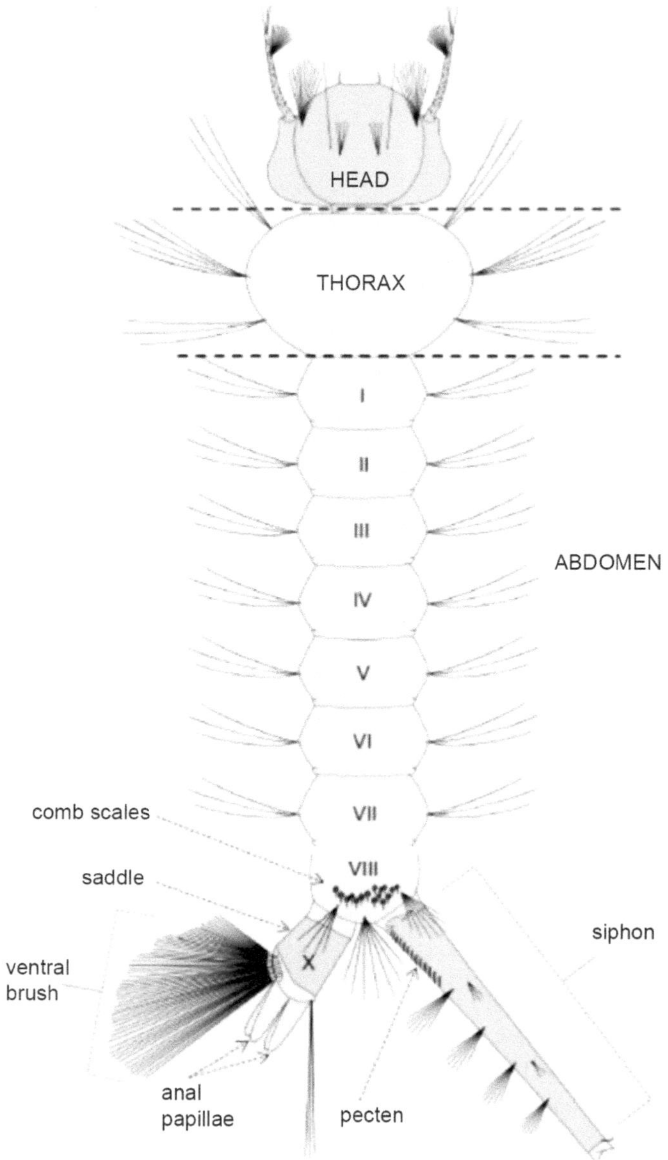

FIGURE 2.2 Diagram of larval mosquito (dorsal view) *Culex*.

2.2 PHYLOGENY

Diptera evolved during the early Permian period (252.2–298.9 million years ago), as presented by the scheme of classification and phylogenetic affinities within the class by Martynov (1938, citing Kiauta, 1975 in Tyagi and Veer, 2016) (Figure 2.3). The order Diptera is among the five most evolved and advanced orders of insects, especially mosquitoes which are genetically equipped with an exceptionally high order of phenotypic plasticity (Edwards, 1932).

Mosquitoes (Culicidae) are regarded as a monophyletic group having evolved from a single common ancestor (Wood and Borkent, 1989; Miller et al., 1997; Harbach and Kitching, 1998), that diverged from its nearest relative, family Chaoboridae Edwards, 1912, about 255 million years ago (mya) (Yeates and Wiegmann, 2005). Currently the two acceptable culicid subfamilies are Anophelinae Grassi, 1900 and Culicinae Meigen, 1818, diverged 229–192 mya (Reidenbach et al.,

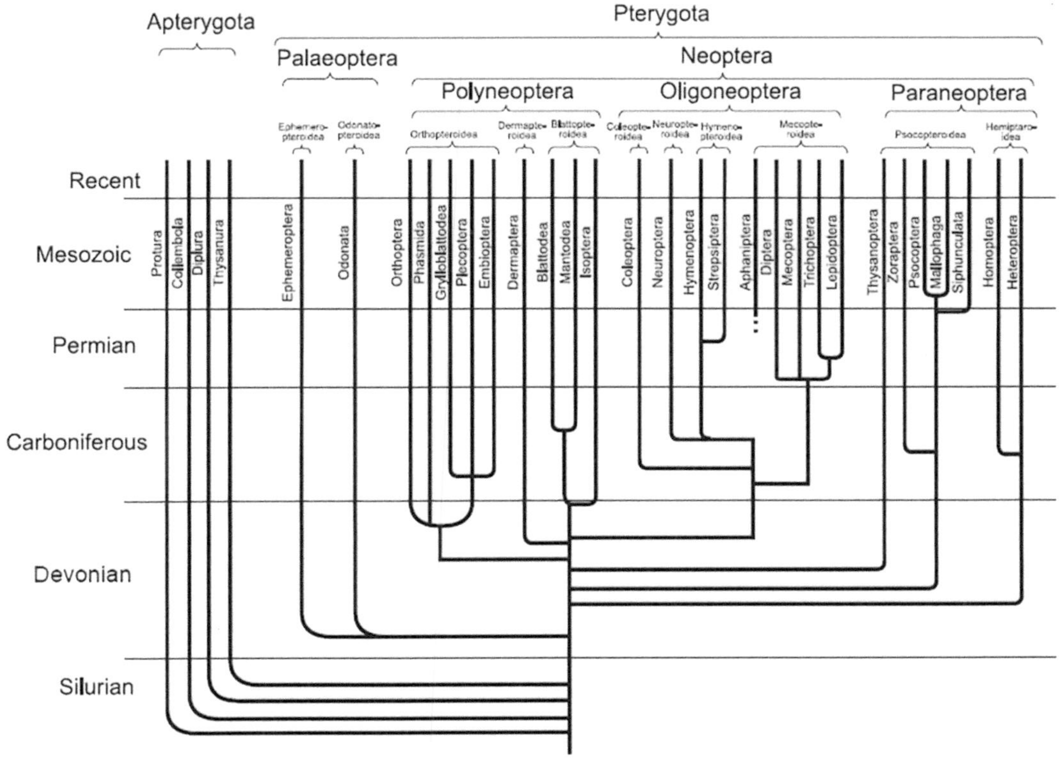

FIGURE 2.3 Martynov's (1938) scheme of insect classification.

***Source*: Tyagi 2019 citing Kiauta 1975.**

2009); the erstwhile third subfamily Toxorhynchitinae having been recently incorporated in the subfamily Culicinae, on the basis of molecular interpretation of phylogeny of the Family which is at the moment quite precarious and needs a continuous research on the matter. At present there are more than 3700 known species, and scores of new ones are being described with each passing year, although, yet again, the taxonomic level of some of these, including *Toxorhynchites* spp., continues to be a matter of personal opinion of the respective research workers and is subject of incessant reorganizations and reclassifications. Interestingly, the modern age tools such as mitochondrial phylogenomics likewise also potentially evidence the geological age of Culicidae (Figure 2.4) (da Silva et al., 2020).

The classification of mosquitoes can be complex but it is very important to group and identify species properly so that they can be studied reliably. The classification of mosquitoes has passed through many changes and with the increased molecular knowledge, tools and processes an increasing number of tribes, genera and subgenera has come to be recognized on regular basis; one such a genus recently erected is *Tewarius* Reinert, 2006 – a new taxon of Aedini (Diptera: Culicidae), created in the honor of an Indian mosquito taxonomist, Satish Chandra Tewari (Tyagi et al., 2015a; Tyagi, 2020). Besides, close to eight dozen new species, including two new taxa under *Toxorhynchites*, have been described from India after Christophers (1933) and Barraud (1934) (*cf.* Tyagi, this work).

2.2.1 The Earliest Mosquito History

In spite of the fact that currently available advanced molecular techniques have dispensed with much of cloud over the phylogeny issues, nonetheless, due to blatant paucity of mosquito fossils, there is no direct indication of the evolutionary history of mosquitoes, i.e., both anophelines and culicines.

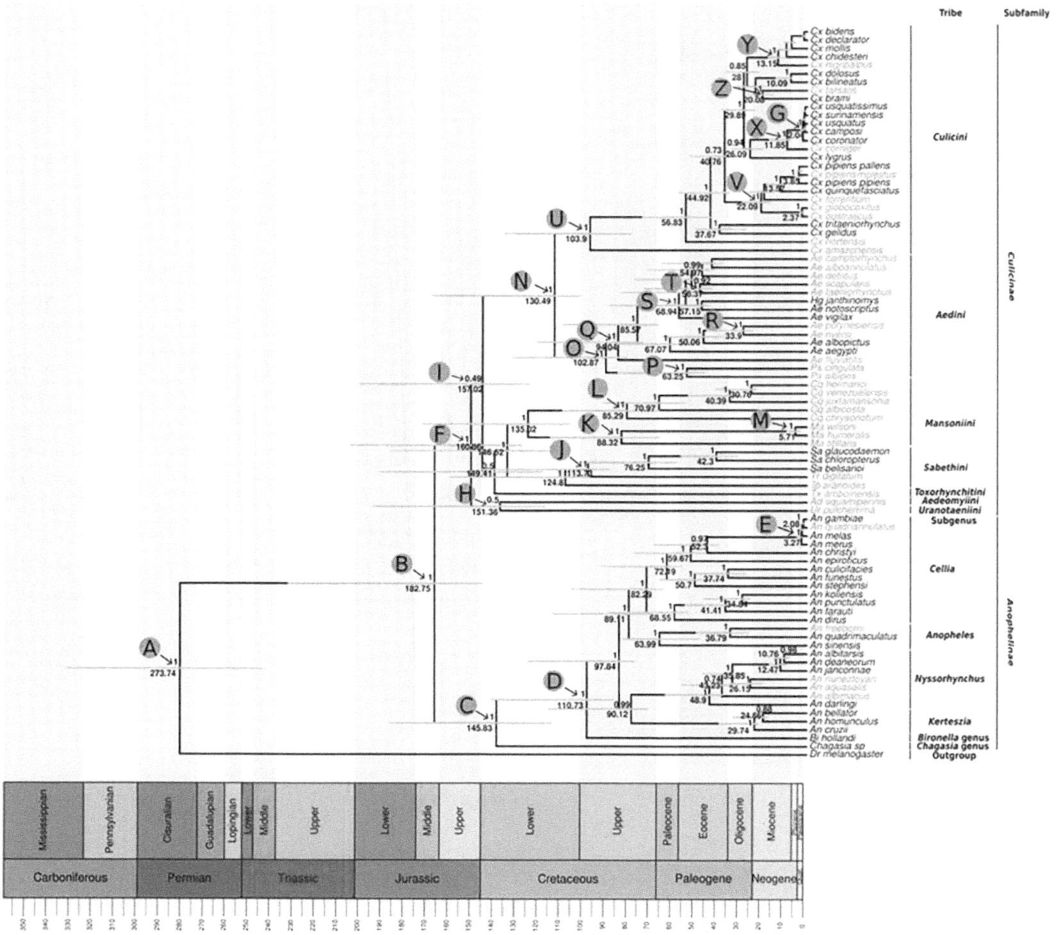

FIGURE 2.4 Evolutionary timescale of Culicidae family. Tree was generated from BEAST using PCGs nucleotide sequences partitioned by gene and codon positions (1st, 2nd, and 3rd separately). Blue bars represent the HPD95%. The numbers above and below the bars show the posterior probability and the predicted median dating respectively for each node. Specific words inside the circles represent the nodes discussed in the text. Light blue tip names represent mitogenomes characterized from SRA data. Orange tip names represent sequenced mitogenomes from this study.

***Source*: da Silva et al., 2020.**

The minuscule fossil records of mosquitoes allude toward a possible evolution of Culicidae, particularly anophelines, in the Jurassic Period (146–200 Mya), along with the early mammals, first birds and first flowering plants (http://www.stratigraphy.org; Borkent and Grimaldi, 2004; Bertone et al., 2008; Lorenz et al., 2021). Chronicling geological age of anopheline and culicine mosquitoes is a major point of debate, though the second oldest fossilized culicine-looking mosquito from the Late Cretaceous (66.0–100.5 Mya), *Paleoculicis minutus*, relates more closer in morphological attributions with culicine mosquitoes than the anophelines, implying that ancestral lineage of Culicinae is much younger than that of Anophelinae duly supported by evidences of fossilized *Anopheles (Nyssorhynchus?) dominicanus* from the mid-Tertiary (about 15–45 Mya) and *An. rottensis* from Late Oligocene of Germany (about 25 Mya) which represent an even younger split than the basal stock of Anophelini. Therefore, the view whether the anophelines are originated earlier than culicines from the main stock, seems to support the argumentation, for reason of convenience in understanding the phylogeny of mosquitoes, that the split would have occurred in the Cenozoic Era (<66.0 Mya), in all probabilities. From divergence times based on sequence data for nuclear protein-coding genes and fossil calibration points, it appears that major mosquito lineages date to the Early Cretaceous (100.5–145.0 Mya), and the ancestral lineage of anophelines may have appeared as early as the Jurassic (~145 Mya) (da Silva et al., 2020).

2.2.2 Roots and Modes of Speciation

The origin of a new species by hybridization between two closely related species is now more frequently encountered among mosquitoes than ever before, implying that speciation due to hybridization in closely evolved or otherwise behaviorally closer mosquitoes is picking growing recognition by clinching evidence across genera. Amidst several species groups, the *Aedes scutellaris* group, comprising more than 30 closely related species having an insular distribution in the South Pacific, provides an exciting evidence for the speciation through hybridization. Most species of this group (distinguished by subtle morphological differences) can easily be inter-crossed with little or no pre-mating barriers and exhibit varying degree of post-zygotic fertility, permitting genetic analyses of species and hybrids. A similar scenario is likely to be encountered with *Aedes albopictus* which is currently evidenced to be formed of eleven closely allied or sibling species. There are evidences that many extant taxa, especially under "Species Complexes," are forging ahead to speciate and evolve under a multitude of factors, especially climate change, global warming, anthropization, and decades of insecticide pressure. One of such groups is the deadly malaria vector, *Anopheles gambiae*, a complex of at least eight species. Recently, several of these closely resembling sibling species have been properly described taxonomically. On the other hand, male *Aedes albopictus* has often been observed in nature to sire the females of *Aedes aegypti*, albeit resultant completely nonviable eggs!

Similarly, unveiling mosquito cryptic species and their reproductive isolation forces members of the group for speciation. Emerging species within the primary malaria vector *Anopheles gambiae* show different ecological preferences and significant prezygotic reproductive isolation, while *Aedes mariae* and *Aedes zammitii* show post-mating reproductive isolation. And, though, none of the Indian complexes' species has so far been properly assigned a scientific name, yet the indication that some of these, viz., *Anopheles culicifacies* and *An. subpictus*, are clearly on way to speciate is more than evident. These features are associated with evolutionary preparations. It is evident that at least in some mosquito cryptic species' evolutionary processes bring about reproductive isolation in the absence of morphological differentiation.

2.3 EVOLUTION

Based on the knowledge derived from the fossils, both the oldest approximately 79 mya (though some believe mosquitoes have been in existence for 226 million years) and the most recent one discovered in 2019, dating back to the mid-Cretaceous Period (145–66 Mya), it has been safely surmised that the fossilized mosquitoes had a number of similarities to the modern-day versions of the mosquito (Figure 2.4). Most notably, the prehistoric mosquitoes strongly resemble today's anopheline mosquitoes by having antennae, abdomens, wing veins, and proboscises that are similar to modern-day species, albeit some minor differences. Likewise, in the light of a fossilized, blood-engorged mosquito – the "oldest direct evidence," it has now become more than evident that mosquitoes were believably already feeding on animal blood meals at least 46 million years ago.

About 10,000 years ago the development of agriculture and dense, sedentary human societies became the new areas for the pest mosquitoes to perpetuate. With the passing time, however, as human populations evolved and began to live and congregate in larger populations, they, with a softer skin and convenient target sites aplenty, possibly provided a new and easy resource for hungry mosquitoes to fulfill their need for blood meals to help complete ovulation in females. And because mosquitoes breed in water, the human tendency to collect water for various different purposes may have caused mosquitoes to depend on "human sources of water for breeding." In the process, many mosquito species evolved to adapt to human behavior for their own blood-feeding and/or breeding necessities (OECD, 2018; Tyagi, 2020, 2023; Tyagi et al., 2006). It appears, thus, that more populations of *Ae. aegypti* will start to prefer humans between now and 2050. For instance, much of Africa and the Great Indian Thar Desert is becoming urbanized; the population boom in India is emerging on a serious note and has already notched up the top position with 1.4 billion people replacing China; and inter-countries' or continents' traveling is increasingly intensifying with every year passing and these factors, including climate change, are expected to have a strong impact on the mosquitoes' evolution such as *Aedes aegypti* and *Anopheles stephensi*, just to mention a few species. These mosquitoes are only distantly related to each other and both have a different life cycle (Christophers, 1960; Rao, 1984; Tyagi 2023). While *Aedes aegypti* has evolved to be strongly anthropophilic (high inclination toward human blood-feeding), at the same time *Anopheles stephensi*, notoriously adapted to transmit intensively the deadly malaria parasite, *Plasmodium falciparum*, has intruded new territories in Africa due to intense urbanization and human congregation coupled with inherent necessity to store water in a variety of modes (Rose et al., 2020; Tyagi, 2023).

2.3.1 Evolution of Anophelinae

The subfamily Anophelinae, consisting of 519 species, with many complex and sibling-species relationships awaiting deeper phylogenetic and taxonomic resolutions mainly in the Anopheles genus, is demonstrably monophyletic – a basal lineage in relation to the family Culicidae (Sallum et al., 2007). To date, the subfamily Anophelinae comprises Anopheles as its representative or nominotypical genus. The subfamily Anophelinae, includes two other genera: Bironella (Australasian) and Chagasia (Neotropical), besides the cosmopolitan Anopheles, but most of the studies focus on the genus Anopheles due to its capacity to transmit malaria parasites, in addition to several arboviruses (Knight and Stone, 1977; Knab, 1913). The unknown medical/veterinary importance of Bironella and Chagasia makes them less attractive genera to be deeply investigated. The monophyletic status for Bironella is either debatable or not supported in most of the studies, but the opposite is found to the genus Anopheles, specifically the subgenus Stethomyia, Lophopodomyia, Kerteszia, and Cellia (Harbach, 2013).

The modern molecular cladistic analyses of morphological data and DNA sequences of various ribosomal (18S, 28S), mitochondrial (COI, COII) and nuclear genes strongly favor the placement of Chagasia in an ancestral relationship to all other anophelines under the subgenera Kerteszia, Nyssorhynchus, Cellia, Lophopodomyia and Stethomyia, which were found to be monophyletic (Sallum et al., 2000, 2002; Collucci and Sallum, 2003). Harbach and Kitching (2005) reworked the above observations with special references to setae of the male gonocoxites, integral to the identification of the subgenera, and inferred based on the parsimony analysis of the data set under implied weighting (IW) that the subgenus Anopheles emerged to belong to a polyphyletic lineage basal to a monophyletic clade consisting of Kerteszia+Nyssorhynchus and Cellia in a sister-group relationship, whereas Bironella, Lophopodomyia and Stethomyia were firmly nested within subgenus Anopheles, which would be paraphyletic even if these taxa were subsumed within it. Also, they supported the monophyly of subgenera Cellia, Kerteszia and Nyssorhynchus, and the sister relationship of Kerteszia + Nyssorhynchus. Subgenus Baimaia, represented by *An. kyondawensis*, was supported as the sister of Bironella + all other Anopheles. Interestingly, contrary to Sallum et al. (2002), Harbach and Kitching (2005) also supported Bironella and Stethomyia to be likely of monophyletic clades separate from subgenus Anopheles. This is the currently admitted anopheline phylogeny and evolutionary relationships based on a greater number of taxonomic groups and homologous characters than all other previous hypotheses (Figure 2.5). However, based on several significant molecular investigations and generated data analyzed using successive approximations character weighting (SACW) and IW, it can be safely surmised that most of the relationships between members of many series and Groups, appearing monophyletic for some and polyphyletic for others, the subgenus Anopheles were

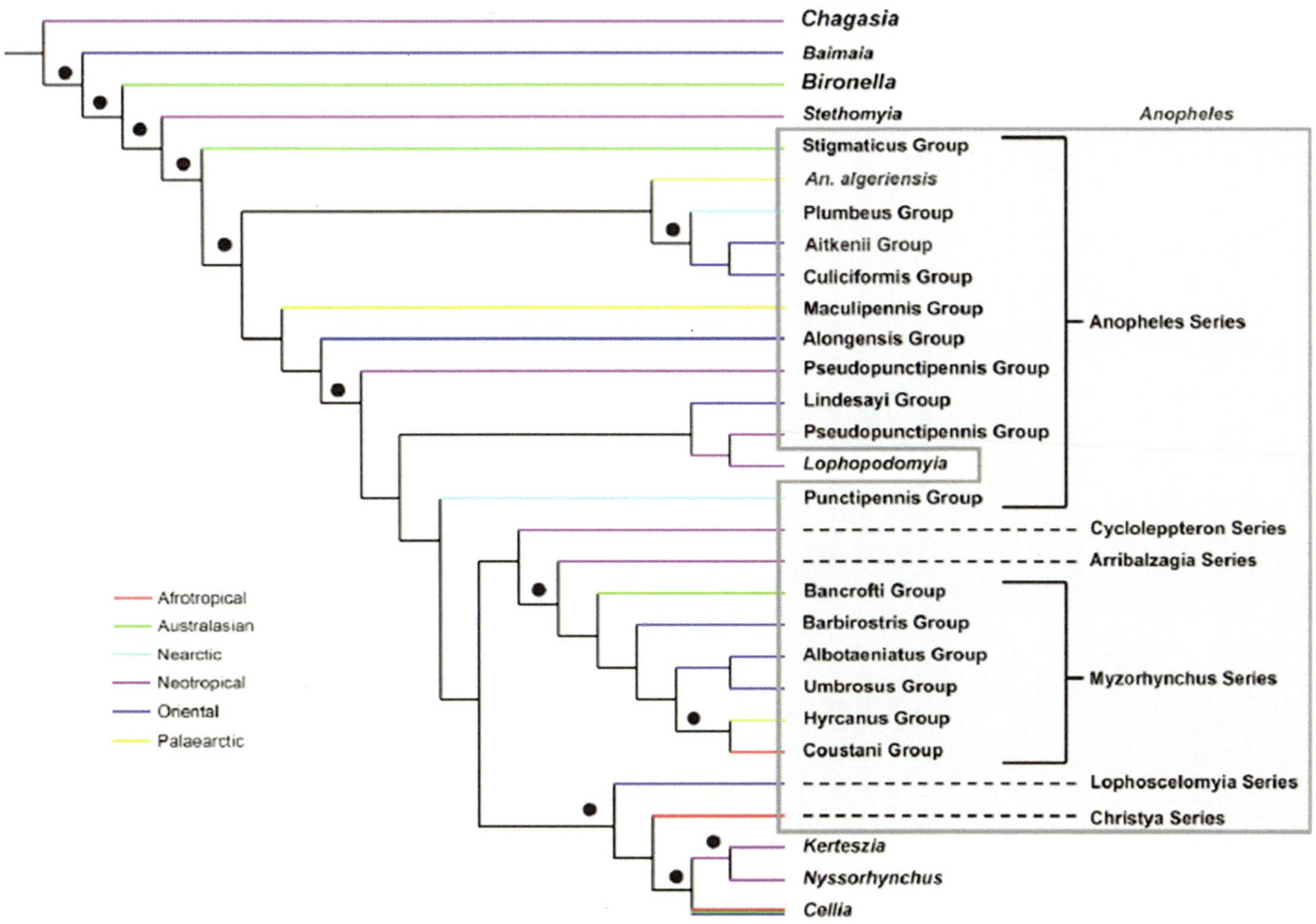

FIGURE 2.5 Phylogeny of subfamily Anophelinae, indicating relationships within subgenus Anopheles. Filled circles indicate Bremer support values greater than 0.8.

Source: **Harbach, 2013.**

either moderately or poorly supported (Sallum et al., 2000, 2002; Harbach and Kitching, 2005; Collucci and Sallum, 2007). According to Sallum et al. (2000), Anopheles series is polyphyletic, but its members are interspersed in a complexity of inter-group relationships rather than arrayed in a pectinate sequence.

Unlike that of subgenus Anopheles where most, if not all, subgenera exhibit monophyly, all the subgenera, viz., Cellia, Kerteszia and Nyssorhynchus, as well as the sister pairing of Kerteszia and Nyssorhynchus are demonstrably monophyletic as evident by a series of phylogenetic studies conducted so far (Harbach, 2013).

2.3.2 Evolution of Culicinae

Like the Anophelinae, the phylogenetic history of the Culicinae subfamily is not completely settled, but some groups have been better studied due to their epidemiological role as potential vectors of microorganisms including helminths, viruses, protozoans etc. to humans as well as animals (OECD, 2018). In sheer contrast to the subfamily Anophelinae where the members of the comparatively lesser number of genera and subgenera are still closely interrelated despite geographical isolations and the consequent diversity, the subfamily Culicinae houses a variety of large number of genera included within a dozen tribes (Meigen, 1818; Theobald, 1901a, 1901b, 1903, 1907, 1910; Knight and Stone, 1977), although rarely cladistic analyses displayed some divergences among the genera, notably the *Wyeomyia* genus where a large paraphyletic taxa were recovered.

The Culicini tribe displays monophyletic status, but because many cryptic species exist there, mainly in the Culex genus, the natural classification of the group is greatly hampered (Komp, 1937; Antunes, 1937; Harbach et al., 2005). For instance, Aedini tribe, which is undoubtedly the most critical, complicated and diverse group in terms of phylogenetic reconstruction, is an enormous group where over time, dozens of subgenera rose to genera status followed by restoration, but still there is hardly any phylogeny consensus toward the tribe. Notwithstanding the popular view that the Aedini tribe, encompassing three genera (Aedes, Haemagogus and Psorophora as per the traditional classification), largely maintains a monophyletic status (Harbach, 2013, 2007; Harbach and Kitching, 1998; Forattini, 2002).

2.3.3 On Absorbance of the Erstwhile Toxorhynchitinae into Culicinae Subfamily

Although, mosquitoes are known to feed on the blood of humans or animals, however exceptions are there and one such group belongs to the genus *Toxorhynchites* which are a paradox among mosquitoes (Donald et al., 2020). *Toxorhynchites* mosquitoes are commonly called "elephant mosquitoes," due largely to their size and their trunk-like proboscis adapted for consuming nectar from plants. Members of this genus are entirely non-blood-feeding and, instead, invariably get their nourishment from plant sources (Figure 2.6). As a result, *Toxorhynchites* mosquitoes are not only completely innocuous to human health, they are actually highly beneficial in controlling the populations of vector and pest species (Pal and Ramalingam, 1981; Collins and Blackwell, 2000; Focks, 2007; Vinogradov et al., 2022). Since the larvae of *Toxorhynchites* mosquitoes are essentially predaceous, they predate on larvae of the harmful mosquitoes and form a potential counter measure against the spread of mosquito-transmitted diseases. The systematic position of the genus *Toxorhynchites* has been dynamically changing between one subfamily to the other and is a regular subject of further investigation.

The Toxorhynchitini tribe is monogeneric, consisting of only *Toxorhynchites* with about 90 species under four subgenera, viz., *Toxorhynchites* (51 spp.), *Lynchiella* (16 spp.), *Afrorhynchus* (19 spp.) and *Ankylorhynchus* (4 spp.) (Donald et al., 2020). They are distributed in the Old World (*Toxorhynchites*), the New World (*Lynchiella* and *Ankylorhynchus*) and Africa (*Afrorhynchus*) and are very similar in appearance making their identification quite confusing at times (Steffan, 1975), not to mention in the least their phylogenic relationships (Steffan and Evenhuis, 1985; Harbach and

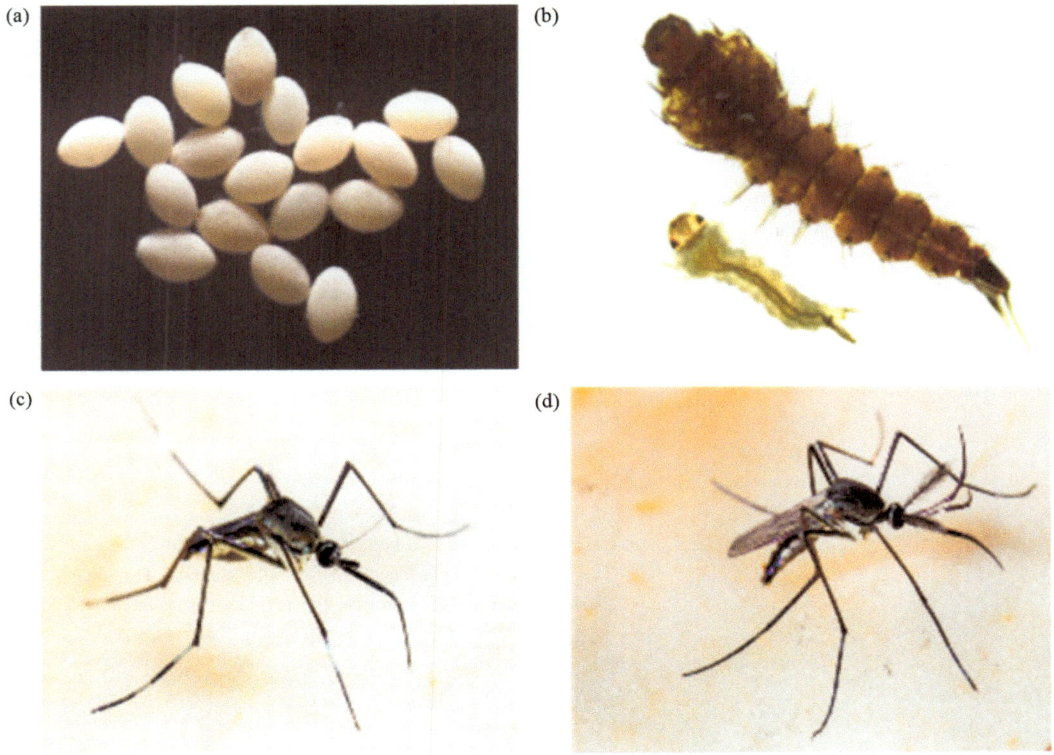

FIGURE 2.6 *Toxorhynchites splendens* life stages. (a) Eggs, (b) comparison between fourth instar larva of *Tx. splendens* and *Cx. quinquefasciatus*, adult (c) female, and (d) male.

***Source*: Donald et al., 2020.**

Kitching, 1998; Ribeiro, 2004; Tyagi et al., 2015b). Of nearly half a dozen species, two species, *Tx. darjeelingensis* Tyagi et al., 2015b and *Tx. tyagii* Krishnamoorthy et al., 2013, are endemic to India.

Although it was initially suggested that Toxorhynchitinae is an independent subfamily within Culicidae (Theobald, 1901), yet, subsequent reports placed it within the Culicinae subfamily (Miller et al., 1997; Shepard et al., 2006; Reidenbach et al., 2009; Da Silva et al., 2019; Mitchell et al., 2002; Belkin, 1962; Harbach, 1977, 1978). It is interesting to remind here that hematophagy, already developed in Anophelinae, has been lost in *Toxorhynchites* or possibly gradually fizzled away with an alternative new adaptation to the needs of survival by these giant mosquitoes that arose in specific lineages after they diverged. This trait is evidenced in many other insects where hematophagy is believed to have been independently acquired and lost on several occasions in course of evolution (Wiegmann et al., 2011). Determining the key genes involved in the evolution of a non-biting life-style of *Toxorhynchites* could provide an alternative approach to eliminating many a disease transmission (Donald et al., 2018).

As discussed earlier, however, the subsuming of Toxorhynchitinae in the subfamily Culicinae appears, at least for the present, arguably acceptable until more information on phylogeny and systematics of all taxa emerges for analysis.

2.4 CHRONOLOGY OF MAJOR CLASSIFICATORY SYSTEMS

Classification of Culicidae has undergone many changes during past nearly one hundred years. The traditional classification of Culicidae is phenetic and, consequently, all levels of classification, underlying broad genus-group concepts exhibiting paraphyletic and polyphyletic taxa, have been

TABLE 2.1
Classification of Culicidae

Subfamily	Tribe	Genus
Anophelinae	Anophelini	*Anopheles* Meigen, *Bironella* Theobald, *Chagasia* Cruz
Culicinae	Aedeomyiini	*Aedeomyia* Theobald
	Aedini	*Aedes* Meigen, *Armigeres* Theobald, *Eretmapodites* Theobald, *Haemagogus* Williston, *Heizmannia* Ludlow, *Opifex* Hutton, *Psorophora* Robineau-Desvoidy, *Udaya* Thurman, *Zeugnomyia* Leicester
	Culicini	*Culex* Linnaeus, *Deinocerites* Theobald, *Galindomyia* Stone & Barreto
	Culisetini	*Culiseta* Felt
	Ficalbiini	*Ficalbia* Theobald, *Mimomyia* Theobald
	Hodgesiini	*Hodgesia* Theobald
	Mansoniini	*Coquillettidia* Dyar, *Mansonia* Blanchard
	Orthopodomyiini	*Orthopodomyia* Theobald
	Sabethini	*Isostomyia* Coquillett, *Johnbelkinia* Zavortink, *Limatus* Theobald, *Malaya* Leicester, *Maorigoeldia* Edwards, *Phoniomyia* Theobald, *Runchomyia* Theobald, *Sabethes* Robineau-Desvoidy, *Shannoniana* Lane & Cerqueira, *Topomyia* Leicester, *Trichoprosopon* Theobald, *Tripteroides* Giles, *Wyeomyia* Theobald
	Uranotaeniini	*Uranotaenia* Lynch Arribalzaga
Toxorhynchitinae		*Toxorhynchites* Theobald

Source: As discussed by Foster and Walker (2002), citing Knight and Stone (1977) (Table modified).

arbitrary groupings based on the subjective interpretation of anatomical similarity. On the contrary, the recent phylogenetic studies based on a melange of morphology and DNA sequence data have deciphered hitherto unknown relationships suggesting major and radical changes in the classification of Culicidae.

Culicidae is among the faunally most fascinating families in the order Diptera, especially due to their hematophagous and disease pathogen-transmitting behavior. With a faunal wealth of >3700 currently recognized species, belonging to 188 subgenera in 41 genera under 11 tribes of mosquitoes worldwide (WRBU, 2018; Rueda and Debboun, 2020), it is being continuously felt that possibly the largest number of taxa is still awaiting to be discovered and brought on record mainly from the tropical rain forests where faunas are more diverse but less well surveyed than temperate regions of the world. Mosquito species which have been known to be disease vectors have also been investigated deeply to often reveal complexes of sibling and/or cryptic species, indicating that many reproductively isolated and niche-specific forms remain to be identified or are undergoing speciation such as, for example, *Anopheles gambiae* in Africa and *An. culicifacies* in India.

Till 1990s, the culicid classification recognized three subfamilies: Anophelinae, Toxorhynchitinae and Culicinae (Knight and Stone, 1977) (Table 2.1), while currently, only two subfamilies are proposed: Anophelinae and Culicinae, with Toxorhynchitinae having got submerged with Culicinae, albeit the fact that both Anophelinae and Toxorhynchitinae are sometimes similarly considered to be relatively primitive groups some kind of specialized derivations of some Culicinae-type ancestor as surmised by other authorities.

2.4.1 Classification by Theobold (1901a, b–1910, Especially 1907)

The first ever global classification of mosquitoes (family Culicidae), that also considered Indian fauna deposited in the Indian Museum, was offered by Theobald (1907; see also 1901–1910). He based his classification on that of Dr A. Lutz (citing "*Mosquitos do Brasil*" by Bourroul, 1904) who

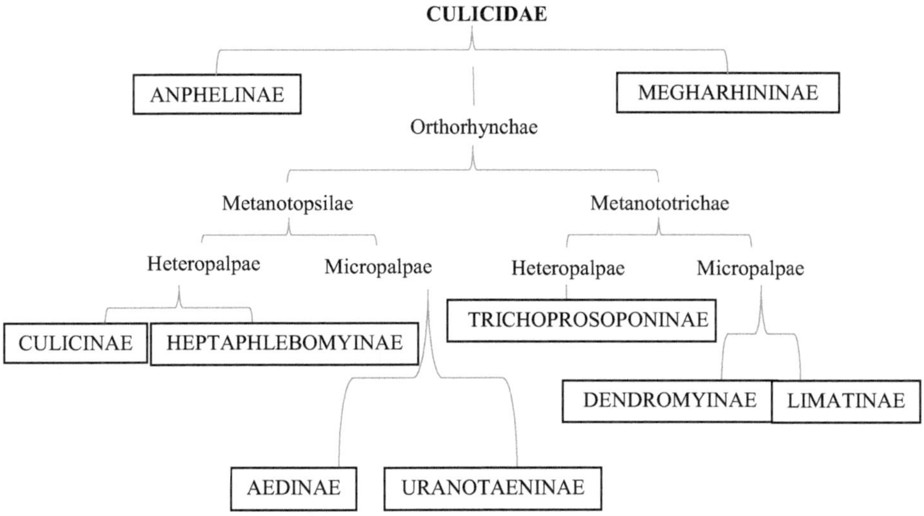

FIGURE 2.7 Classification of Culicidae proposed by Theobald (1907); subfamilies are shown in rectangular blocks.

adopted a new grouping of families and genera on excellent grounds then convinced him. Theobald made several reforms into previous regional proposals and put forth his own based on characteristics of proboscis and palpi. Thus, though clearly recognizing Anophelinae and Megarhininae (Toxorhynchitinae) subfamilies, he had however faltered to directly assign at the same level the position to Culicinae (see Figure 2.7).

Theobald proposed nine subfamilies altogether: Anphelinae, Megarhyninae, Culicinae, Heptaphphlebomyinae, Aedinae, Uranotaeninae, Trichoprosoponinae, Dendromyinae and Limatinae – the last seven subfamilies categorized under a major taxon, Orthorhynchae. Despite being partially correct, in the modern day's light of knowledge on mosquito classification, Theobald's (1907) was nevertheless by far the best general classification yet proposed. However, his system of classification was highly arbitrary and could not stay for long.

2.4.2 Classification by Ross (1951)

Against the backdrop of systems of classification based on anatomical similarities of body parts, Ross (1951) offered a new classification system – a first-ever evolutionary tree of Culicidae – which was constructed based on an intuitive interpretation of comparative bionomics and morphological data (Figure 2.8).

The tree exhibited the traditional division of family Culicidae into three subfamilies: (i) Anophelinae, the basal lineage, (ii) Toxorhynchitinae, the intermediate lineage, and (iii) Culicinae, the large, derived lineage. His evolutionary relationships remained unchallenged and were more or less accepted by subsequent workers until Harbach & Kitching (1998).

2.4.3 Classification by Knight and Stone (1977)

Knight and Stone (1977) offered the first most authentic and duly classified systematic compilation of the world's species of mosquitoes (Culicidae), organizing the family Culicidae, Like Ross (1951), into 3 subfamilies, viz., Anophelinae, Culicinae and Toxorhynchitinae. There are altogether 38 genera (3 genera under one tribe of Anophelinae; 34 genera under 10 tribes of Culicinae; and, 1 in the subfamily Toxorhynchitinae) (Table 2.1). Of all the tribes, the most diverse tribes were Aedini and Sabethini in terms of numbers of genera and species worldwide.

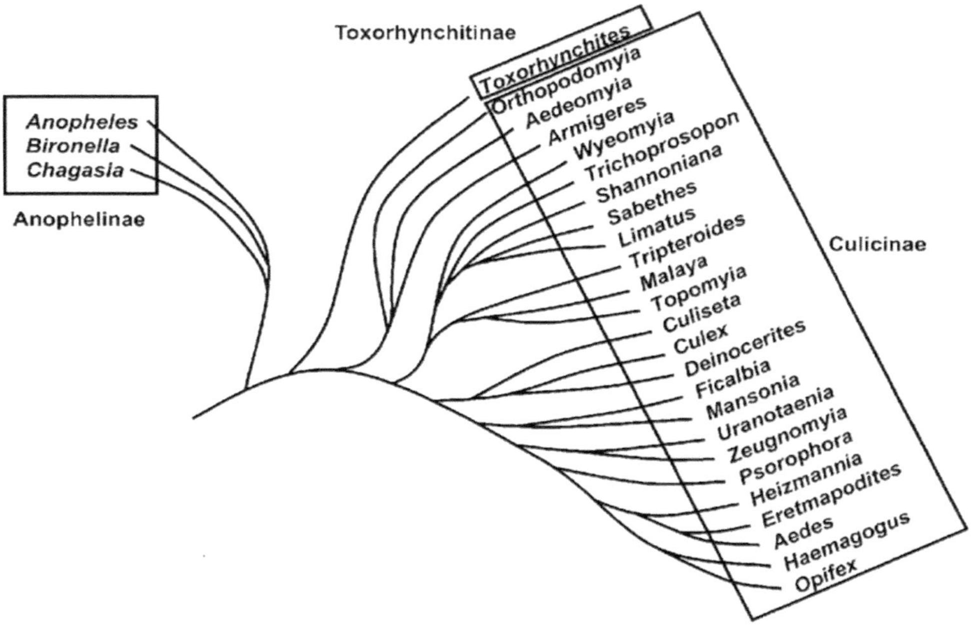

FIGURE 2.8 Intuitive phylogeny of mosquito genera based on "comparative bionomics and morphology", principally larvae.

Source: **Harbach, 2007, quoting Ross, 1951.**

2.4.4 Classification by Harbach & Kitching (1998)

Harbach and Kitching (1998) examined the generic affinities and higher classification of the family based on cladistic analyses of morphological data arising from as many as 73 characters from larvae, pupae and adults from as many as three dozen genera of mosquitoes. Their innovative approach of cladistic analysis of morphological and nucleotide-sequence data supported the concept that the Anophelinae are only distantly related to the other two subfamilies and that the Toxorhynchitinae do not merit subfamily status (Figure 2.9).

As proof, they argued that while, on one hand, anopheline eggs bear characteristic floats, their larvae lack air tubes, and adults have elongate palps in both sexes, the typical culicine and toxorhynchitine larvae, on the other hand, have air tubes, and adult females have short palps. More interestingly, Toxorhynchitines are all predaceous as larvae are unusually large and adults have a curved proboscis suited for feeding on only nectar (see Figure 2.6, and Figures 2.10 and 2.11).

Eggs of mosquitoes carry significant identification characters. In most cases of species, the eggs are elongate, ovoid, spherical or rhomboid, or even spindle-shaped. All eggs are covered in an outermost layer of the eggshell or chorion which generally has complicate surface morphology highlighting intricate structures and patterns quite often diagnostic of the particular species or group of taxa. For instance, the chorions of Anopheles species have air-filled floats which keep the eggs buoyant (Figure 2.10a). It is noteworthy that eggs in species belonging to genera Anopheles, Toxorhynchites, Wyeomyia, Aedes, Ochlerotatus, Psorophora, and Haemagogusspecies are laid individually (Figures 2.10c, 2.10d), albeit with varied maneuvers, whereas those in Culex, Culiseta, Coquillettidia, and Mansonia species, are glued together forming a floating egg raft (Figure 2.11a) or a submerged cluster (Figure 2.11b). Culex eggs are characteristically bearing a cup-shaped corolla at one end (Figure 2.10b), allowing them to align vertically on the water surface while clustered in a raft (Figure 2.11a); the upper ends have apical droplets with a chemical thought to maintain the raft upright.

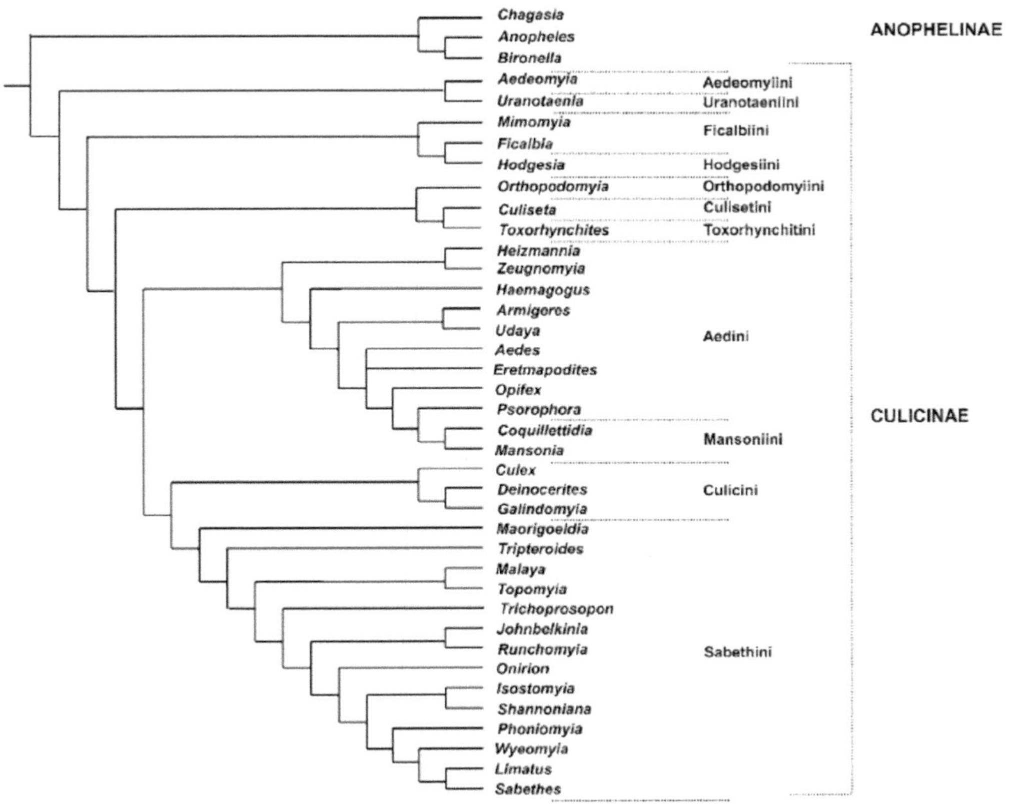

FIGURE 2.9 Phylogeny of mosquito genera based on morphology (after Harbach & Kitching, 1998): MPC from implied-weighting analysis (K=1) of 38 genera and 73 morphological characters.

Source: **Harbach, 2007.**

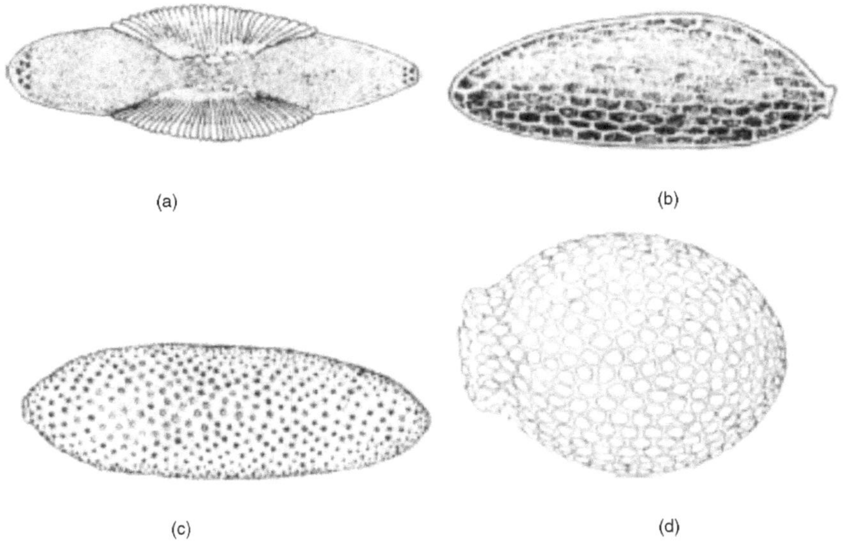

FIGURE 2.10 Eggs of mosquitoes, showing variations in shape and chorionic sculpturing. (a) Anopheles; (b) Culex; (c) *Aedes aegypti*; (d) *Toxorhynchites brevipalpis* (a and b from Ross, 1947; c and d from Harbach and Knight, 1980).

Source: **Foster & Walker, 2002.**

FIGURE 2.11 Mosquito egg rafts and clusters. (a) Floating egg raft of *Culex restuans*; (b) submerged egg cluster of Mansonia, attached to underside of floating leaf.

***Source*: a, from Ross, 1947; b, from Gordon and Lavoipierre, 1962. *Source*: Foster & Walker, 2002.**

They hypothesized monophyly of subfamily Anophelinae, forming the most basal clade of the family, and tribes Culicini and Sabethini. But most diverse was Aedini, due largely to polymorphism, which was recovered as a paraphyletic assemblage with respect to the monophyletic Mansoniini. Other tribes in the family posed much confusion and problems. One of the most radical change was that their analyses did not support the recognition of a separate subfamily for genus Toxorhynchites, which was downgraded merely as a tribe within subfamily Culicinae (Figure 2.9).

2.4.5 Classification by Reinert et al. (2004, 2006, 2008, 2009)

With the advent of new and sophisticated computerized and molecular (COI and COII mitochondrial genes) analyses based methods and application in mosquito taxonomy using comprehensive datasets began a new era to address *de novo* phylogeny and classification of mosquitoes in the late 1990s, albeit numerous problems that usually arise on such subtle occasions to change the mosquito classification and nomenclature largely owing to dependence on a few genes only (Becker et al., 2020). Reclassifications by Reinert et al. (2004, 2006, 2008, 2009) were based on comprehensive morphological studies of all life stages of the tribe Aedini. They published their extensive morphology-cum-phylogenetic studies on tribe Aedini (which contains almost 25% of known species of mosquitoes, including vectors of human disease agents), where they created 74 new, elevated, or resurrected genera from the single genus Aedes, thus increasing about three times the number of genera in the family Culicidae. This classificatory system with an unnecessary large number of unviable taxa did not gain acceptance widely across the world and after due consideration a much revised system with much lesser quantum of subgenera was soon proposed by Harbach (2007) and Wilkerson et al. (2015) who extensively studied Aedini from India and Afghanistan both morphologically and molecularly.

Wilkerson et al. (2015) re-analyzed Reinert et al.'s data on generic groupings and found that their phylogeny was obviously weakly supported and their taxonomic rankings failed priority and other taxon-naming criteria. Wilkerson et al. (2015) proposed simplified Aedine generic designations to restore a more useful classification for the operational community and to correct the unstable classification of tribe Aedini. They reduced the ranks of the genera and subgenera of Reinert et al. to subgenera or informal groups, respectively, and followed the generic structure of tribe Aedini to its status prior to the year 2000 (For more details, visit the online taxonomic catalog of the Walter Reed Biosystematics Unit at www.mosquitocatalog.org).

2.4.6 Classification by Harbach (2007, 2018)

After a well-considered downgrading of Toxorhynchitinae to a tribal status within subfamily Culicinae, the classification of Culicidae before the end of the 20th century comprised 2 subfamilies, 11 tribes, and 38 genera. Subsequent division of Aedes into four genera, i.e., Verrallina, Ayurakitia, Aedes and Ochlerotatus (Reinert, 1999, 2000a,b) and the recent recognition of two additional genera, Borichinda (tribe Aedini) and Kimia (tribe Sabethini), resulted in the modified traditional

TABLE 2.2
Family Culicidae, with 2 subfamilies, 12 tribes, and 44 genera

SUBFAMILY	Tribe	Genus
ANOPHELINAE	Anophelini	*Anophele, Bironella, Chagasia*
CULICINAE	Aedeomyiini	*Aedeomyia*
	Aedini	*Aedes, Armigeres, Ayurakitia, Borichind, Eretmapodites, Haemagogus, Heizmannia,* *Ochlerotatus, Opifex, Psorophora, Udaya, Verrallina, Zeugnomyia*
	Culicini	*Culex, Deinocerites, Galindomyia, Lutzia*
	Culisetini	*Culiseta*
	Ficalbiini	*Ficalbia, Mimomyia*
	Hodgesiini	*Hodgesia*
	Mansoniini	*Coquillettidia, Mansonia*
	Orthopodomyiini	*Orthopodomyia*
	Sabethini	*Isostomyia, Johnbelkinia, Kimia, Limatus,* *Malaya, Maorigoeldia, Onirion, Runchomyia,* *Sabethes, Shannoniana, Topomyia, Trichoprosopon, Tripteroides, Wyeomyia*
	Toxorhynchitini	*Toxorhynchites*
	Uranotaeniini	*Uranotaenia*

Source: Modified from Harbach (2007).

classification indicated in Table 2.2. Based on the intuitive interpretation of morphology, this classification appears to reflect natural groupings, and seem to be backed up molecularly. Later, Harbach (1918) classified Culicidae into 41 genera, 162 subgenera, 3559 species and 1301 subspecies which also comprised 25 fossil species.

2.4.7 Classification by Rueda & Debboun (2020)

These authors followed Harbach (2007) in recognizing two subfamilies, the Anophelinae and Culicinae, under the family Culicidae. While Anophelinae has only one tribe, the subfamily Culicinae comprises 11 tribes. Together, they house a little over 3700 described mosquito species and subspecies, under 41 genera and 188 subgenera of mosquitoes worldwide (WRBU, 2018; Rueda and Debboun, 2020).

2.4.8 Classification by do Nascimento et al. (2021)

All the world's more than 3700 valid species are classified into two subfamilies, viz., Anophelinae and Culicinae, accepted to be monophyletic, more or less completely for the former and for many tribes in the latter (Harbach, 2013; Chen et al., 2023), despite some inner diverse relations that are not entirely understood, and 113 genera (https://mosquito-taxonomic-inventory.info; do Nascimento et al., 2021). The subfamily Anophelinae includes the genera *Anopheles, Bironella* and *Chagasia*, while the subfamily Culicinae includes a variety of genera within the tribes Aedeomyiini, Aedini, Culicini, Culisetini, Ficalbiini, Hoggesiini, Mansoniini, Orthopodomyiini, Sabethini, Toxorhynchitini, and Uranotaeniini, totaling 110 genera (Harbach, 2007, 2013; Harbach and Kitching, 1998; Forattini, 2002).

The Anophelinae subfamily is demonstrably monophyletic, a basal lineage in relation to the Culicidae family. To date, 493 species are recorded, with many complex and sibling-species relationships awaiting deeper phylogenetic and taxonomic resolution, mainly in the Anopheles genus. The other two genera, *Bironella* and *Chagasia*, although better studied phylogenetically, are not occurring in India. Like the Anophelinae, the phylogenetic history of the Culicinae subfamily is not completely settled, but some groups have been better studied due their epidemiological role as potential vectors of microorganisms.

2.4.9 Classification by Wilkerson et al. (2015, 2021)

This is possibly the latest and most acceptable classification of mosquitoes based on detailed molecular and morphological interpretations (Wilkerson et al., 2021), even though greatly influenced by Harbach (2007, 2013, 2018) and Harbach and Kitching (1998). They summarized all the valid subfamilies, tribes, genera, species and subspecies, which are briefly tabulated numerically in the Table 2.3.

Wilkerson et al. (2021) have obviously further simplified on Harbach (2007), in particular, but due to dynamic taxonomy and evolutionary affinities the mosquito classification is apparently still not stabilized and require constant explorations to make it logically concluded and accepted universally.

TABLE 2.3
A Summary of Quantum of Genera, Species, and Subspecies Worldwide

Subfamily	Tribe	Genus	Species (No.)	Subspecies (No.)	Total (No.)
Anophelinae Grassi, 1900	–	***Anopheles*** Meigen, 1818	185	33	509
		Bironella Theobald, 1905	8	–	8
		Chagasia Cruz, 1906	5	–	5
Culicinae Meigen, 1818	Aedeomyiini Theobald, 1901	***Aedeomyia*** Theobald, 1901	7	–	7
		Aedes Meigen, 1818	935	35	970
		Armigeres Theobald, 1901	58	1	59
		Eretmapodites Theobald, 1901	48	3	51
		Haemagogus Williston, 1896	28	–	28
		Heizmannia Ludlow, 1905	40	–	40
		Opifex Hutton, 1902	2	–	2
		Psorophora Robineau-Desvoidy, 1827	49	–	49
		Udaya Thurman, 1954	95	–	95
		Zeugnomyia Leicester, 1908	4	–	4
	Culicini Meigen, 1818	***Culex*** Linnaeus, 1758	769	30	799
		Deinicerites Theobald, 1901	18	–	18
		Galindomyia Stone & Barreto	1	–	1
		Lutzia Theobald, 1903	9	–	9
	Culisetini Belkin, 1962	***Culiseta*** Felt, 1904	39	2	41
	Ficalbiini Belkin, 1962	***Ficalbia*** Theobald, 1903	8	–	8
		Mimomyia Theobald, 1903	45	2	47
	Hodgesiini Belkin, 1962	***Hodgesia*** Theobald, 1904	11	–	11

(Continued)

TABLE 2.3 (Continued)

Subfamily	Tribe	Genus	Species (No.)	Subspecies (No.)	Total (No.)
	Mansoniini Belkin, 1962	***Coquillettidia*** Dyar, 1905	58	–	58
		Mansonia Blanchard, 1901	25	1	26
	Orthopodom-yiini Belkin, Heinemann & Page, 1970	***Orthopodomyia*** Theobald, 1904	35	–	36
	Sabethini Blanchard, 1905	***Isostomyia*** Coquillett, 1906	4	–	4
		Johnbelkinia Zavortink, 1979	3	–	3
		Kimia Vu Duc Huong & Harbach, 2007	5	–	5
		Limatus Theobald, 1901	9	–	9
		Malaya Leicester, 1908	12	–	12
		Maorigoeldia Edwards, 1930	1	–	1
		Onirion Peyton & Harbach, 2000	7	–	7
		Runchomyia Theobald, 1903	8	–	8
		Sabethes Robineau-Desvoidy, 1827	41	–	41
		Shannoniana Lane & Cerqueira, 1942	3	–	3
		Topomyia Leicester, 1908	65	–	65
		Trichoprosopon Theobald, 1901	13	2	15
		Tripteroides Giles, 1904	122	4	126
		Wyeomyia Theobald, 1901	139	1	140
	Toxorhynchi-Tini Lahille, 1904	***Toxorhynchites*** Theobald, 1901	90	10	100
	Uranotaeniini Lahille, 1904	***Uranotaenia*** Lynch ***Arribalzaga***, 1891	271	6	277
Total	**10**	**41**	**3570**	**130**	**3700** (Spp.+ Sub spp.)

Source: Modified from Wilkerson et al. (2021).

2.5 CLASSIFICATION OF INDIAN SPECIES

The Indian mosquitoes were for the first time best dealt with taxonomically in a comprehensive manner by James and Liston (1904), Theobald (1907), Christophers (1924, 1933) and Barraud (1934) during the British Empire and by Puri (1960), Wattal and Kalra (1961), and, to a lesser extent, by restricting to only the vector mosquito species, by Tyagi (2020, 2023) after India's independence in 1947. So far, all these mosquito specialists classified Indian mosquito taxa under three subfamilies (Anophelinae, Toxorhynchitinae and Culicinae) which, following detailed molecular analyses, have now been reorganized into two, Anophelinae and Culicinae, subfamilies while the erstwhile Toxorhynchitinae subfamily having got subsumed with the latter. At present 420 species of mosquitoes (Culicidae) occur in India, the highest number ever inventorized (Tyagi et al., 2015a).

TABLE 2.4
Indian Mosquito Species Arranged under Subfamilies, Tribes, Genera and Subgenera

SUBFAMILY Tribe	No. of Genera	No. of Subgenera	No. of Species
ANOPHELINAE	1	2	64
Anophelini			
CULICINAE	1	1	9
Toxorhynchitini			
Aedeomyiini	1	0	2
Aedini	34	18	181
Culicini	2	8	90
Culisetini	1	2	3
Ficalbiini	2	3	7
Hodgesiini	1	0	2
Mansoniini	2	2	7
Orthopodomyiini	1	0	5
Sabethini	3	4	16
Uranotaeniini	1	2	34
Total	**50**	**42**	**420**

Source: Adapted from Tyagi et al. (2015a).

Due to age old problem of malaria in India by which nearly 100 million people were sickened and about one million died on yearly basis till the first quarter of the 20th century, it was but natural that, compared to Culicinae subfamily, the members of the Anophelinae subfamily were maximally subjected to scientific inquiry into their taxonomy, biology, ecology, behavior, and parasitology (Puri 1949, 1955; Wattal, 1963; Rao, 1984; Das et al., 1990; Nagpal and Sharma, 1995). In the absence of any discovery of the fossilized mosquito from India, the phylogenetic and evolutionary studies are completely wanting so far. In Table 2.4 an updated summary of extant Indian taxa is presented (Bhattacharyya et al., 2014; Tyagi et al., 2015a; Tyagi, 2004, 2020).

2.6 FINAL REMARKS

Although, originally Christophers (1933) and Barraud (1934) had cataloged all the extant Indian taxa and dealt extensively with their taxonomic characters and keys, yet none of their following culicidologists could either comprehensively compile or update the Indian species of mosquitoes (family Culicidae), save for a few sincere but near complete attempts, exclusively for Anophelinae, by Puri (1955, 1960), Wattal and Kalra (1961), Rao (1984) and Nagpal and Sharma (1995). More recently, a couple of inventories of Indian mosquito species were prepared by Bhattacharyya et al. (2014) and Tyagi et al. (2015a); the latter also offered identification keys for the dominant vectors of major mosquito-borne diseases in India, viz., malaria, lymphatic filariasis, dengue and Japanese encephalitis. An updated taxonomic, phylogenetic, evolutionary and geographic treatment of the Indian taxa, especially based on molecular analyses, is still evasive. In modern times. mosquito taxonomists use new methods of computerized and molecular analyses and comprehensive data sets to address phylogeny and classification of mosquitoes. Numerous problems, however, usually arise to change the mosquito classification and nomenclature. The mosquito taxonomy, classification and phylogeny have revealed, with the application of explicit methods of phylogenetic analysis using molecular tools, a number of weaknesses in the traditional phenetic classification system, and despite offering many changes little progress has been made to present a stable and robust classification that reflects comprehensive evolutionary relationships.

Changes to the mosquito classification are unavoidable if the classification is to be presented to reflect evolutionary history as evidenced by molecular interpretations (Harbach, 2007, 2018, Wilkerson et al., 2021). Knowledge of mosquito phylogeny is best derived from a few fossils recovered from Europe and Southeast Asia as well as the various morphological characteristics and their approximations among different taxa under a large number of groups and series, currently analytically rearranged more appropriately molecularly for their placement and evolutionary affinities. Based on robust molecular assays and analyses the revolutionary decision by modern systematists to obliterate Toxorhynchitinae subfamily and reconstitute only two subfamilies, viz., Anophelinae, and Culicinae, with the latter incorporating all the toxorhynchitines in between Anophelinae and Culicinae, is still in a dynamic and fluid state of affairs since there is no consensus on the modern classification yet *hook, line and sinker*, and without doubt more fecund research is required to fill in the gaps in our understanding of Culicidae phylogeny and evolution (Rueda and Debboun, 2020).

REFERENCES

Antunes, P.C.A. (1937). A new *Anopheles* and a new *Goeldia* from Colombia (Diptera: Culicidae). *Bulletin of Entomological Research*, 28, 69–73.

Barraud, P.J. (1934). *The fauna of British India, including Ceylon and Burma. Diptera: Family Culicidae; Tribes Megarhinini and Culicini*. Vol. 5, Taylor & Francis, London, 463 pp.

Becker, N., Dušan, P., Marija, Z., Boase, C., Madon, M.B., Dahl, C. and Kaiser, A. (2020). *Subfamily Anophelinae*. In: *Mosquitoes: Identification, Ecology and Control*. 3rd Edition. Fascinating Life Sciences. Springer, Cham. pp 169–192. https://doi.org/10.1007/978-3-030-11623-1_9

Belkin, J.N. (1962). *The Mosquitoes of the South Pacific (Diptera, Culicidae)*. Volume 1–2. University of California Press; Berkeley/Los Angeles, CA.

Bertone, M.A., Courtney, G.W., Wiegmann, B.M. (2008). Phylogenetics and temporal diversification of the earliest true flies (Insecta: Diptera) based on multiple nuclear genes. *Systematic Entomology*, 33, 668–687.

Bhattacharyya, D.R., Rajavel, A.R., Mohapatra, P.K., Jambulingam, P., Mahanta, J., Prakash, A. (2014). Faunal richness and the checklist of Indian mosquitoes (Diptera: Culicidae). *Check List*, 10(6), 1342–1358.

Borkent, A. and Grimaldi, D.A. (2004). The earliest fossil mosquito (Diptera: Culicidae), in mid-Cretaceous amber. *Annals of the Entomological Society of America*, 97, 882–888.

Bourroul, C. (1904). Mosquitos do Brasil. These, Faculdade de Medicina da Bahia. Officina Typographica, Bahia. [Consisting of Prefacio, (pp. i–viii), Novas especies (pp. 1–6 + 30–32), Quadro das especies encontradas na Bahia (4 pp.) and Proposições (18 pp.) by Bourroul; remainder and bulk of work by Adolpho Lutz]

Chen, D.-H., He, S.-L., Fu, W.-B., Yan, Z.-T., Hu, Y.-J., H. Yuan, Wang, M.-B., Chen, B. (2023). Mitogenome-based phylogeny of mosquitoes (Diptera: Culicidae). https://doi.org/10.1111/1744-7917.13251

Christophers, S.R. (1924). Provisional list and reference catalogue of the Anophelini. *Indian Medical Research Memoirs*, 3, 1–105.

Christophers, S.R. (1933). *The Fauna of British India, including Ceylon and Burma. Diptera: Family Culicidae; Tribe Anophelini*. Vol. 4, Taylor & Francis, London, 271 pp.

Christophers, S.R. (1960). *Aedes aegypti (L.), the Yellow Fever Mosquito: Its Life History, Bionomics and Structure*. London: Cambridge University Press, 739 pp.

Collins, L., and Blackwell, A.B. (2000). The biology of *Toxorhynchites* mosquitoes and their potential as biocontrol agents. *Biocontrol News and Information*, 21, 105–116.

Collucci, E., and Sallum, M.A.M. (2003). Phylogenetic analysis of the subgenus Kerteszia of Anopheles (Diptera: Culicidae: Anophelinae) based on morphological characters. *Insect Systematics and Evolution*, 34, 361–372.

Collucci, E., Sallum, M.A.M. (2007). Cladistic analysis of the subgenus Anopheles (Anopheles) Meigen (Diptera: Culicidae) based on morphological characters. *Memórias do Instituto Oswaldo Cruz*, 102, 277–291.

Da Silva, A., Machado, L.C., De Paula, M.B., Vieira, C.J.D.S.P., de Morais Bronzoni, R.V., de Melo-Santos, M.A.V. and da Luz, W.G. (2019) Time and mode of Culicidae evolutionary history. BioRxiv. https://doi.org/10.1101/871145

Da Silva, A.F., Machado, L.C., de Paula, M.B., da Silva Pessoa Vieira, C.J., de Morais Bronzoni, R.V., de Melo Santos, M.A.V. and Wallau, G.L. (2020) Culicidae evolutionary history focusing on the Culicinae subfamily based on mitochondrial phylogenomics. *Scientific Reports*, 10(1), 18823. https://doi.org/10.1038/s41598-020-74883-3

Das, B.P., Rajagopal, R., and Akiyama, J. (1990). Pictorial key to the species of Indian Anopheline mosquitoes. *Journal of Pure Applied Zoology*, 2, 131–162.

do Nascimento, B.L.S., da Silva, F.S., Nunes-Neto, J.P., de Almeida Medeiros, D.B., Cruz, A.C.R., da Silva, S.P., da Silva, E., Silva, L.H., de Oliveira Monteiro, H.A., Dias, D.D., Vieira, D.B.R., Rosa, J.W. Jr, Brandão, R.C.F., Chiang, J.O., Martins, L.C., da Costa Vasconcelos, P.F. (2021) First description of the Mitogenome and phylogeny of Culicinae species from the Amazon region. *Genes* (Basel), 12(12), 1983. https://doi.org/10.3390/genes12121983

Donald C.L., Varjak M., Aguiar E.R.G.R., Marques J.T., Sreenu V.B., Schnettler E., Kohl A. 2018. Antiviral RNA interference activity in cells of the predatory mosquito. *Viruses*, 10, 694. https://doi.org/10.3390/v10120694

Donald, C.L., Siriyasatien, P., Kohl, A. (2020). *Toxorhynchites* species: A review of current knowledge. *Insects*, 11(11), 747. https://doi.org/10.3390/insects11110747

Edwards, F.W. (1932). *Genera Insectortum. Diptera, Fam. Culicidae. Fascicle 194*. Bruxelles: Desmet-Verteneuil.

Focks, D.A. (2007). Toxorhynchites as biocontrol agents. *Journal of American Mosquito Control Association*, 23(Suppl. 2), 118–127. https://doi.org/10.2987/8756-971X(2007)23[118:TABA]2.0.CO;2

Forattini, O.P. (2002). *Culicidologia Médica: Identificação, Biologia, Epidemiologia*. São Paulo, Brazil: EDUSP.

Foster, W.A. and Walker E.D. (2002). Mosquitoes (Culicidae). *In*: G. Mullen and L. Durden (eds.), *Medical and Veterinary Entomology*, San Diego: Academic Press, pp. 203–262.

Gordon, R. M. and Lavoipierre, M.M.J. (1962). *Entomology for students of medicine*. Oxford: Blackwell Scientific, 353 pp.

Harbach, R.E. (1977) Comparative and functional morphology of the mandibles of some fourth stage mosquito larvae (Diptera: Culicidae) *Zoomorphologie*, 87, 217–236. https://doi.org/10.1007/BF00995821

Harbach, R.E. (1978) Comparative structure of the labiohypopharynx of fourth stage mosquito larvae (Diptera: Culicidae), with comments on larval morphology, evolution and feeding habits. *Mosquito Systematics*, 10, 301–333.

Harbach, R.E. (2007) The Culicidae (Diptera): A review of taxonomy, classification and phylogeny. *Zootaxa*, 1668, 591–638. https://doi.org/10.11646/zootaxa.1668.1.28

Harbach, R.E. (2013) Mosquito Taxonomic Inventory. [(accessed on 3 November 2021)]. Available online: https://www.mosquito-taxonomic-inventory.info/

Harbach, R.E. (2018). *Culicipedia: species-group, genus-group and family-group names in Culicidae (Diptera)*. Wallingford, Oxfordshire, UK: CABI Publishing. 378 pp.

Harbach, R.E., Kitching I.J. (1998). Phylogeny and Classification of the Culicidae (Diptera). *Systematic Entomology*, 23, 327–370. https://doi.org/10.1046/j.1365-3113.1998.00072.x

Harbach, R.E. and Kitching, I.J. (2005). Reconsideration of anopheline phylogeny (Diptera: Culicidae: Anophelinae) based on morphological data. *Systematics and Biodiversity*, 3, 345–374.

Harbach, R.E. and Knight, K.L. (1980). *Taxonomists' Glossary of Mosquito Anatomy*. Marlton, New Jersey: Plexus Publishing, Inc., xi + 415 pp.

Harbach, R.E., Rattanarithikul, R., Harrison, B.A. and Baimaia (2005) a new subgenus for *Anopheles kyondawensis* Abraham, a unique crabhole-breeding anopheline in southeastern Asia. *Proceedings of the Entomological Society of Washington*, 107, 750–761.

Harrison, B.A. and Scanlon, J.E. (1975). Medical Entomology Studies – II. The Subgenus *Anopheles* in Thailand (Diptera: Culicidae). *Contributions of the American Entomological Institute*, 12(1), 308 pp.

James, S.P. and Liston, G. W. (1904). *A monograph of the Anopheles mosquitoes of India*. Calcutta: Thacker, Spink & Co.

Kiauta, B. (1975). Cytotaxonomy of dragonflies, with special reference to the Nepalese fauna. Lectures delivered at the Tribhuvan University, Kathmandu, Vol. 2. Nepal Research Centre, Kathmandu. xiv+78 pp.

Knab, F. (1913). The species of Anopheles that transmit human malaria. *American Journal of Tropical Diseases and Preventive Medicine*, 1, 33–43.

Knight, K.L. and Stone, A.A. (1977). *Catalog of the mosquitoes of the world (Diptera: Culicidae)*. 2nd edition. Thomas Say Foundation 6: 1–611.

Komp, W.H.W. (1937). The species of the subgenus Kerteszia of Anopheles (Diptera, Culicidae). *Annals of the Entomological Society of America*, 30, 492–529.

Krishnamoorthy, R., Munirathinam, A., Dhananjeyan, K.J., Hiriyan, J., Mariappan, T., Samuel P.P. and Venkatesh A. (2013). Description of a new species, *Toxorhynchites (Toxorhynchites) tyagii* (Diptera: Culicidae), from Nilgiri hills, Western Ghats, southern India. *Zootaxa*, 3701(4), 447–459. http://doi.org/10.11646/zootaxa.3701.4.4

Lorenz, C., Alves, J.M.P., Foster, P.G., Suesdek, L. and Sallum, M.A.M. (2021). Phylogeny and temporal diversification of mosquitoes (Diptera: Culicidae) with an emphasis on the Neotropical fauna. *Systematic Entomology*, 46(4), 798–811. https://doi.org/10.1111/syen.12489

Martynov, A.V. (1938). Études sur l'histoire géologique et de phylogénie des ordres des insectes (Pterygota). 1e partie. *Palaeoptera et Neoptera-Polyneoptera, Tray. Inst. paléont. Acad. Sci. U.R.S.S.* 7(4), 150 pp.

Meigen, J.W. (1818) *Systematische Beschreibung der bekannten europäischen zweiflügeligen Insekten*. Volume 1. Aachen.

Miller, B.R., Crabtree, M.B. and Savage, H.M. (1997). Phylogenetic relationships of the Culicomorpha inferred from 18S and 5.8S ribosomal DNA sequences. (Diptera:Nematocera). *Insect Molecular Biology*, 6, 105–114. https://doi.org/10.1111/j.1365-2583.1997.tb00078.x

Mitchell, A., Sperling, F.A. and Hickey, D.A. (2002). Higher-level phylogeny of mosquitoes (Diptera: Culicidae): mtDNA data support a derived placement for Toxorhynchites. *Insects Systematics & Evolution*, 33, 163–174.

Nagpal B.N. and Sharma, V.P. (1995). *Indian Anophelines*. New Delhi: Oxford and IBH Publishing Co. Pvt. Ltd.

OECD (2018). *Safety Assessment of Transgenic Organisms in the Environment, Volume 8: OECD Consensus Document of the Biology of Mosquito Aedes aegypti, Harmonisation of Regulatory Oversight in Biotechnology*. Paris: OECD Publishing, 106 pp. https://doi.org/10.1787/9789264302235-en

Pal, R. and Ramalingam, S. (1981). *Invertebrate predators of mosquitoes*. World Health Organization WHO/VBC/81, p. 799.

Puri, I.M. (1949). *Anophelines of the Oriental region*. Boyd's Malariology. New York: W.B. Saunders 85 Co, pp. 483–505.

Puri, I.M. (1955). *The Distribution of Anopheline Mosquitoes in India*. Health Bulletin No. 17; Malaria Bureau No. 7, Government of India. 104 pp.

Puri, I.M. (1960). Synoptic tables for the identification of the full-grown larvae of the Indian anopheline mosquitoes. Health Bulletin No. 16, Malaria Bureau No. 7, Government of India Press, Calcutta. 104 pp.

Rao T.R. (1984). *The Anophelines of India*. Malaria Research Centre, Indian Council of Medical Research, 518 pp.

Reidenbach, K.R., Cook, S., Bertone, M.A., Harbach, R.E., Wiegmann, B.M. and Besansky N.J. (2009). Phylogenetic analysis and temporal diversification of mosquitoes (Diptera: Culicidae) based on nuclear genes and morphology. *BMC Evolutionary Biology*, 9, 298. https://doi.org/10.1186/1471-2148-9-298

Reinert, J.F. (1999). Restoration of Verrallina to generic rank in tribe Aedini (Diptera: Culicidae) and descriptions of the genus and three included subgenera. *Contributions of the American Entomological Institute (Gainesville)*, 31(3), 1–83.

Reinert, J.F. (2000a). Restoration of Ayurakitia to generic rank in tribe Aedini and a revised definition of the genus. *Journal of the American Mosquito Control Association*, 16, 57–65.

Reinert, J.F. (2000b). New classification for the composite genus Aedes (Diptera: Culicidae: Aedini), elevation of subgenus Ochlerotatus to generic rank, reclassification of the other subgenera, and notes on certain subgenera and species. *Journal of the American Mosquito Control Association*, 16, 175–188.

Reinert, J.F., Harbach, R.E. and Kitching, I.J. (2004) Phylogeny and classification of Aedini (Diptera: Culicidae) based on morphological characters of all life stages. *Zoological Journal of the Linnaean Society*, 142, 289–368.

Reinert, J.F., Harbach, R.E. and Kitching, I.J. (2006) Phylogeny and classification of Finlaya and allied taxa (Diptera: Culicidae: Aedini) based on morphological data from all life stages. *Zoological Journal of the Linnean Society*, 148, 1–101.

Reinert, J.F., Harbach, R.E. and Kitching, I.J. (2008). Phylogeny and classification of *Ochlerotatus* and allied taxa (Diptera: Culicidae: Aedini) based on morphological data of all life stages. *Zoological Journal of the Linnean Society*, 153, 29–114.

Reinert, J.F., Harbach, R.E. and Kitching, I.J. (2009). Phylogeny and classification of Aedini (Diptera: Culicidae). *Zoological Journal of the Linnean Society*, 157, 700–794.

Ribeiro, H.L. (2004). *Toxorhynchites* Theobald de Madagascar (Diptera: Culicidae). *Annales de la Société Entomologique de France* (N.S.) 40, 243–257. https://doi.org/10.1080/00379271.2004.10697423

Rose, N.H., Sylla, M., Badolo, A., White, B.J., Crawford, J.E. and McBride, C.S. (2020). Climate and urbanization drive mosquito preference for humans. *Current Biology*, 30(18), 3570–3579. https://doi.org/10.1016/j.cub.2020.06.092

Ross, H.H. (1947). The mosquitoes of Illinois (Diptera, Culicidae). Ill. *Natural History Survey Bulletin*, 24(1), 1–96.

Ross, H.H. (1951). Conflict with Culex. *Mosquito News*, 11, 128–132.

Rueda, L.M. and Debboun M., (2020). Chapter 1 – Taxonomy, Identification, and Biology of Mosquitoes. (*In*: *Mosquitoes, Communities, and Public Health in Texas*. Eds.: Mustapha Debboun, Martin Reyna Nava, Leopoldo M. Rueda), Academic Press, pp. 3–7. https://doi.org/10.1016/B978-0-12-814545-6.00001-8

Sallum, M.A.M., Foster, P.G., Li, C., Sithiprasasna, R. and Wilkerson R.C. (2007). Phylogeny of the Leucosphyrus Group of Anopheles (Cellia) (Diptera: Culicidae) based on mitochondrial gene sequences. *Annals of the Entomological Society of America*, 100, 27–35.

Sallum, M.A.M., Schultz, T.R., Foster, P.G., Aronstein, K., Wirtz, R.A. and Wilkerson, R.C. (2002) Phylogeny of Anophelinae (Diptera: Culicidae) based on nuclear ribosomal and mitochondrial DNA sequences. *Systematic Entomology*, 27, 361–382.

Sallum, M.A.M., Schultz, T.R. and Wilkerson R.C. (2000). Phylogeny of Anophelinae (Diptera Culicidae) based on morphological characters. *Annals of the Entomological Society of America*, 93, 745–775.

Shepard, J.J., Andreadis, T.G. and Vossbrinck, C.R. (2006). Molecular phylogeny and evolutionary relationships among mosquitoes (Diptera: Culicidae) from the northeastern United States based on small subunit ribosomal DNA (18S rDNA) sequences. *Journal of Medical Entomology*, 43, 443–454. https://doi.org/10.1093/jmedent/43.3.443

Steffan W.A., (1975). Systematics and biological control potential of Toxorhynchites (Diptera: Culicidae) *Mosquito Systematics News Letter*, 7, 59–68.

Steffan, W.A. and Evenhuis, N.L. (1985). Classification of the Subgenus Toxorhynchites (Diptera: Culicidae) I. Australasian, eastern Palaearctic, and Oriental species-groups1. *Journal of Medical Entomology*, 22, 421–446.

Theobald, F.V. (1901a). *A Monograph of the Culicidae or Mosquitoes*. Volume 1. London: Brit-ish Museum (Natural History).

Theobald, F.V. (1901b) *A Monograph of the Culicidae or Mosquitoes*. Volume 2. London: British Museum (Natural History).

Theobald, F.V. (1903). *A Monograph of the Culicidae or Mosquitoes*. Volume 3. London: British Museum (Natural History).

Theobald, F.V. (1907). *A Monograph of the Culicidae or Mosquitoes*. Volume 4. London: British Museum (Natural History).

Theobald, F.V. (1910). *A Monograph of the Culicidae or Mosquitoes*. Volume 5. London: British Museum (Natural History).

Tyagi, B.K. (2004). *The Invincible Deadly Mosquitoes*. Jodhpur: Scientific Publishers (India), 265 pp.

Tyagi, B.K. (2020). *Mosquito Hunters: A history of hostilities against man's deadliest Foe – the mosquito – since 1881*. Scientific Publishers (India), pp. 474.

Tyagi, B.K. (2023). *Desert Malaria: An Emerging Malaria Paradigm and Its Global Impact on Disease Elimination*. Singapore: Springer-Nature, 416 pp.

Tyagi, B.K., Hiriyan, J., Samuel, P.P., Tewari, S.C. and Paramasivan, R. (2006). Dengue in Kerala: A critical review. *ICMR Bulletin*, 36(4–5), 13–22.

Tyagi, B.K., Munirathinam, A. and Venkatesh, A. (2015a). A catalogue of Indian mosquitoes. *International Journal of Mosquito Research*, 2(2), 50–97.

Tyagi B.K., Munirathinam A., Krishnamoorthy R., Baskaran G., Govindarajan R., Krishnamoorthi R., Mariappan T., Dhananjeyanand K.J. and Venkatesh A. (2015b). A revision of genus Toxorhynchites Theobald (1901) in the South-East Asian countries, with description of a new species *Toxorhynchites (Toxorhynchites) darjeelingensis* from West Bengal, India (Diptera, Culicidae). *Halteres*, 6, 13–32.

Tyagi, B.K. and Vijay, V. (2016). *Entomology in the Doon Valley (Garhwal Himalaya): A Stronghold for Insect Research*. Scientific Publishers (India). 300 pp.

Vinogradov, D.D., Sinev, A.Y. and Tiunov, A.V. (2022). Predators as control agents of mosquito larvae in micro-reservoirs (Review). *Inland Water Biology*, 15, 39–53 (2022). https://doi.org/10.1134/S1995082922010138

Wattal, B.L. and N.L. Kalra (1961). Region wise keys to female Indian Anopheles. *Bulletin of Natural Society of India Malaria Mosquito Borne Diseases* 10, 55–138.

Wattal, B.L. (1963). Keys to the Anopheles of the world. 5. Southern and northern Asia. In: *Practical malariology*. 2nd edition. Ed., P.F. Russell, L.S. West, R.D. Manwell and G. MacDonald, London: Oxford University Press, pp. 680–696.

Wiegmann, B.M., Trautwein, M.D., Winkler, I.S., Barr, N.B., Kim, J.W., Lambkin, C., Bertone, M.A., Cassel, B.K., Bayless, K.M., Heimberg, A.M., Wheeler, B.M., Peterson, K.J., Pape, T., Sinclair, B.J., Skevington, J.H., Blagoderov, V., Caravas, J., Kutty, S.N., Schmidt-Ott, U., Kampmeier, G.E., Thompson, F.C., Grimaldi, D.A., Beckenbach, A.T., Courtney, G.W., Friedrich, M., Meier, R. and Yeates, D.K. (2011). Episodic radiations in the fly tree of life. *Proceedings of National Academy of Sciences U.S.A.*, 108(14), 5690–5695. https://doi.org/10.1073/pnas.1012675108

Wilkerson, R.C., Linton, Y.-M., Fonseca, D.M., Schultz, T., Price, D.C. and Strickman, D.A. (2015). Making mosquito taxonomy useful: A stable classification of Tribe Aedini that balances utility with current knowledge of evolutionary relationships. *PLoS ONE* 10(7), e0133602. https://doi.org/10.1371/journal.pone.0133602

Wilkerson, R.C., Linton, Y-M and Strickman, D. (2021). *Mosquitoes of the World.* Vol. I. John's Hopkins University Press. 599 pp.

Wood, D.M. and Borkent, A. (1989). Phylogeny and classification of the Nematocera. In: McAlpine JF, Wood DM, editors. *Manual of Nearctic Diptera*, Vol. 3. Hull: Research Branch Agriculture Canada Monograph No. 32, Canadian Government Publishing Centre.

WRBU (2018). Arthropods as vectors of disease: medically important arthropod species (http://www.wrbu.org/aors/aors_keys.html).

Yeates, D.K. and Wiegmann, B.M. (2005). Phylogeny and evolution of Diptera: recent insights and new perspectives. *In*: Yeates DK, Wiegmann BM, editors. *The evolutionary biology of flies.* Columbia: Columbia University Press.

3 Mosquito Biogeography

B.K. Tyagi
University Institute of Biotechnology, Chandigarh University, Mohali, India

Anil Prakash
Ex-ICMR-National Institute For Research in Environmental Health, Bhopal, India

Ramesh C. Dhiman
Ex-ICMR-National Institute of Malaria Research, New Delhi, India

Vas Dev
Ex-ICMR-National Institute of Malaria Research, Guwahati, India

Abhijit Mazumdar
The University of Burdwan, Golaphag, Burdwan, India

3.1 INTRODUCTION

According to Gressitt (1974), "*Biogeography involves interplay between many disciplines (consilience), both biological and physical, including geology, palaeontology, geography, climate, glaciation etc.*" Biogeography is a reflection of far-reaching actions of life forms analyses factors and causes of past and present events behind distribution in time and space (Bynum, 1994; Ficalbi, 1896, 1893, 1894, 1895; Senthamarai et al., 2015; Senthamarai et al., 2016; Tyagi, 1995). Long-term bio-ecological behaviors where different populations of a species are rendered uncommunicated reproductively can hardly be separated from the process of evolution. The new perspective in biogeography embraces all interacting forces of the environment, including those defining ecology, inclusive of the physical environment and man's use and protection of the environment. Almost everyone has endured the uncomfortable sensation of a mosquito bite. The red mark and itching from mosquito bites are the result of an allergic reaction to the saliva of the insect. The transfer of serious diseases such as malaria, dengue, zika, filaria, and West Nile virus, which can have incapacitating and sometimes fatal effects (such as encephalitis, meningitis, and microcephaly), may be a more significant outcome of some mosquito bites (Ong and Ahmad, 2022). In addition to diseases that affect humans, mosquitoes can also spread a number of illnesses and parasites that dogs and horses are particularly vulnerable to. These include West Nile virus, dog heart worms, and eastern equine encephalitis (Haynes et al., 2022). Due to the movement of people, animals, and things on a worldwide scale as well as climate changes, mosquito-borne diseases are being reported increasingly frequently globally (Weaver and Reisen, 2010). These circumstances also give invasive and frequently alien vector species a chance to spread throughout the world (Becker et al., 2012). Their potential to spread locally amid human populations and their windborne long-distance dispersal over suitable habitats

DOI: 10.1201/9781003326991-4

highlight the need to fill in the knowledge gaps about the distribution of vector species. Monitoring the species composition in each country is one of the most crucial measures in preventing the spread of diseases caused by mosquitoes.

Restricting to mosquitoes, biogeography is still a young discipline due largely to the slow multiplicity of mosquito species and vectors out of them, together with insufficient knowledge of their bioecology, and their distribution and speciation. Mosquito abundance by species is associated with microclimatic variables – pH and temperature. Changes in mosquito distribution and abundance in time and space are predominantly driven by human activities (anthropization), and temperature and rainfall regimes that provide suitable breeding sites for the highly anthropophilic mosquitoes (e.g., *Anopheles stephensi* and *Aedes aegypti*). In recent decades it has been observed that human mobility through air and other ways impacted extension of geographical distribution of certain vector species, e.g., *Ae. albopictus* in context with chikungunya fever (Raharimalala et al., 2012; Dev et al., 2014), and *An. stephensi* in context with malaria (Tyagi, 2023), to new countries and/or regions in the world. In Lakshadweep islands, which was until recently free from dengue and chikungunya, *Aedes* mosquitoes apparently immigrated from mainland India and settled to dominate by breeding in discarded plastic containers and drums (Paramasivan et al., 2009; Samuel et al., 2009; Nihad et al., 2022).

Biogeography of mosquitoes, evidenced from all but one tiny country, Iceland in the Antartica, across the globe and in all conceivable habitats (e.g., forests, prairies, rivers, lakes, marshes, coalfields, alpine ranges, islands, deserts, fresh and polluted water sources, rock holes, caves, phytotelmata etc.), is in a way reflective to their prowess of phenotypic and genotypic plasticity and ecotypic adaptability. In this regard, therefore, Brazil, Indonesia, Malaysia, and Thailand, all of which are richly forested, harbor the highest numbers of species, while Brazil also represents the highest number of type locations and taxonomic output. The fact that 50% of mosquito species are endemic globally possibly only allude toward their specific ecosystem-driven preferences. Although distances between islands do not seem to affect mosquito species composition there, nevertheless, island nations have amassed a relatively higher number of endemic species than mainland countries of similar size, besides total number of species. Interestingly, in island situations, co-existence of multiple species at a micro-habitat level is observed with no evidence of interactions like competition or predation, which indicate their proper use of available space and other vital resources such as food and shelter. Excluding small island countries such as, for example, Panama, French Guiana, Malaysia, and Costa Rica which exhibit the highest densities and endemicity of species, large countries like Brazil, Australia, the Philippines, and Indonesia show the highest numbers of endemic species. The taxonomic output in large regions like Africa and the Middle East is among the lowest (Foley et al., 2007). India is rich in fauna and flora and endemic to many vector species affecting human health particularly the Western and Eastern Ghats regions as well as Andaman and Nicobar Islands, and the northeastern Indian States (Dev, 2020, 2022) (Figure 3.1).

3.2 ZOOGEOGRAPHICAL REGIONS

Philip Sclater and Alfred Wallace's original proposal involved division of Earth's land surface, occupied by plants and animals, into six biogeographic regions: the Nearctic region, Neotropic region, Palaearctic region, Ethiopian region, Oriental region, and Australasian region (*loc. cit.* Rao, 1984).

3.2.1 Nearctic Region

Being a cold and temperate zone, the total number of species, genera, and abundance of mosquitoes is rather restricted, albeit a handsome representation by as many as 168 species is known in this region (Darsie Jr., 1989). Genera like *Anopheles, Culex, Aedes, Psorophora, Melanoconibn, Orthopodomyiu, Deinocentes, Mansonia, Orthopodomyia, Uranotaenia* and *Wyeomyia* are well represented.

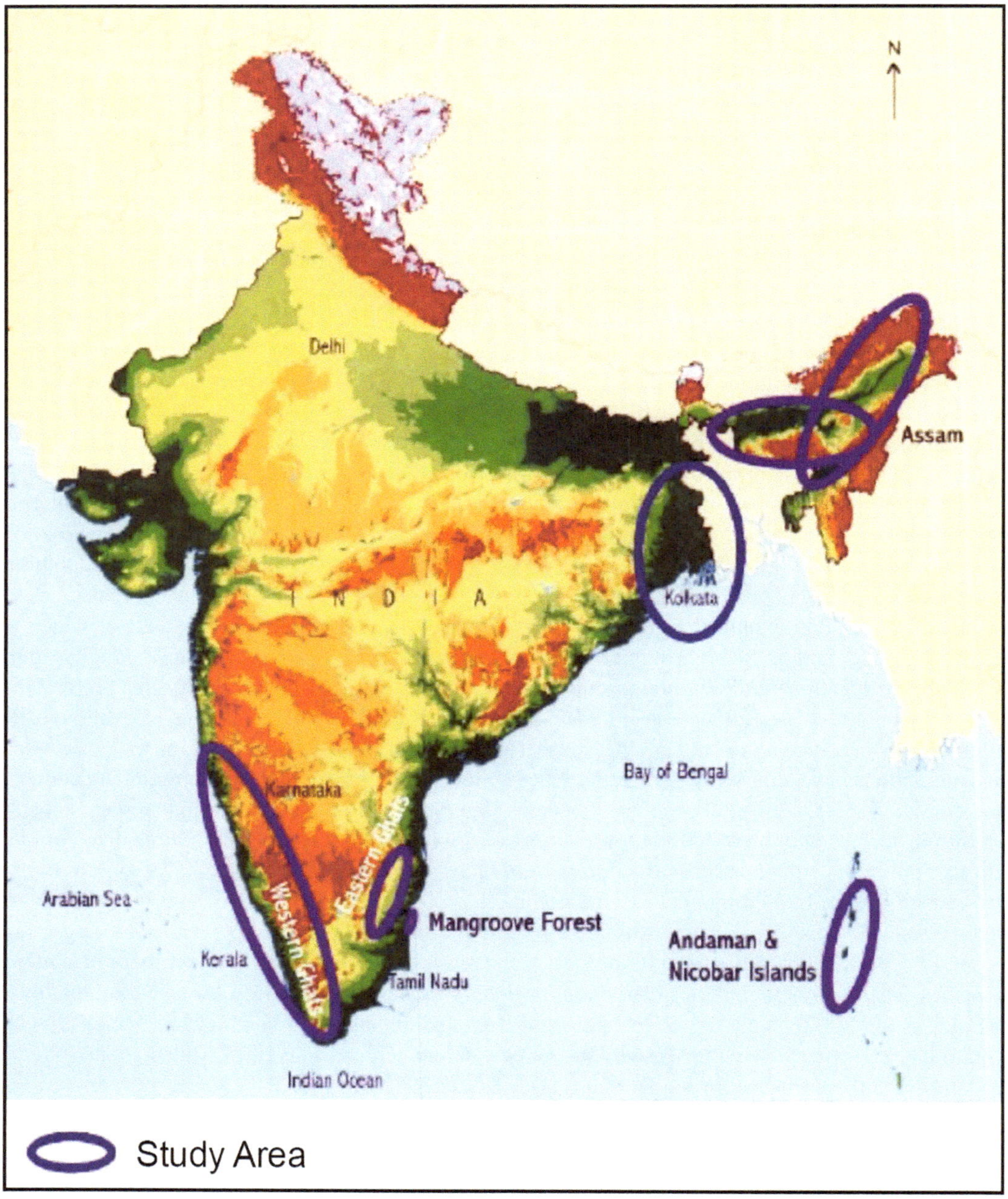

FIGURE 3.1 Important regions in India rich for mosquito fauna with endemic and rare species.

Source: Tyagi et al., 2015b.

3.2.2 Neotropical Region

Physically, this includes the tropical terrestrial ecoregions of the Americas and the entire South American temperate zone. The region is a mosaic with highly variable abiotic and biotic attributions. Temperature and rainfall, in particular, vary considerably from moderately high temperature in central America to appreciably high temperature with high humidity in north-central to the temperate southern South America. Rainfall varies regionally, resulting in a mosaic of habitats controlled by river migration, sea level fluctuations, dense forests, local dryness, and local uplift. Most of the Neotropical countries, such as Brazil, Colombia, Ecuador and Peru, are on the higher ranking of species richness;

particularly Brazil occupies the first position in such rankings and is therefore, considered the most Megadiverse country (Joly, 2008), it is no exaggeration in context with mosquitoes.

Neotropical region boasts of a large number of mosquito genera/species, and new species are regularly being brought on record. Several of these species are cryptic taxa belonging to sibling species complexes. Phylogenetic studies have shown that the genus *Chagasia*, the earliest extant branch in the subfamily Anophelinae, is found only in Latin America. This has led to the quite plausible assertion that genus Anopheles originally evolved in this region of the world (Harbach, 2013). However, the subgenus *Cellia* is characteristically absent.

3.2.3 Palearctic Region

The Palearctic region encompasses Eurasia, including Europe, northern Africa and Asia north of the Oriental region. Ficalbi (1892–1896, 1899) first inventorized the mosquito fauna of the Palearctic region which was later updated by Edwards (1921). Many genera and species are the characteristic feature of the region, but the most common and medically important mosquito for malaria, *Anopheles maculippenis*, a species complex, is highly abundant. Rao (1984) mentioned a total of 26 species under subgenera *Anopheles* and *Cellia* from Palearctic region. He suspected the Ethiopian *An. gambiae is* also present there. Rao (1984) further signifies the Palearctic fauna for the researchers of India as some of the prominent malaria vectors in the Oriental region, such as *An. culicifacies* and *An. stephensi*, are also present in the West Asian/Mediterranean countries.

3.2.4 Ethiopian Region

Ethiopian region, one of the major land areas of the world, encompasses Africa south of the sahara and the southwestern tip of Arabia. Mosquitoes of the Ethiopian region are characterized by their archaic, distinctive and homogenous nature (De Meillon, 1949), although some like *An. pharaoensis* and *An. cousteni* overflow into the southern Mediterranean subregion of the Palearctic region. Although various mosquito genera and species occur, two mosquitoes, namely, *Aedes aegypti*, and, *Anopheles gambiae* (together with *An. arabiensis*) are the deadliest species being the vectors for yellow fever/dengue and malaria, respectively, in Africa and Central and South Americas (Tyagi and Tilak, 2022). Both Indian *Anopheles* subgenera (*Anopheles* and *Cellia*) occur in the Ethiopian region. Mosquito fauna in both larval and adult forms in Ethiopian region has been investigated thoroughly by a number of researchers (Hopkins, 1952; Evans, 1938; Edwards, 1941). A total of 112 anopheline species were cataloged by Rao (1984), although many had since been possibly added to the richness of the fauna.

3.2.5 Australasian Region

The Australasian region includes Australia, New Guinea, New Zealand, eastern of the Indonesian archipelago, Bismarck Archipilago, Vanuatu, the Soloman islands and New Caledonia. Mosquitoes are present throughout Australasian region, due to warmer conditions caused by the combination of heavy rains and increasing temperatures which create the ideal conditions for their populations to boom. All major vector mosquito genera (*Anopheles*, *Culex*, and *Aedes*) occur in Australasian region which has recently been attacked by several kinds of encephalitis fevers, especially Japanese encephalitis, the latter transmitted by *Culex* spp. and likely *Armigeres subalbatus* (Thenmozhi et al., 2016). Common mosquitoes in western Australia are: *Aedes camptorhynchus*, *Ae.notoscriptus*, *Ae. vigilax*, *Anopheles annulipes s.l.*, *Culex annulirostris*, and *Cx. quinquefasciatus*. The mosquito fauna of the Australasia does not strictly show isolation in forms, as both the common Indian Anopheles subgenera, *Anopheles* and *Cellia*, occur in Australia too, notwithstanding high preponderance of the group *Neomyzomyia*. The genus *Bironella* comprised of eight species (seven in New Guinea and one in the Solomon Islands) is exclusively the inhabitants of this region. One of the species, *B. gracillis*, occurs in mainland Australia in its northern part.

3.2.6 Oriental Region

Stone and Delfinado (1973) conceived and organized the Oriental region more or less in consonance with that of Christophers (1933) and Barraud (1934), including within the region the Indian subcontinent (India, Pakistan, Sri Lanka, Bangladesh, Nepal, Bhutan, and Myanmar), Indo-China (Thailand, Laos, Kampuchea, and Vietnam), the Malayan and Indonesian islands, the Philippines, Taiwan, and Southern China (Yunnan, Kwangsi, Kwangtung, and Fukien provinces). However, Rao (1984) felt that, in order to study impact of fauna in neighboring countries, Afghanistan south of Hindukush mountains should also be included in the Oriental region. By far the Oriental region is the richest in mosquito species, abundance and biodiversity represented by a large number of genera and species. Members of all *Anopheles* species have some very clear affinities among themselves, although some certainly overlap with the neighboring Southeastern, Australasian, and West Asia countries and overflow to them and *vice versa*. Puri (1949) initially enlisted 77 anopheline species in the region which were later updated to 116 by Stone and Delfinado (1973). The whole Indian subregion is faunally the most rich and, though, anopheline mosquitoes were updated from time to time after Christophers (1933), yet, no authoritative attempt following Barraud (1934) could be made to deal with culicine fauna in totality despite fragmentary information made now and then.

As far as India is concerned, a gargantuan quantum of mosquito species and subspecies has been recorded (415 which is more than 11% of the world taxa; Tyagi et al., 2016; *This work*). In the context of anopheline mosquitoes, Rao (1984) has classified India into nine geographical zones based on physiography, climate and hydrology. Essentially the same classification has been followed here, albeit certain modifications in the demarcation of a couple of zones and, more importantly, addition of a newly carved out zone – The Great Indian Thar Desert (Figure 3.2). Thus, the ten geographical zones of India include the North Western Zone, Western Zone, Deccan Plateau, Gangetic Valley, East Central India, South Eastern Coastal Areas, Brahmaputra Valley, Himalayan Zone, The Great Indian Thar Desert, and Outlying Island Areas. Thus, while Rao (1984) used nine zones, we have justifiably organized India biogeographically into ten zones.

3.2.6.1 The Great Indian Thar Desert

This zone is characterized by low rainfall, very hot summers and very cold winters, arid and semi-arid features with a vast stretch of desert (app. 100,000 sq. km; 90% of which is in Rajasthan State, covering 62% of the land). The whole Thar Desert in Western Rajasthan state and some northwestern districts of Gujarat State are included in this zone.

Tyagi (1995, 2002, 2003, 2004, 2020, 2023), Tyagi et al. (1995), Tyagi and Yadav (1996), and Tyagi and Chaudhary (1997), while highlighting the impact of canalized irrigation on the xeric Thar's ecosystem transforming it into a malarial inferno with *P. falciparum*-dominated malaria outbreaks, have described a total of 11 species of *Anopheles* from the Thar Desert, as follows;

1. *An. stephensi*
2. *An. culicifacies*
3. *An. subpictus*
4. *An. annularis*
5. *An. vagus*
6. *An. splendidus*
7. *An. barbirostris*
8. *An. fluviatilis*
9. *An. nigerrimus*
10. *An. pulcherrimus*
11. *An. dthali*

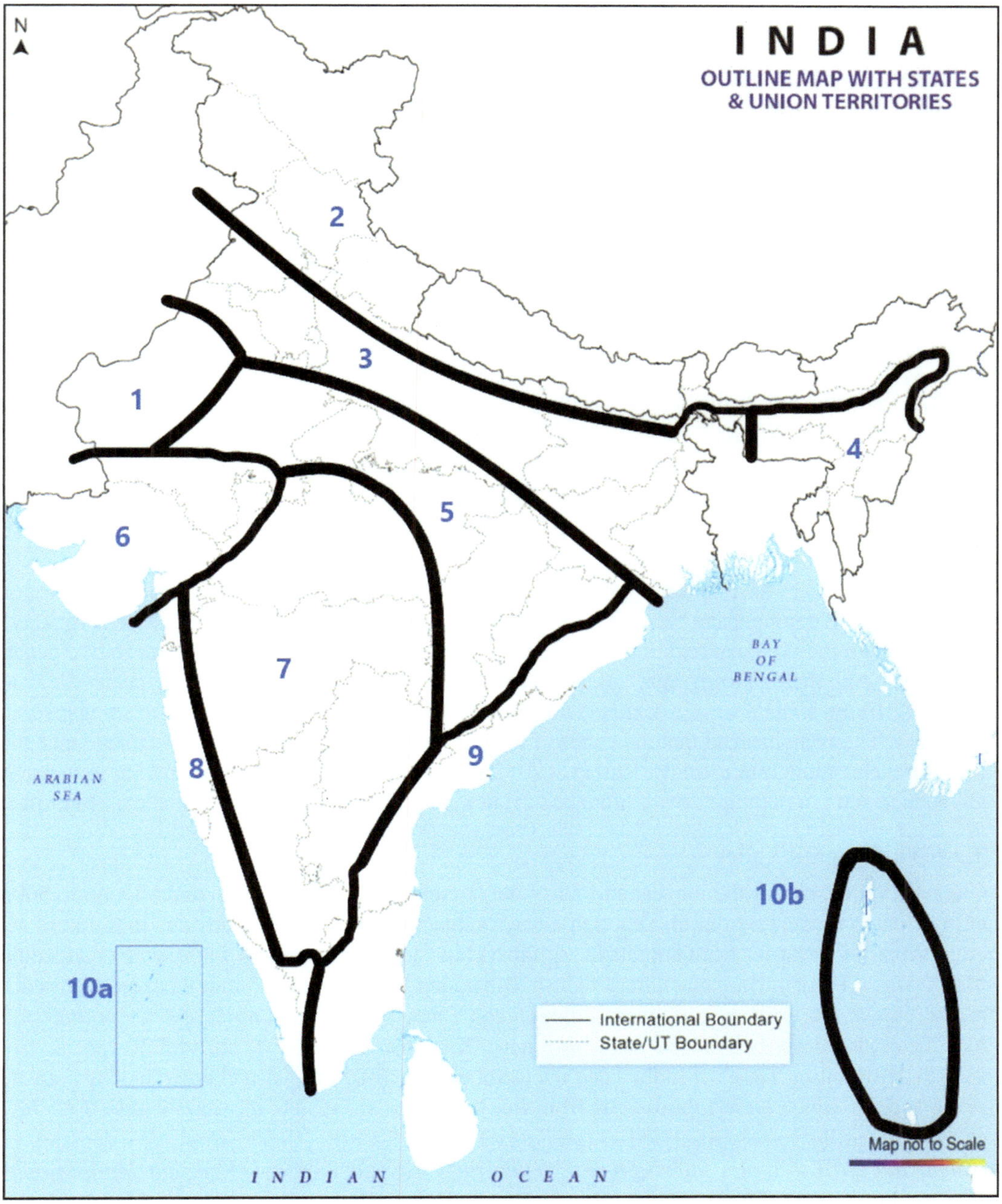

FIGURE 3.2 Indian geographic zones: (1) The Great Indian Thar Desert, (2) Himalayan Zone, (3) Gangetic Valley, (4) Brahmaputra Valley, (5) East Central India, (6) Western Zone, (7) Deccan Plateau, (8) Southwest Zone of Western Ghat, (9) South Eastern Coastal Areas, (10) Outlying Island Areas, and (10a and 10b) The Great Indian Thar Desert.

***Source*: Main author developed this zonal classification taking help from Free Wikipedia for Map of India.**

While *An. stephensi* was regarded by Tyagi (1995) as a true and original desert species, Tyagi (2023) proposed the Great Indian Thar Desert as the cradle of this species. Within Culicinae, a large number of authors have described *Culex quinquefasciatus* from the arid and semiarid areas. Tyagi (2020) also collected *Cx. tritaeniorhynchus* from the Thar Desert, whereas Tyagi and Hiriyan (2004) have reported *Aedes aegypti* from the same xeric ecosystem. While investigating species composition of Barmer district, Chittora et al. (2022) have encountered a total of 15 species (*Aedes (Stegomyia) aegypti, Aedes albopictus, Aedes (Fredwardsius) vittatus, Culex (Culex) quinquefasciatus, Culex (Culex) gelidus, Culex (Culex) vishnui, Culex (Culex) pseudovishnui, Culex (Culex) vagans, Culex (Culex) edwardsi, Culex (Culex) whitei, Anopheles (Cellia) subpictus, Anopheles (Cellia) fluviatilis, Anopheles (Cellia) sundaicus, Anopheles (Cellia) stephensi, Anopheles (Cellia) culicifacie* and *Armigeres (Armigeres) subalbatus*), dominated by *Culex quinquefasciatus* (40.54%), followed closely by *Aedes aegypti* (16.74%).

3.2.6.2 Himalayan Zone

High alpine and subalpine mountain region of the Himalaya including the northern-most state of Jammu and Kashmir, down to Himachal Pradesh, Uttarakhand, parts of West Bengal, Sikkim, and Arunachal Pradesh constitute this zone which is characterized by cold summer to freezingly snowy winter. Both Nepal and Bhutan, countries that brace the Indian borders, fall under this zone. The terai region famous for malaria is categorized within this zone.

Earlier this zone was quite safe for the vector-borne diseases, but during last few decades diseases like malaria and dengue have become rather rampant. Devi and Jauhari (2007) surveyed Himalayan mosquitoes in the Garhwal region and sampled 34 species of mosquitoes belonging to five genera, i.e., Aedes, Anopheles, Armigeres, Culex and Uranotaenia. Some of the prominent species collected included: *Culex mimeticus, Cx. vishnui, Cx. brevipalpis, Anopheles maculatus, An. stephensi*, and *An.vagus*. They emphasized that the zones of lower elevation shared higher species abundance than the higher elevation. Interestingly, Ishtiaq et al. (2017) reiterated that seasonal dynamics in mosquito abundance and temperature do not influence avian malaria prevalence in the Himalayan foothills.

3.2.6.3 Gangetic Valley

Gangetic valley, one of the longest and fairly flat stretches between sea level and hills up to 500m, enjoys low to moderate rainfall, although always subject to heavy inundations due to influx of several torrential tributaries from Himalaya. Summers have high temperatures while winters are cooler. Major part of Punjab, Haryana, Uttar Pradesh, Bihar, and West Bengal fall into this zone. This entire region enjoys most common species, although some new species such as *Anopheles pinjaurensis*, have been also described as new taxa from here. Kirti and Kaur (2004) enlisted 30 species from Punjab. Kirti (citing Times of India dated the October 28th, 2025) has also claimed having collected seven species which were first records from Punjab. Six of these new records include: *Anopheles (Cellia) fluviatilis* James, *An. (Cellia) tesellatus* Theobold, *Culex (Culex) sitiens* Wiedemann, *Cx. (Eumelanomya) brevipalpi* (Giles), *Lutzia (Metalutzia) vorax* Edwards, and *Mansonia (Mansonioides) Indiana* Edward. In the undulating vast stretches of Uttar Pradesh plains intercepted by several major rivers and their distributaries, a large number of mosquito species has been described (Rao, 1984). From the Japanese encephalitis endemic region of Gorakhpur, Kanojia, and Geevarghese (2005) cataloged a total of 67 mosquito species of which 43 were documented for the 1st time; the genus Aedes was represented by 15 species, Anopheles by 15, Culex by 24, Mansonia by 3, Mimomyia, Ochlerotatus, and Verrallina by two each, and the genera Aedeomyia, Armigeres, Coquillettidia, and Uranotaenia were represented by single species each.

3.2.6.4 Brahmaputra Valley

Brahmaputra basin, along with adjoining Barak, Imphal, and Tripura valleys and hill ranges bordering Myanmar and Bangladesh (northeast plateau), forms a distinct physiographic zone and one of the two major biodiversity hot-spots of India. Six of the eight states in the northeast India, viz.,

Assam, Meghalaya, Mizoram, Tripura, Nagaland, and Manipur fall under this zone. Though June-September is the main rainy season under the influence of southwest monsoon usually causing heavy floods, rains generally occurs more or less throughout the year. The climate of Brahmaputra valley is suitable for the transmission of major mosquito-borne diseases such as malaria, filariasis, Japanese encephalitis and dengue while malaria is endemic throughout the zone (Dev et al., 2015). With the availability of fragmented rainforests, the region is extremely rich in mosquito diversity. Review of various past and recent mosquito surveys revealed the presence of a total of 191 species of mosquitoes under 28 genera (subfamily Anophelinae: 1 genus, 45 species; subfamily Culicinae: 27 genera, 146 species) in the zone (Bhattacharyya et al., 2014) including several new species as well as country records such as, for example, *Armigeres mahantai* (Bhattacharyya et al., 2009), *Uranotaenia dibrugarhensis* (Bhattacharyya et al., 2004a) and *Verrallina assamensis* (Bhattacharyya et al., 2004b).

3.2.6.5 East Central India

This zone is characterized by hilly undulations and forests, mainly of Dandakaranya. While receiving downpours of both the monsoons, i.e., southwest and northeast, the temperature ranges between high to moderate during summer. Parts of Chhattisgarh, Jharkhand, Odisha, a few of the eastern most districts of Maharashtra and northern Telangana and Andhra Pradesh barring coastal areas make this zone.

This zone is richly represented by the vectors of malaria, dengue, Japanese encephalitis and lymphatic filariasis. Among malaria vectors, *Anopheles minimus*, which appeared to have disappeared some time ago upon deforestation and the DDT pressure under National Malaria Eradication Program, has recently been reported again from Singhbum hills of east central India (Jambulingam et al., 2005).

3.2.6.6 Western Zone

Almost all of Gujarat, parts of Rajasthan, Madhya Pradesh, Goa, and Maharashtra are included in this zone embracing equally hot and cold seasons with good rains during South West Monsoon, beaches, parts of rugged mountain ranges of Aravalli and Western Ghats etc. Gujarat State is well known for malaria onslaughts for decades. While conducting a major drive to control malaria through environmental methods, Yadav et al. (1989) highlighted the Anopheles of Kheda district. They reported that information on the anopheline fauna of Kheda district in Gujarat dated back to 1950 and in course of time there had been a great number of changes on the species composition in view of the vast ecological changes due to development projects and intensive irrigation. They collected mosquitoes belonging to 16 anopheline species from 164 villages, of which 8 species were recorded for the first time. Maximum number of anopheline species (16) were recorded from canal irrigated area followed by 11 species from riverine area and 10 species from non-canal irrigated area. Four anopheline species namely *An. subpictus*, *An. culicifacies*, *An. annularis*, and *An. stephensi* were predominant and constituted 99.9% of the total anophelines and rest of the 12 species accounted for 0.1%.

Anopheles culicifacies, a complex of five sympatric species (A, B, C, D, and E), drives its "species C" from Songarh Talluk, district Surat (Subbarao et al., 1983). Save for Species D and E, all other sibling species (i.e., A, B, and C) occur in Gujarat (Sharma and Dev, 2015). From the adjoining Goa at a distance, Kumar et al. (2016) came up with a sterling observation regarding *Anopheles subpictus*. They reported, following long-term research in the Thar Desert by Tyagi and Yadav (2001), that *An. subpictus*, which was until then considered a non-vector in Goa and Western India, was found to be a dominant vector (2.8%) for transmission of malaria parasites. Of the 14 *An. subpictus* infections, nested PCR demonstrated three Plasmodium infections in the salivary glands: one *P. vivax* and two mixed infections of *P. falciparum* and *P. vivax*. In addition, ten gut infections (one *P. vivax*, six *P. falciparum*, and three mixed infections) were seen in *An. subpictus* (Figure 3.3).

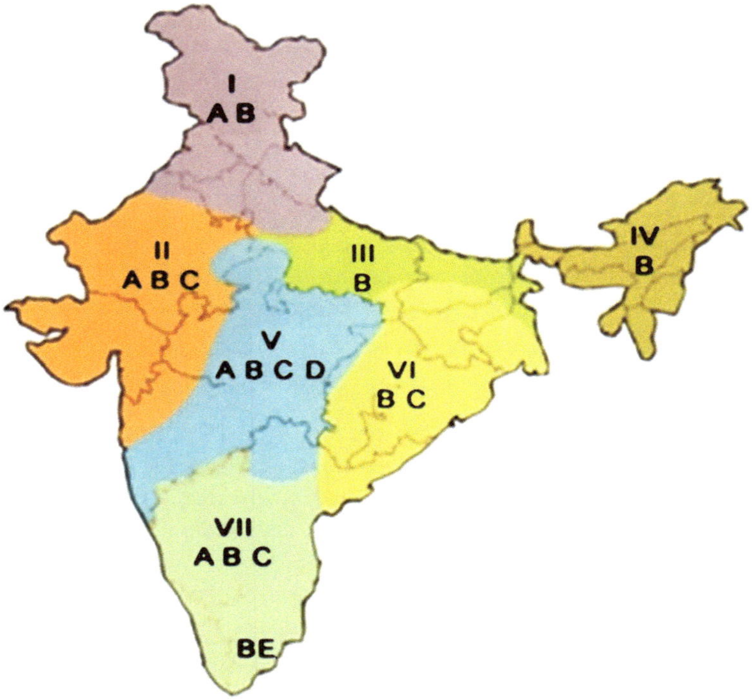

FIGURE 3.3 Map of India showing geographical distribution of *An. culicifacies* complex (sibling species A, B, C, D, and E), and stratification (Divisions I–VII) for vector control options.

Source: **Sharma & Dev, 2015.**

Tyagi (1990), while working in the remote tribal villages such as Serula and Singalkhanch in Songarh Talluk (Surat District), reported 17 species belonging to both subgenera Anopheles (*An. culiciformis*, *An. nigerrimus*, and *An. barbirostris*) and Cellia (*An. tesellatus*, *An. fluviatilis*, *An. culicifacies*, *An. jeyporiensis*, *An. subpictus*, *An. vagus*, *An. turkhudi*, *An. annularis*, *An. jamesii*, *An. moghulensis*, *An. Pallidus*, *An. splendidus*, *An. stephensi* and *An. theobaldi*)). Many of these species were found to be the "first records" for the Gujarat State (Tyagi, 1984).

3.2.6.7 Deccan Plateau

Highlands (500–1000 m above mean sea level) with moderate rainfall from both southwest and northeast monsoons, and equitable temperature characterize this zone. It comprises the highlands of Maharashtra, Madhya Pradesh, Karnataka, and Andhra Pradesh. Generally country lying south of Vindhya mountains and including valleys of Narmada, Tapti, Godavari, and Krishna rivers fall within Deccan Plateau, bounded by the Western and Eastern Ghats.

This zone is richly represented by Culicidae. Many malaria vectors such as *Anopheles culicifacies*, *An. fluviatilis*, and *An. annularis* and dengue/chikungunya vectors prevail here. While *Aedes aegypti* is far more common here, *Ae. albopictus* is also abundantly prevalent. Recently, Shinde and Thete (2014) reported *Aedes scatophagoides* as a new find for Deccan Plateau.

3.2.6.8 Southwest Zone of Western Ghat

Western Ghat mountains and forests make the most of this zone, with a narrow coastal strip of 10–50 km along the Arabian Sea. It houses the southern districts of Gujarat, western districts of Maharashtra, Goa, western districts of Karnataka, the whole of Kerala State, Nilgiris, and Annamalais in Tamil Nadu. The zone enjoys a strong southwest monsoon (May–August) and a weaker northeast monsoon (November–December). Temperature throughout is mild, though moisture is very high particularly along the coast.

This zone is very rich in mosquito fauna, and many surveys have been made in this zone's humongous variety of breeding habitats. Selvan et al. (2016) collected larvae and adults of 23 species (*Culex mimulus*, *Cx. pseudovishnui*, *Cx. quinquefasciatus*, *Cx. vishnui*, *Cx. khazani*, *Cx. uniformis*, *Heizmannia chandi*, *He. grenii*, *He. indica*, *Oclerotatus anureostriatus*, *Oc. albotaeniatis*, *Oc. deccanus*, *Oc. gubernatoris*, *Aedes aegypti*, *Ae. albopictus*, *Ae. edwardsi*, *Ae. krombeini*, *Toxorhynchites minimus*, *Tx. splendens*, *Anopheles aitkenii*, *An. barbirostris*, *An. culiciformis*, and *An. maculatus*), belonging to 6 genera, from cesspits, animal footprints, rock holes, tree holes, drainages at Marudhamalai, Valparai, Mettupalayam in Coimbatore district and Dhottapeta, Coonoor, Gudalur in Nilgiris district, Tamil Nadu, India. Selvan et al. (2016) also sampled *Aedes pseudoalbopicta*, *Ae. aegypti*, *Anopheles elegans*, *Christophersiomyia annularis*, *Ch. thomsoni*, *Downsiomyia albolateralis*, *Do. nivea*, and *Toxorhynchites splendens* specifically breeding in tree holes (phytotelmata) in six hill ranges viz., the Anaimalai hills, Kodaikanal hills, Sitheri hills, Kolli hills, Yercaud hills, and Megamalai hills. Their study revealed high species diversity and relative density associated with different study sites. Shanasree and Sumodan (2019), too, investigated phytotelmata in the Western Ghat's Wayanad district of Kerala during 2017–2018, and collected from 112 tree holes a total of 7 species under 5 genera, viz. *Aedes albopictus*, *Ae. chrysolineatus*, *Armigeres subalbatus*, *Culex brevipalpis*, *Cx. quinquefasciatus*, *Heizmannia chandi*, and *Toxorhynchites splendens*. They found out that *Aedes albopictus* breeding surpassed all other mosquitoes.

Munirathinam et al. (2015) attempted to find out the change in the mosquito biodiversity after a gap of 25 years in the tail-end hill ranges of the Western Ghats, viz., Kanyakumari (Keeriparai) and Tirunelveli districts (Kalakkad-Mundanthurai Tiger Reserve (KMTR) and Courtallam) of Tamil Nadu. Collection of both adults and larvae of mosquitoes from various breeding habitats (slow flowing streams, spring pool, rocky pool, leaf axils, latex cup, tree hole, bamboo stumps, etc.) was done between July 2010 and June 2013. Altogether 4602 immature individuals were collected, reared individually to be identified at the adult stage. A total of 3,583 specimens belonging to 50 species classified under 21 genera and 18 subgenera were recorded. The major vector species found in these hill ranges were *Stegomyia aegypti*, *St. albopicta* (vectors for dengue and chikungunya), *Culex bitaeniorhynchus*, *Cx. tritaeniorhynchus* (vectors for Japanese encephalitis), *Downsiomyia nivea* (vector for diurnally subperiodic filariasis), and *Anopheles mirans* (vector for simian malaria).

This zone has also recently recorded many new species (*Toxorhynchites tyagii*, *Tx. darjeelingensis*, *Anopheles pseudosundaicus*, *Culex katezari*) in addition to erection of a genus, Tewarius.

3.2.6.9 South Eastern Coastal Areas

Mostly built by a communion of low-lying deltas of Mahanadi, Godavari, Krishna and Cauvery in the east of Eastern Ghats, this zone enjoys very heavy northeast monsoon and a moderate southwest monsoon. Summers are hot and the general weather is warm throughout the year. The zone extends from Odisha through coastal Andhra Pradesh and the whole of Tamil Nadu, save for Annamalai and Nilgiri areas. Rajagopalan et al. (1984) demonstrated the significance of *Anopheles culicifacies* in disseminating malaria transmission along with the migratory population within Rameswaram Island as well as between mainland India and Rameswaram Island. Rajavel et al. (2005) reported 74 species belonging to 12 genera and 20 subgenera from Jeypore Hill tracks of Orissa and 43 species belonging to 21 subgenera and 13 genera from mangroves of Bhiterkanika. A total of 55 mosquito species under 12 genera, 17 subgenera, and 3 tribes occurring in Puri and Khurda Districts of Orissa have been documented by Dash (2011). Also, a total of 22 species of mosquitoes (Culicinae 65.6%, Anophelinae 34.4%) belonging to six genera (Anopheles, Aedeomyia, Aedes, Armigeres, Culex and Mansonia) were collected from eight villages in and around the Chilika lake area, Khurda, and Puri Districts from 2006 to 2007 (Dash and Hazra, 2011).

3.2.6.10 Outlying Island Areas

Two major archipelagos, one on each side of the peninsular India, i.e., Andaman and Nicobar in the east and Lakshadweep in the west, and the island of Rameswaram in south characteristically constitute this zone. Sri Lanka, a neighboring country, shares much of the mosquito fauna of southern India.

These islands are full of mosquitoes. While Andaman and Nicobar Islands have experienced outbreaks of dengue and chikungunya transmitted by *Aedes aegypti* and *Ae. albopictus*, the human lymphatic filariasis disease caused by *Wuchereria bancrofti* is subperiodically transmitted by *Downsiomyia nivea*, a day biting mosquito (Shriram, 2015). Lakshadweep islands have already reported chikungunya (Paramasivan et al., 2009; Samuel et al., 2009). Malaria transmitted by *An. culicifacies* in Rameswaram island is deeply investigated in context with human movement (Rajagopalan et al., 1984), although *An. stephensi* has also been occurring there since 2001 (Tyagi, *pers. comm.*). Shriram (2018) studied spatial distribution of *Aedes aegypti* and *Ae. albopictus* in South Andaman and Nicobar Islands (Shriram et al., 2018).

3.3 MEANS AND WAYS OF DISTRIBUTION/DISPERSAL OF MOSQUITOES

Mosquitoes are cosmopolitan (worldwide) in distribution. In nature there are principally three major drivers for spatial distribution of mosquitoes, viz., temperature seasonality, annual precipitation, and mammalian richness, albeit exceptions to this rule, such as the wind. They are found in every land region in the world except Antarctica (Mullen and Durden, 2009), and a few islands with polar or subpolar climates. Iceland is such an island, being essentially free of mosquitoes. Eggs of species of mosquitoes from the temperate zones are more tolerant of cold than the eggs of species indigenous to warmer regions (Hawley et al., 1989; Hanson and Craig, 1995); many even tolerate subzero temperatures. In addition, adults of some species can survive the winter by taking shelter in suitable microhabitats such as buildings or hollow trees (Romi et al., 2006).

Generally mosquitoes live in abundance near people, in forests, marshes, and other similar habitats, but in practice they make use of various kinds of water viz., fast and slow flowing water, fresh and saline water, drinkable and polluted sewage water, water collected in rock holes or in plants axils etc. (phytotelmata), underground and overhead water reservoirs, water coolers and air conditioners' water collecting plates, natural and man-made materials such as tires, vases, *idli*-preparing grinding stone cup, rubber-collecting cup in rubber plantation, squirrel-ate away holes in the fruits of Cocoa pod (*Theobroma cacao*) and, in essence, virtually any imaginable water-bearing structure on Earth (Hiriyan and Tyagi, 2004).

Female mosquitoes need blood for the development of their eggs, and therefore, they have to fly to the site of host availability for blood-feeding soon after emergence. After engorgement on blood they have to, immediately or after a time gap, once again fly back in quest of suitable breeding habitats. Conventional scientific wisdom has long held that mosquitoes have a limited range: they fly low to the ground and typically travel less than 5 km during their brief lifetime depending on the availability of their hosts, on one hand, and the breeding sites, on the other (Rao, 1984). Now, researchers have turned that wisdom on its head. While the yellow fever and dengue vector mosquito, *Aedes aegypti* is confirmed to fly actively up to 840 meters, another major vector of dengue, *Aedes albopictus* (now known as a complex of multiple cryptic species), is reported to fly up to 8 km (Guo et al., 2018). In the Sahel, the semi-desert region just south of the Sahara Desert, malaria-bearing mosquitoes (*Anopheles gambiae sensu stricto*, *An. coluzzii* and *An. squamosus*) are borne on winds that allow them to travel 300 km – and as high as 290 meters above the ground – in a single night (Huestis et al., 2019).

Besides inherent prowess to fly, being true flyers, mosquitoes, can take assistance from not only wind to cover long distances (Dao et al., 2014) but also varieties of transport such as train, ship and airplanes, and the best examples to date are of importation of *An. arabiensis* into South America through slave-ships and intrusion of *Ae. albopictus* into the Europe and United States of America through air-traffic. The recent expansion of *An. stephensi* both within India, the cradle of the species, and extraterritorially southwards into Sri Lanka, on one hand, and westwards into half-a-dozen northeastern African nations is mainly ascribed to anthropization involving erection of towns and cities (Tyagi, 2023).

3.4 INDIAN FAUNA SHARING EXTRATERRITORIAL ELEMENTS

India is strategically located on the world map as its international land-borders are bracing many neighboring countries, and various islands in the Indian Ocean. As a result Indian mosquitoes share a whole lot of taxa with others; while several Indian taxa are common in these neighboring nations, *vice versa* is also true. In other words, impact from all directions could be witnessed on the Indian fauna, viz., (i) Indigenous element, (ii) Northern countries element, (iii) Western countries element, and (iv) South Asian countries element.

With regard to anopheline fauna of India, first Christophers (1933; 42 species and 10 varieties from the Indian subcontinent including India, Pakistan, Bangladesh, Nepal, and Sri Lanka under the then British regime) and subsequently Rao (1984; 51 species and 7 subspecies or varieties) annotated on species' commonality or otherwise based on geographical distribution. Bhattacharyya et al. (2014) prepared a check list of 393 species of mosquitoes, including 61 anophelines, from India. Tyagi et al. (2015b) cataloged 412 species of mosquitoes, of which 63 belonged to the genus *Anopheles*. On the front of culicine mosquitoes, Barraud (1934) described taxonomically and biogeographically all the fauna then available in the subcontinent, but he made a major error by including Chaoboridae (as Chaoborini) and Dixidae (as Dixini) under the family Culicidae while both of these are now recognized as different families altogether. Currently Tyagi (*this work*) has inventorized 416 species and subspecies, although a couple of them have a questionable status of a good species. In the following section the Indian species' specificity and/or generality vis-à-vis many neighboring countries has been highlighted.

3.4.1 Anopheline Mosquitoes

Species like *Anopheles culicifacies*, *An. subpictus*, *An. annularis*, *An. nigerrimus*, *An. splendidus* are having wide geographical distribution in India and beyond into other neighboring nations. *Anopheles stephensi*, having its cradle in India (Tyagi, 2023) and considered an efficient vector of urban malaria, both *Plasmodium falciparum* and *P. vivax*, has during past nearly two decades exhibited a rather enigmatic pan-India migratory trend whereupon it has by now intruded Sri Lanka (2017) in south and several northeastern African nations (Djibouti in 2012, Ethiopia in 2016, Sudan in 2016, Somalia in 2019, Nigeria in 2020 and Yemen in 2021) in the west. WHO (2023) considered the spread of *An. stephensi* to be a major potential threat to malaria control and elimination in Africa and southern Asia, and has recently launched an initiative against the spread of this vector in Africa.

Some species such as, *An. aitkeni*, *An. insulaeflorum*, *An. jamesii*, *An. karwari*, *An. tesselatus*, *An. vagus*, *An. jeyporiensis*, *An. pallidus* are though widely present in India, yet exhibit an eastward movement. *Anopheles culiciformis*, *An. elegans*, *An. sintoni*, *An. pinjaurensis*, *An. pseudosundaicus*, and *An. gigas* are largely localized in character.

Some taxa particularly, *An. dthali*, *An. pulcherrimus*, and *An. multicolor*, are supposed to be western elements.

Similarly, since India is having a vast peninsular region with a vast coastline, species such as, *An. aconitus*, *An. balabacensis*, *An. barbirostris*, *An. umbrosus*, *An. sundaicus*, *An. philippinensis*, *An. sinensis*, *An. minimus*, *An. ramsayi*, *An. kochi*, *An. theobaldi*, and *An. roperi* are impacted by the southeast element. *Anopheles sundaicus* appears to have intruded India from both peninsular coasts.

3.4.2 Culicine Mosquitoes

Like anophelines, many culicine mosquitoes too are impacted by the various trans-India elements. The universal human lymphatic filariasis vector, *Culex quinquefasciatus*is widely distributed in India and overseas, may further penetrate in the areas where unplanned urbanization would mushroom with increased variety of polluted water bodies. The deadly vector for yellow fever and dengue fever, *Aedes aegypti*, whose birthplace is said to be in Africa, is now widely spread in the world (OECD, 2018). It is an exotic mosquito for India settled here many centuries ago. On the other hand, the Asian Tiger mosquito, *Aedes albopictus* – a major vector for dengue and chikungunya – is an Indian mosquito by origin, but has in last few decades spread all over India and the world. The Japanese encephalitis vectors belonging to the *Culex vishnui* subgroup, particularly *Culex tritaeniorhynchus* has a well-defined southeastern element.

India has a good number of *Toxorhynchites* species and those like *Tx. tyagii* and *Tx. darjeelingensis* are endemic to India (Krishnamoorthy et al., 2015; Tyagi et al., 2015a).

REFERENCES

Barraud, P.J. (1934). *The fauna of British India, including Ceylon and Burma. Diptera: Family Culicidae; Tribes Megarhinini and Culicini. Vol. 5*, Taylor & Francis, London, 463 pp.

Becker, N., Pluskota, B., Kaiser, A. and Schaffner, F. (2012). Exotic Mosquitoes Conquer the World. (*In*: Arthropods as Vectors of Emerging Diseases, H. Mehlhorn (ed.), *Parasitology Research Monographs* 3. Springer-Verlag Berlin Heidelberg. https://doi.org/10.1007/978-3-642-28842-5_2

Bhattacharyya, D., Prakash, A., Mohapatra, P. and Sarma D (2009). *Armigeres (Armigeres) mahantai*, a new mosquito species from India. *Journal of the American Mosquito Control Association*, 25, 1–5.

Bhattacharyya, D.R., Rajavel, A.R., Mohapatra, P. K., Jambulingam, P., Mahanta, J. and Prakash, A. (2014). Faunal richness and the checklist of Indian mosquitoes (Diptera: Culicidae). *Check List*, 10(6), 1342–1358. https://doi.org/10.15560/10.6.1342

Bhattacharyya, D., Prakash, A., Mohapatra, P. and Mahanta, J. (2004a). *Uranotaenia dibrugarhensis*, a new species in subgenus Pseudoficalbia from Assam, India. *Journal of the American Mosquito Control Association*, 20, 1–5.

Bhattacharyya, D., Tewari, S., Prakash, A., Mohapatra, P. and Mahanta, J. (2004b). *Verrallina (Neomacleaya) assamensis*, a new species from Assam, India. *Journal of the American Mosquito Control Association*, 20, 115–20.

Bynum, W.F. (1994) An experiment that failed: Malaria control at Mian Mir. *Parassitologia*, 36(1–2), 107–20.

Chittora, S., Johari, S. and Sharma, G. (2022). Species composition and habitat characterization of mosquito fauna in Kota and Barmer region of Rajasthan. *International Journal of Mosquito Research*, 9(5), 57–60. https://doi.org/10.22271/23487941.2022.v9.i5a.633

Christophers, S.R. (1933) *The Fauna of British India, including Ceylon and Burma. Diptera: Family Culicidae; Tribe Anophelini*. Vol. 4, Taylor & Francis, London, 271 pp.

Dao, A., Yaro, A.S., Diallo, M., Timbiné, S., Huestis, D.L., Kassogué, Y., Traoré, A.I., Sanogo, Z.L., Samaké, D. and Lehmann, T. (2014) Signatures of aestivation and migration in Sahelian malaria mosquito populations. *Nature*, 516(7531), 387–90. https://doi.org/10.1038/nature13987

Darsie, R.F. Jr. (1989) Keys to the genera, and to the species of five minor genera, of mosquito pupae occurring in the Nearctic Region (Diptera, Culicidae). *Mosquito Systematics*, 21(1), 1–10.

Dash, S. and Hazra, R.K. (2011). Mosquito diversity in the Chilika lake area, Orissa, India. *Tropical Biomedicine*, 28(1), 1–6.

Dash, S. (2011) Studies on the mosquito fauna inhabiting shoreline habitats of Orissa coast (Culicidae: Diptera). *Records of Zoological Survey of India*, 111(4), 9–24.

De Meillon, B. (1949). The Anophelines of the Ethiopian Region. *In*: Boyd's *Malariology*. W.B. Saunders & Co., New York.

Dev, V. (2020). *Vector Biology and Control: An Update for Malaria Elimination Initiative in India*. The National Academy of Sciences, India, 272 pp.

Dev, V. (2022). *Battling Malaria in North-East India: Targeting Interventions Towards Elimination*. ICMR – National Institute of Malaria Research, New Delhi, India, pp 274.

Dev, V., Khound, K. and Tewari, G.G. (2014). Dengue vectors in urban and suburban Assam, India: Entomological observations. *WHO South-East Asian Journal of Public Health*, 3(1), 51–59

Dev, V., Sharma, V.P., and Barman, K. (2015). Mosquito-borne diseases in Assam, north-east India: Current status and key challenges. *WHO South-East Asian Journal of Public Health*, 4, 20–29.

Devi, N.P. and Jauhari, R.K. (2007). Mosquito species associated within some Western Himalayas phytogeographic zones in the Garhwal region of India. *Journal of Insect Science*, 7, 1–10. https://doi.org/10.1673/031.007.3201

Edwards, F. W. (1921). A Revision of the Mosquitos of the Palaearctic Region. *Bulletin of Entomological Research*, 12(3), 263–351. https://doi.org/10.1017/S0007485300040207

Edwards, F. W. (1941) *Mosquitoes of the Ethiopian Region. III. Culicine Adults and Pupae*. British Museum (Natural History), London, 499 pp.

Evans, A. (1938) *Mosquitoes of the Ethiopian Region. II. Anophelini. Adults and Early Stages*. British Museum (Natural History), London, 404 pp.

Ficalbi, E. (1892) Revisione delle specie europee della fam. delle Zanzare. *Bollettino della Società entomologica italiana*, 24, 257–284.

Ficalbi, E. (1893) Revisione delle specie europee della fam. delle Zanzare. *Bollettino della Società entomologica italiana*, 25, 48–61, 136–144.

Ficalbi, E. (1894) Revisione delle specie europee della fam. delle Zanzare. *Bollettino della Società entomologica italiana*, 26, 66–75, 315–320.

Ficalbi, E. (1895) Revisione delle specie europee della fam. delle Zanzare. *Bollettino della Società entomologica italiana*, 27, 29–38.

Ficalbi, E. (1896) Revisione delle specie europee della fam. delle Zanzare. *Bollettino della Società entomologica italiana*, 28, 108–313.

Ficalbi, E. (1899) Venti specie di Zanzare (Culicidae) Italiane. *Bollettino della Società entomologica italiana*, 31, 46–234.

Foley, D.H., Rueda, L.M. and Wilkerson, R.C. (2007). Insight into global mosquito biogeography from country species records. *Journal of Medical Entomology*, 44(4), 554–67. 10.1603/0022-2585(2007)44[554:iigmbf]2.0.co;2

Gressitt, J. Linsley (1974). Insect biogeography. *Annual Review of Entomology*, 19, 293–321. https://doi.org/10.1146/annurev.en.19.010174.001453

Guo, Y., Song, Z., Luo, L. et al. (2018). Molecular evidence for new sympatric cryptic species of *Aedes albopictus* (Diptera: Culicidae) in China: A new threat from Aedes albopictus subgroup? *Parasites Vectors*, 11, 228. https://doi.org/10.1186/s13071-018-2814-8

Hanson, S.M., and Craig, G.B. (1995). *Aedes albopictus* (Diptera: Culicidae) eggs: Field survivorship during northern Indiana winters. *Journal of Medical Entomology*, 32 (5), 599–604. https://doi.org/10.1093/jmedent/32.5.599

Harbach, R.E. (2013). The phylogeny and classifcation of Anopheles. In: Manguin, S., (Editor), *Anopheles mosquitoes – new insights into malaria vectors*. Rijeka, Croatia: InTech.

Hawley, W.A., Pumpuni, C.B., Brady, R.H., Craig, G.B. (1989). Overwintering survival of Aedes albopictus (Diptera: Culicidae) eggs in Indiana. *Journal of Medical Entomology*, 26(2), 122–9. https://doi.org/10.1093/jmedent/26.2.122

Haynes, Ellen, Cleveland, C.A., Garrett, K.B., Grunert, R.K.A., Bryan II, J.A., Sidouin, M., Oaukou, P.T., Ngandolo, B.N.R., Yabsley, M.J. (2022). Characterization of the genetics and epidemiology of *Brugia* sp. in domestic dogs in Chad, Africa. *Veterinary Parasitology: Regional Studies and Reports*, 35, 1–8. https://doi.org/10.1016/j.vprsr.2022.100784

Hiriyan, J. and Tyagi, B.K. (2004). Cocoa pod (*Theobroma cacao*) – a potential breeding habitat of *Aedes albopictus* (Skuse) in dengue sensitive Kerala State, India. *Journal of American Mosquito Control Association*, 20(3), 323–325.

Hopkins, G.H. (1952). *Mosquitoes of the Ethiopian Region. I. Larval Bionomics of Mosquitoes and Taxonomy of Culicine Larvae*. 2nd ed. Printed by order of the Trustees, British Museum (Natural History), 355 pp.

Huestis, D.L., Dao, A., Diallo, M. et al. (2019) Windborne long-distance migration of malaria mosquitoes in the Sahel. *Nature*, 574, 404–408. https://doi.org/10.1038/s41586-019-1622-4

Ishtiaq, F., Christopher, G.R., Bowden, Y. and Jhala, V. (2017). Seasonal dynamics in mosquito abundance and temperature do not influence avian malaria prevalence in the Himalayan foothills. *Ecology and Evolution*, 7, 8040–8057.

Jambulingam, P., Sahu, S.S. and Manonmani, A. (2005). Reappearance of Anopheles minimus in Singhbum hills of East-Central India. *Acta Tropica*, 96(1), 31–35. https://doi.org/10.1016/j.actatropica.2005.05.013

Joly, C.A. (2008). Biodiversity and climate change in the Neotropical region. *Biota Neotropica*, 8 (1). https://doi.org/10.1590/S1676-06032008000100001

Kanojia, P.C. and Geevarghese, G. (2005). New mosquito records of an area known for Japanese encephalitis hyperendemicity, Gorakhpur District, Uttar Pradesh, India. *Journal of American Mosquito Control Association*, 21(1), 1–4. https://doi.org/10.2987/8756-971X(2005)21[1:NMROAA]2.0.CO;2

Kirti, J.S. and Kaur, J. (2004). Furtrher studies on the mosquito fauna of Punjab, India. *ZOO;s Print Journal*, 19(7), 1523–1524.

Krishnamoorthy R., Munirathinam, A., Dhananjeyan, K.J., Hiriyan, J., Mariappan, T., Samuel, P.P. and Venkatesh, A. (2015). Description of a new species, Toxorhynchites (Toxorhynchites) tyagii (Diptera: Culicidae), from Nilgiri hills, Western Ghats, southern India. *Zootaxa* 3701(4), 447–459.

Kumar, A., Hosmani, R., Jadhav, S., de Sousa, Trelita, Ajeet, M., Milind, N., Adarsh, S., Satyajit, K., Neena, V., Laura, C. and Rathod, P.K. (2016). *Anopheles subpictus* carry human malaria parasites in an urban area of Western India and may facilitate perennial malaria transmission. *Malaria Journal*, 15, 124. https://doi.org/10.1186/s12936-016-1177-x

Mullen, G. and Durden, L. (2009). *Medical and Veterinary Entomology*. London: Academic Press.

Munirathinam A., R. Krishnamoorthi, G. Baskaran, R. Govindarajan, A. Veerapathiran, A. Venkatesh, B. K. Tyagi (2015). Mosquito diversity in Keeriparai and Mundanthurai hill ranges of the Western Ghats, southern India. *Journal of Threatened Taxa*, 7(9), 7552–7556. https://doi.org/10.11609/JoTT.o4193.7552-6

Nihad, M.P.P., Rohini, P. D., Sutharsan G., Anagha Ajith P.K., Sumitha M.K., Shanmuga, P.A., Rahul P., Sasikumar, V., Dasgupta S., Krishnan Jayalakshmi and Ishtiaq, F. (2022). Island biogeography and human practices drive ecological connectivity in mosquito species richness in the Lakshadweep Archipelago. *Scientific Reports* 12, 8060. https://doi.org/10.1038/s41598-022-11898-y

OECD (2018). *Safety Assessment of Transgenic Organisms in the Environment, Volume 8: OECD Consensus Document of the Biology of Mosquito Aedes aegypti*, Harmonisation of Regulatory Oversight in Biotechnology, OECD Publishing, Paris, 106 pp. https://doi.org/10.1787/9789264302235-en

Ong, S.Q. and Ahmad, H. (2022). An annotated image dataset for training mosquito species recognition system on human skin. *Scientific Data*, 9, 413. https://doi.org/10.1038/s41597-022-01541-w

Paramasivan, R., Samuel, P.P., Thenmozhi V., Rajendran R., Victor Jerald Leo S., Dhananjeyan, K.J., Krishnamoorthi, R., Arunachalam, N. and Tyagi, B.K. (2009). Chikungunya virus isolated in Lakshadweep Islands in the Indian Ocean: Evidence of the central/east African Genotype. *Japanese Journal of Infectious Diseases*, 62(1), 67–69.

Puri, I.M. (1949). Anophelines of the Oriental region. *In*: *Boyd's Malariology*. W.B. Saunders 85 Co., New York, pp. 483–505.

Raharimalala, F.N., Ravaomanarivo, L.H., Ravelonandro, P. et al. (2012). Biogeography of the two major arbovirus mosquito vectors, *Aedes aegypti* and *Aedes albopictus* (Diptera, Culicidae), in Madagascar. *Parasites Vectors*, 5, 56. https://doi.org/10.1186/1756-3305-5-56

Rajagopalan, P.K., Jambulingam, P., Sabesan, S., Krishnamoorthy, K., Rajendran, S., Gunasekaran, K. and Kumar, N.P. (1984). Population movement and malaria persistence in Rameswaram Island. *ICMR Bulletin*, 14(11), 125–33.

Rajavel, A.R., Natarajan, R. and Vaidyanathan, K. (2005) Mosquito collections in the Jeypore Hill tracts of Orissa, India, with notes on three new country records, *Culex (Lophoceraomyia) pilifemoralis, Culex (Lophoceraomyia) wilfredi*, and *Heizmannia (Heizmannia) chengi. Journal of American Mosquito Control Association*, 21(2), 121–127.

Rao, T.R. (1984). *The anophelines of India*. Malaria Research Centre, Delhi, 594 pp.

Romi, R., Severini, F. and Toma, L (2006). Cold acclimation and overwintering of female *Aedes albopictus* in Roma. *Journal of the American Mosquito Control Association*, 22(1), 149–51.

Samuel, P.P., Krishnamoorthi, R., Hamzakoya, K.K. and Aggarwal, C.S. (2009). Entomo-epidemiological investigations on chikungunya outbreak in the Lakshadweep islands, Indian Ocean. *Indian Journal of Medical Research*, 129(4), 442–445.

Senthamarai, P.S., Jebanesan, A., Divya, G. and Ramesh, V. (2015) Diversity of mosquitoes and larval breeding preference based on physico-chemical parameters in Western Ghats, Tamilnadu, India. *Asian Pacific Journal of Tropical Disease*, 5(Suppl. 1), S59–S66. https://doi.org/10.1016/S2222-1808(15)60858-1

Senthamarai, P.S., Jebanesan, A. and Reetha, D. (2016). Entomofaunal diversity of tree hole mosquitoes in Western and Eastern Ghats hill ranges of Tamilnadu, India. *Acta Tropica*, 159, 69–82. https://doi.org/10.1016/j.actatropica.2016.03.029

Shanasree, M. and Sumodan, P.K. (2019). Species diversity of tree hole breeding mosquitoes (Diptera: Culicidae) in Wayanad district, Kerala. *Journal of Entomology and Zoology Studies*, 7(3), 324–326.

Sharma, V.P. and Dev, V. (2015). Biology and Control of *Anopheles culicifacies* Giles 1901. *Indian Journal of Medical Research*, 141, 525–536.

Shinde, L. and Thete, K. D. (2014). First time Record of *Aedes scatophagoides* (Theobald, 1934) from Latur (M.S.); Deccan Plateau of India. *Journal of Mosquito Research*, 4, https://doi.org/10.5376/JMR.2014.04.0001

Shriram, A.N. (2015). A bibliometric profile of lymphatic filariasis research in India. *Journal of Vector Borne Diseases*, 52(1), 73–8.

Shriram, A. N., A. Sivan, and A. P. Sugunan (2018). Spatial distribution of Aedes aegypti and Aedes albopictus in relation to geo-ecological features in South Andaman, Andaman and Nicobar Islands, India. *Bulletin of Entomological Research*, 108, 166–174.

Stone, A. and Delfinado, M.D. (1973). *Family Culicidae: Catalog of the Diptera of Oriental Region. Vol. I. Suborder Nematocera.* University Press. Hawaii, Honolulu, 618 pp.

Subbarao, S.K., Vasantha, K., Adak, T. and Sharma, V.P. (1983). *Anopheles culicifacies* complex: Evidence for a new sibling species C. *Annals of Entomological Society of America*, 76, 985–8.

Selvan, P.S., Jebanesan, A. and Reetha, D. (2016). Entomofaunal diversity of tree hole mosquitoes in Western and Eastern Ghats hill ranges of Tamilnadu, India. *Acta Tropica*, 159, 69–82. https://doi.org/10.1016/j

Thenmozhi, V., Balaji, T., Venkatasubramani, K., Dhananjeyan, K.J., Selvam, A., Rajamannar, V. and Tyagi, B.K. (2016). Role of Anopheles subpictus Grassi in Japanese encephalitis virus transmission in Tirunelveli, South India. *Indian Journal of Medical Research*, 144(3), 477–481.

Tyagi, B.K. (1984). A note on the new records of some anopheline mosquitoes from south Gujarat (India). *Geobios new Report*, 1(3), 149–151.

Tyagi, B.K. (1990). Annotated check-list of the anophelines from district Surat (Gujarat). *Journal of Applied Zoological Research*, 1(2), 3–6.

Tyagi, B.K. (1995). Malaria in the Thar Desert. *ICMR Bulletin*, 25, 85–91.

Tyagi, B.K. (2002). *Malaria in the Thar Desert: Facts, Figures and Future.* Agrobios (India), 165 pp.

Tyagi, B.K. (2003). *Handbook of Medical entomology*. Scientific Publishers (India), Jodhpur. 262 pp.

Tyagi, B.K. (2004). *The invincible deadly mosquitoes: India's Health and Economy Enemy No. 1.* Scientific Publishers, India, 266 pp.

Tyagi, B.K. (2020). *Mosquito Hunters: A history of hostilities against man's deadliest Foe – the mosquito – since 1881.* Scientific Publishers (India), 474 pp.

Tyagi, B.K. (2023). *Desert Malaria: An Emerging Malaria Paradigm and Its Global Impact on Disease Elimination.* Springer-Nature, Singapore, 416 pp.

Tyagi, B.K. and Chaudhary, R.C. (1997). Outbreak of falciparum malaria in the Thar Desert with particular emphasis on physiographic changes brought about by extensive canalization and their impact on vector density and dissemination. *Journal of Arid Environments* (UK), 36, 541–555.

Tyagi, B.K., Chaudhary, R.C. and Yadav, S.P. (1995). Epidemic malaria in Thar Desert, India. *The Lancet*, 346, 634–635.

Tyagi, B.K., and Hiriyan, J. (2004). Breeding of dengue vector, *Aedes aegypti* (Linnaeus) in the rural Thar Desert, north western Rajasthan, India. *Dengue Bulletin*, 28, 220–222.

Tyagi, B.K., Munirathinam, A., Krishnamoorthy, R., Baskaran, G., Govindarajan, R., Krishnamoorthi, R., Mariappan, T., Dhananjeyan, K.J. and Venkatesh, A. (2015a). A revision of genus Toxorhynchites Theobald, 1901, in the South-East Asian countries, with description of a new species *Toxorhynchites (Toxorhynchites) darjeelingensis* from West Bengal, India (Diptera, Culicidae). *Halteres*, 6, 13–32.

Tyagi, B.K., Muinirathinam, A. and Venkatesh, A. (2015b). An inventory of the authentically identified and classified mosquito taxa in the "CRME Museum", with annotations on its significance in taxonomic research in India. *International Journal of Mosquito Research*, 3(4), 58–113.

Tyagi, B.K., Munirathinam, A. and Venkatesh, A. (2016). A catalogue of Indian mosquitoes. *International Journal of Mosquito Research*, 3(4), 58–113.

Tyagi, B.K. and Tilak, R. (2022). Which is the world's deadliest animal: A lion, tiger, shark, snake or mosquito? *Journal of Medical Arthropodology & Public Health*, 2(1), 1–7.

Tyagi, B.K. and Yadav, S.P. (1996). Malaria and vector control: Awareness of rural community in the Thar desert following an epidemic. *WHOSEARO Regional Health Forum* 1, 41–45.

Tyagi, B.K. and Yadav, S.P. (2001). Bionomics of malaria vectors in two physiographically difference areas of the epidemic-prone Thar Desert, north-western Rajasthan (India). *Journal of Arid Environments*, 47, 161–172.

Weaver, S.C. and Reisen, W.K. (2010). Present and future arboviral threats. *Antiviral Research*, 85(2), 328–45. https://doi.org/10.1016/j.antiviral.2009.10.008

WHO (2023). *Vector alert*: Anopheles stephensi *invasion and spread in Africa and Sri Lanka.* World Health Organization, Geneva, 8 pp. ISBN 978-92-4-006771-4 (electronic version)

Yadav, R.S., Sharma R.C., Bhatt R.M. and Sharma V.P. (1989). Studies on the anopheline fauna of Kheda district and species specific breeding habitats. *Indian Journal of Malariology*, 26(2), 65–74.

Section II

Taxonomy, Systematics, and Biodiversity

4 Mosquito Inventory of Extant Taxa, National Repositories, and Identification Keys for Dominant Vector Species

B.K. Tyagi
University Institute of Biotechnology, Chandigarh University, Mohali, India

Kailash Chandra
Ex-Zoological Survey of India, Kolkata, India

Devinder Singh
Chandigarh University, Mohali, India

4.1 INTRODUCTION

The family Culicidae, derived from "culex" – the Latin name for "gnat," is a member of one of the main stocks of Nematocera, the infraorder Culicomorpha. It consists of two superfamilies that include all of the piercing/sucking nematocerans, both predators and blood-feeding biters. The superfamily Chironomoidea comprises the families Chironomidae and Thaumaleidae, which have non-piercing mouthparts, and Simuliidae (black Flies) and Ceratopogonidae (biting midges), which pierce either vertebrates or invertebrates. The superfamily Culicoidea comprises the Dixidae, Corethrellidae, Chaoboridae, and Culicidae, the second and fourth of which feed on vertebrate blood. However, among all the culicomorphs, the long proboscis of mosquitoes (Family Culicidae) is distinctive. It is considered the most specialized of biting mouthparts among Nematocera and indicates a long and close association of mosquitoes with vertebrate animals (Wood and Borkent, 1989).

Mosquitoes (Culicidae) have undergone several changes in their classificatory system, and it is still in a fluid state, with emerging advanced technologies. Until recently the culicid classification recognized three subfamilies: Anophelinae, Toxorhynchitinae, and Culicinae; the first two (Anophelinae and Toxorhynchitinae) considered sometimes to be primitive groups, although another school of thoughts regarded them to have been specially derived from some Culicinae-type ancestor. Recent cladistic analysis of morphological and nucleotide-sequence data demonstrated that the Anophelinae are only distantly related to the other two subfamilies, i.e., Culicinae and Toxorhynchitinae, and that the Toxorhynchitinae did not actually merit subfamily status (Harbach, 2007). This phyletic difference is also reflected in the fact that while, on one hand, anopheline eggs bear characteristic floats, their larvae lack air tubes, and adults have elongated palps in both sexes, whereas, typical culicine and toxorhynchitine larvae have, on the other hand, air tubes, and adult females have short palps. Toxorhynchitines are all predaceous in the aquatic larval stage, as also larvae are unusually large and have a curved proboscis suited for feeding on only plant nectar.

DOI: 10.1201/9781003326991-6

TABLE 4.1
Classification of Various Different Taxa from the Higher Most, Phylum, to the Lowest in Hierarch, Species

Phylum – Arthropoda
  Class – Insecta
    Order – Diptera
      Family – Culicidae
        Genus – Anopheles/Aedes/Culex
          Species – *culicifacies/aegypti/quinquefasciatus*

Every mosquito species is assigned, in accordance with the International Code of Zoological Nomenclature, a Latin scientific name under binomial nomenclature system, such as *Anopheles culicifacies* or *Aedes aegypti* or *Culex quinquefasciatus* (Table 4.1). The first word is the "generic" name of a group of closely related mosquito species, while the second word is invariably the "species" name, which denotes a group of individuals or populations or subspecies that are similar in structure and physiology, and capable of interbreeding (vide infra). More specifically, Mayer (2015) defined a species as "*Species ate groups of interbreeding natural populations that are reproductively isolated from other such groups.*"

The name of a mosquito species is, as a rule, invariably written in italics, e.g., *Anopheles culicifacies*, *Aedes aegypti* or *Culex quinquefasciatus*. At times a mosquito name is also observed comprising three, instead of two, words, that is "trinomial," such as, for example, *Aedes aegypti aegypti*, in which situation a nominal subspecies of the main species is differentiated from another subspecies, for example in this context, *Aedes aegypti formosus*. Sometimes, to further characterize the complex nature of species names are instead replaced assigned in English alphabets, until a good specific description is taxonomically made. Thus some mosquitoes, like *Anopheles culicifacies* (in India), *An. gambiae* (in Africa) and *An. maculipennis* (in Europe), have well-characterized sibling species complexes whose individual species though morphologically indistinguishable from other members in the complex (both *sensu stricto* and *sensu lato*) are still good species as demonstrated by cytogenetical and molecular techniques. Such sibling species complexes of certain vector mosquito species carry tremendous biomedical significance as there are some, and not all, species of the complex that are capable of transmitting human disease pathogens such as, for example, malaria parasites. Interestingly while all sibling species belonging to *An. gambiae* complex from Africa were lately taxonomically described as good species (*An. arabiensis, An. bwambae, An. melas, An. merus, An. quadriannulatus, An. gambiae s.s., An. coluzzii* and *An. amharicus*), those belonging to *An. culicifacies*, and many others in India, are still awaiting proper taxonomic description.

4.2 SUMMARY OF EXTANT INDIAN TAXA OF CULICIDAE

There are approximately 3700 species of mosquitoes worldwide (Harbach, 2013), of which 420 species and subspecies occur in India (Bhattacharyya et al., 2015; Rajput and Singh, 1990; da Silva et al., 2020; Tyagi, 2020; Tyagi et al., 2014; Karlekar et al., 2020; Suman et al., 2021). The largest number of species still remaining to be discovered probably inhabits tropical rain forests (e.g., South America's Amazonian region and India's North-East states and Andaman and Nicobar islands), where faunas are more diverse but less well-surveyed than temperate regions. Species that have been studied intensively often reveal that they consist of complexes of closely related sibling species, indicating that many reproductively isolated and niche-specific forms remain to be identified or are undergoing speciation, e.g., *Anopheles gambiae* in Africa and *An. culicifacies* in India.

Indian topophysiography is a world-famous megadiversity, with a gargantuan variety of ecosystems. Mosquitoes occur all over the country, but most of these are present in abundance in the tropical forests such as those in the northeastern regions of lower Himalaya, and those of the Eastern and Western Ghats, as well as the mangroves along the coastline of peninsular India (Tewari et al., 2007; Tyagi et al., 2014; Tyagi, 2020). Both the subfamilies, i.e., Anophelinae and Culicinae, are comprehensively represented by 12 Tribes, 50 genera, 42 subgenera and 420 species which accounts for more than 11% of the world taxa (Figure 4.1). Tribe Aedini (Subfamily Culicinae) with 181 species (including *Aedes aegypti* and *Ae. albopictus* – the deadly vectors of dengue, yellow fever, chikungunya and zika virus, etc.) is the richest among all the culicid tribes. On the other hand, Tribe Culicini (Subfamily Culicinae), the second most species-rich tribe, comprises an extraordinarily diverse range of vectors for different diseases such as human lymphatic filariasis (*Culex quinquefasciatus*), Japanese encephalitis (*Cx. tritaeniorhynhus, Cx. gelidus, Cx. pseovishnui*), West Nile Virus (*Cx. quinquefasciatus, Cx. tritaeniorhynchus, Cx. vishnui, Cx. bitaeniorhynchus* and *Cx. univittatus*). It is interesting that Brugian filariasis (caused by the parasite *Brugia malayi*), rampant mostly in the Kerala State, is transmitted by the mosquitoes belonging to Tribe Mansoniini (Subfamily Culicinae) members of which are vicious biters besides. The Tribe Anophelini has all the human malaria vectors; a total of 11 mosquito species that are capable to transmit the pathogen, though only six of them are primary vectors (*An. culicifacies, An. dirus, An. fluviatilis, An. minimus, An. sundaicus* and *An. stephensi*) (Rao, 1984; Nagpal and Sharma, 1995; Tyagi, 2002, 2023), and the remaining five are of local importance, viz. *An. philippinensis-nivipes, An. varuna, An. annularis, An. subpictus* and *An. jeyporiensis*. In India the subfamily Anophelinae is represented by only single Tribe, Anophelini, comprising two two subgenera, namely, Anopheles, and, Cellia, under the genus Anopheles. All the malaria vector species are classified under Cellia only (Dev and Sharma, 2013).

The exact number of mosquito species relevant to human health is unknown. Yet, a careful global study by Yee et al. (2022) calibrated a total of 331 mosquito species that have the potential to transmit or are able to vectorize some kind of vector-borne or zoonotic infections, which account for more than 17% of all infectious diseases, causing more than 700,000 deaths annually. Further, according to them, a total of 88 of 3578 mosquito species (2.5%) are known vectors for 78 human

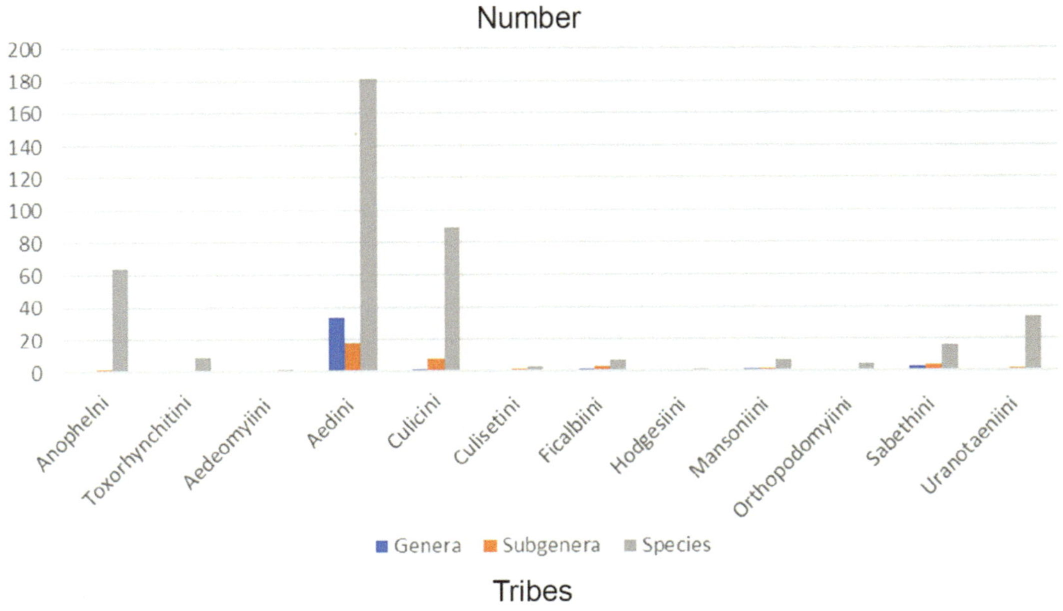

FIGURE 4.1 Indian mosquito species arranged under Subfamilies, Tribes, Genera and Subgenera.

***Source*: Adapted from Tyagi et al., 2014; *This work*.**

disease-causing pathogens; however, an additional 243 species (6.8%) were identified as potential or likely vectors, bringing the total of all mosquitoes implicated in human disease to 331 (9.3%). As far as Indian mosquitoes are concerned, close to 40 species of mosquitoes are vectors for various diseases like malaria, lymphatic filariasis, dengue, chikungunya, zika, Japanese encephalitis, West Nile Virus, etc., while many others are vicious biters that leave behind a characteristic "wheel," which is itchy, painful, and inflammatory, besides a source of allergy and annoyance (Table 4.2).

TABLE 4.2
Public Health Importance of Mosquito-borne Diseases with Their Vectors and Distribution Both Globally and in India

		Distribution	
Diseases (Pathogens)	**Vectors**	**World**	**India**
Malaria *Plasmodium falciparum, P. vivax, P. ovale, P. malariae* and *P. knowlesi*	*Anopheles* culicifacies, *An. stephensi, An. fluviatilis, An. philippinensis, A. dirus, An. sundaicus, An. subpictus, An. minimus, An. jeyporiensis* and *An. varuna*	Mostly Sub-Saharan Africa, Central & South America, Asia	Throughout the country, but more pronouncedly in the North-East, Eastern, Central, Northwestern and Southern Indian states
Lymphatic Filariasis *Wuchereria bancrofti, Brugia malayi*	*Culex quinquefasciatus, Mansonia* spp.	Tropical urban areas	Mostly peninsular states but prevails inland, too; *W. bancrofti* is predominant, while *B. malayi* is restricted mainly to Kerala
Dengue DENV 1–4 (Dengue fever, dengue haemorrhagic fever, and dengue shock syndrome)	*Aedes* aegypti and *Ae. albopictus*	Urban tropics	Throughout India; Prevalent in 35 out of 36 states/UTs
Chikungunya CHKV **Japanese Encephalitis** JEV	*Aedes* aegypti and *Ae. albopictus*	Urban tropics	Throughout India
Yellow Fever YFV	*Aedes aegypti, Haemagogus* spp.	Africa, Central & Southern Americas; recently sporadic imported cases occurred in Asia	Absent
Zika Virus (ZFV)	*Aedes aegypti*; also of lesser importance are species of certain other genera	South America; spread sporadically in many countries (incl. India) across the world during 2015 global outbreaks	Some states only
Japanese encephalitis (JEV)	*Culex* spp. mainly mosquitoes belonging to *Culex vishnui* subgroup, but *Anopheles subpictus* and *Armigeres subalbatus* have been also incriminated with JEV	Southeast Asia-Pacific	Mainly in Uttar Pradesh, southern states like Tamil Nadu, Karnataka, Andhra Pradesh, Kerala, and North-East states

4.3 CHRONOLOGICAL ACCOUNT OF INDIAN MOSQUITO FAUNA

Anopheles stephensi – which was discovered in an Indian State of Maharashtra, by Liston (1901), and shot to fame, even earlier than this, in 1897, when for the first time Dr Ronald Ross, a medical doctor in British Indian Army, demonstrated while stationed in Secunderabad (India) the inextricable relationship of malaria and mosquito, believed to be only *An. stephensi*, is apparently the first Indian anopheline mosquito scientifically discussed in detail (Tyagi, 2020; Tyagi, 2023). It is noteworthy here that all drawings of the original *Anopheles stephensi* publication were created by a noted Indian medical entomologist working in Madras (now Chennai), Dr. D.A. Turkhud, after whose name even a mosquito species was named in his honor, *Anopheles turkhudi* Liston, 1901. Dr. Turkhud, a colleague of Lt. Col. W.G. Liston, worked on anopheline mosquitoes in Kodaikanal near Madurai. While serving as the Director of the famous King Institute of Preventive Medicine at Guindy in Madras, he dedicatedly worked on malaria vectors in south India (Tyagi, 2020). On the other hand, two mosquitoes, namely, *Culex pipiens* Linn. 1758 and *Aedes aegypti* Linn. 1762, which are both found in India but originated from outside it, were already listed, like several other species in many other works. At the same time, *Aedes albopictus*, popularly called the Asian Tiger mosquito and a deadly vector for dengue, chikungunya, zika, etc., originated in Kolkata, West Bengal (India), although described outside of Australia by Skuse (1894).

The first authoritative catalog of Indian mosquitoes was first prepared Theobald (1902), in the course of his *magnum opuses* in five volumes entitled "*A monograph of the mosquitoes of the world*" (Theobald, 1901a, 1901b, 1901c, 1901d, 1903, 1907,1910). Thereafter, several prominent taxonomic works were published in different reputed periodicals particularly Indian Journal of Medical Research. This culminated into a synoptic account of taxonomic description for individual species (totally 294; including 43 Anophelinae and 251 Culicinae) by Christophers (1933) and Barraud (1934), under world-famous *The Fauna of British India* series which, in spite of the fact that they have now become out-of-date with reference to current classification as well as identification keys, besides outright expulsion of the annexed Dixinae and Chaoborinae from Culicidae in case of Volume III: Tribe Anophelini, still continue to be the most authentic reference works on the Indian mosquitoes (Tyagi et al., 2014; Tyagi, 2020). Post-independence as India waged a war against the devastating malaria in the country by launching a major initiative, namely, the National Malaria Control Program, in 1953, later transformed into the National Malaria Eradication Program in 1958, some very useful field guides about mosquito identification and control were produced (Puri, 1949, 1960; Wattal and Kalra, 1961, Das et al., 1990; Nagpal and Sharma, 1995). However, all these monographs dealt with *Anopheles* mosquitoes, largely due to an unprecedented high importance tagged with malaria in the country in those days (Rao, 1984).

Medico-entomological researches between 1950 and 1970 were significantly directed toward the facilitation of malaria eradication, which was almost achieved since less than one million cases were reported without a single death was reported in the mid-1960s (Sharma 1986a,b; Sharma, 1999; Sharma and Mehrotra, 1986). This kind of focus on the anti-malaria campaign, however, overshadowed any appreciable attention to other (non-anopheline) mosquitoes notwithstanding the rise in both intensity and distribution of diseases like *Cx. quinquefasciatus*- transmitted lymphatic filariasis (Tyagi, 2014), *Cx. vishnui* Subgroup-associated Japanese encephalitis (Reuben et al., 1993, 1994; Tyagi, 2011) and *Aedes aegypti*-mediated dengue/chikungunya (Tyagi and Dash, 2006; Tyagi, 2007, 2008). Yet, a few fragmentary but significant publications were brought out on the regional distribution of culicines in the Lower Himalayas, the Western and The Eastern Ghats, Andaman and Nicobar Islands, North-East states, the Thar Desert, the Great Runn of Kutchch, and Malwa region of Punjab (Munirathinam et al., 2014; Tyagi and Verma, 1991; Tyagi et al., 2009; Rao et al., 1973; Reuben et al., 1993; Rajavel et al., 2005a, 2005b, 2005c; Dutta et al., 2003; Sagandeep et al., 1994; Jagbir and Jagdish). A comprehensive check-list of Indian mosquitoes was attempted for the first time by

Tyagi (2004), though it was marginally incomplete and incorrect, and therefore a little while later the catalog was absolutely completely flawlessly reformed (Tyagi et al., 2014; Tyagi, 2020; Bhattacharyya et al., 2015) which is being further updated in this chapter of the book. Some authors have emphasized on ecosystem-based mosquito sampling, e.g., xeric, mangrove, and sylvatic or mountainous areas (Tyagi and Chaudhary, 1997; Tyagi and Yadav, 1996, 2000, 2001; Rajavel et al., 2005a, 2005b; Krishnamoorthy et al., 2013).

4.4 MOSQUITO REPOSITORIES IN INDIA

The oldest museum or repository of mosquito-type specimens is the Indian Museum, Kolkata, much of which is now is in disarray, but at one time even Dr. F.V. Theobald had come from England to this important museum for preparing his classics or *vade mecum* on the world mosquitoes in several editions. More or less parallel to this is the entomology museum of the Government of India's Pusa Institute, established in Bihar in 1905, but due to a devastating earthquake in 1934, this was shifted to New Delhi in 1936. It was also famous for housing types of many insects including mosquitoes. Currently, it is one of the largest insect repositories of agricultural significance and mosquitoes, if at all present, are in a much dilapidated state. A focused and dedicated mosquito museum of various "type specimens" is the few national museums of the present day's National Centre for Diseases Control (erstwhile Central Malaria Bureau, which was established at Kasauli, Himachal Pradesh, India, in 1909. It was renamed the Malaria Institute of India, Delhi in 1938 and further rechristened as the National Institute of Communicable Diseases in 1963, Delhi), where many type specimens discovered in the first half of the 20th century are preserved with a background document (Sharma et al., 2005). The NCDC Museum houses 89,464 specimens of mosquitoes belonging to 31 genera and 539 species, from India and abroad. The oldest specimen is from the United Kingdom, collected in 1902. The museum boasts of housing the oldest mosquito collection of *Anopheles culicifacies* from Karnal, Haryana in 1905. During the past almost 40–60 years, several institutes and centers under the country's premier scientific organization, the Indian Council of Medical Research, have evolved into a stronghold of medico-entomological investigations, particularly in context with mosquito-transmitted diseases (e.g., Centre for Research in Medical Entomology, Madurai (now merged with the ICMR-VCRC), Vector Control Research Centre, Puducherry, National Institute of Malaria Research, Delhi, Regional Medical Research Centre, Dibrugarh, and National Institute of Virology, Pune) which have established exquisite research museums housing type specimens (i.e., holotypes, allotypes, paratypes, etc.) of various different new species described in recent past. Centre for Research in Medical Entomology (now merged with Vector Control Research Centre) housed a total of 1,09,166 fully vouchered specimens exhibiting relevant distributional details. These specimens belong to 242 species (including type specimens of seven new species, viz., *Anopheles pseudosundaicus, Rhinoskusea portonovoensis, Tewarius agastyai, Te. reubenae, Verrallina assamensis, Toxorhynchites tyagii* and *Tx. darjeelingensis*), belonging to 34 subgenera and 39 genera (Tewari et al., 2007; Tyagi et al., 2016; 2015). In addition, the Vector Control Research Centre has in a collection of about 43,388 adult specimens (312 species, 28 new country records, and types of eight new species), of which 36,816 are individually mounted on minute pins, while rests are held in stock vials. It also includes 3979 male and female genitalia mounted on slides, besides 14,131 larvae, larval exuviae, and pupal exuviae (Rajavel et al., 2004, 2005a, 2005b, 2005c, 2005d, 2005, 2005f, 2011).

India's first arthropod museum was established by the Society of Medical Arthropodology, under its newly established International Institute of Medical Arthropodology at Hyderabad under the charge of Prof. Dr. B. Reddya Naik of Zoology Department, Osmania University, Hyderabad, with a major and long-time objective of archiving all arthropodological research titles and publications.

4.5 SALIENT TAXONOMIC FEATURES OF INDIAN MOSQUITOES

4.5.1 Sibling Species Complexes

A species complex tends to encompass a group of closely related, morphologically indistinguishable species, which may occur in sympatry (but not interbreeding), yet still display behavioral differences that could confound any control efforts that ignore their bionomics and epidemiological importance. Moreover, even among those species that are not members, complex, behavioral differences are common depending upon location, such that a species can be considered a primary vector in one area, but of secondary or no importance elsewhere.

Mosquito systematics and phylogeny are often complicated by the excessive nature of species complexes, in both Anophelinae and Culicinae, but more specifically under the genus *Anopheles*. By an estimate, there could be more than 50 species complexes under *Anopheles* which is represented by 465 formally recognized species throughout the world (Sinka, 2013). Nearly 70 of these species have the potential to transmit human malaria parasites and 41 are regarded to be either dominant vector species or complexes of species many of which are capable of transmitting malaria at a level of major concern to public health.

There are as many as five *Anopheles* species complexes in India which are comprised of cryptic or sibling species, viz., *Anopheles culicifacies, An. fluviatilis, An. sundaicusn, An. subpictus* and *An. dirus* (Table 4.3). As regards the various species under the subfamily Culicinae, not much information is available in India (Jose, 2018), although significant research has become available about *Aedes albopictus* from Southeast Asia (Huang, 1972; Minard et al., 2017; Guo et al., 2018). The Scutellaris Subgroup of *Ae. albopictus* is demonstrated to consist of 11 sympatrically cryptic species worldwide on the basis of *Cox-1* (mitochondrial DNA and ribosomal DNA) gene and *ITS-II* gene subunit markers using One Step Multiplex Real Time PCR (RT-PCR).

These species complexes carry immense epidemiological significance since some of these could be serious vectors which only need to be targeted during control campaigns. Unfortunately none of the alphabetically designated members of species complexes in India has so far been deciphered to be properly described taxonomically.

TABLE 4.3
Sibling Species Complex of *Anopheles* Mosquitoes in India

Species	Sibling Species Complex	Designated Species
Anopheles culicifacies	5	A, B, C, D, and E
Anopheles fluviatilis	3	S, T, and U
Anopheles subpictus	4	A, B, C, and D
Anopheles annularis	2	A and B
Anopheles dirus	5 (In India only 3 species occur)	*An. dirus* s.s. (species A), *An. cracens* (species B), *An. scanloni* (species C), *An. baimaii* (species D), *An. elegans* (species E), *An. nemophilous* (species F), *An. takasagoensis*, and a cryptic species tentatively designated as *An.* aff. *Takasagoensis*.
An. sundaicus	4	*An. sundaicus* s.s., *An. epiroticus* (formerly species A), *An. sundaicus* species D and *An. sundaicus* species E.

4.5.1.1 *Anopheles culicifacies* Giles

Anopheles culicifacies, which also has its cradle in India, is the most important vector in India responsible for more than 65% of cases in the country, particularly in the rural environment, although it is absent in the islands of Lakshadweep and Andaman & Nicobar (Sharma and Dev, 2015). The species complex comprises five sibling species, namely A, B, C, D and E; among all siblings Species E, being a potential career for both *Plasmodium vivax* and *P. falciparum* (Subbarao et al., 1988; Surendran et al., 2000), along with Species A, is the most important vector while Species B is a non-vector, being mostly zoophilic (Subbarao, 1988; Vatandoost and Borhani, 2004; Dev, 2020). The range of the *An. culicifacies* complex extends far beyond the Indian subcontinent; it also encompasses large areas of Southeast Asia including Vietnam, Cambodia, Lao PDR, and southern China, and reaches as far as Yemen in the Middle East with a small distribution in Eritrea as well as Nepal, India, Pakistan and Sri Lanka (Sinka et al., 2011). Despite this extensive distribution, the complex has only been studied in detail in India and Sri Lanka (Surendran et al., 2000; Jude et al., 2014; Subbarao, 1988; Subbarao and Sharma, 1997; Amerasinghe and Amerasinghe, 1999; Barik et al., 2009). Of the five species (A, B, C, D, E) of the complex, four are considered vectors in India (A, C, D, and E) (Sharma, 1998; Kar et al., 1999). Sometimes the five sibling species of *An. culicifacies* spatially overlap each other in different proportions and a clear understanding of whether a vector or two among them is predominantly prevalent is to be inevitably determined to understand the intensity of malaria transmission, on the one hand, and to bring about effective vector control, on the other (Figure 4.2).

These cytospecies are best identifiable and differentiable from each other through cytotaxonomic methods involving reading distinguishing characters or markers on the polytene chromosomes from female mosquito's ovarian germline cells or larval salivary glands or even brain cells (Figures 4.3 and 4.4). Though not of much use, sometimes mitotic or even meiotic chromosome complements are also studied to comprehend karyotypic characteristics, including a number of chromosomes, centromere type, and evolution. Generally, mosquitoes have a diploid chromosome number of six

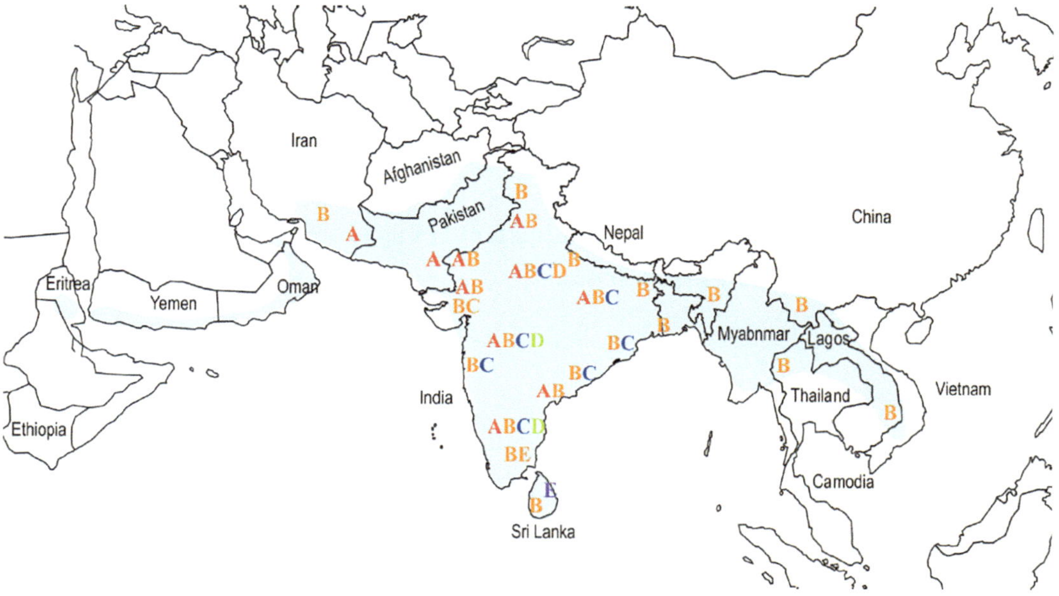

FIGURE 4.2 Distribution of *Anopheles culicifacies* sibling species complex in Asia, and parts of Middle East and Africa.

Source: **Dev, 2020.**

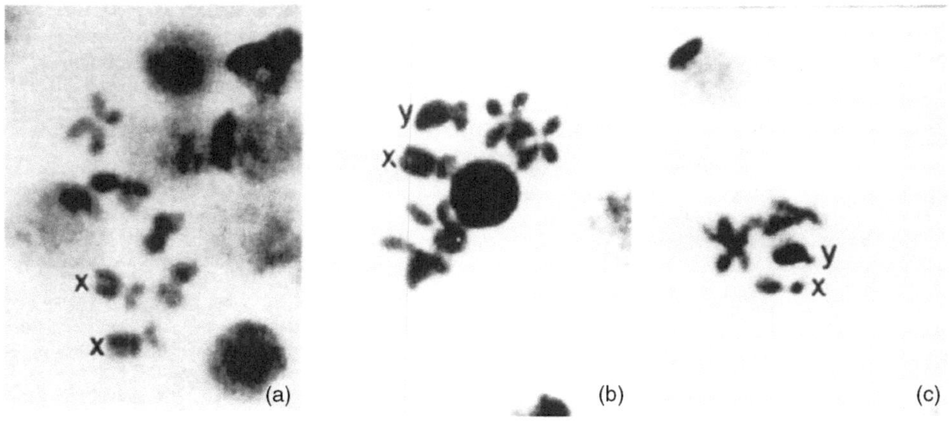

FIGURE 4.3 Mitotic karyotypes of *Anopheles culicifacies*: (a–c) chromosomes prepared from larval brain tissue; (a) Species B female, (b) Species E male with submetacentric Y– chromosome; and (c) Species B male with acrocentric Y–chromosome.

Source: **WHOSEARO, 2007.**

(2n = 6), including two pairs of autosomes and one pair of heterogametic sex chromosomes (Matthews and Munstermann, 1994).

Chromosome number and morphology in mosquitoes is remarkably uniform and chromosomes are invariably metacentric or submetacentric. Obviously, variation in chromosome number does not play any important role in mosquito chromosomal evolution. It appears that, because of the morphological similarity of their chromosomes, structural changes have played little or no role in mosquito karyotypic evolution. Therefore, while Culicidae as a family is robustly stable, signs of evolutionary foresights are nevertheless apparently becoming discernible in a very few species such as those in the *Anopheles gambiae* complex, which comprises at least nine species (Loughlin, 2020). It is interesting to note here that at least two major African malaria vectors in the complex, *An. gambiae* and *An. coluzzii* do hybridize frequently in the wild, and there is also some evidence from genetic markers that species are likely evolving, not to mention in the least that a new member of the complex has very recently been described, *An. fontenillei* (Barron, 2019). In India, too, it can be safely surmised that members of at least some *Anopheles* complexes' species are in the process of evolution, though none of these complexes' species are reported to hybridize. On the other hand, in Culicinae, *Aedes albopictus* males have been documented to be siring the females of *Ae. aegypti*, albeit without viable eggs. It should be mentioned here that both these species are grouped under subgenus *Stegomya*.

Notwithstanding the fact no effort is ever made to morphometrically distinguish the adults of the alphabetized (A, B, C, D, and E) major malaria vector in India, *Anopheles culicifacies*, at least a sound beginning has been recently made by Tyagi et al. (2016) who electron-microscopically differentiated three of the species (A.D and E) on the basis of egg structure (Figure 4.5).

Tyagi et al. (2016) found striking differences particularly in the structure of micropyle disk and micropyle tubercular areas on the polar region (Figures 4.6 and 4.7).

An analysis of morphological differences in floats, micropyle disks, and rosette tubercles in the eggs of *An. culicifacies* sibling species A, D, and E, as observed under a scanning electron microscope, is presented in Table 4.4.

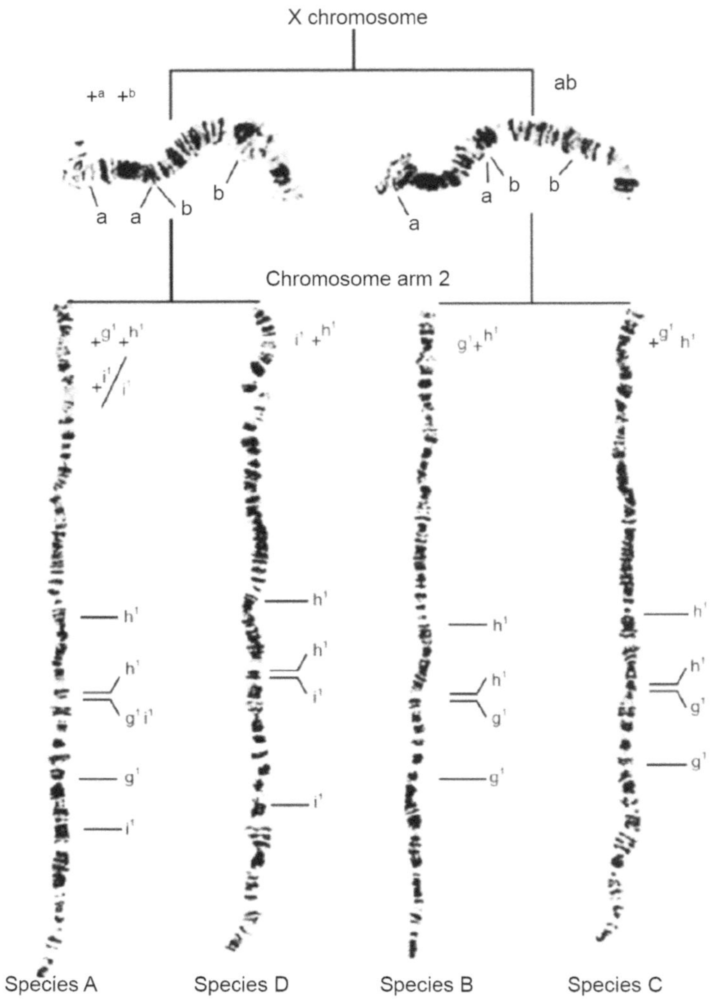

FIGURE 4.4 Schematic representation of polytene chromosomes of *Anopheles culicifacies* sibling species, A, B, C, and D.

Source: WHOSEARO, 2007.

4.5.1.2 *Anopheles dirus* Species Complex

There are eight sibling species in the complex – all forest dwellers and serious vectors of malaria, seven of which have been formally named, i.e., *An. dirus s.s.* (species A), *An. cracens* (species B), *An. scanloni* (species C), *An. baimaii* (species D), *An. elegans* (species E), *An. nemophilous* (species F), *An. takasagoensis*, and a cryptic species tentatively designated as *An.* aff. *takasagoensis*. Among these member species, only *An. baimaii* and *An. elegans* are prevalent in India with a distinct distribution range and epidemiological significance. *Anopheles baimaii* is widely abundant in the northeast states where it pre-eminently transmits malaria.

4.5.1.3 *Anopheles fluviatilis* Species Complex

A dangerous vector for malaria contributing approximately 12–15% of all the annually reported cases, *Anopheles fluviatilis s.l.* is widespread in the mountainous regions and foothill areas of India, such as the terai area in the foothills of northern India as well the Eastern and Western Ghat regions, where the larvae of this complex are associated with slow-flowing water in streams or river

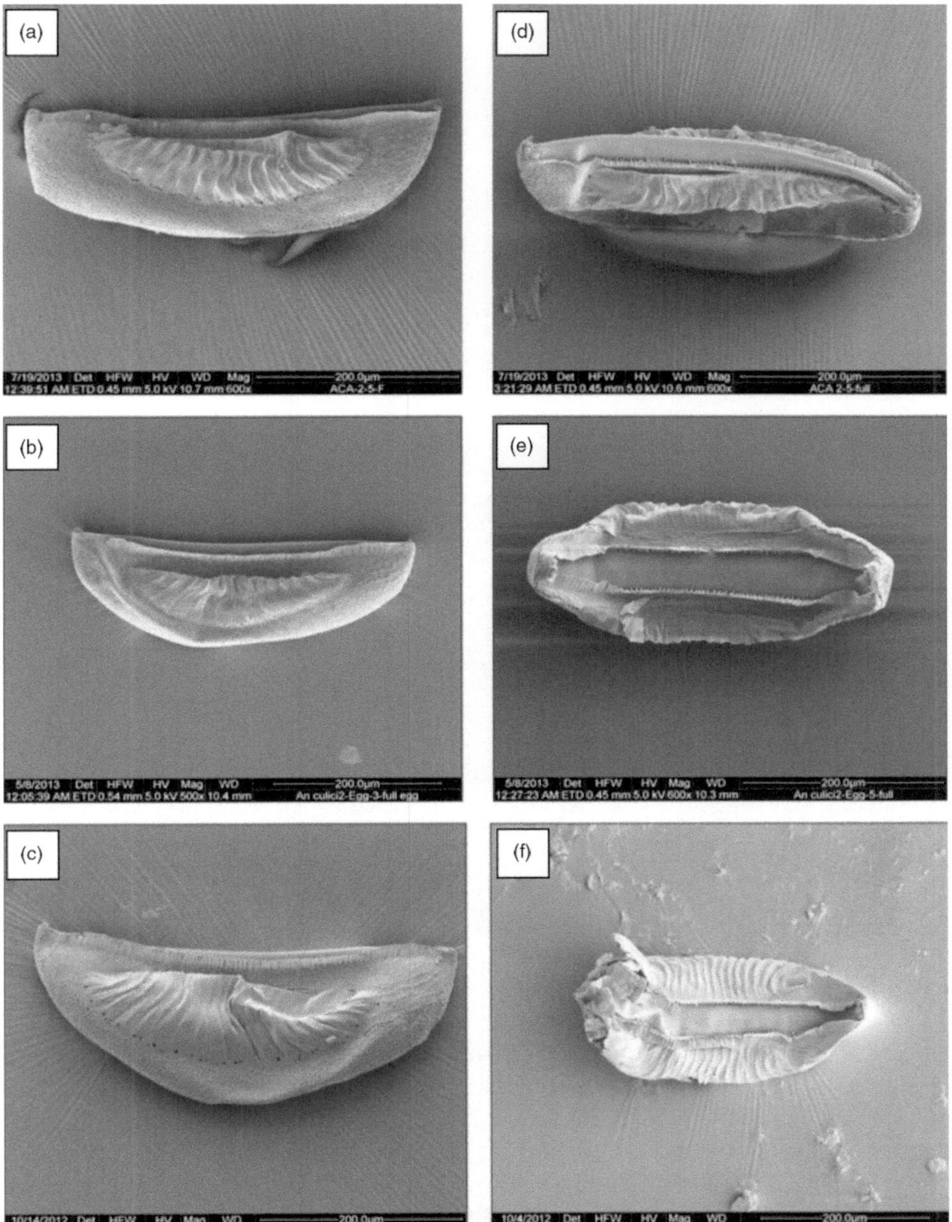

FIGURE 4.5 (a–f). Whole eggs of *An. culicifacies*: (a–c) Lateral aspect showing egg length and floats of Species A, D, and E, respectively; (d–f) Ventral aspect showing deck area of Species A, D, and E.

margins (Figure 4.8). Besides India, the species is also found in Nepal, Bangladesh, Myanmar, Iran, Pakistan, and Afghanistan. The species complex comprises three sibling species, i.e., S, T, and U, and a form "V", based on cytotaxonomic investigations on polytene chromosomes.

Members of the complex are found at a wide range of altitudes, from plains to hilly and mountainous areas such as, for example, the terai area in the foothills of northern India as well as the

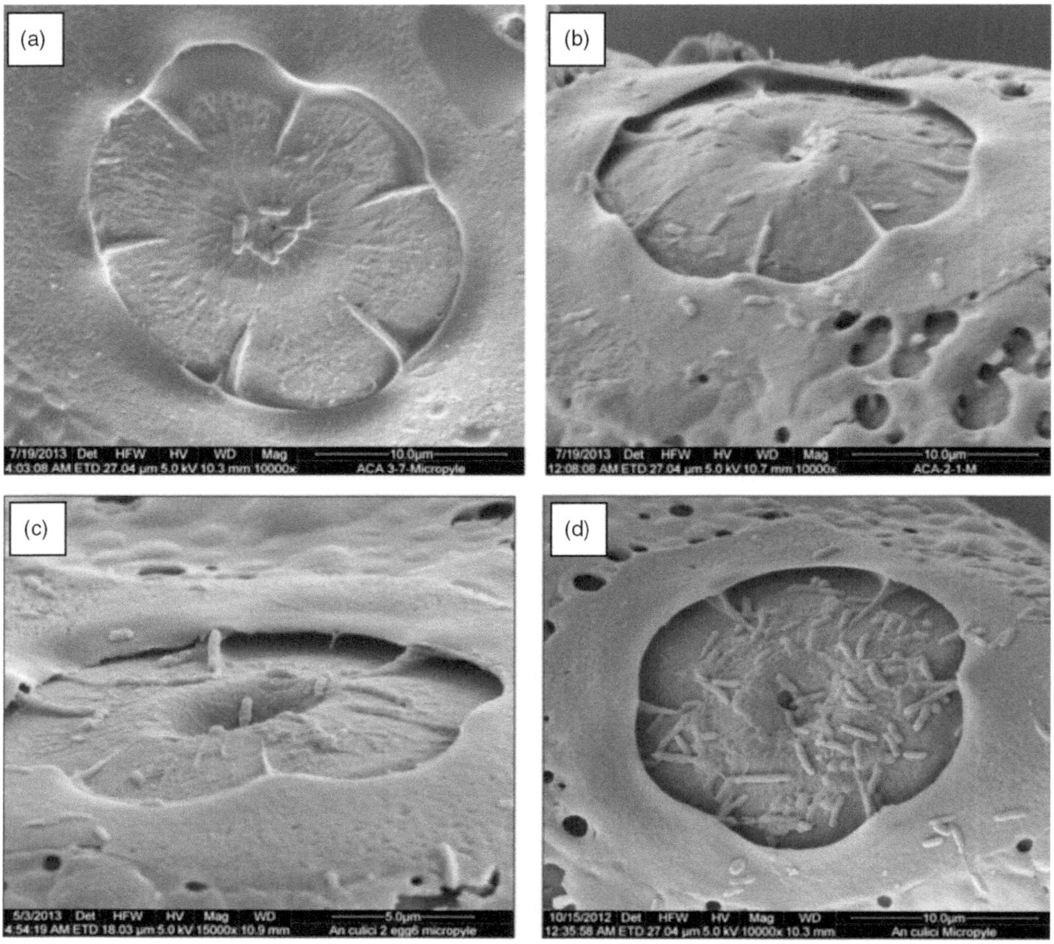

FIGURE 4.6 **(a–d)**. Micropyle disk of eggs of *Anopheles culicifacies* sibling species, A, D, and E.

Eastern and Western Ghat regions, where the larvae of this complex are associated with slow-flowing water in streams or river margins and habitats are especially rich in this species. The habitats they utilize are also varied and include forested and deforested ecotypes and irrigated areas. Consequentially, the larval sites they inhabit are also wide-ranging and include man-made habitats such as irrigation canals, borrow pits, domestic wells, tanks and gutters as well as natural sites such as stream margins and rock pools. Apparently, the specie prefers fresh water sites in comparison to brackish.

Anopheles fluviatilis is mostly investigated both behaviorally and ecologically in India and Iran where the complex occurs. *An. fluviatilis* S, being a strong anthropophilic and endophilic mosquito, is a highly efficient vector in India. Other members, barring V whose status is not yet clear, of the complex Species T and U, both poor or non-vectors, are zoophilic and tend to feed and rest outdoors. In Pakistan, Nepal, and Iran, however, species T is considered an important vector.

4.5.1.4 *Anopheles sundaicus* Species Complex

Besides a small focus in the Kutchch area of Gujarat, this species is a major vector in Andaman & Nicobar islands, besides many other island nations in Southeast Asia. *Anopheles sundaicus* species complex *s.l.* is a complex of four species, i.e., *An. sundaicus s.s.*, *An. epiroticus* Linton & Harbach

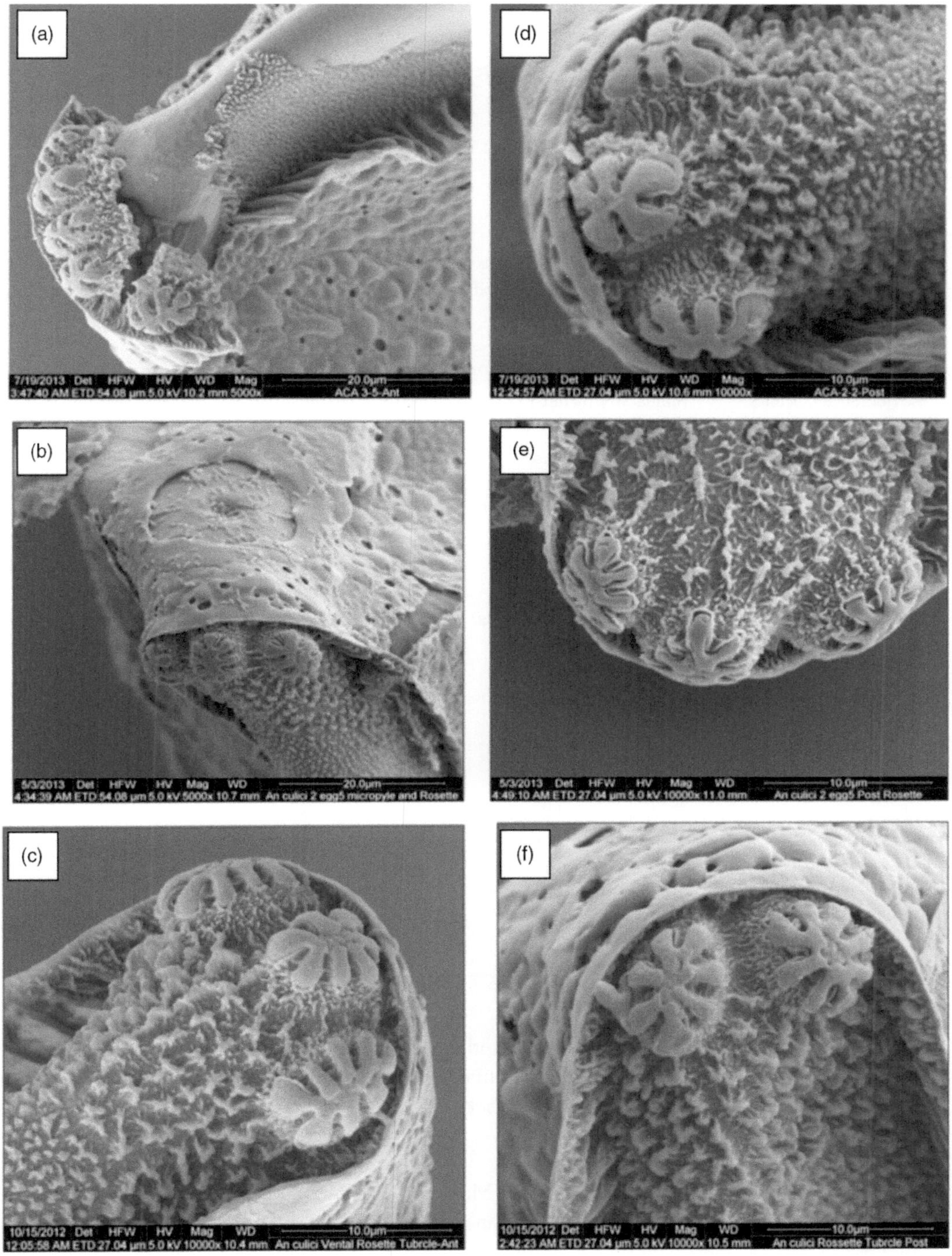

FIGURE 4.7 **(a–f)**. Rosette tubercles of eggs (Scanning Electron Microscope Images) of *Anopheles culicifacies* sibling species A, D, and E.

***Source*: Varun Tyagi et al., 2016.**

TABLE 4.4
Comparative Description of Eggs of *An. culicifacies* Species A, D, and E Showing Morphological Differences

Chasracter	*An. culicifacies "A"*	*An. culicifacies "D"*	*An. culicifacies "E"*
Ventral and dorsal surface	Ventral surface concave, dorsal surface curved	Ventral surface concave, dorsal surface strongly curved	Ventral surface slightly concave, dorsal surface curved
Anterior and posterior ends	Anterior and posterior ends blunt, sometimes pointed	Anterior end blunt, posterior end slightly pointed and sometimes Blunt	Anterior end blunt, posterior end slightly pointed, sometimes blunt
Anterior lobed ventral tubercles	Usually oval	Usually oval or oblong, but occasionally round	Usually oval or oblong
Posterior lobed tubercles	Similar in structure to anterior lobed tubercles	Similar in structure to anterior lobed tubercles	Similar in structure to anterior lobed tubercles but sometimes found to be round
Number of antero-ventral tubercles	3–7	2–3	3–4
Number of postero-ventral tubercles	1–2 No. less from No. of antero-ventral tubercles	2–3 (Similar in No. to No. of antero-ventral tubercles)	Only 1 No. less from No. of antero-ventral tubercles
Micropyle	Outline of micropylar collar irregular in shape and well developed	Micropylar collar irregular in outline, with smooth surface	Outline of micropylar collar irregular in shape, with slight striations
Number of sectors of micropylar disk	Bigger micropyle with 6–7 sectors	Only 6 sectors	Only 6 sectors
Floats	Short and closer to ventral than dorsal surface	Relatively short and narrow in dorso-ventral plane	Relatively long and wide in dorsoventral plane
Deck	Continuous, slightly narrows at middle of float, anterior part of deck usually as wide as posterior part	Continuous, anterior part of deck usually wider than posterior part	Continuous and anterior part of deck usually wider than middle part

Source: Tyagi, Sharma, et al., 2016.

(formerly species A), *An. sundaicus* species D and *An. sundaicus* species E. In Andaman and Nicobar islands, it breeds in brackish water in the populations that have been characterized to be cytotype species D. It is largely a brackish water in a variety of habitats such as swamps, lagoons, creeks, etc.

4.5.1.5 *An. subpictus*

Anopheles subpictus Grassi *s.l.* is a primary (Malaysia, Indonesia) or a secondary malaria vector (India, Sri Lanka) in different Southeastern Asian countries; in Sri Lanka it is secondary to *Anopheles culicifacies* Giles *s.l.* The taxon *An. subpictus* is a complex of four sibling species (A–D) that can be differentiated through polytene chromosome banding patterns and stage-specific morphometric traits (Figure 4.9).

Like India where the species, and the complex, have both been extensively studied (Reuben and Suguna, 1983), all four sibling species of the Subpictus Species complex have been described in the neighboring island nation of Sri Lanka (Jude et al., 2014). Characteristically, sibling species there show distinct bio-ecological characteristics that are important for devising appropriate vector control measures. Species A is present in the interior of mainland India, while Species B, a comparatively stronger vector, exists in coastal areas (Suguna et al., 1994). In Sri Lanka, it is Species C which

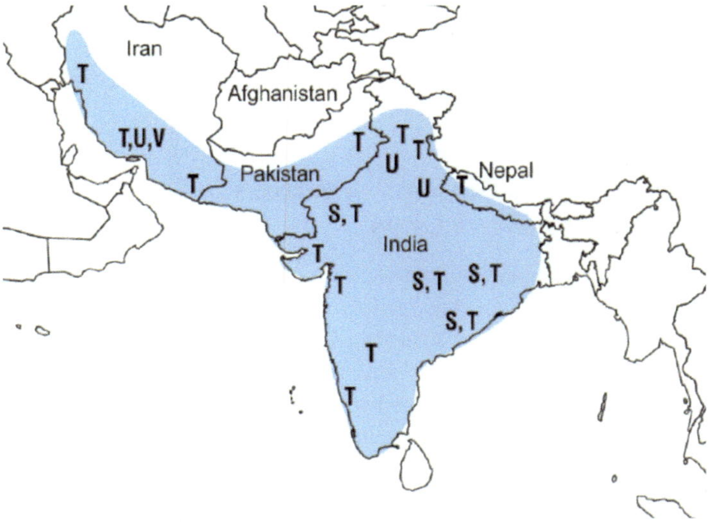

FIGURE 4.8 Distribution of *An. fluviatilis* sibling species complex.

Source: **Dev 2020.**

FIGURE 4.9 Polytene chromosome complement of *An. subpictus* species A. Break-points of inversions, a and b, reported by Suguna et al. (1994) are shown on the X-chromosome.

Source: **WHOSEARO, 2007.**

occurs inland and Species B along the coastline areas. In the Indian Ocean's pearl-island country, Species B, C, and D are both anthropophilic and zoophilic, albeit Species C and D are endophilic and endophagic, respectively, while species B is outdoor-resting with no significant preference for indoor- or outdoor-resting (Jude et al., 2014).

4.5.2 Unresolved Taxonomic Situation of Variants or Subspecies within *Anopheles stephensi* Liston

Anopheles stephensi Liston is a major urban vector for malaria in India, transmitting more than 10% (next only to *An. culicifacies* and *An. fluviatilis*) of all cases annually. It is also a primary vector in the Thar Desert region in the northwestern Rajasthan state. The species is considered to possess three subspecies or variants, viz., *An. stephensi* type species, intermediate, and *An. stephensi* var. *mysorensis*, based on counts of egg ridges and spiracular length (Table 4.5). *Anopheles stephensi* in the Thar Desert has been advocated to be a different ecotype (Tyagi et al., 1991). Interestingly, Firooziyan et al. (2018) hypothesized on the possibility of *Anopheles stephensi* as a species complex based on evidence based on odorant binding protein 1 intron I sequence.

Anopheles stephensi, believed to have its cradle in the Great Indian Thar Desert (Tyagi, 2023), is found across the Indian subcontinent, extending from the Arabian Peninsula, through Iran and Iraq, across to Bangladesh, southern China, Myanmar, and Thailand (Krishnan 1954). It is typically described as an endophilic and endophagic species despite a tendency to bite outdoors during warmer months when people are more active outdoors. Host availability seems to be a driver to a variable anthropophily for this species, and therefore in urban areas, there appears to be a greater tendency for biting humans (Rao 1984), and therefore an increased risk of malaria transmission. The ability of the larval stages of *An. stephensi* to develop in urban areas, making use of artificial containers such as domestic wells, overhead water tanks, room coolers, cisterns, and roof gutters and in water bodies in construction sites and other industrial localities, brings malaria transmission into densely populated areas including the major cities of India such as Delhi, Bangalore, and Hyderabad (Subbarao et al., 2019; Rajagopalan et al., 1987; Tyagi *pers. comm.*). In general, malaria is considered to be a disease confined to rural environments, as a simple consequence of the tendency of anophelines to search for clean and unpolluted larval habitats, and thus the existence of *An. stephensi* in such areas is a defining new characteristic of the species. In India, Sharma et al. (1969) was among the first to have given a clear picture of the polytene chromosomes of *An. stephensi stephensi* (Figure 4.10).

Anopheles stephensi Liston *s.l.* is regarded to have three subspecies or strains, viz., the type subspecies, variety *mysorensis*, and an *intermediate*. These are classically distinguishable through cytogenetic techniques (WHOSEARO, 2007) (Figure 4.10), although recently even more powerful molecular tools have reconfirmed their presence. The type form is a strong vector and occurs in

TABLE 4.5
***Anopheles stephensi* with Three Variants, Forms, or Subspecies Characterized by Distinctive Morphological Features in Egg and Spiracle**

S.No.	*Anopheles stephensi forms*	Ridges on Egg Float	Aver. No. of Ridges Among Eggs Laid	Average Spiracular Length (mm)	Average Spiracular Index (SI)
1	*An. stephensi* Type Form	14–22	16–19	0.11–0.12	8.09–9.23
2	Intermediate Subbarao, 1987	12–17	13–16	–	–
3	*An. stephensi mysorensis* Sweet & Rao, 1937	9–15	10–14	0.09–0.10	6.82–7.60

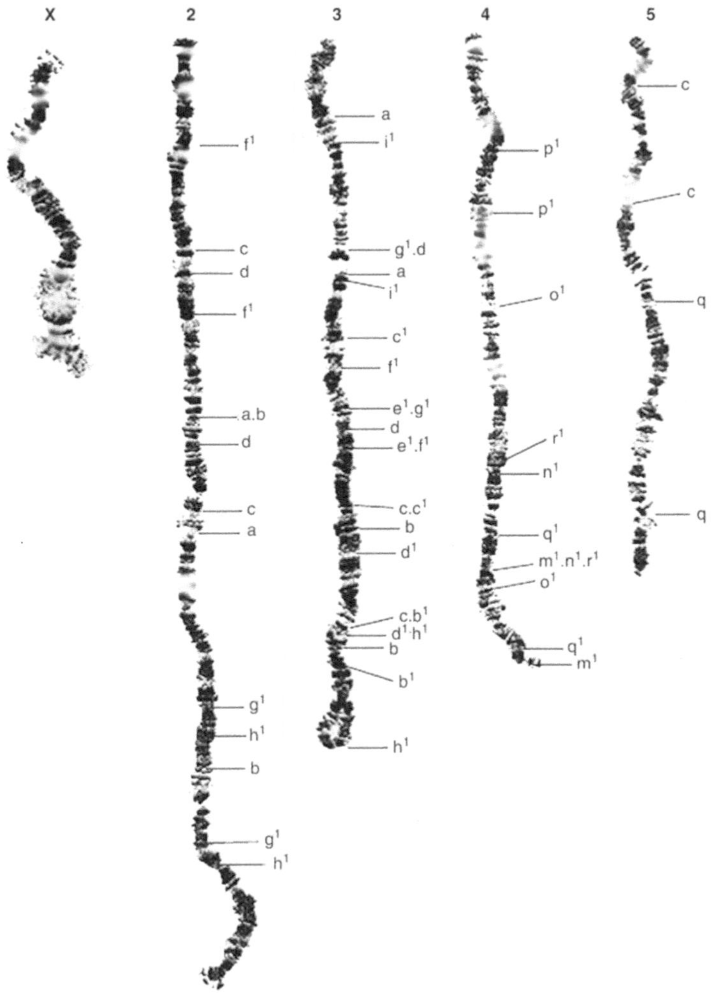

FIGURE 4.10 Photomap of *An. stephensi* polytene chromosomes showing the break-points of the inversions.

Source: **WHOSEARO, 2007.**

urban agglomerations; *mysorensis* is a poor rural vector in certain pockets, while *intermediate* does not show any trait to assert it is a vector of concern.

In the Thar Desert region, *Anopheles stephensi* is the major vector of malaria, and is biologically characterized by the phenomenon of both the aestivation during summer when temperature soars above 48°C and hibernation during winter when the temperature in places dips below 2°C (Tyagi, 2023). In certain parts of Iran, *An. stephensi* var. *mysorensis* was found as the only vector transmitting malaria. In these areas, animal hosts were very low in number or were totally absent. It is most interesting to note here that very recently this species has been speculated to be a complex based on examination of odorant binding protein 1 intron I sequence in *An. stephensi* specimens collected from Iran and Afghanistan (Firooziyan et al., 2018). The three biological species recognized as species A, B, and C correspond to *type* form, intermediate form and var. *mysorensis*, respectively. Singh et al. (2022) have refuted existence of the biological species. According to them, *AsteObp1* cannot serve as a marker for the identification of biological forms of *An. stephensi*. The probable existence of sibling species in *An. stephensi* based on the AsteObp1 intron-1 is thus refuted.

The main strategy to interrupt malaria transmitted by *An. stephensi* in urban areas by the National Vector Borne Disease Control Program (NVBDCP) under the urban malaria scheme is larval control, and in rural areas of Rajasthan where this species is reported resting and biting indoors, indoor residual spraying is used. In Rajasthan, in pre-DDT era this species was found resting on the walls of houses, but recently in the Jodhpur district of Rajasthan this species was found resting on household objects (hanging clothes, furniture, stacked clothes, etc.) avoiding walls both in insecticide sprayed and unsprayed villages, suggesting a change in its resting behavior.

4.5.3 Invasive Alien Species (IAS) in India post-Christophers (1933) and Barraud (1934)

Rao (1984) indicated that several species of Anopheles seem to have transgressed effective natural barriers such as, for example, oceans, deserts, and high mountains and invaded other regions in the past couple of centuries, and cited the example of *An. gamabie* (which actually was *An. arabiensis*, as later confirmed through molecular markers; Parmakelis et al., 2008; Griffing et al., 2015) reported for the first time in Brazil's Natal in 1930. With respect to India, Rao (1984) categorically stated that no recorded evidence was on record for any invasive species, though he suspected *Anopheles sundaicus* had intruded into India from neighboring maritime countries and later spread into the coastal areas of West Bengal, Orissa (now Odisha) and Andhra Pradesh causing severe malaria outbreaks. The species is also a major vector of malaria in the Andamans. Similarly, *An. dirus* is also doubted to have sneaked into the Indian mainland. Before *An. dirus s.l.* (*sensu lato*, i.e., *An. dirus* complex) was discovered, and sibling species were properly named; the closest species recognized in India was *An. balabacensis*. Subsequently *An. baimaii* was also reported from many parts of the country, mainly from the northeastern states of the Assam region. Invariably, all of the *An. dirus* species complex mosquitoes are vectors of malaria in Asian forested zones. Of the three species prevalent in India, at least *An. dirus* Peyton & Harrison, 1979, and *An. baimaii* Sallum & Peyton, 2005, are clearly the invasive alien species to India in recent times. Some species were brought on record of the Indian fauna for the first time, such as *Culex (Lophoceraomyia) pilifemoralis, Culex (Lophoceraomyia) wilfredi, Heizmannia (Heizmannia) Chengi, Aedes fumidus*, and *Aedes amesii* and it needs to be explored if these species are to be regarded invasive alien species. Very recently, *Aedes koreicus* was chanced upon in a village near Bangalore International Airport in 2013 (B.K. Tyagi, *personal archive*).

4.5.4 New Species Descriptions Following Christophers (1933) and Barraud (1934)

Christophers (1933) and Barraud (1934), in their respective *magnum opuses* on Tribes Anophelini and Culicini, published in the *Fauna of British India* series cataloged a total of 336 species and subspecies. In comparison, during subsequent eight-nine decades only close to 100 species could be described *de novo*. It is noteworthy that compared to merely a dozen entomologists active during the first half of the 20th century, well over 500 entomologists engaged in some kind of research on vector-borne diseases were working in government, non-government, universities and medical colleges, but still a number of new species discovered accounted relatively low in quantum. Particularly subfamily Anophelinae was abysmally productive with only very few species added; for instance, *Anopheles pseudosundaicus* (Tyagi et al., 2009) was described after nearly five decades that the last species, i.e., *An. pinjaurensis*, was described, i.e., before the end of the first half of the century. Most species had been described under the subfamily Culicinae, from different parts of the country, e.g., *Culex (Culex) gaugleri* (Suman et al., 2021), *Culex katezari* (Karlekar et al., 2020), *Toxorhynchites darjeelingensis* (Tyagi et al., 2014), and *Tx. tyagii* (Krishnamoorthy et al., 2013).

4.5.5 Species That are Likely "*Nomen nudum*"

At least two examples are readily available of such taxa which are stayed as *nomen nudum* for the want of mandatory taxonomic documentation in support of the new species:

(1) *Anopheles krishnai* (Sathe and Jagtap, 2012): The authors described a purported new species of the *Anopheles maculatus* group of the Neocellia Series, which they named *Anopheles krishnai*. Unfortunately, the paper does not provide credible evidence to support the recognition of a new species. The authors overlooked the extensive morphological and molecular work that has been done on the Neocellia Series in southern and southeastern Asia. Instead of contrasting the purported new species with members of this group, the authors conducted a molecular phylogenetic comparison with distantly related species of subgenera *Nyssorhynchus* in the Neotropical Region and *Cellia* in the Afrotropical Region (also one species, *An. stephensi*, from southern Asia). Needless to say, the comparison does not support the recognition of a new species within the Neocellia Series. Furthermore, in view of the current state of anopheline taxonomy, the immature stages as well as both sexes of the adults should have been described and contrasted with closely related species (in concert with molecular data) to ensure the validity of a new species. Not only have the authors failed to do this, they have not adequately fulfilled the criteria of availability for species names established by the *International Code of Zoological Nomenclature*: (1) they provide a brief description and indicate that the species runs to *An. maculatus* but there is no diagnosis or statement of morphological characters that differentiates the purported new species from related or similar taxa (Article 13.1), and (2) they designate a holotype, but fail to state the name and location of the collection where it is or will be deposited (Article 16.4). For these reasons, the validity of the species is highly doubtful, and the name *krishnai* is hereby regarded as a *nomen nudum*. (*cf.* Thu, 2009-11-26 15:04 – Ralph Harbach).

(2) *Lutzia agranensis* Singh & Prakash, 2008: This species, a *nomen nudum* in all probabilities, was described from larvae collected in Agra, Uttar Pradesh. The description was based on the 4th instar larvae collected from cement tank breeding with those of *Culex quinquefasciatus*, *Anopheles stephensi* and *Aedes aegypti* and the genitalia of male adults subsequently reared. However, authors have failed completely to offer and creditable evidence to support recognition of their purported new species. They did not assign any holotype and lectotype and overlooked any morphological and molecular work s but without extensive explanation. They had often believed their species to be very close to *Lutzia halifaxii* with which a brief, rather inconclusive, comparison was offered. This comparison does not support the recognition of a new species as suggested. Needless to re-emphasize that, in the line of acceptable norms in describing a new species, all the larvae, pupae, and adults should have been described, together with the molecular data, and contrasted with closely related species to ensure the validity of a new species. The authors have not only denied this inevitability, they have failed to fulfill the criteria of availability for species names as per the *International Code of Zoological Nomenclature*. For these reasons, it is nearly impossible to validate their species and, therefore, their proposed new species *agranensis* is here considered but a *nomen nudum*.

4.6 CATALOG OF INDIAN SPECIES OF MOSQUITOES

This inventory is complete and up-to-date as of December 31st, 2023. This includes all species and subspecies irrespective of any taxonomic vagaries or uncertainties associated with any taxa. The sibling species complexes have been avoided as separate taxa, and only assigned with the main species *sensu stricto*. The entire quantum of species has been fully classified between Family and Species/Subspecies levels. All species' authorities are mentioned with the year of publication including the said species.

Family: CULICIDAE
Subfamily: ANOPHELINAE
Tribe: Anophelini

I. Genus *ANOPHELES* Meigen, 1818

(i) **Subgenus *Anopheles* Meigen, 1818**

1. *Anopheles (Anopheles) ahomi* Chowdhury, 1929
2. *Anopheles (Anopheles) aitkenii* James, 1903
3. *Anopheles (Anopheles) annandalei* Prashad, 1918
4. *Anopheles (Anopheles) argyropus* (Swellengrebel, 1914)
5. *Anopheles (Anopheles) baileyi* Edwards,1929
6. *Anopheles (Anopheles) barbirostris* van der Wulp, 1884
7. *Anopheles (Anopheles) barbumbrosus* Strickland & Chowdhury, 1927
8. *Anopheles (Anopheles) barianensis* James, 1911
9. *Anopheles (Anopheles) bengalensis* Puri, 1930
10. *Anopheles (Anopheles) crawfordi* Reid, 1953
11. *Anopheles (Anopheles) culiciformis* Cogill, 1903
12. *Anopheles (Anopheles) gigas* Giles, 1901
13. *Anopheles (Anopheles) hodgkini* Reid, 1962
14. *Anopheles (Anopheles) insulaeflorum* (Swellengrebel & Swellengrebel de Graaf, 1920)
15. *Anopheles (Anopheles) interruptus* Puri, 1929
16. *Anopheles (Anopheles) lesteri paraliae* Sandosham 1959
17. *Anopheles (Anopheles) lindesayi* Giles, 1900
18. *Anopheles (Anopheles) nigerrimus* Giles, 1900
19. *Anopheles (Anopheles) nilgiricus* Christophers, 1924
20. *Anopheles (Anopheles) nitidus* Harrison, Scanlon & Reid, 1973
21. *Anopheles (Anopheles) peditaeniatus* (Leicester, 1908)
22. *Anopheles (Anopheles) pinjaurensis* Barraud, 1932
23. *Anopheles (Anopheles) roperi* Reid, 1950
24. *Anopheles (Anopheles) sinensis* Wiedemann, 1828
25. *Anopheles (Anopheles) sintoni* Puri, 1929
26. *Anopheles (Anopheles) umbrosus* (Theobald, 1903)

(ii) **Subgenus *Cellia* Theobald, 1902**

27. *Anopheles (Cellia) aconitus* Donitz, 1902
28. *Anopheles (Cellia) annularis* van der Wulp, 1884
29. *Anopheles (Cellia) baimaii* Sallum & Peyton, 2005
30. *Anopheles (Cellia) balabacensis* Baisas, 1936
31. *Anopheles (Cellia) culicifacies* Giles, 1901
32. *Anopheles (Cellia) dravidicus* Christophers, 1924
33. *Anopheles (Cellia) dthali* Patton, 1905
34. *Anopheles (Cellia) elegans* (James, 1903)
35. *Anopheles (Cellia) fluviatilis* James, 1902
36. *Anopheles (Cellia) jamesii* Theobald, 1901
37. *Anopheles (Cellia) jeyporiensis* James, 1902
38. *Anopheles (Cellia) karwari* (James, 1903)
39. *Anopheles (Cellia) kochi* Donitz, 1901
40. *Anopheles (Cellia) krishnai* Sathe & Jagtap (2012)

41. *Anopheles (Cellia) maculatus* Theobald, 1901
42. *Anopheles (Cellia) majidi* Young & Majid, 1928
43. *Anopheles (Cellia) minimus* Theobald, 1901
44. *Anopheles (Cellia) mirans* Sallum & Peyton, 2005
45. *Anopheles (Cellia) moghulensis* Christophers, 1924
46. *Anopheles (Cellia) multicolor* Cambouliu, 1902
47. *Anopheles (Cellia) nivipes* (Theobald, 1903)
48. *Anopheles (Cellia) pallidus* Theobald, 1901
49. *Anopheles (Cellia) philippinensis* Ludlow, 1902
50. *Anopheles (Cellia) pseudojamesi* Strickland & Chowdhury, 1927
51. *Anopheles (Cellia) pseudosundaicus* Tyagi et al., 2009
52. *Anopheles (Cellia) pseudowillmori* (Theobald, 1910)
53. *Anopheles (Cellia) pulcherrimus* Theobald, 1902
54. *Anopheles (Cellia) sergentii* (Theobald, 1907)
55. *Anopheles (Cellia) splendidus* Koidzumi, 1920
56. *Anopheles (Cellia) stephensi* Liston, 1901
57. *Anopheles (Cellia) subpictus* Grassi, 1899
58. *Anopheles (Cellia) sundaicus* (Rodenwaldt, 1925)
59. *Anopheles (Cellia) tessellatus* Theobald, 1901
60. *Anopheles (Cellia) theobaldi* Giles, 1901
61. *Anopheles (Cellia) turkhudi* Liston, 1901
62. *Anopheles (Cellia) vagus* Donitz, 1902
63. *Anopheles (Cellia) varuna* Iyengar, 1924
64. *Anopheles (Cellia) willmori* (James, 1903)

Subfamily: CULICINAE
Tribe: Aedeomyiini Theobald, 1901

I. **Genus *AEDEOMYIA* Theobald, 1901**

(i) Subgenus Aedeomyia Theobald, 1901

65. *Aedeomyia (Aedeomyia) catasticta* Knab, 1909
66. *Aedeomyia (Aedeomyia) venustipes* (Skuse, 1889)

Tribe: Aedini Neveu-Lemaire, 1902

I. **Genus *AEDIMORPHUS* Theobald, 1903**

67. *Aedimorphus alboscutellatus* (Theobald, 1905)
68. *Aedimorphus caecus* (Theobald, 1901)
69. *Aedimorphus culicinus* (Edwards, 1922)
70. *Aedimorphus jamesi* (Edwards, 1914)
71. *Aedimorphus lowisii* (Theobald, 1910)
72. *Aedimorphus mediolineatus* (Theobald, 1901)
73. *Aedimorphus nigrostriatus* Barraud, 1927
74. *Aedimorphus pallidostriatus* (Theobald, 1907)
75. *Aedimorphus pampangensis* (Ludlow, 1905)
76. *Aedimorphus pipersalatus* (Giles, 1902)
77. *Aedimorphus punctifemoris* (Ludlow, 1921)
78. *Aedimorphus stenoetrus* (Theobald, 1907)
79. *Aedimorphus syntheticus* (Barraud, 1928)
80. *Aedimorphus taeniorhynchoides* (Christophers, 1911)
81. *Aedimorphus trimaculatus* (Theobald, 1905)
82. *Aedimorphus vexans* (Meigen, 1830)

II. **Genus *ARMIGERES* Theobald, 1901**
  (i) Subgenus Armigeres Theobald, 1901
    83. *Armigeres (Armigeres) aureolineatus* (Leicester, 1908)
    84. *Armigeres (Armigeres) durhami* (Edwards, 1917)
    85. *Armigeres (Armigeres) joloensis* (Ludlow, 1904)
    86. *Armigeres (Armigeres) kesseli* Ramalingam, 1987
    87. *Armigeres (Armigeres) kuchingensis* Edwards, 1915
    88. *Armigeres (Armigeres) mahantai* Bhattacharyya et al., 2009
    89. *Armigeres (Armigeres) pallithorax* Dong, Zhou & Dong, 2004
    90. *Armigeres (Armigeres) subalbatus* (Coquillett, 1898)
    91. *Armigeres (Armigeres) theobaldi* Barruad, 1934
  (ii) Subgenus Leicesteria Theobald, 1904
    92. *Armigeres (Leicesteria) annulipalpis* (Theobald, 1910)
    93. *Armigeres (Leicesteria) annulitarsis* (Leicester, 1908)
    94. *Armigeres (Leicesteria) dentatus* Barraud, 1927
    95. *Armigeres (Leicesteria) digitatus* (Edwards, 1914)
    96. *Armigeres (Leicesteria) dolichocephalus* (Leicester, 1908)
    97. *Armigeres (Leicesteria) flavus* (Leicester, 1908)
    98. *Armigeres (Leicesteria) inchoatus* Barraud, 1927
    99. *Armigeres (Armigeres) joloensis* Ludlow, 1904
    100. *Armigeres (Leicesteria) longipalpis* (Leicester, 1904)
    101. *Armigeres (Leicesteria) magnus* (Theobald, 1908)
    102. *Armigeres (Leicesteria) omissus* (Edwards, 1914)

III. **Genus *AYURAKITIA* Thurman, 1954**
  103. *Ayurakitia peytoni* (Reinert, 1972)

IV. **Genus *BRUCEHARRISONIUS* Reinert, 2003**
  104. *Bruceharrisonius aureostriatus* (Doleschall, 1857)
  105. *Bruceharrisonius christophersi* (Edwards, 1922)
  106. *Bruceharrisonius doonii* (Wattal, Bhatia & Kalra, 1958)
  107. *Bruceharrisonius greenii* (Theobald, 1903)

V. **Genus *CANCRAEDES* Edwards, 1929**
  108. *Cancraedes cancricomes* (Edwards, 1922)
  109. *Cancraedes simplex* (Theobald, 1903)

VI. **Genus *CHRISTOPHERSIOMYIA* Barraud, 1923**
  110. *Christophersiomyia annulirostris* (Theobald, 1905)
  111. *Christophersiomyia gombakensis* (Mattingly, 1959)
  112. *Christophersiomyia ibis* (Barraud, 1931)
  113. *Christophersiomyia thomsoni* (Theobald, 1905)

VII. **Genus *COLLESSIUS* Reinert, Harbach & Kitching, 2006**
  (i) Subgenus *Alloeomyia* Reinert, Harbach & Kitching, 2008
    114. *Collessius (Alloeomyia) pseudotaeniatus* (Giles, 1901)
  (ii) Subgenus *Collessius* Reinert, Harbach & Kitching, 2006
    115. *Collessius (Collessius) elsiae* (Barraud, 1923)
    116. *Collessius (Collessius) macdougalli* (Edwards, 1922)
    117. *Collessius (Collessius) shortti* (Barraud, 1923)

VIII. **Genus *DANIELSIA* Theobald, 1904**
  118. *Danielsia albotaeniata* Leicester, 1904
  119. *Danielsia lepchana* (Barruad, 1923)

IX. **Genus *DENDROSKUSEA* Edwards, 1929**
  120. *Dendroskusea kanarensis* (Edwards, 1934)
  121. *Dendroskusea micropterus* (Giles, 1901)

122. *Dendroskusea periskelata* (Giles, 1902)
123. *Dendroskusea ramachandrai* (Reuben, 1967)
124. *Dendroskusea reginae* (Edwards, 1922)

X. **Genus *DOWNSIOMYIA* Vargas, 1950**

125. *Downsiomyia albolateralis* (Theobald, 1908)
126. *Downsiomyia albonivea* (Barruad,1934)
127. *Downsiomyia mohani* (Knight, 1969)
128. *Downsiomyia nivea* (Ludlow,1903)
129. *Downsiomyia niveoides* (Barruad, 1934)
130. *Downsiomyia novonivea* (Barraud,1934)

XI. **Genus *EDWARDSAEDES* Belkin, 1962**

131. *Edwardsaedes imprimens* (Walkar, 1860)

XII. **Genus *FINLAYA* Theobald, 1903**

132. *Finlaya flavipennis* Giles, 1904
133. *Finlaya poicilia* Theobald, 1903

XIII. **Genus *FREDWARDSIUS* Reinert, 2000**

134. *Fredwardsius vittatus* (Bigot, 1861)

XIV. **Genus *GILESIUS* Reinert, Harbach & Kitching, 2006**

135. *Gilesius pulchriventer* (Giles, 1901)

XV. **Genus *HEIZMANNIA* Ludlow, 1905**

(i) Subgenus *Heizmannia* Ludlow, 1905

136. *Heizmannia (Heizmannia) aureochaeta* (Leicester, 1908)
137. *Heizmannia (Heizmannia) chandi* Edwards, 1922
138. *Heizmannia (Heizmannia) chengi* Lien, 1968
139. *Heizmannia (Heizmannia) complex* (Theobald, 1910)
140. *Heizmannia (Heizmannia) covelli* Barraud, 1929
141. *Heizmannia (Heizmannia) funerea* (Leicester, 1908)
142. *Heizmannia (Heizmannia) greenii* (Theobald, 1905)
143. *Heizmannia (Heizmannia) himalayensis* Edwards, 1922
144. *Heizmannia (Heizmannia) indica* (Theobald, 1905)
145. *Heizmannia (Heizmannia) metallica* Leicester, 1908
146. *Heizmannia (Heizmannia) rajagopalani* Natarajan, Eapen & Jambulingam, 2020
147. *Heizmannia (Heizmannia) reidi* Mattingly, 1957
148. *Heizmannia (Heizmannia) viridis* Barraud, 1929

(ii) Subgenus *Mattinglyia* Lien, 1968

149. *Heizmannia (Mattinglyia) discrepans* (Edwards, 1922)
150. *Heizmannia (Mattinglyia) tripunctata* (Theobald, 1908)

XVI. **Genus *HIMALAIUS* Reinert, Harbach & Kitching, 2006**

151. *Himalaius gilli* (Barraud, 1924)
152. *Himalaius simlensis* (Edwards, 1922)

XVII. **Genus *HOPKINSIUS* Reinert, Harbach & Kitching, 2008**

(i) Subgenus *Yamada* Reinert, Harbach & Kitching, 2008

153. *Hopkinsius (Yamada) albocinctus* (Barraud, 1924)

XVIII. **Genus *HULECOETEOMYIA* Theobald, 1904**

154. *Hulecoeteomyia chrysolineata* (Theobald, 1907)
155. *Hulecoeteomyia formosensis* (Yamada, 1921)
156. *Hulecoeteomyia harveyi* (Barraud, 1923)
157. *Hulecoeteomyia pallirostris* (Edwards, 1922)
158. *Hulecoeteomyia saxicola* (Edwards, 1922)
159. *Hulecoeteomyia cherrapunjiensis* Natarajan et al., 2016

XIX. **Genus *INDUSIUS* Edwards, 1934**
160. *Indusius pulverulentus* (Edwards, 1922)
XX. **Genus *JIHLIENIUS* Reinert, Harbach & Kitching, 2006**
161. *Jihlienius unicinctus* (Edwards, 1922)
XXI. **Genus *KENKNIGHTIA* Reinert, 1990**
162. *Kenknightia dissimilis* (Leicester, 1908)
163. *Kenknightia karwari* (Barraud, 1924)
XXII. **Genus *LORRAINEA* Belkin,1962**
164. *Lorrainea amesii* (Ludlow, 1903)
165. *Lorrainea fumida* (Edwards, 1928)
XXIII. **Genus *MUCIDUS* Theobald, 1901**
Subgenus *Mucidus* Theobald, 1901
166. *Mucidus* (*Mucidus*) *laniger* (Wiedemann, 1820)
167. *Mucidus* (*Mucidus*) *quasiferinus* (Mattingly, 1961)
168. *Mucidus* (*Mucidus*) *scatophagoides* Theobald, 1901
XXIV. **Geunus *NEOMELANICONION* Newstead, 1907**
169. *Neomelaniconion lineatopenne* (Ludlow, 1905)
XXV. **Genus *OCHLEROTATUS* Lynch Arribalzaga, 1891 *Sensu***
Subgenus *Finlaya* Theobald, 1903 *Sensu auctorum*
170. *Ochlerotatus* (*Finlaya*) *auronitens* (Edwards, 1922)
171. *Ochlerotatus* (*Finlaya*) *oreophilus* Edwards, 1916
172. *Ochlerotatus* (*Finlaya*) *sintoni* (Barraud, 1924)
173. *Ochlerotatus* (*Finlaya*) *suffusus* (Edwards, 1922)
174. *Ochlerotatus* (*Finlaya*) *versicolor* (Barraud,1924)
Subgenus uncertain
175. *Ochlerotatus pulcritarsis* (Rondani, 1872)
176. *Ochlerotatus pullatus* (Coquillett, 1904)
XXVI. **Genus *PARAEDES* Edwards, 1934**
177. *Paraedes barraudi* Edwards, 1934
178. *Paraedes chrysoscuta* (Theobald, 1910)
179. *Paraedes jambulingami* Natarajan, 2019
180. *Paraedes menoni* (Mattingly, 1958)
181. *Paraedes ostentatio* (Leicester, 1908)
XXVII. **Genus *PETERMATTINGLYIUS* Reinert, Harbach & Kitching, 2009**
(i) Subgenus *Petermattinglyius* Reinert, Harbach & Kitching, 2009
182. *Petermattinglyius* (*Petermattinglyius*) *franciscoi* (Mattingly, 1959)
183. *Petermattinglyius* (*Petermattinglyius*) *iyengari* (Edwards, 1923)
XXVIII. **Genus *PHAGOMYIA* Theobald, 1905**
184. *Phagomyia assamensis* (Theobald 1908)
185. *Phagomyia cacharana* (Barraud, 1923)
186. *Phagomyia cogilli* (Edwards, 1922)
187. *Phagomyia deccana* (Barraud, 1923)
188. *Phagomyia feegradei* (Barraud, 1934)
189. *Phagomyia gubernatoris* (Giles, 1901)
190. *Phagomyia inquinata* (Edwards, 1922)
191. *Phagomyia khazani* (Edwards, 1922)
192. *Phagomyia lophoventralis* (Theobald, 1910)
193. *Phagomyia prominens* (Barraud, 1923)
194. *Phagomyia stevensoni* (Barraud, 1923)
XXIX. **Genus *RHINOSKUSEA* Edwards, 1929**
195. *Rhinoskusea longirostris* (Leicester, 1908)
196. *Rhinoskusea portonovoensis* (Tewari & Hiriyan, 1992)

197. *Rhinoskusea wardi* (Reinert, 1976)

XXX. **Genus *SCUTOMYIA* Theobald, 1904**

198. *Scutomyia albolineata* Theobald, 1904

XXXI. **Genus *STEGOMYIA* Theobald, 1901**

(i) Subgenus *Actinothrix* Reinert, Harbach & Kitching, 2009

199. *Stegomyia (Actinothrix) edwardsi* Barraud, 1923
200. *Stegomyia (Actinothrix) seampi* (Huang, 1974)

(ii) Subgenus *Heteraspidion* Reinert, Harbach & Kitching, 2009

201. *Stegomyia (Heteraspidion) annandalei* Theobald, 1910
202. *Stegomyia (Heteraspidion) craggi* Barraud, 1923

(iii) Subgenus *Huangmyia* Reinert, Harbach & Kitching, 2009

203. *Stegomyia (Huangmyia) mediopunctata* Theobald, 1905
204. *Stegomyia (Huangmyia) perplexa* Leicester, 1908

(iv) Subgenus *Stegomyia* Theobald, 1901

205. *Stegomyia (Stegomyia) aegypti* (Linnaeus, 1762)

(v) Subgenus *Xyele* Reinert, Harbach & Kitching, 2009

206. *Stegomyia (Xyele) desmotes* Giles, 1904

[subgenus unassigned yet for following]

207. *Stegomyia albopicta* (Skuse, 1894)
208. *Stegomyia flavopicta* (Yamada, 1921)
209. *Stegomyia gardnerii imitator* (Leicester, 1908)
210. *Stegomyia koreicus* (Edwards, 1917)
211. *Stegomyia krombeini* (Huang, 1975)
212. *Stegomyia malayensis* (Colless, 1962)
213. *Stegomyia novalbopicta* (Barraud, 1931)
214. *Stegomyia ostentation* (Leicester, 1908)
215. *Stegomyia patriciae* (Mattingly, 1954)
216. *Stegomyia pseudalbopicta* Borel, 1928
217. *Stegomyia scutellaris* (Walker, 1859)
218. *Stegomyia subalbopicta* (Barraud, 1931)
219. *Stegomyia unilineata* (Theobald, 1906)
220. *Stegomyia w-alba* Theobald, 1905

XXXII. **Genus *TEWARIUS* Reinert, 2006**

221. *Tewarius agastyai* (Tewari & Hiriyan, 1992)
222. *Tewarius nummatus* (Edwards, 1923)
223. *Tewarius reubenae* (Tewari & Hiriyan, 1992)

XXXIII. **Genus *UDAYA* Thurman, 1954**

224. *Udaya argyrurus* (Edwards, 1934)
225. *Udaya subsimilis* (Barraud, 1927)

XXXIV. **Genus *VERRALLINA* Theobald, 1903**

(i) Subgenus *Harbachius* Reinert, 1999

226. *Verrallina (Harbachius) abdita* (Barraud, 1931)
227. *Verrallina (Harbachius) consonensis* (Reinert, 1973)
228. *Verrallina (Harbachius) uniformis* (Theobald, 1910)
229. *Verrallina (Harbachius) yusafi* (Barraud, 1931)

(ii) Subgenus *Neomacleaya* Theobald, 1907

230. *Verrallina (Neomacleaya) agrestis* (Barraud, 1931)
231. *Verrallina (Neomacleaya) andamanensis* (Edwards, 1922)
232. *Verrallina (Neomacleaya) assamensis* Bhattacharyya et al., 2004
233. *Verrallina (Neomacleaya) atria* (Barraud, 1928)
234. *Verrallina (Neomacleaya) cauta* (Barraud, 1928)
235. *Verrallina (Neomacleaya) clavata* (Barraud, 1931)

236. *Verrallina(Neomacleaya) comata* (Barraud, 1931)
237. *Verrallina (Neomacleaya) hirsutipleura* Barraud, 1928
238. *Verrallina (Neomacleaya) indica* (Theobald, 1907)
239. *Verrallina (Neomacleaya) pseudodiurna* (Theobald, 1910)
240. *Verrallina (Neomacleaya) pseudomediofasciata* (Theobald, 1910)
241. *Verrallina (Neomacleaya) rami* (Barraud, 1928)
242. *Verrallina (Neomacleaya) seculata* (Menon, 1950)
243. *Verrallina (Neomacleaya) vallistris* (Barraud, 1928)
244. *Verrallina (Neomacleaya) unca* (Theobald,1901)

(iii) Subgenus *Verrallina* Theobald, 1903

245. *Verrallina (Verrallina) butleri* (Theobald, 1901)
246. *Verrallina (Verrallina) dux* (Dyar & Shannon, 1925)
247. *Verrallina (Verrallina) lugubris* (Barraud, 1928)

Tribe: Culicini Meigen, 1818

I. **Genus *CULEX* Linnaeus, 1758**

(i) Subgenus *Barraudius* Edwards, 1921

248. *Culex (Barraudius) modestus* Ficalbi, 1890

(ii) Subgenus *Culex* Linnaeus, 1758

249. *Culex (Culex) alienus* Colless, 1957
250. *Culex (Culex) annulus* Theobald, 1901
251. *Culex (Culex) barraudi* Edwards, 1922
252. *Culex (Culex) edwardsi* Barraud, 1923
253. *Culex (Culex) fuscifurcatus* Edwards, 1937
254. *Culex (Culex) fuscocephala* Theobald, 1907
255. *Culex (Culex) gaugleri* Suman, De, Sharma, Chandra and Banerjee, 2021
256. *Culex (Culex) gelidus* Theobald, 1901
257. *Culex (Culex) hutchinsoni* Barraud, 1924
258. *Culex (Culex) jacksoni* Edwards, 1934
259. *Culex (Culex) katezari* Karlekar, Andrew and Deshpande, 2020
260. *Culex (Culex) kodaikanalensis* Suman, Veer and Chandra, 2022
261. *Culex (Culex) mimeticus* Noe, 1899
262. *Culex (Culex) mimuloides* Barraud, 1924
263. *Culex (Culex) mimulus* Edwards, 1915
264. *Culex (Culex) murrelli* Lien, 1968
265. *Culex (Culex) nilgiricus* Edwards, 1916
266. *Culex (Culex) perexiguus* Theobald, 1903
267. *Culex (Culex) perplexus* Leicester, 1908
268. Culex (Culex) pipiens Linn.
269. *Culex (Culex) pseudovishnui* Colless, 1957
270. *Culex (Culex) quinquefasciatus* Say, 1823
271. *Culex (Culex) sitiens* Wiedemann, 1828
272. *Culex (Culex) theileri* Theobald, 1903
273. *Culex (Culex) tritaeniorhynchus* Giles, 1901
274. *Culex (Culex) univittatus* Theobald, 1901
275. *Culex (Culex) vagans* Wiedemann, 1828
276. *Culex (Culex) vishnui* Theobald, 1901
277. *Culex (Culex) whitei* Barraud, 1923
278. *Culex (Culex) whitmorei* (Giles, 1904)

(iii) Subgenus *Culiciomyia* Theobald, 1907

279. *Culex (Culiciomyia) bailyi* Barraud, 1934

280. *Culex (Culiciomyia) fragilis* Ludlow, 1903
281. *Culex (Culiciomyia) harrisoni* Sirivanakarn, 1977
282. *Culex (Culiciomyia) nigropunctatus* Edwards, 1926
283. *Culex (Culiciomyia) pallidothorax* Theobald, 1905
284. *Culex (Culiciomyia) ramakrishnii* Wattal & Kalra, 1965
285. *Culex (Culiciomyia) scanloni* Bram, 1967
286. *Culex (Culiciomyia) shebbearei* Barraud, 1924
287. *Culex (Culiciomyia) spathifurca* (Edwards, 1915)
288. *Culex (Culiciomyia) viridiventer* Giles, 1901

(iv) Subgenus *Eumelanomyia* Theobald, 1909

289. *Culex (Eumelanomyia) brevipalpis* (Giles, 1902)
290. *Culex (Eumelanomyia) castrensis* Edwards, 1922
291. *Culex (Eumelanomyia) foliatus* Brug, 1932
292. *Culex (Eumelanomyia) hinglungensis* Chu, 1957
293. *Culex (Eumelanomyia) iphis* Barraud, 1924
294. *Culex (Eumelanomyia) khazani* Edwards, 1922
295. *Culex (Eumelanomyia) malayi* (Leicester, 1908)
296. *Culex (Eumelanomyia) mohani* Sirivanakarn, 1977
297. *Culex (Eumelanomyia) pluvialis* Barraud, 1924
298. *Culex (Eumelanomyia) tenuipalpis* Barraud, 1924

(v) Subgenus *Lophoceraomyia* Theobald, 1905

299. *Culex (Lophoceraomyia) aculeatus* Colless, 1965
300. *Culex (Lophoceraomyia) bengalensis* Barraud, 1934
301. *Culex (Lophoceraomyia) bicornutus* (Theobald, 1910)
302. *Culex (Lophoceraomyia) cinctellus* Edwards, 1922
303. *Culex (Lophoceratomyia) cubitatus* Colless, 1965
304. *Culex (Lophoceraomyia) demissus* Colless, 1965
305. *Culex (Lophoceraomyia) flavicornis* Barraud, 1924
306. *Culex (Lophoceraomyia) gracicornis* Sirivanakarn, 1977
307. *Culex (Lophoceraomyia) inculus* Colless, 1965
308. *Culex (Lophoceraomyia) infantulus* Edwards, 1922
309. *Culex (Lophoceraomyia) lasiopalpis* Sirivanakarn, 1977
310. *Culex (Lophoceraomyia) macdonaldi* Colless, 1965
311. *Culex (Lophoceraomyia) mammilifer* (Leicester, 1908)
312. *Culex (Lophoceraomyia) minor* (Leicester, 1908)
313. *Culex (Lophoceraomyia) minutissimus* (Theobald, 1907)
314. *Culex (Lophoceraomyia) paraculeatus* Sirivanakarn, 1977
315. *Culex (Lophoceraomyia) peytoni* Bram & Rattanarithikul, 1967
316. *Culex (Lophoceraomyia) pholeter* Bram & Rattanarithikul, 1967
317. *Culex (Lophoceraomyia) pilifemoralis* Wang & Feng, 1964
318. *Culex (Lophoceraomyia) quadripalpis* (Edwards, 1914)
319. *Culex (Lophoceraomyia) raghavanii* Rahman, Chowdhury & Kalra, 1968
320. *Culex (Lophoceraomyia) rubithoracis* (Leicester, 1908)
321. *Culex (Lophoceraomyia) seniori* Barraud, 1934
322. *Culex (Lophoceraomyia) singhbhumensis* Natarajan & Rajavel, 2009
323. *Culex (Lophoceraomyia) uniformis* (Theobald, 1905)
324. *Culex (Lophoceraomyia) variatus* (Leicester, 1908)
325. *Culex (Lophoceraomyia) wilfredi* Colless, 1965

(vi) Subgenus *Maillotia* Theobald, 1907

326. *Culex (Maillotia) hortensis* Ficalbi, 1889

(vii) Subgenus *Oculeomyia* Theobald, 1907

327. *Culex (Oculeomyia) bitaeniorhynchus* Giles, 1901

328. *Culex (Oculeomyia) cornutus* Edwards, 1922
329. *Culex (Oculeomyia) epidesmus* (Theobald, 1910)
330. *Culex (Oculeomyia) infula* Theobald, 1901
331. *Culex (Oculeomyia) luzonensis* Sirivanakarn, 1976
332. *Culex (Oculeomyia) sinensis* Theobald, 1903

II. **Genus *LUTZIA* Theobald, 1903**

(i) Subgenus *Metalutzia* Tanaka, 2003

333. *Lutzia (Metalutzia) agranensis* Singh & Prakash, 2008
334. *Lutzia (Metalutzia) fuscana* (Wiedemann, 1820)
335. *Lutzia (Metalutzia) halifaxii* (Theobald, 1903)
336. *Lutzia (Metalutzia) vorax* Edwards, 1921
337. *Lutzia (Metalutzia) tigripes* (de Grandpre and de Charmoy, 1901)

Tribe: Culisetini Belkin, 1962

I. **Genus *CULISETA* Felt, 1904**

(i) Subgenus *Allotheobaldia* Broelemann,1919

338. *Culiseta (Allotheobaldia) longiareolata* (Macquart, 1838)

(ii) Subgenus *Culiseta* Felt,1904

339. *Culiseta (Culiseta) alaskaensis indica* (Edwards, 1920)
340. *Culiseta (Culiseta) niveitaeniata* (Theobald, 1907)

Tribe: Ficalbiini Belkin, 1962

I. **Genus *FICALBIA*** Theobald, 1903

341. *Ficalbia minima* (Theobald, 1901)

II. **Genus *MIMOMYIA* Theobald, 1903**

(i) Subgenus *Etorleptiomyia* Theobald, 1904

342. *Mimomyia (Etorleptiomyia) luzonensis* (Ludlow, 1905)

(ii) Subgenus *Ingramia* Edwards, 1912

343. *imomyia (Ingramia) fusca* (Leicester, 1908)

(iii) Subgenus *Mimomyia* Theobald, 1903

344. *Mimomyia (Mimomyia) aurea* (Leicester, 1908)
345. *Mimomyia (Mimomyia) chamberlaini* Ludlow, 1904 ssp. *clavipalpus* (Theobald, 1908)
346. *Mimomyia (Mimomyia) hybrida* (Leicester, 1908)
347. *Mimomyia (Mimomyia) intermedia* (Barraud, 1929)

Tribe: Hodgesiini Belkin,1962

I. **Genus *HODGESIA* Theobald, 1904**

348. *Hodgesia bailyi* Barraud, 1929
349. *Hodgesia malayi* Leicester, 1908

Tribe : Mansoniini Belkin,1962

I. **Genus *COQUILLETTIDIA* Dyar, 1905**

(i) Subgenus *Coquillettidia* Dyar, 1905

350. *Coquillettidia (Coquillettidia) crassipes* (van der Wulp, 1881)
351. *Coquillettidia (Coquillettidia) novochracea* (Barraud, 1927)
352. *Coquillettidia (Coquillettidia) ochracea* (Theobald, 1903)

II. **Genus *MANSONIA* Blanchard, 1901**
(i) Subgenus *Mansonioides* Theobald, 1907
353. *Mansonia (Mansonioides) annulifera* (Theobald, 1901)
354. *Mansonia (Mansonioides) dives* (Schiner, 1868)
355. *Mansonia (Mansonioides) indiana* Edwards, 1930
356. *Mansonia (Mansonioides) uniformis* (Theobald, 1901)

Tribe: Orthopodomyiini Belkin, Heinemann & Page, 1970

I. **Genus *ORTHOPODOMYIA* Theobald, 1904**
357. *Orthopodomyia albipes* Leicester, 1904
358. *Orthopodomyia andamanensis* Barraud, 1934
359. *Orthopodomyia anopheloides* (Giles, 1903)
360. *Orthopodomyia flavicosta* Barraud, 1927
361. *Orthopodomyia flavithorax* Barraud, 1927

Tribe: Sabethini Blanchard,1905

I. **Genus *MALAYA* Leicester, 1908**
362. *Malaya genurostris* Leicester, 1908
363. *Malaya jacobsoni* (Edwards, 1930)
II. **Genus *TOPOMYIA* Leicester, 1908**
(i) Subgenus *Suaymyia* Thurman, 1959
364. *Topomyia (Suaymyia) cristata* Thurman,1959
(ii) Subgenus *Topomyia* Leicester, 1908
365. *Topomyia (Topomyia) aureoventer* (Theobald, 1910)
366. *Topomyia (Topomyia) bifurcata* Dong,Wang & Lu,1995
367. *Topomyia (Topomyia) hirtusa* Gong,1989
III. **Genus *TRIPTEROIDES* Giles, 1904**
(i) Subgenus *Rachionotomyia* Theobald, 1905
368. *Tripteroides (Rachionotomyia) affinis* (Edwards, 1913)
369. *Tripteroides (Rachionotomyia) aranoides* (Theobald, 1901)
370. *Tripteroides (Rachionotomyia) ceylonensis* (Theobald 1905)
371. *Tripteroides (Rachionotomyia) coonorensis* Mattingly,1981
372. *Tripteroides (Rachionotomyia) edwardsi* (Barraud, 1929)
373. *Tripteroides (Rachionotomyia) serratus* (Barraud, 1929)
(ii) Subgenus *Tripteroides* Giles, 1904
374. *Tripteroides (Tripteroides) indicus* (Barraud, 1929)
375. *Tripteroides (Tripteroides) powelli* (Ludlow,1909)
376. *Tripteroides (Tripteroides) similis* (Leicester, 1908)
377. *Tripteroides (Tripteroides) tarsalis* Delfinado & Hodges, 1968

Tribe: Toxorhynchitini Lahille, 1904

I. **Genus *TOXORHYNCHITES* Theobald, 1901**
(i) Subgenus *Toxorhynchites* Theobald, 1901
378. *Toxorhynchites (Toxorhynchites) albipes* (Edwards, 1922)
379. *Toxorhynchites (Toxorhynchites) edwardsi* (Barraud, 1924)
380. *Toxorhynchites (Toxorhynchites) gravely* (Edwards, 1921)
381. *Toxorhynchites (Toxorhynchites) kempi* (Edwards,1921)
382. *Toxorhynchites (Toxorhynchites) klossi* (Edwards, 1921)

383. *Toxorhynchites (Toxorhynchites) metallicus* Leicester 1904
384. *Toxorhynchites (Toxorhynchites) minimus* (Theobald, 1905)
385. *Toxorhynchites (Toxorhynchites) splendens* (Wiedemann, 1819)
386. *Toxorhynchites (Toxorhynchites) tyagii* Krishnamoorthy et al., 2013

Tribe: Uranotaeniini Lahille, 1904

I. **Genus *URANOTAENIA* Lynch Arribalzaga, 1891**

(i) Subgenus *Pseudoficalbia* Theobald, 1912

387. *Uranotaenia (Pseudoficalbia) atra* Theobald, 1905
388. *Uranotaenia (Pseudoficalbia) bicolor* Leicester, 1908
389. *Uranotaenia (Pseudoficalbia) bimaculata* Leicester, 1908
390. *Uranotaenia (Pseudoficalbia) dibrugarhensis* Bhattacharyya et al., 2004
391. *Uranotaenia (Pseudoficalbia) husaini* Qutubuddin, 1947
392. *Uranotaenia (Pseudoficalbia) luteola* Edwards, 1934
393. *Uranotaenia (Pseudoficalbia) lutescens* Leicester, 1908
394. *Uranotaenia (Pseudoficalbia) maculipleura* Leicester, 1908
395. *Uranotaenia (Pseudoficalbia) mattinglyi* Qutubuddin, 1951
396. *Uranotaenia (Pseudoficalbia) maxima* Leicester, 1908
397. *Uranotaenia (Uranotaenia) micans* Leicester, 1908
398. *Uranotaenia (Pseudoficalbia) nivipleura* Leicester, 1908
399. *Uranotaenia (Pseudoficalbia) novobscura* Barraud, 1934
400. *Uranotaenia (Pseudoficalbia) obscura* Edwards, 1915
401. *Uranotaenia (Pseudoficalbia) ohamai* Tanaka, Mizusawa & Saugstad, 1975
402. *Uranotaenia (Pseudoficalbia) recondita* Edwards, 1922
403. *Uranotaenia (Pseudoficalbia) stricklandi* Barraud, 1926
404. *Uranotaenia (Pseudoficalbia) unguiculata* Edwards, 1913
405. *Uranotaenia (Pseudoficalbia) maikalensis* Natarajan et al., 2017
406. *Uranotaenia (Pseudoficalbia) dandakaranyensis* Natarajan et al., 2017
407. *Uranotaenia (Pseudoficalbia) satpuraensis* Natarajan et al., 2017

(ii) Subgenus *Uranotaenia* Lynch Arribalzaga, 1891

408. *Uranotaenia (Uranotaenia) alboannulata* (Theobald, 1905)
409. *Uranotaenia (Uranotaenia) annandalei* Barraud, 1926
410. *Uranotaenia (Uranotaenia) campestris* Leicester, 1908
411. *Uranotaenia (Uranotaenia) christophersi* Barraud, 1926
412. *Uranotaenia (Uranotaenia) edwardsi* Barraud, 1926
413. *Uranotaenia (Uranotaenia) hebes* Barraud, 1931
414. *Uranotaenia (Uranotaenia) lateralis* Ludlow, 1905
415. *Uranotaenia (Uranotaenia) longirostris* Leicester, 1908
416. *Uranotaenia (Uranotaenia) macfarlanei* Edwards, 1914
417. *Uranotaenia (Uranotaenia) micans* Leicester, 1908
418. *Uranotaenia (Uranotaenia) orientalis* Barraud, 1926
419. *Uranotaenia (Pseudoficalbia) pseudostricklandi* Natarajan, Rajavel & Jambulingam, 2018
420. *Uranotaenia (Uranotaenia) testacea* Theobald, 1905

4.7 IDENTIFICATION KEYS FOR DOMINANT VECTOR MOSQUITOES

Identification of vectors is of prime importance in the field of medical entomology for both operational and research purposes. In the broad sense, vectors encompass any organism involved in the transmission of an infectious agent; more specifically vectors designate any haematophagous arthropod which actively transmits an infectious pathogen from an infected host to a new host (Duvallet et al., 2018). In a given mosquito group, vector species are ordinarily outnumbered by other non-vector species such as, for example, among the nearly 500 Anopheles species, about 60 species only are recognized as vectors of *Plasmodium*, the agent of human malaria (Cohuet et al., 2010). Therefore, identification of arthropod species, both vectors and non-vectors, is a core capacity in the field of medical and veterinary entomology. Accurate identification of a species involved in transmitting a pathogen is essential for a proper understanding of the mechanisms of epidemiology of the disease. Knowing which arthropod species transmit vector-borne diseases is also vital to providing species-specific focused control programs and to correctly identifying disease risk and exposure (Failloux et al., 2017).

Different approaches to mosquito identification are available: morphological, molecular (PCR and nucleic acids sequencing), proteomics tools, isozyme analysis, etc. (Fontenille et al., 2017; Nayduch, 2009). However, for practical reasons, morphological identification, despite three principal limitations, i.e., (i) it relies on expert entomologists performing the identification, especially when many species are present in the collection area, (ii) the level of preservation of the morphological characters used in the sample handling process, and (iii) species complexes and a few morphologically similar species where identification should be based on more than one developmental stage, which is not always accessible, is still the reference method for both research purpose and operational surveillance as it requires little technical equipment, is easy to implement in the field and is inexpensive even when large numbers of individuals need to be identified.

Identification keys are provided for male adults of the dominant mosquito vector species involved in the transmission of major vector-borne diseases, viz., malaria, lymphatic filariasis, dengue/chikungunya/Zika and Japanese encephalitis in India.

4.7.1 Identification Characters

***Head*:** Character and coloration of scaling, and arrangement of bristles; erect, narrow and broad scales on vertex.

***Antenna*:** Coloration of torus, and presence or absence of scales; length of antenna compared with that of proboscis.

***Clypeus*:** Shape; presence or absence of scales.

***Palpi*:** Length, shape, ornamentation, tips of palpi dark or white; palpi with 3 or 4 pale bands; with or without speckling.

***Proboscis*:** Ratio of palpi and proboscis, size and shape, pale rings; pale patches on ventral surface.

***Thorax*:** Shape, with or without pale markings (ornamentation); with or without scale patches in antepronotum and postpronotum; postspiracular area with our without scales or setae, and shape of scutellum.

***Legs*:** With or without speckling, especially presence or absence of pale rings or bands on tarsal segments, if present whether basal or apical, or both; comparative length of tibia and of tarsal segments; ornamentation of the legs is usually in connection with the tarsal segments. They may be uniformly dark, or they may be marked with white or pale bands at the joints; tarsal claws, whether toothed or simple; presence or absence of pulvilli.

***Wings*:** Venation or arrangement of veins; venation and spotting of wing; shape of scales on veins; length of wing from base of costa to tip; presence or absence of hairs or scales on upper calypter and alula.

***Abdomen*:** The abdomen consists of eight visible segments. The first segment usually forms a somewhat transverse bar dorsally carrying long, outstanding hairs. The dorsal surface or dorsum is called as tergites and ventral surface or venter is called as sternites; with or without basal or apical, or both pale bands.

4.7.2 IDENTIFICATION KEY FOR GENERA

1. Scutellum evenly rounded; palpus about equal to the length of proboscis (Fig. 1, 1a); abdomen with sterna and usually terga largely or wholly devoid of scales ....***Anopheles***

   Scutellum trilobed; palpus short (Fig. 2, 2a); abdominal Sterna and terga with dense, uniform covering of scales 2

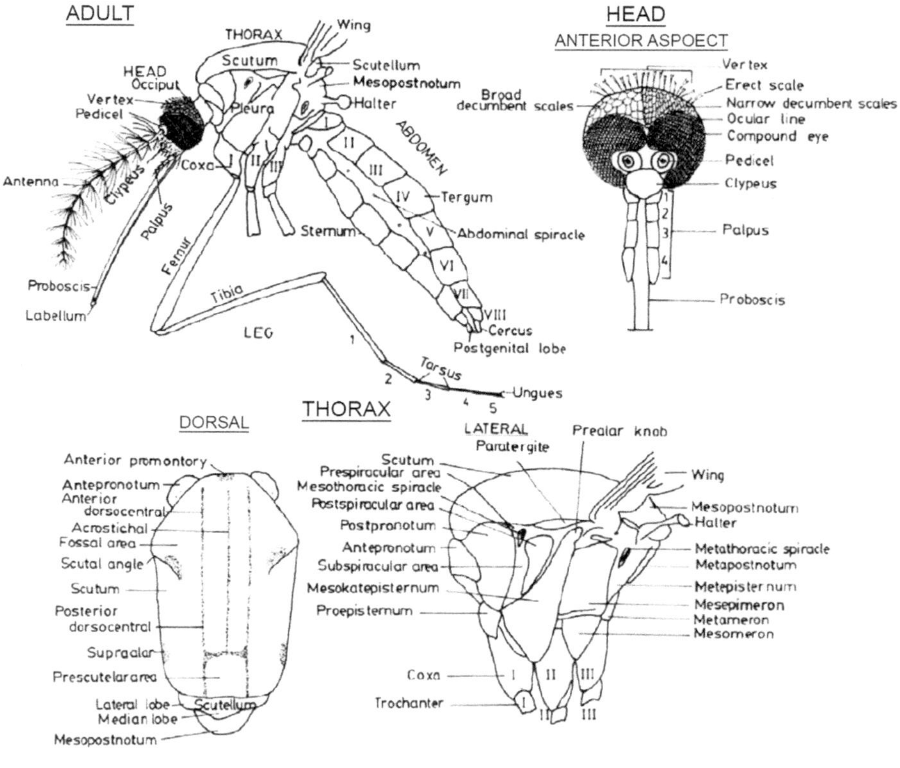

Main body parts

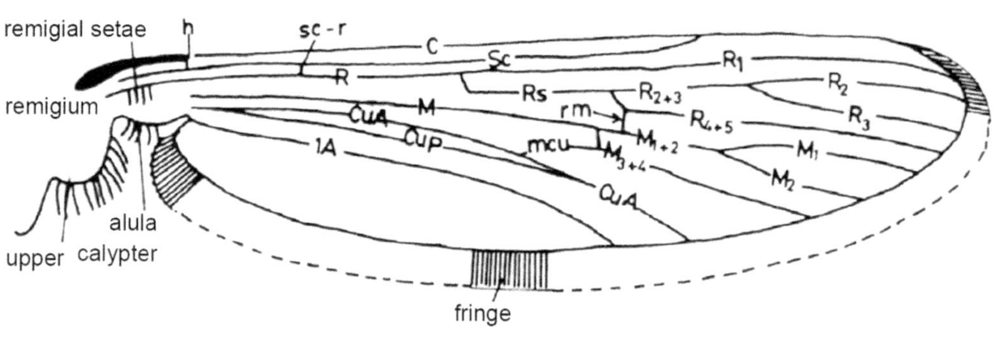

Wing

PLATE I Morphology and diagnostic characters of a female mosquito, C=Costa, CuA=Cubitus Anterior, CuP=Cubitus Posterior, h=Humeral Crossvein, M=Median, M_1=Median One; M1+2=Median One+Two, M_2=Two, M_{3+4}= Median Three+Four, mcu= Mediocubital Crossvein, R=Radius, R_1=Radius One, R_{2+3}=Radius Two+Three, R_{3+4}=Radius Four+Five, Rs=Radial Sector, rm=Radiomedial Crossvein, Sc=Subcosta, sc-r = Subcostal crossvein, 1A=Anal Vein.

2(1). Postspiracular setae present, or fore ungues toothed, or both (Fig. 3, 3a) 3
Postspiracular setae absent; ungues simple (Fig. 4, 4a) 5

3(2). Wing with broad, often asymmetrical scales; dark and pale scales intermixed (Fig. 5)............. ....*Mansonia*

Wing with normal scales 4

4(3). Proboscis curved and laterally compressed (Fig. 6)................*Armigeres*

Proboscis slender and usually straight, never laterally compressed (Fig. 7)...*Aedes*

5(2). Hindtarsi with ungues very small and inconspicuous; pulvilli present (Fig. 8) ....*Culex*

Hindtarsi with ungues not very small; pulvilli absent (Fig. 9); yellow or yellowish brown mosquitoes...............*Coquillettidia*

4.7.2.1 Vectors of Japanese Encephalits

Lower mesepimeral setae absent or 1, 2 weak ones; pleuron with distinct pale scale patches on at least upper and lower mesokatepisternum and anterior mesepimeron (Fig. 10).................subgenus *Culex*

Wing heavily/moderately speckled with pale scale; abdominal terga II-VI with apical bands and/or apicolateral pale patches and basal pale bands (Figs. 24–26) ....(Subgenus *Oculeomyia*)

1. Proboscis and tarsi without pale rings; 1 or 2 lower mesepimeral setae present (Figs. 11, 12)................................2

Proboscis and tarsi with pale rings; lower mesepimeral setae absent (Figs. 13, 14)..4

2(1). Abdominal terga without pale bands, occasionally a few indistinct bands on posterior segments; pleuron with striking pattern of dark and pale stripes (Figs. 15, 16)................*fuscocephala*

Abdominal terga with basal pale bands; pleuron without striking pattern of dark and pale stripes (Figs. 17, 18)...........................3

3(2). Postspiracular area with pale scale patch (Fig. 19).....................*perexiguus*
Postspiracular area without pale scale patch (Fig. 20)................*quinquefasciatus*

4(1). Wing with pale spots on at least 2 areas of costa and 1 area of other veins (Fig. 21)...Mimeticus Subgroup

Wing without distinct pale spots...5

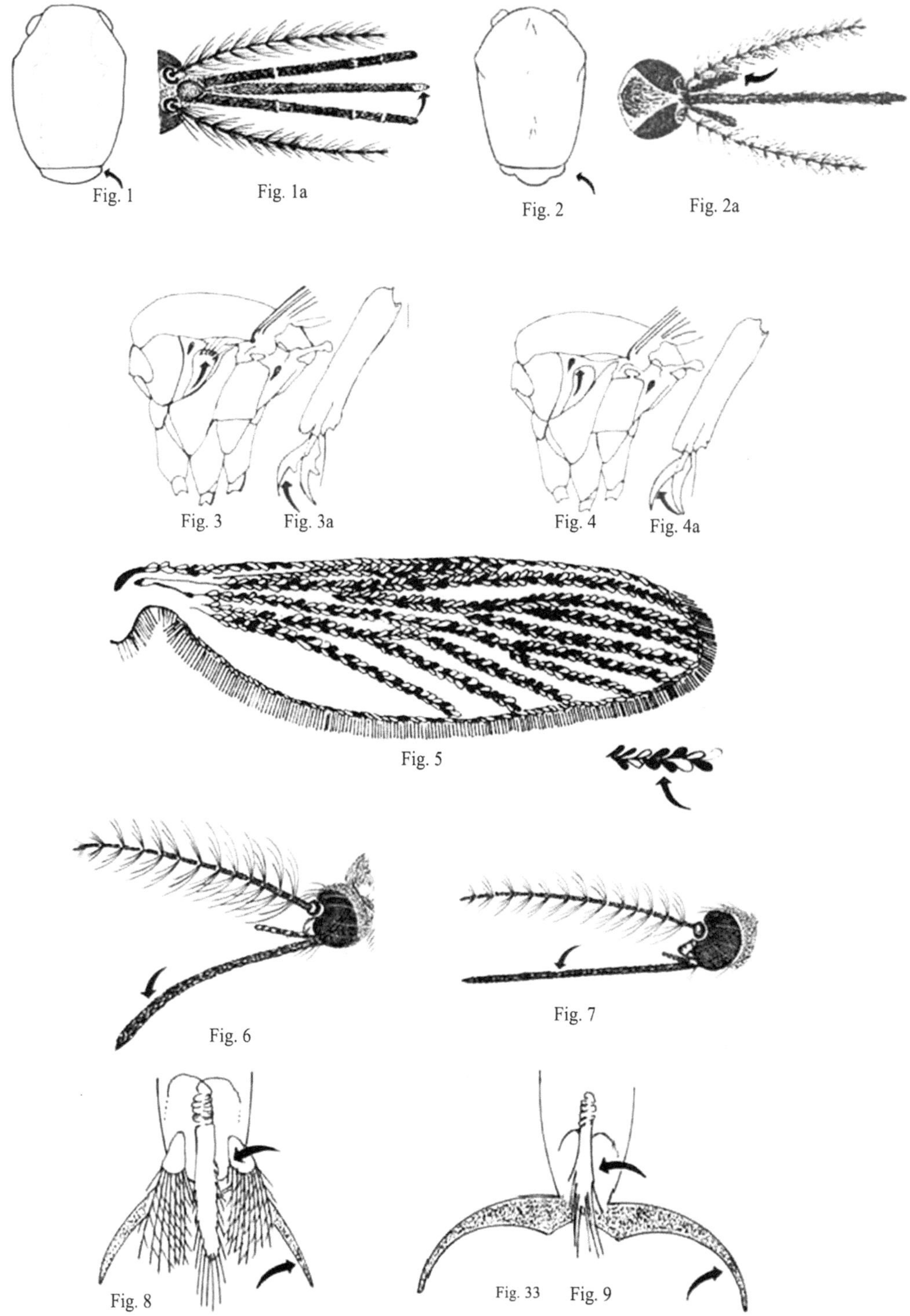

PLATE II **(Figs. 1–9)**: 1 – *Anopheles* scutellum, 1a – *Anopheles* palpi and proboscis, 2a – *Culicine* palpi and proboscis, 3 – pleuron with postspiracular setae, 3a – toothed fore ungues, 4 – pleuron without spiracular setae, 4a – untoothed fore ungues, 5 – wings with asymmetrical scales, 6 – laterally compressed proboscis, 7 – slender proboscis, 8 – pulvilli present, 9 – pulvilli absent.

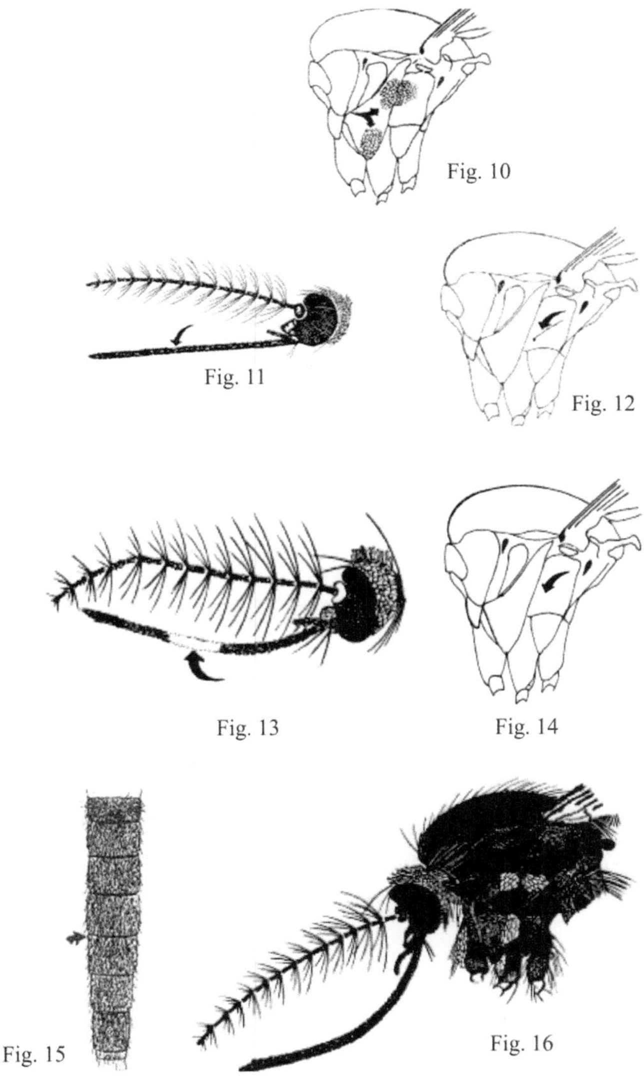

PLATE III (Figs. 10–16): 10 – pleuron with pale scale patches, 11 – proboscis without pale ring, 12 – with lower mesepimeral setae, 13 – proboscis with pale ring, 14 – without lower mesepimeral setae, 15 – abdomen terga without pale band, 16 – pleuron with dark and pale stripes.

5(4). Abdominal terga II- VIII largely yellowish or golden; pale yellow scales on apical portions of wing veins C, R1, R2, R3, and R4+5, forming apical pale area (Figs. 22, 23)................................... *epidesmus*

Abdominal terga II-VIII with dark and pale bands, or completely dark; wing tip without definite pale area...6

6(5). Wing heavily / moderately speckled with pale scale; abdominal terga II-VI with apical bands and/or apicolateral pale patches and basal pale bands.............................7

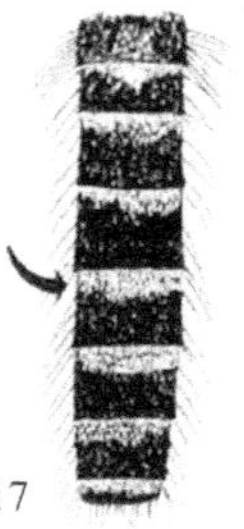

Fig. 17

Fig. 18

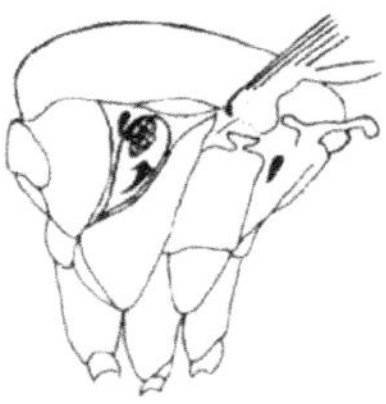

Fig. 19

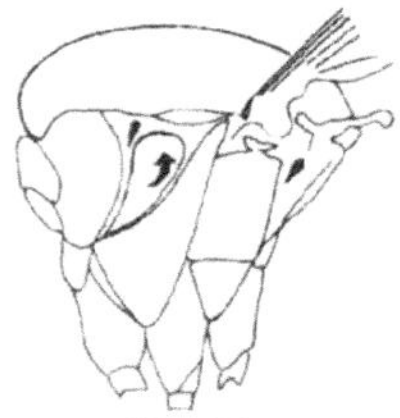

Fig. 20

Fig. 21

PLATE IV (Figs. 17–21): 17 – abdominal terga with basal pale band, 18 – pleuron without dark and pale stripes, 19 – postspiracular area with pale scales, 20 – postspiracular area without pale scales, 21 – wing with pale spots.

Wing without speckling of pale scales; abdominal terga II-VI with basal pale bands..................................8

7(6). Abdominal terga II-VII with evenly broad apical pale bands and without apicolateral pale patches; femora, tibiae, and wings heavily speckled with pale scales (Figs. 24, 25)..*bitaeniorhynchus*

Abdominal terga II-IV largely dark, or with narrow apical bands, apicolateral yellowish patches, and median basal pale bands or spots; terga V-VII with apical and basal bands, basal bands narrower; legs and wings lightly or moderately speckled (Fig. 26)............ *infula*

8(6). Anterior 0.7 of scutum covered with pure white scales......... 9

Anterior 0.7 of scutum covered with beige, yellow, golden or dark scales...10

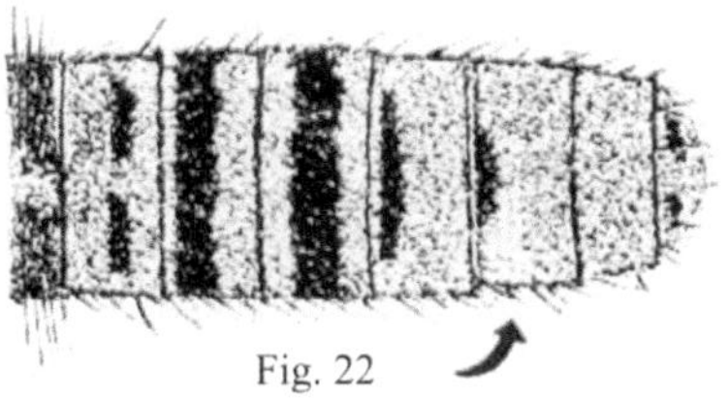

Fig. 22

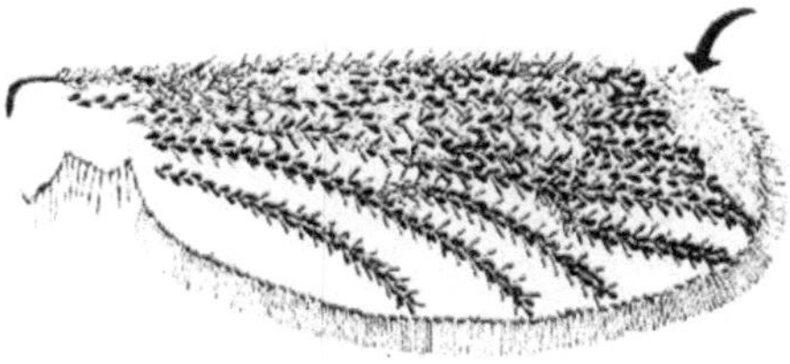

Fig. 23

Fig. 24

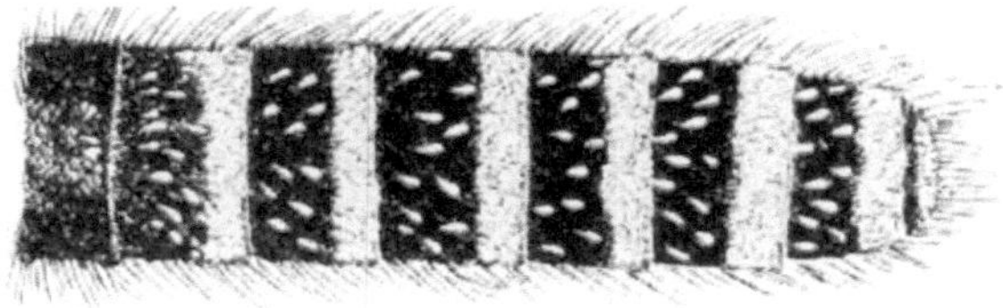

Fig. 25

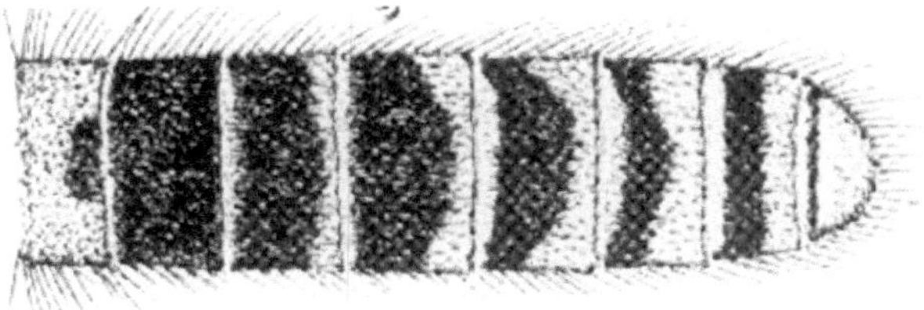

Fig. 26

PLATE V (**Figs. 22–26**): 22 – abdomen terga II–VIII with yellowish scales, 23 – apical portion of wing with pale area, 24 – legs with speckling, 25 – abdomen terga II–VII broad apical band, 26 – abdomen terga II–IV largely dark with narrow apical bands.

9(8). Anterior surface of fore-and midfemora without speckling of pale scales; scales on prescutellar space, behind wing base, and on scutellum entirely dark; pale band on proboscis narrow, less than length of basal dark area (Figs. 27–29)........................*gelidus*

Anterior surface of fore-and midfemora speckled with pale scales; scales on prescutellar space, behind wing base, and on scutellum mainly pale; pale band on proboscis broad, as long as or longer than basal dark area (Figs. 30–32) ....*whitmorei*

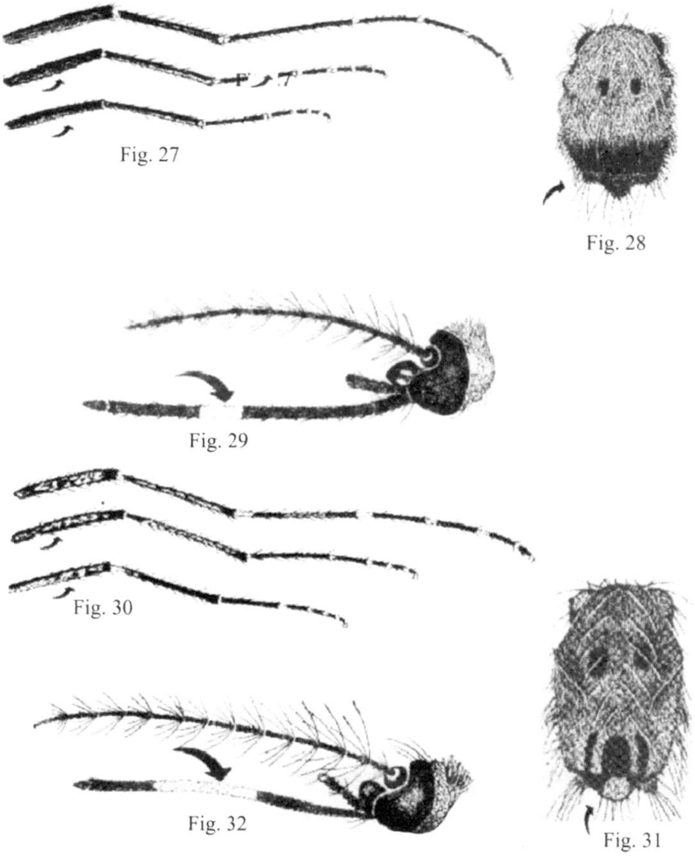

PLATE VI (Figs. 27–32): 27 – legs without speckling, 28 – scutellum entirely dark, 29 – proboscis with narrow pale band, 30 – legs with speckling, 31 – scutellum mainly pale, 32 – proboscis with broad pale band.

10(8). Anterior surface of fore and midfemora speckled with several pale scales, at least on apicodorsal surface......................11

Anterior surface of fore and midfemora entirely dark..................12

11(10). Wing scales mainly dark; scutal integument dark; speckling of pale scales on femora contrasting sharply with dark-scaled area (Fig. 33)..................................*sitiens*

Wing with few to several scattered pale scales; Scutal integument light brown; anterior surface of hind femur with pale stripe not contrasting with dark-scaled area (Fig. 34)..*vishnui*

12(10). Proboscis usually with accessory pale patches or stripe on ventral surface: hindfemur pale with an apical dark ring (Figs. 35, 36); erect scales on vertex all dark..*tritaeniorhynchus*

Fig. 33

Fig. 34

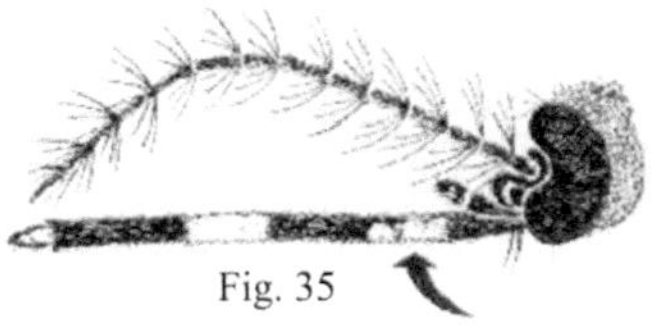

Fig. 35

Fig. 36

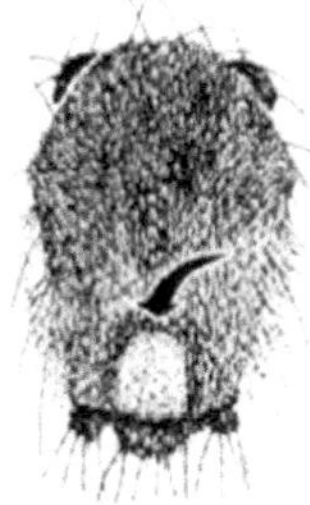

Fig. 37

Fig. 38

PLATE VII (Figs. 33–38): 33 – legs with speckling, 34 – hind femur not contrasting with dark scales, 35 – proboscis with accessory pale patches, 36 – hindfemur with apical dark ring, 37 – prescutellar area with pale scales, 38 – hind femur well contrasting with dark scales.

Prescutellar area entirely covered with pale scales; hind femur with pale stripe on anterior surface contrasting well with dark areas (Figs. 37, 38)................................*pseudovishnui*

Wing with broad, often asymmetrical scales; dark and pale scale intermixed (Fig. 39)..***Mansonia***

1. Mesonotum marked with distinct round spots of white scales...2

Mesonotum not so (though there may be small indefinite white spots or patches) 3

2(1). Yellowish-brown; mesonotum marked with 4 (or more) distinct round white spots; rather broad white scales on mid-lobe of scutellum (Fig. 40)..*annulifera*

Brownish-black; mesonotum marked with 2 (or 3) round white spots; scales on mid-lobe of scutellum narrow (Fig. 41)..*longipalpis*

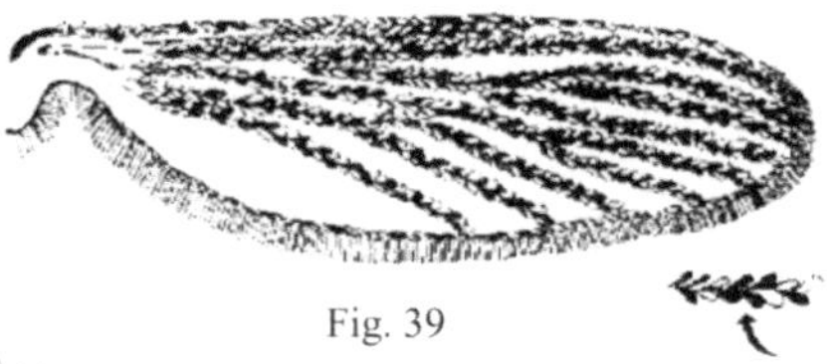

Fig. 39

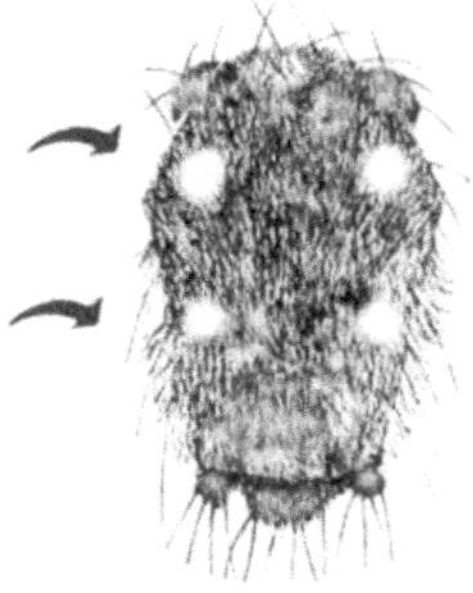

Fig. 40

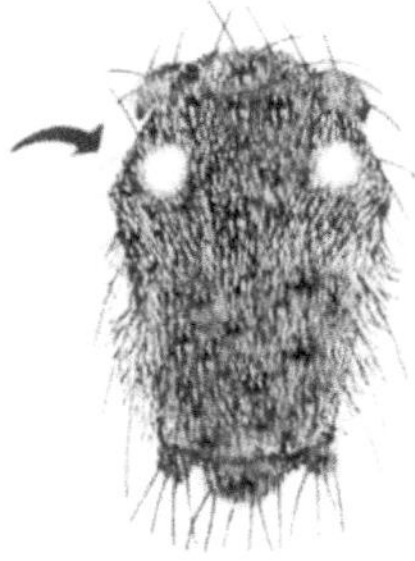

Fig. 41

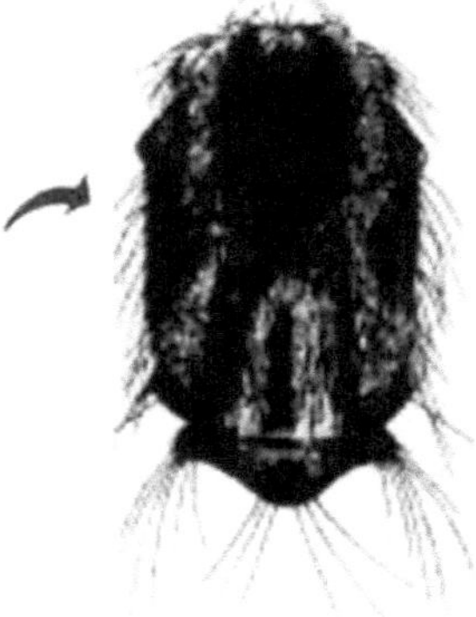

Fig. 42

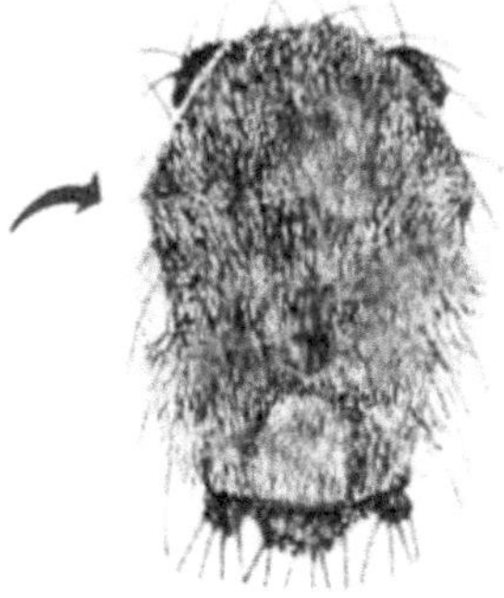

Fig. 43

PLATE VIII (Figs. 39–43): 39 – wing with asymmetrical scales, 40 – mesonotum with four round white spots, 41 – mesonotum with two round white spots, 42 – mesonotum with sublateral greenish stripes, 43 – mesonotum without greenish stripes.

3(1). Mesonotum marked with a pair of sublateral greenish stripes on a brown ground (Fig. 42).. *uniformis*

Mesonotum dark brown, not marked with greenish stripes; some white scales, tending to form indistinct spots or patches, in some specimens (Fig. 43).................................... *indiana*

1. Wings with less than three dark spots on costa, which also involve vein 1 (Fig. 44)..(Subgenus *Anopheles*)

1a. Hind tarsomeres with broad apical pale bands; wing humeral cross vein without dark scales; remigium mostly pale scaled (Figs. 45, 46) ....*peditaeniatus*

1b. A prominent tuft of dark scales on ventral side of abdominal segment VII; inner quarter of costa with scattered pale

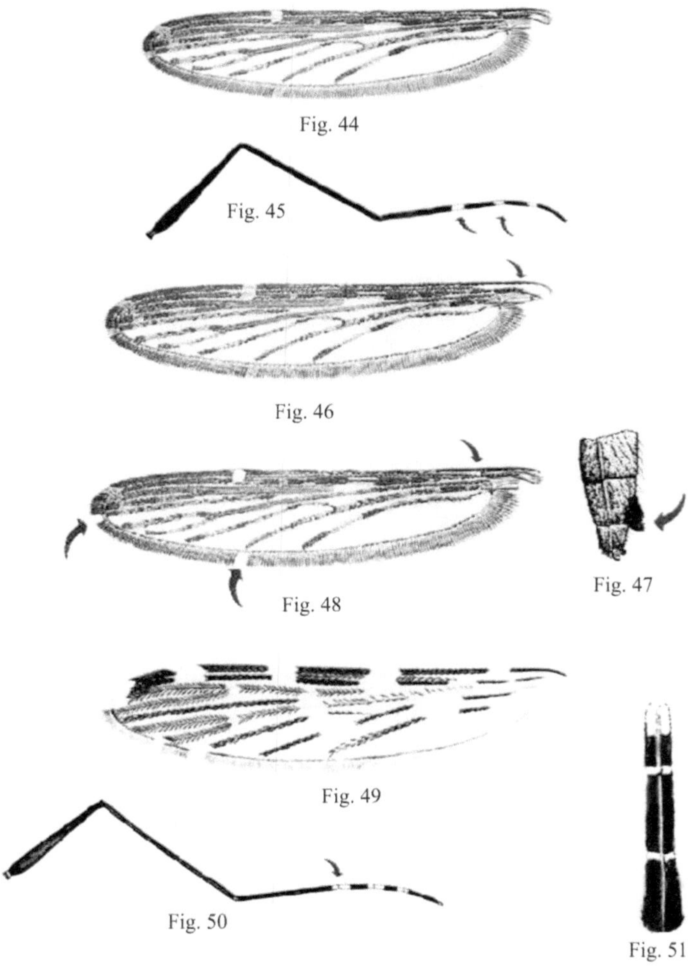

PLATE IX (**Figs. 44–51**): 44 – wings with less than three dark spots, 45 – hind tarsomere with broad apical pale bands, 46 – remigium with pale scales, 47 – abdomen segment VII with tuft, 48- wing with fringe spot at vein 3 and 5.2, 49 – wings with four or more dark spots, 50 – fore tarsi with broad pale bands, 51 – palpi with subapical dark band equal to apical pale band.

scales, usually with a fringe spot at vein 5.2; wing with narrow lower apical fringe spot opposite to vein 3 only (Figs. 47, 48).. *barbirostris*

2. Wings with 4 (or) more dark spots on costa which also involve vein 1 (Fig. 49)..(Subgenus *Cellia*)

2a. Femur and tibia not speckled; hind tarsomere 5 not white; fore tarsi with broad pale bands; palpi with subapical dark band equal to the apical pale band (Figs. 50, 51)..........................*subpictus*

4.7.2.2 Vectors of Dengue/Chikungunya

1. Post spiracular setae present; proboscis slender and usually straight, never laterally compressed .... **Aedes**

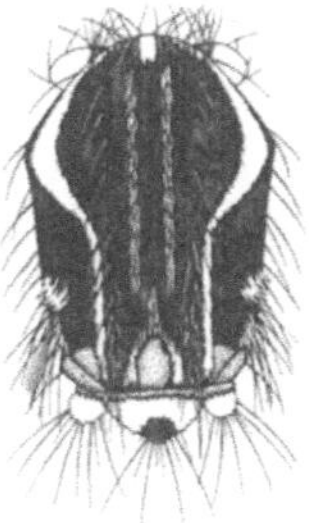

Fig. 52

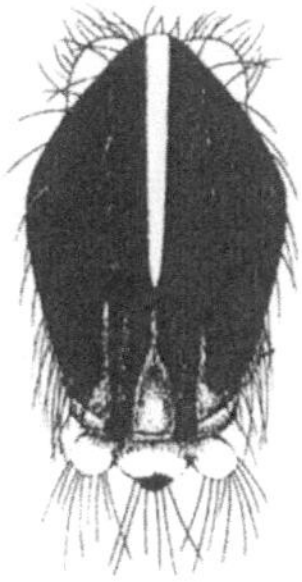

Fig. 53

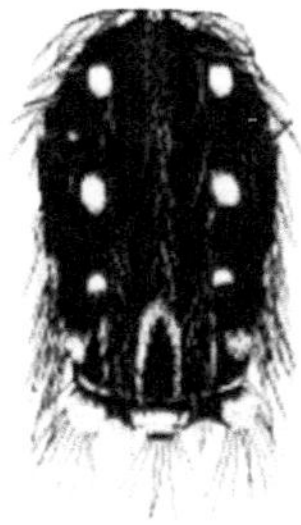

Fig. 54

PLATE X (Figs. 52–54): 52 – mesonotum marked with a pair of lateral curved white lines (*Aedes aegypti*), 53 – mesonotum with narrow median silvery white line (*Aedes albopictus*), 54 – mesonotum marked with 4–6 white spots (*Aedes vittatus*)

1a. Mesonotum marked with a pair of lateral curved white lines, and usually also with a pair of submedian yellowish lines; tibiae without white rings; two dots of white scales on clypeus (Fig. 52).. *aegypti*

1b. Mesonotum with a narrow median silvery white line; white scales on pleurae arranged in irregular patches; tibiae without white rings; abdomen with silvery basal bands on dorsum (Fig. 53)...*albopictus*

1c. Mesonotum marked with 4-6 small white spots; all tibiae with white rings (Fig. 54)...*vittatus*

4.7.2.3 Vectors of Lymphatic Filariasis

1. Lower mesepimeral setae absent or 1, 2 weak ones; pleuron with distinct pale patches on at least upper and lower mesokatepisternum and anterior mesepimeron .... Subgenus Culex

1a. Proboscis and tarsi without pale rings; pleuron without striking pattern of dark and pale stripes; abdominal terga with basal pale bands; (Figs. 55–57)................ *quinquefasciatus*

2. Wing with broad, often asymmetrical scales; dark and pale scale intermixed (Fig. 58).. **Mansonia**

2a. Mesonotum marked with a pair of sublateral greenish stripes on a brown ground (Fig. 59).................................*uniformis*

2b. Yellowish-brown; mesonotum marked with 4 (or more) distinct round white spots; rather broad white scales on mid-lobe of scutellum (Fig. 60)...............................*annulifera*

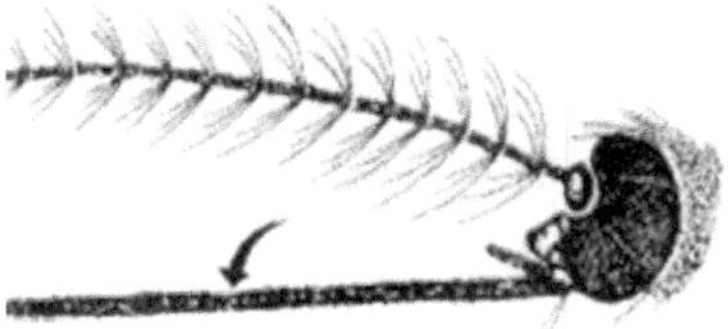
Fig. 55

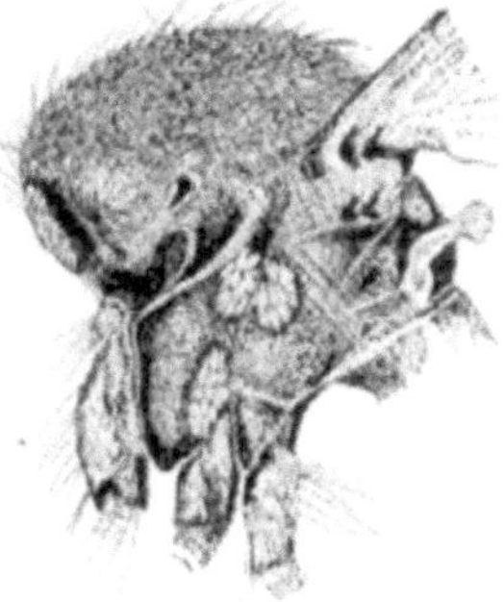
Fig. 56

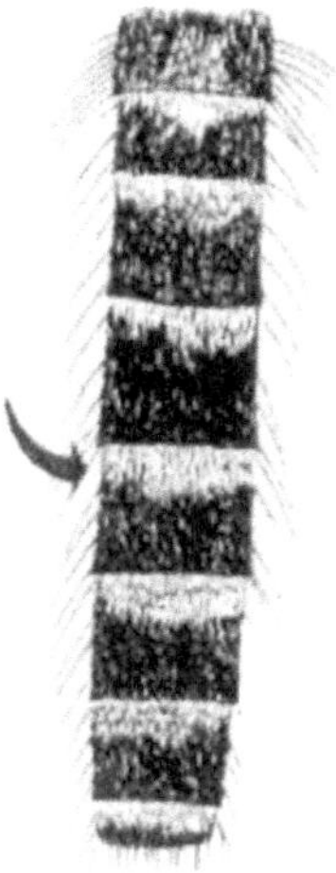
Fig. 57

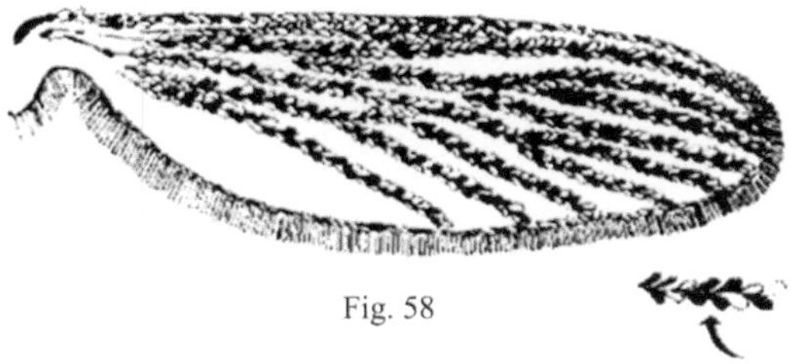
Fig. 58

Fig. 59

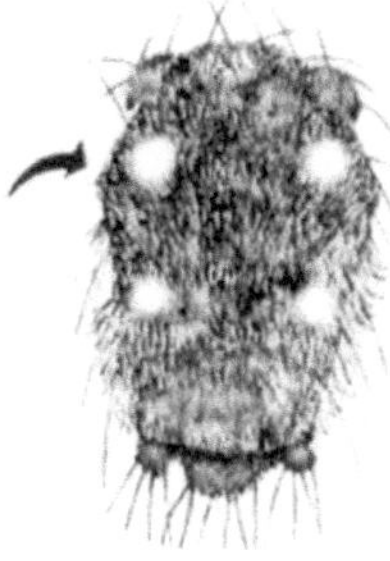
Fig. 60

PLATE XI (Figs. 55–60): 55 – proboscis without pale ring, 56 -pleuron without striking pattern of dark and pale stripes, 57 – abdominal terga with basal pale band, 58 – wing with asymmetrical scales, 59 – mesonotum with sublateral greenish stripes, 60 – mesonotum with four round white spots

***Diurnally Sub-Periodic Filariasis Vector In Andaman & Nicobar Islands*:**

1. Postspiracular setae present; proboscis slender and usually straight, never laterally compressed (Fig. 61)……........................ **Aedes**

1a. Mesonotum with a large snowy white patch in front; wing scales all dark; tarsi entirely dark (Figs. 62, 63)…………………….. *nivea**

----

* *Downsiomyia nivea* (=*Aedes (Finlaya) niveus*)

4.7.2.4 Vectors of Malaria

Wings with four or more dark spots on costa (Fig. 64)…………………….. (Subgenus Cellia)

1. Femur and tibia speckled (Fig. 65)…………………………...............2
   Femur and tibia not speckled (Fig. 66) ……………………………….....7

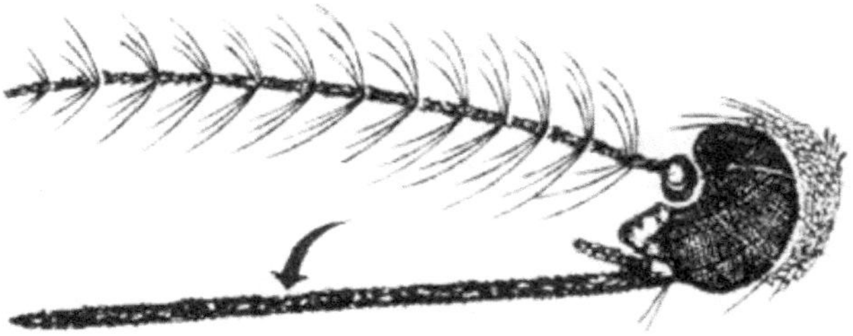

Fig. 61

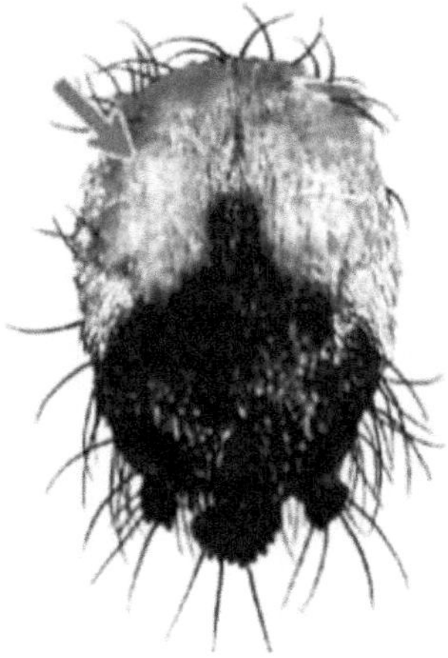

Fig. 62

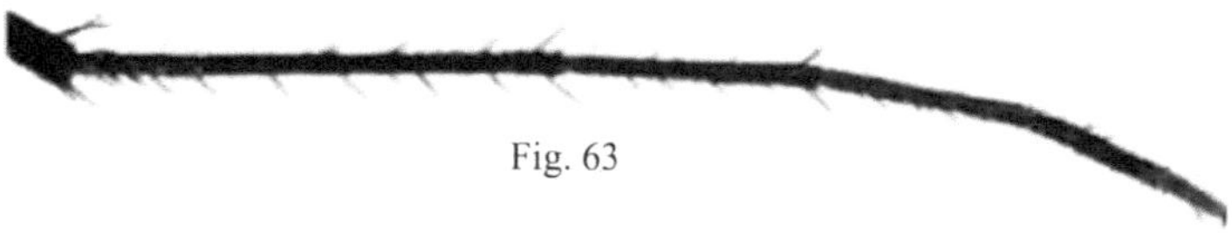

Fig. 63

PLATE XII (Figs. 61–63): 61 – proboscis slender and straight, 62 – mesonotum with snowy white patch, 63 – tarsi entirely dark.

2(1). Hind tarsomere 5 not white (Fig. 66) 3
Hind tarsomere 5 white (Fig. 67) 6

3(2). Palpi with three pale bands (Fig. 68).......................................4
Palpi with four pale bands (Fig. 69).......................................5

4(3). Apical and subapical pale bands of palpi equal; palpi speckled (Fig. 70).. *stephensi*

Apical and subapical pale bands of palpi unequal; palpi not speckled (Fig. 71)...*sundaicus*

5(3). Hind leg with a broad white band at the tibio-tarsal joint; presector dark mark on vein 1 without any pale interruption; apical pale band on hind tibia with

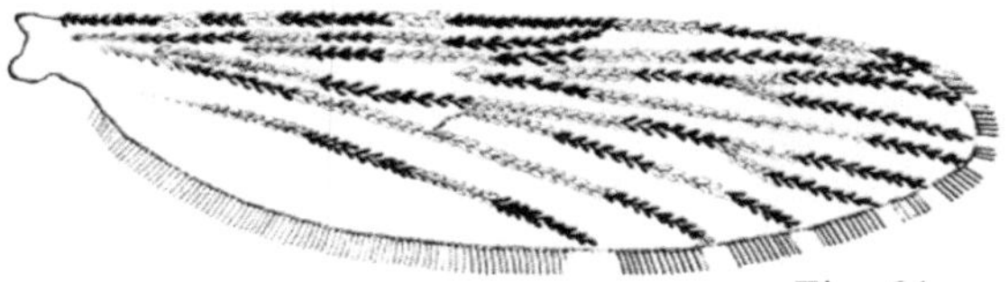

Fig. 64

Fig. 65

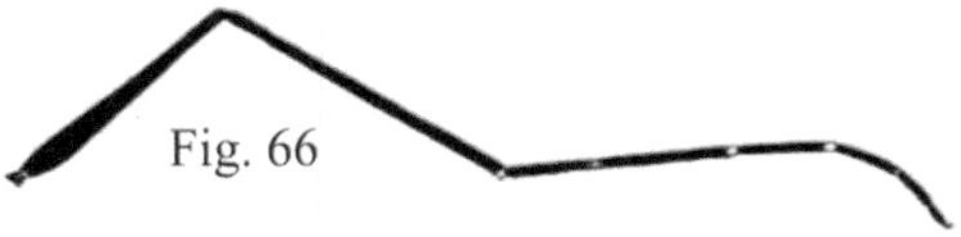

Fig. 66

Fig. 67

Fig. 68

Fig. 69

PLATE XIII (Figs. 64–69): 64 – wings with four or more dark spots on costa; 65 – femur and tibia speckled, 66 – femur and tibia not speckled, 67 – Hind tarsomere 5 white, 68 – palpi with three pale bands, Fig. 69 – palpi with four pale bands.

a longitudinal basal dark stripe on ventral aspect (Figs. 72, 73)..*baimaii* *(=dirus D)*

Hind leg with a broad white band at the tibio-tarsal joint; presector dark mark on vein 1 with one or more pale interruption; apical pale band on hind tibia without a ventral dark stripe (Figs. 74, 75)..............................*elegans*

6(2). Abdominal terga II-VIII shiny, with no pale scales (Fig. 76)...*eudowillmori*
Abdominal terga IV-VIII with a few pale scales (Fig. 77)... *maculatus*

Fig. 70

Fig. 71

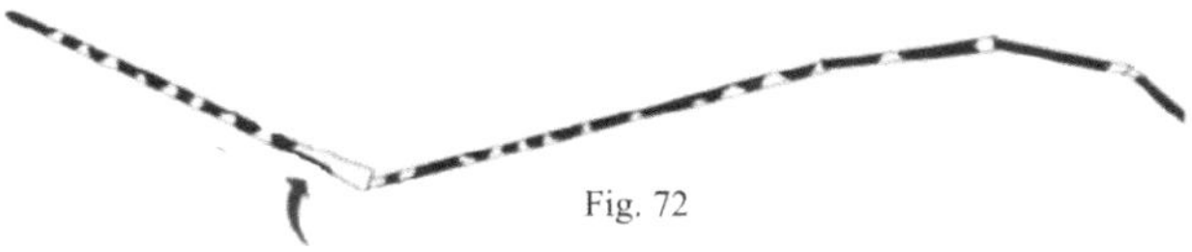

Fig. 72

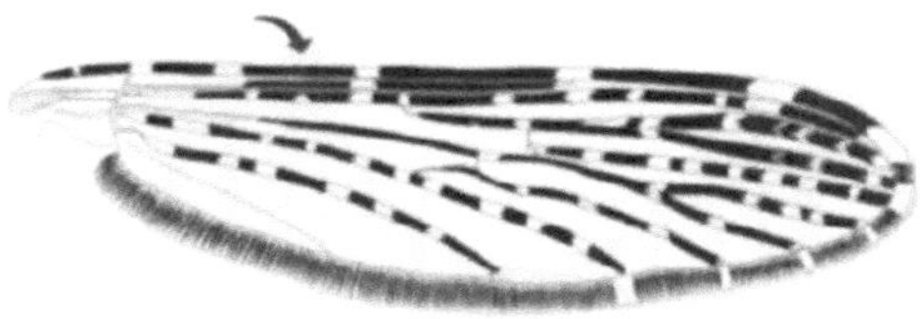

Fig. 73

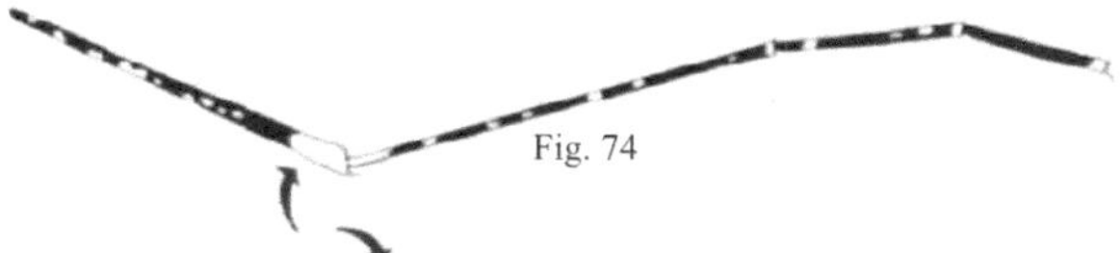

Fig. 74

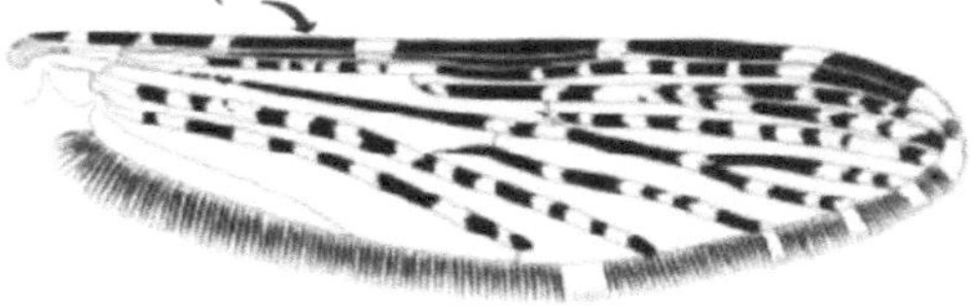

Fig. 75

PLATE XIV (Figs. 70–75): 70 – palpi with speckling, 71 – palpi without speckling, 72 – hindleg with broad white band at tibio-tarsal joint, 73 – presector dark mark on vein 1 without pale interruption, 74 – hind tibia without ventral dark stripe, 75 – presector dark mark on vein 1 with pale interruption.

Fig. 76

Fig. 77

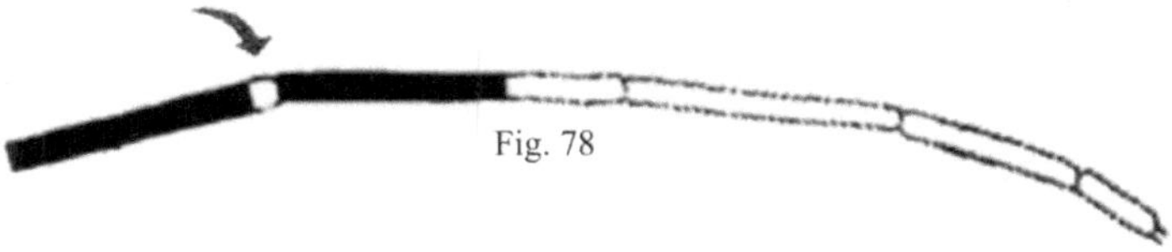

Fig. 78

Fig. 79

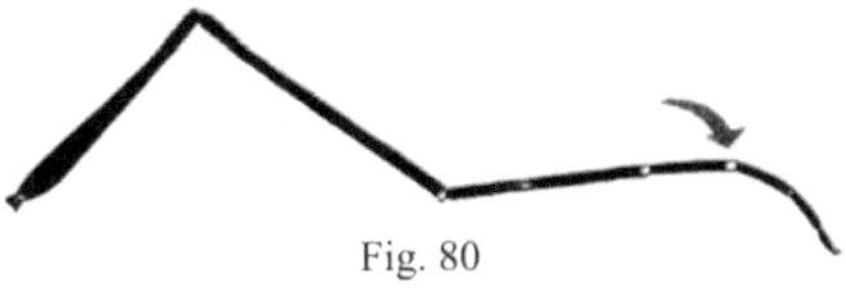

Fig. 80

PLATE XV (Figs. 76–80): 76 – abdominal terga II–VIII without pale scales, 77 – abdominal terga IV–VIII with few pale scales, 78 – hind tarsomere 1 with white band, 79 – wing vein 5 white no dark spot at the point of bifurcation, 80 – fore tarsi with narrow pale band.

7(1). Hind tarsomere 5,4 and 3 completely white; apex of hind tarsomere 1 usually with white band; wing vein 5 mainly white, no dark spot at the point of bifurcation (Figs. 78, ...*philippinensis*

Hind tarsomere not white; fore tarsi with narrow pale bands (Fig. 80)..8

8(7). Wing vein 3 mainly dark; inner quarter costa usually with pale interruption (Fig. 81)... *culicifacies*

Wing vein 3 mainly white (Fig. 82)................................... 9

9(8). Female palpi with both apical and subapical pale bands as broad as or broader than subapical dark band; basal third of costa with white interruption; fore tarsomere 1–4 with very small dorso-apical pale patches (Figs. 82–84).................minimus

Female palpi with subapical pale band, narrow and subapical dark band much broader; vein 6 with two dark spots, distal half mainly dark; hind tarsomeres uniformly dark (Figs. 85–87)..fluviatilis

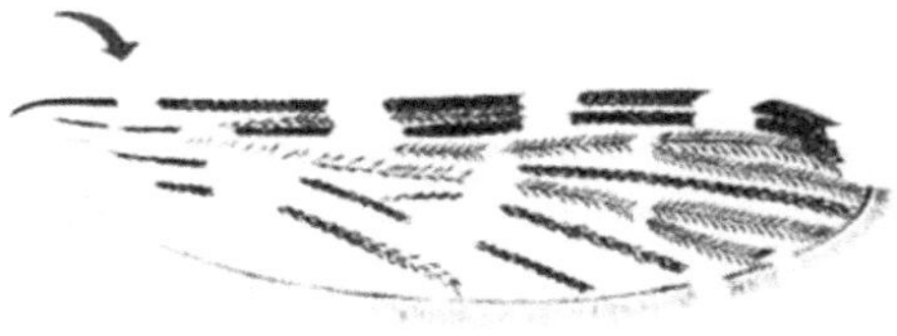

Fig. 81

Fig. 82

Fig. 83

Fig. 84

Fig. 86

Fig. 85

Fig. 87

PLATE XVI **(Figs. 81–87)**: 81 – wing vein 3 mainly dark, 82 – wing vein 3 mainly white, 83 – palpi – apical and subapical pale bands broader than subapical dark band, 84 – fore tarsomere 1–4 with very small pale patches, 85 – palpi – subapical pale band narrow and subapical dark band broader, 86 – wing vein 6 with 2 dark spots, 87 – hind tarsomere uniformly dark.

4.8 CONCLUSION

The mosquito fauna of India is highly rich in species (more than 11% of the world's taxa), with new species being regularly described every year (Tyagi, 2010; Tyagi et al., 2012, 2015). Therefore, mosquito researchers have so far been struggling to calibrate an exact number of extant species and subspecies since the times of *magnum opuses* by Christophers (1933) and Barraud (1934), notwithstanding commendable attempts by Puri (1949, 1960), Wattal and Kalra (1961), Das et al. (1990), Rao (1984), and Nagpal and Sharma (1995), all of whom essentially dealt with *Anopheles* mosquitoes (Subfamily Anophelinae). The vacuum of information about Culicinae, covering all other kinds of mosquitoes including *Toxorhynchites* (Tribe Toxorhynchitini), was removed by an expeditious venture by Tyagi (2004, 2020), Bhattacharyya et al. (2015), and Tyagi et al. (2014). The conundrum needed authentic attention to dispel clouds about the wealth of mosquito fauna in India (420 species), and the same has been done here in this chapter, with a further flip of offering the identification keys of the dominant vector species for malaria, lymphatic filariasis, dengue/chikungunya/zika and Japanese encephalitis, which are the major mosquito-borne diseases in the country (Tyagi, 1994).

REFERENCES

Amerasinghe, P.H. and Amerasinghe, F.P. (1999). Multiple host feeding in populations of *Anopheles culicifacies* and *An. subpictus*. *Medical and Veterinary Entomology*, 13, 124–131.

Barik, T.K., Sahu, B. and Swain, V. (2009). A review on *Anopheles culicifacies*: From bionomics to control with special reference to Indian subcontinent. *Acta Tropica*, 109(2), 87–97. https://doi.org/10.1016/j.actatropica.2008.09.017

Barraud, P.J. (1934). *The fauna of British India, including Ceylon and Burma. Diptera: Family Culicidae; Tribes Megarhinini and Culicini, vol. v*. Taylor & Francis, London, 1934, 463 pp.

Bhattacharyya, D.R., Rajavel, A.R., Natarajan, R., Mohapatra, P.K., Jambulingam, P., Mahanta, J. and Prakash A. (2015). Faunal richness and the checklist of Indian mosquitoes (Diptera: Culicidae) Faunal richness and the checklist of Indian mosquitoes (Diptera: Culicidae). *Check List* 10(6):1342–1358.

Christophers, S.R. (1933). *The fauna of British India, including Ceylon and Burma. Diptera: Family Culicidae; Tribe Anophelini, vol. iv*. Taylor & Francis, London, 371 pp.

Das, B.P., Rajagopal R. and Akiyama, J. (1990). Pictorial key to the species of Indian Anopheline mosquitoes. *Journal of Pure Applied Zoology*, 2, 131–162.

Dev, V. (2020). *Vector Biology and Control – An Update for Malaria Elimination Initiative in India*. The National Academy of Sciences, India. 272 pp.

Dev, V. and Sharma, V. P. (2013). The Dominant Mosquito Vectors of Human Malaria in India. (*In*: *Anopheles mosquitoes – New insights into malaria vectors*. Prof. Sylvie Manguin (Ed.), InTech. https://doi.org/10.5772/55215, pp. 239–271.

Dutta, P., Khan, S.A., Sharma, A.M., Hazarika, C.K. and Mahanta J. (2003). Survey of medically important mosquito fauna in Mizoram. *Entomon*, 28, 237–240.

Firooziyan, S., Dinparast Djadid, N. and Gholizadeh, S. (2018) Speculation on the possibility for introducing *Anopheles stephensi* as a species complex: preliminary evidence based on odorant binding protein 1 intron I sequence. *Malaria Journal*, 17, 366. https://doi.org/10.1186/s12936-018-2523-y

Griffing, S.M., Tauil, P.L., Udhayakumar, V. and Silva-Flannery, L. (2015) A historical perspective on malaria control in Brazil. *Memórias do Instituto Oswaldo Cruz*, 110(6), 701–18. https://doi.org/10.1590/0074-02760150041

Guo, Y., Song, Z., Luo, L., Wang Q., Zhou G., Yang D., Zhong D. and Zheng X. (2018). Molecular evidence for new sympatric cryptic species of Aedes albopictus (Diptera: Culicidae) in China: A new threat from *Aedes albopictus* subgroup? *Parasites Vectors* 11, 228. https://doi.org/10.1186/s13071-018-2814-8

Harbach, R.E. (2007). The Culicidae (Diptera): a review of taxonomy, classification and phylogeny. *Zootaxa*, 1668, 591–638.

Harbach, R.E. (2013). Mosquito Taxonomic Inventory, http://mosquito-taxonomic-inventory.info/, accessed on 26th September, 2020.

Huang, YM. (1972). Contributions to the mosquito fauna of Southeast Asia XIV. The subgenus Stegomyia of *Aedes* in Southeast Asia. I – The Seutellaris group of species. *Contributions of the American Entomological Institute (Ann Arbor)*, 9(l), l–109.

Jose, J. (2018). First Report on *Aedes albopictus* Subgroup Species in India – Short Communication. *International Journal of Life-Sciences Scientific Research*, 4(5), 1983–1985. https://doi.org/10.21276/ijlssr.2018.4.5.4

Jude, P.J., Ramasamy, R. and Surendran, S.N. (2014). Bionomic aspects of the Anopheles subpictus species complex in Sri Lanka. *Journal of Insect Science*, 14, 97. https://doi.org/10.1673/031.014.97

Kar, I., Subbarao, S.K., Eapen, A., Ravindran, J., Satyanarayana, T.S., Raghavendra, K., Nanda, N. and V.P. Sharma (1999). Evidence for a New Malaria Vector Species, Species E, within the *Anopheles culicifacies* Complex (Diptera: Culicidae). *Journal of Medical Entomology*, 36(5), 595–600. https://doi.org/10.1093/jmedent/36.5.595

Karlekar, S., Andrew, R. and Deshpande, M. (2020). *Culex katezari*, A new species of Mimeticus mosquito (Diptera: Culicidae) from the forest of Gadchiroli region of central India. *International Journal of Mosquito Research*, 7(6), 54–58.

Krishnamoorthy R., Munirathinam, A., Dhananjeyan, K.J., Hiriyan, J., Mariappan, T., Samuel, P.P. and Venkatesh, A. (2013) Description of a new species, *Toxorhynchites (Toxorhynchites) tyagii* (Diptera: Culicidae), from Nilgiri hills, Western Ghats, southern India. *Zootaxa*, 3701 (4), 447–459.

Krishnan, K.S. (1954). Malaria vectors of India. IV. Anopheles stephensi Liston, 1901. *Bull National Society of India on Malaria & Other Mosquito-borne Diseases*, 2, 97–109.

Liston, W.G. (1901). A year's experience of the habits of *Anopheles* in Ellichpur. The description of the species found in Ellichpur during the year. *Indian Medical Gazette*, 36, 361–366.

Loughlin, S.O. (2020) The expanding Anopheles gambiae species complex. *Pathogens and Global Health*, 114(1), 1. https://doi.org/10.1080/20477724.2020.1722434

Matthews, T.C., and Munstermann, L.E. (1994). Chromosomal repatterning and linkage group conservation in mosquito karyotypic evolution. *Evolution*, 48(1), 146–154. https://doi.org/10.2307/2410010

Minard, G., Tran Van, V., Tran, F.H., Melaun, C., Klimpel, S., Koch, L.K., Kim, K.L.H., Thuy, T.H.T., Ngoc, H.T., Potier, P., Mavingui, P. and Moro, C.V. (2017). Identification of sympatric cryptic species of Aedes albopictus subgroup in Vietnam: new perspectives in phylosymbiosis of insect vector. *Parasites Vectors*, 10, 276 https://doi.org/10.1186/s13071-017-2202-9

Munirathinam, A., Krishnamoorthi, R., Venkatesh, A. and Tyagi, B.K. (2014). A note on spatial and temporal distribution of anopheline mosquitoes in Western Ghats, Southern India. *International Journal of Mosquito Research*, 1, 14–18.

Nagpal, B.N. and Sharma, V.P. (1995). *Indian Anophelines*, Oxford and IBH Publishing Co. Pvt. Ltd., New Delhi.

Parmakelis, A., Russello, M.A., Caccone, A., Marcondes, C.B., Costa, J., Forattini, O.P., Sallum, M.A.M., Wilkerson, R.C. and Powell, J.R. (2008). Historical analysis of a near disaster: Anopheles gambiae in Brazil. *American Journal of Tropical Medicine and Hygiene*, 78, 176–178.

Puri, I.M. (1949). Anophelines of the Oriental Region. (*In*: *Malariology*. Ed: M.F. Boyd), W.B. Sunders & Co., New York. 483–505 pp.

Puri, I.M. (1960). *Synoptic keys for the identification of the full grown larvae of the Indian anopheline mosquitoes*. Health Bulletin No. 16, Malaria Bureau No. 7, Government of India Press, Calcutta. 104 pp.

Rajagopalan, P.K., Das, P.K., Kalyanasundaram, M., Tyagi, B.K., Arunachalam, N., Somachary, N., Reddy, C.B.S., Reddy, C.M.R. (1987). *Bangalore mosquito control project: Master plan*. Vector Control Research Centre, Pondicherry, 309 pp.

Rajavel, A.R., Natarajan, R. and Vaidyanathan K. (2004). *Aedes fumidus* and *Aedes amesii* in mangrove forests of India – two new country records. *Journal of the American Mosquito Control Association*, 20, 315–316.

Rajavel, A.R., Natarajan, R. and Vaidyanathan, K. (2005a). Mosquito of mangrove forests of India: Part I – Bhitarkanika, Orissa. *Journal of American Mosquito Control Association*, 21, 131–135.

Rajavel, A.R., Natarajan, R. and Vaidyanathan, K. (2005b). Mosquito of mangrove forests of India: Part II – Sundarbans, West Bengal. *Journal of American Mosquito Control Association*, 21, 136–138.

Rajavel, A.R., Natarajan, R. and Vaidyanathan, K. (2005d). Mosquitoes of the mangrove forests of India: Part 1 – Bhitarkanika, Orissa. *Journal of the American Mosquito Control Association*, 21, 131–135. https://doi.org/10.2987/8756-971X(2005)21[131:MOTMFO]2.0.CO;2

Rajavel, A.R., Natarajan, R., Vaidyanathan, K. and Jambulingam, P. (2011). Systematic list of the species added to the mosquito museum at the Vector Control Research Centre, Pondicherry, India. *Journal of the American Mosquito Control Association*, 27, 8–14.

Rajavel, A.R., Natarajan, R., Vaidyanathan, K. and Saniya, N.P. (2005c). Mosquito collections in the Jeypore hill tracts of Orissa, India, with notes on three new country records. *Culex (Lophoceraomyia) pilifemoralis. Culex (Lophoceraomyia) wilfredi*, and *Heizmannia (Heizmannia) Chengi. Journal of American Mosquito Control Association*, 21, 21–127.

Rajavel, A.R., Natarajan, R., Vaidyanathan, K. and Saniya, N.P. (2005f). A list of the mosquitoes housed in the mosquito museum at the Vector Control Research Centre, Pondicherry, India. *Journal of the American Mosquito Control Association*, 21(3), 243–51.

Rajavel, A.R., Natarajan, R., Vaidyanathan, K. and Soniya, V.P. (2005). A list of the mosquitoes housed in the mosquito museum at the Vector Control Research Centre, Pondicherry, India. *Journal of the American Mosquito Control Association*, 21, 243–251.

Rajput. K.B. and Singh, T.K. (1990). Report on the occurrence of *Uranotaenia (Uranotaenia) micans* Leicester, 1908, in the state of Manipur, India. *Mosquito Systematics*, 22(3), 184.

Rao, T.R. (1984). *The Anopheline of India*. Malaria Research Centre, Indian Council of Medical Research, New Delhi, 518 pp.

Rao, T.R., Dhanda, V., Bhat, H.R. and Kulkarni, S.M. (1973). A survey of Haematophagous arthropods in western Himalayas Sikkim and hill districts of West Bengal: A general account. *Indian Journal of Medical Research*, 61, 1421–1461.

Reuben, R. and Suguna, S.G. (1983). Morphological differences between sibling species of the taxon *Anopheles subpictus* Grassi in India with notes on relationship with known forms. *Mosquito News*, 15(2), 117–126.

Reuben, R., Tewari, S.C. and Hiriyan, J. (1993). Studies of the mosquito fauna of south India. *Proceeding of Entomologist Extraordinary: A festschrift in honour of Botha De Meillon*. Centre for Research in Medical Entomology, Madurai, pp. 47–50.

Reuben, R., Tewari, S.C., Hiriyan, J. and Akiyama, J. (1994). Illustrated key to species of Culex (Culex) associated with Japanese encephalitis in Southeast Asia (Diptera: Culicidae). *Mosquito Systematics*, 26, 75–96.

Sharma, G.K. (1986a). *Malaria and its Control in India. I.* Delhi: Directorate of National Malaria Eradication Programme, 347 pp.

Sharma, G.K. (1986b). *Malaria and its Control in India. III.* Delhi: Directorate of National Malaria Eradication Programme. 618 pp.

Sharma, V.P. (1998). Fighting malaria in India. *Current Science*, 75(11), 1127–1140.

Sharma, V.P. (1999). Current scenario of malaria in India. *Parassitologia*, 41(1–3), 349–353.

Sharma, V.P. and Dev V. (2015). Biology and Control of *Anopheles culicifacies* Giles 1901. *Indian Journal of Medical Research*, 141, 525–536.

Sharma, A.K., Lal, S., Saxena, V.K. and Kumar, S. (2005). An inventory of entomological specimens housed in the museum at the National Institute of Communicable Diseases, Delhi, India. *Journal of Communicable Diseases*, 37(4), 333–368 (Supplement).

Sharma, V.P. and Mehrotra, K.N., 1986. Malaria resurgence in India: A critical study. *Social Science & Medicine*, 22: 835–845.

Sharma, G.P., Parshad, R., Narang, S.L. and Kitzmiller J.B. (1969). The Salivary Chromosomes of *Anopheles stephensi stephensi*. *Journal of Medical Entomology*, 6(1), 68–71. https://doi.org/10.1093/jmedent/6.1.68

da Silva, A.F., Machado, L.C., de Paula, M.B., Vieira, C.J. da Silva P., Bronzoni, R.V. de Morais, Santos, M.A.V. de Melo and Wallau, G.L. (2020). Culicidae evolutionary history focusing on the Culicinae subfamily based on mitochondrial phylogenomics. *Scientific Reports*, 10, 18823. https://doi.org/10.1038/s41598-020-74883-3

Singh, O.P., Mishra, S., Sharma, G., Sindhania, A., Kaur, T., Sreehari, U., Das, M.K., Kapoor, N., Gupta, B. (2022). Evaluation of intron-1 of odorant-binding protein-1 of Anopheles stephensi as a marker for the identification of biological forms or putative sibling species. *PLoS One*, 17(7), e0270760. https://doi.org/10.1371/journal.pone.0270760

Singh, G. and Prakash, S. (2008). A new species of *Lutzia* (Diptera: Culicidae) comparable with *halifaxii* Theobald, 1903 in a semiarid zone of India; pp. 195–204 (*In*: B.K. Tyagi, Ed., *Vector Borne Diseases: Epidemiology and Control*. Centre for Research in Medical Entomology, Madurai, India.

Sinka, M.E. (2013). Global distribution of the dominant vector species of malaria. Chapter 4. (Anopheles mosquitoes – New insights into malaria vectors. Ed.: Sylvie Manguin). *Intech. Croatia*, pp. 109–144. http://doi.org/10.5772/54163

Sinka, M.E., Bangs, M.J., Manguin, S., Chareonviriyaphap, T., Patil, A.P., Temperley, W.H., Gething, P.W., Elyazar, I.R.F., Kabaria, C.W., Harbach, R.E. and Hay, S.I. (2011). The dominant *Anopheles* vectors of human malaria in the Asia-Pacific region: occurrence data, distribution maps and bionomic précis. *Parasites Vectors* 4, 89. https://doi.org/10.1186/1756-3305-4-89

Skuse, F. A. A. (1894). The banded mosquito of Bengal. *Indian Museum Notes*, 3(5), 20.

Subbarao, S.K. (1988). The *Anopheles culicifacies* complex and control of malaria. *Parasitology Today*, 4(3), 72–5. https://doi.org/10.1016/0169-4758(88)90199-8

Subbarao, S.K., Nanda, N., Rahi, M. and Raghavendra, K. (2019). Biology and bionomics of malaria vectors in India: existing information and what more needs to be known for strategizing elimination of malaria. *Malaria Journal*, 18, 396. https://doi.org/10.1186/s12936-019-3011-8

Subbarao, S.K. and Sharma, V.P. (1997). Anopheline species complexes and malaria control. *The Indian Journal of Medical Research*, 106, 164–173.

Subbarao, S.K., Vasantha, K., Raghavendra, K., Sharma, V.P. and Sharma, G.K. (1988). *Anopheles culicifacies* siblings species composition and its relationship to malaria incidence. *Journal of the American Mosquito Control Association*, 4(1), 29–33.

Suguna, S.G., Rathinam K.G., Rajavel, A.R. and Dhanda, V. (1994). Morphological and chromosomal descriptions of new species in the *Anopheles subpictus* complex. *Medical and Veterinary Entomology*, 8, 88–94.

Suman, D.S., De, S., Sharma, G., Chandra, K. and Banerjee, D. (2021). *Culex (Culex) gaugleri*, a new species (Diptera: Culicidae) from India. *Records of Zoological Survey of India*, 121(4), 429–439. https://doi.org/10.26515/rzsi/v121/i4/2021/166022

Surendran, S.N., Abhayawardana, T.A., de Silva, B.G.D.N.K., Ramasamy, M.S., Ramasamy, R. (2000). *Anopheles culicifacies* Y chromosome dimorphism indicates the presence of sibling species (B and E) with different malaria vector potential in Sri Lanka. *Medical & Veterinary Entomology*, 14, 437–440.

Tewari, S.C., Hiriyan, J., Ayanar, K., Munirathinam, A., Venkatesh, A., Reuben, R. and Tyagi, B.K. (2007). CRME Mosquito Museum: an annotated checklist of indian mosquito species. *Contributions of the Centre for Research in Medical Entomology, Madurai*, 175 pp.

Theobald, F.V. (1901a). *A monograph on Culicidae mosquitoes*. Brit. Mus. (Nat. Hist.), London, 646 pp.
Theobald, F.V. (1901b). The classification of mosquitoes. *Journal Tropical Medicine*, 4, 229–235.
Theobald, F.V. (1901c). *A monograph of the mosquitoes of the world*. Vol. 1, 424 pp. London.
Theobald, F.V. (1901d). *A monograph of the mosquitoes of the world*. Vol. 2, 391 pp.
Theobald, F.V. (1902). A short description of the Culicidae of India, with descriptions of new species of *Anopheles*. *Proceedings of Royal Society, London, Ser. B*, 69, 367–394.
Theobald, F.V. (1903). *A monograph of the Culicidae of the world*. Vol. 3, 359 pp.
Theobald, F.V. (1907). *A monograph of the mosquitoes of the world*. Vol. 4, 639 pp.
Theobald, F.V. (1910). *A monograph of the mosquitoes of the world*. Vol. 5, 646 pp.
Tyagi, B.K. (1994). Distribution of arthropod vector-borne communicable diseases and control of their vectors in India. *Indian Review of Life Sciences*, 14, 223–243.
Tyagi, B.K. (2002). *Malaria in the Thar Desert: facts, figures and future*. Agrobios, 162 pp.
Tyagi, B.K. (2004). *The invincible deadly mosquitoes: India's Health and Economy Enemy No.1*. Scientific Publishers, India, 266 pp.
Tyagi, B.K. (2007). Chikungunya in India. *Bulletin De La Societe De Pathologie Exotique*, 100(5), 321.
Tyagi, B.K. (2008). *Vector-borne diseases: Epidemiology and control*. Scientific Publishers (India), 298 pp.
Tyagi, B.K. (2010). The Indian mosquito taxonomy at crossroads: Glorious past, dwindling present and promising future roadmap with a preliminary catalogue of Indian mosquitoes. (In: *Advances in Entomology*. Eds. Jagbir Singh Kirti and Ashwani Kumar), Punjabi University, Patiala, pp. 142–160.
Tyagi, B.K. (2011). *Establishment of a Field Station at Vridhachalam, South Arcot district, Tamil Nadu – Two decadal Results on Japanese encephalitis control trial studies established in 1987 at Vridhachalam, Tamil Nadu*. Centre for Research in Medical Entomology, Madurai, 455 pp.
Tyagi, B.K. (2014). *The Tirukoilur Project – A report on the assessment of the added impact of vector control for augmentation of mass drug administration to eliminate lymphatic filariasis in south India & associated studies*. Centre for Research in Medical Entomology, Madurai, 291 pp.
Tyagi, B.K. (2020). *Mosquito hunters: An Indian history of hostilities against man's deadliest foe – the Mosquito – since 1881*. Scientific Publisher (India), 452 pp.
Tyagi, B.K. (2023). *Desert Malaria: An Emerging Malaria Paradigm and Its Global Impact on Disease Elimination*. Springer-Nature, Singapore, 416 pp.
Tyagi, B.K., Bansal S.K. and Verma K.V.S. (1991). Intra-specific morphological variations in two ecotypes of *Anopheles stephensi* Liston in north-western Rajasthan (Diptera: Culicidae). *Annals of Entomology*, 9, 85–89.
Tyagi, B.K. and R.C. Chaudhary (1997). Outbreak of falciparum malaria in the Thar Desert with particular emphasis on physiographic changes brought about by extensive canalization and their impact on vector density and dissemination. *Journal of Arid Environments (UK)*, 36, 541–555.
Tyagi, B.K. and A.P. Dash (2006). Dengue in India with special reference to the interspecific invasive and virus transmission potential of the Asian tiger mosquito, Aedes albopictus (Skuse) in Kerala: An update. (*In*: *Vector Biology*. Ed. V.P. Sharma and Jagbir Singh Kirti). The National Academy of Sciences, Allahabad, India, pp. 142–158.
Tyagi, B.K., Hiriyan, J., Tewari, S.C., Ayanar, K., Samuel, P.P., Arunachalam, N., Paramasivan, R., Krishnamoorthy, R., Dhananjeyan, K.J., Victor Jerald, L.S., Rajendran, R. (2009). Description of a new species of Anopheles pseudosundicus (Diptera : Culicidae) from coastal Kerala. *Zootaxa*, 2219, 49–60.
Tyagi, B.K., Munirathinam, A., Krishnamoorthi, R., Baskaran, G., Venkatesh, A., Kalimuthu, M. and Govindarajan, R. (2014). *A field-based handbook of identification keys to 4th instar larvae of mosquitoes of public health importance in India. Volume 2*. Taxonomy and Biodiversity Cell. Centre for Research in Medical Entomology, 45 pp.
Tyagi, B.K., Munirathinam, A., Krishnamoorthy, R., Baskaran, G., Govindarajan, R., Krishnamoorthi, R., Mariappan, T., Dhananjeyan, K.J., Venkatesh, A. (2015). A revision of genus Toxorhynchites Theobald, 1901, in the South-East Asian countries, with description of a new species *Toxorhynchites (Toxorhynchites) darjeelingensis* from West Bengal, India (Diptera, Culicidae). *Halteres*, 6, 13–32.
Tyagi, B.K., Munirathinam, A., Krishnamoorthy, R., Venkatesh, A. (2012). *A field-based handbook of identification keys to mosquitoes of public health importance in India. Volume 1*. Taxonomy and Biodiversity Cell. Centre for Research in Medical Entomology, 55 pp.
Tyagi, B.K., Munirathinam, A. and Venkatesh, A. (2015). A catalogue of Indian mosquitoes. *International Journal of Mosquito Research*, 2(2), 50–97.
Tyagi, B.K., Munirathinam, A. and Venkatesh, A. (2016). An inventory of the authentically identified and classified mosquito taxa in the "CRME Museum", with annotations on its significance in taxonomic research in India. *International Journal of Mosquito Research*, 3(4), 58–113.

Tyagi, V., Sharma, A.K., Dhiman, S., Srivastava, A.R., Yadav, R., Sukumaran, D., Agrawal, O.P., Veer, V. (2016). Malaria vector *Anopheles culicifacies* sibling species differentiation using egg morphometry and morphology. *Parasite & Vectors*, 9(202), 1–13.

Tyagi, B.K. and Verma, K.V.S. (1991). Anopheline mosquitoes of Sri Ganganagar district (Rajasthan) transmitting malaria parasite. *Journal of Applied Zoological Researches*, 2, 85–91.

Tyagi, B.K. and Yadav, S.P. (1996). Malariological and sociological significance of "Tanka" and "Beri" in the Thar Desert, Western Rajasthan (India). *Journal of Arid Environments (U.K.)*, 33, 497–501.

Tyagi, B.K. and S.P. Yadav (2000). Paddy cultivation and the prospect of malaria in the Thar Desert, India: A review. *Journal of Applied Zoological Researches*, 11, 41–47.

Tyagi, B.K. and Yadav, S.P. (2001). Bionomics of malaria vectors in two physiographically difference areas of the epidemic-prone Thar Desert, north-western Rajasthan (India). *Journal of Arid Environments*, 47, 161–172.

Vatandoost, H. and Borhani, N. (2004). Susceptibility level and irritability of synthetic pyrethroids against main malaria vectors in the endemic areas of Iran. *Acta Medicine Iran*, 42, 247–255.

Wattal, B.L. and N.L. Kalra (1961). Region wise keys to female Indian Anopheles. *Bulletin of Natural Society of India Malaria Mosquito Borne Diseases*, 10, 55–138.

Wood, D.M. and Borkent, A. (1989). Phylogeny and classification of the Nematocera. Manual of Nearctic Diptera, Vol. 3, Research Branch Agriculture Monograph, Vol. 32 (ed. by J. McAlpine), pp. 1333–1370. Canadian Government Publishing, Centre, Hull.

World Health Organization (2007). *Anopheles species complex in South and South East Asia. Anopheles species complexes in South and South East Asia.* SEARO Technical Publication No. 57, 102 pp. http://apps.searo.who.int/pds_docs/B2406.pdf

Yee, D.A., Dean Bermond, C., Reyes-Torres, L.J., Fijman, N.S., Scavo, N.A., Nelsen, J., Yee, S.H. (2022). Robust network stability of mosquitoes and human pathogens of medical importance. *Parasites Vectors*, 15, 216. https://doi.org/10.1186/s13071-022-05333-4

5 Mosquito Biodiversity

B.K. Tyagi
Chandigarh University, Mohali, India

Jagbir Singh Kirti
Ex-Department of Zoology, Punjabi University, Patiala, India

Ritwik Mondal
North Bengal University, Darjeeling, West Bengal, India

S.A. Khan
ICMR-Regional Medical Research Centre, Dibrugarh, India

Vikas Kumar
Zoological Survey of India, Kolkata, India

P.K. Sumodan
Government College, Madappally, India

Sajal Bhattacharya
Asutosh College, Kolkata, India

5.1 INTRODUCTION

"*Biodiversity is a gift of nature. It should not be allowed to get destroyed with man's greed. Biodiversity, by its nature, wants to give a lot for the benefit of human*" extolled the *Bharat Ratna* Dr A.P.J. Kalam, His Excellency the President of India, on 25th October, 2002 during the 72nd National Symposium on "*Biodiversity: A scientific approach – Agenda for the 21st century*" under the auspices of the National Academy of Sciences (Gupta, 2004). Biological diversity was first proposed by Norse and McManus (1980) for two combined concepts involving genetic diversity and ecological diversity. While the former relates to the amount of genetic variability within species, the latter refers to the number of species in a community of organisms. The term "Biodiversity" was accepted at the first planning meeting of the National Forum on Biodiversity in Washington D.C. in 1986 (Wilson and Peters, 1988). In summary, biodiversity includes all life forms, ecosystems, and ecological processes at the genetic, taxonomic, and ecosystem levels (McNeely et al., 1990). Biodiversity assumed a new dimension in the Earth Summit in Rio de Janeiro in 1992 where it was related to scientific, economic, and political scenarios globally – so beautifully epitomized in the ancient Indian aphorism, "*Variety is the spice of life.*" Biodiversity has been differently defined in different times just as the world was impacted differently, and it appears that the phenomenon is a dynamic trait of life. Biodiversity is, in essence, all the different kinds of life such as, for example, animals, plants, fungi, and even microorganisms like bacteria, viruses, etc. in a stupendous variety of form, shape, size, and behavior most suitable to sustain in an area of the natural world. Each of

DOI: 10.1201/9781003326991-7

these species and organisms work together in ecosystems, like an intricate web, to maintain balance and support life. Biodiversity supports everything in nature that we need to survive: food, clean water, medicine, and shelter. However, as human population and anthropization are growing there is an increasing threat of upsetting the balance of ecosystems and losing biodiversity as already, one species, *Culiseta atlantica*, has been brought up to the IUCN Red List of Endangered Species, and more species suggest to be considered for the safety of their genome (Nunes and Borges, 2021).

Flying predatory insects like mosquitoes, which are actually micropredators, are key natural mediators of pathogens such as protozoans, helminths, and viruses resulting in a plethora of deadly and incapacitating diseases that can highly damage human health, economy, and intelligence (Wilkerson et al., 2015). However, there is a lack of knowledge about the diversity in various biological phenomena, e.g., morphological structure, selection of breeding habitats, oviposition, emergence, mating, resting, and flight or movements of flying. In particular, it is still unclear how a mosquito moves in the pitched dark night and still finds a commensurate site on the preferred host's body to feed on blood, essential for maturation of eggs in the female body. Also, is a male mosquito any worth his weight (Tyagi, 2020) and if it plays an essential role in regulating different biological activities of the counterpart, the *femme fatale* (Tyagi, 2023a, 2023b)? The abundance, diversity, distribution, and ecology of mosquitoes (Diptera: Culicidae), especially arbovirus-carrying vectors are important indices for arthropod-borne disease control. Detailed information on the diversity, distribution, and ecology provides a better insight into understanding mosquito population dynamics in relation to vector control measures (Paksa et al., 2019). The diversity of the mosquito species varies among different regions of the world (Singh et al., 2023). The greatest diversity of mosquito species is found in the Neotropical region (31% of total known species), followed by the Oriental (30%), Afrotropical (22%), and Australasian (22%) regions. The Nearctic region (5%), including the United States and Canada, has the lowest species diversity. In the Nearctic region, the greatest number of species in Culicinae is found in the tribe Culicini, followed by Aedini (Rueda, 2004).

Intra-family (Culicidae) variations among over 3600 mosquito species make each of the tribes of mosquitoes specialized to sustain befittingly in a specific ecosystem such as marshes, deserts, mountains, forests, prairies, valleys, rivers, ravines, lakes and phytotelmata etc. (Figure 5.1). Mosquitoes are epitome of both structural and behavioral superiorities among the whole animal kingdom, and their blood-feeding habit through millions of years is a sterling example of their unparalleled sense of innovativeness, adaptability, and survivability. It is no surprise, therefore, that the world's first animal model to show an evolutionary response to climate change comes only from a mosquito, *Wyeomyia smithii* (Emerson et al., 2010; National Science Foundation (2010).

Diversity in mosquitoes is noted not only in the aerial adult life but also in the aquatic larval/pupal life:

(i) Larvae have the power to colonize and live in a wide variety of natural and artificial habitats, such as temporary and permanent water resources, unclean and clean water, large or small water resources, stagnant or stream waters, and even the smallest places where water is gathered, such as buckets of water, pots, tires, bromeliads, animal feet prints, and plant leaf axes;
(ii) Pupae, though bionomically very important as a partially dormant stage since it is not feeding but is highly motile and physiologically active, are highly diverse phenotypically;
(iii) Adult mosquitoes, both male and female sexes, are very diverse bionomically, for example, they are too diverse in host searching, biting behaviors, dispersal and reproductive strategies.

Two factors (abiotic and biotic) affect mosquito life cycle: (i) The biotic factors contain adult sugar and blood meal types, species communications, interactions, and natural enemies, and (ii) Abiotic factors include physicochemical attributes of larval habitats are abiotic factors, which include the type and contents of water, temperature, and rainfall.

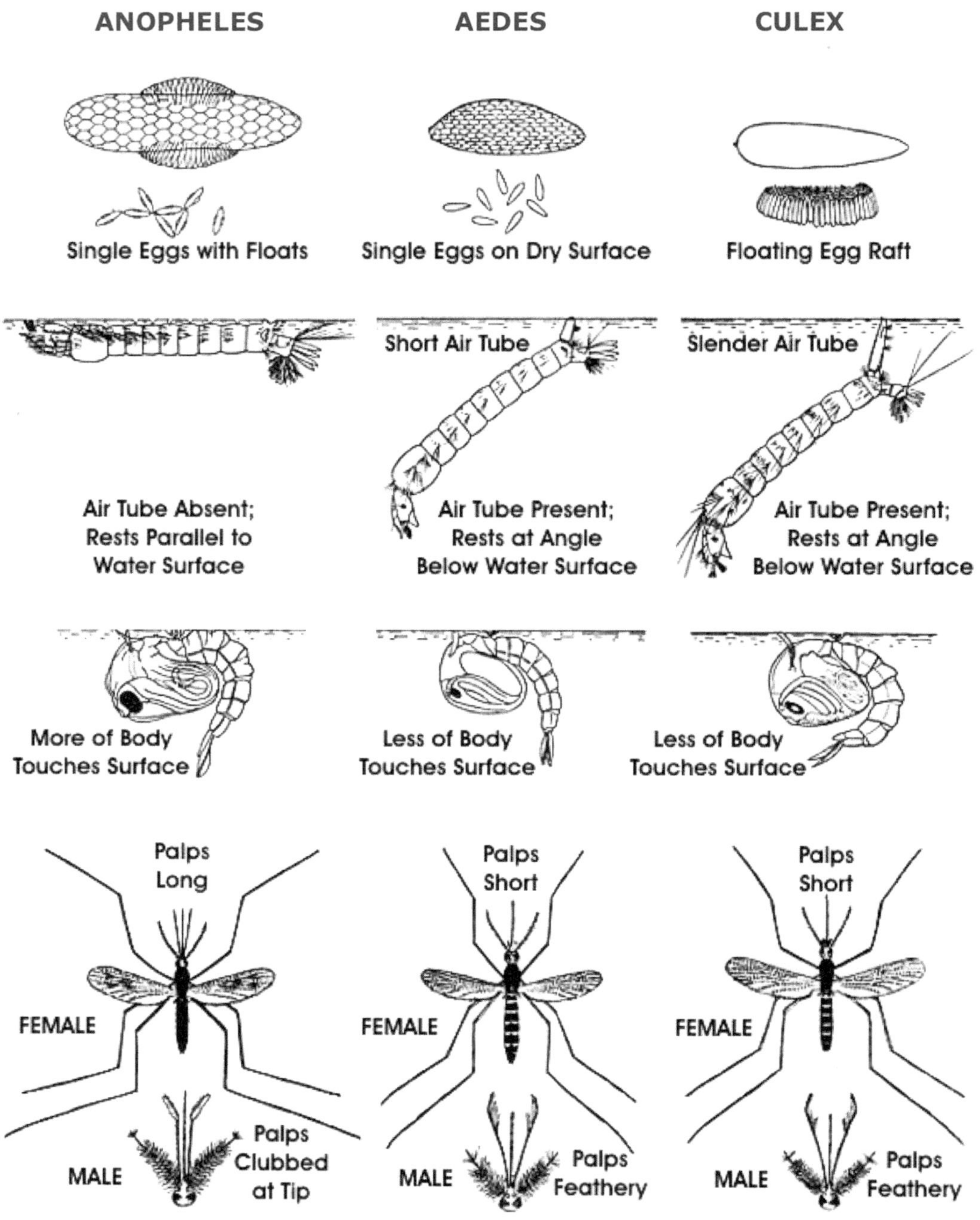

FIGURE 5.1 Diagrammatic summary of fundamental morphological diversity among three major groups of mosquitoes.

There are complex interactions between these factors that significantly affect mosquito ecological adaptability and vectorial capacity for disease transmission with significant concepts for vector management and control at the local and regional levels. Accordingly, investigating biotic and abiotic factors for various mosquito fauna makes it easier to monitor potential modifications of larval habitats affected by rains, global climate change, and man-made activities.

5.1.1 Phenotypic Plasticity

Phenotypic plasticity (i.e., the ability of individual genotypes to produce different phenotypes when exposed to varied environmental conditions; Pigliucci, 2005), is reflective of organismal diversification on a variety of levels of biological organization, evolutionary innovation, new species, and complex life cycles (Moczek, 2010). Mosquitoes offer one of the best examples of phenotypic plasticity highlighting the capacity of a single genotype to exhibit a range of phenotypic outputs and the responsiveness of the underlying developmental process to the environment (nutrition, temperature, and population density). Phenotypic plasticity in mosquitoes has generated novel targets for evolutionary processes to act on, as well as brings about trade-offs during development and evolution, thereby diversifying evolutionary trajectories available to natural populations. Mosquitoes exhibit phenotypic plasticity among species as a way to survive optimally in the surrounding environments. Various attributes such as, for example, size, shape, form, coloration, movement, feeding, egg-laying characterize a species to sustain in its niche.

5.1.2 Polyphenic Development

Polyphenism describes the unique sub-type of phenotypic plasticity where the outputs are not continuous, but rather discrete and multi-stable, resulting in several distinct phenotypes on the same genetic background (Yang and John, 2019). Polyphenism, like phenotypic plasticity, is a major contributor to diversification and evolution in mosquitoes. It is a type of phenotypic plasticity where development exhibits reproducible bifurcations revealing multiple distinct outcomes (e.g., larvae are mobile and able to actively feed vs. pupae which are motile but can not feed and continue with all necessary physiological advancement such as for example, formation of wings, legs, genitalia, etc.). Environmental stimuli are the major external triggers of polyphenisms.

5.1.3 Diversity in Quantum of Species

Globally four countries, namely, Brazil, Indonesia, Malaysia, and Thailand have the highest number of mosquito species, whereas in terms of endemic species, the most notable countries are Brazil, Indonesia, Australia, and the Philippines.

According to Reinert et al. (2009; https://mosquito-taxonomicinventory.myspecies.info/simpletaxonomy/term/6045), mosquitoes (Family Culicidae) are organized under 2 subfamilies and 11 tribes covering 113 genera and 3,614 species, but as per the latest, simplified and more useful classification (vide Harbach, 2007, 2013, 2018; Wilkerson et al., 2015, 2021) mosquitoes, which have two subfamilies Anphelinae and Culicinae, possess currently a total of only 10, tribes, 41 genera and 3715 species and 130 Subspecies. Due to their involvement in transmitting infectious pathogens to humans, the generic triad of *Anopheles, Culex*, and *Aedes* are the most important in biomedical or public health science. The genus *Aedes* have the maximum species in the family, and are also extremely important both as vectors and pests (Lemine et al., 2017). The OECD (2018) has recently discussed *Aedes aegypti* in detail. The Indian mosquito fauna includes 420 species organized into 50 genera and 42 subgenera (Bhattacharyya et al., 2014; Tyagi et al., 2015; *this work*).

5.1.4 Diversity in Cryptic Species

Apart from good species' inter- and intraspecific diversity, there are examples of taxa in Culicidae, particularly among anophelines, which are morphologically indistinguishable but can be identified with the aid of chromosomal and/or molecular characteristics (e.g., *Anoopheles gambiae, An. atroparvus, An. maculipennis, An. culicifacies, An. fluviatilis, An. subpictus*). Subbarao et al. (1983, 1999). Goswami et al. (2006) and WHO SEARO (2007) have not only described the distribution of *An. culicifacies* sibling species in India, along with others in Southeast Asia, but also highlighted

TABLE 5.1
Sibling Species Complex of *Anopheles* Mosquitoes in India

Species	Sibling Species Complex	Designated Species
Anopheles culicifacies	5	A, B, C, D, and E
Anopheles fluviatilis	3	S,T, and U
Anopheles subpictus	4	A, B, C, and D
Anopheles annularis	2	A and B
Anopheles dirus	5 (In India only 3 species occur)	*An. dirus* s.s. (species A), *An. cracens* (species B), *An. scanloni* (species C), *An. baimaii* (species D), *An. elegans* (species E), *An. nemophilous* (species F), *An. takasagoensis*, and a cryptic species tentatively designated as *An.* aff. *takasagoensis*.
An. sundaicus	4	*An. sundaicus* s.s., *An. epiroticus* (formerly species A), *An. sundaicus* species D and *An. sundaicus* species E.

their varied significance to transmit malaria in the country and differential potential to develop resistance against conventional insecticides. Tyagi et al. (2016) have recently attempted to distinguish various cytologically distinguishable cryptic species designated as A, B, C, D, and E under *An. culicifacies* on the basis of SEM-visualized morphometric differences in eggs. Some of these cryptic species are medically important as they bear the potential to transmit pathogens. Recently Tyagi (2021) has summarized taxonomic issues associated with the sibling species complexes under various taxa in India (Table 5.1).

Anopheles stephensi has so been kept outside a species complex, although a hypothesis was first put forth by (Tyagi et al., 1991) and most recently reinforced by (Firooziyan et al., 2018) based on odorant binding protein 1 intron I sequence. The adult populations emerging out of the three subspecies, races, or forms are completely indistinguishable, albeit distinct egg morphometric attributes among them (Table 5.2).

TABLE 5.2
***Anopheles stephensi* with Three Variants, Forms or Subspecies Characterized by Distinctive Features in Morphology, Behavior, and Interaction with Malaria Parasites**

S. No.	*Anopheles Stephensi* Forms	Ridges on the Egg Floats	Aver. Number of Ridges Among the Eggs Laid	Spiracular Length	Spiracular Index (SI)
1	Type form (*An. stephensi* s.s.)*	14–22	16–19	Average Spiracle length0.11–0.12 mm	Average SI 8.09–9.23
2	Intermediate Subbarao, 1987	12–17	13–16	–	–
3	Mysorensis** Sweet & Rao, 1937	9–15	10–14	0.09–0.10 mm	6.82–7.60

Source: (Tyagi, 2021).

* urban vector
** rural vector

Likewise, *Aedes vexans* – a medically highly important mosquito vectoring more than 30 arboviruses including Eastern horse encephalitis, Japanese encephalitis, California encephalitis, Western horse encephalitis, as well as other pathogens such as Tularemia and *Dirofilaria immitis* – is comprised of several subspecies, viz., *Ae. vexans vexans* in Europe, *Ae. vexans nipponii* in East Asia, and *Ae. vexans arabiensis* in Saudi Arabia, Yemen, and Africa. This species is most ubiquitous among mosquitoes, along with *Aedes aegypti* and *Culex quinquefasciatus*.

5.1.5 Geographic Biodiversity in India

Northern India is very rich in mosquito fauna due to sylvatic and mountainous terrains with innumerable varieties of water sources. Jammu & Kashmir, the northern-most state in India, has a known history of malaria, with Anopheles *An. barianensis* collected from the spot (Adie, 1913), although dengue outbreaks have also been notably reported lately from the state (Naik et al., 2022). Dar et al. (2016) archived information on the biodiversity of mosquitoes in the Jammu region of Jammu and Kashmir state and inventorized 41 species under 11 genera and 12 subgenera. Of this, 17 species have been known earlier since 1983, while 24 species were reported for the first time from the region. Six species reported earlier were absent in their collection. Altogether Jammu division reports a total of 47 mosquito species. Faiz and Shahbaz (2020) sampled 2895 specimens of mosquitoes from the POK area in Kashmir, which were identified as *Anopheles barianensis* (Subfamily Anophilinae), *Culex barraudi, Cx. epidesmus, Cx. fuscocephala, Cx. pipirms fatigans, Cx. pipiens pipiens, Cx. pseudovishnui, Cx. vishnui, Aedes aegypti, Ae. micropterus* and *Armigeres subalbatus* (Subfamily Culicinae). The most abundant species was *Armigeres subalbatus*. Devi and Jauhari (2008), who studied mosquito fauna in as many as 87 sites in Dehradun, Pauri, New Tehri, Hardwar, Rudraprayag, Chamoli, and Uttarkashi of Garhwal region of Uttarakhand state between January 2000 and December 2005, enlisted 18 anopheline species, three of which are well-known vectors of malaria, viz., *Anopheles culicifacies, An. fluviatilis* and *An. stephensi*. Haq and Singh (2021) sampled a total of 2822 specimens of 25 species of mosquitoes belonging to 6 genera viz., Anopheles, Culex, Aedes, Armigeres, Toxorhynchites, and Uranotaenia from Dehradun between January 2019 to December 2020 were recovered. Of these, 12 species of Anopheles, 5 species each of Culex and Aedes were the main species, in which Anopheles form 48.12%, Culex 22.89%, and Aedes 18.85% respectively. Shannon and Evenness index for Anopheles (2.212, 0.839) Culex (1.642, 0.746), and Aedes (1.625. 0.657) during the entire study period was obtained. Overall, the highest Shannon index was 1.9358 during September 2020, and the lowest was 1.0781 during January 2020.

Similarly, the northeastern region of India is very rich in Culicidae, and several studies including a checklist from the northeast inventorizing 130 species (37 anophelines and 93 culicines) have been conducted in northeastern states to determine the species composition, diversity, and density of mosquitoes (Malhotra and Mahanta, 1994; Veer and Gopalakrishnan, 2016). Interestingly, some new records of species were made from India for the first time (Devi and Singh (2018) who are credited to have found *Culex (Culex) pipiens* Linn. in Manipur. *Culex (Culex) quinquefasciatus* which is a symapatric to *Cx. pipiens* had been existing in India including Manipur, but not *Cx. pipiens*.

In Assam, entomological surveys in Dibru Saikhowa Biosphere Reserve during pre-monsoon, monsoon, and post-monsoon seasons reported 52 species of mosquitoes under 11 genera. The genus *Anopheles* (18 species) was found to be predominant followed by *Culex*, *Aedes*, *Mansonia*, *Armigeres*, *Mimomyia*, *Ochlerostatus*, *Malaya*, *Toxorhynchites*, *Ficalbia*, and *Aedeomyia* (Bhattacharyya et al., 2014; Dutta et al., 2010). Surveys in Guwahati reported the presence of four genera, namely, *Anopheles*, *Aedes*, *Mansonia*, and *Culex*. The human filariasis vector, *Culex quinquefasciatus*, was found to be the most abundant followed by *Ae. aegypti* and *Ae. minimus* (Alak and Kaushik, 2014). Surveys in Sivasagar and Kamrup districts of Assam reported *Cx. gelidus* and *Cx. tritaeniorynchus* in high densities (Baruah et al., 2018). Manipur is another mosquito-rich state in northeast India, and a total of 111 species of mosquitoes have been reported from Manipur of which 83 species, belonging to 13 genera, are in Culicine. Recently, there were reports of the occurrence of *Lutzia* and *Culex pipiens* from the state (Moirangthem and Singh, 2018a, 2018b). Mosquito surveys during monsoon and post-monsoon seasons in Manipur reported 55 species of mosquitoes belonging to 10 genera. Among the Culicines, nine genera were reported consisting of 17 species of *Culex*, 9 of *Aedes*, three of *Armigeres*, 3 of *Mansonia*, 2 of *Uranotaenia* and 1 each of *Coquillettidia*, *Aediomyia* and *Toxorhynchites*. The study also reported the presence of *An. dirus* and *An. nivipies* among the 17 Anopheles species recorded (Dutta et al., 2005). Recent studies in Manipur revealed the presence of *Cx. quinquefasciatus*, *Cx. mimeticus* and *Culiseta niveitaeniata* (Devi and Chingangbam, 2019).

Among all the seven-sisters states of the northeast, Mizoram and Meghalaya are without doubt a paradise for rare and important taxa, and yet are far more unexplored for Culicidae due largely to a difficult terrain. Previous studies in Mizoram on Anopheles during winter to pre-monsoon season reported 7 species of Anopheles *viz. An. philippinensis*, *An. jamesii*, *An. vagus*, *An. barbirostris-campestris*, *An. theobaldi*, *An. culiciformis* and *An. maculatus*. Both *Ae. aegypti*, and *Ae. albopictus* has been reported from Mizoram (Vanlalruia et al., 2014). On the other hand, earlier studies in Meghalaya have reported the presence of 32 mosquito species belonging to 7 genera (*Anopheles*, *Culex*, *Aedes*, *Armigeres*, *Mansonia*, *Taxorhychites*, and *Uranotaenia*) (Christophers, 1933; Barraud, 1934). Further studies reported the presence of 42 species belonging to six genera (*Anophles*, *Cules*, *Aedes*, *Mansonia*, *Armigeres*, and *Coguillettidia*). The study also reported two new species of Anopheles from this region (Das et al., 1984). *Armigeres (Armigeres) mahantai* a new species of mosquito was first described from Meghalaya (Bhattacharyya et al., 2009). As far as Tripura is concerned, 7 anopheline species have been recorded earlier of which *Anopheles philippinensis* was the most prevalent vector (Das et al., 1991), although recent surveys in Tripura have reported the presence of 11 species of Anopheles out of a total of 31 species belonging to eight genera of mosquitoes (Prakash et al., 1998). It also reported the first incrimination of *An. baimaii* as vector of malaria in Tripura (Sarmah et al., 2019). Strategically Arunachal Pradesh appears more suitable for the prosperity of mosquito taxa. Surveys in three biotypes under Jairampur in Arunachal Pradesh revealed the presence of 35 mosquito species belonging to 8 genera. *Orthopodomyia anopheloides* and *Tripteroides indicus* were recorded for the first time from the state. *Topomyia (Topomyia) hirtusa*, *Topomyia (Topomyia) bifurcata*, and *Topomyia (Suaymyia) cristata* were recorded for the first time in India from Arunachal Pradesh (Bhattacharyya et al., 2007; Prakash et al., 2000). Finally, when Nagaland, the mystique state with fewer explorations for the mosquito taxa, was struck by a dengue outbreak quick survey revealed the presence of three Aedes species, *Ae. albopictus*, *Ae. aegypti* and *Ae. annandalei* (Dutta et al., 2004, Dutta, Prakash, et al., 2010a). Further surveys in Nagaland during monsoon and post-monsoon season reported a total of 58 mosquito species under 11 genera.

Mosquitoes in the central India region exhibit a great degree of diversity commensurate to the surrounding environments (Sinha et al., 2018). Two of the dominant malaria vectors, *Anopheles culicifacies* Giles and *An. stephensi* Liston was originally discovered from this region.

The southern Indian region comprises of five major states, viz., Andhra Pradesh, Telangana, Karnataka, Tamil Nadu, and Kerala, and a Union Territory, Puducherry. The whole of south India is very rich in the species diversity of mosquitoes. The pioneering studies on mosquitoes in Kerala can be attributed to Theobald and Giles (Theobald, 1901; Giles, 1901). The former described three species from Kollam, viz., *Ficalbia minima*, *Mansonia annulifera*, and *Ma. uniformis*. The latter described *Cx. bitaeniorhynchus* and *Cx. tritaeniorhynchus* from Travancore (exact locality not mentioned). The mosquito fauna of the state is amply represented in the two volumes on the mosquito fauna of British India compiled by Christophers (1933) and Barraud (Barraud, 1934). Christophers reported 16 Anopheline species and Barraud 41 species under Culicinae and Toxorhynchitinae. Iyengar reported 72 species of mosquitoes from the state while studying the epidemiology of Filariasis in Travancore (Iyengar, 1938). In the post-independence period, Tewari and Hiriyan described two new species of *Aedes* from the state, viz., *Ae. agastyai* and *Ae. rubenae* (Tewari and Hiriyan, 1992). Hiriyan et al. surveyed a Japanese encephalitis endemic area in Kerala and reported 21 species of mosquitoes (Hiriyan et al., 2003). Subsequently, Arunachalam et al reported 18 species of mosquitoes during a study for determining the vectors of Japanese encephalitis (Arunachalam et al., 2004). Rajavel et al. reported 17 species of mosquitoes from the Mangrove forests of Kannur in North Kerala (2006). In 2009 Tyagi et al described a new species viz., *Anopheles pseudosundaicus* from Kollam (Tyagi et al., 2009). In 1992 Sumodan reported breeding of 12 species of mosquitoes in latex collecting containers in the rubber plantations of Kerala (Sumodan, 2012). So far as many as 144 species of mosquitoes under 18 genera have been reported from the state, viz., *Aedes* (27 species), *Anopheles* (36 species), *Armigeres* (7 species), *Coquillettidia* (1 species), *Culex* (31 species), *Ficalbia* (1 species), *Heizmannia* (5 species), *Hodgesia* (1 species), *Lutzia* (1 species), *Malaya* (1 species), *Mansonia* (3 species), *Mimomyia* (3 species), *Orthopodomyia* (2 species), Uranotaenia (13 species), *Topomyia* (1 species), *Tripteroide* (2 species), *Toxorhynchites* (2 species) and Verrallina (7 species). In Tamil Nadu, a coastal state gifted with a long coastline dotted with deep mangroves infested with a large mosquito fauna (Rajavel and Natarajan, 2009)) where in the mangroves of Muthupet they listed nine species, Munirathinam et al. (2015) collected altogether 4602 immatures from Keeriparai and Mundanthurai hill ranges of the Western Ghats, southern India. They classified a total of 3583 specimens belonging to 50 species under 21 genera and 18 subgenera. The major vector species found in these hill ranges were *Stegomyia aegypti, St. albopicta* (Dengue and Chikungunya), JE vectors *Culex bitaeniorhynchus, C. tritaeniorhynchus*, diurnally sub-periodic filariasis vector *Downsiomyia nivea* and *Anopheles mirans* Simians malaria vectors were recorded. The Karnataka state provides a far heterologous ecosystem to varied mosquitoes particularly in reference to transmission of vector-borne diseases, and was recently on high alert after Zika virus was found in mosquito, *Aedes aegypti* (Kumar et al., 2021). Recently in 2013, an invasive mosquito species, *Aedes koreicus*, was sampled from a village near the airport in Bangalore (*pers. comm.* B.K. Tyagi with S.K. Ghosh). This species is a common species in South Korea, China, Japan, and Russia, but since 2008 this highly invasive alien species has also been recovered repeatedly from several European countries such as Belgium, Italy etc. Like Tamil Nadu, the undivided erstwhile Andhra Pradesh state too has a long coastline where mangroves provide a thriving climate to the mosquitoes. Rajavel et al. (2006) have sampled 12 species, belonging to 5 subgenera and 9 genera (Aedes, Anopheles, Culex, Diceromyia, Lorrainea, Rhinoskusea, Stegomyia, Toxorhynchites, and Verrallina), from tree holes, crab holes, and swamp pools which constituted the larval habitats with *Lo. fumida* being predominant in tree holes. Some species like *Ae. cancricomes, Cx. sitiens, Lo. fumida, Rh. wardi*, and *Ve. lugubris* were involved in daytime feeding on humans. Also, adults were found resting in tree holes, tree trunks, root bases, and crab holes.

5.2 BIODIVERSITY OF DOMINANT VECTOR SPECIES

Many biomedically and/or ecologically significant mosquito species have spread over large regions of the world during past few decades. For convenience of understanding a few major examples from India and the world are offered hereunder involving species *Aedes aegypti, Ae. albopictus, Anopheles gambiae, An. arabiensis, An. stephensi, an. culicifacies, An. sundaicus, Culex quinquefasciatua* and *Wyomiya smithii*, though there are indeed a large number of other examples as well.

5.2.1 *Aedes aegypti* (Linn., 1762)

To date, *Ae. aegypti* remains an invasive tropical species worldwide with a cosmopolitan habitat from 40° N to 40° S latitude. One of the world's most deadly mosquitoes, *Ae. aegypti* is vector for yellow fever, dengue, chikungunya, and Zika etc. The mosquito has two subspecies or races or populations:

(i) The first and foremost, the ancestral *Ae. aegypti formosus*, a synanthropic and much darker form, is adapted to feed on both humans and animals in forests and vegetated ecotones in sub-Saharan Africa. In forests it breeds in tree holes and uses these breeding habitats for egg-laying.

(ii) The second subspecies is *Ae. aegypti aegypti* which is ubiquitous across the tropical and subtropical regions. It is strongly synanthropic and rests, breeds and feeds indoors on humans. The form or subspecies has a characteristic preference for artificial containers provided by humans, but also freely breeds in natural habitats like tree holes. In the Thar Desert (northwestern India) the mosquito has adapted to breed in underground water reservoirs, "Tanka," once the preferred breeding site for malaria vector, *Anopheles stephensi* (Tyagi, 2023b).

A third subspecies, *Ae. aegypti queenslandensis*, described as a light-colored form in the Mediterranean Basin, was recently subsumed with the second subspecies *Ae. aegypti aegypti* as it was genomically indistinguishable.

Aedes aegypti originated from Africa was introduced into South America possibly during the times of slave trade and transport ships resupplied in African ports during the 15th century through to the end of the 17th century. It is possible that through this means the species was introduced to the rest of the world via Ethiopia over to tropical and subtropical areas. The species made an independent introduction to Asia.

5.2.2 *Aedes albopictus* Skuse, 1894

Equally deadly in transmitting a range of pathogens of public and veterinary health importance, *Aedes albopictus*, was described by Skuse in 1894 while working in Australia, from specimens collected in Calcutta, India (Barraud, 1934). Believed to be having its cradle in the tropical forest of Southeast Asia, the species is another medically important vector for many diseases of public and veterinary health importance. It later spread to far north China. Japan, and the Korean peninsula as well as the Palcarctic and Nearclic regions – the northernmost reported distribution points of *Ae. albopictus*.

In Southeast Asia *Ae. albopictus* has extended its range to the coastal cities of Irian Jaya, the Solomon and Santa Cruz Islands, and Papua New Guinea; it has also been found in Brisbane and Darwin, Australia, and recently in New Zealand. Several islands in the Indian Ocean, such as Madagascar, Mauritius, Reunion, the Seychelles, and the Chagos Archipelago, have recently reported *Ae. albopictus*. In addition, following an unprecedented outbreak of chikungunya in the Indian mainland during 2005–2006, the mosquito was also found in Europe (Albania and Italy).

Aedes albopicius – the Asian Tiger mosquito – is believed to be inhabiting, besides the mainland Asian countries, also certain islands of the Indian Ocean and the Hawaiian Islands in the Pacific Ocean since before the early 1980s. It made a hideous entry in August 1985 in the United States of America's Harris County, in and around Houston, Texas where it was found breeding in tire dumps. Later the species was collected from 22 other states in the USA. One of the most important differences between the North American populations of *Ae. albopictus* and *Ae. aegypti* is their latitudinal distribution. While *Ae. aegypti* populations limited to the southern portions of the United States due mainly to their inability to tolerate very low temperatures, *Ae. albopictus* populations have evolved photoperiod-induced egg diapause, which allows them to colonize temperate and northern latitudes. In addition, temperate strains of *Ae. albopictus*, such the North American ones, show egg cold hardiness, which enables the species to survive the suboptimal winter temperatures found in the northern latitudes. Almost simultaneously, in June 1986, *Ae. albopictus* was first reported in Brazil's Rio de Janeiro and Sao Paulo, and within a decade, it spread to many neighboring countries like Panama, and many others far away, e.g., Dominican Republic, Colombia, Suriname, and Mexico. It went on to prevail in continental Africa, as well as Europe, both being huge landmasses.

While the distribution of Aedes albopictus and that of *Aedes aegypti* overlap in the urban environment, *Ae. albopictus* is commonly found in suburban and rural where open, vegetated spaces prevalent. It is generally accepted that *Ae. albopictus* originally sylvan species, breeding and feeding in forest fringes, which began to adapt to the domestic environment throughout several parts of its geographical range by open with plentiful vegetation, surrounded by buildings.

5.2.3 *Anopheles stephensi* Liston, 1901

Considered a native of India (Asia; *cf.* Tyagi, 2023), *Anopheles stephensi* has a geographical range in southern Asia that extends across the Indian subcontinent with a eastward in Bangladesh, southern China, Myanmar, Sri Lanka and Thailand, and a westward extension through Iran and Iraq into the Middle East and Arabian Peninsula and, most recently, a string of northeastern African countries namely Djibouti (2012), Ethiopia (2016), Sudan (2016), Somalia (2019), Nigeria (2020) and Yemen (2021) (WHO, 2022).

Anopheles stephensi, a highly competent vector of *Plasmodium falciparum* and *P. vivax*, is considered an efficient vector of urban malaria, although in the vast stretches of the Great Indian Thar Desert the species there is also a major vector of malaria, besides being the original vector as well (Tyagi, 2002, 2023b). India (species was first incriminated as a vector of malaria in Mumbai in 1911, Gujarat and Madras in 1938, Ahmedabad in 1943 and Broach in 1967), and since then it has been recognized as an important vector of malaria in urban areas all over India, and areas bordering the Persian Gulf, including western and northwestern India.

Anopheles stephensi comprises three egg phenotypes, subspecies, forms or races, namely, (i) mysorensis Sweet & Rao, (ii) typical, and (iii) intermediate. They are distinguishable on the basis of egg dimensions, and the number of ridges on the egg float. Generally speaking, the type form is an efficient vector of urban malaria (the sole exception is the Thar Deseret environments in northwestern India), whereas the *mysorensis*, with a poor vectorial capacity due to its highly zoophilic behavior, is rural in nature, although in Iran both the typical and *mysorensis* forms are important vectors; the sporozoite rates being as high as 47. The intermediate form is typically found in rural villages and peri-urban areas, but very little is known about its vector status.

Larvae of *An. stephensi* breed typically in various different kinds of artificial containers in residential and industrial areas, and in homes (domestic wells, overhead water tanks, room coolers, cisterns, and roof gutters), although in rural areas the species larvae utilize freshwater pools, stream margins, and stream beds, catch basins, seepage canals, wells, and domestic water-storage containers. Greater numbers of larvae are typically found outdoors than indoors. Larvae of the *mysorensis* form appear to exclusively inhabit stone pots and earthenware containers. In the Thar Desert, *An. stephensi* breeds in the man-made underground water reservoirs, "Tanka" and "Beri" (Tyagi, 2002, 2023b).

Anopheles stephensi is both an indoors resting (endophilic) and feeding (endophagic) species even though it bites outdoors during the warmer summer months due to greater outdoor activity of humans and domestic animals, particularly in the Thar Desert. *Anopheles stephensi* is a night biter, typically arising to bite first in the late evening and then in the early morning hours, in two peaks. In the rural Thar Desert's northwestern areas, and in canalized areas of Gujarat, *An. stephensi* is associated with canal-irrigated and non-irrigated villages all year round, but generally in low densities. In urban areas, *An. stephensi* is found throughout the year but is most abundant in the summer months, coinciding with the peak period of malaria transmission.

5.2.4 *Anopheles culicifacies* Giles, 1901

The most malaria-transmitting (app. 65%) species in India, where maximum investigations have been made, *Anopheles culicifacies* is a complex of five cryptic species spread all over India, and many other neighboring Asian countries or areas such as southern China, Pakistan, southern Afghanistan and Iran, with a western extension into the Arabian Peninsula (Yemen) and Ethiopia. The five isomorphic species cytotaxonomically and molecularly (but more sequences need to be analyzed in order to identify those that show more clear species differences among species) designated as A, B, C, D, and E exhibit remarkable biological differences in their behavior, seasonal prevalence, distribution, vectorial capacity and status of insecticide resistance. Although these species await proper taxonomic description in accordance with the International Commission of Zoological Nomenclature, Tyagi (2020) has attempted to distinguish among them based on egg-morphology. Sibling species E, due to its high anthropophilic and endophilic behavior, is the most important and efficient vector of *P. falciparum* and *P. vivax* in southern India and Sri Lanka, although species A, C, and D, with zoophilic and very low anthropophilic rates, also play minor roles in malaria transmission. Species B, with the widest distribution occurring from Iran and Sri Lanka to Southeast Asia, and the only species of the complex being found in the far eastern areas of southern China (Sichuan), Vietnam, Laos, Cambodia, and northwestern Thailand, is considered to be a poor or non-vector on account of its highly zoophilic behavior.

Anopheles culicifacies complex species are abundant in plains, hilly and mountainous areas in the Indian Himalayan region and in Afghanistan' Kabul. It is a highly dynamic species possibly diverging to evolve with a wide spectrum plasticity of resting, breeding and feeding in different ecotypes, ranging from forested areas with perennial streams to deforested riverine ecosystems and irrigated areas. Larvae of the species can be found in freshwater irrigated canals, stream margins, seepages, borrow pits, hoof marks, rock pools, sandy pools near rice fields, rock quarries, newly dug pits, ponds, domestic wells, tanks, and gutters. Immature stages may also develop saline or brackish water. In India, species A has shown to be more abundant in villages with domestic wells, whereas species B was found in higher densities in villages with streams. Studies have shown that adult biting activity occurs during the first half of the night in cooler months (November–March) and during the second and third quarters of the evening in the warmer months (September–October), whereas others reported peak biting activity occurring around 23:00 h to midnight. Post-feeding behavior of the species showed a higher tendency for resting indoors, mainly in cattle sheds, but outdoor resting has also been reported.

5.2.5 *Anopheles sundaicus* (Rodenwaldt, 1925)

Anopheles sundaicus is a complex of at least four sibling species; A (*An. epiroticus* Linton & Harbach), B and C (found in Sumatra, Java, and Thailand), D, and E immatures of which develop in coastal waters with salinity ranging from low, brackish to sea-water concentrations, and they do not exhibit ecological differences such as freshwater/brackish-water preferences. Populations of the species have also been recorded from inland lakes close to seashore (Chilka Lake in Odisha, India), in association with fresh water, particularly in northeastern India, Car Nicobar Island, peninsular Malaysia, Malaysian Borneo (Miri, Sarawak), northern Sumatra and Java, Indonesia. *Anopheles sundaicus* species A occurs most often along the mainland coastal areas from eastern India to Thailand, southern Vietnam and peninsular Malaysia; *An. sundaicus* species D appears to be restricted to the Nicobar and Andaman Islands of India; and *An. sundaicus* species E is found in Sumatra and Java, Indonesia.

Larvae of *An. sundaicus* generally require sunlit habitats containing pooled stagnant water, algae, and non-invasive vegetation, including filamentous floating algae and aquatic plants which are crucial for the development of the larvae as they provide food (micro-algae and bacteria) and protection against predators. Particularly favorable habitats include ponds, swamps, lagoons, open mangroves, rock pools, and coastal shrimp or fishponds (active or abandoned/poorly maintained impoundments such as in Indonesia), as well as irrigated inland sea-water canals.

Anopheles sundaicus females are mainly anthropophilic and exhibit both endophagic and exophagic feeding habits. Peak biting activity typically occurs between 20:00 h and 03:00 h depending on locality. Blood-engorged females can be found resting inside or outside houses. Varying degrees of indoor and outdoor resting occurs, and some members of the complex have been reported to be predominantly endophilic during the gonotrophic cycle.

Depending on location, species of the complex are considered as either major or secondary malaria vectors. They are regarded as the main vectors of malaria along the coastal areas of India, southern Vietnam, and much of Indonesia where they transmit both *P. falciparum* and *P. vivax*, and are responsible for local outbreaks. The ecological and behavioral plasticity of species of the Sundaicus complex poses difficulties for developing efficient and appropriate vector control strategies.

TABLE 5.3
Structural Differences between Male and Female Mosquitoes

Basis of Comparison	Male Mosquito 	Female Mosquito 
Antennae	Male mosquitoes have bushy fine hairs on their long antennae (proboscis) which help them with sensing their potential mate's wing beats.	Female mosquitoes' antennae are sparce and contain several blood odor sensing receptors that help her target blood sources so that they can land on their host's body accurately and feed off their host.
Body Size	Male mosquitoes are relatively smaller than female mosquitoes.	Female mosquitoes are relatively larger than male mosquitoes.
Buzzing Sound	A male mosquito does not make a buzzing sound or do so at a very low note.	A female mosquito makes an audible whining or buzzing sound.
Emergence	Males generally emerge earlier from pupae than females, so as to be able to prepare for consummation with the females to follow.	Females are next to males in emergence.
Lifespan	A male mosquito has a relatively shorter lifespan than the female mosquito. A male mosquito's average lifespan is between one and three weeks.	A female mosquito has a relatively longer lifespan than the male mosquito. A female mosquito's average lifespan is between two and four weeks.
Human Contact	Male mosquitoes do not possess type of mouthparts which can pierce through skin to feed on blood and, therefore, they typically avoid human contact. Instead, they survive on plant juices etc.	The female mosquito has piercing and sucking type of mouthparts so as to be able to feed on human/animal blood and thus carries disease pathogens such as those of malaria, dengue, Japanese encephalitis and yellow fever, etc.

5.3 DIOECIOUS DIVERSITY

Mosquitoes are bisexual (dioecious) organisms, implying male and female mosquitoes have different reproductive organs. Mosquitoes exhibit a remarkable range of diversity in both sexes. Differences between female and male mosquitoes are strikingly clear – in shape, size and form as well as their functions. Mosquitoes are unisexual and are easily distinguishable from each other morphologically, behaviorally and genetically, as succinctly described in Table 5.3.

5.4 STRUCTURAL DIVERSITY

The very first striking feature that puts both sexes apart is their size; apparently, the female is larger, with a needle-like protrusion called a proboscis, by means of which she bestows her itchy, stinging bite through a set of highly sophisticated engineering marvels of six stylets with varied functionalities to probe, serrate through host's integument, bore deep hypodermally, pierce into a blood vessel, and blood-feed with alacrity.

The female mosquito also has sparsely haired antennae. On the other hand, the smaller mosquito boasts a pair of wider, branched, feathery or bushy antennae, and the kind of mouthparts entirely different from those in its fair sex. Most significantly, male mosquitoes typically have clasping genitalia on the venter of the 8th abdominal segment, with a well-developed aedeagus – the structure

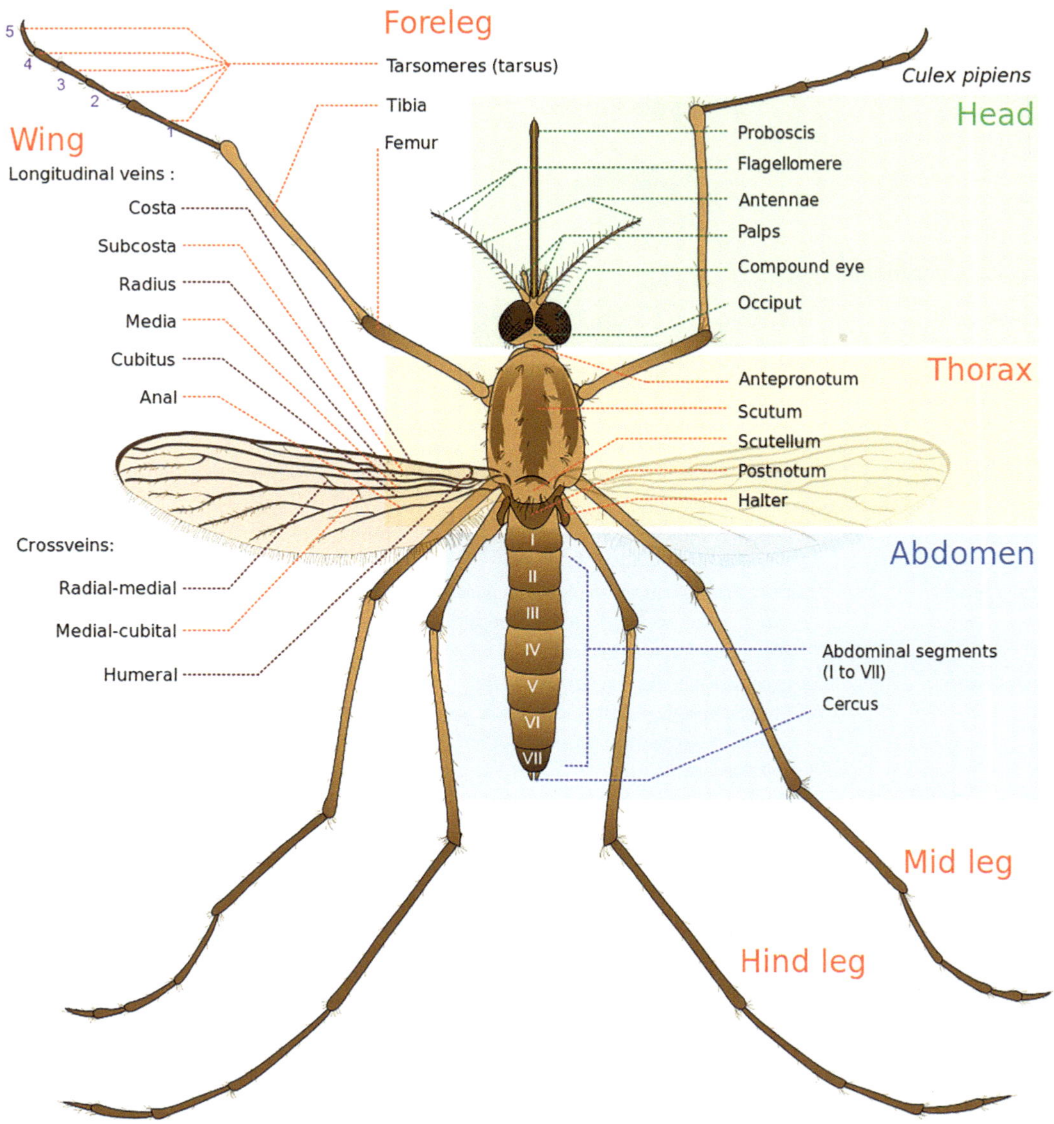

FIGURE 5.2 An illustrated diagram of a culicine female mosquito, *Culex quinquefasciatus*.

loosely analogous to a penis, and through which they secrete sperms from the testes during copulation with a female. These differences in the appearance of reproduction hold significant promise (Figure 5.2).

The anopheline and culicine mosquitoes differ in a number of characteristics. On the head region, however, they differ fundamentally in their mouthparts because while females are blood-sucking (hematophagous), males are essentially surviving on plant secretions such as, for example, nectar, etc. Consequently, males invariably have a pair of highly degenerated mandibles, besides rudimentary maxillae and hypopharynx, among others. They also differ, both, between the two groups and the sexes within the same group of mosquitoes, in the form and shape of their maxillary palps. While in females of culicine mosquitoes, the maxillary palpi are much shorter in length, almost stud-like, and ending close to the head, in males they extend much outwardly beyond the length of the proboscis and are shaped like an angular stick (Figure 5.3).

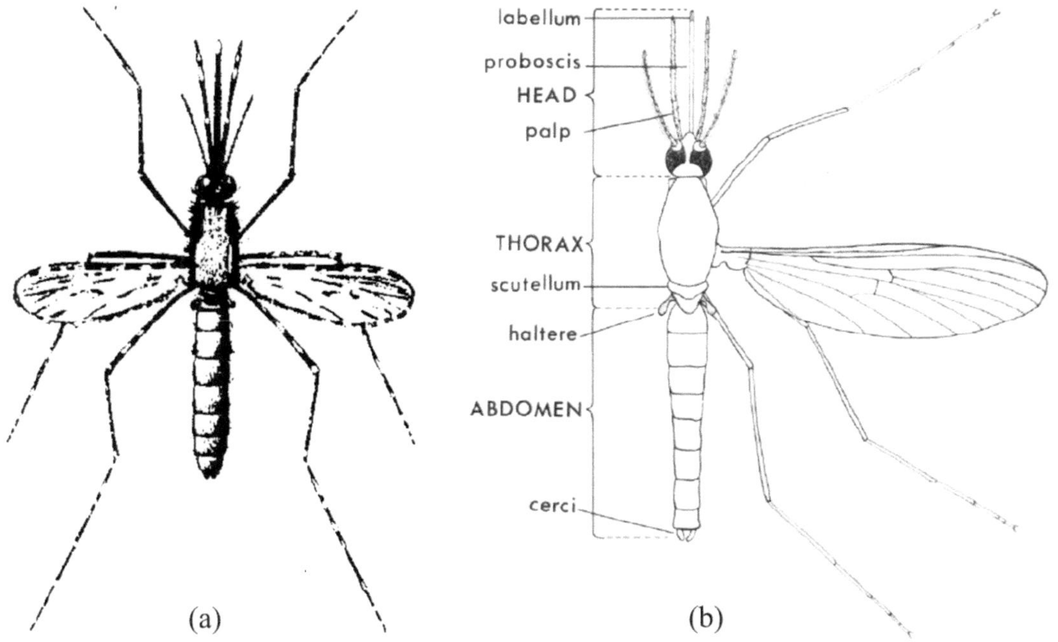

FIGURE 5.3 Anopheline mosquito (dorsal view); (a) unlabeled, (b) labeled, half side of body, and (c) a male mosquito's head, mouthparts, and antennae.

In comparison to culicines, the anopheline palpi in females are curved outwardly more or less similar in length to that of the proboscis; males too have equally or slightly longer palpi which are curved outwardly in a golf-stick-like manner (Figure 5.3).

Males differ from females within both groups of Family Culicidae, i.e., subfamilies Anophelinae and Culicinae, in the form and shape of their antennae. While females essentially have sparce hair on antennae, the males have a markedly bushy pair of antennae (Figure 5.4). The pedicel is basal to the flagellum and appears as a swollen or bulbous segment. Neurosensory cells within the pedicel receive vibratory signals from sensory setae of the flagellum. The scape is the ring-like or cup-like basal segment of the antenna. Below the antennae is the clypeus, which covers the forward projecting portion of the head that gives rise to the paired maxillary palps and the proboscis. The maxillary palps (often referred to simply as the palps), are jointed chemosensory and mechanosensory sensory appendages that flank the proboscis. In most mosquitoes, the palps are shorter in the females than in the males. The proboscis is the conspicuous elongate projecting mouthparts of the adult mosquito. It is composed of a ventral sheath, which holds the styliform (needle-like) elements that pierce host flesh, deliver mosquito saliva and transport blood. At the tip of the proboscis are the labella, two sensory lobes (usually appearing fused) that mosquitoes use to locate host blood vessels.

The thorax, located between the head and the abdomen, bears the legs and wings, and is therefore the locomotory center of the adult mosquito. The major dorsal portion of the thorax is the scutum. The scutum of some mosquitoes is covered in dark and light scales that can form striking patterns. Posterior to the scutum is the scutellum, and posterior to the scutellum is the mesopostnotum. The lateral portion of the thorax is the pleuron. The pleuron has several exoskeletal plates, called sclerites. Two of the larger sclerites are the mesokatepisternum and mesepimeron. The arrangements of setae and scales on the mesokatepisternum and mesepimeron are often used in mosquito identification. The pleuron also bears two large spiracles, openings in the exoskeleton through which the adult mosquito breathes.

FIGURE 5.4 Culicine mosquitoes: Form and shape of antennae and maxillary palps in the male and the female.

Mosquitoes have six legs like other insects. The upper part of the leg is called femur which connects with the thorax through trochanter. Tibia is the middle part of the leg, while tarsus, a 5-segmented structure, is the end part of the leg. The apical tarsomere terminates in a claw.

Mosquitoes technically have four wings, but only the front wings of mosquitoes are used for flying (Figure 5.5). The hind wings, called "halteres," are small and do not resemble true wings at all. The halteres are short and knoblike and used to help maintain balance during flight.

The front wings have long thickenings, called veins, which give the wing rigidity. The veins are covered with scales, which can be dark or light in coloration. There are six major veins, with several subdivisions and cross-veins. The major veins are the costal, subcostal, radial, medial, cubital, and anal veins. The membranous portions of the wing between the veins are called cells, and are named after the vein that they follow, for example, radial cell, costal cell. The apical tip and posterior margins of the wings are bordered with long, narrow setae, called (collectively) the wing fringe.

Adult mosquitoes have six legs, of which the hind legs are the longest. The legs are divided into five segments. The coxa is the basal segment and is followed by the trochanter, the femur, the tibia, and finally the tarsus (plural tarsi). The tarsus is further divided into five subunits, called tarsomeres.

The abdomen, the posterior-most region of the body, is the primary site for digestion, excretion, and reproduction (Figure 5.6). It is divided into ten segments, each composed of a dorsal and ventral plate. The dorsal plates are called tergites, and the ventral plates are called sternites. Tergites and sternites are connected by a membranous exoskeleton that can expand and stretch during feeding.

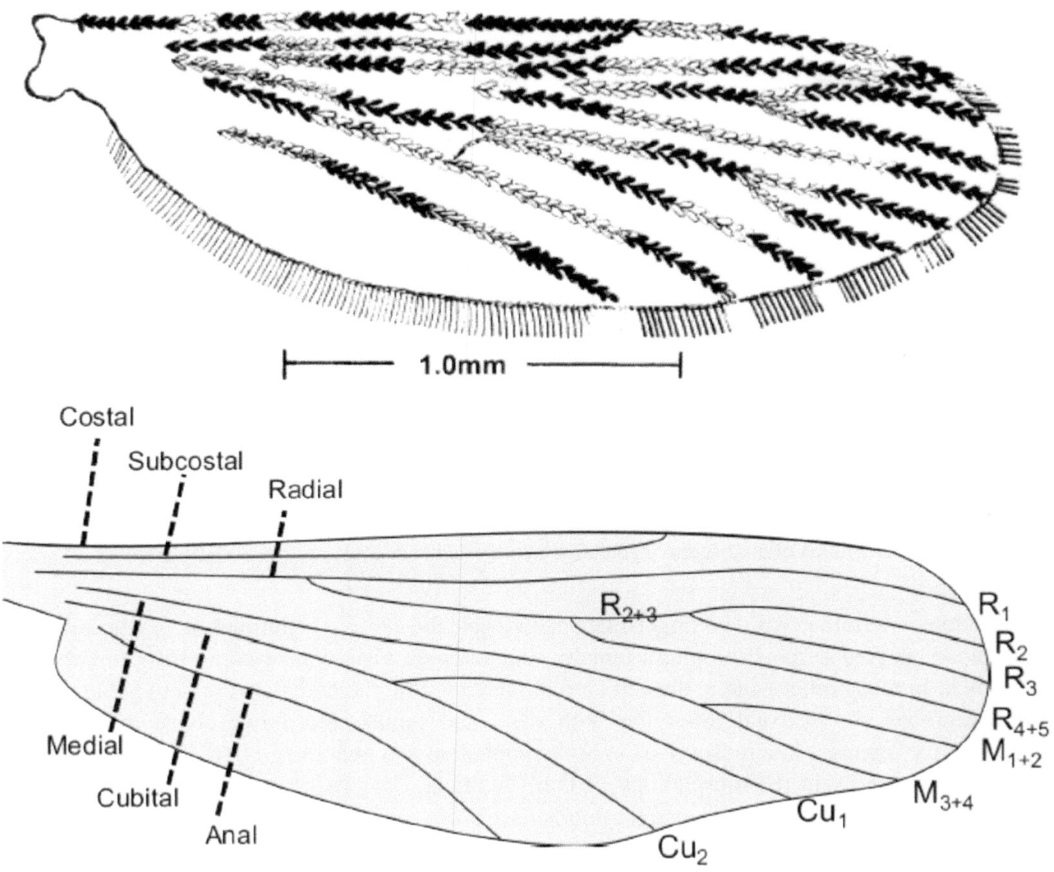

FIGURE 5.5 Diagrams of mosquito's wing, with various taxonomically important structures; (Above) Anopheles mosquito's wing, and (below) Culex mosquito's wing.

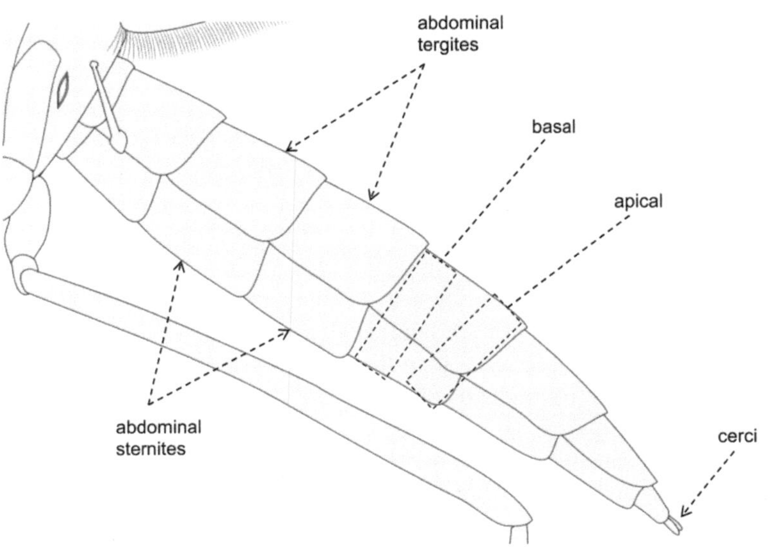

FIGURE 5.6 A diagram of a mosquito's abdomen, with various taxonomically important associated structures.

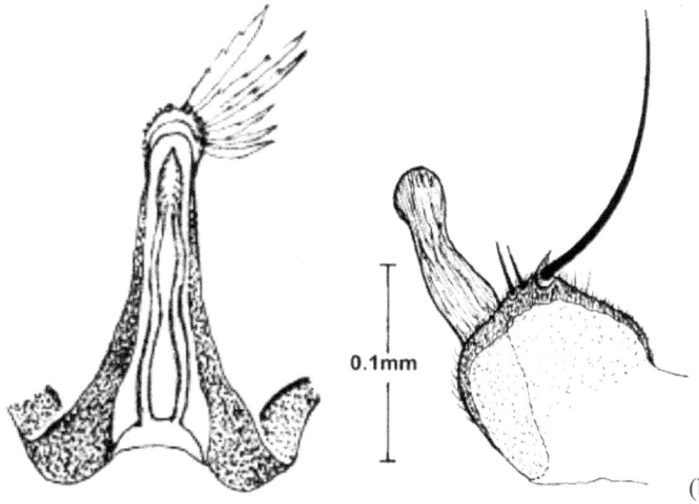

FIGURE 5.7 Diagrams of a mosquito's genital structures and aedeagus; (a) top, and (b) lateral view.

The abdomen terminates in two finger-like appendages, the cerci, which function in egg-laying and copulation. In Aedes and Psorophora females, the cerci are visible, protruding from the tip of the abdomen. In many other genera, the cerci are retracted within the body and are not visible.

Mosquitoes are bisexual, implying both male and female mosquitoes have their different reproductive organs. Male mosquitoes in both anopheline and culicine groups of mosquitoes differ more pronouncedly in the morphology of their genitalia, comprising grabbing structures, paired claspettes, and a penis-like aedeagus (Figure 5.7).

5.5 METAMORPHOSIS AND DEVELOPMENT

Mosquitoes go through a complete metamorphosis to reach adulthood. This means that each stage of the mosquito (from egg to larvae to pupa to adult) looks and acts differently. The egg is laid on a water surface or in a place where water may be after flooding or rain. Eggs hatch once it is in water, and the larvae and pupae live in water their whole life. Male mosquitoes hatch before the females and once they hatch, they wait around the brooding site to mate with the females. Once the female mosquito has mated, she does not need to mate again. Once she gorges on blood, she will return to the brooding site and lay her eggs. The male mosquito does not have blood-sucking mouthparts but instead spends his life gaining nourishment from nectar and plant.

5.6 LIFE CYCLE

Mosquitoes are very slender flies, generally with an elongated piercing proboscis and no ocelli. The palpi is stiff and not pendulous. Legs are long, antennae densely plumose in the males, pilose in females. Wings are fringed with scales along the posterior margin and the veins. The larvae and pupae are aquatic and very active, though only the former feed. The larvae are metapneustic, with an elongated thoracic mass (Figure 5.8).

Mosquitoes have the capability to bite by day or by night, depending on species. They breed in all kinds of water fresh and salt, flowing or stagnant, foul or potable, water in tin cans, tree holes (phytotelmata) and salt marshes. Large coastal areas are occasionally inhabitable because of salt marsh mosquitoes. Mosquitoes may render recreation centers unusable and make normal Living in and around homes extremely unbearable and vexing. Ironically, real estate values may suffer where mosquitoes are present, and losses caused by reduced industrial efficiency can be considerable

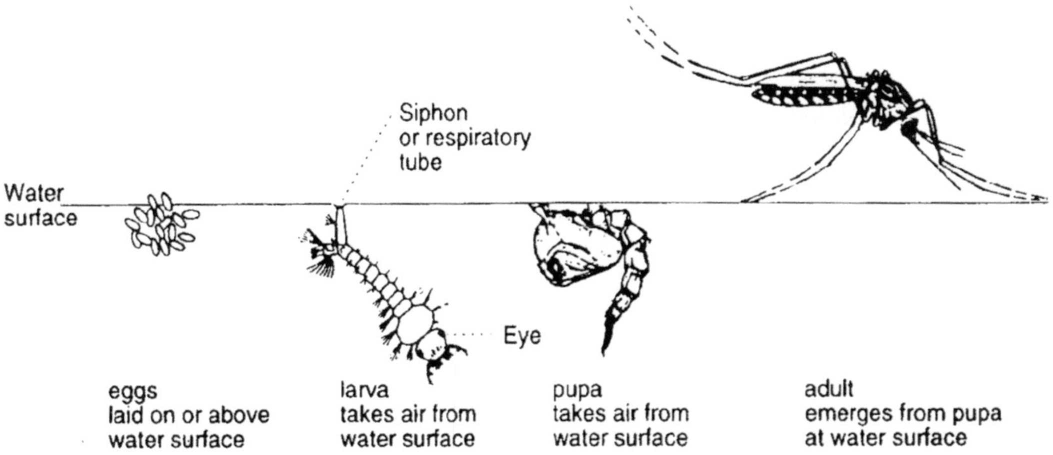

FIGURE 5.8 The life cycle of a mosquito.

because of the mosquito menace. As a principle, this is only the female of the vector mosquito which feeds on the blood of human or animal host, although females of some species are known to supplement their meals with plant sugars, too. Female mosquitoes may feed on the blood of a wide variety of mammals, birds, reptiles and amphibians. Mosquitoes are generally host-specific for their blood-feeding requirements but they may quickly shift in case of any exigency.

Mosquitoes have their body clearly divided in three parts: head, thorax and abdomen. The head bears a pair of large compound eyes, a pair of antennae, and mouthparts which are of the piercing and sucking type, suited primarily for blood-feeding in females, while in males they develop to suck sap from the plants. Antennae also differ, being shorter and sparsely haired in females, while in males they are thickly haired and bushy. There is a pair of fully developed forewings, while the hind wings are reduced to rudimentary halteres or balancers. The males are equipped with external genitalia, called hypopygium. Following mating, males pass spermatozoa to the females where they are stored in spermatheca. Eggs are fertilized as they pass out of the ovary of the female. Females generally lay eggs in mildly moving or stagnant fresh (e.g., *Anopheles*) or polluted water bodies (e.g., *Culex*). Eggs are laid singly directly on water surface (Anopheles), in batches/groups approximated in a boat raft-like shape (*Culex, Culiseta, Mansonia* and *Uranotaenia*) or along the moistened surfaces, like soil etc., close to water or on damp soil that will be flooded by water (*Aedes* and *Ochlerotatus*). Eggs of the genus *Toxorhynchites* are also laid singly on the water surface. Most eggs hatch into larvae within 48 hours, others might withstand subzero winters before hatching. Some mosquitoes (*e.g., Aedes, Toxorhynchites*) are specialized to lay eggs in plant materials or tree holes (phytotelmata). Some species may survive in the egg stage for 1–2 years before flooding occurs. *Aedes*, *Orthopodomyia* and *Psorophora* eggs are commonly laid in this manner. Water is a necessary part of their habitat. In general, the development of mosquitoes is essentially the same.

There are four stages in the life cycle of mosquitoes: egg, larva, pupa, and adult. The development is holometabolous. The immature stages are aquatic in nature. The eggs are usually white when first deposited, becoming dark within an hour or two. Anopheline eggs are oval-shaped and float horizontally. They are usually pointed on one end and have a pair of lateral floats. The larvae of all mosquitoes get their food from the water where they live, but to respire they have to come to the surface. Culicinae larvae have siphon tubes for breathing and hang upside down from the water surface. Anopheles larvae do not have a siphon and lie parallel to the water surface to get a supply of oxygen through a breathing opening or spiracle. Larvae of the genus *Mansonia* and *Coquillettidia* are adapted to take oxygen under water from the air-conducting parts of plants. The larval developmental period includes four instars (i.e., during growth, the larva molts, or sheds its skin four times, and the stages between molts are called instars), which usually take 5–15 days to complete. At the end of the larval development the pupa is formed by discarding the skin of the fourth instar larva. Only larvae feed while pupae do not feed. The larvae feed on microorganisms and organic matter in the water. Larva of *Toxorhynchites* species feed on larvae of other mosquitoes sharing the habitat. *Toxorhynchites* larvae are much larger than other mosquito larvae. The larval body is characteristically divisible in three parts, but the pupa has only two parts; the fused cephalothorax and the curled and segmented abdomen. Pupae, commonly called tumblers, too, occur in water and are actively motile. When disturbed, it dives in a jerking, tumbling motion toward protection and then floats back to the surface. For respiring all pupae, except those of *Mansonia* and *Coquillettidia*, move (tumble) with a flip of their tails (caudal fins) to the surface where with the help of a pair of trumpets atop head region the oxygen is taken in. The pupal stage lasts for one day to several days. The metamorphosis of the mosquito into an adult is completed within the pupal case. When the pupal growth is complete, the pupal skin or exuvium breaks, the adult works its way out, crawls on to the surface of the water for a short time to allow itself to dry and all its body parts to harden. The wings have to spread out and dry properly before the newly emerged adult can fly.

Among all species, the larvae of Aedes species move characteristically emulating a serpentine-like movement (Bar and Andrew, 2013a, 2013b).

Though not a general rule, blood-feeding and mating does not occur for a couple of days after the adult emerges. The adult mosquito in all genera with vector species is a small fragile insect with a slender abdomen, one pair of narrow and delicate wings, covered or thinly so with scales, and three pairs of long slender legs. As the case may be, species may greatly vary in their size varying from slightly over 4 mm to about 12 mm. The three body parts (i.e., head, thorax, and abdomen) are clearly visible.

The egg, larva and pupa stages depend on temperature and species characteristics to determine how long they take for development. For instance, *Culex tarsalis*, a common mosquito in USA might go through its life cycle in 14 days at 70 °F and take 10 days at 80 °F. On the other hand, some species have naturally adapted to go through their entire life cycle in as little as four days or as long as one month (Figure 5.8).

Only female mosquitoes require a blood meal (protein) and bite animals warm or cold blooded. Stimuli that influence biting (blood-feeding) include a combination of carbon dioxide, temperature, moisture, smell, color, and movement. Most mosquitoes are nocturnal in their feeding behavior (e.g., Anopheles), but some are diurnal or crepuscular (e.g., Aedes). When the female finds a suitable host, such as the man, it introduces the piercing mouthparts through skin deep into body and penetrates a capillary, to suck the blood directly from the capillary or from the pool of the blood oozed outside the punctured capillary. Females of certain mosquitoes are single-time feeder while many species tend to feed in shift or multiple occasions on a given night or daytime. The number of eggs formed and developed in the female's ovary is directly dependent on the quantum and nature of the supply of blood meal received by the female mosquito. Depending on a particular species, an engorged female may like to rest inside houses or prefer outside environs. This characteristic of the vector species has for long been exploited for their control through indoor spraying of residual insecticides.

Mosquitoes are the sole vectors of malaria, yellow fever, and dengue, and they participate significantly in transmitting lymphatic filariasis and the encephalitides of various different kinds. Besides, their bites may itch for days, causing much loss of sleep and tranquility, and serious nervous irritation and restlessness. Their saliva contains proteins alien to the human body and repeated bites result in sensitivity to it. Mosquitoes often cause serious economic losses limiting outdoor movements. Male mosquitoes do not bite, but feed on the nectar of flowers or other suitable sugar source. Acquiring a blood meal, *i.e.*, protein, is essential for egg production in females, but mostly both male and female mosquitoes are nectar feeders. Of those female mosquitoes capable of blood-feeding, human blood meals are first or second choices. Horses, cattle, smaller mammals and/ or birds are preferred for blood-feeding. Female *Toxorhynchites* mosquitoes cannot actually obtain a blood meal and are restricted to a nectar diet.

5.7 HEMATOPHAGOUS BEHAVIOR OF FEMALE MOSQUITOES

Because some mosquitoes are transmitters of deadly human and animal diseases, blood-feeding is the most important characteristic of their life cycle, which offers a unique amalgamation of various sophisticated processes. Adult female culicids feed according to their state in the gonotrophic cycle. Following the oviposition, the mosquito will seek a blood meal. Blood ingestion begins and continues until stretch receptors in the abdomen signal that the female mosquito cannot hold anymore. This is referred to as distention-induced host-seeking inhibition. Blood is then digested, and egg maturation begins. Host-seeking is then further inhibited in oocyte-induced host-seeking inhibition. Following the oviposition of this batch of eggs, the cycle begins again. Variations in the cycle occur in mosquito species, which blood-feed more than once during egg maturation. Vertebrate blood contains essential proteins necessary for eggs to mature. Egg production is dependent upon amino acids and does not require the presence of other constituents in the blood. It has been found that female *Aedes aegypti* fed on an artificial diet high in protein developed equal amounts of eggs to those fed on bovine blood (Kogan, 1990). Feeding is also dependent upon the circadian rhythms of the specific mosquito species. Circadian rhythms determine the time of day or night that a mosquito will feed. This is of great importance in determining a means of mosquito control; for example, bed netting can be used to prevent blood-feeding while prospective hosts sleep. Mosquito behavior is also dictated by a species' tendency to blood-feed on humans (anthropophagic) or animals (zoophilic) and whether the mosquito prefers to feed indoors (endophagic) or outdoors (exophagic). *Anopheles gambiae*, a vector of malaria in Africa and *An. culicifacies*, a vector of malaria in India, feed on sleeping hosts inside their homes. Host-seeking behavior is dependent upon chemical and visual cues specific to a desirable host, as well as their internal species-specific cues.

The entire process of mosquito's blood-feeding is a highly intricate activity which is accomplished by a melange of coordinated functions by various different organelles as well as chemical cues. This complexity of their mechanical and chemical systems for finding human host, finding a commensurate site to feed once they land, and then manipulating the body chemistry while they draw their nourishment are all an example of remarkable feat of evolution. To understand the whole *modus operandi* it is inevitable to first understand the feeding apparatus of a mosquito.

The feeding mechanisms in mosquitoes revolve around two main structures, (i) the food canal, and (ii) the salivary duct. Both the food canal and salivary duct are separate, independent and parallelly running structure, never meeting on course. The mosquito squirts saliva before initiating feeding through one passage (and never the remnants of its last blood meal, as one might wrongly infer!) and draws blood up another passageway. As a consequence, the food canal is not flushed out completely like a drug-injecting syringe does and the blood flow is always unidirectional (Figure 5.9).

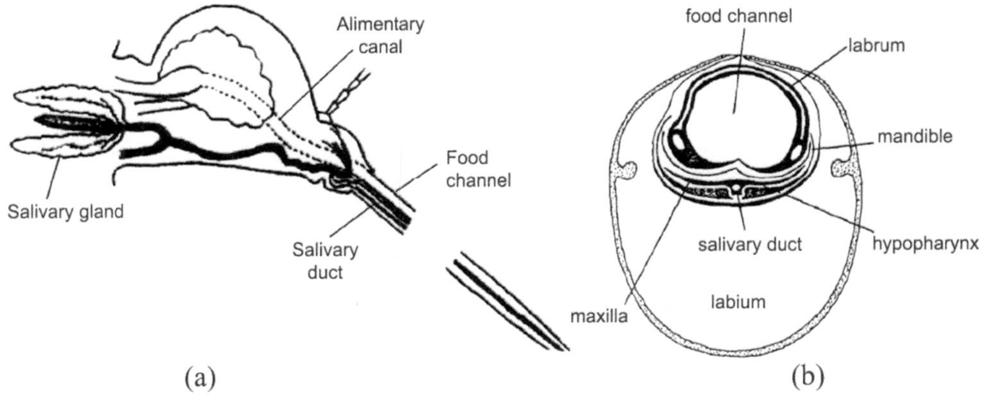

FIGURE 5.9 (a) Anatomy of a mosquito showing separate passageways of each, the food channel and the salivary duct; and (b) structures, including the food channel and the salivary duct.

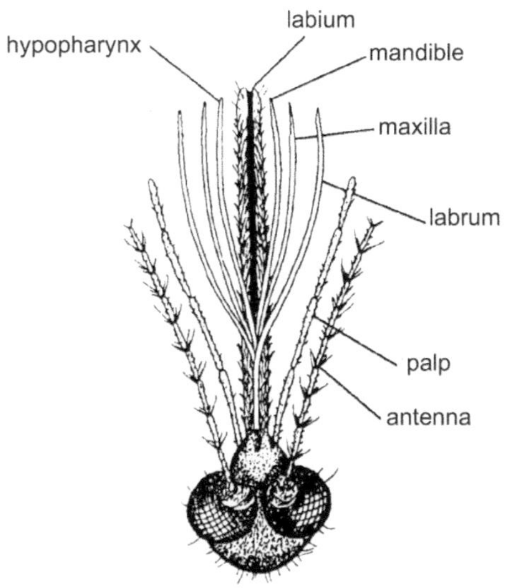

FIGURE 5.10 Piercing-sucking mouthparts of a mosquito (dorsal view).

Mosquito's feeding apparatus then begins to work. It is a complex seven-piece set of devices for doing a variety of actions. One of those parts is a tube that allows the mosquito to secrete its saliva into the body being bitten. That saliva, remarkable in its action, contains a set of chemicals that act like an anesthetic as well as an anticoagulant. On one hand, the mosquito causes numb the biting spot on the skin so that hosts do not sense presence of the mosquito probing around on skin, and, on the other, make the blood flowing intensely so as to enable her continue to suck in a gluttonous manner! Meanwhile, the other parts begin an intricate pattern of coordinated movements that let her hack into man's flesh. The mandibles, on the edges, are serrated on the tip, like tiny steak knives, to carve their way in. Just below the surface, the whole package of savagely armored parts bends and takes a 90° turn, to begin the attack (Figure 5.10).

That set of needle-like parts begins to randomly poke away, anywhere between 12 and 18 times in just several seconds. It is not actually looking for a venule or arteriole just yet. It's indiscriminately poking around to try and poke into anything in the area that may bleed. That tube in the middle of their mouth assembly that does the spitting is actually two straws in one, one inside the other. As they continue to spit out their anesthetic, the other wider tube in the straw is sucking in samples of fluid to see if it has hit pay dirt (Figure 5.11).

Mosquitoes are the only known creatures that can spit and suck at the same time. If they don't find blood, they draw their probing mouth parts back a little, not all the way out, and dig into another spot. This process can last up to two minutes, if somehow the human body remains unaware that the mosquitoes are there.

Selecting a biting site is made on a "trial and error" basis, and not without meticulous planning behind their host-seeking strategy. For zoophilic mosquitoes, host accessibility is dependent upon physical barriers to blood-feeding. Many mosquito species prefer to feed on relatively hairless or featherless areas of a host. These areas are conveniently more vascularized to keep the host warm. Chipmunks and squirrels are more likely to be bitten on the ear (versus the back or feet). As for relatively hairless beings, it seems unlikely that mosquitoes would prefer any one biting site over another. This is not a universal truth since it is proved otherwise. de Jong and Knols (1996) used a test subject inside a bed net, wearing only close-fitting underwear, to determine body regions most likely to be bitten by specific mosquito species. Mosquitoes were released into the bed net, and were recorded if they bit within three minutes. *Aedes aegypti* was found to prefer the shoulder and head region, while *Anopheles gambiae* preferred the feet. *Anopheles gambiae* has also been proven to be

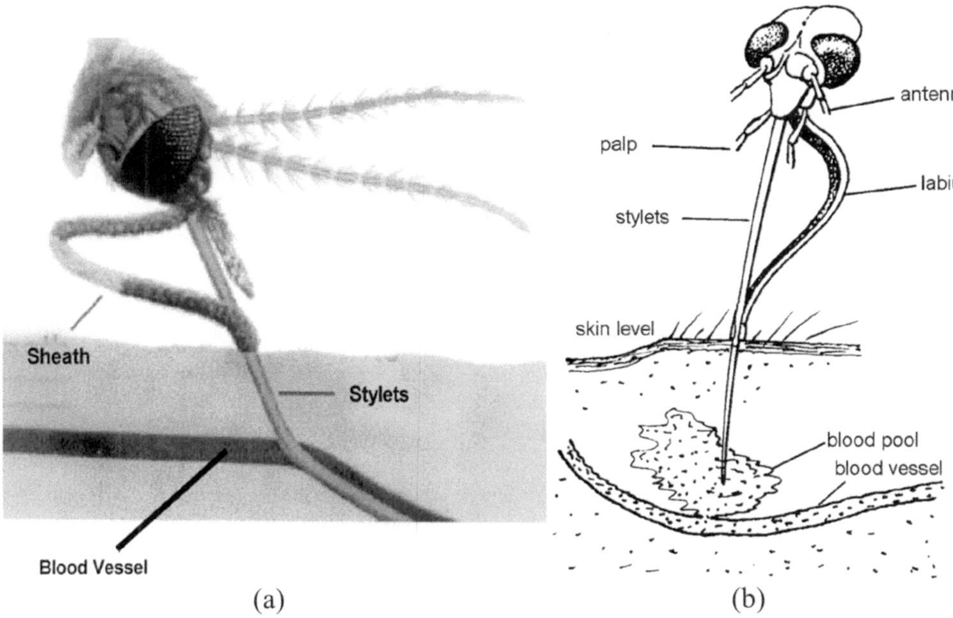

FIGURE 5.11 Head of female mosquito at the commencement of blood-feeding (a) directly from a venule and (b) from a pool of blood formed as a result of strategic puncturing of the blood vessel.

more attracted to Limburger cheese than human breath in wind tunnel experiments (de Jong and Knols, 1996). This suggests that bacterial stimuli may also stimulate host-seeking behavior, and is an interesting possible clue to better understanding host-seeking behavior in mosquitoes. Selection of a biting site can also be further dictated by the presence of a parasite. Host-seeking behavior may also be enhanced by the presence of a parasite in the vector's body. To increase their dispersal, pathogens within a mosquito vector may alter the host's behavior. It is clear that by altering vector behavior to increase hematophagy, a pathogen will gain increased dispersal. Pathogens can affect host attractiveness, increase probing behavior, and reduce engorgement time (Edman et al., 1974), as well as alter the vector's circadian rhythms. Filarial worms, for example, develop according to the activity of the vector and are successfully dispersed. Thus, in mainland India, where the night-biting *Culex quinquefasciatus* is the vector, the microfilarae of *Wuchereria bancrofti* are freely flowing in the man's peripheral blood vascular system in the late evening/early night hours (nocturnal) and in the Andaman & Nicobar ils., where the day-biting *Ochlerotatus (Aedes) niveus* is the vector, the same appears in peripheral blood during the day time (diurnal; subperiodic).

Many factors work in unison to inveigle a hematophagous mosquito toward its host. After orienting toward a host according to chemical cues and using upwind optomotor anemotaxis, a mosquito has to have cues to land. Arresting stimuli, such as sweat, humidity, heat, color, and movement are possible factors which lead to the termination of host-seeking, and the beginning of probing behavior. However, as mentioned above, host-seeking behavior may be prevented by certain stages of the gonotrophic cycle. Some researchers have indicated that the host-seeking inhibitory responses of *Ae. aegypti* result from the neuropeptide known as Head Peptide. Increased amounts of this neuropeptide then act to desensitize sensory receptors to inhibit host-seeking behavior. It has been further determined that nutritional and mating states, and other hormones released during egg maturation control the inhibition of host-seeking behavior. There is a great deal of research yet to be done on hormonal control of host-seeking behavior as understanding of these mechanisms can lead to genetic manipulation for vector control. If however a mosquito does taste blood, it turns on more powerful pumps in its mouth, increasing the suction on its "drinking" straw. Then she draws on the tiny cloud of blood that has formed under the skin. The nearest venule or arteriole, even if it's not the one already wounded, is pulled over to the needle-like tip of the whole apparatus, where it's impaled. That opens up the hole through which the mosquito starts to drink.

Generally, there are two types of blood feeders: vessel feeders and pool feeders. Vessel feeders take blood directly from venules or small veins, while pool feeders take blood from a blood pool resulting from the laceration of blood vessels (James and Harwood, 1969) (Figure 5.11b). The process of a mosquito questing for a blood meal involves a complex, interconnected cascade of behaviors, each probably having its own cadence of cues, be they visual, thermal, or olfactory. The complexity of these questing behaviors may account for the bewildering variations in trapping efficiency noted for certain species of mosquitoes at different times. The finely toothed maxillae of the fascicle begin sawing into the tissue of the skin with fine back-and-forth movements. A fascicle is "a bundle of muscle or tendon fibers." The fascicle is guided into the skin between the labella. As it goes into the skin the labium folds back like a hairpin and the mosquito shifts its legs closer to its body. When about half of the length of the fascicle has been inserted into the skin, the mosquito begins to draw blood. After the mosquito's abdomen is filled, she straightens her front legs to quickly withdraw the fascicle. The fascicle springs upward and forward out of the wound and is fitted back into the deep groove in the labium. Figure 5.11a shows the stylet bundles of mosquitoes piercing, while their sheaths [labium] bend and the stylets come out of the grooves. A stylet is "a long thin pointed instrument."

To better understand the complexity of what is involved in the act of blood-feeding, we need to know something about the anatomy of the mosquito's proboscis and alimentary canal. The proboscis projects directly in front of the insect. It consists of a large, scaly outer lower lip called the labium, which terminates at the tip in two hairy lobes called the labella. The labium forms a deep trough in which is concealed a smaller bundle of long tapering pale yellow feeding stylets that are collectively called the fascicle. The largest stylet is the sharply pointed labrum, which forms an inverted gutter up which the blood is drawn. At the side of the labrum are two long, thin mandibles. Under them are two much larger needle-like maxillae with fine saw-toothed tips. Under them in turn is a flattened stylet, the hypopharynx, down the center of which runs a single salivary canal. The fascicle acts as both a piercing mechanism and a food canal. Then the mosquito is poised for the moment of piercing and it has all six legs on the surface of the victim's skin, its hairy labella touching the skin and its maxillary palps raised. It is believed that after the mosquito spreads its labella the finely toothed maxillae of the fascicle literally saw their way into the tissue of the skin with extremely rapid back-and-forth movements. Next, the large labium is folded back like a hairpin. The skin is actually punctured without any obvious thrusting of the mosquito's body. About half the length of the fascicle is inserted into the skin. As it is inserted, the mosquito gradually shifts its legs closer to its body. It takes an average female mosquito about 50 seconds to insert her fascicle into the human skin, and unless she is interrupted (perhaps by being swatted by the victim) she usually sucks blood for about two and a half minutes. When the female mosquito has drunk her fill, she withdraws the fascicle rather quickly (in about five seconds). The procedure is to straighten her front legs, leaning slightly backward, and moving the labium from side to side. Once the fascicle has been removed from the skin, it springs upward and forward and then returns to the deep groove of the labium (Figure 5.12).

The feeding behavior and habits of mosquitoes vary widely from one species to the next. *Aedes vexans*, the most common plague of Americans, both inside country and city, is different from *Aedes excrucians* of northern wooded areas, as well as from *Aedes tormentor* of the southern woods; *Culex pipiens*, the common house mosquito of the tropics, *Coquillettidia perturbans*, an inland swamp mosquito; *Toxorhynchitis rutilus* mosquitoes, which eat the larvae of *Aedes aegypti*, the yellow fever mosquito, the *Anopheles*, and so on down through more than 3500 different species. To complicate matters even more, the species of mosquitoes that predominate vary widely from area to area, so that the type that hovers around a city bedroom or other parts of the home is likely to be quite different from the one that follows people around on the golf course or in the park.

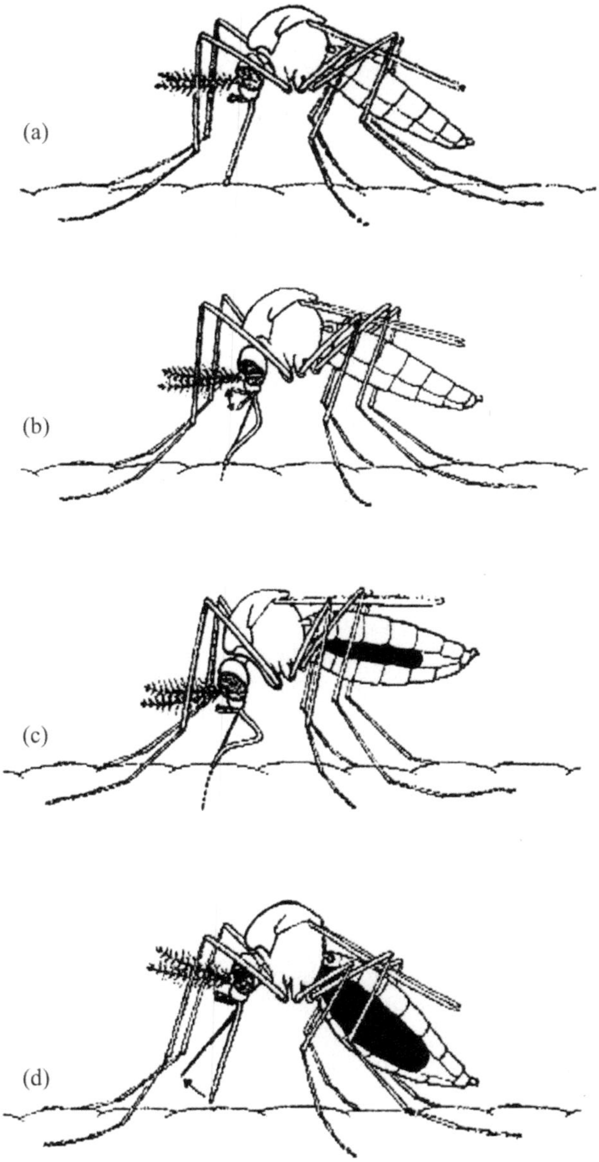

FIGURE 5.12 Feeding mechanism of a mosquito: (a) a female mosquito bends and settles to probe; (b) it forces its mouthparts through the skin; (c) the mouthparts are guided deeper to suck blood with the support of labial sheath sticking outside host's body; and (d) after engorgement.

There are mosquitoes that rest during the day and emerge in the dim light of dawn and evening. Some feed at night, others during bright daylight; still others, like the vector of La Crosse Encephalitis, are most active in late afternoon. After feeding on blood, mosquitoes lay eggs in thousands of carefully selected places: tree holes, discarded tires, decorative fountains, flowerpot trays, elephant footprints, pitcher plants, empty cans, forest ponds, salt marshes, low spots in the yard; anywhere that water is likely to accumulate in the near future. It is said that some species of *Aedes* lay their eggs on dry ground that will be subject to future flooding, and the eggs may survive for as long as two years without water. The whole process from egg to blood-thirsty adult is accomplished in about ten days, and four to five days after her last blood meal, the female is ready to feed and lay again (about 100 at a time).

At the juncture of inserting mouthparts in a host's body a more highly evolved chemistry comes into play. Mosquitoes inject chemicals that act as vasodilators, expanding the blood vessel to increase the flow. They also inject anti-clotting chemicals, to keep the tiny wound from sealing up before they have had its fill. Mixed in with these chemicals are parasites that the mosquitoes had earlier picked up while feeding on the host. The parasites are lodged in the mosquitoes' salivary glands, waiting for the mosquitoes to bite on their next victim. When the mosquitoes squirt their chemical payload into a human body, out come the parasites. This is how mosquitoes spread diseases.

Mosquitoes feed from a wide range of different host animals, including mammals, birds, reptiles, amphibians, fish, and even insects, arachnids, and annelids. Blood-feeding is a risky venture for insects because vertebrates are very large, attempting to blood-feed could be lethal. However, the nutritional materials derived from blood can support present and future life for the mosquito, and all this fascinating and yet intricate activity is accomplished by the saliva present in the salivary glands of mosquitoes (Ribeiro, 1995). The saliva of mosquitoes contains a variety of pharmacologically active substances that counteract the normal hemostatic response to injury in vertebrate hosts. Bioactive salivary compounds exist in the salivary glands of mosquito vectors. These compounds have anticoagulant, antihemostatic, and immunomodulatory properties that facilitate blood-feeding. They can suppress the host's innate and specific acquired immune defenses. The fact that mosquito saliva can enhance pathogen transmission leads to the possibility that vaccinating the host against the components of vector saliva will inhibit pathogen transmission. Mosquitoes produce bioactive molecules in their salivary glands that are inoculated into the host's skin via saliva during feeding. The salivary bioactive molecules have antihemostatic and immunomodulatory functions. Such functions allow mosquitoes to live with and contend with the host immune system. Vasodilators in the saliva of *Aedes aegypti* are in fact tachykinins, which are peptides that have diverse pharmacological actions in the central nervous system, cardiovascular, and glandular tissues (Ribeiro, 1992). This vector secretes an active enzyme called apyrase that inhibits ADP-dependent platelet aggregation. The amino acid sequences obtained match the conceptual translation product of a cDNA clone isolated from an adult female salivary gland library. Sequence comparisons indicate similarities with a ubiquitous family of 5'-nucleotidases. The mosquito protein differs from other members of the family by lacking a carboxyl-terminal hydrophobic domain. The apparent conversion of a gene encoding an enzyme involved in a common metabolic event at the cellular level to a gene involved in the antihemostatic response of mosquitoes illustrates one way this particular insect has adapted to the challenges of blood-feeding. Mosquito anticoagulants seem to possess different potencies, too. For instance, the filariasis vector, *Culex quinquefasciatus*, exhibited an anticoagulant activity twice as high as that of *Aedes aegypti*. New World *Culex* mosquitoes are commonly thought of as bird feeders that have expanded their host range to include humans and other vertebrates. The observed differences between the two mosquito species may reflect the fact that bird hemostasis relies more heavily on coagulation than does mammalian hemostasis.

Bites of some mosquitoes itch sorely, e.g., *Aedes*, *Armigeres*, and *Mansonia* species, etc. A mosquito's bite often leaves behind an imprint, evanescent though, of a slowly developing itching wheal. However, if it is a first-ever mosquito bite, the roundish wheal fails to appear. The immune system in the human body begins thereafter, and develops antibodies to recognize and flag the foreign chemicals. But after just one dose of mosquito spit, the familiar itchy bump left behind by subsequent bites is caused by an allergic reaction to all those chemicals the mosquitoes are squirting into the body. The immune system releases histamines, chemicals that make the cells in the lining of human blood vessels spread apart. That lets fluid leak out into the skin. That fluid carries components of man's immune system that attack and destroy the foreign chemicals from the mosquito. The extra fluid from the leaking blood vessels causes a bump, which irritates other receptors in the skin, leading to itching. Children get bigger bumps from mosquito bites than adults because their immature immune systems respond with larger amounts of histamines. More leaking blood vessels equal bigger bumps, which mean more itching and scratching. However, following repeated mosquito biting, the human immune system also become stronger and exhibits a different reaction to mosquito saliva instead of triggering histamines, another part of the immune system grabs onto the foreign chemicals and drags them away to macrophages, which destroy the chemicals in the mosquito's saliva.

Among mosquitoes, it has been established that only the females desire and extract blood. Male mosquitoes drink only sugary fluids such as flower nectar. Both in the wild and in the laboratory, mosquitoes will visit certain flowers and will feed on fruit placed in their cage. Since they (*Toxorhynchites*) vigorously probe the flowers of some plants, and can distinguish between different types of sugars, they play a certain role in the pollination activity of certain plants. The hematophagous females will also drink sugary fluids, but when hungry, if females are given a choice between sugar water and blood, they will always choose blood. If males are offered the same choice, they will always drink the sugar water as their mouthparts are not suited (or designed) for blood-sucking. Since male mosquitoes do not suck blood, they also do not transmit diseases. Like the males of many other insect families, they are important for just one reason – mating – and then they become superfluous. The female usually needs to mate just once in her life, although she invariably blood feeds following each oviposition which may, under favorable conditions, go up to as many as six or more in her lifetime. She stores sperm in her body and fertilizes her eggs at the moment when she lays them.

Hematophagous mosquitoes do not really sting, in the sense of a hypodermic needle piercing the skin. Instead, they saw their way through, using four cutting stylets kept in a protective sheath along with a duct that carries anticoagulant into the wound and a tube that carries blood out. Mosquitoes fly into the wind, picking up scents that lead them to their victims. Just about everything attracts them, including the carbon dioxide we exhale. Mosquitoes will even suck the blood of birds, frogs, turtles, snakes, and just about every warm-blooded animal. ln fact, some mosquitoes prefer their sources of blood to be other than humans. There are many species of mosquitoes that prefer the blood of birds or other animals (e.g., many culicines). In fact, some species have been observed to feed on numerous mammalian groups, on a variety of reptilian species, and even on other kinds of insects. Anthropophilic mosquitoes are attracted to certain ranges of temperature and humidity, and to carbon dioxide from the exhalation of humans. They seem to prefer dark-colored objects to light. Certain chemical smells may also come into play; for example, components of blood and sweat such as hemoglobin and amino acids. The best explanation of what brings the mosquito to its victim is that she simply flies upwind until she comes upon a potential victim exuding a "host beam" of warm, moist air laden with carbon dioxide. However, this piece of information is far from complete and a lot more research is needed to explore the facts. The mosquito proboscis (long tubular mouth parts used for feeding) consists of six different shafts. Four are cutting and piercing tools; a fifth transports blood from the host; and the sixth transports saliva, thought to act as an anticoagulant for the blood going in the other direction. The saliva also facilitates transportation of the organisms of malaria, yellow fever, dengue, and most of the other diseases for which mosquitoes are notoriously remembered. When a mosquito punctures ("bites") into the flesh, one usually feels an allergic reaction to the saliva, which causes the swelling and the itch. The fact that this reaction is allergic helps explain why some people suffer more than others when their skins are invaded. Some mosquito species are nocturnal, diurnal, or crepuscular (active at dawn and twilight). They also differ in their preferences for altitudes.

5.8 REPRODUCTIVE BEHAVIOR

Mosquitoes exhibit holometabolous metamorphosis and therefore have a life cycle consisting of egg, larvae, pupa, and adult. Ovary ecdysteroidogenic hormone activates egg maturation in female mosquitoes (Figure 5.13).

Adult female culicids feed according to their state in the gonotrophic cycle, and lay in a variety of habitats depending on the species. Following the oviposition, the mosquito will seek a blood meal. Blood ingestion begins and continues until stretch receptors in the abdomen signal that the female mosquito cannot hold anymore. This is referred to as "distention-induced host-seeking inhibition." Blood is then digested and egg maturation begins. Host-seeking is then further inhibited in

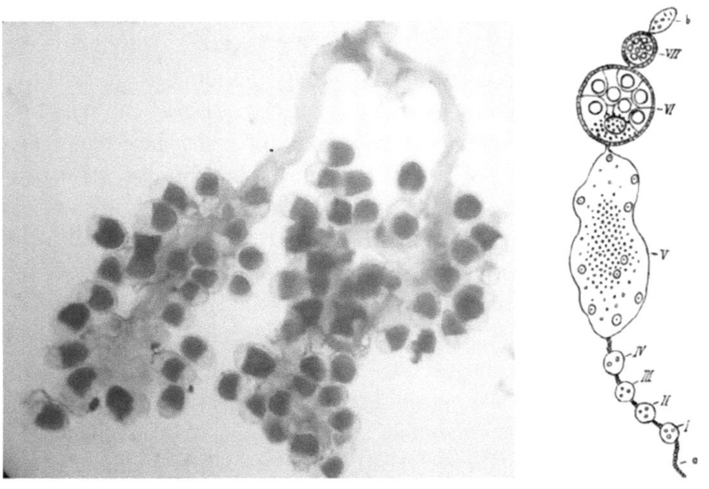

(a) (b)

FIGURE 5.13 (a) Ovaries with developing ova in *Anopheles sundaicus*, a major malaria vector in Andaman and Nicobar islands (India) (*Source*: Vidhya Thiruvoth), and (b) Follicular tube with follicles at fifth gonotrophic cycle of anopheline mosquito (I-IV – pigmented dilatations left by previous four follicles; V – degenerated fifth follicle; VI – sixth follicle about to develop; VII – undifferentiated seventh follicle; a – terminal stalk; b – germinal zone). (Adapted from Beklemishev et al., 1959).

"oocyte-induced host-seeking inhibition". Following the oviposition of this batch of eggs, the cycle begins again. Variations in the cycle occur in mosquito species that blood-feed more than once during egg maturation. Vertebrate blood contains essential proteins necessary for eggs to mature. Egg production is dependent upon amino acids and does not require the presence of other constituents in the blood. It has been found that female *Aedes aegypti* fed on an artificial diet high in protein developed equal amounts of eggs to those fed on bovine blood (Kogan, 1990). Female mosquitoes require the amino acids in blood proteins to mature their eggs. It is during the process of taking blood that the mosquito can transmit diseases such as malaria and dengue fever, which are re-emerging as very significant health problems. It is interesting to know that the female mosquito is not able to digest the proteins at the moment the blood meal is ingested because she has no active trypsin in her gut. Trypsin synthesis requires a series of complex mechanisms following blood-feeding. In the first four–six hours following a blood meal, the female gut produces an insulin-like special type of trypsin called "early trypsin." Although only small amounts of "early trypsin" are made, the activity of "early trypsin" plays a vital role in the regulation of digestive enzyme synthesis by the gut. Somehow "early trypsin," acting on the meal proteins, produces a signal that induces translation of the several proteolytic enzymes required for blood meal protein digestion. The nature of this signal, which is not free amino acids, remains a secret that the mosquito has refused to share despite many attempts to identify it. The synthesis of "early trypsin" itself is regulated in a unique manner.

Transcription of the "early trypsin" gene starts a few hours after the adult emerges from the pupal state. However, the "early trypsin" mRNA is stored in the midgut epithelium and remains untranslated until a blood meal is taken. It seems that free amino acids in the blood are the signal for translation, apparently by providing amino acids to charge tRNAs in the gut, which in the absence of a blood meal are uncharged, thereby limiting translation. Thus, "early trypsin" is part of a fascinating signaling system that conveys the presence of protein in the gut lumen to the epithelial cell. Once the signal is received, the epithelial cells can produce the enzymes required for protein digestion; in the absence of the signal the cell is quiescent. These regulatory mechanisms ensure that proteases are made only when needed. Given the importance of amino acids for producing the proteins of the egg, one might suspect that the mosquito would be quite efficient in using the amino acids. In fact, the contrary is true. A female mosquito uses only 10–15% of the amino acids to make egg proteins. Some of the rest are used to make egg lipids, and some are used to build the energy reserves of the female. But more than 80% of the amino acid carbon is oxidized to CO_2 to provide the energy needed for egg production. Matters related to the partitioning of meal proteins into various products and how the nutritional state of the female affects these processes are a major subject of research now. It is probable that females who had poor larval nutrition will need more than one blood meal to mature a batch of eggs, because the first or even the second meal is needed to build up the energy reserves affecting egg production. Under such circumstances the possibility of disease transmission increases, so it becomes important to understand how the female manages the blood meal. Because the female oxidizes 80% of the amino acids, she has a very large amount of ammonia to deal with. The female mosquito seems to place the ammonia in proline, which she stores in her hemolymph (blood), until she can convert the ammonia to urea or uric acid for secretion. This is a unique process, still not understood adequately, in mosquitoes, and the hemolymph proline concentration can be as high as 80 mM.

BIBLIOGRAPHY

Adie, J.R., (1913). Malaria and mosquitoes in the Kashmir Valley. *The Indian Medical Gazette*, *48*(9), 341–342.

Alak, C. D. and Kaushik, M. (2014). Mosquito vector survey in Guwahati city of Assam, India. *Journal of Entomology and Nematology*, *6*(3), 42–50.

Arunachalam, N., Philip Samuel, P., Hiriyan, J., Thenmozhi, V., and Gajanana, A. (2004). Japanese Encephalitis in Kerala, South India: Can Mansonia (Diptera: Culicidae) play a Supplemental Role in Transmission? *Journal of Medical Entomology*, *41*(3), 456–461.

Bar, A. and Andrew, J. (2013a). Morphology and morphometry of *Aedes aegypti* larvae. *Annual Review and Research in Biology*, *3*, 1–21.

Bar, A. and Andrew, J. (2013b). Morphology and morphometry of *Aedes aegypti* adult mosquito. *Annual Review and Research in Biology*, *3*, 52–69.

Barraud, P. J. (1934). *The Fauna of British India, including Ceylon and Burma. Diptera. Vol. 5. Family Culicildae; Tribes Megarhinini and Culicini.* Taylor & Francis, London.

Baruah, A., Hazarika, R.A., Barman, N.N., Islam, S. and Gulati, B.R. (2018). Mosquito abundance and pig seropositivity as a correlate of Japanese encephalitis in human population in Assam, India. *Journal of Vector Borne Diseases*, *55*(4), 291.

Bhattacharyya, D. R., Prakash, A., Dutta, P., Mohapatra, P. K., & Mahanta, J. (2007). Addition of three species of Topomyia (Diptera: Culicidae) to the mosquito fauna of northeastern India. *Journal of the American Mosquito Control Association*, *23*(1), 76–77.

Bhattacharyya, D.R., Prakash, A., Mohapatra, P.K. and Sarma, D.K. (2009). *Armigeres (Armigeres) mahantai*, a new mosquito species from India. *Journal of the American Mosquito Control Association*, *25*(1), 1–5.

Bhattacharyya, D.R., Rajavel, A.R., Mohapatra, P.K., Jambulingam, P., Mahanta, J. and Prakash, A. (2014). Faunal richness and the checklist of Indian mosquitoes (Diptera: Culicidae). *Check List*, *10*(6), 1342–1358.

Bracco, J.E. et al. (2007). Genetic variability of Aedes aegypti in the Americas using a mitochondrial gene: Evidence of multiple introductions. *Memórias do Instituto Oswaldo Cruz*, *102*(5), 573–580.

Brady, O.J., Golding N., Pigott D.M., Kraemer M.U.G., Messina J.P., Reiner Jr R.C., Scott T.W., Smith D.L., Gething P.W. and Hay S.I. (2014). Global temperature constraints on *Aedes aegypti* and *Ae. albopictus* persistence and competence for dengue virus transmission. *Parasites & Vectors*, *7*, 338. https://doi.org/10.1186/1756-3305-7-338

Brady, O.J., Johansson, M.A., Guerra, C.A., Bhatt, S., Golding, N., Pigott, D.M., Delatte, H., Grech, M.G., Leisnham, P.T., Maciel-de-Freitas R., Styer, L.M., Smith, D.L., Scott, T.W., Gething P.W. and Hay, S.I. (2013). Modelling adult Aedes aegypti and *Aedes albopictus* survival at different temperatures in laboratory and field setting. *Parasites & Vectors*, *6*, 351.

Brown, J.E., McBride, C.S., Johnson, P., Ritchie, S., Paupy, C., Bossin, H., Lutomiah, J., Ildefonso, F.-S., Ponlawat, A., Cornel, A.J., Black W.C., Gorrochotegui-Escalante, N., Urdaneta-Marquez, L., Sylla, M., Slotman, M., Murray, K.O., Walker, C., Powell, J.R. (2011). Worldwide patterns of genetic differentiation imply multiple "domestications" of Aedes aegypti, a major vector of human diseases. *Proceedings of the Royal Society B: Biological Sciences*, *278*, 2446–2454.

Chadee, D.D. (1997). Effects of forced egg-retention on the oviposition patterns of female Aedes aegypti (Diptera: Culicidae). *Bulletin of Entomological Research*, *87*, 649–651.

Chadee, D.D. and P.S. Corbet (1987).Seasonal incidence and diel patterns of oviposition in the field of the mosquito, Aedes aegypti (L.) (Diptera: Culicidae) in Trinidad, West Indies: A preliminary study. *Annals of Tropical Medicine and Parasitology*, *81*, 151–161.

Chadee, D.D., P.S. Corbet and J.J.D. Greenwood (1990). Egg-laying Yellow Fever Mosquitoes avoid sites containing eggs laid by themselves or by conspecifics. *Entomologia Experimentalis et Applicata*, *57*, 295–298.

Christophers, S. R. (1933). *The Fauna of British India, including Ceylon and Burma. Diptera. vol. iv. Family Culicidae. Tribe Anophelini*. Taylor & Francis, London.

Christophers, S.R. (1960). *Aedes aegypti (L.) The Yellow Fever Mosquito. Its Life History, Bionomics and Structure*. Cambridge University Press, Cambridge.

Clements, A.N. (1992). *The Biology of Mosquitoe., vol.I, Development, Nutrition and Reproduction*. Chapman and Hall, London.

Clements, A.N. (2000). *The Biology of Mosquitoes. vol. I: Development, Nutrition and Reproduction*. 2nd Edition, CABI Publishing, Oxford.

Couret, J. and M.Q. Benedict (2014). A meta-analysis of the factors influencing development rate variation in Aedes aegypti (Diptera: Culicidae). *BioMed Central Ecology*, *14*(3), 1–15.

Culicidae Meigen (1818). https://mosquito-taxonomicinventory.myspecies.info/simpletaxonomy/term/6045. Accessed on 23.09.2022

Dar, T.H., Ramalingam, N., Jayakumar, S. and Purusothaman, J. (2016). Mosquitoes (Diptera: Culicidae) of Jammu division of Jammu and Kashmir State, India, with new records. *Journal of Entomology and Zoology Studies*, *4*, 388–393.

Das, S. C., Bhuyan, M., Baruah, I. and Talukdar, P. K. (1991). Mosquito survey in Tripura. *Indian Journal of Malariology*, *28*(2), 129–134.

Das, S. C., Das, N. G. and Baruah, I. (1984). Mosquito survey in Meghalaya. *Indian Journal of Public Health*, *28*(3), 147–51.

de Jong, R. and Knols, B.G. (1996). Selection of biting sites by mosquitoes. *Ciba Foundation Symposium*, *200*, 89–100; discussion 101–8. https://doi.org/10.1002/9780470514948.ch8

Delatte, H., Bagny, L., Brengue, C., Bouetard, A., Paupy, C. and Fontenille, D. (2011). The invaders: Phylogeography of dengue and chikungunya viruses Aedes vectors, on the South West islands of the Indian Ocean. *Infection, Genetics and Evolution*, *11*(7), 1769–1781.

Devi, N.P. and Jauhari, R.K. (2008). Reappraisal on anopheline mosquitoes of Garhwal region, Uttarakhand, India. *Journal of Vector Borne Diseases*, *45*(2), 112–23.

Devi, M.B. and Chingangbam, D. S. (2019). Mosquitoes from senapati district, Manipur. *Indian Journal of Entomology*, *81*(4), 677–681.

Devi, M.B. and Singh, D.C. (2018). New records of *Culex (Culex) pipiens* Linn. from Manipur India. Distt Dehradun, Uttarakhand. *International Journal of Current Microbiology and Applied Sciences*, *4*, 583–589. https://doi.org/10.20546/ijcmas.2021.1004.058

Dutta, P., Khan, S. A., Khan, A. M., Sharma, C. K. and Mahanta, J. (2004). Entomological observations on dengue vector mosquitoes following a suspected outbreak of dengue in certain parts of Nagaland with a note on their susceptibility to insecticides. *Journal of Environmental Biology*, *25*(2), 209–212.

Dutta, P., Khan, S. A., Khan, A. M., Sharma, C. K. and Mahanta, J. (2005). Biodiversity of mosquitoes in Manipur State and their medical significance. *Journal of Environmental Biology*, *26*(3), 531.

Dutta, P., Khan, S. A., Khan, A. M., Sharma, C. K. and Mahanta, J. (2010b). Survey of mosquito species in Nagaland, a hilly state of north east region of India. *Journal of Environmental Biology*, *31*(5), 781.

Dutta, P., Prakash, A., Bhattacharyya, D. R., Khan, S. A., Gogoi, P. R., Sharma, C. K., & Mahanta, J. (2010a). Mosquito biodiversity of Dibru-Saikhowa biosphere reserve in Assam, India. *Journal of Environmental Biology*, *31*(5), 695–699.

Edman, J.D., Webber, L.A. and Schmid, A.A. (1974). Effect of host defenses on the feeding pattern of Culex nigripalpus when offered a choice of blood sources. *The Journal of Parasitology*, *60*(5), 874–883.

Emerson K. J., Merz, C. R., Catchen, J. M., Hohenlohe, P. A., Cresko, W. A., Bradshaw, W. E. and Holzapfel, C. M. (2010). Resolving postglacial phylogeography using high-throughput sequencing. *Proceedings of the National Academy of Sciences*. https://doi.org/10.1073/pnas.1006538107

Faiz, A.-U. and Shahbaz, M. (2020). Biodiversity of Culicidae Mosquitoes in District Bagh, Azad Jammu and Kashmir. *Pakistan Journal of Zoology 43*, 1–4.

Farnesi, L.C., Martins, A.J., Valle, D. and Rezende, G.L. (2009). Embryonic development of Aedes aegypti (Diptera: Culicidae): Influence of different constant temperatures, *Memórias do Instituto Oswaldo Cruz*, *104*(1), 124–126.

Fay, R.W. and Perry, A.S. (1965). Laboratory studies of ovipositional preferences of Aedes aegypti. *Mosquito News*, *25*, 276–281.

Firooziyan, S., Dinparast Djadid, N. and Gholizadeh, S. (2018). Speculation on the possibility for introducing *Anopheles stephensi* as a species complex: preliminary evidence based on odorant binding protein 1 intron I sequence. *Malaria Journal*, *17*, 366. https://doi.org/10.1186/s12936-018-2523-y

Foster, W.A. and Walker, E.D. (2002). Mosquitoes (Culicidae). *In*: G. Mullen and L. Durden (eds.), *Medical and Veterinary Entomology*. Academic Press, San Diego, pp. 203–262.

Giles, G.M. (1901). A plea for the collective investigation of Indian Culicidae, with suggestions as to moot points for enquiry, and a prodromus of species known to the author. *Journal Bombay Natural History Society*, *13*, 592–610.

Gillett, J.D. (1962). Contributions to the oviposition cycle by individual mosquitoes in a population. *Journal of Insect Physiology*, *8*, 665–681.

Goswami, G., Singh, O.P., Nanda, N., Raghavendra, K., Gakhar, S.K., and Subbarao, S.K. (2006). Identification of all members of the *Anopheles culicifacies* complex using allele-specific polymerase chain reaction assays. *American Journal of Tropical Medicine & Hygiene*, *75*(3), 454–460. 1.

Goud, S., Pattnaik, S., Dash, P., Biswal, I., Seth, J.K., Hazra, R.K. and Panda, B.B. (2022). New Record of Mosquito, *Coquillettidia xanthogaster* (Edwards) from India. *Indian Journal of Entomology*, 1–3. https://doi.org/10.55446/IJE.2022.686

Gupta, R.K. (2004). *Advancements in insect biodiversity*. Agrobios (India), 337 pp.

Harbach, R.E. (2007). The Culicidae (Diptera): A Review of Taxonomy, Classification and Phylogeny. *Zootaxa*, *1668*, 591–638. https://doi.org/10.11646/zootaxa.1668.1.28

Harbach, R.E. (2013). Mosquito Taxonomic Inventory. [(accessed on 3 November 2021)]. Available online: https://www.mosquito-taxonomic-inventory.info/

Harbach, R.E. (2018). *Culicipedia: species-group, genus-group and family-group names in Culicidae (Diptera)*. CABI Publishing, Wallingford, Oxfordshire, UK. 378 pp.

Harrington, L. and Edmann, J.D. (2001). Indirect evidence against delayed "Skip-Oviposition" behavior by *Aedes aegypti* (Diptera: Culicidae) in Thailand. *Journal of Medical Entomology*, *38*, 641–645.

Hemme, R.R., Thomas, C.L., Chadee, D.D. and Severson, D.W. (2010). Influence of urban landscapes on population dynamics in a short-distance migrant mosquito: Evidence for the dengue vector *Aedes aegypti*. *PLoS Neglected Tropical Diseases*, *4*(3), e634. http://doi.org/10.11609/JoTT.o4193.xxxx-x

ITIS (2014). *Aedes aegypti*, Integrated Taxonomic Information System (database), www.itis.gov/servlet/SingleRpt/SingleRpt?search_topic=TSN&search_value=126240

Iyengar, M.O.T. (1938). Studies on the epidemiology of filariasis in Travancore. *Indian Medical Research Memoirs*, *30*, 179.

James, M.T. and Harwood, R.F. (1969). *Herm's Medical Entomology*, 6th ed. London: Macmillan Co. 484 pp.

Kogan, P.H. (1990). Substitute blood meal for investigating and maintaining Aedes aegypti (Diptera: Culicidae). *Journal of Medical Entomology*, *27*, 709–712. https://doi.org/10.1093/jmedent/27.4.709

Kraemer, M.U.G., Sinka, M.E., Duda, K.A., Mylne, A.Q.N., Shearer, F.M., Barker, C.M., Moore, C.G., Carvalho, R.G., Coelho, G.E., van Bortel, W., Hendrickx, G., Schaffner, F., Elyazar, I.R.F., Teng, H.-J., Brady, O.J., Messina, J.P., Pigott, D.M., Scott, T.W., Smith, D.L., Wint, G.R.W., Golding, N. and Hay, S. (2015). The global distribution of the arbovirus vectors Aedes aegypti and Ae. albopictus. *eLife*, *4*, e08347.

Kumar, N.P., Kumar, A., Panneer, D., Abidha, S., Muthukumaravel, S., Sankari, T., Ajithlal, P.M., Mathew, J., Koothradan, S., Paramasivan, R., Muniyaraj, M., Singh, H., Saxena, R., Vijayachari, P., Sunish I.P., Shriram A.N., Dutta, P., Patgiri, S.J., Bhattacharyya, D.R., Hoti, S.L., Chattopadhyay, D., Roy, S., Mahapatra, N., Pati, S., Chand, G., Mishra, A.K., Barde, P. and Jambulingam, P. (2021). Nation-wide vector surveillance on Zika and Dengue did not indicate transmission of the American lineage-pandemic Zika virus in India. *International Journal of Infectious Diseases*, *113*, 119–124. https://doi.org/10.1016/j.ijid.2021.09.074

Lemine, A.M., Lemrabott, M.A., Ebou, M.H., Lekweiry, K.M., Salem, M.S., Brahim, K.O., Moukah, M.O., Bouraya, I.N., Brengues, C., Trape, J.F. and Basco, L. (2017). Mosquitoes (Diptera: Culicidae) in Mauritania: a review of their biodiversity, distribution and medical importance. *Parasites & Vectors*, *10*(1), 35.

Lorenz, L., Beaty, B.J., Aitken, T.H., Wallis, G.P., Tabachnick, W.J. (1894). The effect of colonization upon aedes aegypti susceptibility to oral infection with yellow fever virus. *American Journal of Tropical Medicine and Hygiene*, *33*(4), 690–4. https://doi.org/10.4269/ajtmh.1984.33.690

Lounibos, L.P. (1981). Habitat segregation among African treehole mosquitoes. *Ecological Entomology*, *6*, 129–154.

Malhotra, P.R. and Mahanta, H.C. (1994). Check-List of mosquitoes of Northeast India (Diptera: Culicidae), *Oriental Insects*, *28*(1), 125–149. https://doi.org/10.1080/00305316.1994.10432301
Mattingly, P.F. (1967). Taxonomy of *Aedes aegypti* and related species. *Bulletin of the World Health Organization*, *36*(4), 552–554.
McClelland, G.A.H. (1974). A worldwide survey of variation in scale pattern of the abdominal tergum of Aedes aegypti (L.) (Diptera: Culicidae). *Transaction Royal Entomological Society London*, *126*, 239–259.
McNeeley, G.A., Miller, K.A., Reid, W.V., Wittermeir, R.A. and Warner, T.R. (1990). *Conserving the World's Biological Diversity*, International Union for Conservation of Nature & Natural Resources.
Moczek, A.P. (2010). Phenotypic plasticity and diversity in insects. *Philosophical Transactions of the Royal Society of London, B – Journals, Biological Sciences*, *365*(1540), 593–603. https://doi.org/10.1098/rstb.2009.0263
Mogi, M. and Mokry, J. (1980). Distribution of *Wyeomyia smithii* (Diptera: Culicidae) eggs in pitcher plants in Newfoundland, Canada. *Tropical Medicine*, *22*, 1–12.
Moirangthem, B. D. and Singh, D. C. (2018a). New records of *Culex (Culex) pipiens* Linn. from Manipur India. *International Journal of Mosquito Research*, *5*(2), 52–55.
Moirangthem, B. D., and Singh, D. C. (2018b). New records of species of Metalutzia (Diptera: Culicidae) from Manipur, India. *International Journal of Mosquito Research*, *5*(3), 28–31.
Moore, M., Massamba, S., Goss, L., Burugu, M.W., Sang, R., Kamau, L.W., Kenya, E.U., Bosio, C., Munoz, M. de L., Sharakova, M., and Black, W.C. (2013). Dual African origins of global *Aedes aegypti* s.l. populations revealed by mitochondrial DNA. *PLoS Neglected Tropical Diseases*, *7*(4), e2175.
Mousson, L., Dauga, C., Garrigues, T., Schaffner, F., Vazeille, M. and Failloux, A.-B. (2005). Phylogeography of *Aedes (Stegomyia) aegypti* (L.) and *Aedes (Stegomyia) albopictus* (Skuse) (Diptera: Culicidae) based on mitochondrial DNA variations. *Genetical Research* (Camb.), *86*, 1–11.
Munirathinam, A., Krishnamoorthi, R., Baskaran, G., Govindarajan, R., Veerapathiran, A., Venkatesh, A. and Tyagi B.K. (2015). Mosquito diversity in Keeriparai and Mundanthurai hill ranges of the Western Ghats, southern India. *Journal of Threatened Taxa*, *7*(9), 7552–7556. http://doi.org/10.11609/JoTT.o4193.7552-6
Muñoz, M. de L. et al. (2013). Gene flow pattern among Aedes aegypti populations in Mexico. *Journal of the American Mosquito Control Association*, *29*(1), 1–18.
Naik, M., Bhat, T., Wani, A.A., Amin, A. and Jalaalie, U. (2022). Clinical and Laboratory Profile of Dengue in Kashmir Valley. *Journal of the Association of Physicians of India*, *69*(12), 11–12.
National Science Foundation (2010). Mosquitoes: Genetic structure of first animal to show evolutionary response to climate change determined. *Science Daily*. Retrieved August 6, 2023 from www.sciencedaily.com/releases/2010/08/100824132357.htm
Nelson, M.J. (1986). *Aedes aegypti: Biology and Ecology*, Pan American Health Organization, Washington, DC, PNSP/86–63, pp. 50.
Norse, E.A. and McManus, R.E. (1980). Ecology and living resources biological diversity. *In Environmental Quality: The Eleventh Annual Report of the Council on Environmental Quality*. Washington, 31–80.
Nunes, R. and Borges, P.A.V. (2021). *Culiseta atlantica. The IUCN Red List of Threatened Species 2021*, e.T124913500A124930626. https://doi.org/10.2305/IUCN.UK.2021-1
Otero, M., Solari, H.G. and Schweigmann, N. (2006). A stochastic population dynamics model for Aedes aegypti: Formulation and application to a city with temperate climate. *Bulletin of Mathematical Biology*, *68*(8), 1945–1974.
OECD (2018). *Safety Assessment of Transgenic Organisms in the Environment, Volume 8: OECD Consensus Document of the Biology of Mosquito, Aedes aegypti*, Harmonisation of Regulatory Oversight in Biotechnology, OECD Publishing, Paris, 106 pp. https://doi.org/10.1787/9789264302235-en
Paksa, A., Sedaghat, M.M., Vatandoost, H., Yaghoobi-Ershadi, M.R., Moosa-Kazemi, S.H., Hazratian, T., Sanei-Dehkordi, A., Oshaghi, M.A. (2019). Biodiversity of Mosquitoes (Diptera: Culicidae) with Emphasis on Potential Arbovirus Vectors in East Azerbaijan Province, Northwestern Iran. *Journal of Arthropod Borne Diseases*, *13*(1), 62–75.
Pigliucci, M. (2005) Evolution of phenotypic plasticity: where are we going now? *Trends in Ecology and Evolution*, *20*(9), 481–485.
Powell, J.R. and W.J. Tabachnick (2013). History of domestication and spread of Aedes aegypti – A review. *Memórias do Instituto Oswaldo Cruz*, *V108*, 11–17.
Powell, J.R., Tabachnick, W.J. and Arnold, J. (1980). Genetics and the origin of a vector population – Aedes aegypti, a case-study. *Science*, *208*, 1385–1387.
Prakash, A., Bhattacharyya, D.R., Mohapatra, P.K. and Mahanta, J. (1998). Investigation on malaria vectors and mosquito fauna in south Tripura district, Tripura state. *Indian Journal of Malariology*, *35*(3), 151–159.

Prakash, A., Bhattacharyya, D.R., Mohapatra, P.K. and Mahanta, J. (2000). Mosquito fauna and malaria vectors in Jairampur, district Changlang, Arunachal Pradesh. *Indian Journal of Malariology*, *37*(3–4), 74–81.

Rajavel, A.R., Natarajan, R. and Vaidyanathan, K. (2006). Mosquitoes of the mangrove forests of India: Part - VI. Kundapur, Karnataka and Kannur, Kerala. *Journal of American Mosq. Control Association*, *22*(4), 582–585.

Rajavel, A. and Natarajan, R. (2009). Mosquitoes of the Mangrove Forests of India: Part 7—An Overview. *Journal of the American Mosquito Control Association*, *24*, 478–88. https://doi.org/10.2987/08-5762.1

Rašić, G., Filipović, I., Callahan, A.G., Stanford, D., Chan, A., Lam-Phua, S.G., Huat, C.T. and Hoffmann, A.A. (2016). The queenslandensis and the type form of the dengue fever mosquito (Aedes aegypti L.) are genomically indistinguishable. *PLoS Neglected Tropical Diseases*, *10*(11), e0005096.

Reinert J.F., Harbach R. and Kitching I.J. (2009). Phylogeny and classification of tribe Aedini (Diptera: Culicidae). *Zoological Journal of Linnean Society*, *157*(4), 700–794.

Reinert, J.F. and R.E. Harbach (2005). Generic changes affecting European aedine mosquitoes (Diptera: Culicidae: Aedini) with a checklist of species. *Journal of the European Mosquito Control Association, European Mosquito Bulletin*, *19*, 1–4.

Reiter, P. (1998). *Aedes albopictus* and the world trade in used tires, 1988–1995: The shape of things to come? *Journal of the American Mosquito Control Association*, *14*, 83–94.

Ribeiro, J.M.C. (1992). Characterization of a vasodilator from the salivary glands of the yellow fever mosquito Aedes aegypti. *Journal of Experimental Biology*, *165*, 61–71.

Ribeiro, J.M.C. (1995). Blood-feeding arthropods: Live syringes or invertebrate pharmacologists? *Infect. Agents Diseases*, *4*, 143–152.

Rowley, W.A. and C.L. Graham (1968). The effect of temperature and relative humidity on the flight performance of female Aedes aegypti. *Journal of Insect Physiology*, *14*(9), 1251–1257.

Rueda, L. (2004). Pictorial keys for the identification of mosquitoes (Diptera: Culicidae) associated with dengue virus transmission. *ZOOTAXA*, *589*, 60.

Rueda, L.M., Patel, K. J., Axtell, R.C. and Stinner, R. E. (1990). Temperature-dependent development and survival rates of *Culex quinquefasciatus* and *Aedes aegypti* (Diptera. Culicidae). *Journal of Medical Entomology*, *27*(5), 892–898.

Sarmah, N.P., Bhowmik, I.P., Sarma, D.K., Sharma, C.K., Medhi, G.K., Mohapatra, P.K. and Bhattacharyya, D. R. (2019). Role of *Anopheles baimaii*: potential vector of epidemic outbreak in Tripura, North-east India. *Journal of Global Health Reports*, *3*, e2019036. https://doi.org/10.29392/joghr.3.e2019036

Savage, H.M. and G.C. Smith (1995). *Aedes albopictus* y *Aedes aegypti* en las Américas: Implicaciones para la transmisión de arbovirus e identificación de hembras adultas dañadas [*Aedes albopictus* and *Aedes aegypti* in the Americas: Implications for the transmission of arboviruses and identification of damaged adult females]. *Boletin de la Oficina Sanitaria Panamericana*, *118*, 473–487.

Scarpassa, V.M., Cardoza, T.B. and Cardoso Junior, R.P. (2008). Population genetics and phylogeography of Aedes aegypti (Diptera: Culicidae) from Brazil. *The American Journal of Tropical Medicine and Hygiene*, *78*(6), 895–903.

Schaper, S. and F. Hernandez-Charvarria (2006). Scanning electron microscopy of the four larval instars of the Dengue fever vector *Aedes aegypti* (Diptera: Culicidae). *Revista de Biología Tropical*, *54*, 847–852.

Scholte, E., Hartog, W.D., Dik, M., Schoelitsz, B., Brooks, M., Schaffner, F., Foussadier, R., Braks, M. and Beeuwkes, J. (2010). Introduction and control of three invasive mosquito species in the Netherlands, July-October 2010. *Euro Surveillance*, *15*(45), 19710.

Service, M. (2012), *Medical Entomology for Students*, 5th Ed., Cambridge University Press, New York, 303 pp.

Singh, V., Singh, N., and Swami V.P. (2023). Diversity of mosquito in India: Status Report. *Journal of Emerging Technologies and Innovative Research*, *10*(7), 533–545.

Sinha, P., Purwar, R., Mishra, R., Chaurasiya, R. and Bhandari, R. (2018). Biodiversity of mosquitos of Jabalpur city with special reference to *Anopheles, Culex* and *Aedes* Species. *International Journal of Creative Research Thoughts*, *6*(2), 1387–1396.

Subbarao, S.K., Vasantha, K., Adak, T. and Sharma, V.P. (1983). *Anopheles culicifacies*: cytogenetic characterization of rameshwaram island populations. *Journal of American Mosquito Control Association*, *9*(1), 27–31.

Subbarao, S.K., Nanda, N. and Raghavendra, K. (1999). Malariogenic stratification of India using *Anopheles culicifacies* sibling species prevalence. *ICMR Bulletin*, *29*, 75–80.

Tabachnick, W.J. (1982). Geographic and temporal patterns of genetic variation of Aedes aegypti in New Orleans. *American Journal of Tropical Medicine and Hygiene*, *31*, 849–853.

Tabachnick, W.J. (1991). Evolutionary genetics and arthropod-borne disease: The yellow fever mosquito. *American Entomologist*, *37*, 14–24.

Tabachnick, W.J. and Powell, J.R. (1979). A world-wide survey of genetic-variation in the yellow fever mosquito, *Aedes aegypti. Genetical Research*, *34*(3), 215–229.

Tabachnick, W.J., Wallis, G.P., Aitken, T.H., Miller, B.R., Amato, G.D., Lorenz, L., Powell, J.R. and Beaty, B.J. (1985). Oral infection of Aedes aegypti with yellow-fever virus – Geographic variation and genetic considerations. *American Journal of Tropical Medicine and Hygiene*, *34*(6), 1219–1224.

Tewari, S.C. and Hiriyan, J. (1992). Description of two new species of Aedes (Diceromyia) from South India (Diptera: Culicidae). *Mosquito Systematics*, *24*(2), 154–175.

Theobald, F.V. (1901). *A monograph of the Culicidae or mosquitoes*. Vol. *1*, 424 pp.

Tyagi, B.K. (2002). *Malaria in the Thar Desert: Facts, Figures and Future*. Agrobios (India). 165 pp.

Tyagi, B.K. (2020). Are male mosquitoes worth their weight? A mosquito's eye view. *SOMA Newsletter*, *1*(2), 1–27.

Tyagi, B.K. (2021). A synopsis of taxonomic data on the extant species of mosquitoes in India, with an annotated note on additional taxa described after Tyagi (2020). *Journal of Medical Arthropodology & Public Health*, *1*(1), 33–47.

Tyagi, B.K. (2023a). *Male Mosquito: A zaddy and subaltern sex in the disease epidemiology and control*. Scientific Publishers (India), 232 pp.

Tyagi, B.K. (2023b). *Desert Malaria: An Emerging Malaria Paradigm and Its Global Impact on Disease Elimination*. Springer-Nature, Singapore, 416 pp.

Tyagi, B.K., Bansal, S.K. and Verma, K.V.S. (1991). Intra-specific morphological variations in two ecotypes of *Anopheles Stephensi* Liston in north-western Rajasthan (Dipter: Culicidae). *Annals of Entomology*, *9*, 85–89.

Tyagi, B.K., Hiriyan, J., Samuel, P., Tewari, S.C., Ayanar, K. et al. (2009). Description of a new species, *Anopheles pseudosundaicus* (Diptera: Culicidae) from Kerala, India. *Zootaxa*, *2219*, 49–60.

Tyagi, V. (2020a). Can we resolve the mystery of sibling species concept of *Anopheles culicifacies* mosquito using morphological characters? *SOMA Newsletter*, *1*(1), 23–25.

Tyagi, B.K., Munirathinam, A. and Venkatesh, A. (2015). A catalogue of Indian mosquitoes. *International Journal of Mosquito Research*, *2*(2), 50–97.

Tyagi, V., Sharma, A.K., Dhiman, S., Srivastava, A.R., Yadav, R., Sukumaran, D., Agrawal, O. P. and Vijay, V. (2016). Malaria vector Anopheles culicifacies sibling species differentiation using egg morphometry and morphology. *Parasite & Vectors*, *9*(202), 1–13.

Vanlalruia, K., Senthilkumar, N. and Gurusubramanian, G. (2014). Diversity and abundance of mosquito species in relation to their larval habitats in Mizoram, North Eastern Himalayan region. *Acta Tropica*, *137*, 1–18.

Vargas, V.M. (1968). Sexual dimorphism of larvae and pupae of i (Linn.). *Mosquito News*, *28*, 374–379.

Veer, V. and Gopalakrishnan, R. (2016). *Herbal Insecticides, Repellents and Biomedicines: Effectiveness and Commercialization*. Springer, 478 pp.

Wallis, G.P., Tabachnick, W.J. and Powell, J.R. (1984) Genetic-heterogeneity among Caribbean populations of Aedes aegypti. *American Journal of Tropical Medicine and Hygiene*, *33*, 492–498.

Warrell, D.A. and H.M. Gilles (eds.) (2002), *Essential Malariology*, 4th Ed., Hodder Arnold, London, pp. 350.

WHO (2022). Vector alert: *Anopheles stephensi invasion and spread in Africa and Sri Lanka*. World Health Organization, Geneva. ISBN 978-92-4-006771-4 (electronic version)

Wilkerson, R.C., Linton, Y.M., Fonseca, D.M., Schultz, T.R., Price, D.C. and Strickman, D.A. (2015). Making mosquito taxonomy useful: A stable classification of tribe Aedini that balances utility with current knowledge of evolutionary relationships. *PLoS One*, *10*(7), e0133602.

Wilkerson, R.C., Linton, Y-M and Strickman, D. (2021). *Mosquitoes of the World*. Vol. *I*. John's Hopkins University Press. 599 pp.

Wilson, E.O. and Peters, F.M. (1988). *Biodiversity*. National Academy Press, Washington, DC.

WHO SEARO, (2007). Anopheline Species Complexes in South and South-East Asia. World Health Organization. Regional Office for South-East Asia, & ProQuest (Firm).

WRBU (2014). Mosquito Classification Comparison, 2013, The Walter Reed Biosystematics Unit.

Yang, C.-H. and John, A.P. (2019). Polyphenism – A Window Into Gene-Environment Interactions and Phenotypic Plasticity. *Frontiers in Genetics*, *10*, 1–9. https://doi.org/10.3389/fgene.2019.00132

Yang, H.M., Macoris, M.L., Galvani, K.C., Andrighetti, M.T. and Wanderley, D.M. (2009). Assessing the effects of temperature on the population of *Aedes aegypti*, the vector of dengue. *Epidemiology and Infection*, *137*(8), 1188–202. https://doi.org/10.1017/S0950268809002040

6 Mosquitoes as Vectors, Pests, and Allergenics

B.K. Tyagi
University Institute of Biotechnology, Chandigarh University, Mohali, India

Manas Sarkar
Reckitt Benckiser Ltd., Gurugram, India

C. Kandasamy
Consultant – Agrochemicals, Chennai, India

Sajal Bhattacharya
Ex-Department of Zoology, Asutosh College, Kolkata, India

6.1 INTRODUCTION

Mosquitoes and man are literally "inseparable", in their restful and/or playful hours at least because the former always continues to occupy the latter's mind or, say, attention irrespective of time, location or habitat (Harrison, 1978; Spielman and D'Antonio, 2002; Tyagi et al., 2020)! Thus, there are lots of reasons why people should be interested in mosquitoes which are, rhetorically speaking, "ugly, bold and beautiful" – all at the same time! The common perception in the eyes of masses, however, is that mosquitoes are some of the most deadly creatures on the planet. A mosquito is small in size but is giant of a killer. They are responsible for wasting humongous national resources, in particular, by impoverishing human health, reducing economy and emaciating intelligentsia. Customarily, the mosquito is seen as a sort of an evil thing, notwithstanding the fact that there are lots of positive things that they carry in the ecosystem. Yet, the first thought about mosquitoes reminds only of a blood-sucker! The late Professor Felix Amerasinghe, one of the greatest Sri Lankan medical entomologists and mosquito-borne disease experts, has elegantly described the mosquito in the following manner:

> The insect world has always spawned mixed reactions among humankind throughout history – some love them and some abhor them; few are indifferent to them. Many insects are beautiful, others are useful. Some are a nuisance, others a menace. Mosquitoes are not exactly the favourite insects of humankind. The reaction of the average human to a visiting mosquito is to squash it as soon as possible. Or use a bewildering array of killing techniques, ranging from insecticides to electrical currents, to dispose of it. Or, at the very least, drive it away (temporarily, of course) with repellents. Most never see the beauty of these beasts, their wonderful adaptation to what is essentially a piratical life style, their subtle (sometimes brilliant) colouring and scalation that are only fully revealed under a microscope. And not all mosquitoes are harmful – indeed, relatively few of the vast array of mosquito species buzzing around our ears are the deadly disease carriers that we imagine them to be. Most do not even feed on humans, preferring animal blood instead. While their buzzing and singing irritate, their painful bites infuriate – but they are simply going about the business of ensuring that there will be a next generation, something that every self-respecting life form attempts to do.

DOI: 10.1201/9781003326991-8

Contrary to an ordinary man's thinking, mosquitoes are actually an integral and complex part of the terrestrial and freshwater ecosystems with which the future of human being is inextricably linked. They are generally understood by the culicidologists to be the most unique creatures due largely to their remarkable structural and behavioral diversity. Among members of all insect orders and families, or even entire animal kingdom, fewer can match the popularity and importance that these tiny creatures, the mosquitoes, convey to the human life! In fact no animal on earth has affected human lives so directly and profusely as is brought about by the mosquito (Sinka et al., 2020).

Not only mosquitoes trouble people, of both sexes and of all age groups, by their irritating and painful bites, but, in the process, they also mediate transmission of many life-threatening and debilitating diseases, besides a plethora of vexations of varied nature including psychological disturbance (Clements, 1992, 1999, 2011; Eldridge & Edman, 2012; Tyagi, 2004, 2023). A mosquito bite *per se* may not bring instant havoc but it is capable to cause disasters if it carried disease pathogens. Through their bites mosquitoes also levy unbearable uneasiness, ill temperament and loss of sleep, thereby affecting human psychology in attending to important jobs at hand. That is why they are branded objects of deep abhorrence and dislike for all they offer to the mankind – the annoyance, the pain, the nuisance, and the deadly infections such as, for example, malaria, lymphatic filariasis, yellow fever, dengue, chikungunya, zika virus, West Nile virus, Japanese encephalitis, and similar several scores of other kinds of encephalitides. Several mosquitoes are involved, as dominating vector species, in transmitting these diseases. A region-wise summary of the dominating vector species spreading malaria (Sinka et al., 2010, 2011, 2012; Sinka, 2013) filariasis and dengue/chikungunya/yellow fever/zika, and Japanese encephalitis is offered below in Table 6.1.

TABLE 6.1
List of Dominating Vector Species for Malaria, Filariasis, Japanese Encephalitis, and Dengue/Chikungunya/Yellow Fever

Sl. No.	Disease	Region	Dominating Vector Species
1	Malaria	The Americas	*Anopheles albimanus* *An. albitarsis* complex *An. aquasalis* *An. darlingi* *An. freeborni* *An. marajoara* *An. nuneztovari* complex *An. pseudopunctipennis* *An. quadrimaculatus* complex
2		Europe and Middle-East	*An. atroparvus* *An. labranchiae* *An. messeae* *An. sacharovi* *An. sergentii* *An. superpictus*
3		Africa	*An. arabiensis* *An. funestus* *An. gambiae* *An. melas* *An. merus* *An. moucheti* *An. nili complex*

(*Continued*)

TABLE 6.1 (Continued)

Sl. No.	Disease	Region	Dominating Vector Species
4		Asia-Pacific	*An. aconitus* *An. annularis* *An. balabacensis* *An. barbirostris* complex *An. culicifacies* complex *An. dirus* complex *An. farauti* complex *An. flavirostris* *An. fluviatilis* complex *An. koliensis* *An. lesteri* *An. leucospyrus/latens* *An. maculatus* (group) *An. minimus* complex *An. punctulatus* complex *An. sinensis* complex *An. stephensi* *An. subpictus* complex *An. sundaicus* complex
5	Dengue/yellow Fever/ chikungunya/zika	Worldwide	*Aedes aegypti* *Ae. albopictus*
6	Lymphatic filariasis	Worldwide	*Culex quinquefasciatus* *Mansonia* spp.
7	Japanese encephalitis	Mostly Southeast Asia	*Culex tritaeniorhynchus* *Cx. bitaeniorhynchus* *Cx. pseudovishnui* *Cx. gelidus* *Mansonia indiana*

Although a little over 10% of mosquito species only are involved in the transmission of a plethora of deadly and/or incapacitating disease pathogens, viz., protozoans, helminths, viruses, bacteria etc., nonetheless, a large number of species (about 90%) are surprisingly not harmful and in fact, contrary to the common perception, several of them are highly useful in various different ways such as, for example, biocontrol agent, pollinator, source of medicinal product or molecule, human food and animal feed (Panicker and Bai, 1983; Tyagi et al., 2021, 2023). With the growing positive knowledge about mosquitoes the notion that mosquitoes are only villains is gradually shattering. In conclusion, mosquito is an animal which must not always be abhorred; losses and damages are galore due to mosquitoes and so are the benefits and advantages associated with them. Important advice here is to scan their overall performance in nature from an evolutionist-cum-naturalist's viewpoint or, say, look at these creatures from a "mosquito's eye-view"!

6.2 MAN'S EYE-VIEW *VS.* MOSQUITO'S EYE-VIEW

Good or bad, mosquitoes are an important component in our natural ecosystem. Mosquitoes are vital components of a complex ecosystem, just like every other living organism do. Most of the mosquitoes are harmless. But, when they transmit diseases, horrible things can happen. The ones that carry disease have caused human suffering for millennia, e.g., malaria and filariasis. Let us, therefore, take a balanced look on the whole family of mosquitoes, and judge on the basis of their holistic performance rather than merely one trait of being hazardous to human and animal health.

The mosquito is kind of a mixed bag. Many mosquitoes serve an important biological purpose. They can help pollinate plants as they feed on nectar (their usual food source, outside of that crucial blood meal period) and provide a vital source of food for larger animals. Researchers in the Arctic worried that climate change would lead more animals to feast upon local mosquitoes, throwing the food web out of whack and leaving plants unpollinated. As it turns out, warmer water has actually produced an Arctic mosquito boom. The point is that just because humans hate mosquitoes does not mean that they must be wiped out without consequences. That also does not mean that scientists have a softer "live and let live" philosophy about these insects. Many scientists are working on creating genetically modified male mosquitoes that can only father sterile offspring. If released into the population, these males could trick deadly mosquitoes into wasting their blood meals on useless eggs. Consequently researchers are trying to eradicate mosquitoes, by either suppressing or replacing their populations, and they are focusing on the ones that cause harm the most, e.g. (i) *Aedes aegypti*, which transmits the fatal dengue, yellow fever, chikungunya and Zika, and the like, in the absence of any effective drug to treat a patient or a vaccine to prevent onset of the infection to a person (Sebastian et al., 1980, 1990; Venkatesh and Tyagi, 2013, 2015), and (ii) *Anopheles stephensi*, the mosquito that has evolved to live with humans in urban areas and is a major malaria vector in India, is a particularly attractive target due to its recent ingression into several far off countries in Africa and Sri Lanka, and for which WHO (2023) has recently issued an alert for the whole world. In a way, man is on a war with the mosquito and there is no visible end to this conflict in near future.

6.3 WHAT MAKES THE MOSQUITO A SERIOUS PEST OR A DEADLY VECTOR OF DISEASES?

The central point here is the mosquito's hematophagous behavior, though only female mosquitoes bite, and that too only for a very brief portion of their life cycle, i.e., when they need nutrients to create eggs. But during that stage of life, they might bite multiple victims including those of different species, i.e., humans, nonhuman primates, and cattle etc.

6.3.1 Vector Mosquitoes

If a mosquito bites someone infected with an infectious pathogen (e.g., malarial protozoans or arboviruses of dengue, chikungunya, zika, and Japanese encephalitis, etc.) that can replicate inside the mosquito's gut instead of simply being digested, that parasite can then sneak into a new host when the mosquito's saliva slips into a bite (Figure 6.1).

It is not the infected blood itself being passed along, but the whole mechanism of transformation in the pathogen (parasite) biology from merely "infected" to "infective" stage in mosquito's body which is why one cannot get human immunodeficiency virus (HIV) or novel coronavirus (nCoV) from a mosquito's bite. Still, plenty of pathogens have evolved to thrive in mosquito saliva, a perfect sanctuary of some sort. When mosquitoes perform such an act and allow the parasite complete a part of its biology or life cycle in the female mosquito's body then they are regarded as vectors.

6.3.2 Pestilent Mosquitoes

In principle all vector mosquitoes are pest as well since they cause loss of tranquility to people. Theoretically, therefore, there are more pestilent species than the vector species. Although all vector species are also essentially pestilent, the reverse may not be always likely. The hematophagous female mosquitoes leave, depending on the species concerned, behind post-engorgement a reddened "wheel" at the site of bite (Figure 6.2). This causes irritation and inflammation and becomes a reason for scratching the wheel vigorously that may lead to abscesses, occasionally with bleeding. Occasionally this may lead to allergy of some kind at the site of bite.

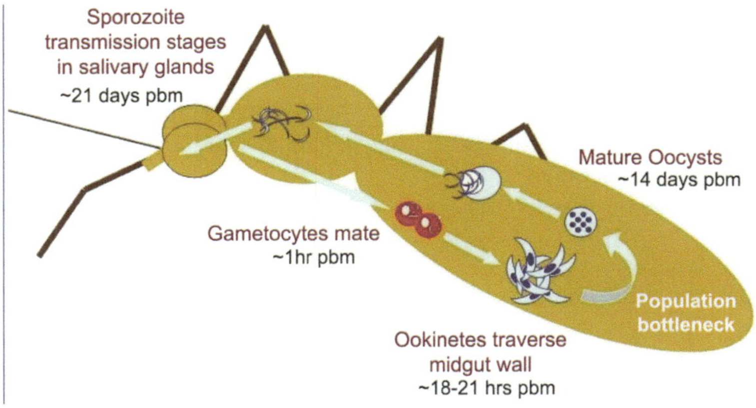

FIGURE 6.1 Graphic representation of development of malaria parasite in mosquito.

Source: **Free Wikipedia.**

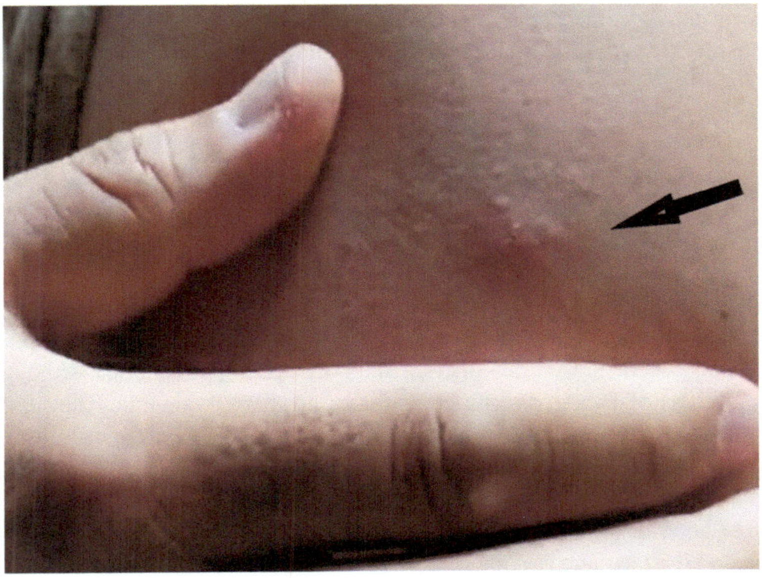

FIGURE 6.2 A mosquito bite on the upper arm of a person. Note the arrow indicating the site of the bite which develops onto a "wheel" later, depending on species.

Mosquito's vicious and painful bite is a very carefully planned activity, with the sole purpose to feed on blood. The bite often leaves behind an imprint, evanescent though, of a slowly developing itchy wheel. However, if it is a first mosquito bite ever, the roundish wheel fails to appear. The immune system in human body begins thereafter, and develops antibodies to recognize and flag the foreign chemicals. But after just one dose of mosquito spit, the familiar itchy bump left behind by subsequent bites caused by an allergic reaction to all those chemicals the mosquitoes are squirting into the body. The immune system releases histamines, chemicals that make the cells in the lining of human blood vessels spread apart. That lets the fluid leak out into the skin. That fluid is carrying components of man's immune system that attacks and destroys the foreign chemicals from the mosquito. The extra fluid from the leaking blood vessels causes a bump, which irritates other receptors in the skin, leading to the itching. Children get bigger bumps from mosquito bites than adults because their immature immune systems respond with larger amounts of histamines. More leaking blood vessels equal bigger bumps, which mean more itching and scratching. However, following repeated mosquito biting, the human immune system also becomes stronger and exhibits a different reaction to mosquito saliva. Instead of triggering histamines, another part of the immune system grabs onto the foreign chemicals and drags them away to macrophages, which destroy the chemicals in the mosquito's saliva.

As an example some of more vicious biters and irritant mosquitoes are briefly described, including a few also prevalent in India.

6.3.2.1 Mansonia and Coquillettidia Mosquitoes

These are large, yellowish-brown to dark-brown mosquitoes with very painful bites. Mansonia occurs in all zoogeographical regions, while Coquillettidia occurs mainly in Afro-tropical region but also in Nearctic, Neotropical, Oriental and Australasian regions (Figure 6.3).

Mansonia mosquitoes are characterized by possessing postspiracular setae, but no prespiracular setae are found. They are in the habit of biting and resting both inside and outside the houses. They breed in the heavily vegetated pond water growing aquatic plants such as *Pistia* spp. etc. Larvae respire on air-conducting parts of the aquatic plants. Coquillettidia mosquitoes have neither pre- or postspiracular setae, but have an overall yellow color. The adult body has a covering of broad scales, which may form fine speckled patterns of color on the wings. The abdomen is blunt ended. The scales on the wings are very broad. The claws are simple and pulvilli are small or even absent. Some species like *Coquillettidia perturbans* breeds in marshes, ponds and lakes having a thick growth of aquatic vegetation. Larval development is generally slow, lasting several months. Just like that in Mansonia species, larvae in this species too breathe by penetrating air-conducting plant tissues with their breathing tubes. They readily enter the house and bite viciously. Some of them, such as *Ma. annullifera* and *Ma. uniformis* are vectors of human lymphatic filariasis caused by *Brugia malayi* In India, this type of lymphatic filariasis is mostly restricted in Kerala state.

6.3.2.2 Hemagogus Mosquitoes

These mosquitoes occur in Neotropical region only where they are of much importance in public health. They mainly occur in Central America and northern South America, although some species inhabit forested areas of Brazil, and range as far as northern Argentina. The female Hemagogus mosquitoes are day-flying and most probably suck human blood. These mosquitoes are identifiable due to certain specific characteristics. Adults are densely scaly mosquitoes, whose general coloration is black with brilliant silvery markings. Scales on their wings are narrow, but they are fairly broad on the body. The anterior pronotal lobes are large, collar like, closely approximated behind the head and nearly meet at the top. The abdomen of these mosquitoes is blunt ended. Large part of dorsum of thorax without bristles; none present even immediately in front of scutellum. Forelegs of the sexes possess toothed claws. There are no prespiracular or postspiracular areas. Proboscis is slender, distinctly longer than fore femur. The palpi not more than 1/6th of length of proboscis, in males these

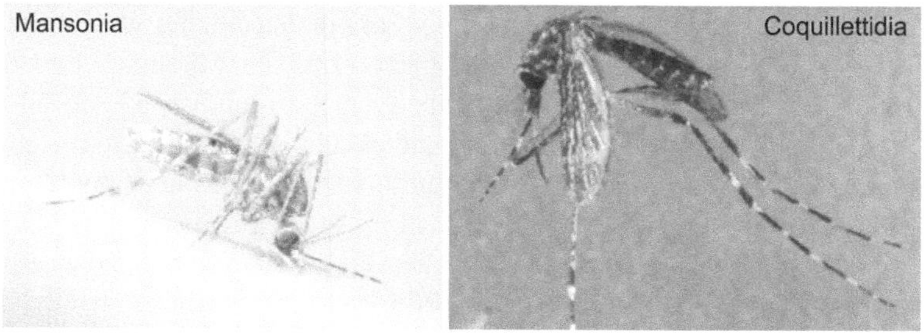

FIGURE 6.3 Mansonia and Coquillettidia mosquitoes.

Source: **Free Wikipedia.**

likewise are shorter than the proboscis. Hemagogus species are nefariously known for their painful bites. Some of these species are serious vectors.

6.3.2.3 Sabethes Mosquitoes

Members of this genus occur only in Neotropical region. Sabethes or Canopy Mosquito are primarily an arboreal genus, breeding in plant cavities. They are identifiable due to scutum and abdomen being covered with scales of a metallic color (Harbach, 1994; Rodaniche and Galindo, 1957). The scales of wings are narrow, but on the body they are fairly broad. The abdomen is blunt ended. The anterior pronotal lobes are very large. The mid or hindlegs have a conspicuous area of long scales in the shape of a paddle (Figure 6.4). Prespiracular bristles are present but there are no postspiracular bristles. There is a group of bristles on the postnotum. The palps of males are shorter than the proboscis. They are doubted to be vectors, particularly yellow fever virus between rhesus macaques.

6.3.2.4 Psorophora Mosquitoes

Members of this genus occur only in the Neartic and Neotropical regions. This genus has several species which are of immense public health importance largely due to their nuisance value particularly when in large numbers. Many species like *P. ciliata* and *P. columbaie* are large with shaggy legs. The scales on body and wings are narrow and without metallic color. Prespiracular and postspiracular bristles are present. The abdomen is pointed at the tip. Both sexes have forelegs with toothed claws. The male palps are longer than the proboscis. They are vicious biter and because of their gigantic appearance they often offer a frightening look. Some species do not lay eggs directly in water but on the ground that is subject to flooding from rainfall, overflow or irrigation. They breed in temporary pools, irrigation ditches and seepage pools, and larvae survive predating on other aquatic insects including mosquito larvae. Like Coquillettidia and Mansonia mosquitoes, the Psorophora mosquitoes too are becoming more pestiferous as an ever-expanding human population invades their natural habitats. They have been found playing a key role in transmitting equine encephalitis virus, West Nile fever virus, and other arboviruses

FIGURE 6.4 A *Sabethes* mosquito.

Source: **Free Wikipedia.**

6.3.2.5 Culiseta Mosquitoes

Found in all zoogeographical regions in the world, they are identifiable due to presence of narrow scales on the body and wings which are non-metallic in color; there are, however, colored patterns on the body. The abdomen is blunt ended. Prespiracular bristles alone are present. Forelegs of males have toothed claws; in female's forelegs they are simple. The stem vein (base of subcostal vein) of the wing usually has bristles on the ventral surface. Most species are moderate to highly aggressive biters, attacking in the evening hours or in the shade during the day. A species named, *Culiseta melanura* is the primary enzootic vector of eastern equine encephalitis (EEEV) throughout eastern North America, although a number of other species serve as bridge vectors.

6.3.2.6 Armigeres Mosquitoes

This is a very painful biter, both during the day and at dusk, and occurs mainly in the Oriental region, but a few species are found in Australian region and one as far as Japan. A medium sized to rather large-sized mosquito it tends to seek blood by attacking and obstinate feeding. Mosquitoes of this genus are not overtly ornamental and resemble in general appearance and structure those of Aedes (Subgenus Stegomyia). Scales of head broad and flat, a few upright scales on nape, at least in some species, scutellar scales are all broad and flat. Proboscis rather stout, slightly curved downwards toward the tip, the curved part somewhat laterally flattened. Palpi of male as long as, or longer than, proboscis, of three segments, the last two of which are long, slender, turned upwards, and without hair-tufts. Palpi of male of variable length, form about 1/2 to 3/4 of length of proboscis. Antennae of male plumose, the plume-hairs projecting fairly evenly round shaft.

The abdomen of the female is somewhat tapering, with 8 partially retractile segments, and short cerci. The coxite of the male hypopygium has a lobe on the inner surface bearing spines. The style is of variable form, but usually with spines along the apical part or at the tip. The phallosome is divided into small lateral plates with teeth. The paraprocts are without teeth or hairs at the crown. The only character distinguishing the genus, as a whole, from Aedes appears to be the form of the proboscis which characteristically arching downwards.

Of the nearly 30 species known under the genus, 13 have so far been recorded in India, of which *Ar. subalbatus* is very common in the Western Ghat ranges in Kerala State (Figure 6.5). All species chiefly occur in forested areas with high rainfall, and the larvae live mainly in bamboos, coconut

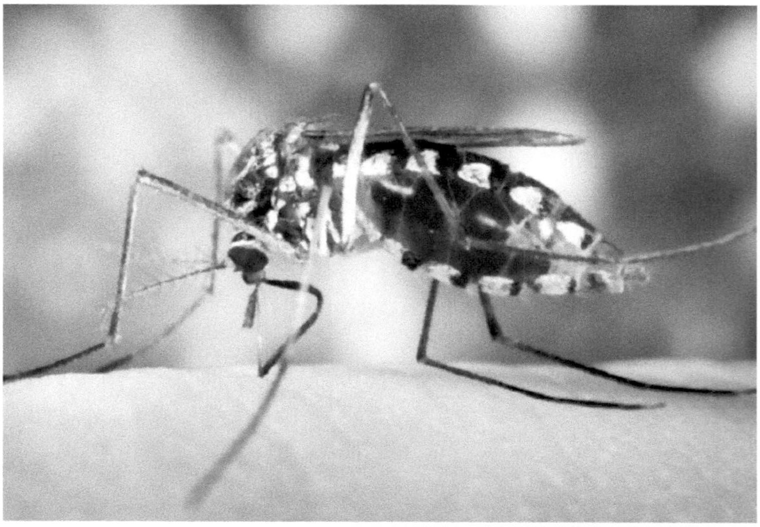

FIGURE 6.5 *Armigeres subalbatus* mosquito.

Source: Free Wikipedia.

shells etc. In Kerala's Western Ghat region where latex producing trees are afforested, *Ar. subalbatus* has often been collected in the latex collecting cup fastened along with the trunk of the tree. Two species, *Ar. subalbatus* and *Ar. flavus*, have been occasionally incriminated with the Japanese encephalitis virus (Tyagi et al., 2014).

6.4 INDIAN PESTILENT AND VECTOR MOSQUITOES

Out of a total of 417 mosquito species occurring in India nearly three dozen are dominant vector species and the rest are supporting or secondary vector species, as illustrated next:

1. Subfamily Anophelinae
   Genus **Anopheles**
   - (i) *Anopheles culicifacies*
   - (ii) *An. stephensi*
   - (iii) *An. fluviatilis*
   - (iv) *An. minimus*
   - (v) *An. baimaii (dirus)*
   - (vi) *An. sundaicus*
   - (vii) *An. philippinensis*
   - (viii) *An. subpictus*
   - (ix) *An. jeyporiensis*
   - (x) *An. annularis*
   - (xi) *An. varuna*

   } Malaria vectors

   - (xii) *An. peditaeniatus*
   - (xiii) *An. barbirostris*
   - (xiv) *An. subpictus*

   } JE vectors
2. Subfamily Culicinae
   Genus **Aedes**
   - (i) *Aedes aegypti*
   - (ii) *Ae. albopictus*

   } Dengue, chikungunya, Zika vectors

   - (iii) *Ae. vittatus*

   } Incriminated with CHKV in lab in India, but in Africa it is a vector for YFV and DENV in nature. A serious pest.

   Genus **Culex**
   - (i) *Culex quinquefasciatus*

   } Lymphatic filariasis vector

   - (ii) *Cx. tritaeniorhynchus*
   - (iii) *Cx. vishnui*
   - (iv) *Cx. pseudovishnui*
   - (v) *Cx. gelidus*
   - (vi) *Cx. epidesmus*
   - (vii) *Cx. bitaeniorhynchus*
   - (viii) *Cx. infula*
   - (ix) *Cx. gelidus*
   - (x) *Cx. whitmorei*
   - (xi) *Cx. fuscocephala*
   - (xii) *Cx. quinquefasciatus*

   } JE vectors

   Genus **Mansonia**
   - (i) *Mansonia uniformis*
   - (ii) *Ma. annularis*
   - (iii) *Ma. indiana*
   - (iv) *Ma. annulifera*

   Genus **Armigeres**
   - (i) *Armegeres subalbatus*

   } JE vector

6.5 HUMAN DISEASES TRANSMITTED BY MOSQUITOES

Mosquitoes are known to transmit diseases to both humans and animals. While in some cases of diseases (e.g., malaria, lymphatic filariasis, dengue, chikungunya, zika) mosquitoes, besides humans, are the only other host for the development and/or growth of the parasites or pathogen, in others (e.g., Japanese encephalitis) they are host to pathogens/parasites which utilize a vertebrate zoonosis as a medium for proliferation without actually harming the zoonotic animal host.

6.5.1 Malaria

Out of a total of 63 *Anopheles* species inventorized in India, only 11 species have been regarded as vectors; of these six are primary vectors, viz., *An. culicifacies, An. fluviatialis, An. stephensi, An. baimaii (dirus), An. sundaicus* and *An. minimus*, and remaining five species are secondary or supportive vectors, i.e., *An. subpictus, An., An. annularis* and *An. philippinensis* and *An. jeyporiensis* and *An. varuna*) (Rao, 1984; Nagpal and Sharma, 1995; Bhattacharyya et al., 2014; Tyagi et al., 2015; Tyagi, 2020; Dev, 2020). While *An. culicifacies* is predominantly a rural vector, transmitting approximately 65% of malaria in India, *An stephensi* is a major urban vector responsible for almost 10% of malaria in the country. *Anopheles fluviatialis* (15%), a major contributor to malaria is mostly found in the mountainous forest and foothill "*terai*" areas. Among the secondary vector species, *An. subpictus* is a significant vector in peninsular coastal region, even though the mosquito has been frequently incriminated with malaria parasite in several northern, western and central states (Panicker and Bai, 1983; Panicker et al., 1981; Tyagi and Yadav, 2001; Kumar et al., 2016). *Anopheles sundaicus*, a brackish water breeder, is mostly a vector in Andaman and Nicobar islands, besides some coastal area in the peninsular India. A likely distribution in India is presented in Figure 6.6.

6.5.2 Dengue, Chikungunya, and Zika

Only two mosquito species, *Aedes aegypti* and *Ae. albopictus*, are dominating vectors for dengue, chikungunya, and zika in the country (Christophers 1960). Contrary to earlier belief, *Ae. albopictus* has emerged as a major vector in recent years (Tyagi et al., 2006; Singh et al. 2019; Kumar et al.,

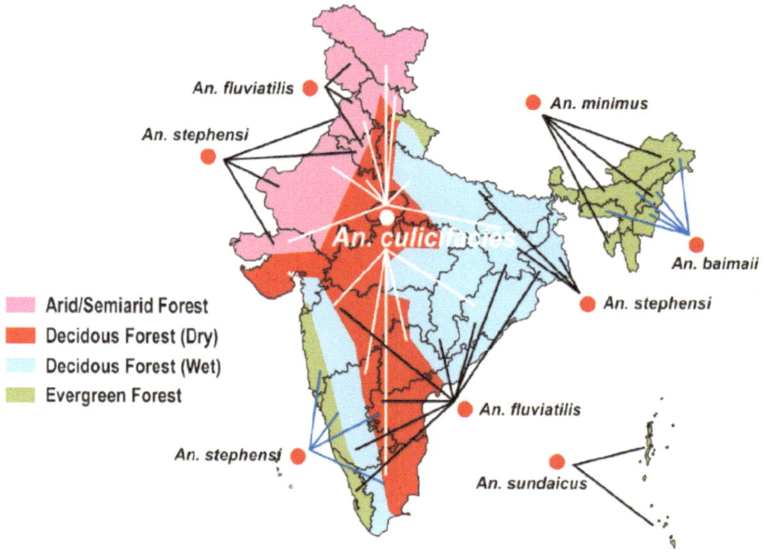

FIGURE 6.6 A likely geographical distribution of malaria vector species in India.

Source: Dev, 2020.

2021). While on one hand, *Ae. aegypti* is strongly anthropophilic, preferring to breed all kinds of man but also natural water-holding containers in close vicinity to human habitation in urban agglomeration, *Ae. albopictus*, on the other hand, is dominantly rural in character, preferring to breed both in natural and artificial containers, and feed and rest outdoors (Tyagi, 2008). A third species, *Ae. vittatus*, is only a rare vector, and has been incriminated with chikungunya virus in a laboratory in India (Mourya and Banerjee, 1987) and with dengue and yellow fever virus in Africa in nature (Sudeep and Shil, 2017). Currently, 35 out of 36 States and Union Territories are reporting dengue (Baruah et al., 2021).

6.5.3 Lymphatic Filariasis

Culex quinquefasciatus, the ubiquitous and highly anthropophilic mosquito, preferring to breed in polluted cesspits, cesspools, wells, drains, laterine pits etc., is the most important vector contributing approximately 96% of the infection by *Wuchereria bancrofti* (Tyagi, 2018; Joshi, 2018; Srivastava et al., 2018; Sunish et al., 2007, 2015; Tyagi, 2014). The remaining 4% is contributed by *Brugian malayi*, largely endemic to Kerala State, is transmitted by many of the species of Mansonia such as *Ma. uniformis*, *Ma. annularis*, etc. (Sabesan et al., 1991; Arunachalam et al., 2004).

6.5.4 Japanese Encephalitis

Mosquito species belonging to *Culex vishnui* subgroup particularly *Cx. tritaeniorhynchus, Cx. pseudovishnui* and *Cx. vishnui* are among the most dangerous vectors in whole of Asia (Arunachalam et al., 2008; Kabilan, 2004; Reuben et al., 1994; Tyagi et al., 2014), although as many as 17 species under four genera have been identified to be involved in JE transmission in nature (Thenmozhi et al., 2015, 2016) (Figure 6.7).

The surprise finding has been the incrimination of *Armigeres subalbatus* from a male specimen. It is noteworthy that this species is an active JEV transmitter in the Torres Strait region of Australia as well as the mainland. The multi-zoonotic and multi-vectorial dynamism of Japanese encephalitis make the disease generally very difficult to control, without vaccination either to the humans, who are the dead-ends in this kind of infection, or zoonosis, e.g., via swine or pigs.

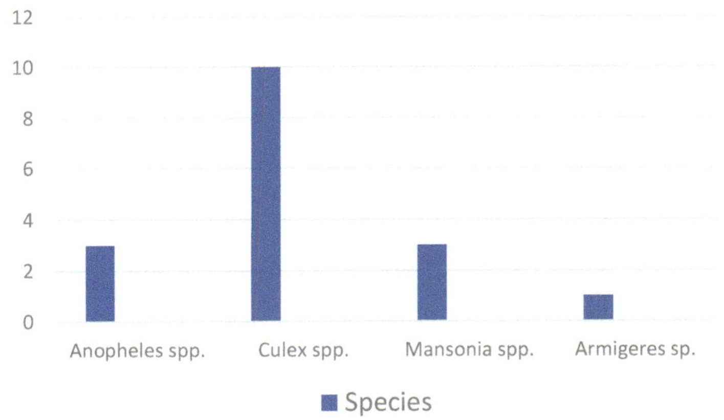

FIGURE 6.7 Various different mosquito vector species incriminated with Japanese encephalitis virus (JEV) in India.

Source: Tyagi, ***this work***.

6.5.5 West Nile Virus

Birds act as a carrier and amplifying hosts of the West Nile Virus (WNV) in nature. Ornithophilic mosquitoes mainly belonging to *Culex* species act as a vector for the transmission of the infection from viraemic birds to a large spectrum of vertebrate hosts. Globally, though isolatedly, *Cx, univittatus* (South Africa, Israel), *Cx. modestus* (France), *Cx. vishnui* complex (India, Pakistan), *Cx. pipiens pipiens* (USA, Romania) act as a vector of the WNV. In India, *Cx. quinquefasciatus Culex vishnui* group mosquitoes have been found to transmit WNV in various southern, central, western and northeast states in India (Banker, 1952; Paramasivan et al., 2003; Khan et al., 2011). The disease of which antibodies in humans were first discovered in Maharashtra (India) in 1952 has since gradually spread in most parts of the country through a variety of mosquitoes.

6.6 ANIMAL DISEASES TRANSMITTED BY MOSQUITOES

6.6.1 Canine Dirofilariasis

Aedes albopictus is the transmitter for dirofilariasis, a parasitic infection in canids caused by *Dirofilaria (Nochtiella) repens* and *D. immitis*. The definitive mammalian hosts for Dirofilaria are primarily domestic dogs, wild canids (such as wolves and foxes), and raccoons. In these hosts, sexually mature worms produce microfilariae that circulate in the blood and are ingested by mosquitoes during a blood meal. Rarely mosquitoes may carry these microfilariae to humans, though human cases are gradually witnessed in more and more places. In India, majority of human dirofilarial cases have been reported from Kerala followed by Karnataka and Maharashtra, although sporadic cases were also reported from Tamil Nadu, Orissa, Bihar, Haryana, Gujarat, Punjab and Assam. A recent case of dirofilariasis was documented from Malappuram District, Kerala (Haroon et al., 2022). Over the past decade, most countries in Southeast Asia and some in Asia have reported the presence of *P. knowlesi*, namely, Singapore, Thailand, Myanmar, Philippines, Indonesia, Vietnam, Cambodia, Brunei, and China.

6.6.2 Avian and Simian Malaria

Avian malaria, caused by a protozoan parasite (*Plasmodium relictum*), is a mosquito-borne disease of birds. *Plasmodium* species are capable of infecting and completing their life cycle in > 400 species of birds. *Plasmodium relictum* reproduces in avian red blood cells and if the parasite load is sufficiently high, the bird loses red blood cells and becomes anemic. Vectors for avian malaria include several mosquitoes from the genera *Culex, Aedes*, and *Culiseta.*

The vectors of *P. knowlesi*, the fifth human malaria parasite, and other Asian simian malarias belong to the Leucosphyrus Group of *Anopheles* mosquitoes which are generally forest dwelling species. So far 33 simian malarial parasites have been documented in several prosimians, New World and Old World monkeys, African and Asian apes (Coatney et al., 1971; Leclerc et al., 2004; Ollomo et al., 2009; Krief et al., 2010; Akter et al., 2015). *Plasmodium knowlesi, P. cynomolgi, P. inui, P. coatneyi*, and *P. fieldi* are the five common species of simian malaria parasites in *Macaca fascicularis* and *M. nemestrina* (Eyles, 196323). Over the past decade, most countries in Southeast Asia and some in Asia have reported the presence of *P. knowlesi*, namely, Singapore, Thailand, Myanmar, Philippines, Indonesia, Vietnam, Cambodia, Brunei, China, and Malaysia (Ng et al., 2008; Jongwutiwes et al., 2004; Jiang et al., 2010; Luchavez et al., 2008; Figtree et al., 2010; Sulistyaningsih et al., 2010; Eede et al., 2009; Khim et al., 2011; Ninan et al., 2012; Zhu et al., 2006). Recently, Tyagi et al. (2013) reported human *P. knowlesi* infections from the Andaman and Nicobar Islands of India which are geographically nearby these Southeast Asian countries.

6.7 MOSQUITOES AS A SOURCE OF ALLERGENS

Mosquito-related allergy occurs worldwide in two major forms: (i) mosquito saliva-based allergies, and (ii) mosquito allergy through body-based *per se*. Mosquitoes being hematophagous bite and blood-feed on humans, and as a result are a source of allergic reactions at the site of piercing. One such severe allergy is called "Skeeter syndrome" – a large local allergic reaction to mosquito bites marked by significant inflammation. The Skeeter syndrome occurs when the mosquito during the process of blood-feeding releases into human bloodstream saliva containing polypeptides to which human immune system is sensitive to produce an array of symptoms: large areas of swelling, skin warmth, skin color or texture changes such as reddening of lighter color skin or darkening, or other appearance changes in darker color skin, and finally, the area getting harder, itching, pain, blisters, and fever. Antibiotics so not help resolve the issue, and having been in full sight for about 8–10 hours after the mosquito bite, the symptoms typically resolve within 3–10 days. However, this can be incapacitating in some highly sensitive people who may have trouble moving (Figure 6.8).

In addition to saliva-driven allergy, mosquito body with parts like scales and wings might also cause some severe allergies like asthma, allergic rhinitis, and probably conjunctivitis. There are about 500 vector mosquito species which are involved in some kind of infectious diseases, although the number of species which are pestilent is even greater. Compared to both pest and vectors,

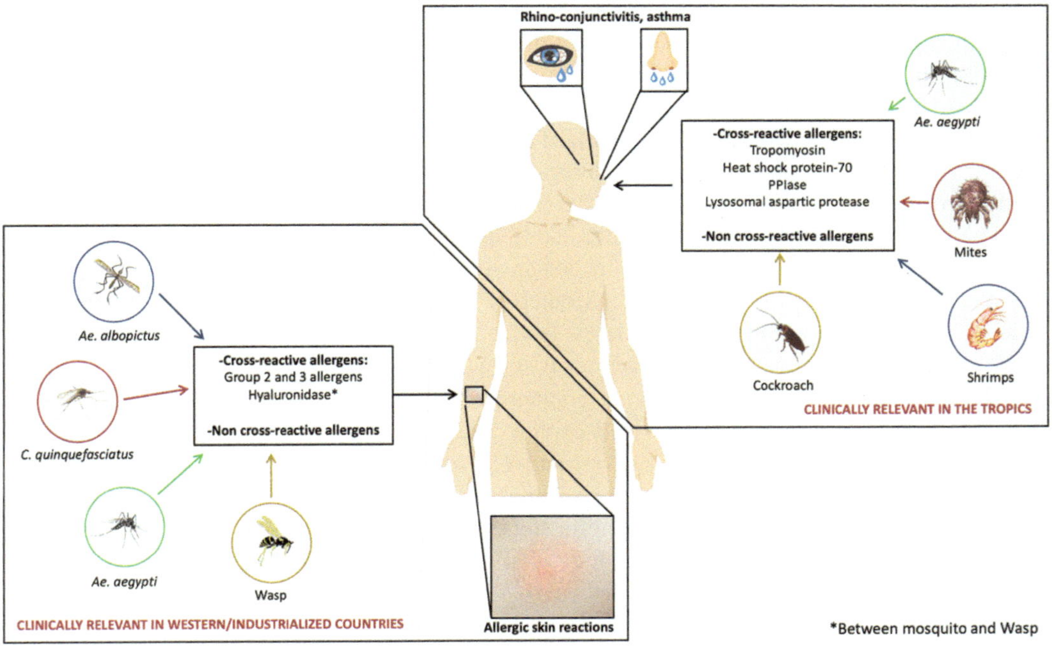

FIGURE 6.8 Unique and cross-reactive mosquito allergens induce different manifestations of mosquito allergy. Mosquito species such as *Ae. aegypti*, *Ae. albopictus*, and *C. quinquefasciatus* are distributed worldwide; however, some studies suggest that two kinds of allergic reactions induced by mosquito allergens are clinically relevant depending on the geospatial location: allergic skin reactions induced by salivary allergens and respiratory reactions induced by body-derived allergens. Skin reactions are common in western and industrialized countries and respiratory reactions are relevant in tropical areas. Cross-reactivity between mosquito species and with several species within Arthropoda may play an important clinical role.

***Source*: Cantillo and Puerta, 2021; Permission received dt 230723.**

mosquitoes that are sources for some kind of allergies surpass them all put together in various different geographical regions. It is noteworthy that mosquito species in different regions may exhibit different levels of sensitivity to humans. It is well known that several species may cause varying degrees of allergies, however, not all of these allergy-causing mosquito species have been studied scientifically, or at least adequately to offer some dependable data. Still, a few of these have been explored region-wise for their role as sources of allergens (Cantillo and Puerta, 2021; Tyagi et al., 2006):

(i) Cosmopolitan: *Anopheles (An.) stephensi, An. minimus, An. sinensis, Ochlerotatus (Oc.) triseriatus, Oc. hendersoni, Culex (Cx.) quinquefasciatus, Cx. tritaeniorhynchus, Cx. pipiens, Cx. pipiens pallens*, and *Cx. tarsalis*.
(ii) Old world: (Africa, Asia, and Europe): *Aedes (Ae.) aegypti, Ae. vexans, Ae. communis, Ae. togoi, Ae. albopictus, Ae. triseriatus* and *Armegeres subalbatus*.
(iii) New world (America): *Culiseta inornate*.

Intensity of human sensitivity appears to vary a great deal to the nature of saliva of different mosquito species, as well as in manifestation of kind of allergy that could be influenced by geographical localization of mosquito species and global warming. Some mosquito species, e.g., *Aedes albopictus* and *Anopheles stephensi* which have recently invaded many countries earlies a *terra incognito*, may pose greater threat of saliva-based allergies to non-immune populations. In such instances this affect spectrum of pathogenesis and the accuracy of the diagnosis of mosquito allergy.

Experimental and clinical data suggest that body allergens mediate different manifestations of allergic reactions such as asthma and rhinitis. In an Indian investigation, 47% of the population with asthma and/or allergic rhinitis were sensitized to mosquito allergens, as determined by skin prick tests (SPT), serum specific IgE antibodies and bronchial provocation tests with whole mosquito body extracts (Agarwal et al., 1991), whereas in a Chinese study in Guangzhou town, in a cohort of 7,047 allergic patients, 4% of them had detectable specific IgE levels to mosquito allergens, ranging from ≥0.35 to <3.5 IU/ml in most of the patients, with peaks of sensitization at age between 15 and 18 years (Sun et al., 2014). Similarly, in Monterrey City, Mexico, a cross-sectional study reported that 82% of patients admitted to the allergy service had specific IgE to mosquitoes, although only 2.5% of them showed positive skin reactions (Gonzalez et al., 2010).

The whole gamut of mosquito allergen response is very complex, since not many species' saliva has so far been extensively and intensively explored for characterization and identification. The best species hitherto examined are *Ae. aegypti* and *Cx. quinquefasciatus*. Cross-reactivity occurring among mosquitoes and several other non-mosquito or even non-insect arthropod species is a serious issue in correct characterization and identification of the components of the saliva. The effects that this phenomenon has on the pathophysiology of allergy diseases is still unknown.

BIBLIOGRAPHY

Agarwal, M.K., Chaudhry, S., Jhamb, S., Gaur, S.N., Chauhan, U.P., Agarwal, H.C. (1991). Etiologic significance of mosquito (*Anopheles stephensi*) in respiratory allergy in India. *Annals of Allergy, Asthma & Immunology*, 67, 598–602.

Akter, R., Vythilingam, I., Khaw, L.T. et al. (2015). Simian malaria in wild macaques: first report from Hulu Selangor district, Selangor, Malaysia. *Malaria Journal*, 14, 386. https://doi.org/10.1186/s12936-015-0856-3

Arunachalam, N., Samuel, P.P., Hiriyan, J., Thenmozhi, V., Gajanana, A. (2004). Japanese encephalitis in Kerala, south India: can Mansonia (Diptera: Culicidae) play a supplemental role in transmission? *Journal of Medical Entomology*, 41(3), 456–61. https://doi.org/10.1603/0022-2585-41.3.456

Arunachalam, N., Samuel, P. Philip, Paramasivan, R., Balasubramanian, A., Tyagi, B.K. (2008). Japanese encephalitis in Gorakhpur Division, Uttar Pradesh. *Indian Journal of Medical Research*, 128(6), 775–777.

Bai, M.G., Panicker, K.N., Rajagopalan, P.K. Biology of the predator mosquito, *Toxorhynchites splendens* (Wiedemann) (Diptera: culicidae). *Indian Journal of Medical Research*, 74, 13–7.

Banker, D.D. (1952). Preliminary observations on antibody patterns against certain viruses among inhabitants of Bombay city. *Indian Journal of Medical Science*, 6, 733–46.

Baruah, K., Arora, N., Sharma, H. and Katewa, A. (2021). Dengue in India: Temporal and spatial expansion in last two decades. *Journal of Medical Arthropodology & Public Health*, 1(1), 15–32.

Bhattacharyya, D., Rajavel, A., Natarajan, R., Mohapatra, P., Jambulingam, P., Mahanta, J. and Prakash, A. (2014). Faunal richness and the checklist of Indian mosquitoes (Diptera: Culicidae). *Check List*, 10(6), 1342–1358.

Cantillo, J.F. and Puerta, L. (2021). Mosquitoes: Important Sources of Allergens in the Tropics. *Front Allergy*, 2, 690406. https://doi.org/10.3389/falgy.2021.690406

Christophers, S.R. (1960). Aedes aegypti *(L.), the Yellow Fever Mosquito: Its Life History, Bionomics and Structure*. Cambridge University Press, London, 739 pp.

Clements, A.N., (1992). *The Biology of Mosquitoes. Vol. 1. Development, Nutrition and Reproduction.* CABI Publishing, 571 pp.

Clements, A.N. (1999). *The Biology of Mosquitoes. Vol. 2. Sensory Reception and Behaviour.* Wallingford: CABI Publishing, 752 pp.

Clements, A.N. (2011). *The Biology of Mosquitoes. Vol. 3: Transmission of Viruses and Interactions with Bacteria.* CABI Publishing, 592 pp.

Coatney, R.G., Collins, W.E., Warren, M. and Contacos, P.G. (1971). *The Primate Malarias.* Washington: U.S. Government Printing Office.

Dev, V. (2020). *Vector Biology and Control – An Update for Malaria Elimination Initiative in India.* The National Academy of Sciences, India. 272 pp.

Eede, P., Van, H., Van Overmeir C, Vythilingam, I., Duc, T., Hung, L., et al. (2009). Human *Plasmodium knowlesi* infections in young children in central Vietnam. *Malaria Journal*, 8, 249.

Eldridge, B.F. and Edman, J.D. (2012). *Medical Entomology: A Textbook on Public Health and Veterinary Problems Caused by Arthropods.* Springer Dordrecht. 659 pp. https://doi.org/10.1007/978-94-007-1009-2

Eyles, D.E. (1963). The species of simian malaria: taxonomy, morphology, life cycle, and geographical distribution of the monkey species. *Journal of Parasitology*, 49, 866–87.

Figtree, M., Lee, R., Bain, L., Kennedy, T., Mackertich, S., Urban, M., et al. (2010). *Plasmodium knowlesi* in Human, Indonesian Borneo. *Emerging Infectious Diseases*, 14, 672–4.

Gonzalez, Diaz S.N., Cruz, A.A., Sedo, Mejia G.A., Rojas, Lozano A.A., Valenzuela, E.A., Vidaurri, Ojeda A.C. (2010). [Prevalence of reactions secundary to mosquito bites Aedes aegypti at en el Regional Center of Allergy and Clinical Immunology, University Hospital, de Monterrey, Nuevo Leon]. *Revista Alergia México*, 57, 37–43.

Harbach, R.E. (1994). The subgenus *Sabethinus* of Sabethes (Diptera: Culicidae). *Systematic Entomology*, 19, 207–234.

Haroon, A., Rajendran, R. and Dilip Kumar, T. (2022). Increasing cases of dirofilariasis in Kerala – an emerging public health concern. *Journal of Medical Arthropodology & Public Health*, 2(1), 9–14.

Harrison, G. (1978). *Mosquitoes, Malaria and Man: A History of Hostilities Since 1880.* London: J. Murray.

Jiang, N., Chang, Q., Sun, X., Lu, H., Yin, J., Zhang, Z. et al. (2010). Co-infections with *Plasmodium knowlesi* and other malaria parasites, Myanmar. *Emerging Infectious Diseases*, 16, 1476–8.

Jongwutiwes, S., Putaporntip, C., Iwasaki, T., Sata, T. and Kanbara, H. (2004). Naturally acquired *Plasmodium knowlesi* malaria in human, Thailand. *Emerging Infectious Diseases*, 10, 2211–3.

Joshi, P.L. (2018). Epidemiology of Lymphatic Filariasis. pp. 1–14. *In*: Tyagi, B.K. (ed.) *Lymphatic Filariasis.* Springer, Singapore. https://doi.org/10.1007/978-981-13-1391-2_1

Kabilan, L. (2004). Control of Japanese encephalitis in India: A reality. *Indian Journal of Pediatrics*, 71, 707–712. https://doi.org/10.1007/BF02730659

Khan, S.A., Dutta, P., Khan, A.M., Chowdhury, P., Borah, J., Doloi, P., Mahanta, J. (2011). West Nile virus infection, Assam, India. *Emerging Infectious Diseases*, 17(5), 947–8. 10.3201/eid1705.100479

Khim, N., Siv, S., Kim, S., Mueller, T., Fleischmann, E., Singh, B., et al. (2011). *Plasmodium knowlesi* infection in humans, Cambodia, 2007–2010. *Emerging Infectious Diseases*, 17, 1900.

Krief, S., Escalante, A.A., Pacheco, M.A., Mugisha, L., André, C., Halbwax, M., et al. (2010). On the diversity of malaria parasites in African apes and the origin of *Plasmodium falciparum* from Bonobos. *PLoS Pathogens*, 6, e1000765.

Kumar, A., Hosmani, R., Jadhav, S. et al. (2016). *Anopheles subpictus* carry human malaria parasites in an urban area of Western India and may facilitate perennial malaria transmission. *Malaria Journal*, 15, 124. https://doi.org/10.1186/s12936-016-1177-x

Kumar, N.P., Kumar, A., Panneer, D., Abidha, S., Muthukumaravel, S., Sankari, T., Ajithlal, P.M., Mathew, J., Koothradan, S, Paramasivan, R., Muniyaraj, M., Singh, H., Saxena, Rekha, Vijayachari, P., Sunish, I.P., Shriram, A.N., Dutta, Prafulla, Patgiri, Saurav Jyoti, Bhattacharyya, D.R., Hoti, S.L., Chattopadhyay, D., Roy, S., Mahapatra, N., Pati, S., Chand, G., Mishra, A.K., Barde, P., Jambulingam, P. (2021). Nationwide vector surveillance on Zika and Dengue did not indicate transmission of the American lineage-pandemic Zika virus in India. *International Journal of Infectious Diseases* 113, 119–124. https://doi.org/10.1016/j.ijid.2021.09.074

Leclerc, M., Hugot, J., Durand, P. and Renaud, F. (2004). Evolutionary relationships between 15 Plasmodium species from New and Old World primates (including humans): a 18S rDNA cladistic analysis. *Parasitology*, 129, 677–84.

Luchavez, J., Espino, F., Curameng, P., Espina, R., Bell, D., Chiodini, P., et al. (2008). Human infections with *Plasmodium knowlesi*, the Philippines. *Emerging Infectious Diseases*, 14, 811–3.

Mourya, D.T. and Banerjee, K. (1987). Experimental transmission of chikungunya virus by *Aedes vittatus* mosquitoes. *Indian Journal of Medical Research*, 86, 269–71.

Nagpal, B.N. and Sharma, V.P. (1995). *Indian Anophelines*, Oxford and IBH Publishing Co. Pvt. Ltd., New Delhi.

Ng, O.T., Ooi, E.E., Lee, C.C., Lee, P.J., Ng, L.C., Pei, S.W. et al. (2008). Naturally acquired human Plasmodium knowlesi infection, Singapore. *Emerging Infectious Diseases*, 14, 814.

Ninan, T., Nalees, K., Newin, M., Sultan, Q., Than, M.M., Shinde, S., et al. (2012). *Plasmodium knowlesi* malaria infection in human. *Brunei International Medical Journal*, 8, 358–61.

Ollomo, B., Durand, P., Prugnolle, F., Douzery, E., Arnathau, C., Nkoghe, D., et al. (2009). A new malaria agent in African hominids. *PLOS Pathogens*, 5(5), e1000446. https://doi.org/10.1371/journal.ppat.1000446

Panicker, K., Bai, M.G., Rao, U.B., Viswam, K. and Murthy, U.S. (1981). *Anopheles subpictus*, vector of malaria in coastal villages of south-east India. *Current Science*, 50, 694–5.

Panicker, K.N. and Bai, M.G. (1983). Field release of *Toxorhynchites splendens* (Diptera; ulicidae) in controlling container breeding mosquitoes in a coastal village. *Indian Journal of Medical Research*, 77, 339–41.

Paramasivan, R., Mishra, A.C. and Mourya, D.T. (2003). West Nile virus: the Indian scenario. *Indian Journal of Medical Research*, 118, 101–8.

Rao, T.R. (1984). *The Anopheline of India* (Revised Edition.). Malaria Research Centre, Indian Council of Medical Research, New Delhi, 518 pp.

Reuben, R., Tewari, S.C., Hiriyan and Akiyama, J. (1994). Illustrated key to species of Culex (Culex) associated with Japanese encephalitis in southeast Asia (Diptera: Culicidae). *Mosquito Systematics*, 26, 75–96.

Rodaniche, E. de and Galindo, P. (1957). Isolation of Ilhéus Virus from Sabethes chloropterus captured in Guatemala in 1956. *American Journal of Tropical Medicine and Hygiene*, 6(4), 686–687.

Sabesan, S., Kumar, N.P., Krishnamoorthy, K. and Panicker, K.N. (1991). Seasonal abundance & biting behaviour of *Mansonia annulifera, M. uniformis & M. indiana* and their relative role in the transmission of Malayan filariasis in Shertallai (Kerala State). *Indian Journal of Medical Research*, 93, 253–8.

Sebastian, A., Sein, M.M., Thu, M.M., Corbet, P.S. (1990). Suppression of *Aedes aegypti* (Diptera: Culicidae) using augmentative release of dragonfly larvae (Odonata: Libellulidae) with community participation in Yangon, Myanmar. *Bulletin of Entomological Research*, 80, 223–232.

Sebastian, A., Thu, M.M., Kyaw, M., Sein, M.M. (1980). The use of dragonfly larvae in the control of Aedes aegypti. *Southeast Asian Journal of Tropical Medicine and Public Health*, 11, 104–107.

Singh, H., Singh, O.P., Akhtar, N., Sharma, G., Sindhania, A., Gupta, N. and Valecha, N. (2019). First report on the transmission of Zika virus by Aedes (Stegomyia) aegypti (L.) (Diptera: Culicidae) during the 2018 Zika outbreak in India. *Acta Tropica* 199, 105114. https://doi.org/10.1016/j.actatropica.2019.105114

Sinka, M., Bangs, M.J., Manguin, S., Rubio-Palis, Y., Chareonviriyaphap, T., Coetzee, M., Mbogo, C.M., Hemingway, J., Patil, A.P., Temperley, W.H., Gething, P.W., Kabaria, CW, Burkot, T.R., Harbach, R.E. and Hay, S.I. (2012). A global map of dominant malaria vectors. *Parasites & Vectors*, 5, 69. https://doi.org/10.1186/1756-3305-5-69

Sinka, M.E. (2013) Global distribution of the dominant vector species of malaria. Chapter 4. (Anopheles mosquitoes – New insights into malaria vectors. Ed. Sylvie Manguin). Intech. Croatia, pp. 109–144; http://doi.org/10.5772/54163

Sinka, M.E., Bangs, M.J., Manguin, S., Chareonviriyaphap, T., Patil, A.P., Temperley, W.H., AGething, P.W., Elyazar, I.R., Kabaria, C.W., Harbach, R.E. and Hay, S.I. (2011) The dominant Anopheles vectors of human malaria in the Asia-Pacific region: occurrence data, distribution maps and bionomic précis. *Parasites & Vectors*, 4, 89. https://doi.org/10.1186/1756-3305-4-89

Sinka, M.E., Bangs, M.J., Manguin, S., Maureen, C., Mbogo, C.M., Hemingway, J., Anand, P.P., Temperley, W.H., Gething, P.W., Kabaria, C.W., Okara, Robi M., Thomas, Van B., Godfray, C.J., Harbach, R.E. and

Simon, I.H. (2010) The dominant Anopheles vectors of human malaria in Africa, Europe and the Middle East: occurrence data, distribution maps and bionomic précis. *Parasites & Vectors*, 3, 117. https://doi.org/10.1186/1756-3305-3-117

Sinka, M.E., Pironon, S., Massey, N.C., Longbottom, J., Hemingway, J., Moyes, C.L., Willis, K.J. (2020). A new malaria vector in Africa: Predicting the expansion range of *Anopheles stephensi* and identifying the urban populations at risk. *Proceedings of National Academy of Sciences*, 117(40), 24900–24908. https://doi.org/10.1073/pnas.2003976117

Spielman, A. and D'Antonio, M. (2002). *Mosquito: The Story of Man's Deadliest Foe*. Hyperion, 272 pp.

Srivastava, P.K., Dhariwal, A.C. and Tyagi, B.K. (2018). The Indian Lymphatic Filariasis Elimination Programme: The Success to Sustain. pp. 193–198. *In*: Tyagi, B.K. (eds) *Lymphatic Filariasis*. Springer, Singapore. https://doi.org/10.1007/978-981-13-1391-2_15

Sudeep, A.B. and Shil, P. (2017). *Aedes vittatus* (Bigot) mosquito: An emerging threat to public health. *Journal of Vector Borne Diseases*, 54, 295–300.

Sulistyaningsih, E., Fitri, L.E., Löscher, T. and Berens-Riha, N. (2010). Diagnostic difficulties with Plasmodium knowlesi infection in humans. *Emerging Infectious Diseases*, 16, 1033.

Sun, B-Q, Zheng, P-Y, Zhang X-W, Huang H-M, Chen D-H, Zeng G-Q. (2014). Prevalence of allergen sensitization among patients with allergic diseases in Guangzhou, Southern China: a four-year observational study. *Multidisciplinary Respiratory Medicine*, 9, 2. https://doi.org/10.1186/2049-6958-9-2

Sunish, I.P., Kalimuthu, M., Rajendran, R., Munirathinam, A., Kumar V.A., Nagaraj, J., Tyagi, B.K. (2015). Decline in lymphatic filariasis transmission with annual mass drug administration using DEC with and without albendazole over a 10 year period in India. *Parasitology International*, 64(1), 1–4.

Sunish, I.P., Rajendran, R., Mani, T.R., Munirathinam, A., Dash, A.P. and Tyagi, B.K. (2007), Vector control complements mass drug administration against bancroftian filariasis in Tirukoilur, India. Vector control in lymphatic filariasis elimination. *Bulletin of the World Health Organization*, 85(2), 138–145.

Thenmozhi, V., Balaji, T., Selvam, A., Venkatasubramani, K. and Dhananjeyan, K.J. (2015). A longitudinal study on abundance and infection frequency of Japanese encephalitis vectors in Tirunelveli district, Tamil Nadu, India. *International Journal of Mosquito Research*, 2(3), 166–169.

Thenmozhi, V., Balaji, T., Venkatasubramani, K., Dhananjeyan, K.J., Selvam, A., Rajamannar, V. and Tyagi, B.K. (2016). Role of *Anopheles subpictus* Grassi in Japanese encephalitis virus transmission in Tirunelveli, South India. *Indian Journal of Medical Research*, 144(3), 477–481. https://doi.org/10.4103/0971-5916.198684

Tyagi, B.K. (2004). *The invincible deadly mosquitoes: India's Health and Economy Enemy No. 1*. Scientific Publishers, India, 266 pp.

Tyagi, B.K. (2008). *Vector-borne diseases: Epidemiology and control*. Scientific Publishers (India), 298 pp.

Tyagi, B.K. (2014). *The Tirukoilur Project*. Centre for Research in Medical Entomology (ICMR), Madurai (TN), India, 202 pp.

Tyagi, B.K. (2018). *Lymphatic Filariasis: Epidemiology, Treatment and Prevention – The Indian Perspective*. Springer, 314 pp.

Tyagi, B.K. (2020). *Mosquito hunters: a history of hostilities against man's deadliest foe – the mosquito – since 1881*. Scientific Publishers, Jodhpur (India), 474 pp.

Tyagi, B.K. (2023). *Desert Malaria: An Emerging Malaria Paradigm and Its Global Impact on Disease Elimination*. Springer-Nature, Singapore, 416 pp.

Tyagi, B.K., Bhattacharya, S. and Reddya, N.B. (2020). *Dr Ronald Ross: mosquito, malaria, India and the Nobel Prize – an untold story of the First Indian Nobel Laureate*. Scientific Publishers (India), 258 pp.

Tyagi, B.K., Hiriyan, J., Samuel, P.P., Tewari, S.C. and Paramasivan, R. (2006). Dengue in Kerala: A critical review. *ICMR Bulletin*, 36(4–5), 13–22.

Tyagi, B.K., Munirathinam, A. and Venkatesh, A. (2015). A catalogue of Indian mosquitoes. *International Journal of Mosquito Research*, 2(2), 50–97.

Tyagi, B.K., Naik, B. Reddya and Bhattacharya, S. (2021). Are mosquitoes only ugly, and not beautiful? *SOMA Newsletter*, 2(1), 1–7.

Tyagi, B.K., Naik, B. Reddya and Bhattacharya, S. (2023) *Mosquito: Ugly, Bold and Beautiful – A Mosquito's-eye view v/s Man's Eye View*. Scientific Publishers (India). (in press).

Tyagi, B.K., Velayutham, T., Selvi, S.K. (2014). Transmission dynamics of Japanese encephalitis, with emphasis on gaps in understanding and priority areas for research on Japanese encephalitis and other acute encephalitis syndromes in India. *Journal of Communicable Diseases*, 46, 24–34.

Tyagi, B.K. and Yadav, S.P. (2001) Bionomics of malaria vectors in two physiographically different areas of the epidemic-prone Thar Desert, north-western Rajasthan (India). *Journal of Arid Environments*, 47, 161–172. https://doi.org/10.1006/jare.2000.0698

Tyagi, R.K., Das, M.K., Singh, S.S., Sharma, Y.D. (2013). Discordance in drug resistance-associated mutation patterns in marker genes of *Plasmodium falciparum* and *Plasmodium knowlesi* during coinfections. *Journal of Antimicrobial Chemotherapy*, 68(5), 1081–1088.

Venkatesh, A. and Tyagi, B.K. (2013). Predatory potential of *Bradinopyga geminata* and *Ceriagrion coromandelianum* larvae on dengue vector *Aedes aegypti* under controlled conditions (Anisoptera: Libellulidae; Zygoptera: Coenagrionidae; Diptera: Culicidae). *Odonatologica*, 42(2), 139–149.

Venkatesh, A. and Tyagi, B.K. (2015). *Bradinopyga geminata* (Anisoptera: Libellulidae) as a predator of *Aedes aegypti* immatures (Diptera: Culicidae). *International Journal of Mosquito Research*, 2(2), 98–105.

World Health Organization (2023). WHO initiative to stop the spread of *Anopheles stephensi* in Africa – 2023 update. WHO/UCN/GMP/2023.06

Zhu, H., Li, J., Zheng, H. (2006). Human natural infection of *Plasmodium knowlesi*. *Chinese Journal of Parasitology and Parasitic Diseases*, 24, 70.

Section III

Sampling Methods

7 Mosquito Sampling
Theory and Practice in India

Rina Tilak and Vivek Anand
Armed Forces Medical College, Pune, India

M. Govindaraju
Bharathidasan University, Tiruchirappalli, India

B.K. Tyagi
University Institute of Biotechnology, Chandigarh University, Mohali, India

7.1 INTRODUCTION

Medical entomologists, by virtue of their work mandate, must assess the risk of infectious diseases affecting public and animal health, and strengthen the capacity for their prevention and control in the country (Silver, 2008; Tyagi, 2003). Vector-borne, or for that matter the mosquito-borne diseases such as, malaria, filariasis, dengue, chikungunya, zika, yellow fever, Japanese encephalitis, and West Nile, as a specific group of highly fatal and debilitating emerging and/or re-emerging infections, pose a serious threat to public health and require particular attention. An important aspect of preparedness for vector-borne diseases is, on the one hand, conducting a regular and systematic survey to determine seasonal variations in density and/or preponderance of both the larval and adult populations, and, on the other hand, monitoring or surveillance on a regular basis to find out the introduction, establishment and spread of the main disease vectors as well as invasive alien species (IAS) which can have a deleterious effect on our living environment. A well-considered, efficient, regular, and systematic sampling strategy together with a surveillance or monitoring campaign is the foundation of a structured vector and/or disease control program.

Sampling methods for mosquitoes can be divided into four main categories: methods for collecting flying adults, resting adults, larvae, and eggs (WHO, 1975a, 1975b). The combined use of these methods depends on the objective of the surveillance or monitoring campaign, the target species, the environmental conditions at the selected sampling sites and the availability of resources. In recent times pupae have also been sampled because of the near calibration of a likely adult population preponderance.

A range of sampling methods is available depending on the target species such as, for example, (i) malaria vectors like *Anopheles culicifacies*, a rural vector, (breeding in open surface running or stagnant water bodies), *An. fluviatilis*, an agro-sylvatic vector (breeding in expansive agriculture/paddy fields in plains and submontane *terai* regions), and *An. stephensi*, an urban vector, (breeding in potable fresh water like wells, underground/overhead water tanks as well as pitcher or containers of variety); (ii) lymphatic filariasis vector *Culex quinquefasciatus* breeding in polluted water bodies, e.g., drains, cess-pits/pools, discarded wells, latrines pits etc., and (iii) dengue/chikungunya vectors like *Aedes aegypti*, an urban vector (breeding in potable/contained water sources of variety) and *Ae. albopictus*, a rural vector (breeding additionally in phytotelmata) (Tyagi, 2008).

DOI: 10.1201/9781003326991-10

Mosquito sampling serves as a fundamental tool for vector surveillance, aiding in the early detection of disease outbreaks and identifying regions at risk. Vector control programs benefit significantly from the insights provided by mosquito sampling. Accurate data on vector abundance and distribution allow for targeted interventions, optimizing resource allocation and reducing the environmental impact of control efforts. Furthermore, the integration of larvicidal and adulticidal measures can be strategically planned based on surveillance results to increase efficacy. However, challenges persist, including the need to improve the sensitivity and specificity of sampling methods, especially in regions with diverse mosquito species. Additionally, ethical considerations and potential unintended ecological consequences necessitate careful evaluation and regulation of sampling techniques. This chapter aims to provide an overview of different mosquito sampling methods, highlighting their principles, advantages, and limitations. By understanding the diverse array of sampling techniques available, researchers, entomologists, and public health professionals can make informed decisions when selecting and implementing mosquito surveillance strategies, contributing to effective vector control programs and the prevention of mosquito-borne diseases.

7.2 SAMPLING STRATEGIES AND OBJECTIVES

Each sampling method is specific to a specific objective, i.e., presence/absence, density of the species, adult population size, seasonality, ecology and behavior, the vector status with pathogen-incrimination, and control options of the target species etc. This information is essential for the design of a successful mosquito control campaign.

7.2.1 Study Design

The study design should meet the main objectives: a cross-sectional design is recommended primarily for distribution and abundance assessments whereas a longitudinal design is recommended for collecting data on population dynamics. Cross-sectional surveys collect data to assess the presence/absence or vectorial capacity (field components) of the target species at randomly selected or risk-based study sites and over a short time period. Cross-sectional surveys may be repeated periodically. Usually, in a repeated cross-sectional survey, sites sampled at one point in time are not intentionally sampled again, although a site of a previous survey could be randomly selected for a subsequent one. However, identical sites can be intentionally re-sampled: hotspots of invasive species, for example, are often sampled several times a year.

Longitudinal surveys are research studies that involve repeated sampling of the same sites (Figure 7.1). A limited number of sites selected by a predefined geographical range, biotope type/s and time frame are monitored over one or more mosquito seasons. In the case of the 'Bangalore Mosquito Control: Master Plan' and the 'Filariasis Control Demonstration Project, Pondicherry,' it was *Culex quinquefasciatus* – the main nuisance mosquito (Rajagopalan et al., 1987, Rajagopalan and Das, 1988).

These surveys were used to assess vector presence/absence and population dynamics (e.g., seasonality, density peaks, number of generations) and vector abundance. Density peak assessments are essential to depict time windows for abundance sampling (Rajagopalan and Das, 1988). These types of data were combined to predict later the risk assessment for lymphatic filariasis infection. Data on presence/absence were collected by several appropriate sampling methods (resting and night biting collection of adults, larval sampling, simultaneous collection of biting mosquitoes from volunteer human bait and animal (a buffalo calf) bait and the estimation of population dynamics and abundance were appropriated based on a single sampling method in accordance with a standardized protocol.

It is invariably important for the team leader to obtain or draw beforehand the landcover maps of the area under survey and its surroundings, including ecological and meteorological information (Figure 7.2). However, this is not possible in all areas and should not be a limiting factor when planning sampling activities.

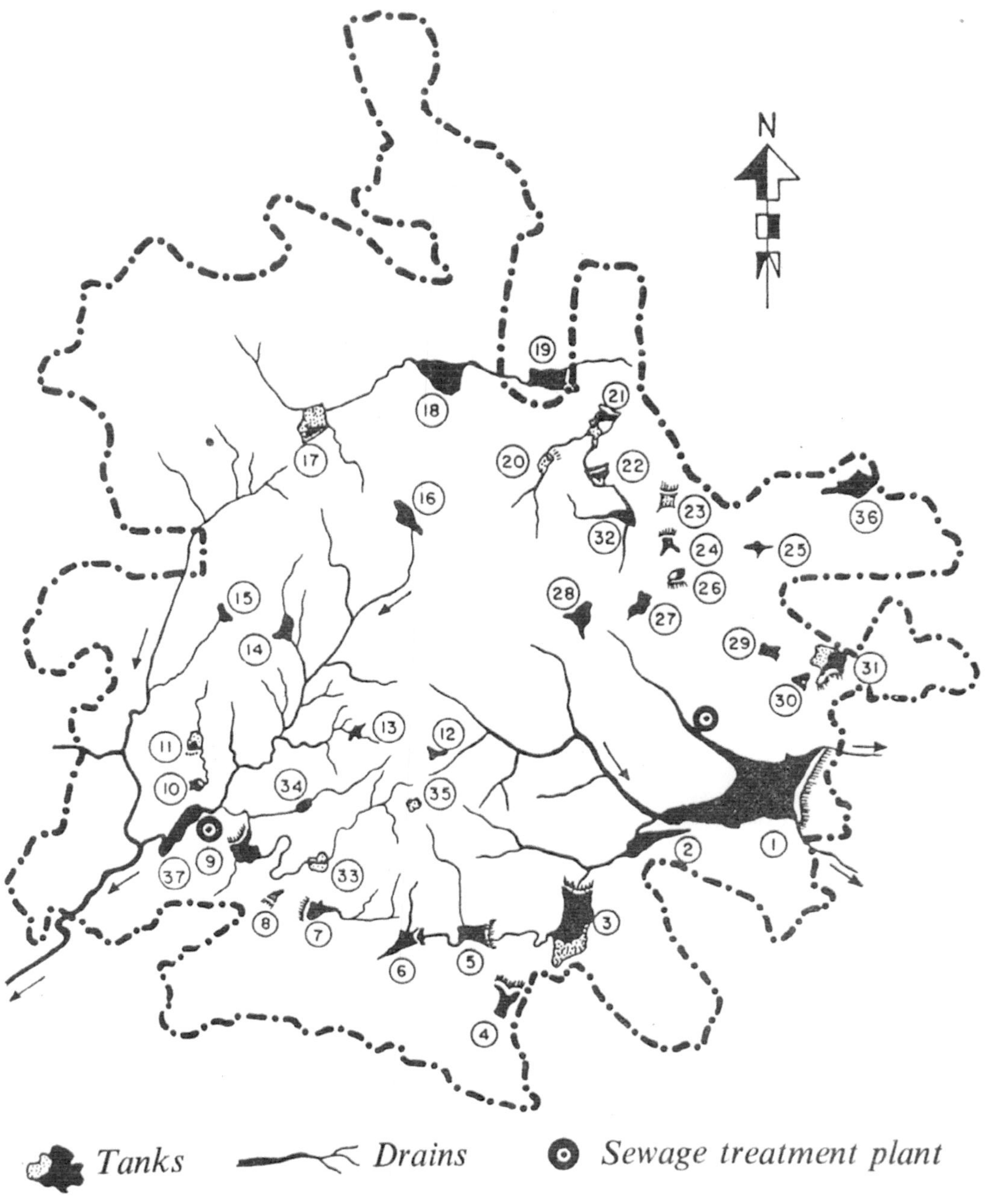

FIGURE 7.1 Bangalore Metropolitan area showing major tanks and drains as mosquito breeding sites; (1) Belandur, (2) Agaram, (3) Madivala, (4)Hulimavu, (5) Duraiswamypalya, (6) Sarakki, (7) Uttrahalli, (8) Arahalli, (9) Hosekerahalli, (10) Nayandaha, (11) Lorrypalya, (12) Lalbagh, (13) Kempabhudi,(14) Dasarahalli, (15)Kamakshipalya, (16) Sankey, (17) Mattikere, (18) Hebbal, (19) Nagavara, (20) Kalagondanahalli, (21) Hennur, (22) Kacherakanahalli, (23) Banasvadi, (24) Naganapalya, (25) Dooravannagar, (26) Byappanihalli, (27) Kodarampalya, (28) Ulsoor, (29) Byrasandra, (30) Annasandrapalya, (31) Doddanekundi, (32) Lingarajapura, (33) Ittamadu, (34) Pappana, (35) Yediyur, (36) Krishnarajapura, (37) Kengeri (*unpublished*; Rajagopalan et al., 1987; B.K. Tyagi is a co-author).

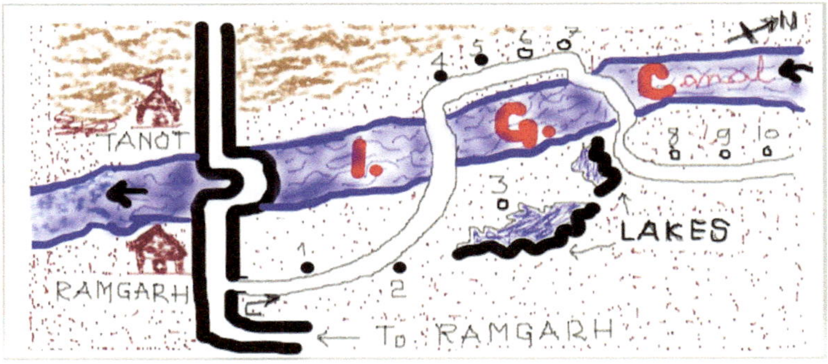

FIGURE 7.2 Spot map of malaria outbreak-hit villages in Ramgarh Talluk, Jaisalmer District in the Thar Desert during April 1999: 1. Raghwa, 2. Seuva, 3. Hada, 4. Raimala, 5. Sadhna, 6. Tejpala, 7. Badha, 8. Nagga Dhani, 9. Nagga village, and 10. Hu;cha. (Note: Villages in dark circles were investigated for the malaria outbreak).

***Source*: Redrawn from Tyagi et al. 2001.**

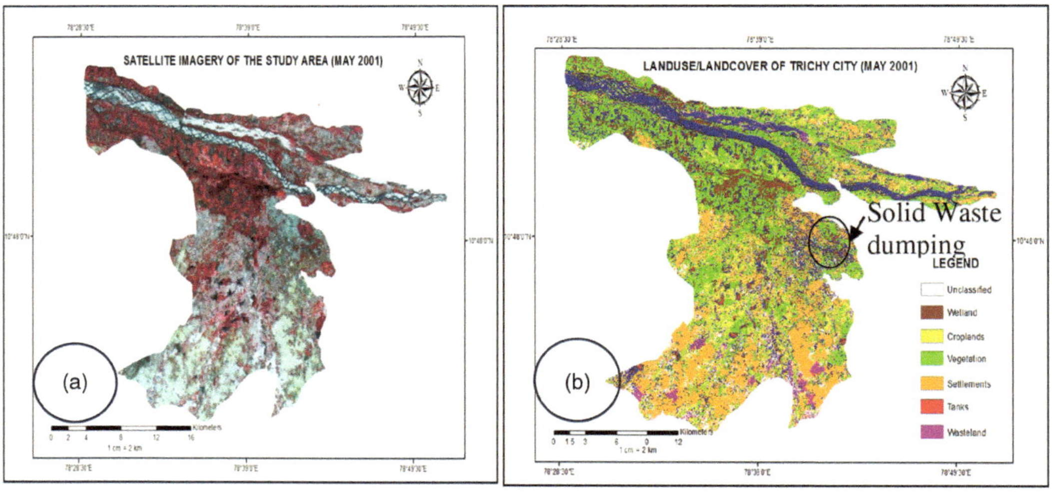

FIGURE 7.3 Sitheri Hills near Tiruchirapalli (Tamil Nadu), India – the study area about vectors of lymphatic filariasis; (a) Satellite imagery, and (b) Land Use/Land Cover Map.

***Source*: Suganthi et al., 2018; B.K. Tyagi is a co-author.**

The conventional mosquito surveillance strategies and sampling have witnessed considerable advancement courtesy the remarkable progress made in the field of robotics, machine learning and artificial intelligence. Integrated software components such as artificial intelligence (AI), Google Earth, and state-of-the-art geographic information systems (GIS) is utilized to facilitate the planning of sampling activities. They allow for the rapid and cost-efficient planning of surveys, which in turn increases the overall success rate (Figure 7.3). These maps are used to identify urban, suburban, and rural zones; floodplains and rice fields; urban 'green and blue islands;' large animal sheds and bunkers or caves etc. Maps also permit to identify resting vegetation (groves, forests) in or nearby flood lands. Unmanned Aerial Vehicles (UAVs), drone-based multispectral imagery and real-time deployable robots using AI are emerging technologies that have been widely exploited to determine potential mosquito breeding areas for identification of hotspots as well as for detection, classification, mapping distribution of mosquitoes in varying eco settings and their control by various researchers globally. UAVs are increasingly being used for identification and detection of larval breeding habitats of malaria vectors using high-resolution imagery and for their targeted control (Carrasco-Escobar et al., 2022; Dias et al., 2018). Commercial drones fitted with a multispectral camera have also been widely used for surveying and mapping of potential mosquito breeding sites in diverse breeding habitats viz. tidal marsh, rice fields, ponds, and in urban and peri-urban areas for prioritizing areas for larval mosquito survey and/or to reduce mosquito oviposition and for control (Haas-Stapleton et al., 2019; Hardy et al., 2017). Robot-assisted mosquito surveillance is gaining popularity as an emerging concept which has the potential to assist in performing automated tasks of mosquito surveillance which is normally challenging for humans to perform over prolonged periods with precision and accuracy (Kim et al., 2019; Semwal et al., 2022). The future of mosquito surveillance and sampling appears promising with the advent of these newer technologies and awaits full exploitation of its potential as a tool for effective management of various mosquito-borne diseases.

7.2.2 Selection of Sampling Methods

A sampling method, choice of trap type and the trapping frequency for a specific objective depends largely on the target species, life stage and status as well as available funding and resources, and the size of the area to be sampled. Sampling adult flying mosquitoes is the preferred method when all immatures have emerged (e.g., after the emergence of a cohort of floodwater *Aedes/Ochlerotatus* species), during swarming (to collect adult males), when sampling larvae is more time-consuming or when breeding sites are scarce because of dry conditions. Some species are difficult to collect with the available adult traps and should therefore be sampled in the larval stage or while resting indoors. The indoor resting mosquitoes have for long been sampled using a mouth-operated aspirator and a torch for the light source (Figure 7.4). The relative density of each mosquito species is represented as per/man/hour, with a minimum single visit per house out of six houses for 10 minutes. Due to human bias and health hazards associated with the mouth-operated aspirator's use, alternatively, a battery-operator aspirator has been in vogue (Figure 7.5). It is mentionable here that to overcome health hazards and improve efficiency in sampling of indoor resting mosquitoes Tyagi (1993) developed a mechanical mosquito sampler.

Tyagi (1993) developed a mechanical mosquito sampler for sampling the indoor resting mosquitoes, particularly those resting at much higher height on the walls or in deep burrows in roofs (Figure 7.6). This mechanical mosquito sampler was specifically developed to sample mosquitoes in the arid environments where it was >2.5 times superior to the conventional mouth-operated aspirator. This is portable, non-hazardous, economical (Rs. 625.00 or US$ 31.25, at 1992 exchange level, per unit) and highly feasible in conducting field sampling. It is also efficient, durable, and reliable for data mining and sustaining physiological fitness of the sampled material. This invention of mechanical mosquito sampler was awarded an Indian Patent (No. 191635) in 2005.

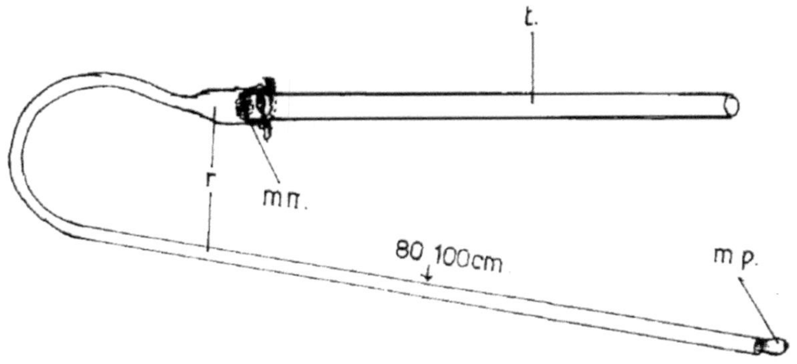

FIGURE 7.4 A mouth-operated aspirator (t = glass or plastic tube, mn = mosquito net, r = rubber tube, mp = mouthpiece).

Source: WHO, 1975b.

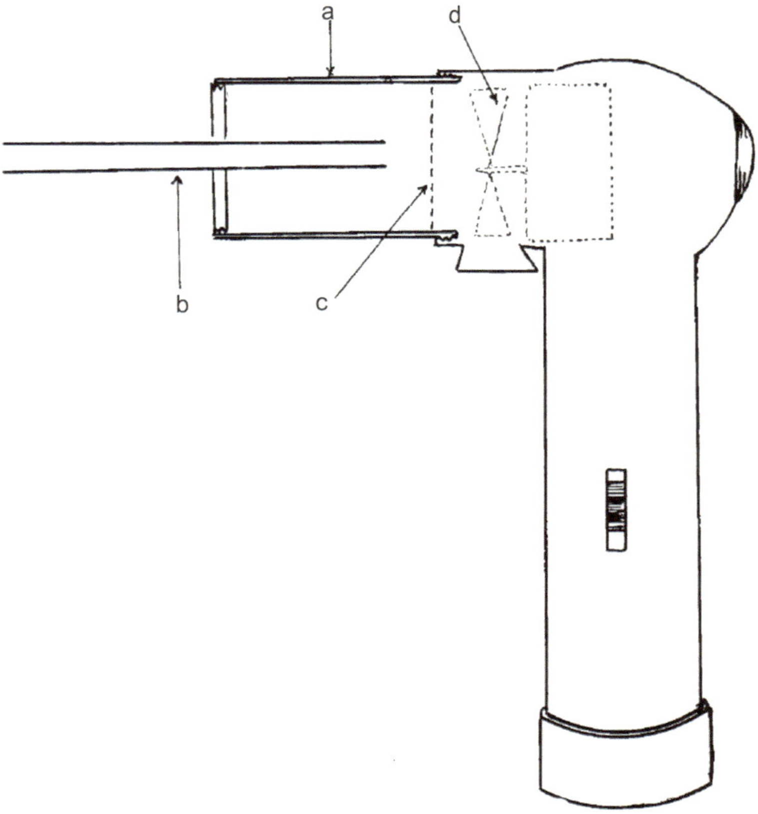

FIGURE 7.5 A battery-operated aspirator for mosquito sampling resting indoors.

Source: **WHO, 1975b.**

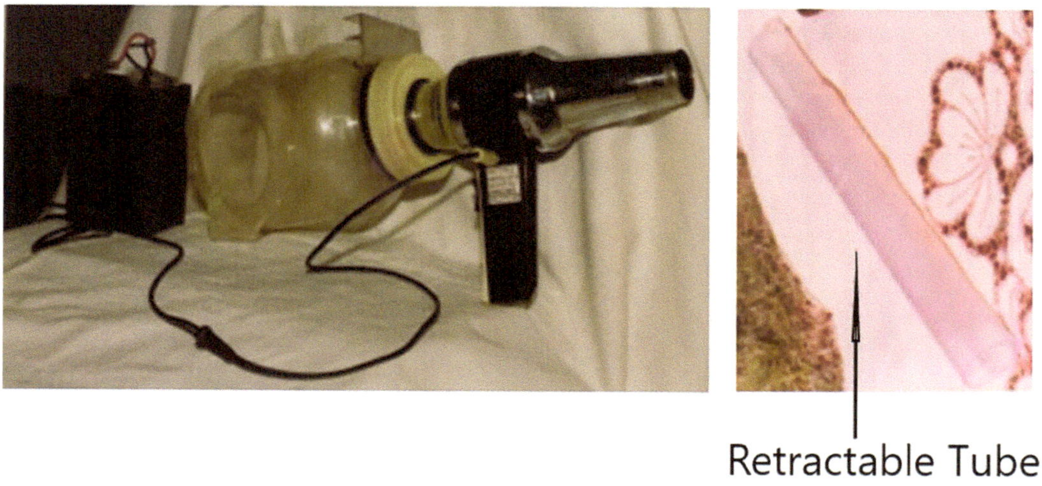

FIGURE 7.6 A mechanical mosquito sampler developed by B.K. Tyagi in the Thar Desert, with an Indian Patent No. 191635.

Source: **Dr B.K. Tyagi archive.**

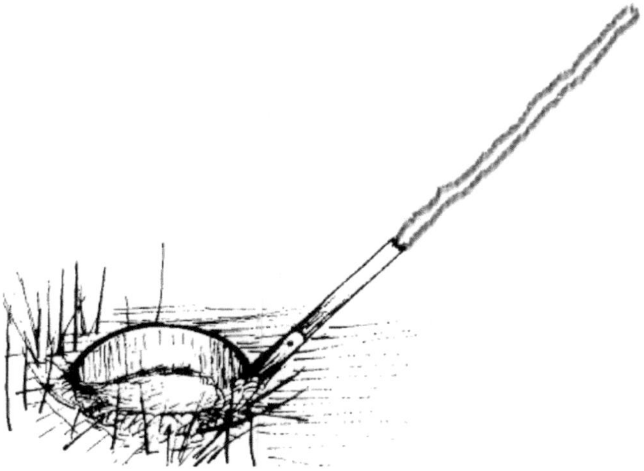

FIGURE 7.7 A long-handle dipper used in sampling mosquito larvae.

Source: **WHO, 1975b.**

Larval sampling may also be required for vector control purposes in case of mosquito nuisance, for example, it is essential to detect the source of the problem, so identification of the larval breeding sites of the species is the key factor. To execute larval sampling – or for that matter also pupal sampling, a long-handled dipper, which is a combination of a 100 ml enamel bowl and a long shaft, is used (Figure 7.7).

Sampling resting mosquitoes is the preferred method to collect adult females that are not attracted to traps (e.g., blood-fed or gravid females which ignore baited traps), to collect adult males (which can facilitate mosquito species identification), and to collect overwintering adult females (Figure 7.8). It should be ensured that traps do not interfere with each other, for example, the range of CO_2, chemical lures, and animal or human bait should be considered when placing traps. In cases of mosquitoes, such as those living in forests, which are difficult to be sampled in the adult stage, larval sampling is the most trusted method to determine their biology, breeding preferences, distribution range and association with humans (Figure 7.9).

7.2.3 Selection of Sampling Sites

The success of sampling of a given mosquito species is dependent on the thoughtful selection of the study sites with crystal clear criteria including the identification of gaps in the knowledge of the distribution of the species, the geographical area to be covered, previous information on earlier field data (lacking altogether, incomplete and/or outdated), passive surveillance in past, and theoretical species distribution models highlighting areas with a high probability of presence where model validation with field data is required. Field maps subsequently developed based on these data facilitate the sampling a great deal (Figure 7.10, also see Figure 7.2).

One of the major malaria vectors' sampling in the xeric and the recently irrigated desert villages was the discovery of superlative dominance of *Anopheles stephensi* in the desert villages, on one hand, while irrigation water savvy *An. culicifacies*, obviously a new entrant in the Thar Desert was far more dominant in the non-desert and irrigated villages (Figure 7.11).

FIGURE 7.8 Sampling of adult mosquitoes resting indoors in (a) open and (b) closed thatched rooms.

Source: **Tyagi, *personal archive.***

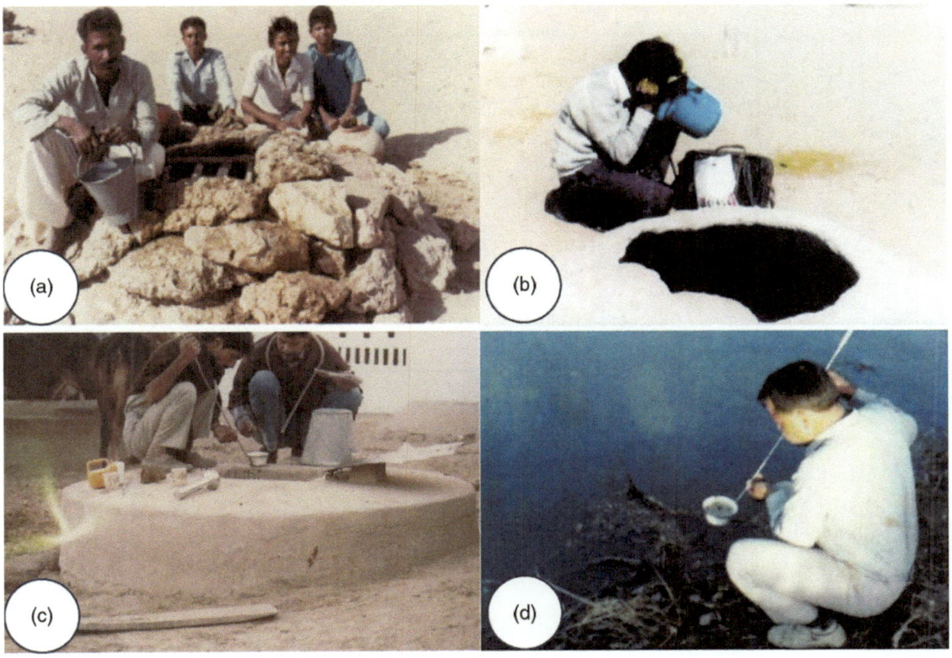

FIGURE 7.9 (a–d). Larval sampling in different kinds of breeding sites.

Source: Tyagi, *personal archive*.

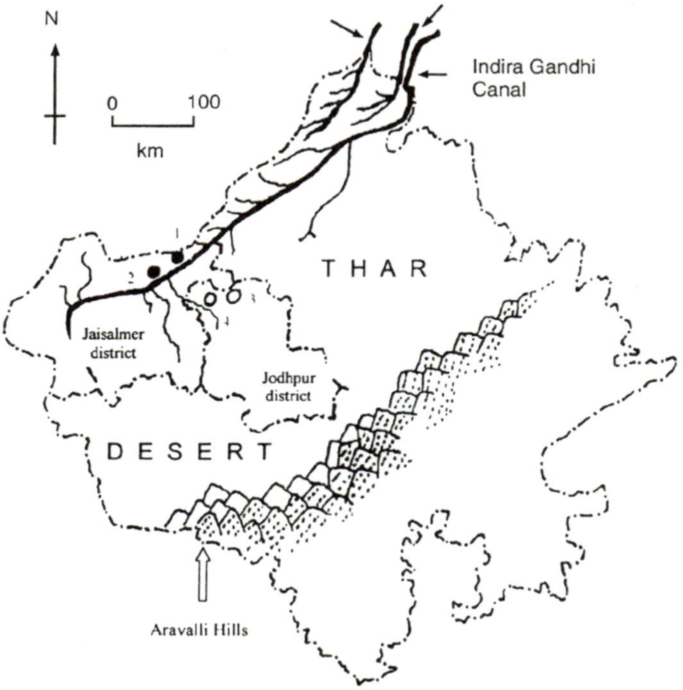

FIGURE 7.10 Location of sites for sampling mosquitoes from villages on either side of the Indira Gandhi Canal across the Thar Desert.

Source: Tyagi, 2002.

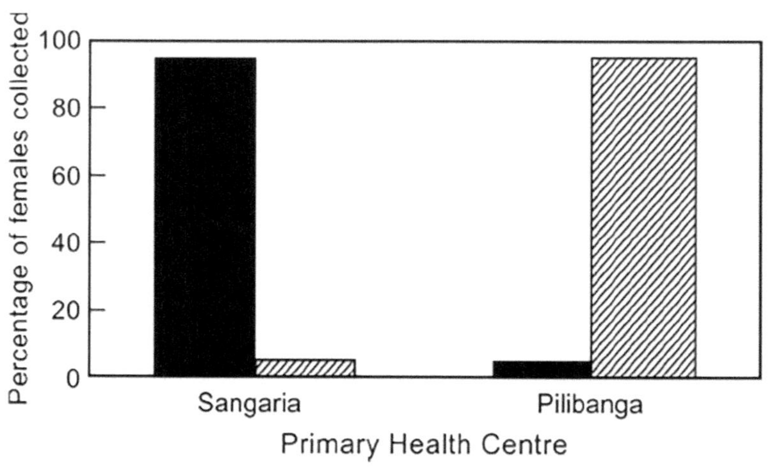

FIGURE 7.11 Dominance of *An. stephensi* (lined pillar) in desert villages, while *An. culicifacies* (black pillar) was dominant in the non-desert and irrigated villages in The Thar Desert.

Source: **Tyagi and Yadav, 1996.**

7.2.4 Sample Size Calculation

Calculating the appropriate sample size for mosquito sampling depends on the specific objectives of the study and the desired level of precision and confidence. The sample size estimation is essential to ensure that the collected data are representative and statistically valid. It ensures the sample size is not too small which can affect the utility of the study while not too large to cause a wastage of precious resources. Steps to calculate sample size for mosquito sampling:

(a) Define objectives: It is important to state the purpose of mosquito sampling study which can be an estimation of mosquito abundance, species diversity, or prevalence of a particular disease transmitted by mosquitoes. The objective will dictate the appropriate sampling methodology to be applied in the study.
(b) Determine the level of precision: How precise do you want your estimates to be? A higher level of precision will require a larger sample size. Precision is often expressed as the margin of error or confidence interval width.
(c) Set the confidence level: The confidence level determines the certainty researcher wants to have in his/her results. Common confidence levels used are 95% or 99% with higher confidence levels requiring larger sample sizes.
(d) Estimate the variability: An estimate of the expected variability (standard deviation) in mosquito counts or other relevant metrics is needed. The previous similar studies form the basis to measure expected variability. In absence of information from previous published literature, a pilot data or conservative estimates from literature can be used.
(e) Choose an appropriate sampling method: Depending on the study objectives, the researcher uses various sampling methods such as simple random sampling, systematic sampling, or stratified sampling. Each method may have different implications for sample size calculations.
(f) Use the appropriate formula: The sample size calculation depends on the statistical test the researcher intends to perform and the sampling method. Commonly used formulas for estimating sample size include those for estimating means, proportions, or comparing groups.

(i) *Estimating mosquito abundance (mean)*:
To estimate the average number of mosquitoes per sampling unit (e.g., per trap, per square meter), sample size estimation for estimating means is given by:

$$n = \frac{z^2\sigma^2}{E^2}$$

Where:
n is the required sample size
Z is the Z-score corresponding to desired confidence level (e.g., 1.96 for 95% confidence)
σ is the estimated standard deviation of mosquito counts
E is the desired margin of error (precision)

(ii) *Estimating mosquito prevalence (proportion)*:
To estimate the proportion of mosquitoes carrying a specific pathogen (e.g., malaria parasites), one can use the formula for sample size estimation for estimating proportions:

$$n = \frac{2Z^2\left(\sigma_1^2 + \sigma_2^2\right)}{d^2}$$

Where:
n is the required sample size
Z is the Z-score corresponding to your desired confidence level (e.g., 1.96 for 95% confidence)
p is the estimated proportion of mosquitoes carrying the pathogen (prevalence)
E is the desired margin of error (precision)

(iii) *Comparing mosquito populations (between two groups)*:
To compare mosquito populations between two different groups (e.g., treated vs. untreated areas), the formula for sample size estimation for comparing two means can be used:

$$n = \frac{2Z^2\left(\sigma_1^2 + \sigma_2^2\right)}{d^2}$$

Where:
n is the required sample size per group
Z is the Z-score corresponding to your desired confidence level (e.g., 1.96 for 95% confidence).
$\sigma_1^2$and σ_2^2 are the variances of mosquito counts in each group
d is the effect size, which represents the minimum difference in means that you want to detect

(g) Usage of statistical software: Statistical software or online sample size calculators may be used to determine the required sample size using all the information available from the previous steps.

Different sample size calculation methodologies have been used by researchers in various studies over the years (Silver, 2008). The variability of mosquito populations necessitates a well-designed sampling strategy to ensure the accuracy and representativeness of results.

7.3 COMMON TRAPS FOR SAMPLING

Based on the objectives of sampling, different types of traps are used, for example, malaise traps, truck traps, electrocution traps, and attractants-based traps. Barring the attractants-based traps, all others can be used to sample the foraging population by placing them between breeding sites and those locations where mosquitoes obtain their blood meals. Such traps are best suited in situations where it is a prerequisite to determine mosquito flight periods and generate information for the correct timing of initiating control measures against the adult mosquitoes. The attractants-based traps are a group of traps which are based on the fact that the bait or host for a blood meal emits olfactory stimuli or kairomones which attract the host-seeking mosquitoes (foraging flight). Female mosquitoes respond to compounds such as carbon dioxide from exhaled breath and host odor components (e.g., lactic acid) derived from sweat. They can also detect water vapor and body heat. The most frequently used outdoor methods include:

(i) collection of females landing on a human (human landing collection, HLC) or inside a drop-net with aspirators (mouth or battery-operated), provided that all ethical issues are fully considered;
(ii) animal bite catches (by removing mosquitoes directly from a tethered animal host with an aspirator or inside a drop-net); it is also possible to use a suction trap baited with small animals;
(iii) suction traps containing attractants like carbon dioxide (CO_2-baited traps), light (UV and incandescent light traps), or combination of both; and
(iv) gravid and sticky ovitraps that attract ovipositing females (Figure 7.12).

Variety of attractants-based traps help in determining the species/sex composition. In addition to the range of conventional mosquito traps used by mosquito specialists, several novel mosquito trapping or monitoring devices have been introduced over the last decades, especially with reference to *Aedes aegypti* (Mackay et al., 2013; Llagas et al., 2016). Some incorporate novel chemical lures, others have alternative airflow systems, while yet others use novel technology to power the fan to sample gravid *Culex quinquefascaiatus* (Figure 7.13) (Mboera et al., 2000).

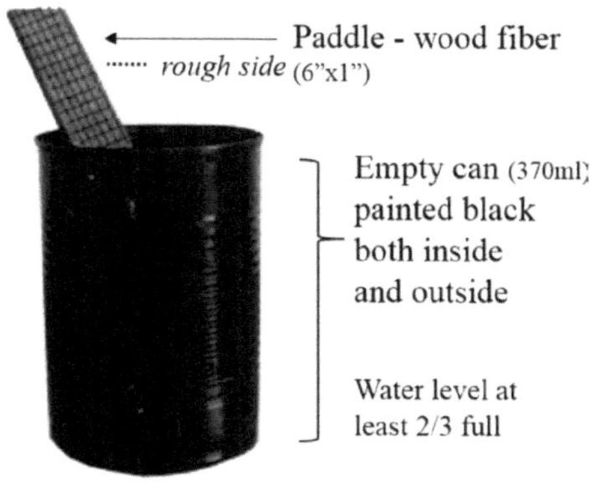

FIGURE 7.12 The Philippine Ovitrap Kit.

Source: **Llagas et al., 2016.**

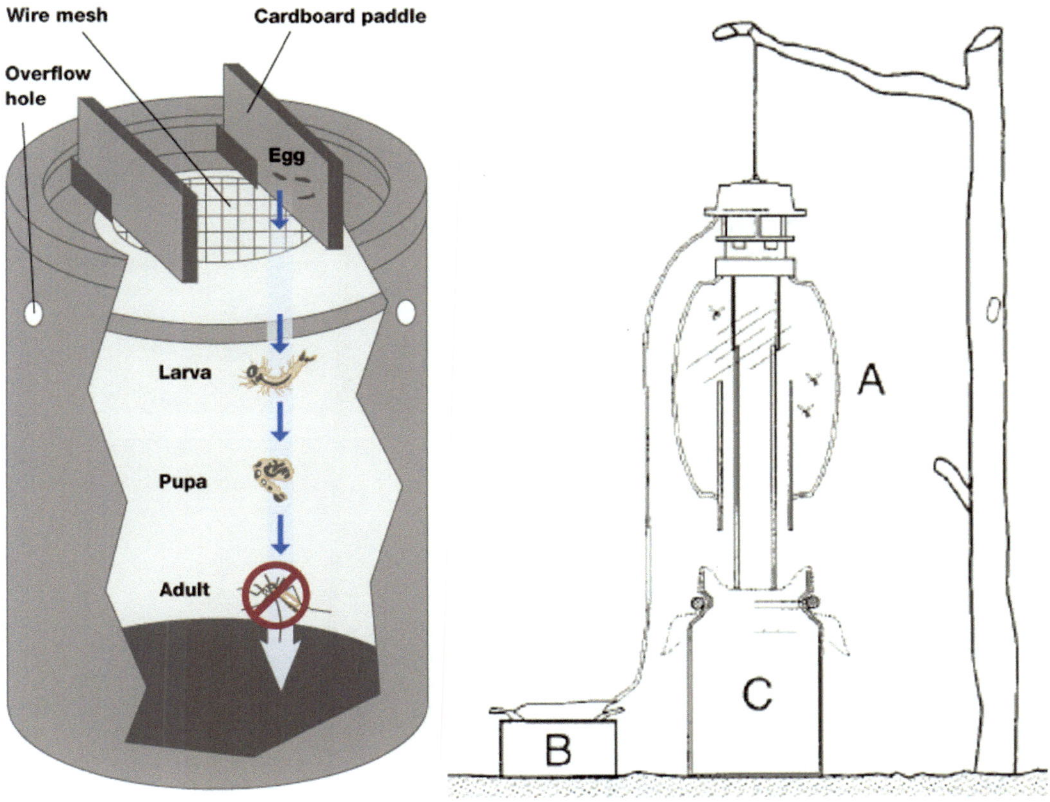

FIGURE 7.13 (a): A lethal autocidal Ovitrap designed for sampling gravid *Aedes aegypti* (*Source*: Wikipedia); (b) Counterflow Geometry (CFG) trap baited with an oviposition attractant; (A) CFG trap, (B) Power source (12-V battery), and (C) Plastic jar containing an oviposition attractant.

Source: **Mboera et al., 2000.**

Many newer and innovative traps have been designed recently and studies have been performed on efficacy of these traps by comparing with the gold standard of HLC.

(i) *Mosquito electrocuting traps (MET)*: It is an exposure-free alternative to HLC and was built on previous work using electrified nets and grids to trap tsetse flies and mosquitoes attracted to hosts or their odors (Figure 7.14(A)).

(ii) *Mosquito Magnet Trap*: This trap emits CO_2 and uses heat to attract mosquitoes. Once attracted, they are sucked into a net by a vacuum fan. The Mosquito Magnet has been used for both residential and commercial mosquito control (Figure 7.14(B)).

(iii) *Biogents Sentinel Trap (BGS)*: This trap is designed to mimic human scent and uses carbon dioxide and heat to attract mosquitoes. It has proven to be effective in capturing *Aedes* mosquitoes which transmit diseases like dengue, chikungunya etc. BG-Sentinel 2 Trap is an improved version of the Biogents Sentinel Trap, with additional features for easier maintenance and more efficient trapping (Figure 7.14(C)).

(iv) *Suna Trap*: This trap uses a combination of attractants, including light, heat, CO_2, and color patterns, to lure mosquitoes and trap them within the device (Figure 7.14(D)).

(v) *In2Care Trap*: This innovative trap combines a resting place with a slow-release larvicide and a fungus that infects and kills adult mosquitoes. The trap is designed to be carried by mosquitoes to their breeding sites, infecting larvae and reducing the mosquito population (Figure 7.14(E)).

FIGURE 7.14 (A) Mosquito electrocuting trap, (B) Mosquito Magnet, (C) BG-Sentinel 2 Trap, (D) SUNA Trap, and (E) In2Care Trap.

Source: **[a] Maliti et al., 2015; [b] Jeyaprakasam et al., 2021.**

7.4 SAMPLING TECHNIQUES AND METHODS

7.4.1 Adult

Vector control is an essential component of nearly all vector/mosquito-borne diseases, but most importantly malaria, lymphatic filariasis and dengue control programs (WHO, 1975a, b). Various entomological indicators and sampling techniques are used to monitor and evaluate the impact of many vector control interventions (Service, 1993). Several methods have been used in sampling mosquitoes, both adults and larvae, to calculate mosquito adult density. Though adult mosquito sampling is inevitably indispensable for the purpose of estimating the entomological inoculation rate (EIR), however, each sampling technique is subject to some bias or shortcomings (Mboera, 2005). Most common adult mosquito sampling techniques in relation to their reliability is the estimation of EIR include human landing catch, light trap, light trap/bednet combination and odor-baited traps. Although the human landing technique is the most reliable, it, however, renders the catcher to likely exposure to mosquito-borne infections. On the other hand, light traps have been found to capture mosquitoes with higher sporozoite rates as compared to those from human bait catch thus leading to an overestimation of EIR. From an epidemiological point of view, the use of a light-trap-bed net combination is an approach that is more meaningful than using a light trap alone because, a light trap functions more efficiently when placed near the normal flight paths of mosquitoes such as inside huts or under the eaves (Tyagi, *unpub. rep.*).

7.4.1.1 Basis of Sampling

Mosquitoes rest either indoors or outdoors. As far as indoors is concerned, only few mosquitoes species rest in man-made shelters such as houses or cattle sheds (e.g., *Culex quinquefasciatus*). However, the dominant vector species of malaria parasites and filarial worms (more rarely arboviruses) are mostly endophilic and therefore should be collected indoors. Collections in the morning are usually more efficient than in the afternoon or evening. The most frequently used methods for indoor sampling include:

(i) collection with tubes;
(ii) collection with aspirators (mouth or battery-operated) after locating the mosquito (with the help of an electric torch) on walls, ceilings, clothing, furniture, under mosquito nets; and
(iii) knock-down pyrethroid spray catches.

Searching indoors for resting adults is particularly efficient for *Anopheles* spp., members of the *Culex pipiens* complex and, to a certain extent, for *Aedes aegypti*.

On the other hand, with respect to outdoors, most mosquito species rest outdoors or in natural shelters (e.g., in vegetation, tree holes, crab holes, animal burrows, caves, rock crevices, and artificial shelters). The most frequently used methods for outdoor sampling include:

(i) collection with hand nets,
(ii) collection with motorized aspirators (large mechanical suction sweepers),
(iii) drop-net cages in the vegetation,
(iv) resting boxes of various types and sizes, an artificial Muirhead–Thomson pit.

Searches for mosquitoes resting outdoors are usually very time-consuming and not very effective. In order to collect a large and representative number of mosquitoes, a major effort must be made.

Sampling methods for mosquitoes can be divided into five main categories and the combined use of these methods depends on the objective of the surveillance or monitoring campaign, the target species, the environmental conditions at the selected sampling sites and the availability of resources.

7.4.1.1.1 Flying Adults

Sampling techniques for flying adults are quite efficient for most mosquitoes of public and veterinary health importance. Adult mosquitoes tend to spend more time on resting than flying, and therefore, sampling adult mosquitoes is generally challenging without the use of attractants. Use of attractants in trapping flying adults is both less time-consuming and more effective than the other trapping methods. During their flight, they are in search of nectar and blood or suitable sites to lay eggs. Mostly these are the females at various stages of the gonotrophic cycle (unfed, fed, half gravid, and gravid) which are targeted, though the fully-fed (blood-engorged) females are preferred for conducting various kinds of experiments such as, determining the status of resistance against a given insecticide, and the preferred natural hosts.

Flying mosquito species (*Aedes aegypti* and *Ae. albopictus*) are most frequently sampled outdoors but miniature traps (with or without light) baited with CO_2 can also be used in semi-open places or indoors (e.g., *Culex pipiens* and *Anopheles culicifacies*). To reveal the blood-feeding behavior of a particular mosquito species, host-baited traps are found useful, albeit with ethical restrictions. However, if ethical agreements on animal testing are obtained beforehand, then the sampling of blood-engorged females using host-baited traps is also possible. In general, however, most flying mosquito sampling techniques are designed to sample outdoor populations (e.g., *Armigeres subalbatus*).

7.4.1.1.2 Resting Adults

This type of sampling is most common and important to know the physiological age of the mosquito and the stage of the gonotrophic cycle. Additionally, such collections also yield male mosquitoes, which are rarely collected with traps that target biting or ovipositing females. The male mosquitoes, besides being very helpful in ascertaining the species identification are also helpful in evaluating the vertical transmission of a pathogen (such as in the case of dengue and chikungunya), and assessing the overwintering of the pathogen in the vector at localities with high virus circulation (Chye et al., 1997; Arunachalam et al., 2008, Thenmozhi et al., 2007; Jain et al., 2016: Yin et al., 2016; Ferreira-de-Lima and Lima-Camara, 2018; Murillo et al., 2019; Goncalves et al., 2020; Heath et al., 2020; Ferreira et al., 2021). The collection of mosquitoes at rest is among the simplest types of collections requiring either, though archaic and health-hazardous - inhaling aspirator or a motorized aspirator. This type of sampling does not require baited traps and one rarely has to resort to traps, save for the use of louver traps in certain hut-based genetic studies (Tyagi, *pers. comm.*). The collection of resting mosquitoes must differentiate between indoor and outdoor sites. Tyagi (1993) invented a battery-operated mechanical mosquito sampler (India Patent No. 191635) which was highly efficient and economical.

7.4.2 Pupa

Pupal sampling is undertaken using the same methods as larval sampling (discussed in detail in next section) depending on the type of aquatic habitat (Service, 1993; Harbison et al., 2018). Pupal sampling, however, provides more accurate information about the likely adult density build-up as compared to larval sampling and is a better tool for density estimation. The steps typically involved in mosquito pupal sampling are:

(a) *Identification of potential breeding sites*: It is important to identify the breeding places like slow-flowing streams or stagnant water bodies like ponds, wetlands, and domestic and peri-domestic breeding sites like buckets, tires, or other artificial or natural containers.
(b) *Choice of appropriate sampling tools*: A dipper or pipette can be used to sample pupa from the breeding area.
(c) *Laboratory rearing*: The pupae collected should be transferred to the nearest laboratory for adult emergence as it is easy and more accurate to identify the adult stage.

(d) *Identification*: Once the adults emerge from the pupa, the adults can be counted, and the species identified. This data provides vital inputs for estimating the likely adult density and aids in the formulation of effective prevention and control strategies for impending mosquito-borne disease outbreaks.

Mosquito pupal populations can vary greatly even within a small area, and throughout the year depending on changes in weather conditions and water availability. It is therefore important to remember that specific methods and tools used for pupal sampling can vary just like in the case of the larval sampling depending on the species of mosquito being sampled as well as the reason for the pupal sampling.

7.4.3 Larvae

Sampling, monitoring, and surveillance of larvae is significant in determining the presence of the vector and, based on the preponderance and density, the risk of adult populations in the area under survey. Thus, larval surveillance is of key importance. Larvae of *Anopheles* spp. can be found in puddles, road tracks, swamp areas, drains, ditches, irrigated croplands, streams, riverbeds, and ponds, while those of *Aedes* spp. can be found naturally in tree holes (phytotelmata), but also in human-made containers such as bird's drinking troughs, discarded waste containers, tires, and tarpaulins (Praveen et al., 2016). In urbanized areas, larval sampling should focus on the available human-made water bodies (both in private and public areas), below in underground tanks and above ground in overhead water tanks (Tyagi, 2023), discarded containers, flower vases, flower pots in gardens and cemeteries, used tires left outdoors, rain water barrels, road drains and catch basins, pits).

Since mosquitoes need water to complete immature life as well as for egg-laying, the aquatic habitats of various kinds assume enormous significance both for determining a possible entry and populational dominance of the vector in an area, on one hand, and for targeting their focused control using environmental, chemical, or biological agents. Because of the stability of most of the large-sized or permanent breeding sites, the sampling of larvae can be made rapidly and on a repetitive basis. The dipping technique is usually used for mosquito larval stages and may differ depending on the type of aquatic habitat being sampled. Water depth, type of vegetation (i.e., emergent, floating or submerged) and density are important factors affecting dipping. Importantly, sampling and identification can be performed on the same day, which ensures a good cost–benefit ratio. Larval sampling is especially helpful when sampling mosquito species whose females are not, or only rarely, attracted to the commonly used trap systems. Larvae of anopheline mosquitoes (e.g., *Anopheles culicifacies, An. subpictus, An. fluviatilis, An. minimus*) prefer fresh stagnant or slow running water of any kind, i.e., man-made or natural, while those of culicine mosquitoes (e.g., *Culex quinquefasciatus*) generally prefer organically polluted water in cess-pools, cess-pits, latrines, ponds, lakes, and drains (Rajagopalan et al., 1987; Rajagopalan and Das, 1988).

Larval sampling is inevitably indispensable in the case of forest-dwelling species (e.g., *Anopheles dirus, Aedes albopictus*). Detection of larvae of univoltine species (e.g., *Ochlerotatus decticus*) can only be made at specific periods while those of multivoltine species (*Aedes sollicitans* and various *Culex* and *Anopheles* species) can be undertaken throughout the entire vegetation period. Therefore, frequency and sites for larval sampling need to be accordingly designed based on a detailed knowledge of mosquito ecology (Service, 1993).

7.4.4 Egg

Egg sampling from different breeding habitats is an efficient method for the prediction of the size of the emerging larval populations and thereby serves as an advance warning of an impending mosquito-borne disease outbreak, Lopp (1957). The ovipositing behavior of mosquitoes varies and depends on the mosquito genus and species. Female mosquitoes of *Anopheles* lay their eggs individually, whereas *Culex*, *Culiseta*, *Coquillettidia*, and *Uranotaenia* females lay their eggs in groups as a raft on the surface of the water. *Mansonia* females in contrast deposit their eggs as star-shaped clusters on the underside of the leaves of aquatic plants like *Pistia stratiotes, Eichhornia, Salvinia*, etc. and the females of the genera *Aedes* lay their eggs on substratum or floating debris while *Ochlerotatus* prefer to lay their small little eggs on the moist surface of areas which get flooded.

Because of the diversity of oviposition sites among the different mosquito genera, different sampling techniques are used. Sampling of *Anopheles* eggs is done by using a dipper (small plastic container). The dipper is modified to have the bottom portion of a different color or alternatively, it is made of fine wire mesh. Artificial oviposition sites like small water-filled borrow pits have also been used for sampling eggs for *Anopheles* species in India (Russell & Rao 1942). Eggs of *Anopheles culicifacies, Anopheles subpictus* and *Anopheles vagus* were collected from the water surface of these artificial pits with the help of wire loops. This technique has also been widely used in West Africa by Muirhead-Thomson (1945). Christie (1958) modified this concept by using a metal box containing pool water as an artificial oviposition site for *Anopheles gambiae*.

For mosquitoes which lay eggs in rafts like *Culex*, *Uranotaenia*, *Coquillettidia*, and *Culiseta*, sampling is done by using pipettes or a small net or a brush. Artificial oviposition sites like that used for *Anopheles* have also been widely used for *Culex* egg sampling using borrow pits as well as metal boxes (Smith & Jones, 1972; De Meillon Soman & Reuben, 1970). Oviposition traps using pots or any other domestic containers filled with water and yeast infusion have also been used for attracting gravid *Culex quinquefasciatus* thus facilitating egg sampling (Southwood et al., 1972; Service, 1970; Yasuno et al., 1973).

In the case of mosquitoes which lay eggs in moist soil, soil samples prove to be an ideal method. Traps have also been developed to sample the eggs of *Culex* (Surgeoner & Helson, 1978; Haeger & O'Meara, 1983; Reiter, 1986; Strickman, 1988), *Toxorhynchites* (Schuler & Beier, 1983) and *Haemagogus* (Chadee et al., 1984). Mansonia eggs have been routinely sampled by examining the leaves of the aquatic plants (Laurence & Samarawickrema, 1970; Lounibos & Dewald, 1989; Gass et al., 1983) and the species identified.

Aedes eggs are usually sampled using ovitraps, which have proven to be a very efficient method of sampling as well as for monitoring *Aedes* populations (Fay & Eliason, 1966; Pratt & Jakob, 1967; Evans & Bevier, 1969; Furlow & Young, 1970; Frank & Lynn, 1982; Chadee & Corbet, 1987, 1990; Nayar, 1981). *Ae. aegypti* egg sampling on a brown hardboard strip has been reported to be successful by many workers (Chadee & Corbet, 1995; Arunachalam et al., 1999). The ovitraps can also be made more attractive using hay infusions, conspecific larval waters, or pheromones (oviposition attractants). Studies on color preferences of oviposition substrates have been undertaken by Panigrahi et al., 2014 wherein they reported that, *Ae. albopictus* and *Ae. aegypti*, preferred to lay eggs on a black substrate in contrast to white, where very few eggs were laid. *Aedes* eggs are routinely sampled from artificial or natural containers in all *Ae aegypti* and *Ae albopictus* surveillance programs. (Service, 1965; Fay & Eliason, 1966; Thaggard & Eliason, 1969; Yates, 1979; Jakob et al., 1970; Subra & Mouchet, 1984; Chadee & Corbet, 1987; Freier & Francy, 1991; Tyagi et al., 2006). *Aedes* eggs from the water samples from these containers can be obtained by sieving and flotation (Service, 1970). Some workers have attempted to quantify the number of eggs laid in a container through models which showed a positive correlation between the number of eggs ($r = +0.70$) with the oviposition surface area. It was expressed by the regression line $y = -14.12 + 0.126x$, where y is the mean number of eggs per sample and x is the surface area in cm^2. The slope of the equation ($b = 0.1266$) means that for each $1cm^2$ of water, the oviposition rate increased by 0.13 eggs, thus an increase of 7.9 cm^2 allows 1 more egg to be laid in the container. A Presence-absence technique has been utilized by Mogi et al. (1990) in Thailand in ovitrap surveillance for predicting a potential outbreak of Aedes-borne diseases by using computer simulation and sequential sampling technique.

7.5 MORTALITY ESTIMATION IN IMMATURES

Immature mosquitoes are the aquatic life stages, such as eggs, larvae, and pupae, that live in water before developing into adult mosquitoes. Mortality estimation in immatures in mosquito sampling involves assessing the proportion of immature mosquitoes that have died within a given population. Monitoring mortality in immatures is essential for understanding mosquito population dynamics, evaluating the mosquito control measures and others like assessing environmental changes, insecticide resistance, etc. A sufficient sample size and multiple sampling events may be required to obtain reliable mortality estimates. The main causes of mortality in mosquitoes include competition for food or space, adverse climatic conditions, pathogens, parasites, and predators (Silver, 2008). Density dependent mortalities have been especially researched over the years. Several researchers have used various laboratory as well as ecological methods to evaluate mortality estimates in mosquitoes.

Some common methods used for mortality estimation in immatures during mosquito sampling include larval/pupal dipping and surveys, emergence traps, mark-release-recapture studies, etc. Larval/pupal dipping involves collecting water samples from breeding sites such as ponds, ditches, or containers and then pouring through a fine mesh net or cone-shaped container to capture the immatures. The captured immatures are counted and categorized as alive or dead based on their movements and appearance. Larval/pupal surveys are used where dipping method is not feasible or in case of larger breeding sites. Field staff visually assess the proportion of living and dead larvae or pupae. Emergence traps are set up near breeding sites to capture emerging adult mosquitoes. By monitoring the number of adult mosquitoes that emerge, researchers can estimate the mortality rate of immatures in the breeding site. In mark-release-recapture studies, a portion of the captured immatures is marked with dyes or tags before being released back into the breeding site. After a certain period, a new sample is taken from the same site, and the proportion of marked and unmarked immatures is compared. By calculating the ratio of marked to unmarked immatures in the second sample, researchers can estimate the mortality rate.

In controlled laboratory settings, researchers can conduct mortality assays on immatures by subjecting them to various environmental factors, insecticides, or pathogens and observing their survival over time. There have been several laboratory studies on the construction of lifetables of mosquitoes, the estimation of survival rates, intrinsic rate of increase, reproduction rates, etc. It is important to note that estimating mortality in immatures can be challenging due to factors like natural mortality, predation, and environmental conditions. Newer methods like remote sensing and GIS, DNA barcoding, biomarker-based assays, and use of nano-sensors and nanomaterials are thus increasingly being researched to improve upon the efficacy of mortality estimation in immatures. The choice of method will depend on the specific objectives of the study, available resources, and the size and characteristics of the mosquito population being sampled.

7.6 CONCLUSION

Sampling mosquitoes is a fundamental component of mosquito monitoring programs, research studies, and vector control initiatives. By understanding the population dynamics, species composition, abundance, and infection rates of mosquitoes, researchers and public health authorities can develop targeted strategies for mosquito control and disease prevention. Sampling mosquitoes involves the use of various methods designed to capture and collect adult mosquitoes, their larvae/pupae or eggs from different habitats. These methods enable the collection of valuable data on mosquito species diversity, distribution, abundance, and potential disease transmission. Moreover, they also provide insights into mosquito behavior, host preferences, and breeding sites, which aid in identifying high-risk areas and implementing appropriate control measures. The technological advancement in modern times with AI, drones and GIS is revolutionizing surveillance and sampling techniques as they facilitate effective and targeted control of the vectors and prevent outbreaks of mosquito-borne diseases. Several mosquito sampling methods have been developed and utilized over the years, each with its advantages, limitations, and specific applications. Some common sampling methods include light traps, CO_2-baited traps, gravid traps, larval/pupal sampling, oviposition traps and human landing catch. The efficacy of many newer traps like mosquito magnet, Biogent sentinel trap, MET, etc. for various mosquito species is being explored regularly. Each method targets different life stages or behavior of mosquitoes, allowing researchers and vector control professionals to gain a comprehensive understanding of the mosquito population within a particular area. In determining the appropriate sampling method, several factors must be considered, including the objectives of the study, target mosquito species, habitat characteristics, available resources, and local regulations.

REFERENCES

Arunachalam, N., Samuel, Philip P., Hiriyan, J. and Gajanana, A. (1999). A comparative study on sampling techniques for *Aedes aegypti* (Diptera: Culicidae) surveillance in Madurai, South India. *Tropical Biomedicine*, 29, 25–29.

Arunachalam, N., Tewari, S.C., Thenmozhi, V., Rajendran, R., Paramasivan, R., Manavalan, R., Ayanar, K., Tyagi, B.K. (2008). Natural vertical transmission of dengue viruses by *Aedes aegypti* in Chennai, Tamil Nadu, India. *Indian Journal of Medical Research*, 127(4),395–397.

Carrasco-Escobar, G., Moreno, M., Fornace, K., Herrera-Varela, M., Manrique, E. (2022) Conn Jan. The use of drones for mosquito surveillance and control. *Parasite Vectors*, 15, 473. https://doi.org/10.1186/s13071-022-05580-5

Chadee, D.D., Connell, N.K., Le Maitre, A. and Ferreira, S.B. (1984). Surveillance for *Aedes aegypti* in Tobago, West Indies (1980–82). *Mosquito News*, 44, 490–492.

Chadee, D.D. and Corbet, P.S. (1990). A night-time role of the oviposition site of the mosquito *Aedes aegypti* (L.) (Diptera: Culicidae). *Annals of Tropical Medicine & Parasitology*, 84,429–433.

Chadee, D. D. and Corbet, P. S. (1987). Seasonal incidence and diel patterns of oviposition in the field of the mosquito, *Aedes aegypti* (L.) (Diptera: Culicidae) in Trinidad, West Indies: a preliminary study. *Annals of Tropical Medicine & Parasitology*, 81, 151–61.

Chadee, D.D., Corbet, P.S. and Talbot, (1995). Proportions of eggs laid by *Aedes aegypti* on different substrates within an ovitrap in Trinidad, West Indies. *Medical and Veterinary Entomology*, 9, 66–70.

Christie, M. (1958). Improved collection of anopheline eggs and analysis of oviposition behaviour in *A. gambiae*. *Journal of Tropical Medicine and Hygiene*, 61, 282–6.

Chye, J. K., Lim, C. T., Ng, K. B., Lim, J. M., George, R., & Lam, S. K. (1997). Vertical transmission of dengue. *Clinical Infectious Diseases: An Official Publication of the Infectious Diseases Society of America*, 25(6), 1374–1377. https://doi.org/10.1086/516126

De Meillon Soman, R. S. and Reuben, R. (1970). Studies on the preferences shown by ovipositing emales of *Aedes aegypti* for water containing immature stages of the same species. *Journal of Medical Entomology*, 7, 485–9.

Dias, T., Alves, V., Alves, H., Pinheiro, L., Pontes, R., Araujo, G., Lima, A., Prego, T. (2018). Autonomous detection of mosquito-breeding habitats using an unmanned aerial vehicle. In Proceedings of the 2018 Latin American Robotic Symposium, 2018 Brazilian Symposium on Robotics (SBR) and 2018 Workshop on Robotics in Education (WRE), João Pessoa, Brazil, 6–10 November 2018; pp. 351–356.

Evans, B. R. and Bevier, G. A. (1969). Measurement of field populations of *Aedes aegypti* with the ovitrap in 1968. *Mosquito News*, 29(3), 347–353.

Fay, R. W. and Eliason, D. A. (1966). A preferred oviposition site as a surveillance method for *Aedes aegypti*. *Mosquito News*, 26, 531–5.

Ferreira, F. C. P. A. D. M., da Silva, A. S. V., Recht, J., Guaraldo, L., Moreira, M. E. L., de Siqueira, A. M., Gerardin, P., Brasil, P. (2021). Vertical transmission of chikungunya virus: A systematic review. *PLOS ONE*, 17(8), e0272761. https://doi.org/10.1371/journal.pone.0272761

Ferreira-de-Lima, V.H., Lima-Camara, T.N. (2018). Natural vertical transmission of dengue virus in *Aedes aegypti* and *Aedes albopictus*: a systematic review. *Parasites Vectors*, 11, 77. https://doi.org/10.1186/s13071-018-2643-9

Frank, J. H. and H. C. Lynn. (1982). Standardizing oviposition traps for *Aedes aegypti* and *Culex quinquefascinnrs*: time and medium. *Journal of Florida Anti-Mosq Association*, 53, 22–27.

Freier, J. E. & Francy, D. B. (1991). A duplex cone trap for the collection of adult *Aedes albopictus*. *Journal of the American Mosquito Control Association*, 7, 73–9.

Furlow, B. M. and Young, W. W. (1970). Larval surveys compared to ovitrap surveys for detecting Aedes aegypti and *Aedes triseriatus*. *Mosquito News*, 30(3), 468–470

Gass, R. F., Deesin, T., Surathin, K., Vutikes, S., Sucharit, S. & Harinasuta, C. (1983). Studies on oviposition characteristics of *Mansonia (Mansonioides)* mosquitoes in southern Thailand. *Annals of Tropical Medical Parasitology*, 77, 605–14.

Goncalves, D.D.S., Hue, K.D.T., Thuy, V.T., Tuyet, N.V., Thi, G.N., Thi Thuy, V.H., Xuan, T.H.T., Thi, D.L., Vo, L. T., Le Anh Huy, H., Van Thuy, N.T., Wills, B.A., Thanh, P.N., Simmons, C.P. and Carrington, L.B. (2020). Assessing the vertical transmission potential of dengue virus in field-reared *Aedes aegypti* using patient-derived blood meals in Ho Chi Minh City, Vietnam. *Parasites Vectors*, 13, 468. https://doi.org/10.1186/s13071–020–04334–5

Haas-Stapleton, E. J., Barretto, M. C., Castillo, E. B., Clausnitzer, R. J., Ferdan, R.L. (2019). Assessing mosquito breeding sites and abundance using an unmanned aircraft. *Journal of the American Mosquito Control Association*, 35, 228–32.

Haeger, J. S. and O'Meara, G. F. O. (1983). Separation of first-instar larvae of four *Culex (Culex)*. *Mosquito News*, 43, 76–7.

Harbison, J. E., Henry, M., Xamplas, C., and Berry, C. (2018). Enhancing Mosquito Identification through Pupal Sampling. *Journal of the American Mosquito Control Association*, 34(2), 131–134.

Hardy, A., Makame, M., Cross, D., Majambere, S., Msellem, M. (2017). Using low-cost drones to map malaria vector habitats. *Parasite Vectors*, 10, 29.

Heath, C.J., Grossi-Soyster, E.N., Ndenga, B.A., Mutuku, F.M., Sahoo, M.K., Ngugi, H.N., Mbakaya, J.O., Siema, P., Kitron, U., Zahiri, N., Hortion, J., Waggoner, J.J., King, C.H., Pinsky, B.A., LaBeaud, A.D. (2020). Evidence of transovarial transmission of Chikungunya and Dengue viruses in field-caught mosquitoes in Kenya. *PLOS Neglected Tropical Diseases*, 14(6), e0008362. https://doi.org/10.1371/journal.pntd.0008362

Jain, J., Kushwah, R. B. S., Singh, S. S., Sharma, A., Adak, T., Singh, O. P., Bhatnagar, R. K., Subbarao, S. K., Sunil, S. (2016). Evidence for natural vertical transmission of chikungunya viruses in field populations of *Aedes aegypti* in Delhi and Haryana states in India-a preliminary report. *Acta Tropica*, 162, 46–55. https://doi.org/10.1016/j.actatropica.2016.06.004

Jakob, W. L., Fay, R. W., von Windeguth, D. L. & Schoof, H. F. (1970). Evaluation of materials for ovitrap paddles in *Aedes aegypti* surveillance. *Journal of Economic Entomology*, 63, 1013–14.

Jeyaprakasam, N. K., Pramasivan, S., Liew, J. W., Van Low, L., Wan-Sulaiman, W.-Y., Ngui, R., Jelip, J., & Vythilingam, I. (2021). Evaluation of mosquito magnet and other collection tools for *anopheles* mosquito vectors of simian malaria. *Parasite Vectors*, *14*(1). https://doi.org/10.1186/s13071–021–04689–3

Kim, K., Hyun, J., Kim, H., Lim, H., Myung, H. (2019). A deep learning-based automatic mosquito sensing and control system for urban mosquito habitats. *Sensors*, 19, 2785. https://doi.org/10.3390/s19122785

Llagas, L.D., Bersales, L.G., Bertuso, A.G., Mistica, M., Ocampo, V.R., Samaniego, J.B., Magsino, E.A., & Alzona, F.D. (2016). Mosquito Larvicidal Trap (MLT) as surveillance and control tool for aedes mosquitoes. *Southeast Asian Journal of Tropical Medicine and Public Health*, 47, 701.

Laurence, B. R. & Samarawickrema, W. A. (1970). Aggregation by ovipositing Mansonioides mosquitoes. *Journal of Medical Entomology*, 7, 594–600.

Lopp, O. V. (1957). Egg sampling as an index of mosquito breeding. *Proceedings of New Jersey Mosquito Extermination Association*, 44, 60–5.

Lounibos, L. P. & Dewald, L. B. (1989). Oviposition site selection by Mansonia mosquitoes on water lettuce. *Ecological Entomology*, 14, 413–22.

Mackay, A. J., Amador, M. & Barrera, R. (2013). An improved autocidal gravid ovitrap for the control and surveillance of *Aedes aegypti*. *Parasite Vectors*, 6, 225. https://doi.org/10.1186/1756–3305–6-225

Maliti, D. V., Govella, N. J., Killeen, G. F., Mirzai, N., Johnson, P. C., Kreppel, K., Ferguson, H. M. (2015). Development and evaluation of mosquito-electrocuting traps as alternatives to the human landing catch technique for sampling host-seeking malaria vectors. *Malaria Journal*, 14, 502. https://doi.org/10.1186/s12936–015–1025–4. Erratum in: *Malaria Journal*. 15;15:558.

Mboera, L. E. (2005). Sampling techniques for adult Afrotropical malaria vectors and their reliability in the estimation of entomological inoculation rate. *Tanzanian Health Research Bulletin*, 7(3):117–24. https://doi.org/10.4314/thrb. v7i3.14248

Mboera, L. E. G., Takken, W., Mdira, K. Y., and Pickett, J. A. (2000). Sampling gravid *Culex Quinquefasciatus* (Diptera: Culicidae) in Tanzania with traps baited with synthetic oviposition pheromone and grass infusions. *Journal of Medical Entomology*, 37(1): 172–176.

Mogi, M., Choochote, W., Khambooruang, C. & Suwanpanit, P. (1990). Applicability of presence-absence and sequential sampling for ovitrap surveillance of Aedes (Diptera: Culicidae) in Chiang Mai, northern Thailand. *Journal of Medical Entomology*, 27,509–14.

Muirhead-Thomson, R. C. (1945). Studies on the breeding places and control of *Anopheles gambiae and A. gambiae var.* melas in coastal districts of Sierra Leone. *Bulletin of Entomological Research*, 36, 185–252.

Murillo, D., Murillo, A., Lee, S. (2019). The Role of Vertical Transmission in the Control of Dengue Fever. *International Journal of Environment Research and Public Health*, 16(5), 803. https://doi.org/10.3390/ijerph16050803

Nayar, J. K. (1981). *Aedes aegypti* (L.) (Diptera: Culicidae): Observations on dispersal, survival, insemination, ovarian development and oviposition characteristics of a Florida population. *Journal of Florida Anti-Mosquito Association*, 52, 24–40.

Panigrahi, S. K., Barik, T. K., Mohanty, S. and Tripathy, N. K. (2014). Laboratory evaluation of oviposition behavior of field collected Aedes mosquitoes. *Journal of Insects*, 2014, 1–8.

Pratt, H. D. and Jakob, W. L. (1967). 'Oviposition Trap Reference Handbook.' *Aedes aegypti* Handbook Series No.6, National Communicable Disease Center, ii + 33 pp.

Praveen, G., Tyagi, B.K., Joe, N., and Thakur, M.B. (2016). An outbreak investigation of dengue fever in the coastal areas of Nagapattinam, Thiruvarur and Thanjavur districts in Tamil Nadu, India during 2012. *International Journal of Mosquito Research*, 3(4), 39–47.

Rajagopalan, P.K., P.K. Das, M. Kalyanasundaram, B.K. Tyagi, N. Arunachalam, N. Somachary, C.B.S. Reddy and C.M.R. Reddy. (1987). *Bangalore Mosquito Control Master Plan*. Vector control Research Centre, Pondicherry, pp. 309.

Rajagopalan, P.K. and P.K. Das. (1988). *Filariasis control Demonstration Project*. Vector Control Research Centre, Pondicherry.

Reiter, P. (1986). A standardized procedure for the quantitative surveillance of certain Culex mosquitoes by egg raft collection. *Journal of American Mosquito Control Association*, 2, 219–21.

Russell, P. F. and Rao, T. R. (1942). On relation of mechanical obstruction and shade to ovipositing of *Anopheles culicifacies*. *Journal of Experimental Zoology*, 91, 303–29.

Schuler, T. C. and Beier, J. C. (1983). Oviposition dynamics of two released species of *Toxorhynchites* (Diptera: Culicidae) and potential prey species. *Journal of Medical Entomology*, 20,371–6.

Semwal, A., Melvin, L. M. J., Mohan, R. E., Ramalingam, B., & Pathmakumar, T. (2022). AI-Enabled Mosquito Surveillance and Population Mapping Using Dragonfly Robot. *Sensors*, 22(13), 4921. https://doi.org/10.3390/s22134921

Service, M. W. (1970). Studies on the biology and taxonomy of *Aedes (Stegomyia) vittatus* (Bigot) (Diptera: Culicidae) in Northern Nigeria. *Transactions of the Royal Society of Entomology, London*, 122, 101–43.

Service, M. W. (1965). The ecology of the tree-hole breeding mosquitoes in the northern guinea savanna of Nigeria. *Journal of Applied Ecology*, 2, 1–16.

Service, M. W. (1993). Mosquito Ecology – Field Sampling Methods. *Springer Dordrecht*, 988pp. https://doi.org/10.1007/978–94–011–1868–2

Silver, J. B. (2008). *Mosquito ecology – Field sampling methods*. 3rd Edition. Springer, The Netherlands, 1494 pp.

Smith, W. W. and Jones, D. W. (1972). Use of artificial pools for determining presence, abundance, and oviposition preferences of *Culex nigripalpus* Theobald in the field. *Mosquito News*, 32, 244–5.

Southwood, T. R. E., Murdie, G., Yasuno, M., Tonn, R. J. & Reader, P. M. (1972). Studies on the life budget of *Aedes aegypti* in Wat Samphaya, Bangkok, Thailand. *Bulletin of the World Health Organisation*, 46, 211–26.

Strickman, D. (1988). Rate of oviposition by *Culex quinquefasciatus* in San Antonio, Texas, during three years. *Journal of American Mosquito Control Association*, 4, 339–44.

Subra, R. and Mouchet, J. (1984). The regulation of preimaginal populations of *Aedes aegypti* (L.) (Diptera: Culicidae) on the Kenya coast. II. Food as a main regulatory factor. *Annals of the Tropical Medicine and Parasitology*, 78, 63–70.

Suganthi, P., Govindaraju, M., Kangabam, R. D., Suganthi, K., Thenmozhi, V., Tyagi, B. K. (2018). Identification and mapping of breeding habitats of the filariasis vector, Culex quinquefasciatus say, 1823, using remote sensing and GIS technologies: a case study from the endemic Tamil Nadu state, India. In: Tyagi BK (ed) *Lymphatic filariasis*. Springer-Nature, Singapore, pp 259–269.

Surgeoner, G. A. and Helson, B. V. (1978). An oviposition trap for arbovirus surveillance in *Culex* sp. mosquitoes (Diptera: Culicidae). *Canadian Entomology*, 110, 1049–52.

Thaggard, C. W. and Eliason, D. A. (1969). Field evaluation of components for an *Aedes aegypti* (L.) oviposition trap. *Mosquito News*, 29, 608–12.

Thenmozhi, V., Hiriyan, J. G., Tewari, S. C., Philip, S. P., Paramasivan, R., Rajendran, R., Mani, T. R., Tyagi, B. K. (2007). Natural vertical transmission of dengue virus in *Aedes albopictus* (Diptera: Culicidae) in Kerala, a southern Indian state. *Japanese Journal of Infectious Diseases*, 60(5): 245–9.

Tyagi, B.K. (1993). An indigenously developed mechanical device for sampling the resting adults of mosquitoes. *Journal of Applied Zoological Researches*, 4, 180–184.

Tyagi, B.K., Yadav, S.P., Sachdev, R. and Dam, P.K. (2001). Malaria outbreak in the IGNP command area in Jaisalmer district, Thar Desert, India. *Journal of Communicable Diseases*, 33(2): 88–95.

Tyagi, B.K. (2002). *Malaria in the Thar Desert: Facts, figures and future. Agrobios* (India), pp. 165.

Tyagi, B. K. (2003). *Medical Entomology: A handbook of medically important insects and other arthropods.* Scientific Publishers (India). 265 pp.

Tyagi, B. K. (2008). *Vector-borne diseases: Epidemiology and control.* Scientific Publishers (India), 298 pp.

Tyagi, B. K., Hiriyan, J., Philip, S. P., Tewari, S. C., Paramasivan, R. (2006). Dengue in Kerala: A critical review. *ICMR Bulletin*, 36(4–5): 13–22.

Tyagi, B. K. (2023). *Desert Malaria: An emerging Malaria Paradigm and its global impact on disease elimination.* Springer-Nature, Singapore, 416 pp.

Tyagi, B.K. and Yadav, S.P. (1996). Malariological and sociological significance of "Tanka" and "Beri" in the Thar Desert, Western Rajasthan (India). *Journal of Arid Environments (U.K.)*, 33, 497–501.

WHO (1975a). Manual of Practical Entomology in Malaria. Part I. Vector Bionomic and Organization of Antimalarial Activities. World Health Organization, WHO Offset Publication, No. 13.

WHO (1975b). Manual of Practical Entomology in Malaria. Part II. Methods and Techniques. WHO Offset Publication No. 13.

Yasuno, M., Kazmi, S. J., LaBrecque, G. C., Rajagopalan, P. K. (1973). Seasonal change in larval habitats and population density of *Culex fatigans* in Delhi Villages. *WHOIVBCI73*, 429, 12 pp.

Yates, M. G. (1979). The biology of the tree-hole breeding mosquito *Aedes geniculatus* (Olivier) (Diptera: Culicidae) in southern England. *Bulletin of Entomological Research*, 69, 611–28.

Yin, X., Zhong, X., Pan, S. (2016). Vertical transmission of dengue infection: the first putative case reported in China. *Revista do Instituto de Medicina Tropical de São Paulo*, 58, 90. https://doi.org/10.1590/S1678–9946201658090

Section IV

Biology

8 Biology of Mosquitoes

B. Reddya Naik
University College of Science, Osmania University, Hyderabad, India

Diwakar Singh Dinesh
Ex-ICMR-RMRIMS, Patna, India

M. Siddaiah
Nizam College (OU), Hyderabad, India

Sambashiva Daravath
Nizam College (OU), Basheerbagh, India

B.K. Tyagi
Chandigarh University, Mohali, India

8.1 INTRODUCTION

Arthropods form the largest among animal phyla with over 75% of all living creatures classified under it many species carry enormous medical or veterinary importance including some dangerous diseases for humans, such as malaria, Chagas disease, leishmaniasis, African trypanosomiasis, lymphatic filariasis, onchocerciasis, dengue, yellow fever, chikungunya and Zika. These diseases prosper when there is a suitable eco-bio-social matrix of environment, particularly driven by climate change and anthropization. Environmental changes are driven by humans, with special reference to urbanization. One of the major and integral factors to the emergence or re-emergence of several mosquito-borne diseases in areas where they were not previously recorded is superlative preponderance of dominant vector species with higher degree of inclination toward anthropophagy and, in some cases, synanthropy (Figure 8.1). Mosquitoes are the main vectors for arboviruses (dengue, yellow fever, chikungunya) and parasites (Plasmodium spp., Filariasis spp., etc.) and contribute to their global expansion.

Arthropods, which comprise mosquitoes (Family Culicidae) as one of their members, are the predominantly abundant animals on the earth. An appreciable quantum of arthropods are underestimated predators of vertebrates. Fossil records show that some of the archaic arthropods were giant-sized potential macropredators. The great majority of modern time's arthropods of public health importance are micropredators (Benelli et al., 2022). These micropredators, in a narrow sense, are interpreted as free-living organisms that prey on host blood or other tissues, usually for an ephemeral time and without forming an obligatory relationship with their host (Benelli et al., 2020). Consequently, micropredators usually show a scarce specificity, and differently from "true" parasites, they feed on multiple host individuals per generation, with hosts potentially belonging to one or more species. This category includes all the blood-sucking Diptera that attack humans (e.g., mosquitoes, sandflies, biting midges, blackflies, tumbu flies, tsetse flies, louse flies, stable flies), as well as insects belonging to the orders Hemiptera (kissing bugs, bed bugs, and bat bugs), Anoplura (Human lice) and Siphonaptera (fleas), and arachnids of the types of ticks and mites. Mosquitoes are

DOI: 10.1201/9781003326991-12

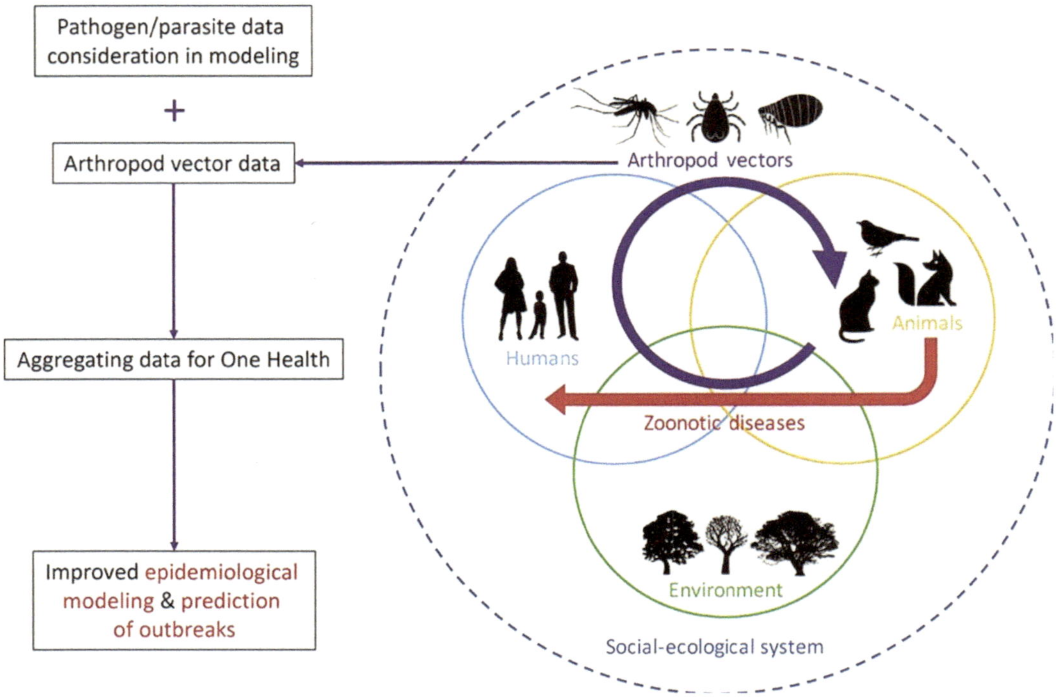

FIGURE 8.1 Arthropod-borne diseases in a matrix of eco-bio-social environment systems with various ramifications.

***Source*: Benelli and Duggan, 2018 (Permission obtained).**

the main vectors for arboviruses (dengue, yellow fever, chikungunya) and parasites (malaria, filariasis etc.) and contribute to their global expansion (Berenger and Parola, 2017).

Biology of mosquitoes, an insect and arthropod, follows a typical holometabolous pathway, albeit several novelties specific to them particularly hematophagy restricted to females only (Clements, 1992, 1999, 2011; Tyagi, 2003). Between the two sexes, mosquito females have always attracted more attention of humans in the past, due largely to their inherent trait of hematophagy, whereby they evolve the opportunity to also infect their hosts (e.g., human and other vertebrates) with pathogens that can cause serious illness (Table 8.1; Figure 8.2). In the entire history of humankind, no creature has touched directly the lives of more human beings than the mosquito which has often been branded as the humankind's deadliest foe ever (Spielman and D'Antonio, 2001). Like other arthropods, female mosquitoes too need to feed on the protein-rich source of blood necessary for their own advantage of maturing eggs. Blood is the typical resource of female mosquitoes – the vessel-feeding micropredators, like other hematophagous insects such as, for example, kissing bugs, bed bugs, and fleas, and in the process of evolutionary fitment have modified their mouthparts for cannulating host dermal blood vessels (Benelli and Mehlhorn, 2018). Contrarily, other blood-feeding insects like blackflies, biting midges, horse flies, stable flies, tsetse flies, and sandflies are pool-feeders and suck blood through micro-lacerations opened on the host skin (Lane and Crosskey, 1993).

Mosquitoes are cosmopolitan in distribution and are found in all climatic zones and zoogeographical regions. Each kind of mosquito has a slightly different secret for surviving cold, drought, predators and other dangers, and different types of mosquitoes carry different diseases. By classifying them into species and studying their habits, scientists can figure out the best way to get rid of the ones that carry the worst diseases. Although, there are about 3500 species (under 112 genera) all over the world, however only fewer than 100 species under 8 genera are vectors of

TABLE 8.1
Structural Differences between Males and Female Mosquitoes

Basis of Comparison between Males and Female Mosquitoes	 ♀	 ♂
Antennae	Male mosquitoes have bushy fine hairs on their long antennae (proboscis) which help them with sensing their potential mate's wing beats.	Female mosquitoes' antennae are sparce and contain several blood odor sensing receptors that help her target blood sources so that they can land on their host's body accurately and feed off their host.
Body Size	Male mosquitoes are relatively smaller than female mosquitoes.	Female mosquitoes are relatively larger than male mosquitoes.
Buzzing Sound	A male mosquito does not make a buzzing sound or do so at a very low note.	A female mosquito makes an audible whining or buzzing sound.
Emergence	Males generally emerge earlier from pupae than females, so as to be able to prepare for consummation with the females to follow.	Females are next to males in emergence.
Lifespan	A male mosquito has a relatively shorter lifespan than the female mosquito. A male mosquito's average lifespan is between one and three weeks.	A female mosquito has a relatively longer lifespan than the male mosquito. A female mosquito's average lifespan is between two and four weeks.
Human Contact	Male mosquitoes do not possess type of mouthparts which can pierce through skin to feed on blood and, therefore, they typically avoid human contact. Instead they survive on plant juices, etc.	The female mosquito has piercing and sucking type of mouthparts so as to be able to feed on human/animal blood and thus carries disease pathogens such as those of malaria, dengue, Japanese encephalitis and yellow fever, etc.

diseases (Service, 2008). Worldwide the medically important genera are: Anopheles, Culex, Aedes, Hemagogus, Sabethes, Bartonella and Mansonia, while Coquillettidia and Psorophora are of lesser importance (Service, 2008; Rudolf et al., 2020). In India there are a little over 400 species (Tyagi et al., 2015). Listed next are the major genera of mosquitoes found in India, with some of the important vector species within each genus indicated.

1. **Anopheles**
    i. *An. culicifacies*
    ii. *An. stephensi*
    iii. *An. fluviatilis*
    iv. *An. minimus*
    v. *An. baimaii*

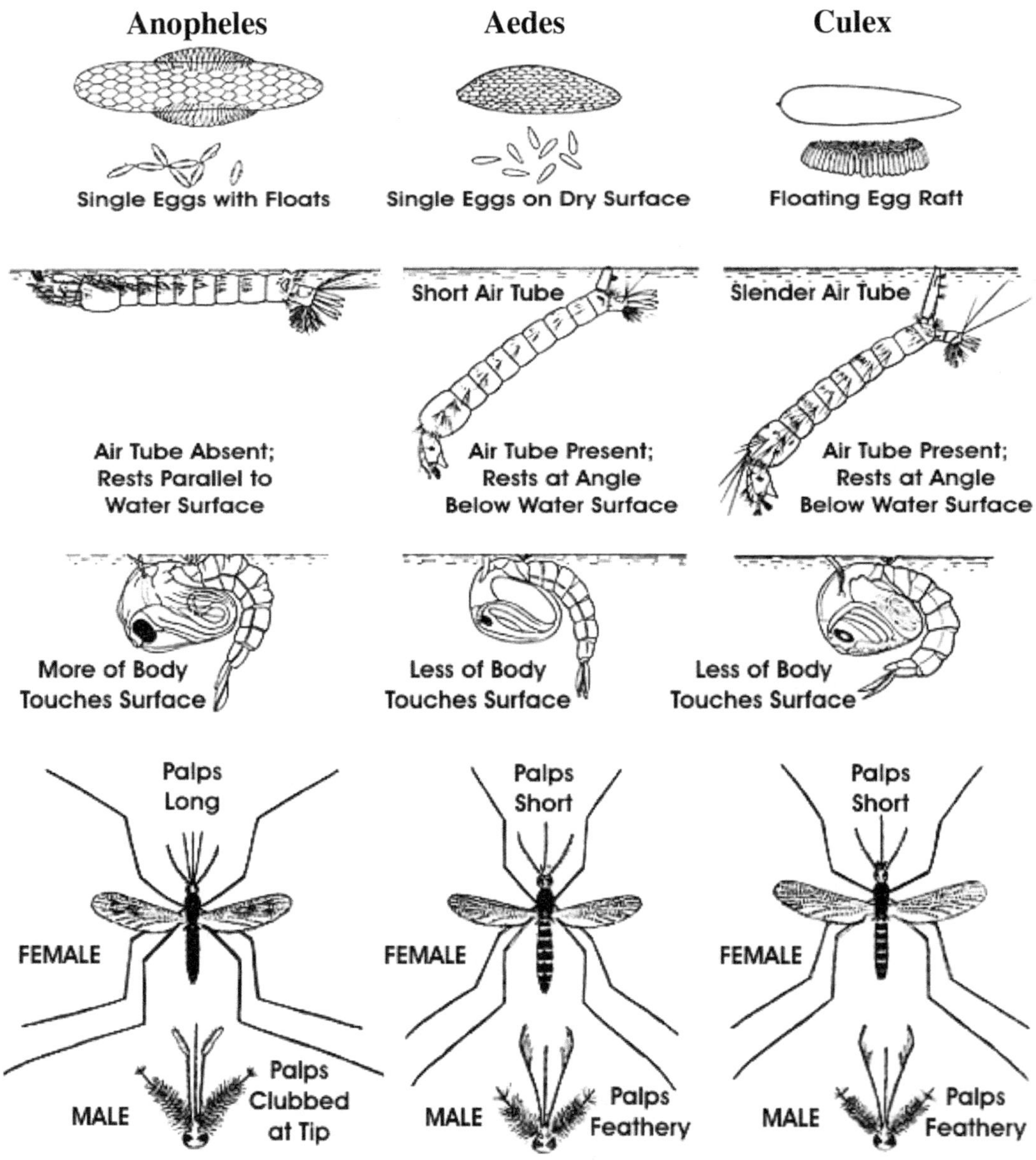

FIGURE 8.2 Morphological and functional differences among three major genera of mosquitoes in each phase of development of life cycle.

    vi. *An. sundaicus*
    vii. *An. philippinensis*
2. **Aedes**
    i. *Ae. aegypti*
    ii. *Ae. albopictus*
3. **Culex**
    i. *Cx. quinquefasciatus*
    ii. *Cx. tritaeniorhynchus*
    iii. *Cx. vishnui*
    iv. *Cx. pseudovishnui*
    v. *Cx. gelidus*

4. **Mansonia**
   i. *Ma. uniformis*
   ii. *Ma. annularis*
   iii. *Ma. indiana*
5. **Armigeres**
   i. *Ar. Subalbatus*

8.2 ECONOMICS OF THE MOSQUITO LIFE CYCLE

It is estimated that there are possibly 200 trillion mosquitoes at any point in time. If Nature had not controlled population of mosquitoes, this universe might have been buried under manifold layers of the critters. From a schematic diagram of the total quantum of adults expected to be borne from a small batch pair of mosquitoes, it can be easily understood (Figure 8.3).

Mosquitoes are holometabolous organisms and, therefore, exhibit all the four inexorable stages of development: egg, larva, pupa, and adult. Illustrations of the main differences between the eggs, larvae, pupae and adults of three common genera, *Anopheles*, *Aedes*, and *Culex* found in India.

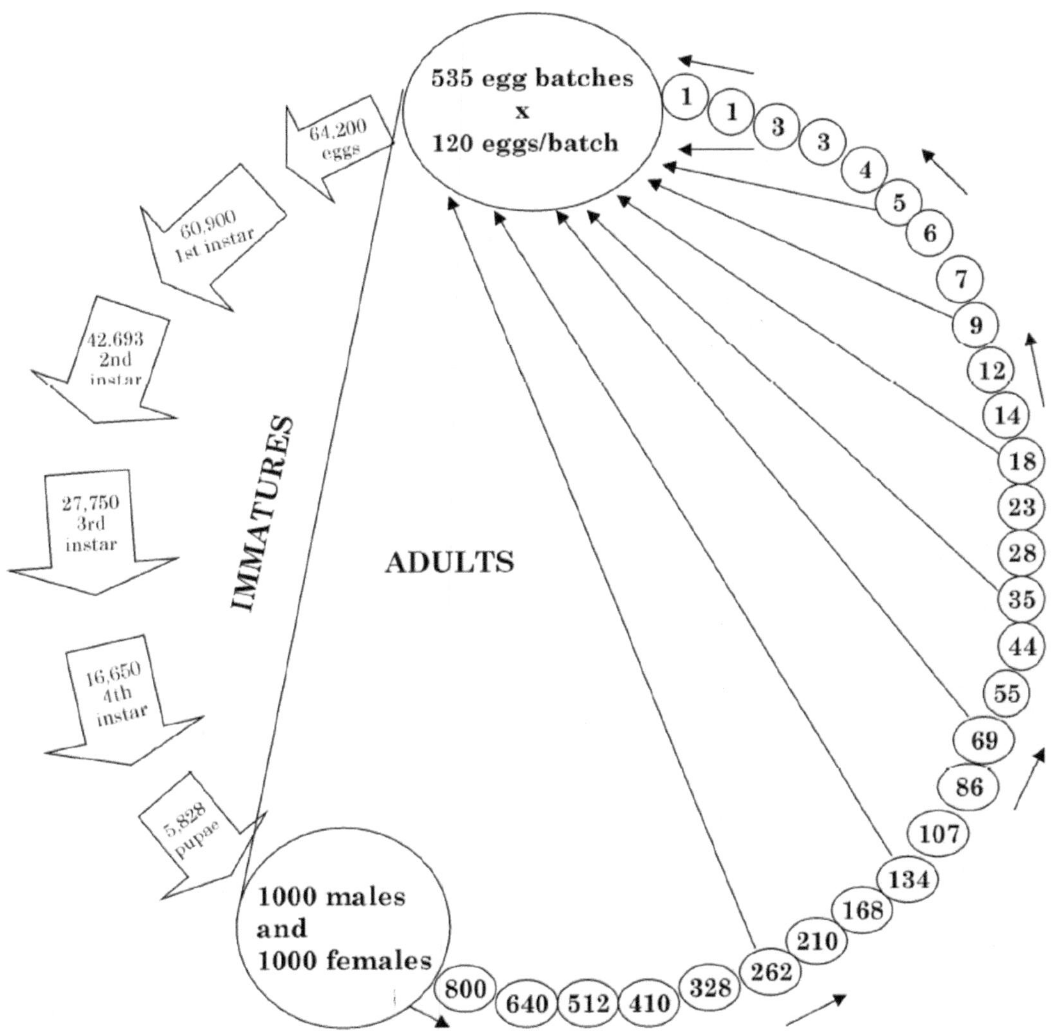

FIGURE 8.3 Relationship of survival and egg-deposition to number of mosquitoes in different development stages.

8.3 ANOPHELES MOSQUITOES

From a medical standpoint, this genus is the most important of all genera of mosquitoes. While many species are serious vectors of the deadly infection, malaria, to humans, several others are painful biters as well as nuisance mosquitoes. Only the Anopheles species can transmit human malaria parasites to humans, i.e., *Plasmodium vivax, P. falciparum, P. ovale, P. malariae* and *P. knowlesi.* There are about 500 species of Anopheles globally, of which nearly 65 species occur in India (Nagpal & Sharma, 1995; Tyagi et al., 2015). Species of this genus are found in all geographical regions of the world, and the Oriental Region is the richest (Rao, 1984).

Important taxonomic and biological features of each larval stage, pupa and adult are described below (Figure 8.4):

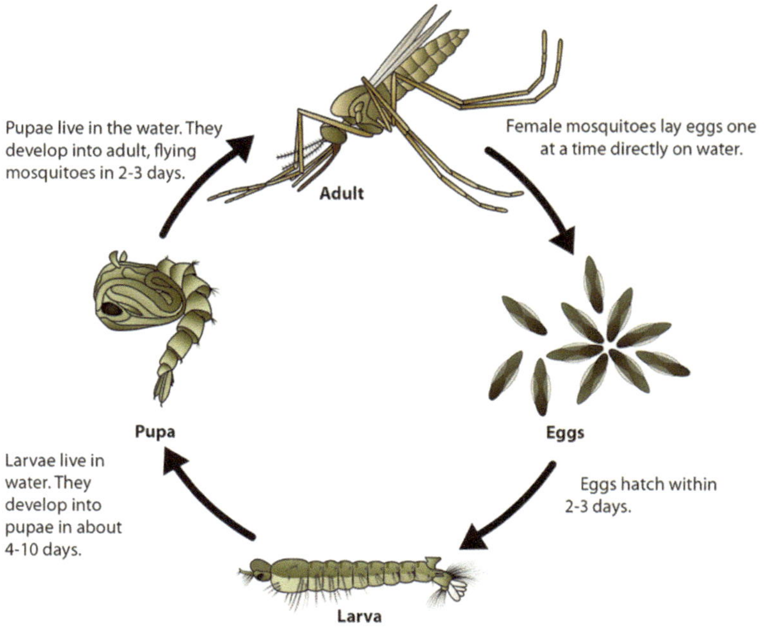

FIGURE 8.4 Life stagers and cycle of Anopheles mosquitoes.

Source: CDC, 2012.

8.3.1 The Egg

Eggs of Anopheles mosquitoes are boat-shaped with an upper deck having a micropyle at one end and a fringe all around, and usually with lateral floats with transverse striations called ridges. Eggs are laid singly. In optimal temperature it takes about 36–48 hours for eggs to hatch (Figures 8.5 and 8.6).

Tyagi et al. (2016) have successfully distinguished all the five sibling species of *Anopheles culicifacies*, viz., A, B, C, D, and E, on the basis of egg morphology under electron microscopy, although he did not designate any name to a species yet which remains a major task in future.

8.3.2 The Larva

This is very agile from the time it hatches out. There are four larval instars before it converts into the pupal stage. It rests characteristically parallel to the surface of water. The head of a larva is capable to rotate at 180° to enable it feed more intensely and vigorously. Characteristic palmate hairs, present

FIGURE 8.5 Singly laid eggs of Anopheles mosquitoes.

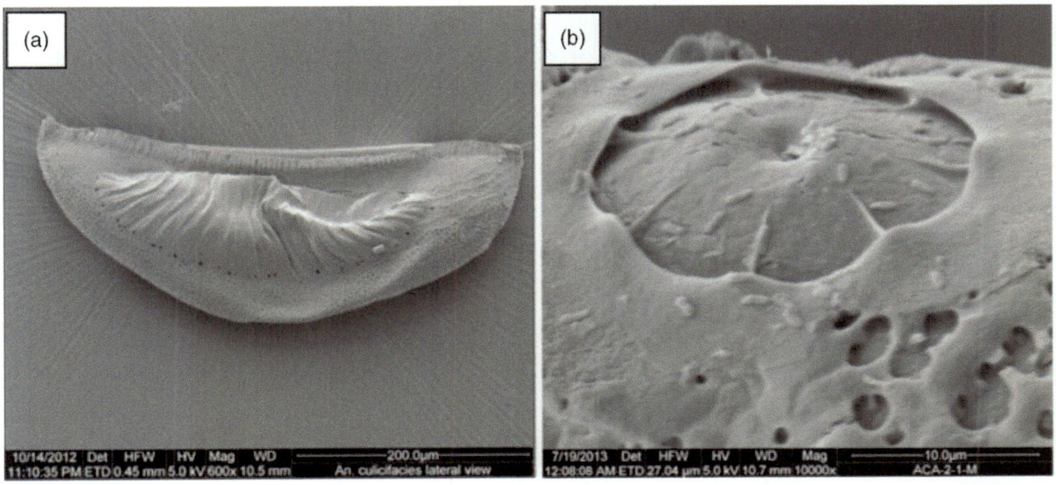

FIGURE 8.6 Size, shape and form of the egg of *Anopheles culicifacies*, with specific features of (a) boat-like shape of egg with lateral floats, and (b) micropyle.

Source: **Tyagi et al., 2016.**

in many species, occur on both thorax and abdomen. Two (one on each side) reversible organs, called Nuttal's organs, occur on the first segment of the thorax. These are best seen in the live specimens as they are fleshy and protrusible. Larvae have a dorsal spiracular plate supporting two spiracles meant for bringing about respiration. The total duration taken by larvae under normal conditions of temperature ranges from 5 to 7 days.

8.3.3 The Pupa

This has a peculiar "comma" form with highly active up and down vertical movements. The paddle is oval in outline. It may be either undivided or only slightly invaginated, with the tip having a pair of setae one of which is ventral and sub-apical. On the 3rd to 7th abdominal segments, there are located spine-like seta No. 9 (Belkin's system) at the postero-lateral corners usually with branches on the 8th abdominal segment. The trumpet, located on the top of the cephalothorax, often bears at least one cleft. Pupal development completes in about 48 hours.

8.3.4 The Adult

Soon after the adult emerges from pupae it flies away to take shelter in some suitable place, where it rests for some time before taking its first blood meal. Males generally emerge earlier than females, and have shorter lifespan than their counterparts (Figure 8.7).

Anopheline adults are identifiable from the rest of the mosquitoes due to their very characteristic color patterns mostly on the palpi, wings, and legs, beside other prominent traits as seen in their posture at rest, etc. The palpi of both sexes are of similar length, almost equal to that of the proboscis. In males, however, the two distal segments in palpi are enlarged giving them a club-like appearance. The posterior edge of the scutellum is only slightly curved and is continuous without any marked division into lobes.

Hairs on these lobes are in continuous series. The axes of the head, thorax, and abdomen are nearly in a straight line. Anopheline mosquitoes rest on an object in a characteristic posture. The angle of resting, varying from species to species, is close to 60° with the tip of the abdomen directed away from the surface (e.g., *An. hyrcanus*, *An. barbirostris*). In a few species like *An. culicifacies*, *An. culiciformis*, *An. sintoni*, etc. the body is almost parallel to the surface, giving the adult the semblance of a Culex-like posture.

Of the three acini of the salivary glands of each side, the middle is saclike instead of being tubular as is noticed in culicines. Wings are spotted or unspotted; the former condition being very common, though. On the other hand, members of subgenus Anopheles, like *An. aitkenii*, have completely dark wings. Males have only one claw in the fore tarsi. Females have only one spermatheca.

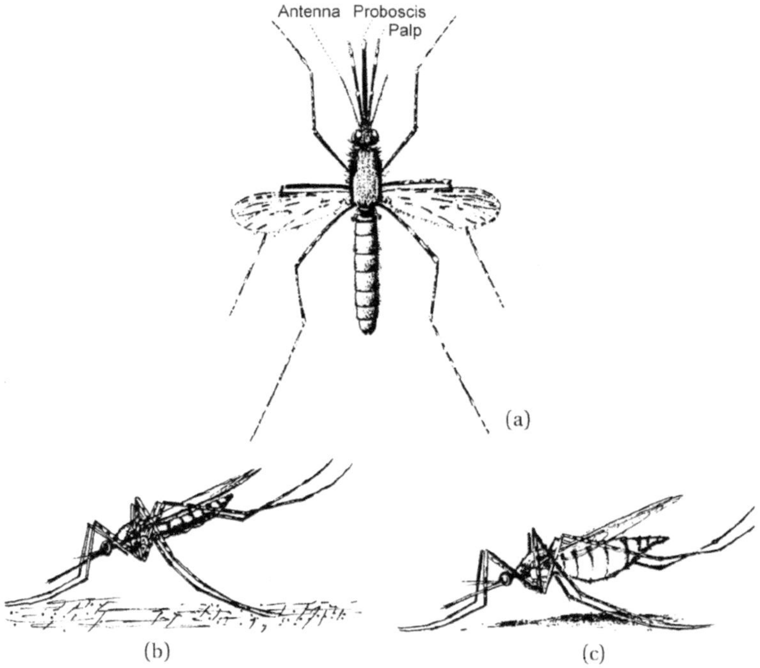

FIGURE 8.7 Adult of Anopheles mosquito: (a) general format, (b) at rest, (c) after engorgement.

Source: Tyagi, 2020.

TABLE 8.2
Sibling Species Complexes among Indian Anophelines

Species	No. of Sibling Species in the Complex	Nomenclature Assigned to Sibling Species
An. culicifacies	5	A, B, C, D, and E
An. fluvciatilis	3	S, T, and U
An. subpictus	4	A, B, C, and D
An. annularis	2	A and B

Source: Subbarao et al., 1999.

Most vector species are endophilic and endophagic in behavior (*An. culicifacies, An. subpictus, An. stephensi*), although a few are sylvatic in nature, too (*An. dirus, An. pulcherrimus*). Although, there is no direct proof of hibernation exhibited by any Indian species, there are species in colder temperate regions elsewhere in the world that manifest overwintering. In India, yet, *An subpictus* is suspected to possess the potential to overwinter. Some species exhibit "sibling species complexes," a phenomenon that carries tremendous medical significance in the sense that while some of them may be vectors, others are either less vectorial or are not able to transmit the pathogen at all. The most widely examined species complexes belong to *An. gambiae* in Africa and *An. culicfacies* in India (Subbarao et al., 1999).

In India as many as eight Anopheles species complexes exist (Table 8.2). Strangely, *An. stephensi* has two races, forms or subspecies but not sibling species. Tyagi and Verma (1991) has advocated presence of two "forms," bigger and smaller, in the Thar Desert ecosystem – the original home in India for the species. In the Thar Desert in northwestern Indian State of Rajasthan, *An. stephensi* has been observed to overwinter at temperatures as low as 0°C.

Knowledge of species complexes is highly vital particularly to malaria control strategy, especially when target-specific selective and sustainable vector control is urgently needed. It is important that vector control, as a part of the global as well as the regional malaria control strategy, should succeed. Its success would depend on our total knowledge of biology, ecology, and behavior, as also governed by genetics, of the vector species involved, and of the vector control options and their selective use. Most of Anopheles known to transmit malaria in India have been determined as species complexes. Members of a species complex, commonly known as sibling species, are reproductively isolated evolutionary units with distinct gene pools and, therefore, differ in biological characteristics.

Differences in biological characteristics of members of the complexes have an important bearing on malaria transmission dynamics. It is therefore imperative to determine sibling species composition and their bionomics as well as their roles in the transmission of malaria. At the cellular level, sibling species among mosquitoes were first determined through cyotaxonomic techniques where as a rule either chromosomes under mitotic and/or meiotic divisions in ova of the developing ovaries in females or the testes in males became object (Figure 8.8), but more characteristically, these were the polytene chromosomes from the germinal layer of the developing ovaries or the salivary glands which exhibited more clearly all the necessary features of identification (Figure 8.9). Later molecular techniques have replaced the cytotaxonomic techniques because they are quicker and more accurate (Figure 8.10). The time taken to arrive at a decision on the correct determination of a sibling species by any of these methods, however, is still a crucial factor.

8.4 AEDES MOSQUITOES

Several species of the genus Aedes are of tremendous medical and public health importance, e.g., *Ae. aegypti, Ae. albopictus, Ae. sollicitans, Ae. nigramaculis, Ae. vexans* etc. As far as Southeast Asia, which includes India as well, the most important vector species are *Ae. aegypti* and *Ae.*

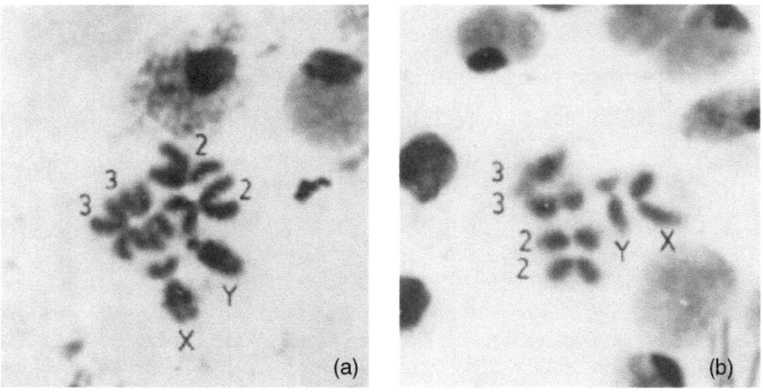

FIGURE 8.8 Male mitotic karyotypes of *An. culicifacies* from Rameswaram island (a) with acrocentric Y-chromosome (species B); (b) with sub-centric Y-chromosome (Species E).

Source: **WHOSEARO, 2007.**

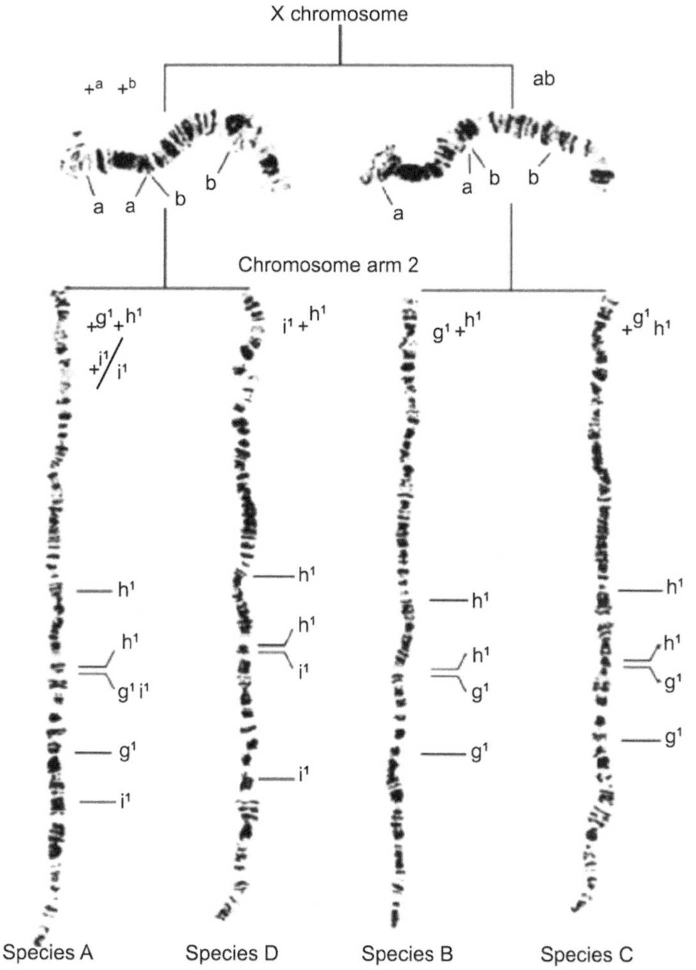

FIGURE 8.9 Schematic representation of polytene chromosomes of *Anopheles culicifacies* sibling species.

Source: **WHOSEARO, 2007.**

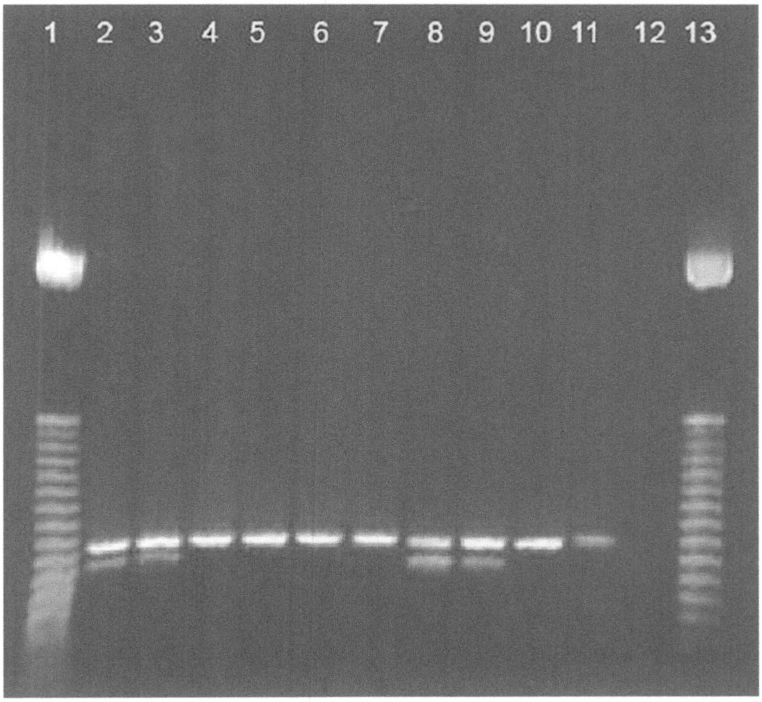

FIGURE 8.10 PCR assay using primers designed from D2 region of 28S rDNA which differentiates species A and D from species B,C and E of *An. culicifacies*. Lanes 1 & 13: DNA ladder; Lanes 2 & 3: species A; Lanes 4 & 5: species B, Lanes 6 & 7: species C; Lanes 8 & 9: species D; Lanes 10 & 11: species E, Lane 12: negative control.

Source: **WHOSEARO, 2007.**

albopictus, due to their involvement in transmission of dengue, chikungunya, and zika fever (Dash et al., 2012). In Africa and South America, where deadly yellow fever prevails, the vector is *Ae. aegypti* (Christophers, 1960).

Biology and life cycle of *Aedes aegypti* are described below:

Main biological and taxonomical features are described for the two important vector species (i.e., *Ae. aegypti* and *Ae. albopictus*). In general, the biology of *Ae. aegypti* follows the pattern of other culicid mosquitoes. The life cycle is completed through four distinct stages: egg, larva, pupa, and adult. The larval development passes through four instars, each undergoing a progressive increase in size to the pupal stage. Pupae and adults do not show any growth. Only female adults of *Ae. aegypli* feed on blood, sugar, and water for the maturation of eggs, energy requirements, and maintenance of the osmotic balance of the body (Figure 8.11).

8.4.1 The Egg

The eggs are small, about 1 mm in length, torpedo-shaped, and deep black, although when freshly laid they are translucent and pure white. The eggs are capable of remaining viable for weeks or even months and can withstand prolonged desiccation. Like eggs of Ochlerotatus mosquitoes, Aedes mosquito eggs, too, are more resistant to drying out (in fact, some complete drying out before the eggs will hatch) and hatch only when flooded with water (salt water high tides, irrigated pastures, tree holes flooded by rains, flooded stream bottoms). Eggs house fully developed larvae ready to hatch. The head of the larva occupies nearly one-third of the egg-volume, while the siphon and other terminal parts fill most of the remaining posterior pole. The egg has very characteristic polygonal

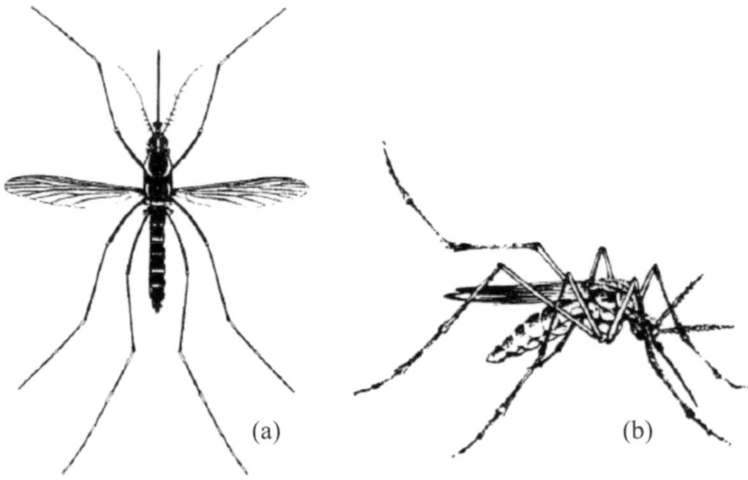

FIGURE 8.11 Female *Aedes aegypti* mosquito: (a) dorsal, and (b) lateral views.

Source: **Free Wikipedia.**

markings, lined with a network of milky white membranous channels filled with air at the base of the exochorion. Beneath exochorion is a dense, tough, and hard endochorion membrane, which gives the egg its shape and rigidity.

8.4.2 The Larva

As in all other mosquitoes, there are four larval instars, each instar ending in a molt or ecdysis (Table 8.3). The first instar larva hatches out from the egg by first widening the transverse crack at the dorsal side anteriorly. Finally, the larva breaks the eggs by means of a thorn-shaped egg breaker. Initially, the freshly hatched larva is unable to move but shortly the young larva wriggles its way to the water surface to make contact with the free air. It takes about 2–3 days at 27–29°C for the larva to hatch out from the egg. However, the larva may take a few minutes to several hours to emerge at the surface. The first instar larva is identifiable on the basis of the presence of the egg breaker on the head. The larva carries a characteristic siphon. The pectin spine arises on the soft membrane numbers 4–5. The siphon carries the trachea. The comb teeth, which are delicate, number only five.The second and third instars can be identified by their size and general appearance. Little change occurs

TABLE 8.3
Some Identification Characteristics of Various Larval Instars as Well as Their Time of Development

Larval Instar	Transverse Diameter of Head (mm)	No. of Pectin Teeth	Period for Completion of Four Instars (in hrs)
1st	0.3	6	24
2nd	0.45	7–10	24
3rd	0.65	12–15	24
4th	0.95	12–20	96–240

Source: (Tyagi et al., 2014).

until the fourth instar when the development of the future pupal structure begins. The compound eyes or the optic lobes begin to form in the late third instar. The imaginal buds at the bases of the pleural hairs become apparent in the second instar, and are quite conspicuous in the third, though their development is a feature of the fourth instar. The gonads are already visible in the sixth abdominal segment in the third instar, when also the buds of the developing gonoxites become apparent. During the fourth instar period, a condition is reached in which the larva is little more than the roughly sculptured body of the future adult disguised in the larval skin. Such a condition has been termed a prepupa. The duration of the larval stages is influenced by the water temperature. The most common period lies between 6 and 8 days (Figure 8.10).

8.4.3 The Pupa

The pupa of *Ae. aegypti* resembles most of other culicines. However, it differs from the anopheline pupa in having a more rounded body and cylindrical breathing trumpets. The main factors used to identify the pupa of *Ae. aegypti* from other culicine pupae are: (a) the shape of the respiratory trumpets; (b) the arrangement and characters of the hairs, particularly those in the abdominal region; and (c) the characteristics of the tail fins or paddles. The pupal body consists of a large globular anterior region, the cephalothorax, and a narrower abdomen which is normally kept flexed under the former to propel the insect when swimming (Figure 8.12).

8.4.4 The Adult

The adult is a small dark structure easily identifiable by the lyre-shaped, silvery-white lines on the thorax and the white bands on the tarsal segments. The adult is both a vector as well as a pest when present in large numbers and therefore deserves special attention in medical entomology. This is a thoroughly domesticated mosquito and breeds almost exclusively in artificial containers in and around human habitations. The females lay their eggs singly on the water just at the margin, or on the sides of the container above the water line. Given the opportunity, the adult prefers human blood to that of other animals. Adults readily enter houses to find suitable hosts. The adult quietly hails to settle for feeding, preferring to bite around the ankles, under shirtsleeves, or in the back of the neck. Males can be easily distinguished from females on the basis of antennae and maxillary palps (Figure 8.13).

Like Ochlerotatus, the Aedes mosquitoes are also painful and persistent biters. They search for a blood meal early in the morning, at dusk (crepuscular feeders), and into the evening. Some are diurnal (daytime biters), especially on cloudy days and in shaded areas. They are strong flyers and are known to fly many miles from their breeding sources. The adult of *Ae. aegypti* resembles markedly to that of *Ae. albopictus*, commonly called the Asian Tiger mosquito, another mosquito of great medical significance, except that the lyre-shaped markings on the thorax are replaced by a stripe of silvery-white scales on the dorsum of thorax.

Aedes albopictus is a forest dwelling species and has adequately adapted itself to urban environments as well. A native of Asia, it has spread to most tropical parts of the globe. In Kerala State, southern India, it breeds abundantly in the latex collecting cups associated with rubber plantations, besides cocoa and coconut shells (Hiriyan and Tyagi, 2004; Tyagi et al., 2006). It has the nasty habit of displacing other species, including *Ae. aegypti*, from their breeding strongholds and has quickly adapted to breed in all types of discarded containers, including tires and littered plastic tea cups (Hiriyan et al., 2003).

8.5 CULEX MOSQUITOES

The genus Culex comprises several medically important species, such as *Cx. pipiens, Cx. quinquefasciatus, Cx. tarsalis, Cx. nigripalpus, Cx. tritaeniorhynchus, Cx. bitaeniorhynchus, Cx.*

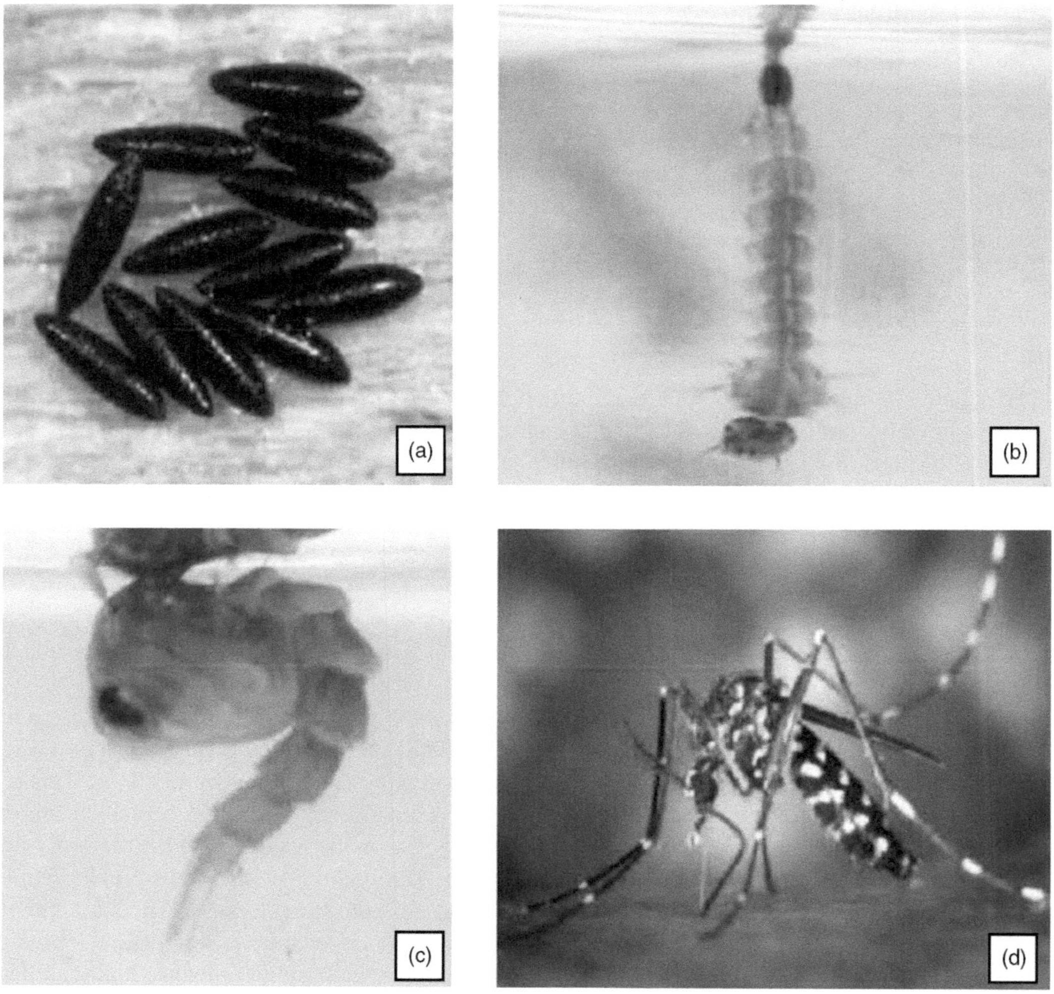

FIGURE 8.12 Different developmental stages of an aedine mosquito, *Aedes albopictus*: (a) eggs, (b) larva, (c) pupa, and (d) adult female.

Source: **Free Wikipedia.**

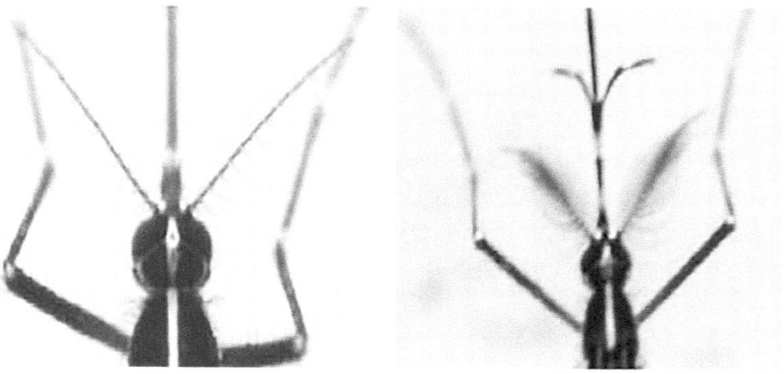

FIGURE 8.13 Mouthparts and antennal differences between female and male mosquitoes of *Ae. albopictus*.

Source: **Free Wikipedia.**

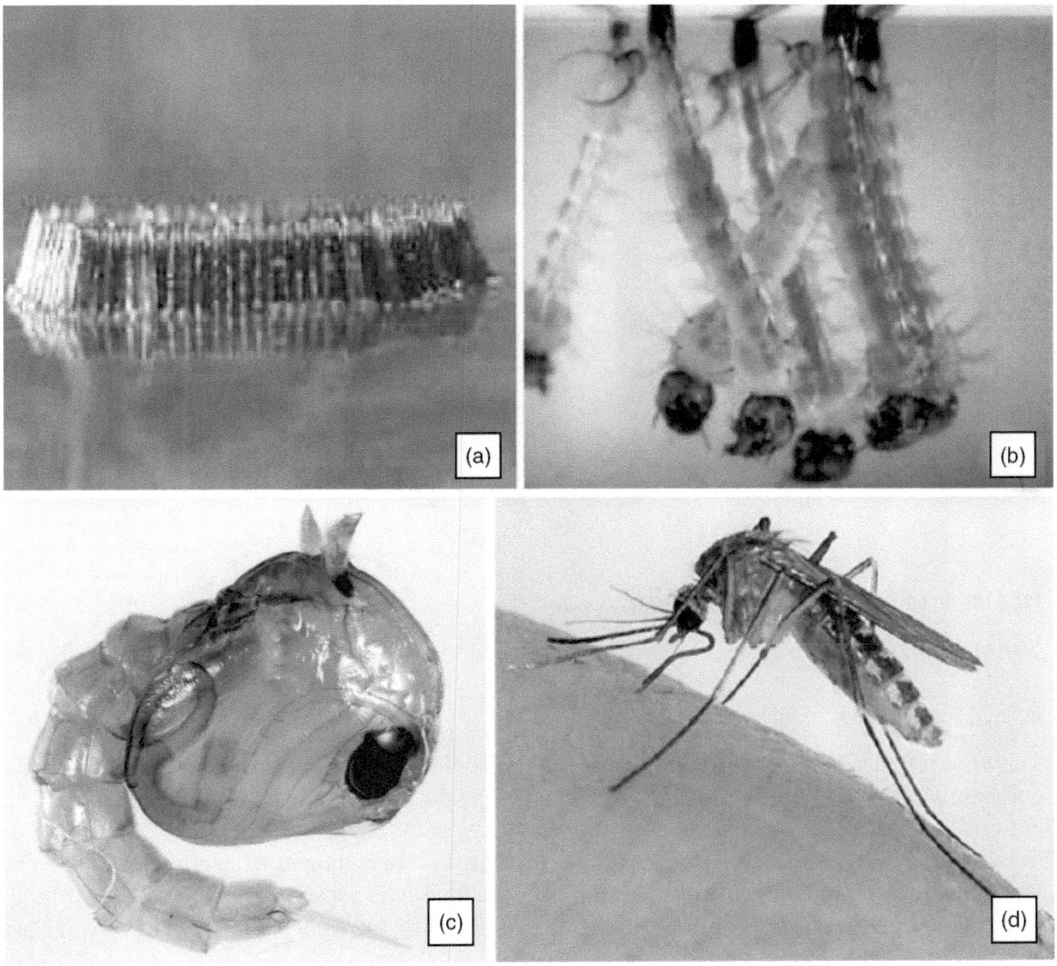

FIGURE 8.14 The four stages of life cycle of a Culex mosquito: (a) eggs (arranged in an egg-raft), (b) larva, (c) pupa, and (d) adult (in feeding process).

Source: **Free Wikipedia.**

pseudovishnui, Cx. restans and *Cx. salinarius* etc. In Indian context, of greatest importance are *Cx. quinquefasciatus* and *Cx. tritaeniorhynchus* which, besides being serious pests, are in the first place responsible for transmitting bancroftian filariasis and Japanese encephalitis, respectively. These are brown-colored mosquitoes of medium size and like the rest of their group, these have four stages of development in their life cycle (Figures 8.14 and 8.15).

8.5.1 The Egg

Many mosquitoes, such as *Culex quinquefasciatus*, lay their eggs on the surface of fresh or stagnant water. Culex mosquitoes usually lay their eggs at night over a period of time, sticking them together to form a raft of 100 to 300 eggs. A raft of eggs looks like a speck of soot floating on the water and is about 1/4 inch long and 1/8 inch wide. A female mosquito may lay a raft of eggs every third night during its life span. Like the eggs of Anopheles and Mansonia mosquitoes, Culex eggs are also more susceptible to long periods of drying out.

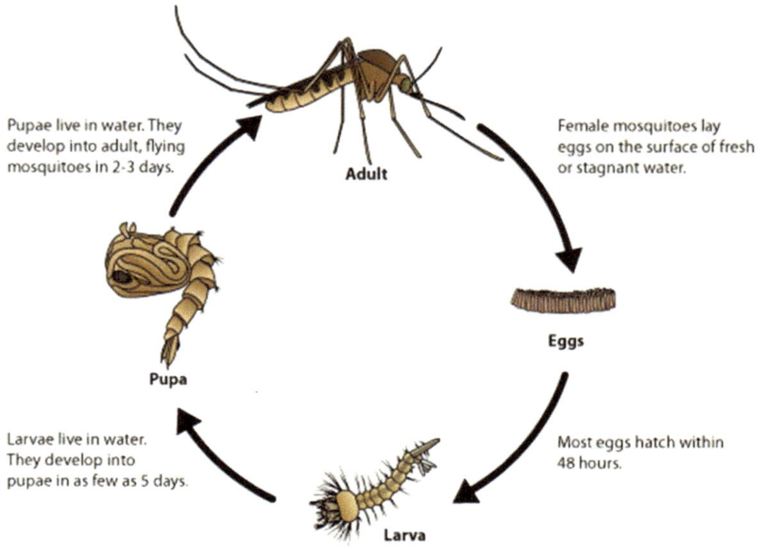

FIGURE 8.15 The life cycle of a Culex mosquito.

Source: **Free Wikipedia.**

8.5.2 The Larva

Tiny mosquito larvae emerge from the eggs within 24–48 hours almost in unison. Mosquito larvae, commonly called "wrigglers," live in water from 4 to 14 days depending on water temperature. They come to the surface at frequent intervals to obtain oxygen through a breathing tube called a siphon. They tend to feed constantly since maturation requires a huge amount of energy and food. They hang upside down at the water surface with the breathing tube up and the brushes by their mouth filtering anything small enough to be eaten toward their mouths to nourish the growing larva (Figure 8.16).

They feed on algae, plankton, fungi, and bacteria and other microorganisms.

8.5.3 The Pupa

A typical culicine pupa (*cf.* Aedes mosquito pupa).

8.5.4 The Adult

Culex quinquefasciatus is fondly referred to as the house mosquito due to its abundant occurrence in houses. It breeds in cesspools, storm-sewer catch basins, street gutters, open septic tanks, and effluence from sewage disposal plants. Adults are active only at night. When disturbed, they can be easily spotted flying out of their hideouts in and around houses, outbuildings and other shelters.

Culex tritaeniorhynchus breeds mainly in rice fields and is both an exophilic and exophagic mosquito. Culex mosquitoes are painful and persistent biters but prefer to attack at dusk and after dark. They readily enter dwellings for blood meals. Domestic and wild birds usually are preferred over humans, cows, and horses. *Culex nigripalpus* is known to transmit St. Louis encephalitis to humans in the USA.

Culex mosquitoes are generally weak flyers and do not move far from home, although they have been known to fly up to 4 km. They live only a few weeks during the warm summer months. Those females that emerge in late summer search for sheltered areas where they "hibernate" until spring.

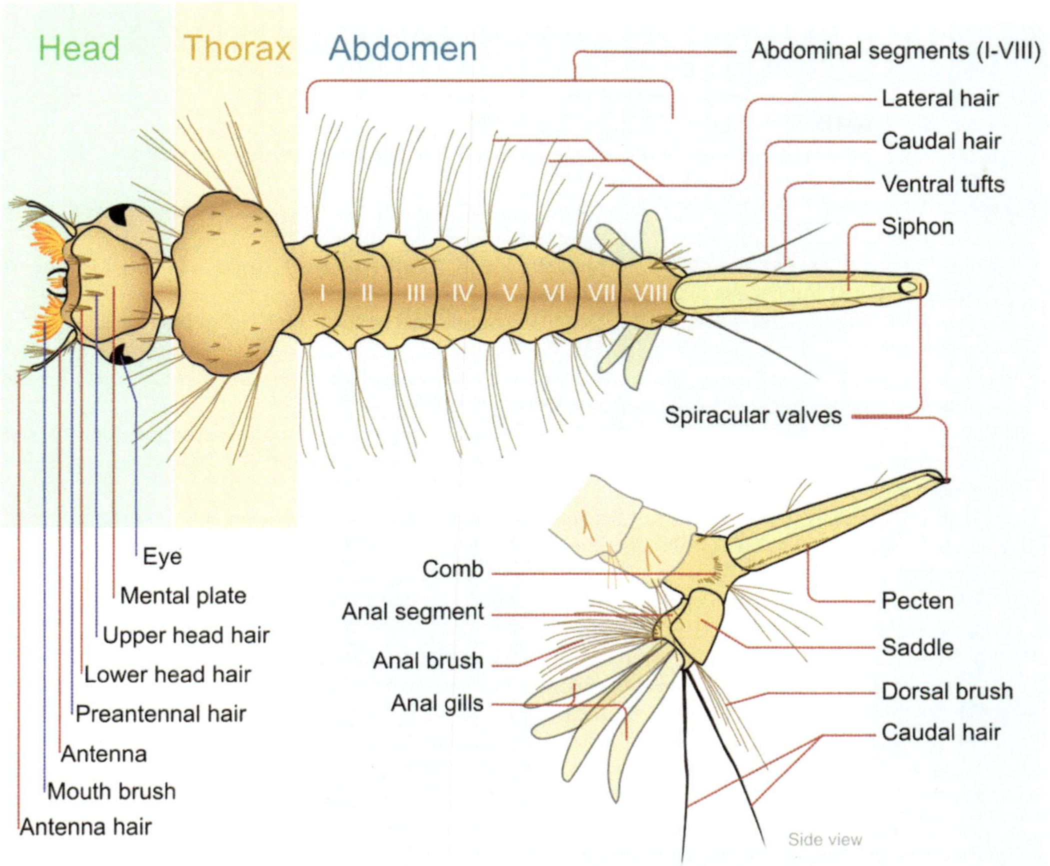

FIGURE 8.16 Diagram of larval mosquito (dorsal view) Culex.

***Source*: Free Wikipedia.**

Warm weather brings them out again in search of water, on which to lay their eggs. *Culex pipiens* and *Cx. quinquefasciatus* are found in most of the tropical countries. All culicines generally rest with their entire body more or less close to the surface. The scales are narrow and not metallic colored. There are no distinct patterns or markings on the body. The abdomen ends bluntly. Prespiracular and postspiracular areas are without bristles. Forelegs in males bear toothed claws. In females, the claws are there but simple. The tarsi in females have large pulvilli.

REFERENCES

Benelli, G. and Duggan, M.F. (2018). Management of arthropod vector data – Social and ecological dynamics facing the One Health perspective. *Acta Tropica*, 182, 80–91. https://doi.org/10.1016/j.Actatropica.2018.02.015

Benelli, G. and Mehlhorn, H. (2018). *Mosquito-borne diseases implications for public health*. Springer, Berlin. https://doi.org/10.1007/978-3-319-94075-5

Benelli, G., Wilke, A.B. and Beier, J.C. (2020). *Aedes albopictus* (Asian tiger mosquito). *Trends in Parasitology*, 36(11), 942–943. https://doi.org/10.1016/j.pt.2020.01.001

Benelli, G., Canale, A., Lucchi, A. and Giovanni, F.D. (2022). Insects and Mites of Medical and Veterinary Importance: A Broad Overview. *In*: *Encyclopedia of Infection and Immunity*. Nima Rezaei (Ed.). Elsevier, 793–800 pp. https://doi.org/10.1016/B978-0-12-818731-9.00040-9

Berenger, J.-M. and Parola, P. (2017). Arthropod vectors of medical importance. (*In: Infectious Diseases* (4th Edition), Jonathan Cohen, William G. Powderly, Steven M. Opal (Eds.), Elsevier, 104–112 pp. https://doi.org/10.1016/B978-0-7020-6285-8.00012-5

Christophers, S.R. (1960). *Aedes aegypti* (L.), *the yellow fever mosquito: Its life history, bionomics and structure*. London, UK: Cambridge University Press.

Clements, A.N. (1992). *The Biology of Mosquitoes. Vol. 1. Development, Nutrition and Reproduction.* CABI Publishing, 571 pp.

Clements, A.N. (1999). *The Biology of Mosquitoes. Vol. 2. Sensory Reception and Behaviour.* Wallingford: CABI Publishing, 752 pp.

Clements, A.N. (2011). *The Biology of Mosquitoes. Vol. 3. Transmission of Viruses and Interactions with Bacteria.* CABI Publishing, 592 pp.

Dash, A.P., Bhatia, R. and Kalra, N.L. (2012) Dengue in South-East Asia: an appraisal of case management and vector control. *Dengue Bulletin*, 36, 1–13.

Hiriyan, J., Tewari, S.C. and Tyagi, B.K. (2003). Plastic cups inveigling *Aedes albopictus* (Skuse) breeding around tea vendor spots in Ernakulam city, Kerala State. *Dengue Bulletin*, 27, 195–196.

Hiriyan, J. and Tyagi, B.K. (2004). Cocoa pod (*Theobroma cacao*) – a potential breeding habitat of *Aedes albopictus* (Skuse) in dengue sensitive Kerala State, India. *Journal of American Mosquito Control Association*, 20(3), 323–325.

Lane, R.P. and Crosskey, R.W. (1993). *Medical Insects and Arachnids.* London: Chapman and Hall.

Nagpal, B.N. and Sharma, V.P. (1995). *Indian Anophelines.* New Delhi: Oxford and IBH Publishing Co. Pvt. Ltd.

Rao, T.R., 1984. *The Anopheline of India* (Revised Edition.), New Delhi: Malaria Research Centre, Indian Council of Medical Research, 518 pp.

Rudolf, I., Blažejová, H., Mendel, J., Straková, P., Šebesta, O., Rettich, F., Čabanová, V., Miterpáková, M., Betášová, L., Peško, J., Barbušinová, E., McKee, C., Osikowicz, L., Šikutová, S., Hubálek, Z. and Kosoy, M. (2020) Bartonella species in medically important mosquitoes, Central Europe. *Parasitology Research*, 119(8), 2713–2717. https://doi.org/10.1007/s00436-020-06732-1

Service, M.W. (2008). *Medical Entomology for Students.* Cambridge University Press, 306 pp.

Spielman, A. and M. D'Antonio, 2001. *Mosquito: The Story of Man's Deadliest Foe.* London: Faber & Faber, vi+247 pp.

Subbarao, S.K., Nanda, N. and Raghavendra, K. (1999). Malariogenic stratification of India using *Anopheles culicifacies* sibling species prevalence. *ICMR Bulletin*, 29(7), 75–80.

Tyagi, B.K. (2003). *Handbook of Medical entomology.* Jodhpur: Scientific Publishers (India), pp. 262.

Tyagi, B.K., Hiriyan, J., Samuel P. Philip, Tewari, S.C. and Paramasivan R. (2006). Dengue in Kerala: A critical review. *ICMR Bulletin*, 36(4–5), 13–22.

Tyagi, B.K., Munirathinam, A. and Venkatesh, A. (2015). A catalogue of Indian mosquitoes. *International Journal of Mosquito Research*, 2(2), 50–97.

Tyagi, B.K., and Verma, K.V.S. (1991). Anopheline mosquitoes of Sri Ganganagar district (Rajasthan) transmitting malaria parasite. *Journal of Applied Zoological Researches*, 2, 85–91.

Tyagi, V., Sharma, A.K., Dhiman, S. et al. (2016) Malaria vector *Anopheles culicifacies* sibling species differentiation using egg morphometry and morphology. *Parasites & Vectors*, 9, 202. https://doi.org/10.1186/s13071-016-1478-5

Tyagi, B.K., Thenmozhi V. and Selvi, S. Karthigai (2014). Transmission dynamics of Japanese encephalitis, with emphasis on gaps in understanding and priority areas on Japanese encephalitis and other acute encephalitis syndromes in India, *Journal of Communicable Diseases*, 1(1): 1–10.

Section V

Ecology and Behavior

9 Ecology and Behavior

Ritwik Mondal
University of North Bengal, India

Syed Afrin Azmi
Ex-DH & FWS, CMOH Office, Purba Bardhaman, India

Shakya Sinha, Chandrima Bose, Tanuka Ghosh, and Sajal Bhattacharya
Asutosh College (Affiliated to Calcutta University), Kolkata, India

B.K. Tyagi
University Institute of Biotechnology, Chandigarh University, Mohali, India

9.1 INTRODUCTION

While Spielman and D'Antonio (2001) and Winegardl (2019) have branded the mosquito as the man's deadliest foe and predator on Earth, Tyagi (2004) has termed it as invincible as the *numero uno* enemy to the country's economy. Evolved in the mid-Mesozoic period, mosquitoes are responsible for several infectious diseases to man and animals worldwide, mostly in tropical environment (Tyagi, 2020). Thus, it is more than clear that most of the vectors and pathogens as well as diseases are environmentally sensitive (Reiter, 2001). It is quite obvious that in the process of survival for such a long time, they have encountered several ecological changes and have been subjected to natural selection. It is well established that mosquitoes are ecologically sensitive insects which means several environmental factors, both biotic and abiotic, can modulate the life-history traits of mosquitoes (Chandrasegaran et al., 2020). The variation of dispersal of mosquitoes is also dependent on ecological parameters. Mosquitoes may use wind speed and wind direction as environmental cues for their non-oriented flights (Verdonschot and Besse-Lototskaya, 2014), while some also use the wind-assisted dispersal, where the direction and distance of dispersal are heavily affected by the wind (Surendran et al., 2020). Hitherto, there are about over 3500 identified mosquito species in existence, and about a couple of hundred are known to be vectors of several diseases (Bhattacharya et al., 2016). In the context of vector-borne diseases, mosquitoes remain the forerunners, causing millions of deaths globally every year and naturally the cause of severe public health maladies (WHO, 2003; Ewing et al., 2019). An interplay of a range of factors including preferential blood meal, oviposition, flight, and other biotic and abiotic components concomitantly serve to delineate the behavioral adaptation and ecological niche of mosquitoes. Further, exploring the driving force behind these behavioral patterns underscores the influence of both species complex-specific and abiotic factors in defining the temporal dynamics of mosquito abundance (Whittaker et al., 2021). The variability in mosquito abundance is strongly associated with temperature and rainfall combined with high anthropogenic activities providing breeding grounds for their survival. In addition, as a result of rampant deforestation and rapid urbanization have expanded their range to the peridomestic region and enhanced preferential biting toward humans (Afolabi et al., 2019). Changes in

DOI: 10.1201/9781003326991-14

vector abundance and species dominance are coupled to processes affecting two-winged insects' population dynamics determined by the interaction of density-independent and density-dependent processes affecting mosquito survival and fecundity. The character and action of the density-dependent processes are notably important because they set the mean abundance concerning that population to fluctuate. On the opposite hand, behavioral resistance was found to exist in natural mosquito populations which can play a major role in checking the potential trade-offs and associations between these different protection strategies. Physiological resistance may occur through single mutations, whose expression levels will be modulated by variation in several genes, creating a quantitative genetic attribute (Ranson et al., 2004; Wondji et al., 2009). Almost like quantitative behavioral resistance, tolerance may also occur through behavioral thermoregulation if mosquitoes rest at some specific temperatures (Abram et al., 2017) or by self-medication if mosquitoes feed on specific diets like nectars (De Roode and Hunter, 2019). Additionally, activity resistance is also rising as a very important topic for further analysis (Gatton et al., 2013).

In this chapter we attempt to develop a proper understanding of the ecology and behavior of the vector mosquitoes that may be crucial for establishing regulatory strategies for their control.

9.2 OVIPOSITIONAL BEHAVIOR OF MOSQUITOES

Ovipositional behavior of mosquitoes varies within species concerning egg-laying pattern, ovipositional site selection, etc. to avoid interspecific competition and optimization of egg and larval survival. These variations in ovipositional behavior are primarily dependent on several biotic and abiotic factors, environmental cues along with the feeding behavior, flying ability, and other physiological parameters of the mosquitoes.

9.2.1 Environmental Cues

Ovipositional flight and selection of ovipositional sites by mosquitoes are largely determined by meteorological conditions such as rainfall, relative humidity, temperature, and wind. Mosquitoes in general prefer warm, humid, moist, and calm conditions for oviposition (Bidlingmayer, 1974). Two of the most important environmental cues that gravid females rely on to detect potential ovipositional sites during their ovipositional flight are the visual and olfactory cues (Bidlingmayer, 1974, 1975). The visual cues and the olfactory cues, along with the tactile cues, also help female mosquitoes to assess the quality of the detected possible ovipositional sites. Vision predominantly acts as a long-range cue for oviposition site location and flight orientation by many mosquito species. Visual cues essentially help identify and separate different aquatic habitats, *viz.*, ponds, streams, bogs, marshes, and flooded agricultural land, along with natural and artificial containers (Kennedy, 1942). Dependence of *Aedes aegypti* and *Aedes albopictus* on the texture and reflectance of the water-holding containers accompanied by the cues from other substrates for them in-water oviposition behavior has been reported by O'Gower (1963). Apprehensions have also been made that nocturnal mosquitoes possibly depend on the reflection of light from natural and artificial containers for the determination of ovipositional sites as they have been found to oviposit mostly at twilight or during moon-lit nights (O'Gower, 1963; Belton, 1967; Birley and Charlwood, 1989). Nevertheless, experimentally it has been demonstrated that *Aedes aegypti* and *Aedes triseriatus* mosquitoes get attracted by colored water (Dark water) more than clear water while searching for their preferential ovipositional site (Williams and DeLong, 1961) whereas *Culex quinquefasciatus* has a preference for polluted water for breeding (Tyagi, 1994). Moreover, Williams and DeLong (1961) also concluded that dark water may increase the oviposition rate of *Aedes aegypti*.

Apart from the visual cues, olfaction plays a key role in defining the oviposition behavior of mosquitoes (Bentley and Day, 1989). The range of olfactory signals from the ovipositional sites as diverse as tree holes, flooded ditches, wastewater retention ponds, cattle hoof prints, and flooded artificial containers in some cases get detected by mosquitoes from distances of up to several meters while they are still in flight. Olfactory receptors located on the antennae, palps, labrum, and tarsi of mosquitoes help them to sense and evaluate the volatile chemical emanations from the oviposition sites. For *Culex* mosquitoes, the main olfactory organ is found to be the antenna, which is adorned with thousands of sensilla housing the olfactory receptor neurons (ORNs), that respond to various types of volatile chemicals (Syed and Leal, 2009; Hill et al., 2009; Seenivasagan et al., 2009; Dhanalakshmi et al., 2018). In *Culex* mosquitoes, the apical droplet factor is considered an attractant for the mosquitoes for the selection of oviposition sites (Bruno and Laurence, 1979). In *Culex quinquefasciatus*, the apical droplet factor is erythro-6-acetoxy-5-hexadecanolide, commonly referred to as Mosquito Oviposition Pheromone (MOP) (Laurence and Pickett, 1982). During ovipositional behavioral studies, involving the natural egg rafts/apical droplet factors or synthetic MOP, it was demonstrated that the gravid female mosquitoes in the semi-filed and field setups got attracted to the ovipositional sites when they were manifested to the mentioned chemicals as olfactory cues (Otieno et al., 1988; Mboera et al., 2000; Olagbemiro et al., 2004; Braks et al., 2007; Wachira et al., 2010). In several other experiments, MOP has also been shown to be behaviorally active in inducing increased oviposition of gravid females, on its own or synergistically with other substrates (Pickett and Woodcock, 1996; Michaelakis et al., 2009; Navarro-Silva et al., 2009).

It has been seen that anautogenous female mosquitoes, i.e., the mosquitoes which must blood-feed before oviposition to acquire the protein necessary to develop a batch of eggs, show temperature dependency for oviposition. This dependency is because blood-meal digestion is temperature-dependent. In the tropics, it takes 2–3 days for a mosquito to oviposit while in temperate environments it takes 5–8 days. Nonetheless, mosquitoes having ready access to sugar before and after a blood meal may display delayed oviposition, possibly due to reduced flight activity or reduced attraction to oviposition sites (Klowden and Dutro, 1990).

9.2.2 Circadian Control of Oviposition

Distinct circadian timings and behaviors have been developed by different mosquito species through evolution according to their temporal/ecological niches. Some mosquito species such as *Aedes aegypti* are diurnal, while some others are nocturnal such as *Anopheles coluzzii*. Numerous behavioral variations of mosquitoes with regards to flight activity, mating, and oviposition have been observed depending on the time of day (Jones et al., 1967; Taylor and Jones 1969; Das and Dimopoulos, 2008; Sawadogo et al., 2013; Rund et al., 2012; Araripe et al., 2018; Sumba et al., 2004). Initiation of oviposition searching flights by gravid females generally corresponds with their normal daily activity periods. It has been observed that many mosquito species oviposit during the dawn and dusk crepuscular periods, depending on their daily activity period. Similar links of environmental synchronization are visible in some species, such as *Cx. nigripalpus*, the key to the flooding of oviposition sites following periods of drought is a signal to oviposit (Day and Curtis, 1994). Gillett et al. (1959), Gillett (1962), and Haddow et al. (1961) reported that in experimental conditions, for *Aedes aegypti* the oviposition occurred at nearly the same time each afternoon, regardless of ambient temperature. In their experiment they concluded there appeared an overall circadian activity as oviposition, blood-feeding, and sugar-feeding occurred at about the same time each day as well, which encompassed all of the major behaviors of *Aedes aegypti* females. However, when the mosquitoes were maintained under conditions of constant light or constant darkness, all cyclical behavior ceased. In addition, the same group apprehended that the mosquito's circadian clock was reset daily by exposure to light. Justifying this observation, when tested in the laboratory where gravid mosquitoes were maintained under constant darkness, it was observed that oviposition was random (Gillett et al., 1959).

9.2.3 Ovipositional Site Selection

The selection of a favorable oviposition site is indispensable for mosquitoes in their reproductive process. In the reproductive process of mosquitoes, one of the crucial steps is the selection of a suitable oviposition site that is both environmentally and chemically beneficial for larval survival and egg development. Therefore, gravid females must seek oviposition sites that are aquatic or supposed to be aquatic in the future. Some of these sites include tree and crab holes, temporary woodland pools, man-made containers (such as phytotelmata, moist soil in low-lying areas, abandoned automobile tires) that readily become flooded as well as a great variety of other aquatic environments (Weber and Tipping, 1993). However, different mosquito species may exhibit a preference for different ovipositional sites according to their requirements. Container-breeding behavior is a strategy adapted by several mosquito species with time. Larvae of container-breeding mosquitoes are found to develop in a wide range of container habitats. Mosquitoes demonstrate preference concerning container types and sizes (Vijayakumar et al., 2014). As mentioned earlier, oviposition choice by female mosquitoes is critical for the survival of the resulting larvae, and in this regard, the choice of larger containers potentially reduces the impact of density-dependent competition (Grill and Juliano, 1996; Griswold and Lounibos, 2006; Kesavaraju et al., 2008; Juliano 2009). However, due to phenological differences, species composition in container habitats may change over time. Females of species including, *Aedes triseriatus*, *Aedes aegypti*, and *Toxorhynchites amboinensis, Toxorhynchites rutilus*, typically prefer natural and artificial containers and may use visual search images to identify these oviposition sites (Focks et al., 1983). For domestic species, like *Aedes aegypti*, visual cues likely help them orient in and around buildings and locate the artificial and natural water-holding containers that they prefer for oviposition.

Water-filled tree holes are an important temporal ecosystem for several different aquatic and semiaquatic organisms (Tyagi, 2012; Munirathinam et al., 2014). Tree holes and rot holes may invite inspection by gravid mosquitoes by providing visual cues that contrast with the background. Mosquitoes belonging to genera *Aedes*, *Culex*, and *Anopheles* were recovered from tree holes during studies in southern India (Selvan and Jebanesan, 2014). Some mosquitoes such as *Culex quinquefasciatus* prefer to oviposit in water with a high organic content like that found in dairy and human wastewater holding ponds and runoff from agricultural treatment plants (Rajagopalan et al., 1987). Mosquitoes sometimes use ovipositional site selection and behaviors as a bet-hedging strategy. For *Aedes albopictus*, the deposition of eggs directly on the water is viewed as a bet-hedging strategy (Ripa et al., 2010) that allows the eggs to develop faster, and allows development during dry periods, avoids mortality associated with delayed hatching, and avoids predators (Madeira et al., 2002). It is considered bet-hedging because the eggs of *Aedes aegypti* are less vulnerable to desiccation than those of the *Aedes albopictus* (Juliano et al., 2002) and requires early development and emergence. Table 9.1 offers differences in ovipositional behavior of mosquitoes.

9.2.4 Egg Types and Ovipositional Patterns

There are two basic types of eggs deposited by mosquitoes during oviposition, *viz*., rapid-hatch (RH) eggs and delayed-hatch (DH) eggs. The RH eggs are deposited directly into water or on the water surface or substrate close to the water. These eggs usually hatch within 48 hrs of the deposition if all environmental conditions are suitable for hatching. The RH eggs can be laid individually, in small groups, or in rafts that contain up to several hundred such eggs. *Culex* female mosquitoes are unique among major insect vector species due to their habit of laying eggs in clusters of over 100 eggs or more, called rafts (Clements, 1999). On the contrary, the DH eggs are usually deposited singly or in small groups. The DH eggs are usually drought-resistant, have a long survival time out of the water, and hatch soon after being re-flooded. Sometimes these DH eggs enter a photoperiod-induced diapause to survive temperate and arctic winters.

So far, four common egg-laying strategies have been observed among different mosquito species. In the first case, several mosquitoes that lay RH eggs, i.e., the eggs hatch soon after oviposition, usually within 28 to 36 h. For RH egg-laying mosquitoes, aquatic habitats must be permanent enough to ensure complete larval and pupal development. Secondly, mosquito species depositing

TABLE 9.1
Ovipositional Behaviors of Some Important Vector Mosquitoes

Mosquito Species	Main Findings	References
Aedes albopictus	For *Aedes albopictus*, the most common breeding sites are water-holding containers found (outdoor) of plastic, followed by coconut shells, and tires. A variety of tree hole breeding is also evident in the case of these mosquitoes.	Hiriyan and Tyagi, 2004; Tyagi et al., 2006; Vijayakumar et al., 2014; Selvan and Jebanesan, 2014
Aedes aegypti	In India, cement tanks and plastic containers got identified as breeding habitats of *Aedes aegypti*. In several surveys, the overhead tanks and curing tanks were identified as key containers of *Aedes* breeding. Opportunistic tree hole breeding is also evident in their case.	Balakrishnan et al., 2006; Arunachalam et al., 2010; Mondal et al., 2014; Nagpal et al., 2016; Praveen et al., 2016
Culex quinquefasciatus	The mosquito prefers to breed in polluted water in ponds, pools, drains, laterine pits etc.	Rajagopalan et al., 1987; Okiwelu and Noutcha, 2012; Emidi et al., 2017
Culex pipiens	Relatively permanent water bodies such as water bodies along river fringes, ponds, and puddles, and water bodies with no predators are favorable sites for the mosquito for their breeding.	Liu et al., 2019
Culex tritaeniorhynchus	*Culex. tritaeniorhynchus* prefers low-lying water bodies with abundant plants.	Fakoorziba and Vijayan, 2008; Liu et al., 2018
Culex vishnui	Their preferred breeding sites are rice fields and ponds, and their dynamics are closely related to the phases of rice growth. In the Asia-Pacific region, rice fields are often located at the intersection of human dwellings and natural areas.	Maquart et al., 2022
Anopheles gambiae	They breed in small, open, sunlit, freshwater bodies. However, there is a growing body of evidence that these larvae may, in fact, breed in any available water, even in dirty and polluted habitats.	Imam and Deeni, 2014; Zogo et al., 2019
Anopheles stephensi	They breed mainly in clean/clear water containers such as overhead tanks, wells, cisterns, barrels, or drums, sumps (underground tanks), roof gutters, curing pits in construction sites, fountains, and ornamental tanks.	Thomas et al., 2016
Anopheles dirus	Wells, path pools, rock pools, stream systems, and slow-moving streams are the larval habitats of *Anopheles dirus*.	Obsomer et al., 2007
Anopheles culicifacies	The preferred breeding habitats for *Anopheles culicifacies* are wastewater, built wells, rainwater pools, earth wells, burrow pits, and tank margins.	Tyagi and Verma, 1991; Gunathilaka et al., 2013
Anopheles sundaicus	The breeding habitats of *Anopheles sundaicus* include sunlit bodies of stagnant water, viz. ponds with vegetation and floating algae, swamps, mangrove, wells, and rockpools, along with the coast or irrigated by inland sea water canals.	Dusfour et al., 2004

desiccation-resistant eggs prefer moist substrates for oviposition. These eggs may hatch soon after the completion of embryonic development or they may dry with the habitat substrate and DH eggs for periods as long as two years. Thirdly, some species lay mixed clutches of RH and DH eggs. Apart from that in the fourth case, some species of mosquitoes lay eggs that are protected (brooded) by the female until they hatch.

Depending on the requirement of blood, oviposition can be grouped into two groups, *viz.*, autogenous and anautogenous. Anautogenous female mosquitoes must be blood-fed to acquire the protein necessary to develop a batch of eggs, while autogenous mosquitoes can mature eggs without feeding on blood. In autogenous mosquitoes, the nutritional status of a blood-fed mosquito is important in determining the ultimate number of eggs that she will develop. It has been observed that sugar feeding, i.e., the natural feeding behavior of mosquitoes, increases previtellogenic follicular development in autogenous as well as anautogenous species. However, a sugar meal that immediately precedes a blood meal may have the opposite effect by decreasing clutch size (Day et al., 1990).

9.3 ROLE OF ABIOTIC AND BIOTIC COMPONENTS ON THE ECOLOGY AND BEHAVIOR OF MOSQUITOES

"Ecology" stems from the Greek word "*oikos*" which translates to home, the immediate environment of an organism, and the study of ecology deals with the relationship of the organism with one another as well as their surroundings (Ricklefs and Miller, 2000). The spatial distribution of mosquitoes is varied with seasonal fluctuations and the availability of host organisms (Golding et al., 2015). Taking into account the perspective of species richness and diversity, India holds the fifth position in terms of mosquito biodiversity (Foley et al., 2007; Bhattacharya et al., 2014). Since multiple mosquito species are responsible for mosquito-borne diseases, it is undeniable that several biotic and abiotic factors together determine the survival and disease competence of such mosquitoes. An interplay of such biotic and abiotic factors promulgates the assemblage and consequent ecology and behavior of mosquitoes in that region and unraveling these complex interactions will aid in the prevention and control of mosquito invasions (Hunt et al., 2017). The ecology of mosquitoes in different physiographic regions of India has been investigated in the urban/peri-urban Delhi (Ansari and Razdan, 1998), arid ecosystem of the Thar Desert (Tyagi and Yadav, 2001), mountainous altitudes of Jeypore, mangroves of Sunderbans and the Coromandel region mangrove and Bhitarkanika National Park (Rajavel et al., 2005a, 2005b, 2005c), and expansive rubber plantation fields in Kerala (Tyagi et al., 2006).

9.3.1 Abiotic and Biotic Factors Critical to Mosquitoes

With the rapidly changing global environmental conditions, several abiotic parameters such as water, temperature, humidity, and wind among others are thought to be the key players in exposing mosquitoes to a plethora of different physical surroundings even affecting their life-history traits (Clements, 1999; Ramasamy and Surendran, 2012; Chandrasegaran et al., 2018; Chandrasegaran and Juliano, 2019; Chandrasegaran et al., 2020). Biotic interactions of mosquitoes like predation and competition, vegetation cover, and host population density are also pivotal in understanding the vector mosquito dynamics (Gouagna et al., 2012; Golding et al., 2015; Kolimenakis et al., 2021). Climate change has been strongly proposed to provide a conducive environment for the increase in geographical distribution and epidemic potential of vector-borne diseases especially in the tropical highlands over the Indian subcontinent (Chaturvedi and Dwivedi, 2020; Couper et al., 2021). Although non-climatic factors like biotic interaction and their effect on the responses of mosquitoes to climatic factors such as temperature remain hitherto unexplored, yet climate-induced alterations affect several aspects of the biotic community of mosquitoes by modifying their ecological structure and function (Hunt et al., 2017; Couper et al., 2021).

9.3.2 Abiotic Factors

Globalization and urbanization have a widespread impact on disease transmission as a result of intensifying climate change due to global warming (Sutherst, 2004; Ramasamy and Surendran, 2012; Couper et al., 2021). Malaria is a significant public health malady, especially in the tropical parts of the world. There is an increased concern regarding a paradigm switch of certain diseases across India, influenced by climatic variables (Chaturvedi and Dwivedi, 2020; Tyagi, 2023). Temperature, rainfall, and humidity, among other factors, are the climate change parameters which are considered to have a major impact on mosquitoes. However, other factors such as pollution and wind may also have a significant impact (Ramasamy and Surendran, 2012).

9.3.2.1 Temperature

Insects being poikilotherms, temperature is in fact one the limiting factors determining the presence or absence of these vectors and has a direct influence over mosquito life cycle, behavior, disease competence and physiology and is therefore linked with mosquito vector-borne disease dynamics (Ramasamy and Surendran, 2012; Cator et al., 2020; Couper et al., 2021). Their internal temperature varies with respect to changes in the surrounding environment (Bellone and Failloux, 2020). Studies have noted that 22 to 32°C being the optimum growth range, extremely low (below 10°C) and extremely high (above 34°C) temperatures can escalate mosquito biting, blood digestion, reproducing and even egg-laying capacity (Lyons et al., 2013; Reinhold et al., 2018; Ahmed et al., 2019; Agyekum et al., 2021; Kumar et al., 2022). For example, the biting activity of mosquitoes may shift to the cooler night-time reducing disease transmission as human populations are more seemingly protected by bed nets (Pates and Curtis, 2005; Moiroux et al., 2012; Thomsen et al., 2017; Carrasco et al., 2019; Couper et al., 2021). The eggs of the anthropophilic vector *Aedes aegypti* can tolerate lower temperatures than that of *Aedes albopictus* expanding its geographical distribution to both tropical and temperate regions. Interestingly, increased temperature leads to a decreased duration of gonotrophic cycle of *Aedes aegypti* (Hanson and Craig Jr, 1994; Kraemer et al., 2015; Bellone and Failloux, 2020). Malaria incidents in the cooler belts of the African highlands are thought to be repercussions of shifting mosquito habitats in order to survive in soaring temperatures (Ko et al., 1999; Ahmed et al., 2019). In fact, current research focused on the "unimodal relationship" of mosquitoes with temperature speculate that future warming may decline transmission risk in certain zones where the risk is presently high, as temperatures begin to exceed the thermal optima and limits of mosquitoes. However, it should be noted that if mosquitoes acclimatize to the changing climate, these speculations are likely to underestimate future disease risk if any (Gething et al., 2010; Ryan et al., 2015, 2019; Mordecai et al., 2019, 2020; Couper et al., 2021).

9.3.2.2 Water

Water in the form of rainfall has always been a conservative determinant of mosquito temporal dynamics as it increases the surface water and creates new breeding sites allowing adult female mosquitoes to thrive and breed in them (Okwa and Adetutu, 2018; Ahmed et al., 2019). A study conducted by Prasad et al. (2021) in the Dakshina Kannada district of Karnataka in India, found that abundant rainfall combined with tropical temperature and humidity created several temporary water bodies alongside permanent ones, providing conducive habitats for notable mosquito vectors such as *Culex tritaeniorhynchus*, *Culex vishnui*, *Anopheles barbirostris*, *Culex gelidus* among others. In fact, despite substantial deforestation, urbanization, and industrialization, the Western Ghats receives good rains throughout and has seen a humongous population of over 30 *Anopheles* species between the years 2000–2012 (Munirathinam et al., 2014). Moreover, a positive indication of a steep rise in the larval density is given by a water pollution index. In an experiment designed by Ghosh et al. (2020) in West Bengal, India, several water parameters such as pH, hardness of water, and chemical oxygen demand (COD) were indicated to have a positive correlation with respect to the mosquito vectors *Culex vishnui*, *Aedes albopictus*, and *Anopheles stephensi*. Since oviposition is closely linked to mosquito breeding grounds, an interesting insight has been noted in the case of *Aedes aegypti* females who show a particularly evasive 'skip oviposition' behavior. *Aedes aegypti* generally lay eggs on the oviposition site wall, which is a few inches above their breeding water level (Fay and Perry 1965; Reiter 2007; Abreu et al., 2015). Absence of their preferred oviposition site could insinuate the dispersal of the vector in order to find a new site, thereby increasing the chances of vector-borne diseases in a region that have been hitherto unaffected (Reiter et al., 1995; Edman et al., 1998; Abreu et al., 2015). On the other hand, high precipitation could also potentially decrease the number of mosquitoes by flushing out their larvae from the breeding sites altogether (Tian et al., 2015; Jemal and Al-Thukair, 2018; Drakou et al., 2020).

9.3.2.3 Relative Humidity

Mosquitoes are especially sensitive to desiccation as an outcome of their increased surface area to volume ratio (Parham and Michael, 2010; Paaijmans and Thomas, 2011; Lunde et al., 2013; Santos-Vega et al., 2022). An increased humidity is therefore correlated to high precipitation favoring the survival of adult mosquitoes and their abundance while mosquito longevity and survival is hindered in the absence of humidity (Yamana and Eltahir 2013; Chandrasegaran et al., 2020). Experiments have shown that a relative humidity (RH) of at least 60% is an ideal and favorable condition for especially maintaining the transmission of *An. stephensi* (Gaaboub et al., 2009; Kessler and Guerin, 2008; Santos-Vega et al., 2022). Similarly, studies have shown that exceptionally trifling levels of RH are detrimental to mosquitoes and ultimately reflect upon their abundance in arid zones (Yamana and Eltahir, 2013; Santos-Vega et al., 2022). *Anopheles stephensi*, a potent urban malaria vector, is not known to rely explicitly on rainfall for larval development. In the context of the Indian subcontinent, the relevance of humidity becomes cardinal in its transmission within cities (Santos-Vega et al., 2022). Other studies have reported that a range of humidity between 44% and 69% is well suited for stimulating mosquito flight activity, with 65% being the precise percentage (Khan et al., 2018; Jemal and Al-Thukair, 2018; Drakou et al., 2020). The vapor content in the air was also identified as one of the most crucial parameters in explaining the dynamics of oviposition, given the choice for mosquitoes to locate oviposition sites in moist surfaces (Day et al., 1990, Edman et al., 1998; Mwingira et al., 2020). An increase in temperature is consistent with the rise of the moisture content in the proximity of potential larval habitations. In effect, this environmental variable influences gravid mosquitoes to soundly locate these sites (Okal et al., 2013; Mwingira et al., 2020). Hence, interestingly water vapor serves as a long-range cue for mosquitoes and thus an important abiotic factor in determining their ecology and behavior (Amarakoon et al., 2008; Mwingira et al., 2020).

9.3.2.4 Wind

Much like many other insect vectors, wind seems to be one of the most predominant factors in fine tuning their dispersal for their gene flow and disease spread (Chapman et al., 2015; Huestis et al., 2019; Sedda et al., 2022). However, mosquito behavior with respect to wind has not yet been completely understood. Research has revealed that in order to migrate to larger distances, these insects show a downwind flight while the reverse is true while host-seeking for blood meal guided by a CO_2 plume emanating from humans (Cardé and Willis, 2008; Lehmann et al., 2014; Endo and Eltahir, 2018). Conversely, it has been studied that flight activities seem to have a species-specific impairment at high wind velocities especially while seeking nectar and/or blood. In fact, wind speeds above 1.67 ms^{-1} have shown to plummet to the incidences of dengue fever in New Delhi, India (Cardé, 2015; Adeleke et al., 2022). The flight range of *Aedes albopictus* was experimentally estimated to be within 250 m to 1 km, although some others suggest it may range up to 8.6 kms. This variation in the range is thought to be a result of wind direction (Kaufmann et al., 2013; Adeleke et al., 2022). Higher wind speeds around breeding reservoirs of *Anopheles* mosquitoes in particular have shown a negative interrelationship. The effect of high wind velocity corresponds to increase in height of waves created near the shoreline of such water bodies. As a result, the height of the waves keeps amplifying with increasing depth of the water body and results in a fatal outcome on the aquatic stages of these insects (Endo and Eltahir, 2018). Having said that, a study by Klassen and Hocking (1964) also reported that the newly emerged adults of *Aedes cataphylla* were interestingly well adapted to take-off even in windy weather conditions (Service, 1980).

9.3.3 Biotic Factors

Mosquitoes are the bearers of vector-borne diseases that lead to social and medical damage to the human population (Ellwanger et al., 2021). As pointed out by Juliano and Lounibos (2005), mosquitoes are among the most prominent invasive species that create an impact on human and animal health through biotic interactions that include predation and parasitism, competition and counting (Williamson, 1996; Lounibos et al., 2002). In addition, anthropogenic alterations of surface land and vegetation have intensified the availability of mosquito larval habitat. These changes have a direct effect on mosquito distribution, providing them with suitable habitats and indirectly altering the composition of the biotic community within these modified habitats (Petric et al., 2014; Hunt et al., 2017).

9.3.3.1 Microbiota Interactions

Consistent with the notion that mosquito ideally follow their well-known ecology, aquatic habitats are the zones of larval production. The habitat preferences of mosquitoes with special emphasis on breeding sites have shown skewness toward aquatic ditches, water, ponds, marshy lakes, and similar other places. However, this environment also possesses a considerable threat to the growing mosquito larvae. An extensive study on *Anopheles maculipennis* and *Culex modestus* have revealed that certain fishes, ditch shrimps, and some amphipods are voracious feeders and predators of their larva (Jeffries, 1988; Roberts, 1995; Medlock and Snow, 2008; Golding et al., 2015). Consequently, a negative correlation between these two species interaction forces mosquitoes to choose oviposition sites free of predators. In this regard, indirect evidence of dispersal of *Aedes aegypti* for oviposition was observed through a "release and capture" method highlighting their preference toward indoor water containers free from such predators (Ritchie and Laidlaw-Bell, 1994; Pates and Curtis, 2005). The effect induced by the release of synomones, a type of allelochemicals, has provided excellent insights on the habitat selection of mosquitoes particularly pertaining to gravid individuals. This volatile chemical emitted from bacteria acts as cues to detect superior nutrient quality and habitat availability of a location (Sumba et al., 2004; Coon et al., 2014; Mwingira et al., 2020). In fact, several species of mosquitoes either consume a number of bacterial species or are in a mutualistic association as gut microbiota (Guegan et al., 2018; Mwingira et al., 2020; Ranasinghe and Amarasinghe, 2020). Current studies on Geosmin have uncovered its role as an info chemical produced by the cyanobacteria *Kamptonema* sp. is used to lure *Aedes aegypti* for egg-laying (Gouagna et al., 2012; Mwingira et al., 2020). Competition of larvae with naturally occurring microcrustaceans, phytoplanktons and similar other microbiota in terms of resource availability has proven to be another important benchmark in contributing to the selection of breeding and oviposition sites. Studies have highlighted Cladocera competing with *Cx. pipiens* for similar functional food and in turn reduce the possibility of invasion of *Cx. pipiens* (Kroeger et al., 2014; Ranasinghe and Amarasinghe, 2020).

9.3.3.2 Human Dependence

Humans are the single most prominent hosts of mosquitoes promising them of a potentially nutritious blood meal to support their egg production. Of primary focus is the shift in mosquitoes from generalists to majorly anthropophilic (Stone and Gross, 2018; Potter, 2019). As an instance, when we talk of *Culex pipiens* and its hybrid North American ancestry, they show host preference at a class level, *i.e.*, human preference over birds (Kilpatrick et al., 2007; Stone and Gross, 2018). It has been well documented that a complex mixture of exhaled carbon dioxide, lactic acid, and other odorants such as ketones, acetates, and alcohols emitted from the human body, also referred to as "kairomones" are perceived by gustatory receptors along with odorant receptor (OR) having an obligate corecepted called Orco (Sbarbati and Osculati, 2006; McMeniman et al., 2014; Potter, 2019; Ellwanger et al., 2021). Although recent studies have highlighted a genetic basis for *An. gambiae (s.s.)* and *Ae. aegypti* being opportunistic feeders preferring humans over other vertebrates while *Culex quinquefasciatus* and *An. arabiensis* being a generalist for taking equal proportions of blood meals from both, the availability of host in that particular location must be taken into consideration (Kilpatrick et al., 2007; Takken and Verhulst, 2013; Bakker et al., 2020; Ellwanger et al., 2021). On the other hand, indiscriminate felling of trees, rapid urbanization, and irrigation projects have brought mosquitoes such as *An. stephensi*, an established vector of urban malaria even closer to the human periphery and more potent in invading residential complexes to take their blood meal (Dev and Sharma, 2013; Cator et al., 2013; Santos-Vega et al., 2016; Bhattacharya et al., 2020). In fact, the movement of humans across the globe has passively dispersed native species into foreign lands (Petric et al., 2014). Justifiable mosquitoes are particularly aggregating near urban areas with a pronounced and ongoing increase in population (Keiser et al., 2004; UN, 2014; Santos-Vega et al., 2016). A study conducted in Ahmedabad, India, reiterated how the tropical climate combined with urbanization and a huge repertoire of hosts acts synergistically effectively promoting the transmission dynamics of several mosquitoes borne diseases (Robert et al., 2003; Donnelly et al., 2005; Reiner et al., 2015; Santos-Vega et al., 2016).

9.3.3.3 Faunal Interactions

Although humans are the primary hosts of mosquitoes, the final blood-meal host of mosquitoes depends upon other parameters such as the abundance and availability of hosts (Takken and Verhulst, 2013; Mburu et al., 2021). In such a scenario, the rearing and/or presence of cattle close to human dwellings could either reduce or increase the burden of mosquito-borne diseases (Bouma and Rowland, 1995; Ghebreyesus et al., 2000; Mayagaya et al., 2015; Mburu et al., 2021). The presence of cattle may prove to be effective in diverting blood-seeking mosquito vectors from human hosts through zooprophylaxis or by enhancing the abundance of secondary hosts (WHO, 1982; Mburu et al., 2021). Besides, the introduction of natural enemies of mosquitoes, such as hydras, planaria, fish, birds, bats, and similar other organisms into the habitat of mosquitoes can potentially deteriorate their population (Collins et al., 2019). This is consistent with the findings that for oviposition, female gravid mosquitoes are likely to make a choice between sites without aquatic macro-organisms and predatory fishes (McCrae, 1984; Bentley and Day, 1989; Munga et al., 2005; Gouagna et al., 2012). The general predilection of *Ae. tritaeniorhynchus* is to normally oviposit on salt marshes and mangrove topsoil. It was then observed to avoid ovipositing near those water bodies with high fish concentrations (Ritchie and Laidlaw-Bell, 1994; Pates and Curtis, 2005). Nonetheless, during periods of heavy rainfall, the dilution of fish densities as a result of increased water levels allowed prompt hatching of mosquito eggs thereby proliferating a presumably diminished mosquito population (Carlson and Vigliano, 1985; Pates and Curtis, 2005).

9.3.3.4 Plant-Flora Association

Apart from the blood meal, plant nectar is a vital source of nutritional supplementation for mosquitoes in order to gain optimum energy in their natural environment (Nikbakhtzadeh et al., 2014; Bassene et al., 2020). The consumption of floral nectar has been confirmed to help vitalize males to fertilize females and also provide a supplementary food in dire times of need thereby sustaining the lifespan of mosquitoes (Stone et al., 2009; Nikbakhtzadeh et al., 2014; Bhattacharya et al., 2021). Evidence suggests that procurement and consumption of sugar is absolutely critical for *Anopheles sergentii* to be able to allow the maturation of sporozoites within their salivary glands (Gu et al., 2011; Bhattacharya et al., 2021). Indeed, rampant deforestation destroys the

FIGURE 9.1 An unusual scene of a mosquito (culicine, possibly a *Culex* sp.) feeding on the Trumpet Tail male dragonfly, *Acisoma panorpoides* (see also inset).

***Source*: Nazneen Siddique.**

nectar source from mosquitoes forcing them to choose an alternate form of food and giving them access to human-mosquito interaction which altogether impacts the mosquito disease ecology (Cutler et al., 2010; Stone and Gross, 2018; Bhattacharya et al., 2021). Rice is a popular staple in Indian diets, and also for 60% of the population of the world. *Anopheles culicifacies* is infamously known to breed in rice fields and mosquito-borne disease transmission is enhanced in fields committed to paddy cultivation (Russell and Rao, 1942; Singh et al., 1989). Highly synanthropic mosquito populations are also not likely to be able to persist without phytophagy, even if blood from a vertebrate host is made readily available to them. Thus, phytophagy is crucial to mosquito ecology and behavior (Peach and Gries, 2020; Stone et al., 2009; Bhattacharya et al., 2021). A very unusual attempt to feed on another insect (a dragonfly, *Acosoma panorpoides*) was recently observed (Figure 9.1) by Tyagi and Nazneen (2021).

9.4 FLIGHT MECHANISMS AND DISPERSAL OF MEDICALLY AND ECOLOGICALLY IMPORTANT MOSQUITOES

The average distance of a mosquito's flight can be as small as 50 m, and it gets as long as 50 km, depending on the mosquito species and prevailing environmental conditions of that particular habitat (Rao, 1984; OECD, 2018). For instance, *Aedes aegypti* mosquitoes are short flyers in their attempt of ensuring that they remain close to humans and proximal to blood-meal sources, harborage areas, and oviposition sites (Ritchie, 2014). Some mosquitoes like *Aedes taeniorhynchus, Aedes vigilax* engages in long dispersive flights of up to tens of kilometers from their larval developmental sites (Vlach et al., 2006; Ritchie, 1993). The range of non-random short daily flight varies and is typically associated with foraging, mating, oviposition, and resting requirements (Service, 1997). The non-random daily flights get influenced by several environmental factors and vary according to the need of the mosquito species. Service (1963) defined mosquito dispersal as the full range of mosquito movement, whether intentional and self-powered or passive.

9.4.1 Mosquito Dispersal Types

Service (1997) has distinguished mosquito dispersal and migration into three primary categories, *viz.*, unintentional long-distance migration through human transportations, wind-assisted long-distance passive migrations of mosquitoes away from breeding sites, and short daily flights (Bailey et al., 1965; Smith and Carter, 1984; Services, 1997). Adding to that, depending on the distance of dispersal, Southwood and Henderson (2000), classified mosquito dispersal as the short-distance movement of mosquitoes to adjacent areas, namely the "neighborhood dispersal," and long-distance dispersal, that is the "jump dispersal."

Nevertheless, considering the previous classification of mosquito dispersal types by various researchers, the flight and dispersal of mosquitoes may be categorized broadly into two groups, *viz.*, "intentional dispersals" and "unintentional dispersal." Among intentional dispersal, the most studied and well-described is the "non-oriented flight," while the unintentional dispersals can be "wind-assisted dispersal" and "dispersal through human transportations."

9.4.1.1 Non-Oriented Dispersal of Mosquitoes

For host-seeking behavior and ovipositional flights, the non-oriented dispersal increases the chances of mosquitoes reaching their potential destinations, i.e., the hosts and ovipositional sites, respectively. The non-oriented dispersal takes place just after the emergence of the adults from the larva, as described by Nayar (1985), and is predominantly dependent on various environmental factors (Verdonschot and Besse-Lototskaya, 2014). The meteorological conditions influencing the non-oriented flight are wind velocity, relative humidity, temperature; illumination levels, local topography, vegetation type, host presence, and density, also the availability of suitable oviposition sites (Schäfer et al., 1997; Verdonschot and Besse-Lototskaya, 2014). Nevertheless, the non-oriented short flights generally do not get influenced by the wind direction. Vegetation is of special significance in maintaining non-oriented flight as it directly influences the humidity, wind pressure, and temperature in the flight path. Mosquitoes, even in an attempt to avoid the meteorological alterations in their dispersal path, fly close to the ground or just above the top of the vegetation (Verdonschot and Besse-Lototskaya, 2014). Becker et al. (2010) classified mosquitoes into four groups based on their non-oriented flights viz. weak flyers, which are urban domestic and container-breeding species breeding and resting close to their hosts' habitats; weak-moderate flyers, which occur in woodlands; moderate-strong flyers, which generally occur along field and forest edges; and strong flyers, occupying the open areas as habitat and dispersing over long distances. The non-oriented flight is followed by the "appetitive flight" or "goal-oriented flight" of mosquitoes. In these goal-oriented flights, the cessation of flight happens in response to different stimuli, and the mosquito participates in other activities such as nectar feeding, blood-feeding, mating, or oviposition, etc. (Silver, 2008). As a method of host searching behavior, the non-oriented flyers that are ornithophilic such as *Culex pipiens pipiens* and *Culiseta morsitans* disperse in a vertical direction toward trees, higher than species that feed on mammals such as *Aedes* and *Ochlerotatus*, etc. (Verdonschot and Besse-Lototskaya, 2014).

9.4.1.2 Wind-Assisted Dispersal

Long-range wind-assisted dispersal or wind-borne dispersal of mosquitoes often gets the status of accidental dispersal as they are not actively regularly performed by mosquitoes. Nevertheless, in recent studies, it has been suggested by several groups of researchers that mosquitoes frequently get migrated using long-range wind-assisted dispersal methods (Dao et al., 2014; Huestis et al., 2019). Simulation models predicted that the range of the wind-borne mosquito dispersal can range hundreds of kilometers (Huestis et al., 2019). During the wind-assisted dispersal, mosquitoes show non-appetential flights, which means this flight is involuntary and adult mosquitoes do not control the direction or distance of the dispersal. This flight is considered part of mosquito population biology. Mosquitoes possibly achieve wind-assisted flight by attaining a higher altitude using the non-oriented methods, as this increases the chances of traveling long distances. The long-distance migration of vector mosquitoes such as *Anopheles coluzzii*, the primary vector of malaria in the tropical African region, depicts the importance of wind-assisted migration in the disease dynamics of mosquito-borne diseases. In a study by Huestis et al. (2019) it was found that 90% of the mosquito samples using wind-assisted dispersal were gravid females after having a blood meal.

9.4.1.3 Unintentional Dispersal through Human Transportations

Another involuntary or passive method of mosquito dispersal is the dispersal through human transportation such as ships, aircraft, vehicles, etc. Today, the world is more interconnected than ever with a huge acceleration in global trade and travel sectors. It has been reported by several groups of researchers that mosquitoes, especially the invasive mosquitoes, can exploit human transportation to facilitate their invasion in newer places. The transport of *Anopheles* mosquitoes as non-paying passengers in aircraft to different European countries and their subsequent establishment resulting in malaria outbreaks is well studied (Silver, 2008). Eritja et al. (2017) reported the intercity dispersal of *Aedes albopictus* using cars as a medium of involuntary transportation (Table 9.2).

9.4.2 Flight Mechanism of Mosquitoes

The flight mechanism of mosquitoes is unique among the insects with regards to wing kinematics especially their stroke amplitude and flap frequency. Their wings are long, slender, and flap at remarkably high frequencies for their size and with lower stroke amplitudes compared to other insects (Simões et al., 2016; Bomphrey et al., 2017). Depending on their size, mosquitoes typically beat their wings at a high flapping frequency in the range of 600–800 Hz. This in turn also lets them conceal their presence and helps them remain aloft. Their wings navigate between the most anterior and posterior wing positions through a stroke amplitude of about 40 to 45°, which is quite less compared to other flying insects having a stroke amplitude >120° (Singh et al., 2021). Most insects depend on the leading-edge vortex for lift; however, mosquitoes have additional flying mechanisms such as; rotational drag, wake capture reinforcement of the trailing-edge vortex, and added mass effect (Singh et al., 2021). From the mechanical perspective, the high angular rate and exquisite timing of stroke reversal are crucial for mosquitoes to generate forces large enough to support their weight in flight. In insect flight lift due to rotational drag is proportional to the square of the pitching angular rate, but equally important is the precise axis of rotation. In mosquitoes, the pitching rotational axis of the wing moves from leading to the trailing edge at the end of the upstroke. By rotating first around an axis close to the leading edge, low pressure develops close to the trailing edge which creates a component of aerodynamic force that supports the mosquito's weight. The great benefit of lift mediated by rotational drag is that the aerodynamic force is not dependent on the radial position. It is therefore equally effective along the entire wingspan (Bomphrey et al., 2017). During flight, the pair of balancing organs, called halteres, are essential elements. Halteres, small ball-like structures, help mosquitoes to maintain balance during flight (Singh et al., 2021). Nevertheless, the unusual development of kinematic patterns beyond the boundaries used by other insects remains a field to explore from an evolutionary perspective.

9.4.3 Factors Influencing Mosquito Flight and Dispersal

Environmental factors influence mosquito flights in different ways and their effect may vary depending on the species also. As flight is the key behavior to support oviposition, host-seeking, mating, etc. mosquitoes rely on environmental factors related to these for determination of the distance, direction, and duration of their flight and dispersal. The key environmental parameters influencing mosquito flight are temperature, relative humidity, wind speed, and direction, along with the visual and olfactory cues.

TABLE 9.2
Dispersal Patterns of Mosquitoes

Mosquito Species	The Average Non-oriented Flight Distance	Main Findings	References
Aedes aegypti	Relatively short distance flyer, up to 100 meters, but has been observed to fly more than 800 meters when vital requirements are absent or poorly present.	Wind speed and wind direction can influence the spatial dispersal pattern along with different climatic factors.	Honório et al., 2003; Harrington et al., 2005; OECD, 2018
Aedes albopictus	Different studies revealed different dispersal range of these mosquitoes. It got reported that the average dispersal range is 250 m for them; however, they were reported to fly even more than 8 km possibly due to the influence of wind.	Wind speed and wind direction can determine the flight range of *Aedes albopictus*.	Adeleke et al., 2022
Culex tritaeniorhynchus	The average distance of *Culex tritaeniorhynchus* dispersal is within the range of 2.2 km – 7.5 km. However, the flight after taking blood meal was found to be around 300 m in experimental setup.	Wind-assisted dispersal is well reported for *Culex tritaeniorhynchus* mosquitoes and is related with their vector potential.	Longbottom et al., 2017; Xie et al., 2021.
Cx. vishnui	The average flight distance of *Culex vishnui* mosquitoes is around 1.5–2.0 km.	Flight is typically dependent on environmental cues most of the times.	Ciota et al., 2012
Cx. annulirostris	The average dispersal of *Culex annulirostris* is between 4.5 and 5 km. In studies the mosquito was found to fly more than 8.5 km even.	The rate of dispersal is about 2.2 km per day for *Culex annulirostris*.	Bryan et al., 1986
Cx. quinquefasciatus	The females of *Culex quinquefasciatus* fly more than 1–2 km for their oviposition and host seeking. However, ovipositional flight more than 2 km was also reported.	The distance of flight for ovipositional flight and host-seeking flight are variable for these mosquitoes.	Medeiros et al., 2017
Anopheles stephensi	In mark-recapture studies the average flight distance was estimated to be about 150 m for *Anopheles stephensi*.	Wind-assisted migration and invasion of *Anopheles stephensi* is well evident. The possible migration can be more than 100 km.	Verdonschot and Besse-Lototskaya, 2014; Surendran et al., 2020
Anopheles culicifacies	The average dispersal is around 250 m for them, but they are found up to 500 m from their larval habitat.	Wind-assisted migration has got reported for them.	Curtis and Rawlings, 2008; Verdonschot and Besse-Lototskaya, 2014
Anopheles fluviatilis	The breeding females of *Anopheles fluviatilis* flies up to 1.7 km.	The flight of immature females and host-seeking females are shorter in comparison to the ovipositional flight taking females.	Sadanandane et al., 1993
Anopheles sundaicus	In coastal region the flight distance was found to be about 300 m.	Wind and boat-borne dispersal has been reported in case of *Anopheles sundaicus*.	Nixon et al., 2014
Mansonia uniformis	Average dispersal range is 170 m for *Mansonia uniformis*.	Dispersal of these mosquitoes are influenced by habitat types.	Verdonschot and Besse-Lototskaya, 2014

9.4.3.1 Temperature and Relative Humidity

Arthropods are poikilothermic animals (i.e., they cannot keep the body's internal temperature constant) and that makes them dependent on environmental temperature from behavioral perspectives (Bellone and Failloux, 2020). The influence of temperature fluctuations on the dispersal of *Aedes aegypti* mosquitoes has been documented extensively by Rowley and Graham (1968). They concluded that the optimum temperature for the dispersal of *Aedes aegypti* is between 15°C and 32°C. They also showed the lower and higher extremes of the temperature at which dispersal occurs to be 10°C and 35°C, respectively (Rowley and Graham, 1968). The same group confirmed that relative humidity also has a role in mosquito flight. The optimum humidity for dispersal of *Aedes aegypti* is 30–90% and the flight stopped below the 30% level (Rowley and Graham, 1968). The flight speed and wing beating frequency of *Aedes aegypti* are also gets influenced by temperature (Reinhold et al., 2018). Flight of another medically important mosquito *Culex tarsalis* is studied by Bailey et al. (1965) and they concluded that *Culex tarsalis* can fly within a range of temperature of 13–33°C, with the optimum flight being recorded at 18–24°C.

9.4.3.2 Wind

The wind is an essential component to discuss while talking about mosquito flight for host-seeking and oviposition. The wind is believed to reduce mosquito host finding primarily by countering upwind progress. When winds increase to intolerable levels, mosquitoes are reported to alight or be closer to the ground to take advantage of lower wind velocities in the boundary layer (Snow 1982; Hoffmann and Miller, 2003). For example, *Aedes aegypti* prefers low-level flight as it prevents the adult mosquito from flying above the wind boundary layer above which wind speed can increase dramatically (Srygley and Dudley, 2008), potentially sweeping s all insects away to unsuitable habitats. The wind velocities may also dilute the stimulus such as an attractant or repellent on which mosquitoes rely for their host-seeking or ovipositional behaviors, resulting in loss of goal-oriented flight (Hoffmann and Miller, 2003).

9.4.3.3 Olfactory Cues

Many behavioral expressions of mosquitoes are mediated by olfaction (Takken and Knols Bart, 1999; Takken and Knols, 2007). An insect's navigational decisions during plume tracking must be made in response to sparsely distributed odor encounters that provide incomplete information about the source location (Cardé and Willis, 2008). Therefore, one can understand that mosquito host-seeking and ovipositional flights are directly influenced by the olfactory cues they receive through their olfactory receptors. Upon receiving the olfactory cues mosquitoes perform a "cast and surge" flight to locate the host or source of the cue (Cooperband and Cardé, 2006). One of the most prominent cues that influence mosquito host-seeking flight is the presence of CO_2. In a broad range of mosquito species, CO_2 elicits a suite of context-dependent effects (Gillies, 1980) that are thought to promote host-seeking behavior. Exposure to CO_2 alone strongly activates mosquito flight, increasing the probability of flight take-off and locomotor activity in the absence of directional air currents (Erias and Jepson, 1991). Filamentous intermittent plumes of CO_2 (McMeniman et al., 2014) elicit stereotyped and sustained patterns of upwind flight toward the CO_2 source (Dekker and Cardé, 2011). In *Aedes aegypti*, CO_2 increases both the kinetics and fidelity of flight responses to host odor plumes, thereby enhancing source-finding (Dekker and Cardé, 2011).

9.5 HOST PREFERENCES OF MOSQUITOES

Mosquitoes are the most significant cluster of arthropods involved in the transmission of several deadly diseases to humans and animals (Silver, 2008). Among the diseases, malaria, filariasis, dengue, and Japanese encephalitis account for appreciable morbidity and mortality across the world (WHO, 2003, 2021; Silver 2008). Even with the many methods adopted together with prophylactic measures for dipterous insect control, mosquito-borne diseases are still massive and customary in several tropical and semitropic countries, including India (WHO, 2021). Among the varied potential reasons, failure to sustain personal and community hygiene, convenience of the breeding sites of mosquitoes beneath anthropogenic and natural influences, and ineffective mosquito regulation methods seem to be the most prominent. Integrated regulative strategies consisting of a mix of physical, chemical, and biological approaches are framed for the intervention of the mosquito population, and thus reducing the likelihood of the sickness (Ghosh et al., 2008; Floore 2007). A proper understanding of the biology and ecology of the vector mosquito may be a requirement for establishing the regulatory strategies for the social control of mosquitoes (Silver, 2008; Becker et al., 2010).

9.5.1 Seasonal Prevalence

The study concerning the seasonal prevalence of *Culex* mosquitoes was started in the middle of the 20th century. Everywhere the world reports expressed that the height density of the resting population of *Culex* mosquitoes was in the rainy and late summer season and also the density was minimal in winter (Heisch et al., 1959; Service, 1963; Hayes and Downs, 1980; Choi et al., 2017; Kim et al., 2018). However, in low land and flood-prone areas, the density of *Culex pipiens fatigans* were significantly lower in the rainy season because larval breeding sites were flooded (Lie and Amaliah, 1960; De and Chandra, 1994). Ogunba (1971), during his work in Western Nigeria, found that *Culex pipiens fatigans* was rich throughout the year with high population density throughout the dry months; similar observations were recorded in coastal areas of Digha, India wherever the population of *Culex quinquefasciatus* was higher in summer (Azmi et al., 2015). Throughout the summer season, large ponds and aquatic marshy-lands are converted into shallow aquatic bodies that act as a parcel of land for these two-winged insect species. The temperature in summer in all probability influenced the fast metamorphosis of this vector species (Azmi et al., 2015). Lower wetness and lower temperature below the optimum limit in winter were not favorable for the breeding of two-winged insect species. *Aedes* and *Armigeres* mosquitoes were found most prevailing in human habitations throughout monsoon (Jun–Aug) among all the seasons in India (Mondal et al., 2021).

In an extensive study in Orissa, the density of *Anopheles fluviatilis* and *Anopheles minimus* vector species in human dwellings peaked throughout the rainy and winter seasons followed by summer (Sahu et al., 2011). In the case of *Anopheles*, the rise in mosquito density was ascertained during rainy and post-rainy months which can flow from increased breeding in paddy fields as reported by Rao (1984). In Assam, the peak abundance of *Anopheles minimus* was noticed throughout the rainy season (Dev, 1996).

9.5.2 Host Selection by Adult Mosquitoes

Generally, mosquitoes sustain by feeding upon vertebrate blood during which they also mediate transmission of diverse pathogens, including viruses, protozoa, and nematodes causing various deadly and/or debilitating diseases such as malaria, lymphatic filariasis, dengue, chikungunya, yellow fever, West Nile virus, and Japanese encephalitis (Clements, 1992; Tyagi, 2003). Feeding on vertebrate blood is characteristic of mosquitoes of all genera, with few exceptions. All species of *Toxorhynchites* (89 species) and *Malaya* (12 species), and possibly others (e.g., *Topomyia, Maorigoeldia*) do not require a blood meal to complete egg development (Foster and Walker, 2009). Adults of species of these genera feed exclusively on plant-derived sugars, either directly or from the carbohydrate-rich solution regurgitated by ants. Autogenous, or partially autogenous, mosquito species are also found in a number of genera the females of which are otherwise blood-feeders on vertebrates. More recently the mosquito *Uranotaenia sapphirine* had been discovered to specialize for blood-feeding on the non-vertebrate hosts, the annelids, for the time in the annals of mosquito or annelid science (Reeves et al., 2018). The first instance of a mosquito feeding on an insect other than an ant was found out by Tyagi and Nazneen (2020) (Figure 9.2).

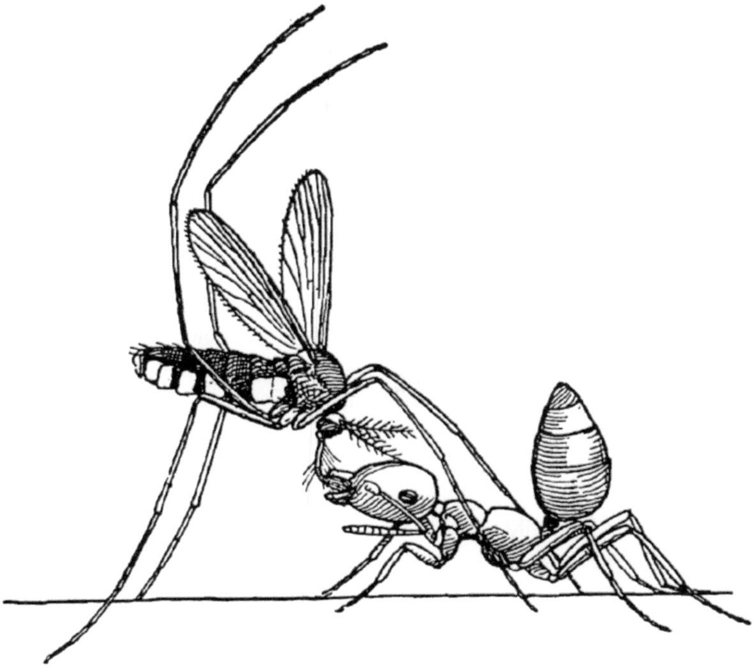

FIGURE 9.2 *Malaya* sp. taking food from mouth of an ant (*Cremastogaster* sp.).

Source: **Free Wikipedia.**

Anophelines exhibit a large variety of host preferences, admiring humans, livestock, birds, and reptiles (Subbarao, 1998). *Anopheles philippines*, *Anopheles minimus s.l.*, *Anopheles baimai*, and *Anopheles annularis* prefer a human host. Wharton (1953) reported that the Malayan *Anopheles maculatus* feeds uniformly on animals, though it exhibits a small preference for humans even once animals are accessible. *Anopheles minimus*, *Anopheles annularis*, *Anopheles maculatus*, and *Anopheles pallidus* showed opportunist blood-feeding behavior, indicating that they selected either human or bovine hosts looking at conditions (Bashar et al., 2012). Host-choice studies within the western world have shown that *Culex quinquefasciatus* mosquitoes are ornithophilic but will feed on a substantial range of hosts, as well as mammals and reptiles (Janssen et al., 2015, Stenn et al., 2019). In an exceedingly recent study on provender analysis (Santos et al., 2019), it has been found that the majority of the blood meals of *Culex quinquefasciatus* originated from chickens, followed by dogs, humans, and passerine birds. All of the blood meals known as human blood were further detected as chicken blood in females captured inside human habitation. The remaining vertebrates were identified from the blood meals of females *Culex* sp. collected from the adjacent human housing. The genus *Culex* solely explored wild birds as their blood source, apart from the taxonomic group *Microculex*, which had a bird and two amphibians as hosts. The choice of birds as hosts was additionally determined in other species of the genus *Culex* mosquitoes. Completely different species *Culex mollis*, is thought of as the main ornithophilic (Forattini et al., 1987), with different species exhibiting variation in their host selection (Stein et al., 2013). On the opposite hand, *Culex intrincatus* is considered a generalist for choosing a range of hosts as well as birds, mammals, amphibians, and reptiles (Forattini et al., 1987). The ornithophilic habit was significantly common in females of *Culex* mosquitoes. Within the recent provender study by microsatellite analysis (Kothera et al., 2020), it has been determined that over 99% of bloodmeals are known from *Culex pipiens* complicated mosquitoes were taken from birds, and <1% was taken from mammals, advocating the results of Fritz et al. (2015) wherever it had been found that *Culex pipiens* females were most well-liked and *Culex molestus* females preferred human blood. Though *Culex pipiens* prefers humans over alternative mammals exhibiting a preference toward a human host (Hamer et al., 2009). In coastal areas of West Bengal, *Cx. quinquefasciatus* disclosed 63.56% anthropophilic index and 36.44% zoophilic (cattle, pig, poultry birds) (Azmi et al., 2015). This observation is additionally just like the findings of Fitzpatrick et al. (2019) and Janssen et al. (2015) who also determined high rates of human-derived bloodmeals by *Culex quinquefasciatus*. *Culex* mosquitoes are observed to change their feeding preferences supported by host accessibility and abundance (Asigau et al., 2019).

The ultimate selection of the host by a hungry mosquito could rely on extraneous or intrinsic factors. Extraneous factors, such as the availability of the host, its variety of flight, and resting habits, etc., may influence the selection. However, the mosquito continues to be able to exert a free selection once extraneous factors lead it to many attainable hosts (e.g., in huts wherever humans and animals live together). Other extraneous factors, such as the sleeping habits of the hosts, the odors given out by them, the massive exposed space of heat skin, the accessibility of the capillary blood vessels, the palatableness of the blood, or probably straightforward random selection, which causes the ultimate choice. Nevertheless, intrinsic factors, a genetic demand for blood from a definite host or a particular capability to retort to some impulses only-may override these tendencies. Additionally, to this, certain characteristics of behavior could also be because of the interaction of each intrinsic and extraneous factor. This is often the case of the distinction between indoor and outside biting behaviors (Swami and Srivastava, 2012).

9.5.3 Blood Meal Analysis

The life cycle of the genus *Culex* is completed in sequence through the egg, larva, insect and adult stages involving each aquatic and terrestrial habitat (Clements, 1992). The procreative immature stages are tailored to the aquatic atmosphere whereas the adults are adapted to the various sorts of the terrestrial environment. Throughout the larval stages, the resources within the kind of detritus from both animal and plant origin and microbes are utilized as food, which is further assimilated in somatic and reproductive tissues. Within the adult stage, the food necessities vary with the sex for the vector mosquitoes. The female needs associate obligatory provender for egg maturation and development (Clements 1992; Harrington et al., 2001; Silver, 2008; Becker et al., 2010), except for the fluid obtained from plant sap. So, blood meal is important for the feminine mosquitoes to attain procreative success and fitness (Harrington et al., 2001). During a study on Andaman and Nicobar Islands (Sivan et al., 2015) *Aedes aegypti* and *Aedes albopictus* shows a primarily anthropophagic nature, which indicates that *Aedes albopictus* most popular to forage on a wider vary of hosts, viz., cow (18.3%), goat (9.6%), fowl (4.8%), pig (3.8%), and rat (1.9%). However, during a study at Kolkata, each species showed a preference for one host than multiple hosts (Tandon and Ray, 2000). Among the *Aedes* mosquitoes found within the Northern part of India, the Human Blood Predominance (H.B.P.) and Pure Human Blood generality (P.H.B.P.) were found highest for *Aedes albopictus* (Mondal et al., 2021). In the case of *Anopheles* mosquitoes, studies disclosed that *Anopheles subpictus* had a preference toward bovine blood as compared to a human, whereas *Anopheles culicifacies* and *Anopheles stephensi* most popular human blood and *Anopheles annularis* gut content showed the presence of solely bovine blood (Swami and Srivastava, 2012). Bhatt et al. (2008) found that the general anthropophilic index of *Anopheles culicifacies* were 11.9%. As per the investigation of Joshi et al. (1988), the provender results of *Anopheles culicifacies* indicated the dipterous insect species to be preponderantly zoophagic. Roy et al. (1991) ascertained the proportion of mixed feeds, i.e., mosquito feeding each on human and bovine hosts, was only 3–4%. Further, it was conjointly found that mosquitoes collected from human dwellings did not show a better proportion of human feeding than bovine feeding. *Anopheles dirus* was rendered to be extremely anthropophilic by Dutta et al. (1996). Thapar et al. (1998) reported a high proportion of *Anopheles culicifacies* between 76% and 100% to feed on bovine hosts followed by 0.32–24.2% on humans, which indicates *Anopheles culicifacies* as zoophagic.

9.5.3.1 Methods for Blood-Meal Analysis

A variety of methodologies are tailored to screen the blood meals of mosquitoes, admire serological and molecular ways (Santos et al., 2019). In the pre-PCR primarily based analysis period, common assays like precipitin assays and ELISA were seeming to indicate low specificity, as a result of that identification supported those tests, solely indicated teams of vertebrates as potential hosts instead of single species (Alencar et al., 2015; Mucci et al., 2015). In recent times, species-level identification is accomplished with the assistance of molecular methods based on the DNA within the ingested blood by mosquitoes (Kirchgatter et al., 2014; Hernandez-Triana et al., 2017). DNA sequencing using the PCR technique is among the foremost definite and targeted approaches for the study of hematophagous arthropods that choose numerous hosts, notably in the case of untamed animals (Kent, 2009).

In the case of contemporary blood meals, the impact of temperature on the amplification of host polymer for short-run temperature storage conditions of full feminine mosquitoes is negligible. Over the past few decades, it has been shown that DNA detection and amplification by PCR are a lot of accurate and economical once engorged mosquitoes are maintained at temperatures up to or below 70°C (Coulson et al., 1990).

9.5.4 Biting Rhythm of Vector Mosquitoes

Mosquito exhibited a nocturnal biting pattern, during which biting continued throughout midnight and showed a declining trend once the midnight peak. In addition, a gentle biting activity was conjointly ascertained by Birley and Rajagopalan (1981), and Pandian and Manoharan (1994). Three forms of pulsating biting patterns, namely nocturnal, diurnal, and crepuscular, were ascertained and a couple of mosquito species were found to exhibit a restricted pattern of biting behavior. Moreover, it was found that a temporal variation within the biting pattern of mosquitoes and therefore the adaptational feature of avoiding competition among themselves by sharing the offered vertebrate hosts in the rural areas throughout the various periods in the diel cycle, is extremely vital for implementing vector management ways to forestall man vector contact for the hindrance of vector-borne diseases (Paramasivan et al., 2015). *Culex tritaeniorhynchus* showed a typical nocturnal biting pattern and therefore the biting rhythm reached a peak at midnight. Once the hour peaks, a decrease in the density of biting mosquitoes was ascertained up to dawn (Rohani et al., 2013). *Culex gelidus*, a zoophilic mosquito, exhibited a nocturnal biting pattern with a distinguished post-nightfall peak of activity. The biting activity was restricted throughout the first part of the night, i.e., before midnight. A typical nocturnal uniform biting activity was exhibited by *Culex infula* and therefore the biting was restricted uniformly during the dark section. *Culex vishnui* and *Culex pseudovishnui* showed a nocturnal biting rhythm and showed post-dusk biting peak like *Culex gelidus* (Paramasivan et al., 2015). *Anopheles hyrcanus* cluster exhibited a typical nocturnal biting pattern followed by the looks of a pointy peak within the premid-night period. Once midnight, the speed of biting was reduced. The presence of nocturnal biting activity with the appearance of a significant peak throughout the post-nightfall section was shown by *Anopheles subpictus* mosquitoes, which were projected to play a secondary role in animal disease transmission of virus infections (Klinkenberg et al., 2008). However, *Anopheles barbirostirs* showed a typical nocturnal biting pattern with two peaks of activity; one major biting peak during the hour and another minor peak during the pre-dawn period (Paramasivan et al., 2015). *Aedes vexans* and *Aedes vittatus* showed a nocturnal biting activity with the post-dawn peak and these mosquitoes exhibited a restricted nocturnal biting pattern. The poor attraction of *Aedes vexan* toward men within the geographic area was detected (Gingrich and Williams, 2005). Asian tiger mosquito showed a scattered diurnal biting activity principally during the morning and evening hours. *Armigeres subalbatus* showed a typical dark biting pattern with a significant peak during the nightfall period that was slightly extended up to midnight. A minor biting peak was ascertained throughout the dawn amount (Pandian and Chandrashekaran, 1980).

9.6 BEHAVIORAL ADAPTATION OF MOSQUITOES

Behavioral variations may end up from essential or inducible traits that are generally thought of as complex, quantitative genetic traits. However, there are instances of a single mutation in mosquitoes having a big impact on behavior. Phenotypical plasticity, the pliability of a given genotype to supply utterly completely different phenotypes in response to different environmental conditions, can therefore alter organisms to quickly answer environmental changes by producing higher matching phenotypes. To identify the genetic determinants of these traits, at intervals of a previous number of years, genomic analysis searching for signatures of alternatives within the genomes of dipteron vectors and their connected species are sequenced (Neafsey et al., 2015; Carrasco et al., 2019) (Figure 9.3).

9.6.1 Adaptation Against Chemical Strain

9.6.1.1 Indoor Behavioral Adaptation of Mosquitoes

Indoor behavioral modifications were ascertained in changes in host predilection. Mosquito species are often delineated as being either generalists or specialists reckoning on the fidelity of their host preference (Wolff and Riffell, 2018). The preference of mosquitoes for one specific host over another could also be regulated by genes concerned with the chemosensory detection of hosts (McBride, 2016). At the single species level, some behavioral resistance traits can be fixed (e.g., early biting) whereas others are malleable; and at the population level, some individuals could show constituent resistance traits while others display plastic resistance traits. The potential behavioral responses for qualitative and quantitative resistance, behavioral tolerance, and exploitation may result from the expression of either constituent or induced behavioral traits (Carrasco et al., 2019).

It has been reported that there is an association between chromosomal polymorphism and differential endophily and endophagy in *An. arabiensis* (Coluzzi et al., 1977), however the preference of individual mosquitoes for a given resting location is not consistent (Lines et al., 1986). Meanwhile,

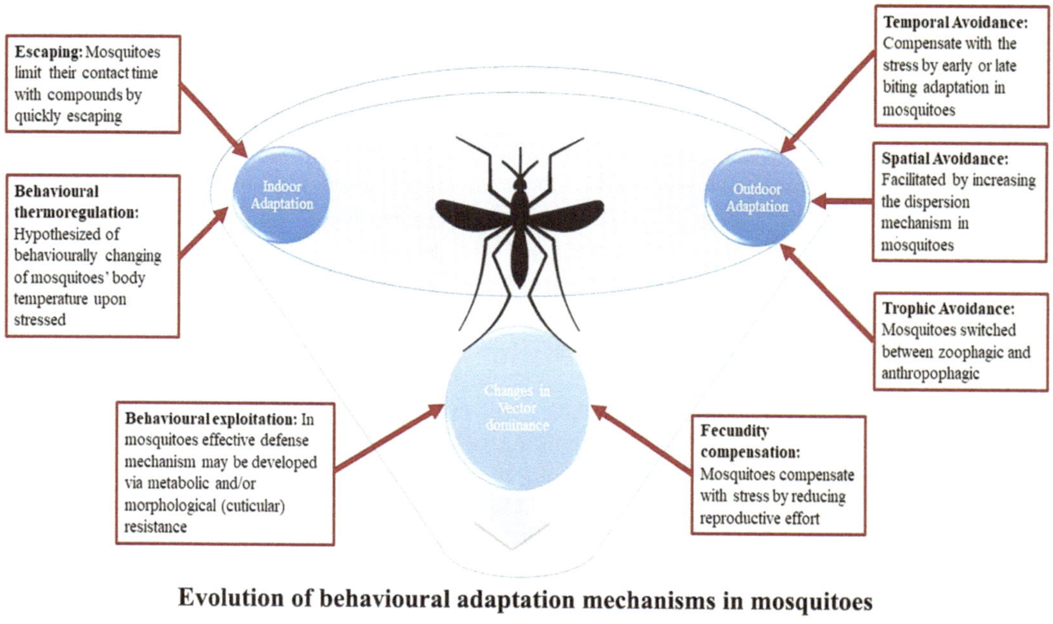

FIGURE 9.3 Schematic presentation of different adaptive features related to mosquito adaptation. (Adapted and redrawn from Carrasco et al., 2019).

An. arabiensis shows site fidelity (returning to the situation of feeding) instead of host fidelity (McCall et al., 2001). During this scenario, mosquitoes retain their inherent feeding preferences (e.g., location, host, and time) and ideally can feed consistently with these preferences. Mosquitoes might not succeed in feeding indoors, being discomfited by net barriers or repelled by insecticides from domestic dwellings. A number of these mosquitoes might reach their hunt for provender elsewhere, whereas others rest outside until the subsequent evening at which point they recommence their search. These vectors may initiate their search before long when crepuscule and feed opportunistically on any outdoor hosts they encounter on the way to an additional most well-liked indoor feeding location (Figure 9.1). The potential impact of insecticide resistance, each activity and physiological, on protozoal infection transmission was assessed utilizing two independent mathematical models: Imperial (Griffin et al., 2010) and OpenMalaria (Smith et al., 2008; Chitnis et al., 2012).

In contrast, there also are studies showing host preference modulation because of changes in environmental conditions (e.g., host availability) (Lefèvre et al., 2009). It has been shown that mosquitoes will modify their activity responses to visual and sense modality stimuli acknowledging both appetitive and aversive associative learning, which is a grace of phenotypical plasticity and might play a crucial role in restructuring host selection preferences in response to insecticide exposure (Bernáth et al., 2016). The evolution of resistance usually entails fitness prices to the mosquito, which are most likely to manifest once the insect is stressed, specifically when it is subject to density-dependent mortality (Kraaijeveld and Godfray, 1997). There have been conjointly massive variations within the foretold impact of resistance, particularly behavioral resistance, relating to the structure of the mosquito population, specifically whether or not there is one homogeneously mixed population or distinct populations of indoor and outside feeding mosquitoes (Menda et al., 2013).

Residual house spraying is currently seldom used as a sway technique against *Culex* or *Aedes* mosquitoes. This can be part of a result of several vector species have out of door resting habits or resting inside on unsprayed objects akin to garments or curtains. *Culex* mosquitoes exhibit an innate tolerance to residual pesticide deposits, presumably because of the pulvilli at lower place *Culex tarsi* (Brown and Pal, 1971).

9.6.1.2 Outdoor Behavioral Adaptation of Mosquitoes

Populations composed of some systematically endophilic individuals and different consistently exophilic ones led to non-uniform exposure of the dipteron population to the insecticide; the exposed fraction of the population could also be so affected that it nearly disappears. However, the unexposed fraction of the population persisted to confirm that sufficient transmission continued (Pates and Curtis, 2005).

Temporal dodging within the variety of activity shifts from night to evening aggressiveness has been determined in *Anopheles farauti* on the Solomon Islands (Taylor, 1975) and Papua island (Thomsen et al., 2017), *Anopheles funestus* in an African country (Russell et al., 2011), and in *Anopheles funestus* and *Anopheles arabiensis* in Kenya (Bayoh et al., 2014). *Anopheles gambiae* and *Anopheles funestus* were found to display multiplied doors host-seeking following the implementation target specific substances (Russell et al., 2011). Biological process avoidance with a shift toward increased zoophilic behaviors was reported in *Culex quinquefasciatus*, *Anopheles funestus* (Bøgh et al., 1998), *Anopheles gambiae* (Bøgh et al., 1998; Ndenga et al., 2016), *Anopheles fluviatilis* and *Anopheles culicifacies* (Waite et al., 2017). Activity quantitative resistance might embrace escape reactions that scale back the contact period with the insecticide (Gerold and Laarman, 1964), resistance-promoting behavioral thermoregulation (Abram et al., 2017), and/or curative self-medication (De Roode and Hunter, 2019). The same as biochemical metabolic resistance, which lowers the quantity of insecticide reaching target sites, these behaviors might limit the insecticide's direct adverse effects following non-lethal exposures (Carrasco et al., 2019). If mosquitoes cannot avoid exposure (qualitative resistance) or scale back the direct effects of the pesticide once exposure will occur (quantitative resistance), then they may still limit their fitness loss through activity tolerance (Carrasco et al., 2019). This might occur by altering behaviors related to offspring production that increase their current procreative efforts, appreciate maximizing nutrient intake quality (e.g., blood-feeding choice), minimizing energy expenditure (e.g., feeding rates and resting times), or adjusting egg production and allocation patterns (Cutler, 2013).

Whether the propensity for door biting by individuals of a given species is increasing or there is simply a residual population of outdoor-biting vectors is debatable, however, the results of this transformation are important. Outside biting is tough to counter with obtainable management methods. Larval source management, spatial repellents, transgenic mosquitoes, and enticing deadly sugar bait may well be used or developed for protozoal infection control, but are either difficult to proportion altogether locations or are tools which will like rather more exploration before they will be successfully deployed (Gatton et al., 2013). Inflicting vectors to feed a lot of often outdoors might represent new opportunities for control. Blood-feeding vectors are captured in odor-baited traps (Okumu et al., 2010), killed by insecticide-treated cows (Rowland et al., 2001), or when feeding on enticing noxious sugar bait (Müller et al., 2010), whereas large females can be targeted by developing oviposition traps (Harris et al., 2011; Devi et al., 2013). These spatiotemporal shifts are usually discovered due to the contradictory impact that humans are usually outdoors in the evenings and mornings whereas they are inside at night, making it difficult to differentiate the underlying reason behind the activity shift (Figure 9.1). Besides it has been established that circadian activity in dipteron species, regulated by genetic expressions, (Leming et al., 2014) with olfactive functions are also essential for host-seeking, being adjusted by daily rhythms (Eilerts et al., 2018) indicates *Anopheles culicifacies* as zoophagic.

9.6.2 Changes in Vector Dominance

Phenotypes caused by single cistron mutations typically demonstrate an exponential increase in frequency. In contrast, phenotypes supported by standing genetic variation in multiple genes generally have a special dynamic; generally exhibited by a direct and sustained modification in their constitution distributions (Gatton et al., 2013).

Reductions in mosquito abundance will have additional effects on malady transmission mediated through density dependence. Lyimo and Koella (1992) among others have pointed out that not only lower larval densities increase survivability and adult size, but also lower development rates have a significant pertinent to disease transmission (Figure 9.1). If larger individuals live longer then is more possible to survive through the disease's latent period.

The existence within the population of variation in the ability of species to notice and perceive chemicals as baneful will verify their level of exposure. Thus, the adjusted evolution of neuroscience skills to detect pesticides, either upon contact or from a definite distance, would considerably scale back the impact of insecticides through qualitative or quantitative behavioral resistance or maybe behavioral exploitation, and supply a method for the expression of each constitutive (e.g., escaping) and evoked behavioral responses (e.g., different host choice, etc.) (Carrasco et al., 2019) indicates *Anopheles culicifacies* as zoophagic.

9.6.3 Adaptation Associated with Dehydration

Water loss has been directly concerned with altered dipteran behavior with foreseen will increase in West Nile River virus (WNV) transmission (Hagan et al., 2018), however, it is conjointly coupled to reduced survival and oviposition (Canyon et al., 1999), reduced nutritionary reserves, egg production (Benoit et al., 2010), and shifted geographic and microgeographic distributions (Reidenbach et al., 2014). Mosquitoes are extremely liable to water loss as a result of high rates of evaporation through both respiratory spiracles and the cuticle (Piermarini, 2016). Though, water loss will be countered by alterations in body size, metabolism, cuticular hydrocarbon composition, and several other behavioral adaptations (Yang et al., 2017). These behavioral variations can present in all mosquitoes, however, can be intensified in adverse conditions, resembling cold, heat, or drought.

Over-wintering (diapausing) mosquitoes show an enlarged desiccation resistance that confers a reduced water loss rate (Rinehart et al., 2006), whereas some Anophelinae mosquitoes have tailored to survive hot and arid climates via long-distance migration or aestivation over the summer (Dao et al., 2014; Mamai et al., 2016). Interestingly, some mosquito species are liable to a lot of water loss, while different species have developed relatively superior desiccation resistance (Sinclair et al., 2007). Dehydration stress has resulted in genetic alterations and behavioral variations (both long- and short-term) that are integrated into dipteran physiology, survival, and distribution (Benoit et al., 2010). Dao et al. (2014) postulated that the flexibility to exploit arid environments might have driven the divergence between several mosquito species, though species-specific adaptations are apparent. Mosquitoes encounter an excess of osmotic pressures that are related to blood-feeding as a result of inflated sodium content from blood plasma, and increased potassium, and iron content from red blood cells, at the side of an enormous inflow of water (Drake et al., 2010).

Dry heat has been attributed to the activation of mosquitoes, however, dampish heat has been involved in augmented searching behavior (Khan and Maibach, 1971). Because the time void of water increases, mosquitoes discriminate less on the supply of provender (Khan and Maibach, 1970), and, once a host is found, the feeding rates of mosquitoes in higher temperatures likewise increase (Grossman and Pappas, 1991). Interestingly, dehydration phenotypes might possess sensitivity to humidness and wet hosts, each of which is a necessary host cue (Van Breugel et al., 2015).

Mosquitoes utilize an extremely economical urinary organ discharge system to control osmolality and ion imbalances (Piermarini, 2016). Following a feed, enlarged rates of excretion through symptoms are ascertained as a method of eliminating non-essential parts of a blood meal. The suitable regulation of water volume and ion concentration permits blood-meal retention/utilization, diminished mortality, and an increased desiccation resistance between blood meals (Holmes and Benoit, 2019).

BIBLIOGRAPHY

Abram, P. K., Boivin, G., Moiroux, J., and Brodeur, J. (2017). Behavioural effects of temperature on ectothermic animals: Unifying thermal physiology and behavioural plasticity. *Biological Reviews, 92*(4), 1859–1876.

Abreu, F. V. S. D., Morais, M. M., Ribeiro, S. P., and Eiras, Á. E. (2015). Influence of breeding site availability on the oviposition behaviour of *Aedes aegypti*. *Memórias do Instituto Oswaldo Cruz, 110*, 669–676.

Adeleke, E. D., Shittu, R. A., Beierkuhnlein, C., and Thomas, S. M. (2022). High wind speed prevents the establishment of the disease vector mosquito *Aedes albopictus* in its climatic Niche in Europe. *Frontiers in Environmental Science*, 508, 1–13.

Afolabi, O. J., Akinneye, J. O. and Igiekhume, A. M. (2019). Identification, abundance, and diversity of mosquitoes in akure south local government area, Ondo state, Nigeria. *The Journal of Basic and Applied Zoology, 80*, 1–7.

Agyekum, T. P., Botwe, P. K., Arko-Mensah, J., Issah, I., Acquah, A. A., Hogarh, J. N., Dwomoh, D., Robins, T. G., and Fobil, J. N. (2021). A systematic review of the effects of temperature on *Anopheles* mosquito development and survival: Implications for Malaria control in a future warmer climate. *International Journal of Environmental Research and Public Health, 18*(14),7255.

Ahmed, T., Hyder, M. Z., Liaqat, I., and Scholz, M. (2019). Climatic conditions: Conventional and nanotechnology-based methods for the control of mosquito vectors causing human health issues. *International Journal of Environmental Research and Public Health, 16*(17), 3165.

Alencar, J., Mello, C. F. D., Gil-Santana, H. R., Giupponi, A. P. D. L., Araújo, A. N., Lorosa, E. S., Guimarães, A. É., and Silva, J. D. S. (2015). Feeding patterns of mosquitoes (Diptera: Culicidae) in the Atlantic Forest, Rio de Janeiro, Brazil. *Journal of Medical Entomology, 52*(5), 783–788.

Amarakoon, D., Chen, A., Rawlins, S., Chadee, D. D., Taylor, M., and Stennett, R. (2008). Dengue epidemics in the Caribbean-temperature indices to gauge the potential for onset of dengue. *Mitig Adapt Strat Global Change, 13*, 341–357.

Ansari M. A. and Razdan, R. K. (1998) Seasonal Prevalence of *Aedes aegypti* in Five Localities of Delhi, India. *World Health Organisation Dengue Bulletin, 22*, 28–34.

Araripe, L. O., Bezerra, J., Rivas, G., and Bruno, R. V. (2018). Locomotor activity in males of *Aedes aegypti* can shift in response to females' presence. *Parasites and Vectors, 11*(1), 254.

Arunachalam N., Tana, S., Espino, F., Kittayapong, P., Abeyewickreme, W., Wai, K. T., Tyagi, B. K., Kroeger, A., Sommerfeld, J., and Petzold, M. (2010). Dengue vectors at the dynamic interface of environmental, biological and social ("eco-bio-social") determinants: A multi-country study in urban and peri-urban Asia. *Bulletin of World Health Organization, 88*, 173–184.

Asigau, S., Salah, S., and Parker, P. G. (2019). Assessing the blood meal hosts of *Culex quinquefasciatus* and *Aedes taeniorhynchus* in Isla Santa Cruz, Galápagos. *Parasites and Vectors, 12*(1), 1–10.

Azmi, S. A., Das, S. and Chatterjee, S. (2015). Seasonal prevalence and blood meal analysis of filarial vector *Culex quinquefasciatus* in coastal areas of Digha, West Bengal, India. *Journal of Vector Borne Diseases, 52*(3), 252.

Bailey, S. F., Eliason, D. A., Hoffman, B. L. (1965). Flight and dispersal of the mosquito *Culex tarsalis* Coquillett in the Sacramento valley of California. *Hilgardia, 37*, 73–113.

Bakker, J. W., Loy, D. E., Takken, W., Hahn, B. H., and Verhulst, N. O. (2020). Attraction of mosquitoes to primate odours and implications for zoonotic Plasmodium transmission. *Medical and Veterinary Entomology, 34*, 17–26.

Balakrishnan, N., Venkatesh, S., and Lal, S. (2006). An entomology study on the dengue vector during outbreak of dengue in Tirupur town and its surroundings, Tamil Nadu, India. *Journal of Communicable Diseases, 38*, 164–168.

Bashar, K., Tuno, N., Ahmed, T. U. and Howlader, A. J. (2012). Blood-feeding patterns of Anopheles mosquitoes in a malaria-endemic area of Bangladesh. *Parasites & Vectors, 5*(1), 1–10.

Bassene, H., Niang, E. H. A., Fenollar, F., Doucoure, S., Faye, O., Raoult, D., Sokhna, C., and Mediannikov, O. (2020). Role of plants in the transmission of *Asaia sp.*, which potentially inhibit the *Plasmodium* sporogenic cycle in Anopheles mosquitoes. *Scientific Reports, 10*(1), 7144.

Bayoh, M. N., Walker, E. D., Kosgei, J., Ombok, M., Olang, G. B., Githeko, A. K., Killeen, G. F., Otieno, P., Desai, M., Lobo, N. F., and Vulule, J. M. (2014). Persistently high estimates of late night, indoor exposure to malaria vectors despite high coverage of insecticide treated nets. *Parasites and Vectors, 7*(1), 1–13.

Becker, N., Petric, D., Zgomba, M., Boase, C., Madon, M., Dahl, C., and Kaiser, A. (2010). *Mosquitoes and their control*, Springer Science and Business Media.

Bellone, R., and Failloux, A. B. (2020). The Role of Temperature in Shaping Mosquito-Borne Viruses Transmission. *Frontiers in Microbiology, 11*, 584846.

Belton, P. (1967). Effect of illumination and pool brightness on oviposition by *Culex restuans* (Theo) in the field. *Mosquito News, 27*, 66–68.

Benoit, J. B., Patrick, K. R., Desai, K., Hardesty, J. J., Krause, T. B. and Denlinger, D. L. (2010). Repeated bouts of dehydration deplete nutrient reserves and reduce egg production in the mosquito *Culex pipiens. Journal of Experimental Biology, 213*(16), 2763–2769.

Bentley, M. D., and Day, J. F. (1989). Chemical ecology and behavioral aspects of mosquito oviposition. *Annual Review of Entomology, 34*, 401–421.

Bernáth, B., Anstett, V. and Guerin, P. M. (2016). *Anopheles gambiae* females readily learn to associate complex visual cues with the quality of sugar sources. *Journal of Insect Physiology, 95*, 8–16.

Bhatt, R. M., Srivastava, H. C., Rajnikant and Yadav, R. S. (2008). Dynamics of *Anopheles culicifacies*-transmitted malaria in the absence of effective zooprophylaxis in a riverine settlement in Gujarat, India. *Current Science*, 82–87.

Bhattacharya, S., Pal, S. and Acharyya, D. (2016) Are mosquitoes "A Necessary Evil"? *International Journal of Fauna and Biological Studies, 3*(1), 124–129

Bhattacharya, S., Sinha, S., Ghosh, T., and Bose, C. (2021). Mosquito-plant interaction with reference to eco-epidemiological importance: An exploratory review. *Journal of Medical Arthropodology and Public Health, 1*(1), 67–78.

Bhattacharya, T., Newton, I. L. and Hardy, R. W. (2020). Viral RNA is a target for *Wolbachia*-mediated pathogen blocking. *PLoS Pathogens, 16*(6), e1008513.

Bhattacharya, D. R., Rajavel, A. R., Mohapatra, P. K., Jambulingam, P., Mahanta, J., and Prakash, A. (2014). Faunal richness and the checklist of Indian mosquitoes (Diptera: Culicidae). *Check List, 10*(6), 1342–1358.

Bidlingmayer, W. L. (1974). The influence of environmental factors and physiological stage on flight patterns of mosquitoes taken in the vehicle aspirator and truck, suction, bait and New Jersey light traps. *Journal of Medical Entomology, 11*(2), 119–146.

Bidlingmayer, W. L. (1975). Mosquito flight paths in relation to the environment. Effect of vertical and horizontal visual barriers. *Annals of the Entomological Society of America, 68*, 51–57.

Birley, M. H., and Charlwood, J. D. (1989). The effect of moonlight and other factors on the oviposition cycle of malaria vectors in Madang, Papua New Guinea. *Annals of Tropical Medicine and Parasitology, 83*(4), 415–422.

Birley, M. H. and Rajagopalan, P. K. (1981). Estimation of the survival and biting rates of *Culex quinquefasciatus* (Diptera: Culicidae). *Journal of Medical Entomology*, *18*(3), 181–186.

Bøgh, C., Pedersen, E. M., Mukoko, D. A., and Ouma, J. H. (1998). Permethrin impregnated bednet effects on resting and feeding behaviour of lymphatic filariasis vector mosquitoes in Kenya. *Medical and Veterinary Entomology*, *12*(1), 52–59.

Bomphrey, R. J., Nakata, T., Phillips, N., and Walker, S. M. (2017). Smart wing rotation and trailing-edge vortices enable high frequency mosquito flight. *Nature*, *544*(7648), 92–95.

Bouma, M., and Rowland, M. (1995). Failure of passive zooprophylaxis: Cattle ownership in Pakistan is associated with a higher prevalence of malaria. *Transactions of the Royal Society of Tropical Medicine and Hygiene*, *89*, 351–353.

Braks, M. A., Leal, W. S., and Cardé, R. T. (2007). Oviposition responses of gravid female *Culex quinquefasciatus* to egg rafts and low doses of oviposition pheromone under semifield conditions. *Journal of Chemical Ecology*, *33*(3), 567–578.

Brown, A. W. A., and Pal, R. (1971). Insecticides resistance in arthropods. *WHO*, Monograph Series No. 38, 2nd edit.

Bruno, D. W. and Laurence, B. R. (1979). The influence of the apical droplet of *Culex* egg rafts on oviposition of *Culex pipiens faticans* (Diptera: Culicidae). *Journal of Medical Entomology*, *16*(4), 300–305.

Bryan, J. H., O'Donnell, M. S., Berry, G., and Carvan, T. (1986). Dispersal of adult female *Culex annulirosris* in Griffith, new south wales, Australia: A further study. *Journal of the American Mosquito Control Association*, *8*(4), 398–403.

Canyon, D. V., Hii, J. L. K. and Müller, R., (1999). Adaptation of *Aedes aegypti* (Diptera: Culicidae) oviposition behaviour in response to humidity and diet. *Journal of Insect Physiology*, *45*(10), 959–964.

Cardé, R. T. (2015). Multi-Cue integration: How female mosquitoes locate a human host. *Current Biology*, *25*(18), R793–R795.

Cardé, R. T., and Willis, M. A. (2008). Navigational strategies used by insects to find distant, wind-borne sources of odor. *Journal of Chemical Ecology*, *34*(7), 854–866.

Carlson, D. B., and Vigliano, R. R. (1985). The effects of two different water management regimes on flooding and mosquito production in a salt marsh impoundment. *Journal of American Mosquito Control Association*, *1*, 203–211

Carrasco, D., Lefèvre, T., Moiroux, N., Pennetier, C., Chandre, F., and Cohuet, A. (2019). Behavioural adaptations of mosquito vectors to insecticide control. *Current Opinion in Insect Science*, *34*, 48–54.

Cator, L. J., Johnson, L. R., Mordecai, E. A., Moustaid, F. E., Smallwood, T. R. C., LaDeau, S. L., Johansson, M. A., Hudson, P. J., Boots, M., Thomas, M. B., Power, A. G., and Pawar, S. (2020). The role of vector trait variation in vector-borne disease dynamics. *Frontiers in Ecology and Evolution*, *8*, 189.

Cator, L. J., Thomas, S., Paaijmans, K. P., Ravishankaran, S., Justin, J. A., Mathai, M. T., Read A. F., Thomas, M. B., and Eapen, A. (2013). Characterizing microclimate in urban malaria transmission settings: A case study from Chennai, India. *Malaria Journal*, *12*, 84.

Chandrasegaran, K., and Juliano, S. A. (2019). How do trait-mediated non-lethal effects of predation affect population-level performance of mosquitoes? *Frontiers in Ecology and Evolution*, *7*, 25.

Chandrasegaran, K., Kandregula, S. R., Quader, S., and Juliano, S. A. (2018). Context-dependent interactive effects of non-lethal predation on larvae impact adult longevity and body composition. *Plos One*, *13*(2), e0192104.

Chandrasegaran, K., Lahondère, C., Escobar, L. E., and Vinauger, C. (2020). Linking mosquito ecology, traits, behavior, and disease transmission. *Trends in Parasitology*, *36*(4), 393–403.

Chapman, J. W., Reynolds, D. R., and Wilson, K. (2015). Long-range seasonal migration in insects: Mechanisms, evolutionary drivers and ecological consequences. *Ecology Letters*, *18*(3):287–302.

Chaturvedi, S., and Dwivedi, S. (2020). Estimating the malaria transmission over the Indian subcontinent in a warming environment using a dynamical malaria model. *Journal of Water and Health*, *18*(3), 358–374.

Chitnis, N., Hardy, D., and Smith, T. (2012). A periodically-forced mathematical model for the seasonal dynamics of malaria in mosquitoes. *Bulletin of Mathematical Biology*, *74*(5), 1098–1124.

Choi, M. B., Lee, W. G., Kang, H. J., Yang, S. C., Song, B. G., Shin, E., and Kwon, O. (2017). Seasonal prevalence and species composition of mosquitoes and chigger mites collected from Daegu, Gunwi and Sangju in South Korea, 2014. *Journal of Ecology and Environment*, *41*(1), 1–7.

Ciota, A. T., Drummond, C. L., Ruby, M. A., Drobnack, J., Ebel, G. D., and Kramer, L. D. (2012). Dispersal of *Culex* mosquitoes (Diptera: Culicidae) from a wastewater treatment facility. *Journal of Medical Entomology*, *49*(1), 35–42.

Clements, A. N. (1999). *The biology of mosquitoes. Volume 2: Sensory reception and behaviour.* CABI Publishing.

Clements, A. N. (1992). *The biology of mosquitoes*, Volume 3 Transmission of viruses and interactions with bacteria (Vol. 3). *CABI*.

Collins, C. M., Bonds, J. A. S., Quinlan, M. M., and Mumford, J. D. (2019). Effects of the removal or reduction in density of the malaria mosquito, *Anopheles gambiae* s.l., on interacting predators and competitors in local ecosystems. *Medical Veternary Entomology*, *33*(1), 1–15

Coluzzi, M., Sabatini, A., Petrarca, V., and M. A. Di Deco. (1977). Behavioural divergences between mosquitoes with different inversion karyotypes in polymorphic populations of the *Anopheles gambiae* complex. *Nature*, *266*(5605), 832–833.

Coon, K. L., Vogel, K. J., Brown, M. R., and Strand, M. R. (2014). Mosquitoes rely on their gut microbiota for development. *Molecular Ecology*, *23*, 2727–2739.

Cooperband, M. F., and Cardé, R. T., (2006). Orientation of *Culex* mosquitoes to carbon dioxide-baited traps: Flight manoeuvres and trapping efficiency. *Medical Veternary Entomology*, *20*, 11–26.

Coulson, R. M., Curtis, C. F., Ready, P. D., Hill, N. and Smith, D. F. (1990). Amplification and analysis of human DNA present in mosquito bloodmeals. *Medical and Veterinary Entomology*, *4*(4), 357–366.

Couper, L. I., Farner, J. E., Caldwell, J. M., Childs, M. L., Harris, M. J., Kirk, D. G., Nova, N., Shocket, M., Skinner, E. B., Uricchio, L. H., Exposito-Alonso, M., and Mordecai, E. A. (2021). How will mosquitoes adapt to climate warming? *Elife*, *10*, e69630.

Curtis, C. F., and Rawlings, P. (2008). A preliminary study of dispersal and survival of *Anopheles culicifacies* in relation to the possibility of inhibiting the spread of insecticide resistance. *Ecological Entomology*, *5*(1), 11 – 17.

Cutler, G. C. (2013). Insects, insecticides and hormesis: Evidence and considerations for study. *Dose Response*, *11*(2), 154–177.

Cutler, S. J., Fooks, A. R., and van der Poel, W. H. M. (2010). Public health threat of new, reemerging, and neglected zoonoses in the industrialized world. *Emerging Infectious Diseases*, *16*(1), 1–7.

Dao, A., Yaro, A. S., Diallo, M., Timbiné, S., Huestis, D. L., Kassogué, Y., Traoré, A. I., Sanogo, Z. L., Samaké, D., and Lehmann, T. (2014). Signatures of aestivation and migration in Sahelian malaria mosquito populations. *Nature*, *516*(7531), 387–390.

Das, S., and Dimopoulos, G. (2008). Molecular analysis of photic inhibition of blood-feeding in *Anopheles gambiae*. *BMC Physiology*, *8*, 23.

Day, J. F., Ramsey, A. M., and Jin-Tong, Z. (1990). Environmentally mediated seasonal variation in mosquito body size. *Environmental Entomology*, *19*, 469–473

Day, J., and Curtis, G. A. (1994). When it rains they soar—and that makes *Culex nigripalpus* a dangerous mosquito. *American Entomologists*, *40*, 162–167.

De Roode, J. C., and Hunter, M. D. (2019). Self-medication in insects: When altered behaviours of infected insects are a defense instead of a parasite manipulation. *Current Opinion in Insect Science*, *33*, 1–6.

De, S. K., and Chandra, G. (1994). Studies on the filariasis vector--*Culex quinquefasciatus* at Kanchrapara, west Bengal (India). *The Indian Journal of Medical Research*, *99*, 255–258.

Dekker, T., and Cardé, R. T. (2011). Moment-to-moment flight manoeuvres of the female yellow fever mosquito (*Aedes aegypti L.*) in response to plumes of carbon dioxide and human skin odour. *The Journal of Experimental Biology*, *214*(20), 3480–3494.

Dev, V. (1996). *Anopheles minimus*: Its bionomics and role in the transmission of malaria in Assam, India. *Bulletin of the World Health Organization*, *74*(1), 61.

Dev, V., and Sharma, V. P. (2013). The Dominant Mosquito Vectors of Human Malaria in India. In (Ed.), *Anopheles* mosquitoes – New insights into malaria vectors. *IntechOpen*.

Devi, N. P., Jauhari, R. K. and Mondal, R. (2013). Ovitrap surveillance of *Aedes* mosquitoes (Diptera: Culicidae) in selected areas of Dehradun district, Uttarakhand, India. *Global Journal of Medical Research Diseases*, *13*, 53–57.

Dhanalakshmi, P., Sankaraiyah, K., Kavipriya, J., Vijayalakshmi, M. N. V. (2018). Distribution of sensory structures on antennae of female *Culex* mosquito: A SEM study. *Journal of Entomology and Zoology Studies*, *6*(2), 1900–1903.

Donnelly, M. J., McCall, P., Lengeler, C. Bates, I., D'Alessandro, U., Barnish, G., Konradsen, F., Klinkenberg, E., Townson, H., Trape, J. F., Hastings, I. M., & Mutero, C.(2005). Malaria and urbanization in sub-Saharan Africa. *Malaria Journal*, *4*, 12.

Drake, L. L., Boudko, D. Y., Marinotti, O., Carpenter, V. K., Dawe, A. L., and Hansen, I. A. (2010). The Aquaporin gene family of the yellow fever mosquito, *Aedes aegypti*. *PloS One*, *5*(12), e15578.

Drakou, K., Nikolaou, T., Vasquez, M., Petric, D., Michaelakis, A., Kapranas, A., Papatheodoulou, A., and Koliou, M. (2020). The Effect of Weather Variables on Mosquito Activity: A Snapshot of the Main Point of Entry of Cyprus. *International Journal of Environmental Research and Public Health*, *17*(4), 1403.

Dusfour, I., Harbach, R., and Manguin, S. (2004). Bionomics and systematics of the Oriental *Anopheles sundaicus* complex in relation to malaria transmission and vector control. *The American Journal of Tropical Medicine and Hygiene*, *71*, 518–24.

Dutta, P., Bhattacharyya, D. R., Khan, S. A., Sharma, C. K. and Mahanta, J. (1996). Feeding patterns of *Anopheles dirus*, the major vector of forest malaria in north east India. *The Southeast Asian Journal of Tropical Medicine and Public Health*, *27*(2), 378–381.

Edman, J. D., Scott, T. W., Costero, A., Morrison, A. C., Harrington, L. C. and Clark, G. G. (1998). *Aedes aegypti* (Diptera: Culicidae) movement influenced by availability of oviposition sites. *Journal of Medical Entomology*, *35*(4), 578–583.

Eilerts, D. F., Vandergiessen, M., Bose, E. A., Broxton, K., and Vinauger, C. (2018). Odor-specific daily rhythms in the olfactory sensitivity and behaviour of *Aedes aegypti* mosquitoes. *Insects*, *9*(4), 147.

Ellwanger, J. H., Cardoso, J. D. C., and Chies, J. A. B. (2021). Variability in human attractiveness to mosquitoes. *Current Research in Parasitology and Vector-borne Diseases*, *1*, 100058.

Emidi, B., Kisinza, W. N., Mmbando, B. P., Malima, R., and Mosha, F. W., (2017). Effect of physicochemical parameters on *Anopheles* and *Culex* mosquito larvae abundance in different breeding sites in a rural setting of Muheza, Tanzania. *Parasites Vectors*, *10*, 304.

Endo, N., and Eltahir, E. A. B. (2018). Modelling and observing the role of wind in Anopheles population dynamics around a reservoir. *Malaria Journal*, *17*, 1–9.

Erias, A. E., and Jepson, P. C. (1991). Host location by *Aedes aegypti* (Diptera: Culicidae): A wind tunnel study of chemical cues. *Bulltin of Entomological Research*, *81*, 151–160.

Eritja, R., Palmer, J. R. B., Roiz, D., and Bartumues, F. (2017). Direct evidence of adult *Aedes albopictus* Dispersal by Car. *Scientific Reports*, *7*, 14399.

Ewing, D. A., Purse, B. V., Cobbold, C. A., Schäfer, S. M. and White, S. M. (2019). Uncovering mechanisms behind mosquito seasonality by integrating mathematical models and daily empirical population data: *Culex pipiens* in the UK. *Parasites & Vectors*, *12*(1), 1–19.

Fakoorziba, M. R., and Vijayan, A. (2008). Breeding habitats of *Culex tritaeniorhynchus* (Diptera: Culicidae), A Japanese Encephalitis Vector, and Associated Mosquitoes in Mysore, India. *Journal of Entomological Research Society*, *10*(3), 1–9.

Fay, R. W., and Perry, A. S. (1965). Laboratory studies of oviposition preferences of Aedes aegypti. *Mosquito News*, *25*(3), 276–281.

Fitzpatrick, D. M., Hattaway, L. M., Hsueh, A. N., Ramos-Niño, M. E., and Cheetham, S. M. (2019). PCR-based bloodmeal analysis of *Aedes aegypti* and *Culex quinquefasciatus* (Diptera: Culicidae) in St. George Parish, Grenada. *Journal of Medical Entomology*, *56*(4), 1170–1175.

Floore, T. G. (2007). Biorational control of mosquitoes. *American Mosquito Control Assoc.*

Focks, D. A., Sackett, S. R., Dame, D. A., and Bailey, D. L. (1983). Ability of *Toxorhynchites amboinensis* (Doleschall) (Diptera: Culicidae) to locate and oviposit in artificial containers in an urban environment. *Environmental Entomology*, *12*, 1073–1077.

Foley, D. H., Rueda L. M., and Wilkerson, R. C. (2007). Insight into mosquito biogeography from country species records. *Journal of Medical Entomology*, *44*, 554–567.

Forattini, O. P., Gomes, A. D. C., Natal, D., Kakitani, I. and Marucci, D. (1987). Feeding preferences of Culicidae mosquitoes in the Ribeira Valley, S. Paulo State, Brazil. *Revista de saude publica*, *21*, 171–187.

Foster, W. A., and Walker, E. D. (2009). Mosquitoes (Culicidae), 207–259. In: G. R. Mullen & L. A. Durden (eds.), *Medical and Veterinary Entomology*, 2nd edn., Academic Press, San Diego, California.

Fritz, M. L., Walker, E. D., Miller, J. R., Severson, D. W., and Dworkin, I. (2015). Divergent host preferences of above and below ground *Culex pipiens* mosquitoes and their hybrid offspring. *Medical and Veterinary Entomology*, *29*(2), 115–123.

Gaaboub, I. A., El-Sawaf, S. K., and El-Latif, M. A. (2009). Effect of different relative humidities and temperatures on egg-production and longevity of adults of *Anopheles* (Myzomyia) pharoensis Theob.1. *Z. Angew Entomol. 67*, 88–94.

Gatton, M. L., Chitnis, N., Churcher, T., Donnelly, M. J., Ghani, A. C., Godfray, H. C. J., Gould, F., Hastings, I., Marshall, J., Ranson, H., and Rowland, M., (2013). The importance of mosquito behavioural adaptations to malaria control in Africa. *Evolution: International Journal of Organic Evolution*, *67*(4), 1218–1230.

Gerold, J. L., and Laarman, J. J. (1964). Selection of some strains of *Anopheles atroparvus* with different behavioural responses to contacts with DDT. *Nature*, *204*(4957), 500–501.

Gething, P. W., Smith, D. L., Patil, A. P., Tatem, A. J., Snow, R. W., and Hay, S. I. (2010). Climate change and the global malaria recession. *Nature*, *465*, 342–345.

Ghebreyesus, T. A., Haile, M., Witten, K. H., Getachew, A., Yohanne, M., Lindsay, S. W., and Byass, P. (2000). Household risk factors for malaria among children in the Ethiopian highlands. *Transactions of the Royal Society of Tropical Medicine and Hygiene*, *94*(1), 17–21.

Ghosh, A., Das, B. K., Roy, A., Mandal, B., and Chandra, G. (2008). Antibacterial activity of some medicinal plant extracts. *Journal of Natural Medicines*, *62*(2), 259–262.

Ghosh, S. K., Podder, D., Panja, A. K., and Mukherjee, S. (2020). In target areas where human mosquito-borne diseases are diagnosed, the inclusion of the pre-adult mosquito aquatic niches parameters will improve the integrated mosquito control program. *PLoS Neglected Tropical Diseases*, *14*(8), e0008605.

Gillett, J. D. (1962). Contributions to the oviposition cycle by individual mosquitoes in a population. *Journal of Insect Physiology*, *8*, 665–681.

Gillett, J. D., Corbet, P. S., and Haddow, A. J. (1959). Observations on the oviposition-cycle of *Aedes* (Stegomyia) *aegypti* (Linnaeus). III. *Annals of Tropical Medicine and Parasitology*, *53*, 132–136.

Gillies, M. T. (1980). The role of carbon dioxide in host finding by mosquitoes (Diptera: Culicidae): A review. *Bulletin of Entomological Research*, *70*, 525–532.

Gingrich, J. B. and Williams, G. M. (2005). Host-feeding patterns of suspected West Nile virus mosquito vectors in Delaware, 2001–2002. *Journal of the American Mosquito Control Association*, *21*(2), 194–200.

Golding, N., Nunn, M. A., and Purse, B. V. (2015). Identifying biotic interactions which drive the spatial distribution of a mosquito community. *Parasites and Vectors*, *8*(1), 1–10.

Gouagna, L. C., Rakotondranary, M., Boyer, S., Lempérière, G., Dehecq, J. S., and Fontenille, D. (2012). Abiotic and biotic factors associated with the presence of *Anopheles arabiensis* immatures and their abundance in naturally occurring and man-made aquatic habitats. *Parasites and Vectors*, *5*(1), 1–12.

Griffin, J. T., Hollingsworth, T. D., Okell, L. C., Churcher, T. S., White, M., Hinsley, W., Bousema, T., Drakeley, C. J., Ferguson, N. M., Basáñez, M. G. and Ghani, A. C., (2010). Reducing *Plasmodium falciparum* malaria transmission in Africa: A model-based evaluation of intervention strategies. *PLoS medicine*, *7*(8), e1000324.

Grill, C. P., and S. A. Juliano. (1996). Predicting speices interactions based on behavior: Predation and competition in container-dwelling mosquitoes. *Journal of Animal Ecology*, *65*, 63–76.

Griswold, M. W., and Lounibos, L. P. (2006). Predator identity and additive effects in a treehole community. *Ecology*, *87*(4), 987–995.

Grossman, G. L., and Pappas, L. G. (1991). Human skin temperature and mosquito (Diptera: Culicidae) blood feeding rate. *Journal of Medical Entomology*, *28*(3), 456–460.

Gu, W. D., Muller, G., Schlein, Y., Novak, R. J., and Beier, J. C. (2011). Natural plant sugar sources of *Anopheles* mosquitoes strongly impact malaria transmission potential. *PLoS One*, 6, e15996.

Guegan, M., Zouache, K., Demichel, C., Minard, G., Tran Van, V., Potier, P., Mavingui, P., and Valiente Moro, C. (2018). The mosquito holobiont: Fresh insight into mosquito microbiota interactions. *Microbiome*, *6*(1), 1–17.

Gunathilaka, N., Fernando, T., Hapugoda, M., Wickremasinghe, R., Wijeyerathne, P., and Abeyewickreme, W. (2013). *Anopheles culicifacies* breeding in polluted water bodies in Trincomalee District of Sri Lanka. *Malaria Journal*, *12*, 285.

Haddow, A. J., Gillett, J. D., and Corbet, P. S. (1961). Observations on the oviposition-cycle of *Aedes* (Stegomyia) *aegypti* (Linnaeus), V. *Annals of Tropical Medicine and Parasitology*, *55*, 343–356.

Hagan, R. W., Didion, E. M., Rosselot, A. E., Holmes, C. J., Siler, S. C., Rosendale, A. J., Hendershot, J. M., Elliot, K. S., Jennings, E. C., Nine, G. A., and Perez, P. L. (2018). Dehydration prompts increased activity and blood feeding by mosquitoes. *Scientific Reports*, *8*(1), 1–12.

Hamer, G. L., Kitron, U. D., Goldberg, T. L., Brawn, J. D., Loss, S. R., Ruiz, M. O., Hayes, D. B., and Walker, E. D. (2009). Host selection by *Culex pipiens* mosquitoes and West Nile virus amplification. *The American Journal of Tropical Medicine and Hygiene*, *80*(2), 268–278.

Hanson, S. M. and Craig Jr G. B. (1994) Relationship between cold hardiness and supercooling point in *Aedes albopictus* eggs. *Journal of the American Mosquito Control Association 11*, 35–38.

Harrington, L. C., Edman, J. D., and Scott, T. W. (2001). Why do female *Aedes aegypti* (Diptera: Culicidae) feed preferentially and frequently on human blood? *Journal of Medical Entomology*, *38*(3), 411–422.

Harrington, L. C., Scott, T. W., Lerdthusnee, K., Coleman, R. C., Costero, A., Clark, G. G., Jones, J. J., Kitthawee, S., Kittayapong, P., Sithiprasasna, R., and Edman, J. D. (2005). Dispersal of the dengue vector *Aedes aegypti* within and between rural communities. The *American Journal of Tropical Medicine and Hygiene*, *72*(2), 209–220.

Harris, C., Kihonda, J., Lwetoijera, D., Dongus, S., Devine, G., and Majambere, S. (2011). A simple and efficient tool for trapping gravid *Anopheles* at breeding sites. *Parasites and Vectors*, *4*(1), 1–5.

Hayes, J. and Downs, T. D. (1980). Seasonal changes in an isolated population of *Culex pipiens quinquefasciatus* (Diptera: Culicidae): A time series analysis. *Journal of Medical Entomology*, *17*(1), 63–69.

Heisch, R. B., Nelson, G. S. and Furlong, M. (1959). Studies in filariasis in East Africa. 1. Filariasis on the Island of Pate, Kenya. *Transactions of the Royal Society of Tropical Medicine and Hygiene*, *53*(1),41–53.

Hernandez-Triana, L. M., Brugman, V. A., Prosser, S. W. J., Weland, C., Nikolova, N., Thorne, L., De Marco, M. F., Fooks, A. R., and Johnson, N. (2017). Molecular approaches for blood meal analysis and species identification of mosquitoes (Insecta: Diptera: Culicidae) in rural locations in southern England, United Kingdom. *Zootaxa*, *4250*(1), 067–076.

Hill, S. R., Hansson, B. S., and Ignell, R. (2009). Characterization of antennal trichoid sensilla from female southern house mosquito, *Culex quinquefasciatus* Say. *Chemical Senses*, *34*(3), 231–252.

Hiriyan, J. and Tyagi, B. K. (2004). Cocoa pod (*Theobroma cacao*) – a potential breeding habitat of *Aedes albopictus* (Skuse) in dengue sensitive Kerala State, India. *Journal of American Mosquito Control Association*, *20*(3), 323–325.

Hoffmann, E. J., and Miller, J. R. (2003). Reassessment of the Role and Utility of Wind in suppression of Mosquito (Diptera: Culicidae) Host Finding: Stimulus Dilution Supported over Flight Limitation. *Journal of Medical Entomology*, *40*(5), 607–614.

Holmes, C. J., and Benoit, J. B. (2019). Biological adaptations associated with dehydration in mosquitoes. *Insects*, *10*(11), 375.

Honório, N. A. et al. (2003). "Dispersal of *Aedes aegypti* and *Aedes albopictus* (Diptera: Culicidae) in an urban endemic dengue area in the state of Rio de Janeiro, Brazil", *Memórias do Instituto Oswaldo Cruz*, *98*, 191–198.

Huestis, D. L., Dao, A., Diallo, M., Sanogo, Z. L., Samake, D., Yaro, A. S., Ousman, Y., Linton, Y. M., Krishna, A., Veru, L., Krajacich, B. J., Faiman, R., Florio, J., Chapman, J. W., Reynolds, D. R., Weetman, D., Mitchell, R., Donnelly, M. J., Talamas, E., Chamorro, L., Strobach, E., and Lehmann, T. (2019). Windborne long-distance migration of malaria mosquitoes in the Sahel. *Nature*, *574*(7778), 404–408.

Hunt, S. K., Galatowitsch, M. L., and McIntosh, A. R. (2017). Interactive effects of land use, temperature, and predators determine native and invasive mosquito distributions. *Freshwater Biology*, *62*(9), 1564–1577.

Imam, A. A., and Deeni, Y. (2014). Common types of *Anopheles gambiae* breeding habitats in north western Nigeria. *Journal of Innovative Research in Engineering and Science*, *4*, 496–504.

Janssen, N., Fernandez-Salas, I., Díaz González, E. E., Gaytan-Burns, A., Medina-de la Garza, C. E., Sanchez-Casas, R. M., Börstler, J., Cadar, D., Schmidt-Chanasit, J., and Jöst, H. (2015). Mammalophilic feeding behaviour of *Culex quinquefasciatus* mosquitoes collected in the cities of Chetumal and Cancun, Yucatán Peninsula, Mexico. *Tropical Medicine & International Health*, *20*(11), 1488–1491.

Jeffries, M. (1988). Individual vulnerability to predation: The effect of alternative prey types. *Freshwater Biology*, *19*(1),49–56.

Jemal, Y., and Al-Thukair, A. A. (2018). Combining GIS application and climatic factors for mosquito control in Eastern Province, Saudi Arabia. Saudi. *Journal of Biological Sciences*, *25*, 1593–1602.

Jones, M. D., Hill, M., and Hope, A. M. (1967). The circadian flight activity of the mosquito *Anopheles gambiae*: Phase setting by the light régime. *The Journal of Experimental Biology*, *47*(3), 503–511.

Joshi, H., Vasantha, K., Subbarao, S. K. and Sharma, V. P. (1988). Host feeding patterns of *Anopheles culicifacies* species A and B. *Journal of the American Mosquito Control Association*, *4*(3), 248–251.

Juliano, S. A. (2009). Species interactions among larval mosquitoes: Context dependence across habitat gradients. *Annual Review of Entomology*, *54*, 37–56.

Juliano, S. A., and Lounibos, L. P. (2005). Ecology of invasive mosquitoes: Effects on resident species and on human health. *Ecology Letters*, *8*(5), 558–74.

Juliano, S. A., O'Meara, G. F., Morrill, J. R., and Cutwa, M. M. (2002). Desiccation and thermal tolerance of eggs and the coexistence of competing mosquitoes. *Oecologia*, *130*(3), 458–469.

Kaufmann, C., Collins, L., and Brown, M. (2013). Influence of Age and Nutritional Status on Flight Performance of the Asian Tiger Mosquito *Aedes albopictus* (Diptera: Culicidae). *Insects*, *4*(3), 404–412.

Keiser J., Utzinger J., De Castro M. C., Smith T. A., Tanner, M., and Singer, B. H. (2004). Urbanization in sub-Saharan Africa and implication for malaria control. *The Intolerable Burden of Malaria II: What's New, What's Needed: Supplement to Volume 71 (2) of the American Journal of Tropical Medicine and Hygiene*.

Kennedy, J. S. (1942). On water-finding and oviposition by captive mosquitoes. *Bulletin of Entomological Research*, *32*, 279–301.

Kent, R. J. (2009). Molecular methods for arthropod bloodmeal identification and applications to ecological and vector-borne disease studies. *Molecular Ecology Resources*, *9*(1), 4–18.

Kesavaraju, B., Damal, K., and Juliano, S. A. (2008). Do natural container habitats impede invader dominance? Predator-mediated coexistence of invasive and native container-dwelling mosquitoes. *Oecologia*, *155*(3), 631–639.

Kessler, S., and Guerin, P. M. (2008). Responses of *Anopheles gambiae*, *Anopheles stephensi*, *Aedes aegypti*, and *Culex pipiens* mosquitoes (Diptera: Culicidae) to cool and humid refugium conditions. *Journal of Vector Ecology*, *33*, 145–149.

Khan, A. A., and Maibach, H. I. (1970). A Study of the Probing Response of *Aedes aegypti* Effect of Nutrition on Probing. *Journal of Economic Entomology*, *63*(3), 974–976.

Khan, A. A., and Maibach, H. I. (1971). A study of the probing response of *Aedes aegypti*. 4. Effect of dry and moist heat on probing. *Journal of Economic Entomology*, *64*(2), 442–443.

Khan, M. A., Elhossary, S., Khan, I. A., Al Zahrani, M. H., Al Zahrani, F. S., and Al Bashri, F. M. (2018). The impact of climatic variables with GIS application on the abundance of medically important mosquitoes (Diptera: Culicidae) in Jeddah, Saudi Arabia. *International Journal of Mosquito Research*, *5*, 12–18.

Kilpatrick, A. M., Kramer, L. D., Jones, M. J., Marra, P. P., Daszak, P., and Fonseca, D. M. (2007). Genetic influences on mosquito feeding behavior and the emergence of zoonotic pathogens. *The American Journal of Tropical Medicine and Hygiene*, *77*(4), 667–671.

Kim, D. M., Noh, B. E., Heo, J., Lee, W. G., Yang, S. C., and Lee, D. K. (2018). Seasonal prevalence of mosquitoes collected from light traps in Gyeongsangnam province, Republic of Korea (2013–2014). *Entomological Research*, *48*(5), 437–445.

Kirchgatter, K., Tubaki, R. M., Malafronte, R. D. S., Alves, I. C., Lima, G. F. M. D. C., Guimarães, L. D. O., Zampaulo, R. D. A., and Wunderlich, G. (2014). *Anopheles* (Kerteszia) *cruzii* (Diptera: Culicidae) in peridomiciliary area during asymptomatic malaria transmission in the Atlantic Forest: Molecular identification of blood-meal sources indicates humans as primary intermediate hosts. *Revista do Instituto de Medicina Tropical de São Paulo*, *56*, 403–409.

Klassen, W., and Hocking, B. (1964). The influence of a deep river valley system on the dispersal of *Aedes* mosquitos. *Bulletin of Entomological Research*, *55*(2), 289–304.

Klinkenberg, E., McCall, P. J., Wilson, M. D., Amerasinghe, F. P. and Donnelly, M. J. (2008). Impact of urban agriculture on malaria vectors in Accra, Ghana. *Malaria Journal*, *7*, 1–9.

Klowden, M. J. and Dutro, S. M. (1990). Effects of carbohydrate ingestion on the pre-oviposition behavior of the mosquito *Aedes aegypti* (L.). *Bulletin of the Society for Vector Ecology*, *15*(1), 59–62.

Ko, A. I., Galvão Reis, M., Ribeiro Dourado, C. M., Johnson, W. D., Jr., and Riley, L. W. (1999). Urban epidemic of severe leptospirosis in Brazil. Salvador Leptospirosis Study Group. *Lancet*, *354*, 820–825.

Kolimenakis, A., Heinz, S., Wilson, M. L., Winkler, V., Yakob, L., Michaelakis, A., Papachristos, D., Richardson, C., & Horstick, O. (2021). The role of urbanisation in the spread of *Aedes* mosquitoes and the diseases they transmit—A systematic review. *PLoS Neglected Tropical Diseases*, *15*(9), e0009631.

Kothera, L., Mutebi, J. P., Kenney, J. L., Saxton-Shaw, K., Ward, M. P., and Savage, H. M. (2020). Bloodmeal, host selection, and genetic admixture analyses of *Culex pipiens* complex (Diptera: Culicidae) mosquitoes in Chicago, IL. *Journal of Medical Entomology*, *57*(1),78–87.

Kraaijeveld, A. R., and Godfray, H. C. J. (1997). Trade-off between parasitoid resistance and larval competitive ability in *Drosophila melanogaster*. *Nature*, *389*(6648), 278–280.

Kraemer, M. U., Sinka, M. E., Duda, K. A., Mylne, A. Q., Shearer, F. M., Barker, C. M., Moore, C. G., Carvalho, R. G., Coelho, G. E., Van Bortel, W. and Hendrickx, G. (2015). The global distribution of the arbovirus vectors *Aedes aegypti* and *Ae. albopictus*. *elife*, *4*, e08347.

Kroeger, I., Liess, M., and Duquesne, S. (2014). Temporal and spatial habitat preferences and biotic interactions between mosquito larvae and antagonistic crustaceans in the field. *Journal of Vector Ecology*, *39*(1), 103–111.

Kumar, P., Pisudde, P., and Sarthi, P. P. (2022). Meteorological linkage of Malaria cases in the eastern state of India. *The Journal of Climate Change and Health*, *5*, 100064.

Laurence, B. R. and Pickett, J. A. (1982). Erythro-6-acetoxy-5-hexadecanolide, the major component of a mosquito oviposition attractant pheromone. *Journal of the Chemical Society, Chemical Communications*, (1), 59–60.

Lefèvre, T., Gouagna, L. C., Dabiré, K. R., Elguero, E., Fontenille, D., Renaud, F., Costantini, C., and Thomas, F. (2009). Beyond nature and nurture: Phenotypic plasticity in blood-feeding behaviour of *Anopheles gambiae s.s.* when humans are not readily accessible. *The American Journal of Tropical Medicine and Hygiene*, *81*(6), 1023–1029.

Lehmann, T., Dao, A., Yaro, A. S., Huestis, D. L., Diallo, M., Timbiné, S., Kassogué, Y., Adamou, A., Traoré, A. I., and Samaké, D. (2014). Phenotypic divergence among the members of the African malaria mosquitoes and strategies of persistence throughout the dry season. *Malaria Journal*, *13*(1), 1–1.

Leming, M. T., Rund, S. S., Behura, S. K., Duffield, G. E., and O'Tousa, J. E. (2014). A database of circadian and diel rhythmic gene expression in the yellow fever mosquito *Aedes aegypti*. *BMC Genomics*, *15*(1), 1–9.

Lie, K. J., and Amaliah, S. (1960). Wuchereria bancrofti infection in Djakarta, Indonesia, a study of some factors influencing its transmission. *Indian Journal of Malariology*, *14*(4), 339–352.

Lines, J. D., Lyimo, E. O., and Curtis, C. F. (1986). Mixing of indoor-and outdoor-resting adults of *Anopheles gambiae* Giles sl and *A. funestus* Giles (Diptera: Culicidae) in coastal Tanzania. *Bulletin of Entomological Research*, *76*(1), 171–178.

Liu, B., Gao, X., Ma, J., Jiao, Z., Xiao, J., and Wang, H. (2018). Influence of host and environmental factors on the distribution of the Japanese Encephalitis Vector *Culex tritaeniorhynchus* in China. *International Journal of Environmental Research and Public Health*, *15*(9), 1848.

Liu, X., Baimaciwang, Wu H., Pengcuociren, Guo Y., Cirenwangla, Ren D., Danzenggongga, Dazhen, Yang, J., Zhaxisangmu, Li J., Cirendeji, Zhao N., Sun, J., Li, J., Wang, J., Cirendunzhu, and Liu, Q. (2019). Breeding site characteristics and associated factors of *Culex pipiens* Complex in Lhasa, Tibet, P. R. China. *International Journal of Environmental Research and Public Health*, *16*(8), 1407.

Longbottom, J., Browne, A. J., Pigott, D. M., Sinka, M. E., Golding, N., Hay, S. I., Moyes, C. L., and Shearer, F. M. (2017). Mapping the spatial distribution of the Japanese encephalitis vector, *Culex tritaeniorhynchus* Giles, 1901 (Diptera: Culicidae) within areas of Japanese encephalitis risk. *Parasites and Vectors*, *10*(1), 148.

Lounibos, L. P., Suárez, S., Menéndez, Z., Nishimura, N., Escher, R. L., O'Connell, S. M., and Rey, J. R. (2002). Does temperature affect the outcome of larval competition between *Aedes aegypti* and *Aedes albopictus*? *Journal of Vector Ecology*, *27*, 86–95.

Lunde, T. M., Bayoh, M. N., and Lindtjørn, B. (2013). How malaria models relate temperature to malaria transmission. *Parasites and Vectors*, *6*(1), 1–10.

Lyimo, E. O. and Koella, J. C. (1992). Relationship between body size of adult *Anopheles gambiae* sl and infection with the malaria parasite *Plasmodium falciparum*. *Parasitology*, *104*(2), 233–237.

Lyons C. L., Coetzee, M., and Chown, S. L. (2013). Stable and fluctuating temperature effects on the development rate and survival of two malaria vectors, *Anopheles arabiensis* and *Anopheles funestus*. *Parasites and Vectors*, *6*, 1–9.

Madeira, N. G., Macharelli, C. A., and Carvalho, L. R. (2002). Variation of the oviposition preferences of *Aedes aegypti* in function of substratum and humidity. *Memorias do Instituto Oswaldo Cruz*, *97*(3), 415–420

Mamai, W., Simard, F., Couret, D., Ouédraogo, G. A., Renault, D., Dabiré, K. R., and Mouline, K. (2016). Monitoring dry season persistence of *Anopheles gambiae* sl populations in a contained semi-field system in south-western Burkina Faso, West Africa. *Journal of Medical Entomology*, *53*(1), 130–138.

Maquart, P. O., Chann, L., & Boyer, S. (2022). *Culex vishnui* (Diptera: Culicidae): An overlooked vector of arboviruses in South-East Asia. *Journal of Medical Entomology*, *59*(4), 1144–1153.

Mayagaya, V. S., Nkwengulila, G., Lyimo, I. N., Kihonda, J., Mtambala, H., Ngonyani, H., Russell, T. L., and Ferguson, H. M. (2015). The impact of livestock on the abundance, resting behaviour and sporozoite rate of malaria vectors in southern Tanzania. *Malaria Journal*, *14*, 1–14.

Mboera, L. E. G., Knols, B. G. J., Braks, M. A. H. and Takken, W. (2000). Comparison of carbon dioxide-baited trapping systems for sampling outdoor mosquito populations in Tanzania. *Medical and Veterinary Entomology*, *14*(3), 257–263.

Mburu, M. M., Zembere, K., Mzilahowa, T., Terlouw, A. D., Malenga, T., van den Berg, H., Takken, W., & McCann, R. S. (2021). Impact of cattle on the abundance of indoor and outdoor resting malaria vectors in southern Malawi. *Malaria Journal*, *20*(1), 1–11.

McBride, C. S. (2016). Genes and odours underlying the recent evolution of mosquito preference for humans. *Current Biology*, *26*(1), 41–46.

McCall, P. J., Mosha, F. W., Njunwa, K. J., and Sherlock, K. (2001). Evidence for memorized site-fidelity in *Anopheles arabiensis*. *Transactions of the Royal Society of Tropical Medicine and Hygiene*, *95*(6), 587–590.

McCrae A. W. (1984). Oviposition by African malaria vector mosquitoes. II. Effects of site tone, water type and conspecific immatures on target selection by freshwater Anopheles gambiae Giles, sensu lato. *Annals of Tropical Medicine and Parasitology*, *78*(3), 307–318.

McMeniman, C. J., Corfas, R. A., Matthews, B. J., Ritchie, S. A., and Vosshall, L. B. (2014). Multimodal integration of carbon dioxide and other sensory cues drives mosquito attraction to humans. *Cell*, *156*(5), 1060–1071.

Medeiros, M. C., Boothe, E. C., Roark, E. B., and Hamer, G. L. (2017). Dispersal of male and female *Culex quinquefasciatus* and *Aedes albopictus* mosquitoes using stable isotope enrichment. *PLoS Neglected Tropical Diseases*, *11*(1), e0005347.

Medlock, J. M., and Snow, K. R. (2008). Natural predators and parasites of British mosquitoes – a review. *European Mosquito Bulletin*, *25*, 1–11.

Menda, G., Uhr, J. H., Wyttenbach, R. A., Vermeylen, F. M., Smith, D. M., Harrington, L. C., and Hoy, R. R. (2013). Associative learning in the dengue vector mosquito, *Aedes aegypti*: Avoidance of a previously attractive odour or surface colour that is paired with an aversive stimulus. *Journal of Experimental Biology*, *216*(2), 218–223.

Michaelakis, A., Papachristos, D., Kimbaris, A., Koliopoulos, G., Giatropoulos, A. and Polissiou, M. G. (2009). Citrus essential oils and four enantiomeric pinenes against *Culex pipiens* (Diptera: Culicidae). *Parasitology research*, *105*, 769–773.

Moiroux, N., Gomez, M. B., Pennetier, C., Elanga, E., Djènontin, A., Chandre, F., Djègbé, I., Guis, H., and Corbel, V. (2012). Changes in *Anopheles* funestus biting behavior following universal coverage of long-lasting insecticidal nets in Benin. *The Journal of Infectious Diseases*, *206*, 1622–1629.

Mondal, R., Devi, N. P., and Jauhari, R. K. (2014). Occurrence of *Aedes* mosquitoes (Diptera: Culicidae) in urban areas of Doon valley, (Uttarakhand), India. *Modern Parasitology*, 28, 255–262.

Mondal, R., Devi, N. P. and Bhattacharya, S. (2021). Seasonal prevalence and host preference of some medically important *Aedes* species of Doon Valley, India. *Journal of Communicable Diseases (E-ISSN: 2581-351X & P-ISSN: 0019-5138)*, *53*(3), 96–103.

Mordecai, E. A., Caldwell, J. M., Grossman, M. K., Lippi, C. A., Johnson, L. R., Neira, M., Rohr, J. R., Ryan, S. J., Savage, V., Shocket, M. S., Sippy, R., Stewart Ibarra, A. M., Thomas, M. B., and Villena, O. (2019). Thermal biology of mosquito-borne disease. *Ecology Letters*, *22*, 1690–1708.

Mordecai, E. A., Ryan, S. J., Caldwell, J. M., Shah, M. M., and LaBeaud, A. D. (2020). Climate change could shift disease burden from malaria to arboviruses in Africa. *The Lancet Planetary Health*, *4*, e416–e423.

Mucci, L. F., Cardoso, R. P., Paula, M. B. D., Scandar, S. A. S., Pacchioni, M. L., Fernandes, A., and Consales, C. A. (2015). Feeding habits of mosquitoes (Diptera: Culicidae) in an area of sylvatic transmission of yellow fever in the state of São Paulo, Brazil. *Journal of Venomous Animals and Toxins including Tropical Diseases*, *21*, 1–10.

Müller, G. C., Beier, J. C., Traore, S. F., Toure, M. B., Traore, M. M., Bah, S., Doumbia, S., and Schlein, Y. (2010). Successful field trial of attractive toxic sugar bait (ATSB) plant-spraying methods against malaria vectors in the *Anopheles gambiae* complex in Mali, West Africa. *Malaria Journal*, *9*(1), 1–7.

Munga, S., Minakawa, N., Zhou, G., Barrack, O. J., Githeko, A. K., and Yan, G. (2005). Oviposition site preference and egg hatchability of *Anopheles gambiae*: Effects of land cover types. *Journal of Medical Entomology*, *42*,993–997.

Munirathinam, A., Krishnamoorthi, G., Baskaran, R., Govindarajan, A., Venkatesh, A., and Tyagi, B. K. (2014). Mosquito species biodiversity in Phytotelmata from Western Ghats, South India. *Halteres*, *5*, 56–3.

Mwingira, V., Mboera, L. E., Dicke, M., and Takken, W. (2020). Exploiting the chemical ecology of mosquito oviposition behavior in mosquito surveillance and control: A review. *Journal of Vector Ecology*, *45*(2), 155–179.

Nagpal, B. N., Gupta, S. K., Shamim, A., Vikram, K., Srivastava, A., Tuli, N. R., Saxena, R., Singh, H., Singh, V. P., Bhagat, V. N., Yadav, N. K., and Valecha, N. (2016). Control of *Aedes aegypti* Breeding: A novel intervention for prevention and control of dengue in an endemic zone of Delhi, India. *PloS One*, *11*(12), e0166768.

Navarro-Silva, M. A., Marques, F. A. and Duque, L. J. E. (2009). Review of semiochemicals that mediate the oviposition of mosquitoes: A possible sustainable tool for the control and monitoring of Culicidae. *Revista Brasileira de Entomologia*, *53*, 1–6.

Nayar K. (1985). Bionomics and physiology of *Aedes taeniorhynchus* and *Aedes sollicitians* the salt marsh mosquitoes of Florida. Bulletin 852: Agricultural Experiment Station, Institute of Food and Agricultural Sciences, University of Florida.

Ndenga, B. A., Mulaya, N. L., Musaki, S. K., Shiroko, J. N., Dongus, S. and Fillinger, U. (2016). Malaria vectors and their blood-meal sources in an area of high bed net ownership in the western Kenya highlands. *Malaria Journal*, *15*(1), 1–10.

Neafsey, D. E., Waterhouse, R. M., Abai, M. R., Aganezov, S. S., Alekseyev, M. A., Allen, J. E., Amon, J., Arcà, B., Arensburger, P., Artemov, G. and Assour, L. A. (2015). Highly evolvable malaria vectors: The genomes of 16 *Anopheles* mosquitoes. *Science*, *347*(6217), 1258522.

Nikbakhtzadeh, M. R., Terbot, J. W., Otienoburu, P. E., and Foster, W. A. (2014). Olfactory basis of floral preference of the malaria vector Anopheles gambiae (Diptera: Culicidae) among common African plants. *Journal of Vector Ecology*, *39*(2), 372–383.

Nixon, C. P., Nixon, C. E., Arsyad, D. S., Chand, K., Yudhaputri, F. A., Sumarto, W., Wangsamuda, S., Asih, P. B., Marantina, S. S., Wahid, I., Han, G., Friedman, J. F., Bangs, M. J., Syafruddin, D., and Baird, J. K. (2014). Distance to *Anopheles sundaicus* larval habitats dominant among risk factors for parasitemia in meso-endemic Southwest Sumba, Indonesia. *Pathogens and Global Health*, *108*(8), 369–380.

O'Gower, A. K. (1963). Environmental stimuli and the oviposition behavior of *Aedes aegypti* var. Queenslandensis Theobald (Diptera: Culicidae). *Animal Behavior*, *11*, 189–197

Obsomer, V., Defourny, P. and Coosemans, M. (2007). The *Anopheles dirus* complex: Spatial distribution and environmental drivers. *Malaria Journal*, *6*(1), 1–16.

OECD. (2018). Safety Assessment of Transgenic Organisms in the Environment, Volume 8: OECD Consensus Document of the Biology of Mosquito *Aedes aegypti*. Harmonisation of Regulatory Oversight in Biotechnology, OECD Publishing, Paris.

Ogunba, E. O. (1971). Observations on *Culex pipiens* fatigans in Ibadan, western Nigeria. *Annals of Tropical Medicine & Parasitology*, *65*(3), 399–402.

Okal, M. N., Francis, B., Herrera-Varela, M., Fillinger, U., and Lindsay, S. W. (2013). Water vapour is a pre-oviposition attractant for the malaria vector *Anopheles gambiae* sensu stricto. *Malaria Journal*, *12*(1), 1–8.

Okiwelu, S., and Noutcha, A. E. (2012). Breeding sites of Culex quinquefasciatus (Say) during the Rainy Season in Rural Lowland Rainforest, Rivers State, Nigeria. *Public Health Research*, *2*, 64–68.

Okumu, F. O., Govella, N. J., Moore, S. J., Chitnis, N. and Killeen, G. F. (2010). Potential benefits, limitations and target product-profiles of odour-baited mosquito traps for malaria control in Africa. *PLoS One*, *5*(7), e11573.

Okwa, O. O., and Adetutu, S. A. (2018). Oviposition and breeding water sites preferences of mosquitoes within Ojo area, Lagos state, Nigeria. *Biomedical Journal*, *1*, 7.

Olagbemiro, T. O., Birkett, M. A. and Pickett, J. A. (2004). Laboratory and field responses of the mosquito, *Culex quinquefasciatus*, to plant-derived *Culex* spp. oviposition pheromone and the oviposition cue skatole. *Journal of Chemical Ecology*, *30*, 965–976.

Otieno, W. A., Mutinga, M. J. and Laurence, B. R. (1988). Response of an East African mosquito, *Culex pipiens fatigans* Wied to synthetic oviposition attractant pheromone in the laboratory at Mbita Point, Kenya, East Africa. *International Journal of Tropical Insect Science*, *9*(2), 261–262.

Paaijmans, K. P., and Thomas, M. B. (2011). The influence of mosquito resting behaviour and associated microclimate for malaria risk. *Malaria Journal*, *10*(1), 1–7.

Pandian, R. S. and Chandrashekaran, M. K. (1980). Rhythms in the biting behaviour of a mosquito Armigeres subalbatus. *Oecologia*, *47*, 89–95.

Pandian, R. S. and Manoharan, A. C. (1994). Survey of the mosquitoes of Avanyapuram & Sewage fodder farm area of Madurai Corporation in Tamil Nadu, India. *Geobios New Reports*, *13*, 109–113.

Paramasivan, R., Philip, S. P. and Selvaraj, P. R. (2015). Biting rhythm of vector mosquitoes in a rural ecosystem of South India. *International Journal of Mosquito Research*, *2*(3), 106–113.

Parham, P. E. and Michael, E. (2010). Modeling the effects of weather and climate change on malaria transmission. *Environ Health Perspec*, *118*, 620–626.

Pates, H. and Curtis, C. (2005). Mosquito behavior and vector control. *Annual Reviews Entomology*, *50*, 53–70.

Peach, D. A., and Gries, G. (2020). Mosquito phytophagy–sources exploited, ecological function, and evolutionary transition to haematophagy. *Entomologia Experimentalis et Applicata*, *168*(2), 120–136.

Petric, D., Bellini, R., Scholte, E.-J., Rakotoarivony, L. M., and Schaffner, F. (2014). Monitoring population and environmental parameters of invasive mosquito species in Europe. *Parasites and Vectors*, *7*(1), 1–14.

Pickett, J. A. and Woodcock, C. M. (1996). The role of mosquito olfaction in oviposition site location and in the avoidance of unsuitable hosts, pp. 109–123, *in* R. G. Bock and G. Cardew (eds.). *Olfaction in Mosquito-Host Interactions*. Wiley, Chichester, UK.

Piermarini, P. M. (2016). Chapter 13: Renal excretory processes in mosquitoes. In *Advances in Insect Physiology*, *51*, 393–433. https://doi.org/10.1016/bs.aiip.2016.04.003

Potter, C. J. (2019). Olfaction: Mosquitoes love your acid odors. *Current Biology*, *29*(8), 282–284.

Prasad, K. I., Govindarajan, R., and Sreepada, K. S. (2021). Seasonal diversity of mosquito species in Dakshina Kannada district, Karnataka, India. *Journal of Vector Borne Diseases*, *58*(2), 119.

Praveen, G., Tyagi, B. K., Joe, N., and Thakur, M. B. (2016). An outbreak investigation of dengue fever in the coastal areas of Nagapattinam, Thiruvarur and Thanjavur districts in Tamil Nadu, India during 2012. *International Journal of Mosquito Research*, *3*(4), 39–47.

Rajagopalan, P. K., Das, P. K., Kalyanasundaram, M., Tyagi, B. K., Arunachalam, N., Somachary, N., Reddy, C. B. S., and Reddy, C. M. R. (1987). *Bangalore Mosquito Control Master Plan*. Vector Control Research Centre, Pondicherry, 309.

Rajavel, A. R., Natarajan, R., and Vaidyanathan, K. (2005b). Mosquito of mangrove forests of India: Part I-Bhitarkanika, Orissa. *Journal of American Mosquito Control Association*, *21*, 131–135.

Rajavel, A. R., Natarajan, R., and Vaidyanathan, K. (2005c). Mosquito of mangrove forests of India: Part II-Sundarbans, West Bengal. *Journal of American Mosquito Control Association*, *21*, 136–138.

Rajavel, A. R., Natarajan, R., Vaidyanathan, K., and Saniya, N. P. (2005a). Mosquito collections in the Jeypore hill trcts of Orissa, India, with notes on three new country records. *Culex (Lophoceraomyia) pilifemoralis. Culex (Lophoceraomyia) wilfredi*, and *Heizmannia (Heizmannia) Chengi. Journal of American Mosquito Control Association, 21*, 121–127.

Ramasamy, R., and Surendran, S. N. (2012). Global climate change and its potential impact on disease transmission by salinity-tolerant mosquito vectors in coastal zones. *Frontiers in Physiology, 3*, 198.

Ranasinghe, H. A. K. and Amarasinghe, L. D. (2020). Naturally occurring microbiota associated with mosquito breeding habitats and their effects on mosquito larvae. *BioMed Research International, 2020*, 4065315. https://doi.org/10.1155/2020/4065315

Ranson, H., Paton, M. G., Jensen, B., McCarroll, L., Vaughan, A., Hogan, J. R., Hemingway, J., and Collins, F. H. (2004). Genetic mapping of genes conferring permethrin resistance in the malaria vector, Anopheles gambiae. *Insect Molecular Biology, 13*(4), 379–386.

Rao, T. R., (1984). *The Anopheline of India* (Revised Edition.), Malaria Research Centre, Indian Council of Medical Research, New Delhi, 518 pp.

Reeves, L. E., Holderman, C. J., Blosser, E. M., Gillett-Kaufman, J. L., Kawahara, A. Y., Kaufman, P. E. and Burkett-Cadena, N. D. (2018). Identification of Uranotaenia sapphirina as a specialist of annelids broadens known mosquito host use patterns. *Communications Biology, 1*(1), 1–8.

Reidenbach, K. R., Cheng, C., Liu, F., Liu, C., Besansky, N. J., and Syed, Z. (2014). Cuticular differences associated with aridity acclimation in African malaria vectors carrying alternative arrangements of inversion 2La. *Parasites and Vectors, 7*(1), 1–13.

Reiner, R. C., Smith, D. L., and Gething, P. W. (2015). Climate change, urbanization and disease: Summer in the city. *Transactions of the Royal Society of Tropical Medicine and Hygine, 109*, 171–172.

Reinhold, J. M., Lazzari, C. R., and Lahondère, C. (2018). Effects of the environmental temperature on *Aedes aegypti* and *Aedes albopictus* mosquitoes: A review. *Insects, 9*(4), 158.

Reiter, P. (2001). Climate change and mosquito-borne disease. *Environmental Health Perspective, 109* Suppl 1, 141–61.

Reiter, P. (2007). Oviposition, dispersal and survival in *Aedes aegypti*: Implications for the efficacy of controls strategies. *Vector-Borne Zoonot Diseases, 7*, 261–273.

Reiter, P., Amador, M. A., Anderson, R. A., and Clark, G. G. (1995). Dispersal of *Aedes aegypti* in the urban area after blood feeding as demonstrated by rubidium-marked eggs. *American Journal of Tropical Medicine and Hygine, 52*, 177–179.

Ricklefs, R. E., and Miller, G. L. (2000). *Ecology*. Macmillan.

Rinehart, J. P., Robich, R. M. and Denlinger, D. L. (2006). Enhanced cold and desiccation tolerance in diapausing adults of *Culex pipiens*, and a role for Hsp70 in response to cold shock but not as a component of the diapause program. *Journal of Medical Entomology, 43*(4), 713–722.

Ripa, J., Olofsson, H., and Jonzén, N. (2010). What is bet-hedging, really? *Proceedings Biological Sciences, 277*(1685), 1153–1154.

Ritchie, S. A., and Laidlaw-Bell, C. (1994). Do fish repel oviposition by *Aedes taeniorhynchus*? *Journal of American Mosquito Control Association, 10*, 380–84.

Ritchie, S. A. (1993) *Aedes vigilax*: The taeni of Oz. *Arbovirus Research in Australia 6*, 52–58.

Ritchie, S. A. (2014). Dengue Vector Bionomics: Why Aedes aegypti is Such a Good Vector, (In eds. eds D. J. Gubler et al.) *CAB International Dengue and Dengue Hemorrhagic Fever*, 2nd Edition.

Robert V., Macintyre, K., Keating, J., Trape, J. F., Duchemin, J. B., Warren, M., and Beier, J. C. (2003). Malaria transmission in urban sub-Saharan Africa. *American Journal of Tropical Medicine and Hygine, 68*, 169–176.

Roberts, G. (1995). Salt-marsh crustaceans, *Gammarus duebeni* and *Palaemonetes varians* as predators of mosquito larvae and their reaction to *Bacillus thuringiensis* subsp. *israelensis*. *Biocontrol Science and Technology, 5*(3), 37–41.

Rohani, A., Zamree, I., Ali, W. N. W. M., Hadi, A. A., Asmad, M., Lubim, D., Nor, Z. M. and Lim, L. H. (2013). Nocturnal man biting habits of mosquito species in Serian, Sarawak, Malaysia. *Advances in Entomology, 1*(2), 42–49.

Rowland, M., Durrani, N., Kenward, M., Mohammed, N., Urahman, H., and Hewitt, S. (2001). Control of malaria in Pakistan by applying deltamethrin insecticide to cattle: A community-randomised trial. *The Lancet, 357*(9271), 1837–1841.

Rowley, W. A., and Graham, C. L. (1968). The effect of age on the flight performance of female *Aedes aegypti* mosquitos. *Journal of Insect Physiology, 14*(5), 719–728.

Roy, A., Ansari, M. A. and Sharma, V. P. (1991). Feeding behavior patterns of anophelines from Uttar Pradesh and Gujarat states of India. *Journal of the American Mosquito Control Association, 7*(1), 11–15.

Rund, S. S., Lee, S. J., Bush, B. R., and Duffield, G. E. (2012). Strain- and sex-specific differences in daily flight activity and the circadian clock of *Anopheles gambiae* mosquitoes. *Journal of Insect Physiology*, *58*(12), 1609–1619.

Russell, P. F., and Rao, T. R. (1942). On relation of mechanical obstruction and shade to oviposition of *Anopheles culicifacies*. *Journal of Experimental Zoology*, *91*, 303–329.

Russell, T. L., Govella, N. J., Azizi, S., Drakeley, C. J., Kachur, S. P., and Killeen, G. F. (2011). Increased proportions of outdoor feeding among residual malaria vector populations following increased use of insecticide-treated nets in rural Tanzania. *Malaria Journal*, *10*(1), 1–10.

Ryan, S. J., Carlson, C. J., Mordecai, E. A., and Johnson, L. R. (2019). Global expansion and redistribution of *Aedes*-borne virus transmission risk with climate change. *PLOS Neglected Tropical Diseases*, *13*, e0007213.

Ryan, S. J., McNally, A., Johnson, L. R., Mordecai, E. A., Ben-Horin, T., Paaijmans, K., and Lafferty, K. D. (2015). Mapping physiological suitability limits for malaria in Africa under climate change, *Vector-Borne and Zoonotic Diseases*, *15*, 718–725.

Sadanandane, C., Gunasekaran, K., Jambulingam, P., and Das, P. K. (1993). Studies on dispersal of malaria vectors in a hilly tract of Koraput District, Orissa State, India. *The Southeast Asian Journal of Tropical Medicine and Public Health*, 24(3), 508–512.

Sahu, S. S., Gunasekaran, K., Vanamail, P. and Jambulingam, P. (2011). Seasonal prevalence & resting behaviour of *Anopheles minimus* Theobald & An. fluviatilis James (Diptera: Culicidae) in east-central India. *The Indian Journal of Medical Research*, *133*(6), 655.

Santos, C. S., Pie, M. R., da Rocha, T. C., and Navarro-Silva, M. A. (2019). Molecular identification of blood meals in mosquitoes (Diptera, Culicidae) in urban and forested habitats in southern Brazil. *PloS One*, *14*(2), e0212517.

Santos-Vega, M., Bouma, M. J., Kohli, V., and Pascual, M. (2016). Population density, climate variables and poverty synergistically structure spatial risk in urban malaria in India. *PLoS Neglected Tropical Diseases*, *10*(12), e0005155.

Santos-Vega, M., Martinez, P. P., Vaishnav, K. G., Kohli, V., Desai, V., Bouma, M. J., and Pascual, M. (2022). The neglected role of relative humidity in the interannual variability of urban malaria in Indian cities. *Nature Communications*, *13*(1), 1–9.

Sawadogo, S. P., Costantini, C., Pennetier, C., Diabaté, A., Gibson, G., and Dabiré, R. K. (2013). Differences in timing of mating swarms in sympatric populations of *Anopheles coluzzii* and *Anopheles gambiae* s.s. (formerly *An. gambiae* M and S molecular forms) in Burkina Faso, West Africa. *Parasites and Vectors*, *6*, 275.

Sbarbati, A., and Osculati, F. (2006). Allelochemical communication in vertebrates: Kairomones, allomones and synomones. *Cells Tissues Organs*, *183*, 206–219.

Schäfer, M., Storch, V., Kaiser, A., Beck, M., and Becker, N. (1997). Dispersal behavior of adult snow melt mosquitoes in the Upper Rhine Valley, Germany. *Journal of Vector Ecology: Journal of the Society for Vector Ecology*, *22*(1), 1–5.

Sedda, L., McCann, R. S., Kabaghe, A. N., Gowelo, S., Mburu, M. M., Tizifa, T. A., Chipeta, M. G., van den Berg, H., Takken, W., van Vugt, M., Phiri, K. S., Cain, R., Tangena, J. A., and Jones, C. M. (2022). Hotspots and super-spreaders: Modelling fine-scale malaria parasite transmission using mosquito flight behaviour. *PLoS Pathogen*, *6*, 18(7), e1010622.

Seenivasagan, T., Sharma K. R.,Shrivastava, A, Parashar, B. D., Pant, S. C., and Prakash S. (2009). Surface morphology and morphometric analysis of sensilla of Asian tiger mosquito, *Aedes albopictus* (Skuse): An SEM investigation. *Journal of Vector Borne Disease*, *46*, 125–135.

Selvan, S., and Jebanesan, A. (2014). Survey and epidemiology of tree hole breeding mosquitoes in Annamalai University campus, Tamil Nadu, India. *International Journal of Current Research*, *6*, 6462–6465.

Service, M. W. (1963). The ecology of the mosquitos of the northern Guinea savannah of Nigeria. *Bulletin of Entomological Research*, *54*, 601–632.

Service, M. W. (1980). Effects of wind on the behaviour and distribution of mosquitoes and blackflies. *International Journal of Biometeorology*, *24*(4), 347–353.

Service, M. W. (1997). Mosquito (Diptera: Culicidae) dispersal--the long and short of it. *Journal of Medical Entomology*, *34*(6), 579–588.

Silver, J. B. (2008). Measuring adult dispersal. In Service, M. W. and Silve, J. B., eds. *Mosquito Ecology Field Sampling Methods*, 3rd Edition. Springer, Netherlands: 1377–1424.

Simões, P. M., Ingham, R. A., Gibson, G., and Russell, I. J. (2016). A role for acoustic distortion in novel rapid frequency modulation behaviour in free-flying male mosquitoes. *The Journal of Experimental Biology*, *219*, 2039–2047.

Sinclair, B. J., Gibbs, A. G., and Roberts, S. P. (2007). Gene transcription during exposure to, and recovery from, cold and desiccation stress in *Drosophila melanogaster. Insect Molecular Biology*, *16*(4), 435–443.

Singh, B., Yidris, N., Basri, A. A., Pai, R., and Ahmad, K. A. (2021). Study of mosquito aerodynamics for imitation as a small robot and flight in a low-density environment. *Micromachines*, *12*(5), 511.

Singh, N., Singh, O. P., and Soan, V. (1989). Mosquito breeding in rice fields and its role in malaria transmission in Mandla district, M.P. *Indian Journal of Malariology*, *26*(4), 191–198.

Sivan, A., Shriram, A. N., Sunish, I. P. and Vidhya, P. T. (2015). Studies on insecticide susceptibility of *Aedes aegypti* (Linn) and *Aedes albopictus* (Skuse) vectors of dengue and chikungunya in Andaman and Nicobar Islands, India. *Parasitology Research*, *114*, 4693–4702.

Smith, A., and Carter, I. D. (1984). International transportation of mosquitoes of public health importance. In: Laird M (ed) *Commerce and the spread of pests and disease vectors. Praegar Scientific*, New York, 1–17.

Smith, T., Maire, N., Ross, A., Penny, M., Chitnis, N., Schapira, A., Studer, A., Genton, B., Lengeler, C., Tediosi, F., and De Savigny, D. (2008). Towards a comprehensive simulation model of malaria epidemiology and control. *Parasitology*, *135*(13), 1507–1516.

Snow, W. F. (1982). Further observations on the vertical distribution of ßying mosquitoes (Diptera: Culicidae) in West African savanna. *Bulletin Entomology Research*, *72*, 695–708.

Southwood, T. R. E., and Henderson, P. A. (2000). *Ecological methods*. 3rd edn. Blackwell Science, Oxford.

Spielman, A., and D'Antonio, M. (2001). *Mosquito: The story of man's deadliest foe*. London: Faber & Faber, VI247.

Srygley, R. B. and Dudley, R. (2008). Optimal strategies for insects migrating in the flight boundary layer: Mechanisms and consequences. *Integrative and Comparative Biology*, *48*, 119–133.

Stein, M., Zalazar, L., Willener, J. A., Almeida, F. L., and Almirón, W. R. (2013). *Culicidae* (Diptera) selection of humans, chickens and rabbits in three different environments in the province of Chaco, Argentina. *Memórias do Instituto Oswaldo Cruz*, *108*, 563–571.

Stenn, T., Peck, K. J., Rocha Pereira, G., and Burkett-Cadena, N. D. (2019). Vertebrate hosts of *Aedes aegypti*, *Aedes albopictus*, and *Culex quinquefasciatus* (Diptera: Culicidae) as potential vectors of Zika virus in Florida. *Journal of Medical Entomology*, *56*(1), 10–17.

Stone, C. M., Taylor, R. M., Roitberg, B. D., and Foster, W. A. (2009). Sugar deprivation reduces insemination of *Anopheles gambiae* (Diptera: Culicidae), despite daily recruitment of adults, and predicts decline in model populations. *Journal of Medical Entomology*, *46*(6), 1327–37.

Stone, C., and Gross, K. (2018). Evolution of host preference in anthropophilic mosquitoes. *Malaria Journal*, *17*, 257.

Subbarao, S. K. (1998). *Anopheles species complexes in South East Asia Region*. Technical Publication No.18. New Delhi: World Health Organization South East Asia Regional Office.

Sumba, L. A., Okoth, K., Deng, A. L., Githure, J., Knols, B. G., Beier, J. C., and Hassanali, A. (2004). Daily oviposition patterns of the African malaria mosquito *Anopheles gambiae Giles* (Diptera: Culicidae) on different types of aqueous substrates. *Journal of Circadian Rhythms*, *2*(1), 6.

Surendran, S. N., Jayadas, T., Tharsan, A., Thiruchenthooran, V., Santhirasegaram, S., Sivabalakrishnan, K., Raveendran, S., and Ramasamy, R. (2020). Anopheline bionomics, insecticide resistance and transnational dispersion in the context of controlling a possible recurrence of malaria transmission in Jaffna city in northern Sri Lanka. *Parasites and Vectors*, *13*(1), 156.

Sutherst, R. W. (2004). Global change and human vulnerability to vector-borne diseases. *Clinical Microbiology Reviews*, *17*(1), 136–173.

Swami, K. K. and Srivastava, M. (2012). Blood meal preference of some anopheline mosquitoes in command and non-command areas of Rajasthan, India. *Journal of Arthropod-Borne Diseases*, *6*(2), 98.

Syed, Z. and Leal, W. S. (2009). Acute olfactory response of Culex mosquitoes to a human-and bird-derived attractant. *Proceedings of the National Academy of Sciences*, *106*(44), 18803–18808.

Takken, W., and Knols Bart, G. J. (1999). Odor-mediated behavior of afrotropical malaria mosquitoes. *Annual Review of Entomology*, *44*, 131–157.

Takken, W., and Knols, B. G. J. (2007). *Emerging pests and vector-borne diseases in Europe. Ecology and control of vectorborne diseases Vol. 1*. Wageningen Academic Publishers, Wageningen, the Netherlands, 499.

Takken, W., and Verhulst, N. O. (2013). Host preferences of blood-feeding mosquitoes. *Annual Review of Entomology*, *58*, 433–53.

Tandon, N. and Ray, S. (2000). Host Feeding Pattern of *Aedes aegypti* and *Aedes albopictus* in Kolkata India. *Dengue Bulletin*, *23*, 117–120.

Taylor, B. (1975). Changes in the feeding behaviour of a malaria vector, *Anopheles farauti* Lav. following use of DDT as a residual spray in houses in the British Solomon Islands Protectorate. *Transactions of the Royal Entomological Society of London*, *127*(3), 277–292.

Taylor, B., and Jones, M. D. (1969). The circadian rhythm of flight activity in the mosquito *Aedes aegypti (L.)*. The phase-setting effects of light-on and light-off. *The Journal of Experimental Biology*, *51*(1), 59–70.

Thapar, B. R., Sharma, S. N., Dasgupta, R. K., Kaul, S. M., Bali, A., Chhabra, K. and Lal, S. (1998). Blood meal identification by using Microdot ELISA in vector mosquitoes. *The Journal of Communicable Diseases*, *30*(4), 283–287.

Thomas, S., Ravishankaran, S., Justin, J. A., Asokan, A., Mathai, M. T., Valecha, N., Thomas, M. B., and Eapen, A. (2016). Overhead tank is the potential breeding habitat of Anopheles stephensi in an urban transmission setting of Chennai, India. *Malaria Journal*, *15*, 274.

Thomsen, E. K., Koimbu, G., Pulford, J., Jamea-Maiasa, S., Ura, Y., Keven, J. B., Siba, P. M., Mueller, I., Hetzel, M. W., and Reimer, L. J. (2017). Mosquito behaviour change after distribution of bednets results in decreased protection against malaria exposure. *The Journal of Infectious Diseases*, *215*(5), 790–797.

Tian, H. Y., Bi, P., Cazelles, B., Zhou, S., Huang, S. Q., Yang, J., Pei, Y., Wu, X. X., Fu, S. H., Tong, S. L., Wang, H. Y., and Xu, B. (2015). How environmental conditions impact mosquito ecology and Japanese encephalitis: An eco-epidemiological approach. *Environmental International*, *79*, 17–24.

Tyagi, B. K. (1994). Effect of the choice of oviposition medium on the breeding behaviour of *Culex quinquefasiatus* Say. *Geobios New Reports 13*, 28–30.

Tyagi, B. K. (2003). *Medical Entomology: A handbook of medically important insects and other arthropods*. Scientific Publishers, India. 265.

Tyagi, B. K. (2004). *The invincible deadly mosquitoes: India's Health and Economy Enemy No. 1*. Scientific Publishers, India, 266.

Tyagi, B. K. (2012). Culicide phytotelmata (Pitcher plant dwelling mosquitoes): Evolutionary link to Earth's changing climate – a review. *International Conference on Entomology*, Patiala (February 17–19, 2012), 17–18.

Tyagi, B. K. (2020). Mosquito Hunters: A history of hostilities against man's deadliest Foe – the mosquito – since 1881. *Scientific Publishers* (India), 474.

Tyagi, B. K. (2023). *Desert Malaria – An emerging malaria paradigm and its global impact on disease. Springer – Nature*, Singapore, 416.

Tyagi, B. K., and Verma K. V. S. (1991). Anopheline mosquitoes of Sri Ganganagar district a(Rajasthan) transmitting malaria parasite. *Journal of Applied Zoological Researches*, *2*, 85–91.

Tyagi, B. K., and Yadav, S. P. (2001) Bionomics of malaria vectors in two physiographically difference areas of the epidemic-prone Thar Desert, north-western Rajasthan (India). *Journal of Arid Environments*, *47*, 161–172.

Tyagi, B. K., Hiriyan, J., Philip Samuel, P., Tewari S. C., and Paramasivan, R. (2006). Dengue in Kerala: A critical review. *ICMR Bulletin*, *36*(4–5), 13–22

United Nations. (2014). *World Urbanization Prospects: The 2014 Revision, Highlights (ST/ESA/SER.A/352)* New York: United Nations DoEaSA, Population Division.

Van Breugel, F., Riffell, J., Fairhall, A., and Dickinson, M. H. (2015). Mosquitoes use vision to associate odour plumes with thermal targets. *Current Biology*, *25*(16), 2123–2129.

Verdonschot, P. F. M., and Besse-Lototskaya, A. A. (2014). Flight distance of mosquitoes (Culicidae): A metadata analysis to support the management of barrier zones around rewetted and newly constructed wetlands. *Limnologica – Ecology and Management of Inland Waters*, *45*, 69–79.

Vijayakumar, K., Sudheesh Kumar, T. K., Nujum, Z. T., Umarul, F., and Kuriakose, A. (2014). A study on container breeding mosquitoes with special reference to *Aedes (Stegomyia) aegypti* and *Aedes albopictus* in Thiruvananthapuram district, India. *Journal of Vector Borne Diseases*, *51*(1), 27–32.

Vlach, J. J., Hall, K. J., Day, J. F., Curtis, G. A., Hribar, L. J., and Fussell, E. M. (2006). Interisland dispersal of the black salt marsh mosquito, Ochlerotatus taeniorhynchus (Diptera: Culicidae) in the Florida Keys. *Journal of the American Mosquito Control Association*, *22*(4), 615–621.

Wachira, S. W., Ndung'u, M., Njagi, P. G. N. and Hassanali, A. (2010). Comparative responses of ovipositing *Anopheles gambiae* and *Culex quinquefasciatus* females to the presence of *Culex* egg rafts and larvae. *Medical and Veterinary Entomology*, *24*(4),369–374.

Waite, J. L., Swain, S., Lynch, P. A., Sharma, S. K., Haque, M. A., Montgomery, J., and Thomas, M. B. (2017). Increasing the potential for malaria elimination by targeting zoophilic vectors. *Scientific Reports*, *7*(1), 1–10.

Weber, R. G. and Tipping, C. (1993). Preoviposition drinking by *Culex restuans* (Diptera: Culicidae). *Journal of Insect Behavior*, *6*, 343–349.

Wharton, R. H. (1953). The habits of adult mosquitoes in Malaya: III.—Feeding preferences of Anophelines. *Annals of Tropical Medicine & Parasitology*, *47*(3), 272–284.

Whittaker, C., Winskill, P., Sinka, M., Pironon, S., Massey, C., Weiss, D. J., Nguyen, M., Gething, P. W., Kumar, A., Ghani, A. and Bhatt, S. (2021). The ecological structure of mosquito population dynamics: Insights from India. *medRxiv*, 2021–01.
Williams, R. E., and DeLong, D. M. (1961). Increasing the rate of egg productivity in *Aedes aegypti. Journal of Economical Entomology*, 54(6), 1265–6.
Williamson, M. (1996). *Biological Invasions*. Chapman and Hall; New York, USA. p. 244.
Winegardl, T. C. (2019). The Mosquito: A Human History of Our Deadliest Predator.
Wolff, G. H., and Riffell, J. A. (2018). Olfaction, experience and neural mechanisms underlying mosquito host preference. *Journal of Experimental Biology*, *221*(4), jeb157131.
Wondji, C. S., Irving, H., Morgan, J., Lobo, N. F., Collins, F. H., Hunt, R. H., Coetzee, M., Hemingway, J., and Ranson, H. (2009). Two duplicated P450 genes are associated with pyrethroid resistance in *Anopheles funestus*, a major malaria vector. *Genome Research*, *19*(3), 452–459.
World Health Organization (2003). *Malaria entomology and vector control* (No. WHO/CDS/CPE/SMT/2002.18 Rev. 1). World Health Organization.
World Health Organization (2021). Global insecticide use for vector-borne disease control: A 10-year assessment (2010–2019).
World Health Organization (1982). *Manual of environmental management for mosquito control, with special emphasis on malaria vectors*. Geneva: World Health Organization;
Xie, G. L., Ma, X. R., Liu, Q. Y., Meng, F. X., Li, C., Wang, J., and Guo, Y. H. (2021). Genetic structure of *Culex tritaeniorhynchus* (Diptera: Culicidae) based on COI DNA barcodes. *Mitochondrial DNA B Resources*, 6(4), 1411–1415.
Yamana, T. K., and Eltahir, E. A. (2013). Incorporating the effects of humidity in a mechanistic model of Anopheles gambiae mosquito population dynamics in the Sahel region of Africa. *Parasites and Vectors*, *6*, 235.
Yang, L., Denlinger, D. L., and Piermarini, P. M. (2017). The diapause program impacts renal excretion and molecular expression of aquaporins in the northern house mosquito, *Culex pipiens. Journal of Insect Physiology*, *98*, 141–148.
Zogo, B., Koffi, A. A., Alou, L., Fournet, F., Dahounto, A., Dabiré, R. K., Baba-Moussa, L., Moiroux, N., and Pennetier, C. (2019). Identification and characterization of Anopheles spp. breeding habitats in the Korhogo area in northern Côte d'Ivoire: A study prior to a Bti-based larviciding intervention. *Parasites and Vectors*, *12*(1), 146.

Section VI

Mosquito Genomes, Genomics, and Proteomics

10 Mosquito Genome

B.K. Tyagi
Chandigarh University, Mohali, India

Varun Tyagi
Eurofins AgroSciences Services Pvt. Ltd, Bengaluru, India

Vas Dev
Ex-ICMR-National Institute of Malaria Research, Dwarka, New Delhi, India

10.1 INTRODUCTION

Mosquitoes (Family Culicidae), comprising over 3600 species and subspecies (Harbach, and Besansky, 2014; Tyagi, *this work*), occupy a unique status among the world's fauna, largely because of their role in the transmission of a large number of vector-borne and zoonotic diseases such as, for example, malaria, filariasis, yellow fever, dengue, zika, West Nile, and several encephalitides. Members of both subfamilies, viz., Anophelinae and Culicinae, are vectors of infections to humans and animals, although human malaria is invariably transmitted by *Anopheles* mosquitoes. On the other hand, certain culicine mosquitoes (several genera, including *Culex* and *Aedes*) are carriers of bird, reptile, and/or amphibian malaria. Mosquito-borne infections take a major chunk (17%) of global illnesses, with close to a million deaths annually. Mosquito is not only man's deadliest foe and predator on the planet, it is also branded as economy enemy No. 1 (Spielman and D'Antonio, 2001; Tyagi, 2004; Winegard, 2019). Mosquitoes have unequivocally impacted human life more than any other animal on Earth so much so that even it has been extolled to have dictated a lot of human history; in battle-field (e.g., Spanish-American War about Panama's freedom), in giant projects (e.g., Panama Canal construction where it was said that more men had fallen to the mosquito bites and sicknesses it spreads like malaria, yellow fever and dengue than probably any other reason), and retreat of forces (Alexander The Great was constrained that the enemy to crossover Sindhu river in Punjab (India) and while returning through the ravines died due to *Plasmodium falciparum* malaria). Mosquito-borne diseases not only weaken humans by health, they also impoverish nations' economies and intelligentsia. Despite a history of more than 150 years of intensive and extensive investigations into its biology and behavior, we have actually made abysmal progress in both our understanding of the mosquito and its control. The pointed advice of Rao (1984) comes very true on the subject: "*... All my fellow workers who gave their best to conquer malaria but found to their disappointment that the enemy had outwitted them. Let better results attend the efforts of those who follow us fortified by an intimate knowledge of the enemy and his strength and weaknesses*." Cracking genome of mosquitoes, particularly those which are vectors of diseases and serious pests of humans and animals, and thereby knowing deeply the various biological functions will be an important arsenal in man's armory against this enemy invincible hitherto.

DOI: 10.1201/9781003326991-16

10.2 WHAT IS A GENOME?

A genome can be defined as the complete set of genes or DNA (deoxyribonucleic acid or the genetic material) in a cell or living thing. Like other animals, in mosquitoes, too, almost every cell in the body contains a complete copy of the genome, which is comparable to an encyclopedia or Wikipedia that contains all of the genetic information needed for an organism to develop and grow. More than a dozen mosquitoes' genomes have been decoded so far, the most significant of which being the dominant disease vectors are given in Table 10.1. Genome size for a species is subject to variation due to different populations treated.

Among all mosquitoes *Anopheles gambiae* (PEST strain from Africa) is the first species whose initial sequencing, assembly, and annotation were completed (Holt et al., 2002) and this provided the very first opportunity to explore the use of comparative genomics approaches to investigate insect biology, genetics, and evolution. As seen from the Table 10.1, the genome size of the malaria vector, *Anopheles stephensi* is the smallest among the dominant disease vector species (Dev and Sharma, 2013), which debatably points out its rather archaic, yet so highly evolved, nature of limited breeding preferences (Tyagi, 2004, 2023; Neafsey et al., 2015). *Anopheles stephensi* has three major gene-rich chromosomes (X, 2, 3) and a gene-poor, heterochromatic Y chromosome (Mittal and Dev, 1977; Chakraborty et al., 2021). Notably, variation in prowess for transmission of human malaria pathogens among *Anopheles* species is determined by many factors, including ecology and behavior, reproductive plasticity, immunity, and biology – traits that correlate with the genomic attributions being a direct pointer to a species' vectorial capacity and exploring new avenues for its control. Some determinants of vectorial capacity, such as chemosensory genes, do not show elevated turnover but instead diversify through protein-sequence changes. This dynamism of anopheline genes and genomes may contribute to their flexible capacity to take advantage of new ecological niches, including adapting to humans as primary hosts.

Generally, the genome size for the genus *Anopheles* appears to be smaller (at least for the present) than the other two prominent genera, *Culex* and *Aedes*. The genus *Aedes* is arguably the most weighted taxon, which is also certainly more recent, advanced, and adaptive to the environment.

10.3 LARGEST MOSQUITO GENOME

To date, the mosquito in which the genome size outweighs the rest of the Culicidae members is the Asian Tiger, *Aedes albopictus* (1,967 Mb), the mosquito which is a serious disease vector and a pest on both humans and animals (Cancrini et al., 2003; Pietrobelli, 2008) (Figure 10.1); followed by all the rest five dominant disease vector species such as, for example, *Ae. aegypti* (1,376 Mb), *Culex quinquefasciatus* (540 Mb), *Anopheles gambiae* (280 Mb), *An. stephensi* (221 Mb), and *An. darlingi* (174 Mb), respectively (Nene et al., 2007; Arensburger et al., 2010; Marinotti et al., 2013; Neafsey

TABLE 10.1
Genome Size of Some Dominant Vector Species

S.No.	Mosquito Species	Genome Size	Authority
1	*Aedes aegypti*	1,376 Mb	Nene et al. (2007)
2	*Aedes albopictus*	1,967 Mb	Chen et al. (2015)
3	*Culex quinquefasciatus*	540 Mb	Arensburger et al. (2010)
4	*Anopheles gambiae*	280 Mb	Holt et al. (2002)
5	*Anopheles stephensi*	221 Mb	Jiang et al. (2014)
6	*Anopheles darlingi*	174 Mb	Marinotti et al. (2013)

FIGURE 10.1 *Aedes albopictus* Skuse (1894), regarded at present the most invasive mosquito in the world.

Source: **Free Wikipedia.**

et al., 2015). A relatively higher or highest genome size also quite justifies the mosquito's power of agility to expand in distribution, adaptability to environment, and reproductive plasticity.

Aedes albopictus, having its origin in Asia, is one of the versatile and highly adaptive species in tropical and temperate zones. During last two decades, with sole exception of Antarctica, it has expanded so strikingly into new continental regions, hitherto *terra incognito* for the species, that it has justifiably earned a sobriquet of "most invasive mosquito" in the world at present. Its widespread geographic distribution, coupled with its potential to vectorize arboviruses such as, for example, dengue, chikungunya, and zika, besides being implicated as a vector of filarial nematodes of veterinary and zoonotic significance, markedly highlight its global public health risks. A complete genome sequence and transcriptome data for the *Ae. albopictus* Foshan strain was collected from its historical origin in Asia. The large genome (1,967 Mb) comprises an abundance of repetitive DNA classes and expansions of the numbers of gene family members involved in insecticide resistance, diapause, sex determination, immunity, and olfaction. Expansion of the number of members in gene families also contributes to the larger size of the species. In addition, portions of integrated flavivirus-like genomes support a shared evolutionary history of association of these arboviruses with their vector, *Ae. albopictus*. Thus, this large genome repertory and plasticity seem to likely contribute to its success as a highly penetrative, intrusive and, above all, invasive species.

10.4 SIGNIFICANCE OF CRACKING GENOME OF A SPECIES

The fact that not all mosquitoes feed on blood and/or have the potential to transmit devastating disease pathogens such as arboviruses, filarial worms, or malaria parasites alludes to a great degree of behavioral variation among the more than 3,500 documented mosquito species. Exploring this diversity in the ecology and behavior as well as the biology through application of omics technologies is fetching increasingly greater resolution that enhances our understanding to characterize mosquito biology and evolution. The advanced omics technology, or mozomics knowledge, resources and their application will facilitate our understanding of the putative links between gene and genome dynamism and mosquito diversity.

10.5 GENOME DETERMINATION AND INSIGHTS INTO MOSQUITO BIOLOGY

The three major culicine mosquitoes' (*Anopheles gambiae, Aedes aegypti*, and *Culex quinquefasciatus*) larger genomes revealed high levels of gene turnover with notably expanded gene repertoires of zinc-finger and insect pheromone-binding proteins, olfactory and gustatory receptors, as well as genes involved in immune responses and resistance to insecticides. Their comparative evolutionary genomics, supported by genomic sampling with a focus on the Anophelinae and in particular on those most closely related to *An. gambiae*, therefore, founded the fundamental expectation for conserved or divergent and common or unique features that characterize mosquito genome biology. Notwithstanding small sampling of mosquito diversity, these resources have nevertheless facilitated detailed characterizations of the evolutionary dynamics of mosquito genomes and their encoded genes, which offer insights into their adaptability to changing environments as well as their co-evolution with the pathogens they transmit.

10.6 FUTURE IMPLICATIONS TOWARD CONTROLLING DOMINANT VECTOR SPECIES

Mosquitoes are widely abhorred for their exasperating buzzing and irritating bites, and more poignantly because the female mosquitoes, if already infected, may transmit pathogens that cause deadly and/or incapacitating infections. However, all hematophagous mosquitoes do not transmit disease pathogens, and those that do (less than 500 species of mosquitoes actually can spread infections of some kind to humans) vary a great deal in transmitting such pathogens like viruses, filarial worms, or malaria parasites. It is therefore inevitably indispensable to determine this diversity among mosquitoes, which is now possible through omics technologies. Such modern tools can sample and explore the biology underlying these variations, and bring greater resolution of the factors that enhance our understanding of mosquito evolution. Application of recent advances in various available resources to a string of mosquito omics or "mozomics" has helped to characterize mosquito biology and evolution, with a focus on the intersection of evolutionary and functional genomics to understand the putative links between gene and genome dynamism and mosquito diversity (Ruzzante et al., 2019) (Table 10.2).

Chakraborty et al. (2021) have recently discovered the thrilling contribution of certain "hidden elements in the chromosome-level genome assembly" of the invasive malaria vector, *Anopheles stephensi*. They discovered 29 previously hidden members of insecticide resistance genes, uncovering new candidate genetic elements for the widespread insecticide resistance observed in *An. stephensi*. Their results and resources underscore the significance of previously hidden genomic elements in the biology of malaria mosquitoes and will accelerate the development of genetic control strategies for malaria transmission.

Genomic sampling with the initial sequencing, assembly, and annotation of the first deadly African malaria vector, *Anopheles gambiae*, lent support to delve deeply into mosquito biology and evolution by characterizing the conserved and divergent features underscoring mosquito diversity and hematophagy (Ruzzante et al., 2019) (Figure 10.2). Compared to the relatively small-sized genomes in Anophelines, the generally obtained larger genomes of the culicine mosquitoes such as, for example, *Aedes aegypti* and *Culex quinquefasciatus*, demonstrate larger pools of genes exhibiting vast potentialities of zinc-finger and insect pheromone-binding proteins, olfactory, and gustatory receptors, as well as genes involved in immune responses and resistance to insecticides.

Comparative evolutionary genomics across the three major mosquito genera (viz., Anopheles, Aedes and Culex) therefore established the baseline expectation for conserved or divergent and common or unique features that characterize mosquito genome biology.

TABLE 10.2
Sequenced Mosquito Genomes to Date from VectorBase (VB-2018-06)

Genus	Subgenus	Species	Notes	References
Aedes	*Stegomyia*	*aegypti*	Liverpool strain, now chromosomal-level	Nene et al. 2007; Dudchenko et al. 2017; Matthews et al. 2018; Whitfield et al. 2017
Culex	*Stegomyia*	*albopictus*	Foshan (Chinese) and Rimini (Italian) strains, and cell line C6/36 assembly	Chen et al. 2015; Dritsou et al. 2015; Miller et al. 2018
	Culex	*quinquefasciatus*	Johannesburg strain, now chromosomal-level assembly	Arensburger et al. 2010; Dudchenko et al. 2017
Anopheles	*Nyssorhynchus*	*albimanus*	STECLA strain (El Salvador), now chromosomal-level assembly	Neafsey et al. 2015; Artemov et al. 2017
		darlingi		
	Anopheles	*atroparvus*	EBRO strain (Spain) now chromosomal-level assembly	Neafsey et al. 2015; Artemov et al. 2018
		sinensis	China and SINENSIS (Korea) strains	Zhou et al. 2014; Neafsey et al. 2015
	Cellia (Neomyzomyia)[a]	*dirus*	WRAIR2 strain (Thailand)	Neafsey et al. 2015
		farauti	FAR1 strain (Papua New Guinea)	
		koliensis	AKwgs3 strain (Papua New Guinea), contig-level assembly	Logue et al. 2015
		punctulatus	APwgs2 strain (Papua New Guinea), contig-level assembly	
		nili	Dinderesso strain (Burkina Faso), contig-level assembly	Peery et al. 2011
	Cellia (Myzomyia)[a]	*culicifacies*	A-37 strain (Iran)	
		funestus	FUMOZ strain (Mozambique)	Neafsey et al. 2015
		minimus	MINIMUS1 strain (Thailand)	
	Cellia (Neocellia)[a]	*maculatus*	maculatus3 strain (Malaysia)	
		stephensi	Indian and SDA-500 (Pakistan) strains	Jiang et al. 2014; Neafsey et al. 2015
	Cellia (Pyretophorus)	*coluzzii*	Formerly gambiae M (Mali-NIH) strain	Lawniczak et al. 2010
		gambiae	PEST strain, chromosomal-level assembly Pimperena (Mali) strain	Holt et al. 2002; Lawniczak et al. 2010
		arabiensis	Dongola (Sudan) strain	
		christyi	ACHKN1017 (Kenya) strain	
		epiroticus	Epiroticus2 (Vietnam) strain	Neafsey et al. 2015
		melas	CM1001059_A (Cameroon) strain	
		merus	MAF (South Africa) strain	
		quadriannulatus	SANGWE (South Africa) strain	

[a] Anopheline Cellia species are grouped by series.
Source: Ruzzante et al., 2019.

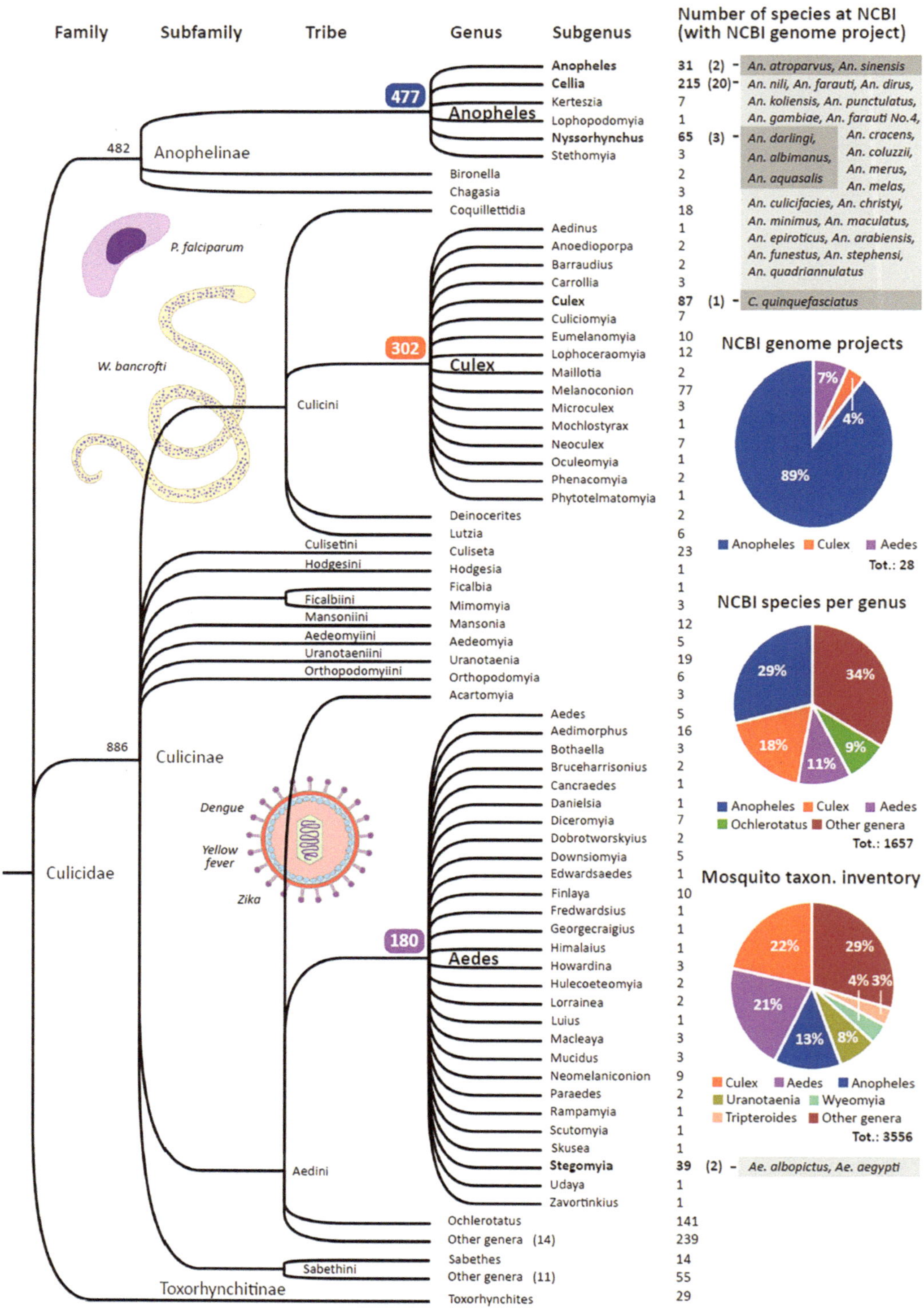

FIGURE 10.2 Current status of genomic sampling and view of mosquito taxonomic diversity.

Source: Ruzzante et al., 2019.

REFERENCES

Arensburger, P., et al. (2010) Sequencing of *Culex quinquefasciatus* establishes a platform for mosquito comparative genomics. *Science*, 330(6000), 86–88. https://doi.org/10.1126/science.1191864

Artemov, G.N., Bondarenko, S.M., Naumenko, A.N., Stegniy, V.N., Sharakhova, M.V., Sharakhov, I.V. (2018). Partial-arm translocations in evolution of malaria mosquitoes revealed by high-coverage physical mapping of the Anopheles atroparvus genome. *BMC Genomics*, 19, 278. https://doi.org/10.1186/s12864-018-4663-4

Artemov, G.N., Peery, A.N., Jiang, X., Tu, Z., Stegniy, V.N., Sharakhova, M.V., Sharakhov, I.V. (2017). The physical genome mapping of *Anopheles albimanus* corrected scaffold misassemblies and identified interarm rearrangements in genus *Anopheles*. *G3 (Bethesda)*, 7(1), 155–164. https://doi.org/10.1534/g3.116.034959

Cancrini, G., Regalbono, A.F., Ricci, I., Tessarin, C., Gabrielli, S., Pietrobelli, M. (2003). *Aedes albopictus* is a natural vector of *Dirofilaria immitis* in Italy. *Veterinary Parasitology*, 118, 195–202.

Chakraborty, M., et al. (2021). Hidden genomic features of an invasive malaria vector, *Anopheles stephensi*, revealed by a chromosome-level genome assembly. *BMC Biology*, 19, 28. https://doi.org/10.1186/s12915-021-00963-z

Chen, X.-G., et al. (2015). Genome sequence of the Asian Tiger mosquito, *Aedes albopictus*, reveals insights into its biology, genetics, and evolution. *Proceedings of the National Academy of Sciences*, 112(44), E5907–E5915. https://doi.org/10.1073/pnas.1516410112

Dev, V., & Sharma, V.P. (2013). The dominant mosquito vectors of human malaria in India. *In*: Manguin S, (Ed.): *Anopheles Mosquitoes: New Insights into Malaria Vectors*. *Croatia*: InTech Open Access; 2013. pp. 239–271. https://doi.org/10.5772/55215

Dritsou, V., et al. (2015). A draft genome sequence of an invasive mosquito: an Italian *Aedes albopictus*. *Pathogens and Global Health*, 109(5), 207–20. https://doi.org/10.1179/2047773215Y.0000000031

Dudchenko, O., et al. (2017). De novo assembly of the Aedes aegypti genome using Hi-C yields chromosome-length scaffolds. *Science*, 356(6333), 92–95. https://doi.org/10.1126/science.aal3327

Harbach R.E., & Besansky, N.J. (2014). Mosquitoes. *Current Biology* 24, R14–R15.

Holt, R.A., et al. (2002). The genome sequence of the malaria mosquito Anopheles gambiae. *Science*, 298(5591), 129–149. https://doi.org/10.1126/science.1076181

Jiang, X., et al. (2014). Genome analysis of a major urban malaria vector mosquito, *Anopheles stephensi*. *Genome Biology*, 15(9), 459. https://doi.org/10.1186/s13059-014-0459-2

Lawniczak, M.K., et al. (2010). Widespread divergence between incipient Anopheles gambiae species revealed by whole genome sequences. *Science*, 330(6003), 512–4. https://doi.org/10.1126/science.1195755

Logue, K. et al. (2015). Whole-genome sequencing reveals absence of recent gene flow and separate demographic histories for Anopheles punctulatus mosquitoes in Papua New Guinea. *Molecular Ecology*, 24, 1263–1274.

Marinotti, O., et al. (2013). The genome of Anopheles darlingi, the main neotropical malaria vector. *Nucleic Acids Research*, 41(15), 7387–400. https://doi.org/10.1093/nar/gkt484

Matthews, B.J., et al. (2018). Improved reference genome of Aedes aegypti informs arbovirus vector control. *Nature*, 563, 501–507. https://doi.org/10.1038/s41586-018-0692-z

Miller, J.R., et al. (2018). Analysis of the Aedes albopictus C6/36 genome provides insight into cell line utility for viral propagation. *Gigascience*, 7(3), 1–13.

Mittal, O.P. and Dev, V. (1977). Chromosomal basis of dosage compensation in Anopheles stephensi Liston (Culicidae: Diptera). *Microbios Letters*, 4, 79–8.

Neafsey, D.E., et al. (2015). Mosquito genomics. Highly evolvable malaria vectors: The genomes of 16 Anopheles mosquitoes. *Science*, 347(6217), 1258522.

Nene, V., et al. (2007). Genome sequence of *Aedes aegypti*, a major arbovirus vector. *Science* 316(5832), 1718–1723. https://doi.org/10.1126/science.1138878

Peery, A., Sharakhova, M.V., Antonio-Nkondjio, C., Ndo, C., Weill, M., Simard, F., Sharakhov, I. V. (2011). Improving the population genetics toolbox for the study of the African malaria vector *Anopheles nili*: microsatellite mapping to chromosomes. *Parasites & Vectors*, 4, 202. https://doi.org/10.1186/1756-3305-4-202

Pietrobelli, M. (2008). Importance of *Aedes albopictus* in veterinary medicine. *Parassitologia*, 50, 113–115.

Rao, T.R. (1984). *The Anopheline of India* (Revised Edition.). Malaria Research Centre, Indian Council of Medical Research, New Delhi, 518 pp.

Ruzzante, L., Reijnders, M.J.M.F., Waterhouse, R.M. (2019). Of genes and genomes: Mosquito evolution and diversity. *Trends in Parasitology*, 35(1), 32–51. https://doi.org/10.1016/j.pt.2018.10.003

Spielman, A., D'Antonio, M. (2001). *Mosquito: the story of man's deadliest foe*. London: Faber & Faber, vi+247 pp.

Tyagi, B.K. (2004). *The invincible deadly mosquitoes: India's Health and Economy Enemy No. 1*. Scientific Publishers, India, pp. 266.

Tyagi, B.K. (2023). *Desert Malaria: An Emerging Malaria Paradigm and Its Global Impact on Disease Elimination*. Springer-Nature, Singapore, 416 pp.

Whitfield, Z.J., et al. (2017). The diversity, structure, and function of heritable adaptive immunity sequences in the Aedes aegypti Genome. *Current Biology*, 27(22), 3511–3519.e7. https://doi.org/10.1016/j.cub.2017.09.067

Winegard, T.C. (2019). *The Mosquito: A Human History of Our Deadliest Predator*. Dutton, 496 pp.

Zhou, D., et al. (2014). Genome sequence of Anopheles sinensis provides insight into genetics basis of mosquito competence for malaria parasites. *BMC Genomics* 15, 42. https://doi.org/10.1186/1471-2164-15-42

11 Indian Malarial Vectors
Tissue-Specific Molecular Complexity and Integrative Biology

Tanvi Singh
ICMR-National Institute of Malaria Research, Delhi, India; AcSIR-Academy of Scientific & Innovative Research, Ghaziabad, India

Seena Kumari
ICMR-National Institute of Malaria Research, Delhi, India

Vaishali Saini and Pooja Rohilla
ICMR-National Institute of Malaria Research, Delhi, India; AcSIR-Academy of Scientific & Innovative Research, Ghaziabad, India

Sanjay Tevatiya
ICMR-National Institute of Malaria Research, Delhi, India

Sanjeev Kumar
Chaudhary Bansi Lal University, Bhiwani, India

Rajnikant Dixit
ICMR-National Institute of Malaria Research, Delhi, India; AcSIR-Academy of Scientific & Innovative Research, Ghaziabad, India

11.1 INTRODUCTION

Among several vector-borne diseases, malaria alone accounts for 0.69 million deaths worldwide (WHO, 2019). Its transmission is heavily influenced by seasonal fluctuations in the vector population. Though effective control of these mosquito vectors by chemical means such as the indoor residual spray (IRS) and long-lasting insecticidal nets (LLIN) is still popular, the alarming rate of vector resistance to almost all insecticides is a serious concern. Similar to African countries, India alone contributes about 76% of the malarial burden in the Southeast Asian region (Kumar et al., 2007). Of a total of 58 anopheline species reported from India (Subbarao et al., 2019), about 10 species comprise both primary (6 species) and secondary (4 species) vectors for malaria (Rao, 1984; Nagpal & Sharma 1995; Tyagi et al., 2015). Among the six primary Indian vectors only *An. stephensi* and *An. culicifacies* have been recognized as dominant urban and rural malarial vectors, respectively (Subbarao et al., 2019; Tyagi, 2023). Field studies demonstrate that both mosquito species are equally susceptible to *Plasmodium falciparum* and *P. vivax*, but carry distinct ecological adaptations. Investigation of insectary-maintained mosquitoes further confers a significant variation in their physiology such as feeding, mating, breeding, immunity, and insecticide resistance/tolerance. Interestingly, the genetic

DOI: 10.1201/9781003326991-17

basis of this species-specific variation is yet to be deciphered. Designing economical and eco-friendly new alternative tools is now a priority in the fight against malaria. Advanced genomic technologies have geared up the understanding to uncover the molecular basis of mosquito mysteries, and provide a new lead to successfully control the vectors in diverse habitats.

Comparative genomic studies of 16 clinically important anopheline mosquito strains originating from geographically distinct areas of malaria-affected countries clearly highlight that inherent genetic variation is the key driving force to feeding, breeding, and adapting distinct ecologies (Neafsey et al., 2015). Many international laboratories are focusing on uncovering the molecular and physiological relevance of genetic factors that influence the complex nature of mosquito population dynamics, and hence malaria transmission. Among the newly sequenced draft genomes, some of the anopheline vectors are common to the Indian region and provide a valuable resource for a basic understanding of their genomic complexity. However, in Africa, comparative genome analysis of more than 1000 individual mosquito genomes of *Anopheles gambiae*, collected from different localities reinforces us to decode and understand the molecular complexity of the most dangerous local vector species (*An. gambiae* 1000 Genomes Consortium). With a collaborative strategy, the efforts of mosquito genome sequencing and fine-scale assembly using combinations of transcriptomics, proteomics, and other relevant homology models are in progress to improve pre-existing draft genomes for the Indian malaria vectors (Jiang et al., 2014; Prasad et al., 2017).

With this reference knowledge, a few Indian laboratories took the initiative to carry forward high throughput functional genomic studies involving large-scale analysis of ESTs, RNA-Seq.-based transcriptomic, and proteomic studies in two Indian species, viz., *An. stephensi* and *An. culicifacies*. In this chapter, we briefly update and highlight important molecular studies conceptualizing and correlating their potential role in tissue-specific biologies, carried out so far in Indian vectors of malaria.

11.2 FEEDING-ASSOCIATED MOLECULAR COMPLEXITY OF ADULT FEMALE MOSQUITOES

Newly emerged adult male and female mosquitoes seek sugar sources for their routine activities. After a certain age which may vary from 2 to 6 days among different species, the adult female mosquitoes start to show host-seeking behavior to feed on the blood of vertebrate hosts. The evolutionary basis and molecular mechanism of this shift/switch from sugar-to-blood-feeding are yet unknown. Most of the molecular studies on female-specific host-seeking and blood-feeding behavior in the mosquitoes *An. gambiae* highlights the important role of individual tissues, such as the impact of olfactory organs in the navigation and salivary glands in the blood-meal uptake, and parasite transmission, but the role of the neuro-regulation is still unclear.

11.3 SALIVARY GLANDS

Mosquitoes' salivary glands play an important role in nectar/sugar meal acquisition in both sexes. However, unlike the male, adult female mosquitoes' salivary gland also facilitates rapid blood-meal uptake and parasite transmission. Due to its unique role in blood-feeding and disease transmission, a number of transcriptomic and proteomic studies have been carried out in the *An. gambiae*, and other non-Indian anopheline mosquitoes. However, research investigation in Indian laboratories dates back to 2009, covering the first small-scale catalog of *An. stephensi* salivary transcripts in response to *Plasmodium berghei* infected blood meal (Dixit et al., 2009). Using subtractive hybridization sequencing protocol several new putative salivary transcripts that encode the diverse nature of salivary-specific secretory proteins, immune proteins, and other salivary physiology-associated proteins have been identified, which may likely prove to be necessary for blood-meal acquisition and parasite transmission. A detailed comparative RNA-Seq. analysis of uninfected and *P. vivax*-infected salivary glands demonstrates that the survival of sporozoites is favored by the suppression of metabolic machinery of the salivary gland (Kumari et al., 2021).

Likewise, a subsequent study further captured molecular signatures of sugar-feeding-associated salivary transcripts, and expression and suppression of sugar transporter, and a cAMP-generating DAPIT, signifies that adult female mosquitoes' salivary expression may undergo an observable modulation when switching from the sugar-to-blood-feeding stage (Dixit et al., 2011). While a comparative analysis of salivary proteomic insecticide susceptible and insecticide-resistant strains of *An. stephensi* documents the upregulation of several proteins and enzymes such as cytochrome 450, short-chain dehydrogenase reductase, and phosphodiesterase, which may likely have an impact on insecticide resistance and xenobiotic detoxification (Vijay et al., 2015).

Clarifying the molecular basis of differential regulation of salivary gland under physiologically distinct feeding states, a detailed comparative RNA-Seq. analysis of naïve sugar-fed and immediate blood-fed *An. culicifacies* confers that the adult female mosquito evolved with the unique ability of salivary gene-expression switching, possibly to manage meal-specific responses (Sharma et al., 2015). Another comparative proteomic analysis of salivary proteins examines the differential expression of at least 38 proteins, further confirming that uncovering the nature and mechanism of gene-expression switching regulation may clarify how salivary glands manage dual feeding-associated behavioral physiology (Rawal et al., 2015).

Furthermore, in an unusual but surprising finding of a large pool of Plant Like Transcripts (PLTs) only from sugar-fed salivary glands, followed by strong validation of the Indigenous nature of mosquitoes' PLTs origin, authors provided the first molecular evidence of Horizontal Gene Transfers (HGTs) from plants to mosquitoes, though the mechanism remains unknown (Sharma et al., 2015). Through a comprehensive analysis of a cluster of putative PLTs encoding photosynthetic pathway-related proteins and the finding of salivary microbiome having photosynthetic properties (Sharma et al., 2014), the authors also predicted that unlike other anopheline species such as *An. stephensi* and, *An. gambiae*, the biology of *An. culicifacies* is more complex and much to be decoded yet.

The molecular complexity is further heightened after the recognition of at least five sibling species, *viz*., A, B, C, D, and E for *An. culicifacies* that have diverse geographical distribution patterns. Observation of the sibling species B as a poor vector is indicative of the fact that it is important to decipher the molecular nature and functions of the genetic factors enabling this species to be refractory toward Plasmodium infection. A comparative salivary proteomic analysis of susceptible (sp. A) and refractory (sp. B) identifies a pool of at least 82 differentially regulated proteins that are dominantly associated with immunity, digestion, energy, and carbon metabolic pathways (Rawal et al., 2018).

Summarily, the available transcriptomic/proteomic studies clearly demonstrate that mosquito salivary glands are evolved with a unique ability to manage meal-specific responses (Figure 11.1). A comprehensive review of data also proposes that identified salivary-specific factors that express uniquely under the distinct pathophysiological status of feeding, i.e., sugar or blood and *Plasmodium* infection, may serve as a key target for future functional characterization. Additional findings of PLTs in the *An. culicifacies* not only provide unique evidence of HGTs in higher eukaryotes but also correlate with species-specific salivary complexity. Further decoding species-specific salivary complexity among anopheline vectors may clarify many unresolved pending questions on mosquito evolution and adaptation to diverse ecologies.

11.4 OLFACTION

Mosquitoes' olfactory system is an integrated organ of three important sub-organs including antennal lobes, proboscis, and maxillary palp. It serves as an essential peripheral tissue system to detect and discriminate among millions of chemical cues present in the environment that enables mosquitoes to navigate and perform diverse activities such as food-source detection, host-seeking, blood-feeding, mate partner finding, oviposition site finding, etc. To achieve these goals, the olfactory system encodes a diverse nature of proteins, broadly including family members of odorant-binding proteins (OBPs) and odorant receptor proteins (ORs). Research for more than two decades in the

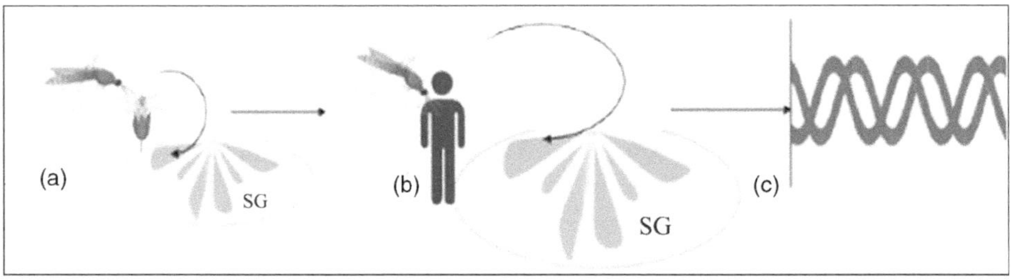

FIGURE 11.1 Molecular complexity of salivary gene expression: Schematic demonstration that when adult female mosquitoes switch from sugar-feeding (a) to blood-feeding (b), the blood-meal not only alters the morphological architecture, resulting in the swelling, widening, and extension of the salivary lobes, also switch gene expression of the salivary transcript to assist the optimal blood-meal acquisition from the vertebrate host. The evolution of this unique ability of gene-expression switching is key to managing meal-specific responses, and hence the adaptation to different feeding habits, though the detailed mechanism is yet to unravel.

***Source*: Sharma et al., 2015.**

mosquitoes *An. gambiae*, *Aedes aegypti*, and some *Culex* spp. highlight that mosquitoes are able to control the complex behavioral coordinates with a limited pool of OBPs and ORs, (Manoharan et al., 2013), about which the scientific community has no clue regarding the mechanism of OBPs & ORs coordination.

However, the molecular physiology of olfaction in Indian malarial vectors remains un-decoded, until the recent initiative of tracing the possible correlation of the olfactory system in the regulation of salivary gland gene-expression-switching. A comparative RNA-Seq. analysis identifies at least ~63 putative OBPs, and 50 ORs of the olfactory system in *An. culicifacies* (De et al., 2018). Notably, in a large-scale spatial/temporal expression analysis of selected OBPs, and ORs, the authors conclude that synergistic coordination of olfactory encoded OBPs/ORs is essential to drive the complex host-seeking and blood-feeding associated behavioral activities (Figure 11.2). Compared with *An. stephensi*, enriched expression of transcripts encoding the chemosensory/sensory appendage proteins (CSP/SAP) in the legs of the adult female mosquito *An. culicifacies* proposes that this species may have more sensitive appendages to detect outer environments such as insecticide, which could be an evolutionary benefit for survival and adaptation in diverse ecologies. Uncovering the regulatory mechanism and strategy to disrupt this synergistic coordination of the OBPs/ORs could pave a new way to design a novel disorientation strategy for vector control. Unfortunately, compared to other regions of the world, ongoing Indian studies in the *An. stephensi* olfactory system have not yet arrived at a definitive conclusion.

11.5 BRAIN

The brain plays a central role in the management of body homeostasis. Comparative to the vertebrate, the insect/mosquito brain harbors nearly a thousand neurons, and the mystery of how this miniature brain achieves a lifestyle associated with complex regulatory behavioral decisions such as sugar-feeding, swarming, mating, host-seeking, blood-feeding, gonotrophic cycle, oviposition, etc. continues to be unsolved yet? Therefore, decoding the molecular nature and function of the neuro-encoded factors such as neuro-signaling, neuro-transmitters, neuro-peptides, and hormones is one of the fast-emerging areas of research in the field of insect neurobiology (Heinze et al., 2021). Relative to a handful of investigations into different mosquito species such as *Aedes aegypti* (Matthews et al., 2016; Zhao et al., 2022), and *An. gambiae* (Dwivedi et al., 2014), the neurobiology of any other mosquito species remains elusive.

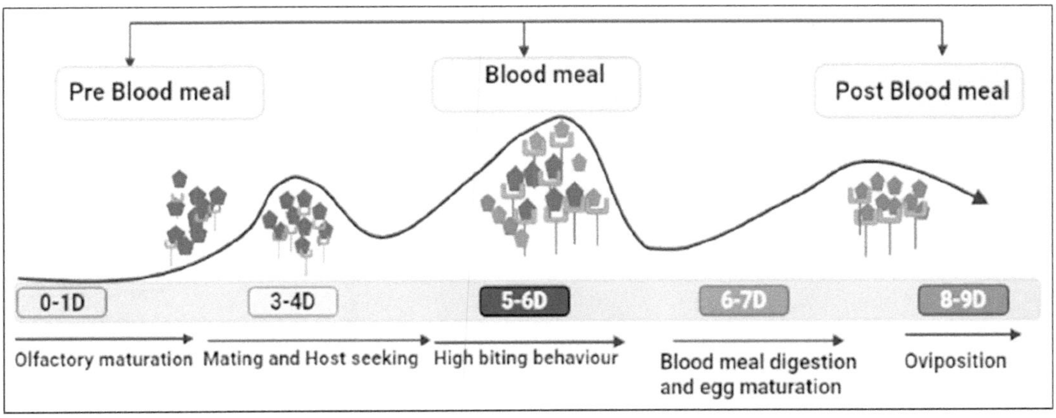

FIGURE 11.2 Schematic overview showing olfactory response modulation in the adult female mosquitoes, when switching from one physiological status to another physiological status: Newly emerged adults are exclusively sugar feeders. Gradual learning and maturation of the olfactory system with exposure to diverse environmental cues enables successful mating and host-seeking. Blood-feeding manipulates the energy in blood-meal digestion in the gut and egg maturation in the ovary. After successful blood digestion, olfactory responses are recovered and reactivated to find the oviposition site. Current studies highlight that the coordinated action of olfactory encoded molecular factors (OBPs & ORs) is necessary to drive all complex behavioral activities.

Source: **De et al., 2018.**

Considering previous findings on the modulation of olfactory (De et al., 2018) and salivary gene expression (Sharma et al., 2015), further, attempts were made to clarify the possible role of brain engagement during switching from pre-to-post-blood-meal in *An. culicifacies*. Interestingly, the authors hypothesize (i) after a blood-meal olfactory-guided neuro action must be arrested temporarily, at least for the first 30 hrs, (ii) during this phase active gut-metabolic switching must modulate the brain functioning to internal coordination for distant organs such as malpighian tubules, midgut, fat body, ovary, and (iii) after blood-meal digestion and egg maturation, brain engagement gradually re-switch to perform external communication for the oviposition, and next gonotrophic cycle-associated behavioral activities.

To establish and validate the possible correlation of brain function, in their recent study (De et al., 2022), authors initially, compared and analyzed naïve sugar-fed and blood-fed mosquito brain RNA-Seq data and noticed that enriched expression of mitochondrial function and energy metabolism-related transcripts may likely result in the shift of the brain's engagement to manage body homeostasis. Further profiling of transcript encoding neuro-signaling and neuromodulatory proteins in both the gut and brain after blood-feeding correlates an active gut-brain communication. Surprising observation of an unusual shift in the level of neurotransmitters (NTs), from the brain to the gut after blood-feeding, further ensured that the gut also carries a unique ability to serve as a "second brain." The dsRNA-mediated histamine receptor silencing reveals moderate enrichment of the microbial population, and gut-immunity alteration endorses the gut microbiome's role in gut-brain–axis communication. While evaluating the gut-associated bacterial population dynamics alteration, the authors noted that blood-feeding enriches the family members of the Morganellaceae and Pseudomonadaceae bacterial communities. And a rapid proliferation of *Pseudomonas* bacterial species and tryptophan enrichment in the gut, and altered NTs dynamics of naïve and aseptic mosquitoes, highlight that gut-endosymbionts are key modulators for the synthesis of major neuroactive molecules. Taken together these data establishes a new conceptual understanding of microbiome-gut-brain–axis communication in mosquitoes (Figure 11.3).

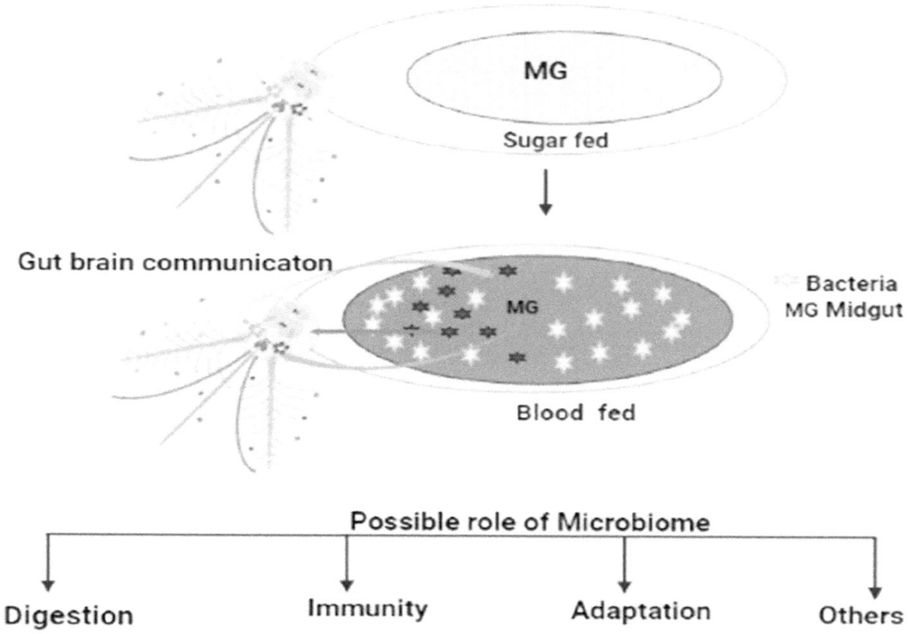

FIGURE 11.3 The gut microbiome's role in gut-brain-axis communication and other mosquito physiologies: periodic ingestion of a protein-rich blood meal by adult female mosquitoes causes a drastic metabolic change in their innate physiological status. This metabolic switch promotes multi-fold enrichment of gut microbiota, which not only helps in the establishment of bi-directional gut-brain-axis communication, but engages and manages multi-organ communication via the release of neuro-modulators for proper digestion, and also ensures immune activation (see below section), adaptation, and many other biological functions, such as egg maturation, larval development, etc.

Source: De et al., 2022.

11.6 GUT-MICROBE INTERACTION AND IMMUNE PHYSIOLOGY

Mosquito midgut is one of the major sites for blood-meal digestion and turns it into seed material for ovary development and egg maturation. Immediately after blood-meal, gut-metabolic machinery activation is rapidly followed by a thousand-fold enrichment of gut bacteria within 24 hrs. Together this contributes to supplying and meeting the nutritional requirement via the fat body and hemocyte, during the vitellogenesis process, though the mechanism is yet unclear.

A few studies on microbiome diversity in wild-caught larvae, and the gut of *An. stephensi* identifies several classes of bacterial species that may likely have a key role in mosquito physiology and adaptation (Rani et al., 2009; Pereira et al., 2021). Interestingly, a gut-salivary microbiome comparison of naïve sugar-fed adult female mosquitoes highlights that the salivary gland harbors a more diverse bacterial community than the gut in *An. culicifacies* (Sharma et al., 2014). Although valuable, bacterial influence on mosquito immunity and physiology is yet to be established.

Mapping tissue-specific alteration of antimicrobial peptides (AMPs) expression after exogenous (by thorax-injection) or endogenous (by oral supplementation) of gram-negative as well as gram-positive bacteria, clarify that an interorgan immune communication among gut-fat-body and hemocyte is managed by the modulation of NOS and REL immune signals in the mosquito *An. stephensi* (De et al., 2018). This study also highlighted that each tissue has its own ability to mount immune action, but hemocytes are more specialized than the fat body and gut of the mosquitoes (Figure 11.4). Follow-up RNA-Seq. analysis of naïve mosquito hemocytes in *An. stephensi* (Thomas

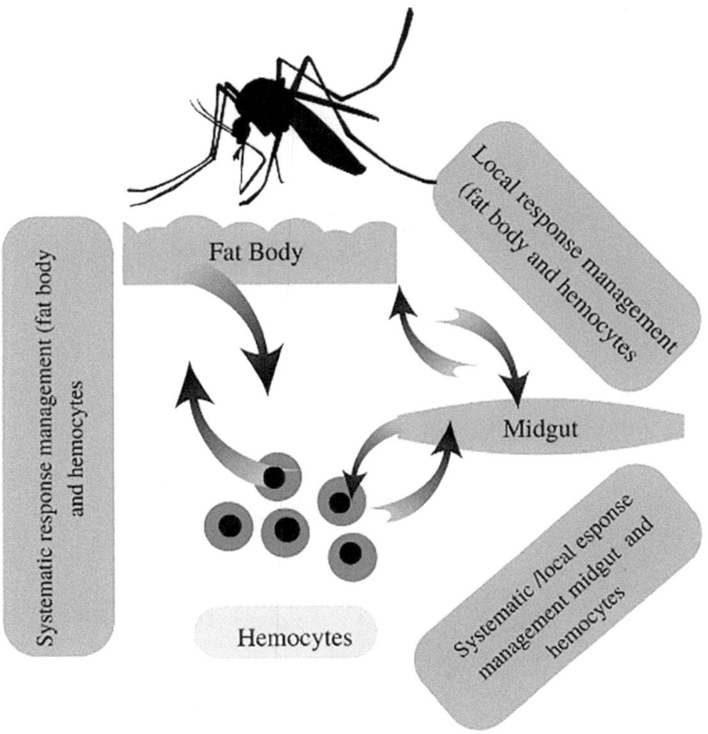

FIGURE 11.4 Interorgan immune communication in the mosquito: During any altered pathophysiology (endogenous and/or exogenous microbial infection), fat body and midgut activated antimicrobial peptides (AMPs), mount a local immune response, and hemocytes play an important role to mount a systemic immune response to clear off the infection remnants. Although the exact mechanism of immune signaling is yet unclear, the current hypothesis suggests each tissue evolved with a unique ability to maintain the flow signals during infection.

***Source*: De et al., 2018.**

et al., 2016), and further comparison with *An. culicifacies* (Rani et al., 2021), confers that hemocytes encode more diverse immune factors controlling cellular immunity.

Since the sexual cycle of the *Plasmodium* development in the mosquito host is key to malaria transmission, and because *Plasmodium* infection also has a negative impact on the nutritional supply, and reproductive outcome, thus decoding the nature and function of genetic factors, and linked molecular mechanism is worth important, but remains poorly understood in Indian malarial vectors. A gut-specific ESTs project comparing uninfected and rodent parasite *P. yoelli* infected *An. stephensi* mosquitoes, provide the first catalog of gut transcripts (Patil et al., 2009). Likewise, evaluation of a molecular network of oxidative stress responses, and dynamics expression alteration of miRNA during rodent malarial parasite *P. vinckei petteri* infection, predicts several putative oxidative stress-related genes, and miRNAs influencing the gonotrophic cycle and Plasmodium development in the mosquito *An. stephensi* (Shrinet et al., 2014; Jain et al., 2014).

However, due to high species-specific genetic variation, and natural compatibility among the Anopheline-Plasmodium, more studies need to be focused on human malarial parasites such as *P. falciparum* and *P. vivax* (Su et al., 2019). Though a handful of investigations defining individual tissues, *e.g.*, midgut and/or salivary glands, on Anopheles-*P. falciparum* originating from international laboratories is available (Shaw et al., 2022), but no single study exists decoding the tissue-specific molecular complexity of the Anopheles-*Plasmodium vivax* interaction during development inside the mosquito host in Indian malarial vectors.

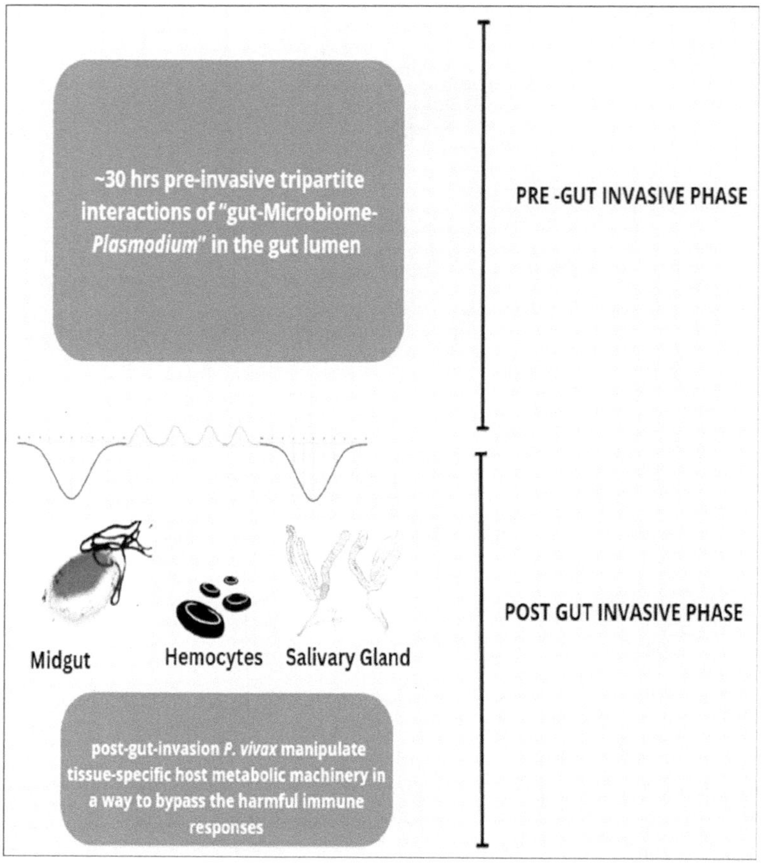

FIGURE 11.5 Experimental demonstration and schematic presentation of ***P. vivax*** survival strategy inside the mosquito host: During its pre-invasive phase of tripartite gut-microbe-parasite interaction in the lumen, ***P. vivax*** follows suppression of gut-microbes, and immune alteration (Sharma et al., 2020) strategy to escape and invade the gut; and post-gut-invasion ***P. vivax*** manipulate tissue-specific host metabolic machinery in a way to bypass the harmful immune responses.

Source: **Kumari et al., 2021.**

Very recently, two published studies detailing ~30 hrs pre-invasive tripartite interactions of "gut-microbiome-*Plasmodium*" in the gut lumen (Sharma et al., 2020), and post-invasive interaction of Plasmodium with gut-hemocyte-salivary glands, together provide a unique nature of relationship during *P. vivax* development in the mosquito *An. stephensi*. Notably, these findings define how *P. vivax* smartly suppresses the gut microbiota in the lumen, allowing it to bypass the gut immunity, and successfully alters the immune physiology of the directly interacting tissues after the invasion of the gut-epithelia and hemocyte and salivary glands (Figure 11.5). Relevant details of respective tissue-specific interactions have also been covered in Chapter 13, and therefore the readers may refer to it.

11.7 CONCLUSION

Studies on understanding the population genetic structure, and evaluation of markers for monitoring insecticide resistance network in the Indian malarial vector have been well under-taken (Singh et al., 2010; Gakhar et al., 2013; Sunil et al., 2013; Sharma et al., 2014; Dykes et al., 2015, 2016; Lather et al., 2022). However, the functional genomic studies correlating tissue-specific physiology of the Indian malarial vectors have just begun in the last decade. Here, we updated the important molecular studies, focused on decoding, integrating, and conceptualizing the fundamentals of the tissue-specific biology of the Indian malarial vectors. Through this update on the conceptual knowledge, we believe readers would be able to easily follow-up in the next chapter "Indian Malarial Vectors: Functional Genomics," detailing the ongoing efforts on the characterization of new putative targets, and their crucial functions.

BIBLIOGRAPHY

Anopheles gambiae 1000 Genomes Consortium (2020). Genome variation and population structure among 1142 mosquitoes of the African malaria vector species *Anopheles gambiae* and *Anopheles coluzzii. Genome Research*, 30(10), 1533–1546. https://doi.org/10.1101/gr.262790.120

De, T.D., Sharma, P, Tevatiya, S., Chauhan, C., Kumari, S., Yadav, P., Singla, D., Srivastava, V., Rani, J., Hasija, Y., Pandey, K.C., Kajla, M., Dixit, R. (2022). Bidirectional Microbiome-Gut-Brain-Axis Communication Influences Metabolic Switch-Associated Responses in the Mosquito *Anopheles culicifacies. Cells*, 11(11), 1798. https://doi.org/10.3390/cells11111798

De, T.D., Thomas, T., Verma, S., Singla, D., Chauhan, C., Srivastava, V., Sharma, P., Kumari, S., Tevatiya, S., Rani, J., Hasija, Y., Pandey, K.C., Dixit, R. (2018). Synergistic Transcriptional Regulation of Olfactory Genes Drives Blood-Feeding Associated Complex Behavioral Responses in the Mosquito *Anopheles culicifacies. Frontiers in Physiology*, 9, 1–15.

Dixit, R., Rawat, M., Kumar, S., Pandey, K. C., Adak, T., Sharma, A. (2011). Salivary Gland Transcriptome Analysis in Response to Sugar Feeding in Malaria Vector *Anopheles stephensi. Journal of Insect Physiology*, 57(10), 1399–1406.

Dixit, R., Sharma, A., Mourya, D., Kamaraju, R., Patole, M., Shouche, Y. (2009). Salivary Gland Transcriptome Analysis during Plasmodium Infection in Malaria Vector *Anopheles stephensi. International Journal of Infectious Diseases*, 13(5), 636–46.

Dixit, R., Sharma, P., De, T. D., Sharma, S., Mishra, A. K., Thomas, T., Verma, S., et al. (2015). Deep Sequencing Revealed Molecular Signature of Horizontal Gene Transfer of Plant like Transcripts in the Mosquito Anopheles culicifacies: An Evolutionary Puzzle. F1000Research, 4(December).

Dwivedi, S. B., Muthusamy, B., Kumar, P., Kim, M. S., Nirujogi, R. S., Getnet, D., Ahiakonu, P., De, G., Nair, B., Gowda, H., Prasad, T. S., Kumar, N., Pandey, A., Okulate, M. (2014). Brain Proteomics of Anopheles gambiae. *OMICS*, 18(7), 421–437. https://doi.org/10.1089/omi.2014.0007

Dykes, C. L., Das, M. K., Eapen, A., Batra, C. P., Ghosh, S. K., Vijayan, V. A., Mishra, S., Singh, O. P. (2016). Knockdown Resistance (kdr) Mutations in Indian Anopheles stephensi (Diptera: Culicidae) Populations. *Journal of Medical Entomology*, 53(2), 315–320. https://doi.org/10.1093/jme/tjv242

Dykes, C. L., Kushwah, R. B., Das, M. K., Sharma, S. N., Bhatt, R. M., Veer, V., Agrawal, O. P., Adak, T., Singh, O. P. (2015). Knockdown Resistance (kdr) Mutations in Indian Anopheles culicifacies Populations. *Parasites & Vectors*, 8, 333. https://doi.org/10.1186/s13071-015-0946-7

Gakhar, S. K., Sharma, R., Sharma, A. (2013). Population Genetic Structure of Malaria Vector Anopheles stephensi Liston (Diptera: Culicidae). *Indian Journal of Experimental Biology*, 51(4), 273–279.

Heinze, S., El Jundi, B., Berg, B. G., Homberg, U., Menzel, R., Pfeiffer, K., Hensgen, R., Zittrell, F., Dacke, M., Warrant, E., Pfuhl, G., Rybak, J., Tedore, K. (2021). A Unified Platform to Manage, Share, and Archive Morphological and Functional Data in Insect Neuroscience. *eLife*, 10, e65376. https://doi.org/10.7554/eLife.65376

Jain, S., Rana, V., Shrinet, J., Sharma, A., Tridibes, A., Sunil, S., Bhatnagar, R. K. (2014). Blood Feeding and Plasmodium Infection Alters the miRNome of Anopheles stephensi. *PLoS ONE*, 9(5), e98402. https://doi.org/10.1371/journal.pone.0098402

Jiang, X., Peery, A., Hall, A. B., Sharma, A., Chen, X. G., Waterhouse, R. M., Komissarov, A., Riehle, M. M., Shouche, Y., Sharakhova, M. V., Lawson, D., Pakpour, N., Arensburger, P., Davidson, V. L., Eiglmeier, K., Emrich, S., George, P., Kennedy, R. C., Mane, S. P., Maslen, G., Oringanje, C., Qi, Y., Settlage, R., Tojo, M., Tubio, J. M., Unger, M. F., Wang, B., Vernick, K. D., Ribeiro, J. M., James, A. A., Michel, K., Riehle, M. A., Luckhart, S., Sharakhov, I. V., Tu, Z. (2014). Genome Analysis of a Major Urban Malaria Vector Mosquito, *Anopheles stephensi. Genome Biology*, 15(9), 459. https://doi.org/10.1186/s13059-014-0459-2

Kumar, A., Valecha, N., Jain, T. and Dash, A.P. (2007). Burden of Malaria in India: Retrospective and Prospective View. In: Breman, J.G., Alilio, M.S., White, N.J., editors. *Defining and Defeating the Intolerable Burden of Malaria III: Progress and Perspectives: Supplement to Volume 77(6) of American Journal of Tropical Medicine and Hygiene.* Northbrook (IL): American Society of Tropical Medicine and Hygiene; Dec. Available from: https://www.ncbi.nlm.nih.gov/books/NBK1720/De

Kumari, S., Chauhan, C., Tevatiya, S., Singla, D., Das De, T., Sharma, P., Thomas, et al. (2021). Genetic Changes of Plasmodium vivax Tempers Host Tissue-Specific Responses in Anopheles stephensi. *Current Research in Immunology*, 2(February), 12–22.

Lather, M., Mallick, P. K., Sharma, D., Kale, S., Dang, A. S., Adak, T., Singh, O. P. (2022). Population Genetic Structure of the Malaria Vector Anopheles fluviatilis Species T (Diptera: Culicidae) in India. *Medical and Veterinary Entomology*, 36(2), 194–202. https://doi.org/10.1111/mve.12566

Manoharan, M., Chong, M. N. F., Vaïtinadapoulé, A., Frumence, E., Sowdhamini, R., Offmann, B. (2013). Comparative Genomics of Odorant Binding Proteins in *Anopheles gambiae*, *Aedes aegypti*, and *Culex quinquefasciatus*. *Genome Biology and Evolution*, 5(1), 163–180. https://doi.org/10.1093/gbe/evs131

Matthews, B. J., McBride, C. S., DeGennaro, M., Despo, O., Vosshall, L. B. (2016). The Neurotranscriptome of the Aedes aegypti Mosquito. *BMC Genomics*, 17, 32. https://doi.org/10.1186/s12864-015-2239-0

Nagpal, B. N., and Sharma, V. P. (1995) *Indian Anophelines*. Science Publishers, Inc., USA.

Neafsey, D. E., Waterhouse, R. M., Abai, M. R., Aganezov, S. S., Alekseyev, M. A., Allen, J. E., et al. (2015). Highly Evolvable Malaria Vectors: The Genomes of 16 Anopheles Mosquitoes. *Science*, 347(6217), 1258522. https://doi.org/10.1126/science.1258522

Patil, D. P., Atanur, S., Dhotre, D. P., Anantharam, D., Mahajan, V. S., Walujkar, S. A., Chandode, R. K., Kulkarni, G. J., Ghate, P. S., Srivastav, A., Dayananda, K. M., Gupta, N., Bhagwat, B., Joshi, R. R., Mourya, D. T., Patole, M. S., Shouche, Y. S. (2009). Generation, Annotation, and Analysis of ESTs from Midgut Tissue of Adult Female Anopheles stephensi Mosquitoes. *BMC Genomics*, 10, 386. https://doi.org/10.1186/1471-2164-10-386

Pereira, M. H., Mohanty, A. K., Garg, S., Tyagi, S., Kumar, A. (2021). Characterization of Midgut Microbiome of Anopheles stephensi Liston. *Journal of Vector Borne Diseases*, 58(1), 74–84. https://doi.org/10.4103/0972-9062.289392

Prasad, T. S., Mohanty, A. K., Kumar, M., Sreenivasamurthy, S. K., Dey, G., Nirujogi, R. S., et al. (2017). Integrating Transcriptomic and Proteomic Data for Accurate Assembly and Annotation of Genomes. *Genome Research*, 27(1), 133–144. https://doi.org/10.1101/gr.201368.115

Rani, A., Sharma, A., Rajagopal, R., Adak, T., Bhatnagar, R. K. (2009). Bacterial Diversity Analysis of Larvae and Adult Midgut Microflora using Culture-Dependent and Culture-Independent Methods in Lab-reared and Field-Collected Anopheles Stephensi-an Asian Malarial Vector. *BMC Microbiology*, 9, 96. https://doi.org/10.1186/1471-2180-9-96

Rani, J., Chauhan, C., Das De, T., Kumari, S., Sharma, P., Tevatiya, S., Patel, K., Mishra, A. K., Pandey, K. C., Singh, N., Dixit, R. (2021). Hemocyte RNA-Seq analysis of Indian malarial vectors Anopheles stephensi and Anopheles culicifacies: From similarities to differences. *Gene*, 798, 145810.

Rao, T. R. (1984). *The Anopheline of India* (Revised Edition). Malaria Research Centre, Indian Council of Medical Research, New Delhi, 518 pp.

Rawal, R., Vijay, S., Kadian, K., Adak, T., Pande, V., Sharma, A. (2015). Functional Annotation of Salivary Gland Proteins of Anopheles culicifacies Refractory and Sensitive Sibling Species. *Tropical Medicine & International Health*, 20(1, SI), 404.

Rawal, R., Vijay, S., Kadian, K., Adak, T., Pande, V., Sharma, A. (2018). Comparative Proteomics of Salivary Glands of Anopheles culicifacies Mosquitoes Using Tandem Mass Tag (TMT) Mass Spectrometry. *Journal of Vector Borne Diseases*, 55(2), 98–110.

Sharma, A., Deshmukh, A., Sharma, R., Kumar, A., Mukherjee, S., Chandra, G. C., Gakhar, S. K. (2014). Population genetic structure of malaria vector Anopheles stephensi using mitochondrial cytochrome oxidase II gene in Indian populations. *Indian Journal of Experimental Biology*, 52(10), 996–1002.

Sharma, P., Rani, J., Chauhan, C., Kumari, S., Tevatiya, S., Das De, T., Savargaonkar, T., Pandey, K. C., Dixit, R. (2020). Altered Gut microbiota and immunity defines Plasmodium vivax survival in Anopheles stephensi. *Frontiers in Immunology*, 11(May), 609.

Sharma, P., Sharma, S., Mishra, A. K., Thomas, T., Das De, T., Rohilla, S. L., Singh, N., Pandey, K. C., Valecha, N., Dixit, R. (2015). Unravelling Dual Feeding Associated Molecular Complexity of Salivary Glands in the Mosquito Anopheles Culicifacies. *Biology Open*, 4(8), 1002–15.

Shaw, W. R., Marcenac, P., Catteruccia, F. (2022). Plasmodium Development in Anopheles: A Tale of Shared Resources. *Trends in Parasitology*, 38(2), 124–135. https://doi.org/10.1016/j.pt.2021.08.009

Shrinet, J., Nandal, U. K., Adak, T., Bhatnagar, R. K., Sunil, S. (2014). Inference of the Oxidative Stress Network in *Anopheles stephensi* Upon Plasmodium Infection. *PLoS One*, 9(12), e114461. https://doi.org/10.1371/journal.pone.0114461

Singh, O., Dykes, C., Adak, T. (2010). kdr-like Mutations in the Voltage Gated Sodium Channel of a Malaria Vector *Anopheles stephensi* and Development of PCR-based Assays for Their Detection. *Malaria Journal*, 9(Suppl 2), P67.

Sinka, M. E., Bangs, M. J., Manguin, S., Chareonviriyaphap, T., Patil, A. P., Temperley, W. H., Gething, P. W., Elyazar, I. R., Kabaria, C. W., Harbach, R. E., Hay, S. I. (2011). The Dominant Anopheles Vectors of Human Malaria in the Asia-Pacific Region: Occurrence Data, Distribution Maps and Bionomic Précis. *Parasites & Vectors*, 4, 89. https://doi.org/10.1186/1756-3305-4-89

Sinka, M., Bangs, M. J., Manguin, S., Rubio-Palis, Y., Chareonviriyaphap, T., Coetzee, M., Mbogo, C. M., Hemingway, J., Patil, A. P., Temperley, W. H., Gething, P. W., Kabaria, C. W., Burkot, T. R., Harbach, R. E., Hay, S. I. (2012). A Global Map of Dominant Malaria Vectors. *Parasites & Vectors*, 5, 69. https://doi.org/10.1186/1756-3305-5-69

Su, X. Z., Lane, K. D., Xia, L., Sá, J. M., Wellems, T. E. (2019). Plasmodium Genomics and Genetics: New Insights into Malaria Pathogenesis, Drug Resistance, Epidemiology, and Evolution. *Clinical Microbiology Reviews*, 32(4), 1–29.

Subbarao, S. K., Nanda, N., Rahi, M., Raghavendra, K. (2019). Biology and bionomics of malaria vectors in India: existing information and what more needs to be known for strategizing elimination of malaria. *Malaria Journal*, 18(1), 396.

Sunil, S., Singh, O. P., Nanda, N., Raghavendra, K., Reddy, B. P., Subbarao, S. K. (2013). Analysis of Population Genetic Structure of Indian *Anopheles culicifacies* species A Using Microsatellite Markers. *Parasites & Vectors*, 6, 166. https://doi.org/10.1186/1756-3305-6-166

Thomas, T., De, T. D., Sharma, P., Lata, S., Saraswat, P., Pandey, K. C., Dixit, R. (2016). Hemocytome: deep sequencing analysis of mosquito blood cells in Indian malarial vector Anopheles stephensi. *Gene*, 585(2), 177–190.

Tyagi, B.K. (2023). *Desert Malaria: An emerging malaria paradigm and its global impact on disease*. Springer-Nature, Singapore, 416 pp.

Tyagi, B.K., Munirathinam, A., Venkatesh, A. (2015). A Catalogue of Indian Mosquitoes. *International Journal of Mosquito Research*, 2(2), 50–97.

Tyagi, B.K., Munirathinam, A., Krishnamoorthy, R., Venkatesh, A. (2012). *A field-based handbook of identification keys to mosquitoes of public health importance in India. Volume 1*. Taxonomy and Biodiversity Cell. Centre for Research in Medical Entomology, 55 pp.

Tyagi, B.K., Munirathinam, A., Krishnamoorthi, R., Baskaran, G., Venkatesh, A., Kalimuthu, M., Govindarajan, R. (2014). *A field-based handbook of identification keys to 4th instar larvae of mosquitoes of public health importance in India. Volume 2*. Taxonomy and Biodiversity Cell. Centre for Research in Medical Entomology, 45 pp.

Vijay, S., Rawal, R., Kadian, K., Raghavendra, K., Sharma, A. (2015). Annotated Differentially Expressed Salivary Proteins of Susceptible and Insecticide- Resistant Mosquitoes of Anopheles stephensi. *PLoS ONE*, 10(3), e0119666.

WHO. (2019). *Malaria report 2019*. World Health Organization, Geneva.

Zhao, Z., Zung, J.L., Hinze, A., et al. (2022). Mosquito brains encode unique features of human odour to drive host seeking. *Nature*, 605, 706–712. https://doi.org/10.1038/s41586-022-04675-4

12 Functional Genomics
Indian Malarial Vectors

Tanvi Singh
ICMR-National Institute of Malaria Research, Delhi, India; AcSIR-Academy of Scientific & Innovative Research, Ghaziabad, India

Seena Kumari
ICMR-National Institute of Malaria Research, Delhi, India

Nirmala Sankhala and Pooja Yadav
ICMR-National Institute of Malaria Research, Delhi, India; AcSIR-Academy of Scientific & Innovative Research, Ghaziabad, India

Sanjay Tevatiya
ICMR-National Institute of Malaria Research, Delhi, India

Lalita Gupta
Chaudhary Bansi Lal University, Bhiwani, India

Rajnikant Dixit
ICMR-National Institute of Malaria Research, Delhi, India; AcSIR-Academy of Scientific & Innovative Research, Ghaziabad, India

12.1 INTRODUCTION

Mosquitoes spread pathogens that have critical and rampant consequences. They transmit malaria, chikungunya, dengue hemorrhagic fever, Japanese encephalitis, lymphatic filariasis, etc. Among these the infection with the highest mortality is malaria. For the year 2020, the World Health Organization reported 241 million new cases. Malaria deaths increased by 12% compared to 2019, to an estimated 627,000 (Monroe et al., 2022).

Anopheles mosquitoes are responsible for the transmission of malaria pathogens. Among 58 recognized anopheline species involved in malaria transmission, *Anopheles stephensi* and *An. culicifacies* are dominant vectors in India (Subbarao et al., 2019). Therefore, understanding these species' molecular physiology and genetics can aid effective vector control. A clue to our understanding of the complex biology of these anophelines lies in their genomes. Comprehending fundamental biological processes at the genome level among these species is vital and holds potential for the prevention of malaria transmission. Additionally, the knowledge of how genomes hold the transformation to ecotypes and eventual speciation is the key to laying solid groundwork for further research in this area.

Classical studies have provided valuable insights regarding genome-size variability and complexity among anophelines, even though these studies still find utility in guiding genome-sequencing

DOI: 10.1201/9781003326991-18

efforts. Nevertheless, the relevance of this genomic data lies in the functional aspect. Translation of these resources into knowledge regarding underlying physiology and mechanisms is of utmost importance. Here we strive to compile and review the current status of functional genomics of the Indian malarial vectors.

12.2 FROM GENOME TO BIOLOGY

Rapid advances in genome technologies have disclosed meaningful information on phenotype-specific transcriptomics and proteomics. These resources allow for detailed comparative analyses across species. There are multiple protocols for establishing a correlation between genotype and phenotype. Figure 12.1 summarizes them and provides insights into the twisty turvy path from genome to biology.

12.2.1 Comparative Genomics

Comparative genomics is a comprehensive approach that compares two or more genomes to analyze the similarities and differences to investigate the molecular aspects of the functional biology of the individual genomes. Concerning different levels of the genomes, comparative studies have multiple standpoints about the organisms. Broadly three analyses provide significant biological insights (Wei et al., 2002). These include the analyses:

(1) at genome structure level – for repeats, genome rearrangement at the gene level, and synteny;
(2) at the coding region level for gene content, transcript content, protein content, orthologs, and paralogs; and
(3) at the noncoding region level that includes a prediction of regulatory elements.

12.2.2 Transcriptional Profiling and Proteomics

Transcriptomic profiling is also known as "expression profiling." It involves quantitative analysis of the expression of many genes in cells/tissue samples at the transcription (RNA) level. It begins with the collection of biological samples, followed by RNA extraction (in most cases, total RNA) at fixed time points in a time series, thereby creating a "photograph" of expression patterns. This photograph is studied using techniques like real-time PCR or RNA-Seq. Similarly, mass spectroscopy can take a "snapshot" of protein expression.

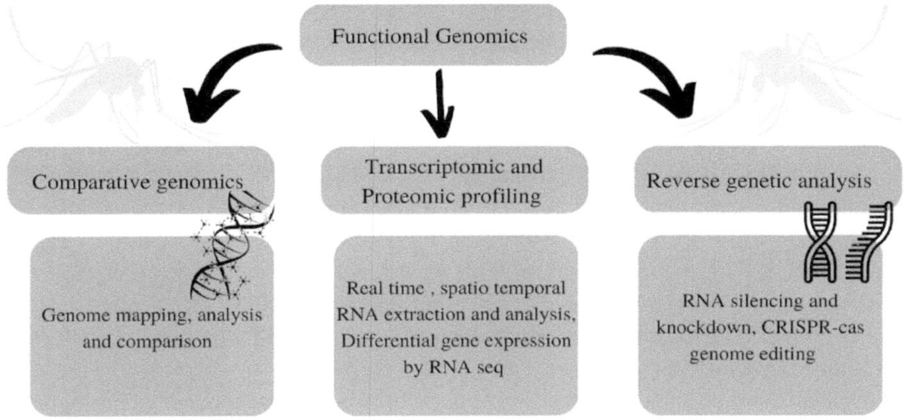

FIGURE 12.1 Various molecular biology techniques for functional genomics.

12.2.3 Reverse Genetic Analysis

One way to establish a correlation between genotype and phenotype is to study the effect during the absence of that trait. A technique to silence gene expression is RNA interference using microinjection.

This gene knockdown technique utilizes a double-stranded siRNA molecule that is a sequence complementary to the gene of interest. Once the RNAi cascade initiates, the Dicer enzyme starts to process siRNA. This event leads to degraded mRNA and destroys instructions needed to build particular proteins (Agrawal et al., 2003). In India, various studies have been conducted based on this protocol and accumulated significant data and knowledge to link the mosquito genome with a specific phenotype essential for mosquito survival and all-around fitness.

For better understanding, we divided the overview of studies into two broad categories: (i) gene-specific studies, and (ii) tissue-specific studies (Tables 12.1 and 12.2).

12.3 FACTORS AFFECTING OLFACTION

The collaborative actions of the environmental and genetic factors are crucial to shaping host-seeking behavior in mosquitoes. During evolution, anautogeny proved to be an adaptive advantage for mosquitoes. There is a clear correlation between host-seeking and adaptive success. Processing olfactory cues followed by salivary actions facilitates locating the host and eventual blood meal uptake, by an adult female mosquito. Therefore, elucidating the underlying molecular mechanism of odor sensing through the detection machinery (olfactory system) might unravel molecular pathways that can be potential targets to disrupt feeding behavior in the mosquito (Das De and Dixit, 2020).

For a successful blood-feeding event, an adult female mosquito manages various behavioral coordinates such as foraging, navigating, and landing over a suitable host. Host-seeking is followed by probing for a site to pierce and siphon blood within two minutes. Later, a blood-fed mosquito is required to enter a new habitat that accommodates successful oviposition. Before blood acquisition, the mosquito's senses heighten, which gradually undergoes a suppressive effect after the blood-feeding. This shift of physiological and metabolic changes, together ensures the success of behavioral events, though the underlying molecular mechanism remains ambiguous.

Mosquitoes interact with the outer environment using their peripheral appendages: antennae, maxillary palp, and labium equipped with hair-like structures known as sensilla. Sensilla are responsible for the reception of environmental cues such as odors. Sensilla consists of a diverse range of receptors. Chemosensory proteins such as sensory appendage protein (SAP) and odorant binding proteins (OBP) with olfactory receptors (OR) are principal molecules for host-seeking and feeding.

A study meticulously traced olfaction-associated molecular events in *An. culicifacies* to apprehend the underlying genetics behind these events. The authors conducted a comprehensive RNA-Seq analysis of *An. culicifacies* olfactory tissues before, and after blood meal uptake. This research revealed a change in the olfactory genes that might be vital in administering diverse behavioral responses.

Age-dependent transcriptional profiling depicted that the adult female mosquito's chemosensory system gradually developed to drive the host-seeking and blood meal uptake from the timepoint of 5–6 days old. Time scale expression analysis of OBPs demonstrated a strong association with the peak biting time – dusk to midnight. Moreover, blood meal-induced switching of various sets of *OBP* genes and OR expression overlapped with the change in the innate physiological status of the mosquitoes. A functional correlation that coherent actions of olfactory encoded factors enabled adult female mosquitoes to administer the blood-feeding associated with complex behavioral activities is evident from the study.

The article also reported the dominant expression of two sensory appendage proteins (SAP-1 and SAP2) in the legs of *An. culicifacies*, suggesting that this genus might have the advantage of having more sensitive appendages than *An. stephensi* (Das De et al., 2018).

TABLE 12.1
List of Gene-specific Studies on Indian Malarial Vectors

S.No	Associated Physiology	Biological Process	Gene Studied	Vector Study Model	Title of the Study
1	Behavioral	Olfaction	SAP	*An. culicifacies*	A synergistic transcriptional regulation of olfactory genes drives blood-feeding associated complex behavioral responses in the mosquito *Anopheles culicifacies*
2			OBP	*An. culicifacies*	
3		Olfaction and stress response	Attractin	*An. culicifacies*	Transcriptional responses of *attractin* gene in the mosquito *Anopheles culicifacies*: a synergistic neuro-olfactory regulation
4		Mating	HPX12	*An. stephensi*	A testis-expressing heme peroxidase *HPX12* regulates male fertility in the mosquito *Anopheles stephensi*
5			QTC	*An. culicifacies*	Sex-specific molccular responses of quick-to-court protein in Indian malarial vector *Anopheles culicifacies*: conflict of mating versus blood-feeding behavior
6		Gut- microbiome -brain homeostasis	HR -Histamine Receptor NRY-Neuropeptide Y Leukokinin	*An. Culicifacies*	Bidirectional microbiome-gut-brain-axis communication influences metabolic switch-associated responses in the mosquito *Anopheles culicifacies*
7	Nutrition	Sugar metabolism	As-TreH	*An. stephensi*	Molecular and functional characterization of trehalase in the mosquito *Anopheles stephensi*
8	Reproduction	Oviposition	Transferrin	*An. culicifacies*	Functional disruption of transferrin expression alters reproductive physiology in *Anopheles culicifacies*
9	Immunity	Immune response	FREP13	*An. stephensi*	Hemocyte-specific FREP13 abrogates the exogenous bacterial population in the hemolymph and promotes midgut endosymbionts in *Anopheles stephensi*
10			HPX15	*An. stephensi*	Characterization and expression analysis of gene encoding heme peroxidase HPX15 in major Indian malaria vector *Anopheles stephensi* (Diptera: Culicidae)
11			APN1	*An. stephensi*	Identification, molecular characterization, and expression of aminopeptidase N-1 (APN-1) from *Anopheles stephensi* in SF9 cell line as a candidate molecule for developing a vaccine that interrupt malaria transmission
12		Gut- microbiome homeostasis	HPX2	*An. stephensi*	Heme peroxidase 2, a peroxinectin-like gene, regulates bacterial homeostasis in *Anopheles stephensi* midgut
13		Immune response	AsDuox	*An. stephensi*	*Anopheles stephensi* dual oxidase silencing activates the thioester-containing protein 1 pathway to suppress *Plasmodium* development
14			SOCS	*An. stephensi*	Molecular characterization of *SOCS* gene and its expression analysis on *Plasmodium berghei* infection in *Anopheles culicifacies*
15			AsDBLOX	*An. culicifacies*	Identification, characterization, and expression analysis of *Anopheles stephensi* double peroxidase

TABLE 12.2
Tissue-Specific Studies on Indian Malarial Vector

S. No.	Tissue Studied	Vector	Title
1	Fat body	*An. stephensi*	Response to blood meal in the fat body of *Anopheles stephensi* using quantitative proteomics: toward new vector control strategies against malaria
2	Salivary gland	*An. stephensi*	Salivary gland transcriptome analysis in response to sugar feeding in malaria vector *Anopheles stephensi*
			Unraveling dual feeding associated molecular complexity of salivary glands in the mosquito *Anopheles culicifacies*
3	Hemocyte	*An. stephensi*, *An. culicifacies*	Hemocyte RNA-Seq analysis of Indian malarial vectors *Anopheles stephensi* and *Anopheles culicifacies*: from similarities to differences
4	Salivary gland, midgut, and hemocyte of infected mosquito	*An. stephensi*	Genetic changes of *Plasmodium vivax* tempers host tissue-specific responses in *Anopheles stephensi*
5	Brain, midgut	*An. culicifacies*	Bidirectional microbiome-gut-brain-axis communication influences metabolic switch-associated responses in the mosquito *Anopheles culicifacies*

12.4 GENETIC ELEMENTS: SALIVARY GLANDS

Adult female salivary glands are specialized pairs of tri-lobed single-layered epithelial tissues. Salivary glands release diverse biomolecules essential for blood-feeding, e.g., anticoagulants, vasodilators, sialo kinins, etc. For the last decade, multiple transcriptomic and proteomic studies identified various salivary factors in Indian Malarial Factors.

A study observed increased transcripts in the blood-fed salivary glands compared to sugar-fed mosquitoes. These differentially expressed transcripts were autophagy (APHAGs), clip domain serine proteases (CLIPs), peroxidase (PRDX), fibrinogen-related proteins (FREPs), relish-like proteins (RELs), inhibitors of apoptosis (IAPs), peptidoglycan recognition proteins (PGRPs), and scavenger receptor (SCRs) family of proteins. The transcripts from these families were higher in response to blood-feeding. (Sharma et al., 2015).

The CLIP family transcripts were upregulated, while downregulation of antimicrobial peptides (AMPs) in response to blood-feeding was detected. A real-time PCR analysis verified these observations. It further strengthened the notion that immune components participate in physiological responses during blood meal uptake by the mosquito. It is likely that these components released by the salivary gland during the process of blood-feeding aid the evasion of complement proteins in host blood which have a damaging effect otherwise.

This comprehensive transcriptomic analysis for hematophagy provided preliminary evidence that mosquito salivary glands encode meal-specific diverse biomolecules. Comparative molecular analysis of blood meal impact supported the hypothesis that adult female salivary glands evolved with the ability to manage meal-specific responses. The same study provided another valuable insight regarding meal-specific cellular changes, such as increased apoptosis. Another study validated these findings by analyzing differential protein expression in sugar-fed and blood-fed in *An. culicifacies* salivary gland using mass spectroscopy. The differential expression reported a 1.79-fold increase in protein with the caspase-like domain and a 1.5-fold increase in the beclin-1 protein involved in autophagy (Rawal et al., 2016). Therefore, the data suggests mosquitoes adopted an elaborated mechanism to tackle unwanted microorganisms or pathogens and stress enhanced during the blood-feeding process in their salivary glands.

Another salivary factor, HPX12, was characterized in a study (Kumari et al., 2022a). dsRNA-mediated silencing experiments depicted that salivary *AsHPX12* might regulate pre-blood meal-associated behavioral properties such as probing time, probing propensity, and host attraction.

The above-mentioned studies are indicative of the fact that salivary-specific genetic elements are essential for the survival of mosquito species.

12.5 A-TELL-TALE BRAIN

Mosquitoes have well-developed neuroendocrine systems, and their brains consist of neurosecretory cells that regulate growth and development. Brain studies of the Anopheles species identified several classes of genetic elements associated with neuronal functioning, synapses, and sensory reception, which might affect the survival and transmission ability of the mosquito (Dwivedi et al., 2014).

A recent study highlighted the role of neuropeptides in feeding behavior. The study focused on establishing the concept of the gut-brain axis in mosquitoes, simultaneously accentuating the brain factors responsible for suppressed host-seeking. An extensive RNA-Seq pattern of neuro-signaling and neuro-modulatory genes alteration in the brain for *An. culicifacies* was analyzed. A transient increase in Neuropeptide Y and leukokinin expression was observed in brain tissue after blood meal uptake, and the expression peaked 30 hours post-blood meal again. It can be speculated that the transient increase in expression of these neuropeptides might be due to gut distension and then again 30 hours after blood meal uptake to suppress host-seeking (Das De et al., 2017). Satiety signals in mosquitoes tend to suppress host-seeking as the food requirements are already fulfilled, and host-seeking otherwise would be an energetically wasteful process. Leukokinins share a significant amount of homology with tachykinins in vertebrates. Tachykinins are known to regulate the hunger-satiety cycle in higher animals.

The same study provided insights regarding another genetic factor in the brain – the histamine-gated-chloride channel (HR) gene. The knockdown data for this factor likely disrupted the histamine signaling, consequently decreasing the expression of genes responsible for immunity, such as defensins. Due to this decrease in immunity, increased microbial proliferation was observed in mosquitoes with silenced HR.

This is how the master control canter – the brain – controls vital mechanisms of survival like immunity, microbial proliferation, and blood-feeding.

12.6 THE EMERGING TRIO: NUTRITION, REPRODUCTION, AND OMICS

Reproductive fitness is the by-product of proper nutrition in all species. Anthropophilic female mosquitoes are famous for their preference for human hosts, but plant nectar is a regular energy source in their diets. On the other hand, male mosquitoes are exclusively dependent on plant nectar or alternative sugar sources. Disrupting this flow from nutrition to reproductive fitness is a new-era model for vector control.

12.6.1 From Genes to Nutrition

Many unique physiological features play a part in the incredible evolutionary success of mosquitoes. These functions also tend to place high-energy demands on mosquitoes. Thereby carbohydrate metabolism (CM) must be synchronized with these needs. (Hou et al., 2015). Trehalose, a non-reducing disaccharide, is one of the significant blood sugars in insects. It is an instant source of energy for mosquitoes and hence, vital for their survival. Trehalose is cleaved into glucose or soluble trehalose by an enzyme trehalase. A recent study characterized trehalase-2 homolog protein from the mosquito *Anopheles stephensi*. The study reported a consistent upregulation until 42 hours of starvation, which was restored to normal levels after sugar supplementation, suggesting a crucial role of the trehalase enzyme in stress tolerance. Additionally, the enrichment of expression in the midgut and salivary glands of the *Plasmodium vivax*-infected mosquitoes revealed that *As-TreH* might favor pathogen development and survival in the mosquito vector. Furthermore, the study implied a potential correlation between trehalase expression and the reproductive physiology of adult female mosquitoes. This correlation is strengthened by functional knockdown by dsRNA-mediated silencing that demonstrated a significant loss in the egg numbers, possibly by impairing energy metabolism in the developing oocytes (Tevatiya et al., 2020).

The famous saying "We are what we eat" implies to mosquitoes, too, and thus "mosquitoes are what they eat." Therefore, exploring their nutritional habits and physiology in detail will open paths toward developing efficient vector control tools (Kumar et al., 2017).

12.6.2 From Genes to Fecundity

The proliferation and survival of species depend on reproductive fitness and hence, the reproductive fitness of the anophelines is an apt target to prevent malaria transmission. In mosquitoes, iron metabolism is crucial for vitellogenesis (Zhou et al., 2007).

It would not be an overstatement that iron metabolism equates to reproductive fitness. Transferrin is a family of extracellular proteins which contains a high-affinity iron-binding site in their two homologous lobes. A 2021 Indian study characterized and conducted a functional analysis for transferrin protein (AcTrf1a) in *An. culicifacies. AcTrf1a* has a putative N-linked glycosylation site, a characteristic feature of certain mammalian transferrin and non-blood-feeding insects. The N-terminal lobe of the transferrin was further confirmed by structural modeling prediction. Spatial and temporal expression analysis under multiple pathophysiological conditions demonstrated that *AcTrf1a* was expressed in the fat body and ovary. An increased expression for AcTrf1a was observed after blood meal uptake. Further, non-heme iron supplementation of FeC13 at 1mM concentration augmented the *AcTrf1a* expression in the fat body. Additionally, it also increased the reproductive fecundity of gravid adult female mosquitoes. RNAi-mediated silencing of *AcTrf1a* caused a significant reduction in egg count, indicating the role of *AcTrf1a* in oocyte maturation (Rani et al., 2022).

Another study analyzed the role of heme peroxidase, a member of the antioxidant enzyme family proteins, in the male fertility of the mosquito *An. stephensi.* It was observed that heme peroxidase 12 (*HPX12*) is a vital factor in protecting the sperms from oxidative stress and maintains semen quality in male mosquitoes. A knockdown of the *HPX12* impaired the sperm's motility and viability and also caused a significant downregulation of male accessory gland transcripts expression, which likely encodes for putative accessory gland proteins. Mating with *HPX12* knockdown male mosquitoes resulted in a ~ 50% reduction in egg-laying and decreased larval hatchability for the gravid female mosquitoes. The data further indicated increased ROS levels in the *HPX12* mRNA-depleted mosquitoes, establishing a strong correlation between HPX12 and sperm health (Kumari et al., 2022b). Since there is a direct correlation between high reproductive capacity and mosquito population, manipulating the male mosquito's reproductive physiology could effectively combat vector-borne diseases.

12.7 THE ALLEGORY OF MOSQUITO'S IMMUNE FACTORS

The mosquito defends itself by recruiting physical and physiological barriers that restrain the entry of the pathogen and the eventual establishment of the pathogen within the mosquito. However, in pathogen entry, the insect ascends an innate cellular and humoral immune response against the pathogen, restricting the pathogen's proliferation to non-pathogenic levels (Weber et al., 2020). Here, we enlist the genetic factors associated with immune response in mosquitoes.

12.7.1 Immune Responses: The Underpinnings of Vector-pathogen Co-evolution

An evolutionary contest emerged due to counter-adaptations for survival and infections; owing to this natural selection, only genus Aedes mosquitoes can transmit the dengue fever virus, and only Culex mosquitoes transmit the Japanese encephalitis virus. Likewise, amid almost 3,000 known species of mosquitoes, only a subdivision of species from the genus Anopheles can transmit Malaria (Riehle et al., 2006). As it happens among the susceptible Anopheles subset, the idiosyncratic genetic factors come into play (genetic factors for tackling the pathogen), rendering some individuals resistant to infection, giving rise to refractory phenotype (Niaré et al., 2002). The genetic model of refractory phenotypes signifies the importance of immune mechanisms in this genus.

12.7.2 Genetic Basis of Immune Mechanisms

Pathogen elimination by mosquitoes is carried out by three primary mechanisms: cell-mediated phagocytosis, melanization, and lysis. These pathways are initiated by pattern recognition receptors (PRRs). The elements responsible for killing can be further divided into cellular and humoral components (Agrawal et al., 2003; Bartholomay et al., 2004). The two required courses of action by insect innate immune response are:

- The humoral response involves a complement-like system accompanied by the transcriptional upregulation of AMPs and various immune effectors.
- A cell-mediated response comprised phagocytosis and melanization of the parasite by elaborated signaling pathways – primarily – Toll, Imd, and JAK-STAT. Regardless of this compartmentalization, the line between cellular and humoral immunity is hazy as many humoral components induced by hemocytes participate in cellular immune responses (Hillyer, 2010; Rani et al., 2021).

12.7.2.1 Pathogen Recognition Receptors

The FREPs are a PRR family exhibiting anti-*Plasmodium* activity. A recent comparative RNA-Seq analysis study of *P. vivax* of infected and non-infected mosquito hemocytes in the Indian *An. stephensi* strain identified at least 13 FREPs (Chauhan et al., 2020). Transcriptional profiling of FREP transcripts depicted distinct responses against various pathophysiological conditions, whereas exclusive expression of FREP12 after ten days of *P. vivax* infection was observed. The same study reported a significant delay of 24 h in the enrichment of gut endosymbionts in the FREP13-silenced mosquitoes. An Indian infectivity study in *An. stephensi* (wild caught) demonstrated that the total microbiota of the mosquitoes could negatively impact parasite development. Thus, reducing the competence of mosquitoes to disease transmission (Kalappa et al., 2018).

12.7.2.2 Reactive Oxidative Immune Mediators

Heme peroxidase (HPX2) is a crucial regulator of midgut epithelial nitration and antiplasmodial immunity that intensifies nitric oxide toxicity in *Anopheles gambiae* (Oliveira et al., 2012). Nitric oxide is a free radical formed during the oxidation process of L-Arginine to L-Citrulline by the nitric oxide synthase enzyme A study divulged a negative correlation between the expression of the *AsHPX2* gene (laboratory-reared Indian strain *An. stephensi*) and gutted bacterial growth (Kajla et al., 2016; Kakani et al., 2020). Furthermore, it was observed that blood-feeding drastically reduced *AsHPX2* gene expression leading to enhanced growth of midgut microbiota (Kumar et al., 2018).

Induction of the STAT (Signal Transducer of Activation) pathway leads to nitric oxide synthase transcription (Gupta et al., 2009). In mosquitoes, activation of the STAT pathway is regulated by a negative feedback loop monitored by the suppressor protein-SOCS. In *An. culicifacies*, SOCS mRNA is significantly induced in Plasmodium-infected midgut tissue. This could be indicative of the fact that SOCS suppress the STAT pathway leading to a reduced Nitric Oxide Pathway (Dhawan et al., 2015).

12.7.2.3 Phagocytosis

Phagocytosis is an immune process used for killing and eliminating tiny microorganisms. Thioester-containing proteins (TEPs) are involved in the phagocytosis process. Induction of the Duox gene was observed in Plasmodium-infected mosquito midguts. RNA interference-mediated silencing of this gene hindered parasite development via activation of the TEP1 pathway, ultimately leading to phagocytosis (Kakani et al., 2020).

12.8 CONCLUSION: THE MALARIA WARFARE

A corner so dark and deep, even after the combined effort to sweep the buzz of mosquitoes still hinders sleep.

In this war against malaria, it is essential to understand the enemy's nature. Therefore, a detailed and accurate map of tactics adopted by our foe: the mosquito, is vital to end this disease. One way to achieve this dream is by creating tools to block their survival in every sphere of their existence. Thereby we ought to base our understanding of biology on the functional genomics of the vector. This would help target the physiological process efficiently (Figure 12.2). Functional genomics would allow us to outsmart anopheline survival mechanisms at each step – olfaction and host-seeking, deriving nutrition, mating, blood-feeding, pathogen infection, and reproduction.

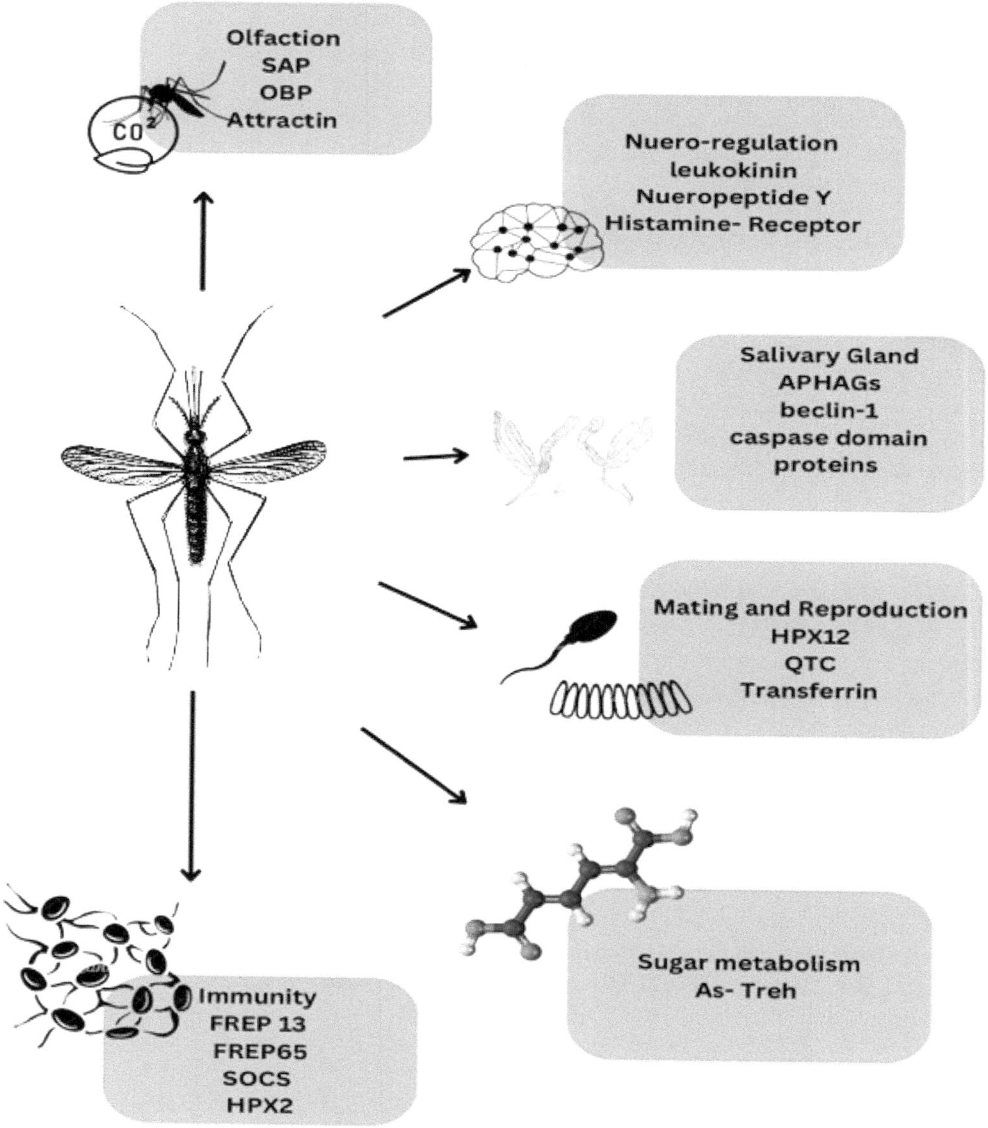

FIGURE 12.2 Molecular targets in Indian malarial vectors.

REFERENCES

Agrawal, N., Dasaradhi, P.V., Mohmmed, A., Malhotra, P., Bhatnagar, R.K., and Mukherjee, S.K. (2003). RNA interference: biology, mechanism, and applications. *Microbiology and Molecular Biology Reviews: MMBR*, *67*(4), 657–685.

Bartholomay, L.C., Cho, W.L., Rocheleau, T.A., Boyle, J.P., Beck, E.T., Fuchs, J.F., Liss, P., Rusch, M., Butler, K.M., Wu, R.C., Lin, S.P., Kuo, H.Y., Tsao, I.Y., Huang, C.Y., Liu, T.T., Hsiao, K.J., Tsai, S.F., Yang, U.C., Nappi, A.J., Perna, N.T., Chen, C.C. and Christensen, B.M. (2004). Description of the transcriptomes of immune response-activated hemocytes from the mosquito vectors *Aedes aegypti* and *Armigeres subalbatus*. *Infection & Immunology*, 72(7), 4114–26. doi: 10.1128/IAI.72.7.4114-4126.2004

Chauhan, C., Das De, T., Kumari, S., Rani, J., Sharma, P., Tevatiya, S., Pandey, K. C., Pande, V., & Dixit, R. (2020). Hemocyte-specific FREP13 abrogates the exogenous bacterial population in the hemolymph and promotes midgut endosymbionts in *Anopheles stephensi*. *Immunology and cell biology*, 98(9), 757–769.

Choudhury, T.P., Gupta, L., and Kumar, S. (2019). Identification, characterization and expression analysis of *Anopheles stephensi* double peroxidase. *Acta Tropica*, 190, 210–219.

Das De, T., and Dixit, R. (2020). Neuro-Olfactory Regulation and Salivary Actions: A Coordinated Event for Successful Blood-Feeding Behaviour of Mosquitoes. *IntechOpen*. doi: 10.5772/intechopen.90768

Das De, T., Hasija, Y. & Dixit, R. (2018). Transcriptional responses of attractin gene in the mosquito Anopheles culicifacies: A synergistic neuro-olfactory regulation. *Journal of Vector Borne Diseases*, 55(2), 89–97.

Das De, T., Sharma, P., Rawal, C., Kumari, S., Tavetiya, S., Yadav, J., Hasija, Y., & Dixit, R. (2017). Sex-specific molecular responses of quick-to-court protein in Indian malarial vector Anopheles culicifacies: conflict of mating versus blood feeding behaviour. *Heliyon*, 3(7), e00361.

Dhawan, R., Gupta, K., Kajla, M., Kumar, S., Gakhar, S. K., Kakani, P., Choudhury, T. P., & Gupta, L. (2015). Molecular characterization of SOCS gene and its expression analysis on Plasmodium berghei infection in Anopheles culicifacies. *Acta Tropica*, 152, 170–175.

Dwivedi, S. B., Muthusamy, B., Kumar, P., Kim, M. S., Nirujogi, R. S., Getnet, D., Ahiakonu, P., De, G., Nair, B., Gowda, H., Prasad, T. S., Kumar, N., Pandey, A., & Okulate, M. (2014). Brain proteomics of Anopheles gambiae. *OMICS: A Journal of Integrative Biology*, 18(7), 421–437.

Gupta, L., Molina-Cruz, A., Kumar, S., Rodrigues, J., Dixit, R., Zamora, R. E., & Barillas-Mury, C. (2009). The STAT pathway mediates late-phase immunity against Plasmodium in the mosquito Anopheles gambiae. *Cell Host & Microbe*, 5(5), 498–507.

Hillyer, J. F. (2010). Mosquito immunity. *Advances in Experimental Medicine and Biology*, 708, 218–238.

Hou, Y., Wang, X. L., Saha, T. T., Roy, S., Zhao, B., Raikhel, A. S., & Zou, Z. (2015). Temporal coordination of carbohydrate metabolism during mosquito reproduction. *PLoS Genetics*, 11(7), e1005309.

Kajla, M., Kakani, P., Choudhury, T. P., Gupta, K., Gupta, L., & Kumar, S. (2016). Characterization and expression analysis of gene encoding heme peroxidase HPX15 in major Indian malaria vector Anopheles stephensi (Diptera: Culicidae). *Acta Tropica*, 158, 107–116.

Kakani, P., Gupta, L., & Kumar, S. (2020). Heme-Peroxidase 2, a Peroxinectin-Like Gene, Regulates Bacterial Homeostasis in Anopheles stephensi Midgut. *Frontiers in Physiology*, 11, 572340.

Kalappa, D.M., Subramani, P.A., Basavanna, S.K. et al. (2018). Influence of midgut microbiota in Anopheles stephensi on Plasmodium berghei infections. *Malaria Journal*, 17, 385.

Kumar, A., Srivastava, P., Sirisena, P., Dubey, S. K., Kumar, R., Shrinet, J., & Sunil, S. (2018). *Mosquito Innate Immunity, Insects*, 9(3), 95.

Kumar, M., Mohanty, A. K., Sreenivasamurthy, S. K., Dey, G., Advani, J., Pinto, S. M., Kumar, A., & Prasad, T. (2017). Response to blood meal in the fat body of Anopheles stephensi using quantitative proteomics: Toward new vector control strategies against Malaria. *OMICS: A Journal of Integrative Biology*, 21(9), 520–530.

Kumari, S., Chauhan, C., Tevatiya, S., Singla, D., De, T. D., Sharma, P., Thomas, T., Rani, J., Savargaonkar, D., Pandey, K. C., Pande, V., & Dixit, R. (2021). Genetic changes of Plasmodium vivax tempers host tissue-specific responses in Anopheles stephensi. *Current Research in Immunology*, 2, 12–22.

Kumari, S., De, T. D., Chauhan, C., Rani, J., Tevatiya, S., Sharma, P., Pande, V., & Dixit, R. (2022a). Salivary AsHPX12 influences pre-blood meal associated behavioral properties in Anopheles stephensi. *Journal of Vector Borne Diseases*, 59(3), 206–215.

Kumari, S., Tevatiya, S., Rani, J., Das De, T., Chauhan, C., Sharma, P., Sah, R., Singh, S., Pandey, K. C., Pande, V., & Dixit, R. (2022b). A testis-expressing heme peroxidase HPX12 regulates male fertility in the mosquito Anopheles stephensi. *Scientific Reports*, 12(1), 2597.

Monroe, A., Williams, N.A., Ogoma, S., Karema, C., and Okumu, F. (2022). Reflections on the 2021 World Malaria Report and the future of malaria control. *Malaria Journal*, 21(1), 154.

Niaré, O., Markianos, K., Volz, J., et al. (2002). Genetic loci affecting resistance to human malaria parasites in a West African mosquito vector population. *Science*, 298(5591), 213–216.

Riehle, M. M., Markianos, K., Niaré, O., et al. (2006). Natural malaria infection in Anopheles gambiae is regulated by a single genomic control region. *Science*, 312(5773), 577–579.

Oliveira, G., Lieberman, J., & Barillas-Mury, C. (2012). Epithelial nitration by a peroxidase/NOX5 system mediates mosquito antiplasmodial immunity. *Science*, 335(6070), 856–859.

Rani, J., Chauhan, C., Das De, T., Kumari, S., Sharma, P., Tevatiya, S., Patel, K., Mishra, A. K., Pandey, K. C., Singh, N., & Dixit, R. (2021). Hemocyte RNA-Seq analysis of Indian malarial vectors Anopheles stephensi and Anopheles culicifacies: From similarities to differences. *Gene*, 798, 145810.

Rani, J., De, T. D., Chauhan, C., Kumari, S., Sharma, P., Tevatiya, S., Chakraborti, S., Pandey, K. C., Singh, N., & Dixit, R. (2022). Functional disruption of transferrin expression alters reproductive physiology in Anopheles culicifacies. *PLoS ONE*, 17(3), e0264523.

Rawal, R., Vijay, S., Kadian, K., Singh, J., Pande, V., & Sharma, A. (2016). Towards a proteomic catalogue and differential annotation of salivary gland proteins in blood fed malaria vector Anopheles culicifacies by Mass Spectrometry. *PLoS ONE*, 11(9), e0161870.

Sharma, P., Sharma, S., Mishra, A. K., Thomas, T., Das De, T., Rohilla, S. L., Singh, N., Pandey, K. C., Valecha, N., & Dixit, R. (2015). Unravelling dual feeding associated molecular complexity of salivary glands in the mosquito Anopheles culicifacies. *Biology Open*, 4(8), 1002–1015.

Subbarao, S. K., Nanda, N., Rahi, M., & Raghavendra, K. (2019). Biology and bionomics of malaria vectors in India: existing information and what more needs to be known for strategizing elimination of malaria. *Malaria Journal*, 18(1), 396.

Tevatiya, S., Kumari, S., Sharma, P., Rani, J., Chauhan, C., Das De, T., Pandey, K. C., Pande, V., & Dixit, R. (2020). Molecular and Functional Characterization of Trehalase in the Mosquito Anopheles stephensi. *Frontiers in Physiology*, 11, 575718.

Weber, J. J., Kanost, M. R., & Gorman, M. J. (2020). Iron binding and release properties of transferrin-1 from Drosophila melanogaster and *Manduca sexta*: Implications for insect iron homeostasis. *Insect Biochemistry and Molecular Biology*, 125, 103438.

Wei, L., Yueyi, L., Inna, D., John, S., John, P. (2002). Comparative genomics approaches to study organism similarities and differences. *Journal of Biomedical Informatics*, 35(2), 142–150.

Zhou, G., Kohlhepp, P., Geiser, D., Frasquillo, M. del C., Vazquez-Moreno, L., & Winzerling, J. J. (2007). Fate of blood meal iron in mosquitoes. *Journal of Insect Physiology*, 53(11), 1169–1178.

13 Oleuropein-Induced Autophagy

A Plausible Mechanism for Anti-Malarial Drug Resistance

Nikunj Tandel
Institute of Science, Nirma University, Ahmedabad, India

Neil Roy Chaudhary
CSIR-Institute of Microbial Technology, Chandigarh, India

Tejaswini Vaidya and Mihir Makwana
Institute of Science, Nirma University, Ahmedabad, India

Rajeev K. Tyagi
CSIR-Institute of Microbial Technology, Chandigarh, India

13.1 INTRODUCTION

13.1.1 Malaria: A Systemic Inflammatory and Infectious Disease

Malaria, a serious global health problem, is responsible for greater morbidity and mortality. India substantially adds to the overall burden in Southeast Asia (WHO, 2022). From 2000 to 2019, the number of malaria deaths decreased sizably from 897,000 in 2000 to 577,000 in 2015 and 568,000 in 2019. However, an estimated 625,000 people died from malaria in 2020 and saw a 10% rise from 2019. The deaths were slightly decreased in 2021 to 6,19,000 (WHO, 2022). The percentage of total malaria deaths among children under the age of five fell from 87% in 2000 to 76% in 2015. It is declared an endemic disease in majority of the African and Asian countries, and 40% of world's populations reside in these endemic areas (WHO, 2021). Malaria-endemic countries, especially underdeveloped African countries, are facing economic barriers as large populations are falling prey to this disease (WHO, 2021). Between 2000 and 2021, there were an anticipated 2 billion cases of malaria worldwide and 11.7 million fatalities from the disease (WHO, 2021, 2022).

The causative agent of malaria is a protozoan parasite of *Plasmodium* species. The life cycle of the malaria parasite is quite complex since it spawns its cycle in the primary (arthropod vector: mosquitoes)and intermediate (humans) host (Milner, 2018). An arthropod vector is a female *Anopheles* mosquito. *Plasmodium* is usually host- and vector-specific, and it infects a limited range of hosts and vectors (Sato, 2021). Human malaria is mostly caused by four distinct *Plasmodium* spp., *P. vivax*, *P. falciparum*, *P. malariae*, and *P. ovale*. The parasite *P. falciparum* has been the deadliest in terms of causing greater mortality and morbidity (Sato, 2021). The higher parasite load associated with *P. falciparum* infection is largely responsible for the malaria pathology (Milner, 2018; Phillips et al., 2017). Malaria symptoms including fever, chills, headache, aches in the

DOI: 10.1201/9781003326991-19

muscles, and exhaustion. The most vulnerable groups to catching malaria are young children, pregnant women, and tourists from non-endemic regions (Ashley, Pyae Phyo, & Woodrow, 2018; Phillips et al., 2017).

13.1.2 Understanding the Host-Parasite Interaction

Malaria infection is transmitted by a female *Anopheles* mosquito bite that releases the sporozoites while taking blood meal through the proboscis into the subcutaneous tissues (Phillips et al., 2017). These sporozoites travel from the skin to the liver via the bloodstream and render infection to the hepatocytes by invading Kupffer cells, followed by forming schizont (liver-stage/pre-erythrocytic stage). Schizogony is a replicative stage where the parasite undergoes nuclear division, but not cytoplasmic division to form progeny called merozoites (Phillips et al., 2017; Sato, 2021). The primary maturation of merosomes takes place within the liver, and these merosomes rupture and give rise to merozoites. Sporozoite does not undergo immediate asexual replication in the case of *P. ovale* and *P. vivax* instead it enters a dormant stage known as hypnozoites (Antonelli et al., 2020; Santos-Reis & Nina, 2022). These merozoites enter the bloodstream and infect the erythrocytes to start an asexual blood-stage infection. The parasite breaks down hemoglobin into amino acids, and forms a malarial pigment called hemozoin (erythrocytic/blood-stage) (Milner, 2018; Sato, 2021). The asexual replication of the malaria parasite continues for indeterminate generations and then a few parasites exit the asexual multiplication and transform into gametocytes and start sexual stages. Blood-staged parasites develop the clinical manifestations of malaria which are detectable through various means (Milner, 2018; Phillips et al., 2017).

The gametocytes are then taken up by the mosquitoes during a blood meal. The second cycle begins and is termed the "sporogonic cycle." The microgametes present within the stomach of mosquitoes penetrate the macrogametes, which then generates zygotes (Venugopal, Hentzschel, Valkiūnas, & Marti, 2020). These zygotes move to the midgut of the mosquito and develop into oocysts. Following this, the oocysts develop into the sporozoites and move to the salivary glands of mosquitoes. The mosquito then inoculates these infected sporozoites into a new human host commencing the new cycle (Meibalan & Marti, 2017; Sato, 2021; Venugopal et al., 2020).

13.2 DRUG RESISTANCE IN MALARIA

Quinine (derived from the bark of the chinchona tree) was the first medicine to demonstrate efficacy in the prevention of disease (Slater, 2009). Atabrine (quinacrine) was clinically approved for the treatment of malaria in 1983 (Slater, 2009). Each drug follows a different mechanism of action against stages of the malaria parasite. The majority of drugs target the blood-stage infection, and some disrupt hypnozoites in the liver or gamete production by killing the asexual blood-stage infection (Fong & Wright, 2013; Tse, Korsik, & Todd, 2019). There are several aryl amino alcohol molecules, anti-folate compounds, including artemisinin and its derivatives available for malaria therapy (Belete, 2020; Tse et al., 2019). One of the first drugs to be effectively administered, chloroquine, became potent due to its effectiveness, stability, and low toxicity. During the 1960s and 1970s, it was the recommended drug for malaria treatment (Belete, 2020; Fong & Wright, 2013). The chloroquine inhibits the polymerization of heme to form hemozoin, causing *Plasmodium* to die irreversibly (Fong & Wright, 2013). However, the evolution of drug resistance has been a key limiting factor of anti-malarial drugs.

13.2.1 Anti-Malarial Drug Resistance

Drug resistance is a major challenge in the treatment and control of malaria, and understanding the underlying mechanisms is crucial for developing effective therapies (Su, Lane, Xia, Sá, & Wellems, 2019). The first incidence of resistance was discovered in Thailand in 1957, and it quickly spread throughout South and Southeast Asia, Sub-Saharan Africa, and South America until the end of 1970 (Packard, 2014). Molecular investigations revealed that a few genes, including *pfmdr1* and *2*, and *pfcrt*, are associated with multidrug resistance in *P. falciparum* (Ikegbunam et al., 2019; Somé et al., 2016). These genes encode for a protein linked with a transmembrane area in the digestive vacuole of the parasite, that possibly exports the drug from the food vacuole (Wicht, Mok, & Fidock, 2020). Mefloquine (Larium) attracted attention for malaria prophylaxis worldwide in the late 1970s. Unfortunately, the parasite showed an evolution of resistance against mefloquine, and copy number and polymorphism of the *pfmdr1* gene is shown to be associated with the resistance (Gupta et al., 2018). Later on, pyrimethamine with a combination of chloroquine was widely used in prophylaxis. However, different mutations in this gene responsible for resistance were identified in different geographical locations (Rasmussen, Alonso, & Ringwald, 2022; Roux et al., 2021). Furthermore, another class of compound artemisinin has shown a rapid drug-activated killing of both the erythrocyte stage and early sexual gametocyte forms of *P. falciparum* parasites within hours (Rasmussen et al., 2022). However, the first resistance case against artemisinin was reported in western Cambodia later followed by Vietnam, Myanmar, Thailand, and Laos (Imwong et al., 2017; Rasmussen et al., 2022). The increasing anti-malarial drug resistance was the cause of discovery of novel therapeutic drugs. From 2001 WHO recommended four Artemisinin combination therapy (ACT) artemether-lumifantrine (Coartem), artesunate-mefloquine, artesunate-amodiaquine and artesunate-sulfadoxine/Pyrimethamine (Lover, Baird, Gosling, & Price, 2018; Reyburn, 2010; van der Pluijm, Amaratunga, Dhorda, & Dondorp, 2021). ACTs have been the best therapeutic treatment for endemic malaria countries. Although ACTs have been shown effective in clearing parasite infection, the parasite is reportedly shown to develop resistance against the partner drugs (Nsanzabana, 2019). Resistance in partner drugs is causing ACT treatment failure in several malaria-endemic regions. Delay in parasite clearance even after treatment with ACT characterizes resistance in artemisinin (Nsanzabana, 2019; Tyagi et al., 2018). Parasite clearance is also affected by differences in host immunity, drug efficacy, and many other genetic factors. The general mechanism of resistance is a point mutation or gene amplification that renders decreased drug sensitivity and promotes survival under drug pressure. The use of sub-optimal therapeutic medication, total parasite load, treatment compliance, and other variables could contribute to the resistance (Wicht et al., 2020). The delayed parasite clearance time (DPCT) suggests the reduced sensitivity of the ring stage is affected by mutations in the kelch-13 (Tyagi et al., 2018). Artemisinin resistance evolved due to mutations in the *K13* gene (kelch protein) which codes the kelch propeller region of the parasite *P. falciparum*. Low levels of ubiquitinated proteins are observed in kelch-mutated parasites, which relate to UPR upregulation (Mbengue et al., 2015; Suresh & Haldar, 2018). Various studies have found that artemisinin targets the *P. falciparum* *Pf*PI3K through reduced ubiquitination, where *Pf*PI3K levels and its lipid product PI3P were increased, concluding reduced artemisinin sensitivity suggesting artemisinin resistance (Mbengue et al., 2015; Ouattara et al., 2015; Suresh & Haldar, 2018). The emergence of drug resistance begets the essentiality of devising novel drugs. Vaccination is an effective malaria-control method that confers long-term immunity. However, there have been many challenges in the development of vaccines including antigen diversity, complicated parasite biology, and low immunogenicity (Nlinwe, Kusi, Adu, & Sedegah, 2018; Tandel & Dalai, 2021; Vekemans, Schellenberg, Benns, O'Brien, & Alonso, 2021).

13.2.2 Malaria Vaccine: A Viable Approach to Treat Malaria

Several vaccines targeting different stages of the parasite's life cycle have been developed, but RTS,S/AS01 is the first malaria vaccine to have completed all clinical trials and is commercially accessible in the endemic areas (Duffy & Patrick Gorres, 2020). The RTS,S/AS01 vaccine contains the circumsporozoite protein (CSP) protein, which is co-expressed with the HBsAg viral peptide, and the adjuvant AS01(Neafsey et al., 2015). From 2019, it has been implemented in endemic countries. In phase-3 studies, vaccine effectiveness was 36% after four years of follow-up among children aged 5–17 months. A seven-year follow-up of 5–17-month-old infants who got three doses of vaccination in Kilifi, Kenya and Korogwe, Tanzania revealed a declining impact of effectiveness with time (Neafsey et al., 2015; RTS, S. C. T. P, 2014). The vaccination decreases both child fatalities and the frequency with which a child contracts malaria, especially severe, life-threatening malaria. RTS, S is the only malaria vaccine currently approved by WHO (RTS, S. C. T. P, 2015). According to WHO, the vaccine can be used as an alternative strategy for malaria (of *P. falciparum*) prevention in children living in regions with moderate to high malaria transmission (Vekemans et al., 2021). The results showed that not only is RTS, S comparable to seasonal malaria vaccination preventing malaria, but that combining the two interventions is markedly superior to either intervention alone (Cairns et al., 2022; Chandramohan et al., 2020). Using both together resulted in an approximately 70% reduction in malaria mortality and morbidity and a 60% reduction in uncomplicated malaria. Research to increase the efficacy of the vaccine is still needed. One recent breakthrough was the introduction of the R21 vaccine, which contains a higher density of CSP protein and matrix adjuvant and achieved 76% effectiveness against malaria over a 12-month period (Datoo et al., 2022). Many vaccine options targeting different stages/phases (of the *Plasmodium* life cycle) are currently under trial, but an effective vaccine that may effectively neutralize the malaria infection and confer long-lasting immunity is still needed (WHO, 2021).

13.3 OLEUROPEIN: AN IMPORTANT NATURAL PHYTOCHEMICAL

Phytochemicals are naturally occurring molecules found in plants that are not essential for growth or reproduction. These chemicals have been extensively investigated for their possible health advantages, including antioxidant, anti-inflammatory, and anti-cancer properties (Bansal & Priyadarsini, 2021; Zhang et al., 2015). Phytochemicals are substances that differ in structure and function and include flavonoids, carotenoids, phenolic acids, and alkaloids (Tungmunnithum, Thongboonyou, Pholboon, & Yangsabai, 2018). Each of these chemicals has distinct characteristics and health advantages. Olive oil has been consumed by mankind throughout history, from Palestine, where the first crop was grown, to South America (Scotece et al., 2015). These oils are composed of two fractions: roughly 97-98% saponifiable oil and a minority of non-saponifiable oil, which accounts for 2% of the total oil (Blanco-Benítez et al., 2022; Scotece et al., 2015). The minority fraction contains about 200 chemicals, including phenolic compounds such as phenolic acids, phenolic alcohols, lignans, flavonoids, and secoiridoids, as well as secoiridoid-related phenolics such as oleuropein (OLP), hydroxytyrosol (HTy), and oleacein (OLCN), among many more. Iridoids and secoiridoids are glycosidically bound and are produced from the secondary metabolism of terpenes (Andrewes, Busch, de Joode, Groenewegen, & Alexandre, 2003; Blanco-Benítez et al., 2022). OLP, a secoiridoid, is abundant in *Oleaceas*, *Gentianales*, and *Cornales*, as well as many other plants (Nediani, Ruzzolini, Romani, & Calorini, 2019; Wichers, Soler-Rivas, & Espı, 2000).

Oleuropein (OLP) is a polyphenol compound derived from the olive plant, present in the leaves and oil of its fruit. Ole (*Olea europaea L.*) is the most abundant phenol present in olives and constitutes up to 14% of the olive fruit's dry weight (Amiot, Fleuriet, & Macheix, 1986). Ole is commonly known as zaytoon in the Mediterranean region and is an important part of the Mediterranean diet (Ahamad et al., 2019). OLP has a number of health benefits including antioxidant activity and also protects from cardiovascular and metabolic disorders like diabetes. It is known to regulate blood pressure (Ahamad et al., 2019; Sun, Frost, & Liu, 2017). It is also associated with proliferation and apoptosis by modulating many signaling pathways such as the induction of autophagy (Rigacci et al., 2015). OLP is a glucoside of an ester of elenoic acid and HTy, with an oleosidic skeleton, which is common to secoiridoid glycosidase of plants belonging to the *Oleaceae* family. OLP and HTy are loaded with health benefits. Ole is heterogeneous in nature and its bioavailability, and its metabolism is gender dependent (Ahamad et al., 2019).

Our group, as others (Hassen, Casabianca, & Hosni, 2015) has also been investigated the OLP for its anti-inflammatory properties. It has been studied for its possible anti-cancer qualities in addition to its antioxidant and anti-inflammatory characteristics. OLP is shown to suppress the development and multiplication of many cancer cells (Goulas et al., 2009; Hamdi & Castellon, 2005; Rishmawi, Haddad, Dokmak, & Karaman, 2022). OLP possesses the following biological activities:

(1) **Antioxidant activity**: OLP has potent antioxidant properties as it protects cells from the damage caused by free radicals (Hassen et al., 2015).
(2) **Anti-inflammatory activity**: OLP has been found to inhibit the production of pro-inflammatory cytokines and enzymes and hence embellishes it a useful natural agent for reducing inflammation and associated health problems (Nediani et al., 2019).
(3) **Anti-cancer activity**: OLP has been found to inhibit the growth and proliferation of various types of cancer cells, including breast cancer and prostate cancer cells (Rishmawi et al., 2022).
(4) **Cardiovascular benefits**: OLP has been found to help lower blood pressure, reduce the risk of heart disease, and improve overall cardiovascular health (Piroddi et al., 2017).
(5) **Anti-diabetic activity**: OLP has been found to improve glucose metabolism and insulin sensitivity, potentially making it useful for managing diabetes (Carnevale et al., 2018; Wainstein et al., 2012).
(6) **Neuroprotective activity**: OLP is found to have neuroprotective effects, potentially protects against neurodegenerative diseases like Alzheimer's and Parkinson's (Butt, Tariq, Iahtisham Ul, Naz, & Rizwan, 2021; Cordero, García-Escudero, Avila, Gargini, & García-Escudero, 2018; S. H. Omar, 2010a).
(7) **Antimicrobial activity**: OLP is shown to have antimicrobial properties and fight against bacterial and viral infections (Y. Liu, McKeever, & Malik, 2017; S. H. Omar, 2010b).

Overall, OLP is a diverse natural chemical with a variety of biological functions that have the potential to be beneficial to the human health. These benefits, however need to be explored, and the therapeutic uses should be confirmed with the experimental evidence.

13.3.1 Anti-Inflammatory Effects of OLP

OLP has also been investigated for its anti-inflammatory properties and demonstrated to inhibit the production of pro-inflammatory cytokines such IL-6 and TNF-α. Hence, it could be a valuable natural agent to suppress inflammation and associated issues (Hassen et al., 2015). Our group has explored the anti-inflammatory and signaling potential of OLP in the human THP-1 macrophage model of inflammation (Devi, Negi, Sharma, Tandel, & Tyagi, 2024; Negi et al., 2024; Sharma et al., 2024). The anti-inflammatory effects of OLP were studied using RAW 264.7 cells and a zebrafish model, and IL-6 and IL-1β production was seen suppressed by the OLP treatment. These results suggest that OLP inhibit TLR and MAPK signaling mediated reduction of the inflammatory response (Ryu et al., 2015). OLP-treated PBMC showed low secretion of TNF-α level. In an *ex-vivo* study, OLP showed anti-inflammatory effects at a concentration of 20 μg/ml when stimulated PBMCs were treated with OLP (Miles, Zoubouli, & Calder, 2005).

OLP is known to suppress the activity of metabolic enzymes responsible for eliciting inflammation such as cyclooxygenase-2 (COX-2) and lipoxygenase (LOX). These enzymes are responsible for eliciting pro-inflammatory prostaglandins and leukotrienes, and contribute to mount the chronic inflammation (Yahfoufi, Alsadi, Jambi, & Matar, 2018). Moreover, OLP inhibits the activation of NF-κB, a key regulator of the inflammatory response. NF-κB is involved in the expression of genes that promote inflammation, and thus inhibition of this pathway could reduce inflammation (Nediani et al., 2019). OLP has been shown to affect the activity of other immune effecters such as macrophages which are crucial players in eliciting the inflammatory response (Magrone et al., 2018). OLP regulates the inflammation by downregulating the pro-inflammatory immune markers and upregulating the anti-inflammatory mediators (Vezza et al., 2017).

Our research is consistent with others (Parra-Perez et al., 2022; Yan, Chai, Cai, Miao, & Ma, 2015) wherein OLP may have anti-inflammatory properties via the PI3K-AKT pathway. This has been involved in a variety of cellular functions and associated with a number of inflammatory conditions. The PI3K-AKT signaling pathway is important in a variety of biological activities, including inflammation and its dysregulation has been linked to a variety of inflammatory illnesses, including rheumatoid arthritis and inflammatory bowel disease (Acosta-Martinez & Cabail, 2022; Stark, Sriskantharajah, Hessel, & Okkenhaug, 2015). The mechanism of how OLP could affect this pathway is shown as:

(i) **Binding to PI3K**: OLP may binds to the p110 subunit of phosphatidylinositol 3-kinase (PI3K), which is responsible for PI3K-AKT pathway activation. This binding inhibits PI3K activity, which in turn inhibits downstream signaling (Sain, Kandasamy, & Naskar, 2022).

(ii) **Inhibition of AKT activation**: OLP has been demonstrated to suppress AKT activation, a PI3K downstream effector. AKT is involved in many biological processes, including inflammation (M. Liu, Wang, Huang, Chen, & Li, 2016).

(iii) **Inhibition of NF-κB activation**: The PI3K-AKT pathway is also involved in the activation of NF-κB, a transcription factor that increases the expression of inflammation-related genes, and OLP has been demonstrated to block NF-κB activation, resulting in to decrease the inflammation (L. Liu et al., 2019).

(iv) **Modulation of inflammatory cytokines**: This pathway is also involved in the synthesis of pro-inflammatory cytokines including IL-6 and TNF-α. OLP has been demonstrated to reduce the production of these cytokines, perhaps by the inhibition of PI3K-AKT pathway (Nediani et al., 2019; Rishmawi et al., 2022; Yan et al., 2015).

Overall, the anti-inflammatory actions of OLP are likely attributable to its capacity to target various inflammatory pathways. OLP could be a valuable natural agent to suppress the inflammation induced due to arthritis, cardiovascular disease, and neurological illnesses (Ding et al., 2023).

13.3.2 Anti-Malarial Activity of OLP

OLP is suggested to have anti-malarial activity and a crude extract of OLP was tested using the rodent malaria parasite (*P. berghei*) infected mice (Misganaw, Engidawork, & Nedi, 2019). The treated animals showed significant inhibition in the parasite load. Moreover, two other compounds extracted from the crude extract showed significant anti-malarial activity with no toxicity seen with 2 g/kg dose (Hailesilase, Rajeshwar, Hailu, Sibhat, & Bitew, 2020). Misganaw *et al.* also observed a decrease in parasitemia after administration of *Olea europaea* leaf extract to *P. berghei*-infected mice following Peter's four-day suppressive test and Ran's test aiming at determining the therapeutic efficacy of OLP. It has been proposed that the temperature of the extract stabilizes the activity and causes a drop in metabolic rate and hypothermia leading to the parasite's death. The polyphenolic substances may protect RBCs from oxidative stress and help survive uninfected and parasitized RBCs (Misganaw et al., 2019).

The mechanism of action to confer anti-malarial activity is yet to be dissected. Nonetheless, several mechanisms have so far been proposed. One possible explanation is that OLP interferes with the heme detoxifying process of the parasite, which is crucial for the survival of the parasite. Heme is a poisonous byproduct of the parasite's hemoglobin digestion, and its detoxification is critical for the survival of the parasite. Another proposition is that OLP prevents the parasite from penetrating and multiplying within red blood cells. Collectively, present data advocates for the anti-malarial activity of oleuropein.

13.4 MALARIA PARASITE EMPLOYS AUTOPHAGY TO CIRCUMVENT ANTI-*PLASMODIAL* DEFENSE

A cellular mechanism involved in the maintenance of cellular homeostasis known as autophagy is associated with the development of drug resistance in malaria. Interestingly, autophagy is seen upregulated in the drug-resistant *Plasmodium* parasites (Ray, Mathur, Choubey, Karmodiya, & Surolia, 2022). The chloroquine-resistant parasites have shown increased autophagy as compared to that with the chloroquine-sensitive parasites (Kamil et al., 2022). Employment of autophagy by the drug-resistant parasite enhances survival by promoting the breakdown of damaged organelles and proteins, reduces oxidative stress, and improves food absorption (Coppens, 2011; Ray et al., 2022). The increasing drug resistance issues with the *P. falciparum* increases; therefore, it is critical to understand the mechanism underlying the degradation pathway of the parasite, that in turn may lead to the discovery of new anti-malarial drugs (Belete, 2020). In eukaryotic cells, the autophagosome-lysosome pathway controls protein breakdown and organelle recycling. Autophagy consists of many processes, including cargo selection and tagging, transport to proteolytic machinery, cargo destruction, and polypeptide recycling. There are about 30 autophagy-related genes (ATG) in the yeast. ATG8 is required for the development of autophagosomes.

The anti-malarial drugs target autophagy to kill the parasite, but they are ineffective against *P. falciparum* has acquired tolerance (Hain & Bosch, 2013). *P. falciparum* has downstream proteins rather than upstream kinases. *Pf*ATG was identified as the protein responsible for autophagosome production, retrieval, and vesicle disintegration (Cervantes et al., 2014). *Pf*ATG8 is a phagophore and autophagosome marker identified in the apicoplast membrane of late schizonts and merozoites. Apicoplasts are non-photosynthetic plastids that originated through endosymbiosis. *Pf*ATG8 is a phosphatidylethanolamine (PE) conjugated protein, and *Pf*ATG8-coated vesicles were seen throughout the whole sexual and asexual erythrocytic stages. Also, *Pf*ATG8 was found in host RBC (Cervantes et al., 2014). Parasites have expanded the classic secretory channel to transfer parasite protein to the host RBC by exporting machinery that modulates the vesicle-mediated transport to the host RBC.

Autophagy was observed in the mice infected with *P. berghei ANKA* through the LC3-mediated autophagy pathway (LAP), and independent of Ulk1. This transfer and degrades the *P. berghei ANKA*-microvesicles (*Pb*A-MVs) inside astrocytes (Leleu, Alloo, Cazenave, Roland, & Pied, 2022). LAP promotes the breakdown of cellular components, which are subsequently released into the extracellular environment, leading to the release of microvesicles. The work suggests that LC3-II is involved in the formation of the double-membrane structure that surrounds the cargo in the LAP pathway. This structure eventually fuses with lysosomes to promote degradation and the release of MVs (Leleu et al., 2022). Further, these microvesicles (containing parasite-derived proteins and nucleic acids) may contribute to the pathogenesis of malaria by favoring the inflammation and activation of the immune system (Leleu et al., 2022). Further research is needed to completely understand the function of LC3-mediated autophagy in the formation of MVs and in influencing host response to the infection.

Recent research suggests that targeting autophagy pathways might be a potential strategy for combating drug resistance in malaria (Cervantes et al., 2014; Joy et al., 2018; Walczak, Ganesan, Niles, & Yeh, 2018). The suppression of the autophagy in the artemisinin-resistant parasites made them susceptible to the treatment (Ray et al., 2022). Combining the autophagy inhibitor with standard anti-malarial drugs has shown increased parasite clearance both in the routine culture and *in vivo*. However, the role of autophagy in drug resistance is complex and context-dependent. While upregulation of autophagy has been observed in some drug-resistant parasites, other studies have shown that autophagy is downregulated in response to drug treatment. Furthermore, the effects of autophagy inhibition on drug-resistant parasites can be variable, and the optimal timing and dosage of autophagy inhibitors could depend on the specific drug and parasite strain being targeted (Chakraborty, 2016).

The release of MVs by *Plasmodium* is implicated in the disease pathogenesis and host immune responses (Mantel & Marti, 2014). The process of MVs' formation is closely linked to autophagy, one of the important cellular homeostatic processes. Recently Leleu and colleagues have discussed the role of autophagy in the formation of *Plasmodium*-derived MVs and its implications for malaria pathogenesis (Leleu et al., 2022). It discusses how autophagy is important in the formation of *Plasmodium*-derived MVs. The parasite undergoes autophagy to digest hemoglobin and obtain nutrients for growth and replication during asexual blood-stage infection of the malaria parasite. This process leads to the accumulation of autophagosomes, which can fuse with the host cell membrane to release MVs containing parasite-derived material (Leleu et al., 2022). The study addressed how *Plasmodium*-derived MVs might affect malaria pathogenesis in both positive and negative ways.

On one hand, MVs can exacerbate disease severity by inducing inflammation and suppressing the host immune response (Leleu et al., 2022). On the other hand, MVs may induce protective immune responses and serve as a source of antigens for vaccine development. Autophagy and MV release are tightly regulated and can be altered by extrinsic stimuli such as nutrition availability and oxidative stress (Chiaradia et al., 2021). However, further investigations are required to comprehend the complicated relationship between autophagy, MV release, and malaria clinical outcomes. In brief, further investigations are required to understand the mechanisms underlying autophagy-mediated drug resistance.

13.4.1 OLP-Induced Autophagy During Malaria Infection

Autophagy is a lysosomal degradation process that clears proteins, cytotoxic aggregates, and damaged organelles. It is necessary for maintaining homeostasis, preventing nutritional deficiency, and preventing pathogenesis induced by its dysregulation. Autophagy starts with the formation of a phagophore that grows and terminates in auto-phagosome completion. The typical markers of autophagy are Beclin-1 and LC3 which are essential for phagophore formation (Hurley & Young, 2017). Autophagy dysfunction is evident in disorders including cancer and neurological disease. OLP is known to modulate autophagy. Since there are several forms of autophagy based on how the cargo is removed, including chaperone-mediated autophagy, microautophagy, and macroautophagy (Hale et al., 2013; Parzych & Klionsky, 2014). Heat shock proteins, such as heat shock cognate 60 and 70 (HSC60 and HSC70), have a role in chaperone-mediated autophagy. Microautophagy is a non-selective breakdown characterized by cytosolic component sequestration by lysosomal membrane invagination (Blanco-Benítez et al., 2022; Parzych & Klionsky, 2014). The cytosolic cargo is gathered by macroautophagy, also known as autophagy, into double-membrane vesicles known as autophagosomes, which eventually merge with the lysosomes to produce autophagolysosome/autolysosome (Hale et al., 2013; Parzych & Klionsky, 2014).

Oleuropein aglycone (OA), a compound found in olive leaves, induces autophagy in cells. The researcher discovered that OA triggered autophagy in cells by activating the AMP-activated protein kinase (AMPK) and inhibiting the mammalian target of the rapamycin (mTOR) signaling pathway (Rigacci et al., 2015). AMPK is a cellular energy sensor that governs metabolism, whereas mTOR is a protein kinase regulating cell growth and proliferation. Furthermore, $Ca2^{+}$-CAMKKβ (calcium/calmodulin-dependent protein kinase kinase β)-AMPK axis plays a critical role in the mechanism by which OA induces autophagy in cells, where increased intracellular $Ca2^{+}$ concentration leads to the activation of the CAMKKβ (Rigacci et al., 2015). CAMKKβ, in turn, activates AMPK by phosphorylation. Moreover, the inhibition of CAMKKβ or AMPK activity significantly attenuated OA-induced autophagy, indicating that the $Ca2^{+}$-CAMKKβ-AMPK axis is necessary for OA-induced autophagy (Rigacci et al., 2015).

In brief, $Ca2^{+}$-CAMKK-AMPK axis is a critical component of the process through which OA promotes autophagy in cells, and targeting this route could be an approach to develop autophagy-mediated treatments. Porcu et al. aimed to investigate the effect of OLP, a phenolic compound found in olive leaves and oil, on non-alcoholic fatty liver disease (NAFLD) and its underlying mechanisms

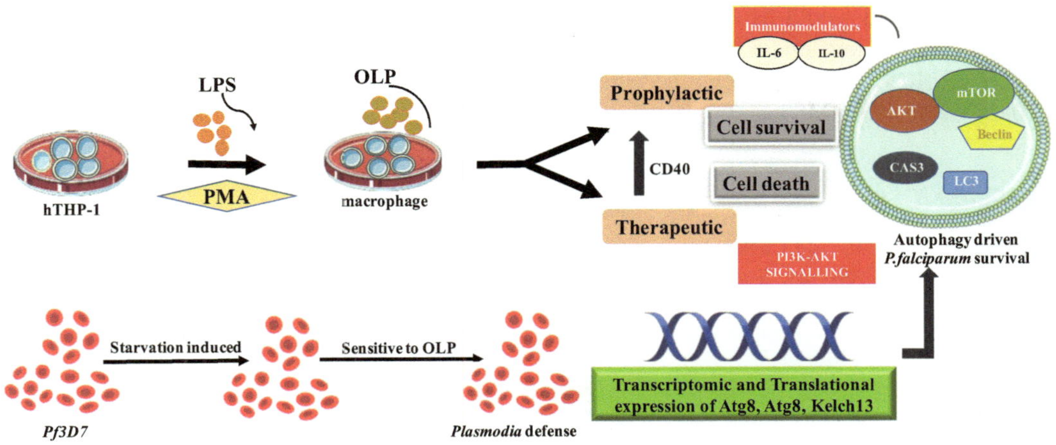

FIGURE 13.1 Proposed model to study the oleuropein-induced autophagy to circumvent anti-inflammatory and anti-plasmodial defense. The signaling and anti-inflammatory potential of OLP in the human THP-1 (hTHP-1) macrophages model of inflammation (upper panel) and detection of the autophagy employed by the *P. falciparum* at transcript and protein level expression of the growth and development and autophagy markers.

(Porcu et al., 2018). Specifically, the study focused on the role of autophagy and AMPK signaling. The study was conducted using male and female mice that had been fed a high-fat diet to induce NAFLD followed by treatment with OLP for 8 weeks to assess the effects of the treatment (Porcu et al., 2018). Results showed that OLP treatment significantly reduced liver fat accumulation and improved the liver function irrespective of gender in the NAFLD mice. The treatment also increased the expression of autophagy-related genes and proteins (Akt, LC3B, mTOR, Beclin-1, and ULK1), indicating the induction of autophagy in the liver tissue of the mice. Additionally, OLP treatment stimulated the AMPK signaling pathway in mouse liver tissue, which is known to control the autophagy. Besides, blocking the AMPK pathway with a particular inhibitor partially prevented the oleuropein-induced autophagy, indicating that OLP-induced autophagy is in part mediated by AMPK signaling (Porcu et al., 2018). Overall, these findings indicate that oleuropein may activate autophagy and confer the extended survival of the diseased mice which is mediated by the signaling pathway (Figure 13.1).

13.5 CONCLUSION

Autophagy could be implicated during the evolution of the anti-malarial resistance. This is confirmed by the fact that drug-resistant parasites have been shown to upregulate autophagy for long-lasting survival of the parasite. Autophagy is also known to enhance drug sensitivity in drug-sensitive parasites, suggesting that regulation of autophagy could be a promising approach for overcoming drug resistance. The induction of autophagy by OLP is thought to enhance parasite susceptibility to anti-malarial drugs by promoting the degradation of damaged organelles and proteins, reducing oxidative stress, and enhancing nutrient uptake in the first phase of anti-malarial drug treatment. Overall, these studies suggest that inducing autophagy through the use of compounds like OLP could provide a promising strategy for enhancing the efficacy of anti-malarial drugs and overcoming drug resistance by the induction and regulation of autophagy.

ACKNOWLEDGMENTS

Rajeev Tyagi would like to express his gratitude to DBT, New Delhi, Ramalingaswami Re-entry Fellowship Project (No. BT/RLF/Re-entry/27/2018) and Indian Council of Medical Research (ICMR), New Delhi extramural grant (35/1/2020-Nano/BMS) for generously supporting this study.

Rajeev K. Tyagi would like to express his thanks to the central MIL facility of CSIR-IMTECH, Chandigarh. Nikunj Tandel would like to thank Nirma University and the Indian Council of Medical Research (ICMR) for providing the fellowship to carry out his research (ICMR award letter No.: 2020-7623/CMB-BMS).

REFERENCES

Acosta-Martinez, M. and Cabail, M. Z. (2022). The PI3K/Akt pathway in meta-inflammation. *International Journal of Molecular Sciences*, *23*(23). doi:10.3390/ijms232315330

Ahamad, J., Toufeeq, I., Khan, M. A., Ameen, M. S. M., Anwer, E. T., Uthirapathy, S., Mir, S. R. and Ahmad, J. (2019). Oleuropein: A natural antioxidant molecule in the treatment of metabolic syndrome. *Phytotherapy Research*, *33*(12), 3112–3128. doi:10.1002/ptr.6511

Amiot, M. J., Fleuriet, A. and Macheix, J. J. (1986). Importance and evolution of phenolic compounds in olive during growth and maturation. *Journal of Agricultural and Food Chemistry*, *34*(5), 823–826.

Andrewes, P., Busch, J. L., de Joode, T., Groenewegen, A. and Alexandre, H. (2003). Sensory properties of virgin olive oil polyphenols: identification of deacetoxy-ligstroside aglycon as a key contributor to pungency. *Journal of Agricultural and Food Chemistry*, *51*(5), 1415–1420. doi:10.1021/jf026042j

Antonelli, L. R., Junqueira, C., Vinetz, J. M., Golenbock, D. T., Ferreira, M. U. and Gazzinelli, R. T. (2020). The immunology of Plasmodium vivax malaria. *Immunological Reviews*, *293*(1), 163–189. doi:10.1111/imr.12816

Ashley, E. A., Pyae Phyo, A., & Woodrow, C. J. (2018). Malaria. *Lancet*, *391*(10130), 1608–1621. doi:10.1016/s0140-6736(18)30324-6

Bansal, A., & Priyadarsini, C. (2021). Medicinal properties of phytochemicals and their production. In: El-Shemy, H. A., ed. *Natural drugs from plants*. IntechOpen, London, UK.

Belete, T. M. (2020). Recent progress in the development of new antimalarial drugs with novel targets. *Drug Design, Development and Therapy*, *14*, 3875–3889. doi:10.2147/dddt.s265602

Blanco-Benítez, M., Calderón-Fernández, A., Canales-Cortés, S., Alegre-Cortés, E., Uribe-Carretero, E., Paredes-Barquero, M., Gimenez-Bejarano, A., Duque González, G., Gómez-Suaga, P., Ortega-Vidal, J., Salido, S., Altarejos, J., Martínez-Chacón, G., Niso-Santano, M., Fuentes, J. M., González-Polo, R. A., & Yakhine-Diop, S. M. S. (2022). Biological effects of olive oil phenolic compounds on mitochondria. *Molecular & Cellular Oncology*, *9*(1), 2044263. doi:10.1080/23723556.2022.2044263

Butt, M. S., Tariq, U., Iahtisham Ul, H., Naz, A., & Rizwan, M. (2021). Neuroprotective effects of oleuropein: Recent developments and contemporary research. *Journal of Food Biochemistry*, *45*(12), e13967. doi:10.1111/jfbc.13967

Cairns, M., Barry, A., Zongo, I., Sagara, I., Yerbanga, S. R., Diarra, M., Zoungrana, C., Issiaka, D., Sienou, A. A., Tapily, A., Sanogo, K., Kaya, M., Traore, S., Diarra, K., Yalcouye, H., Sidibe, Y., Haro, A., Thera, I., Snell, P., Grant, J., Tinto, H., Milligan, P., Chandramohan, D., Greenwood, B., Dicko, A., & Ouedraogo, J. B. (2022). The duration of protection against clinical malaria provided by the combination of seasonal RTS,S/AS01(E) vaccination and seasonal malaria chemoprevention versus either intervention given alone. *BMC Medicine*, *20*(1), 352. doi:10.1186/s12916-022-02536-5

Carnevale, R., Silvestri, R., Loffredo, L., Novo, M., Cammisotto, V., Castellani, V., Bartimoccia, S., Nocella, C., & Violi, F. (2018). Oleuropein, a component of extra virgin olive oil, lowers postprandial glycaemia in healthy subjects. *British Journal of Clinical Pharmacology*, *84*(7), 1566–1574. doi:10.1111/bcp.13589

Cervantes, S., Bunnik, E. M., Saraf, A., Conner, C. M., Escalante, A., Sardiu, M. E., Ponts, N., Prudhomme, J., Florens, L., & Le Roch, K. G. (2014). The multifunctional autophagy pathway in the human malaria parasite, Plasmodium falciparum. *Autophagy*, *10*(1), 80–92. doi:10.4161/auto.26743

Chakraborty, A. (2016). Understanding the biology of the Plasmodium falciparum apicoplast; an excellent target for antimalarial drug development. *Life Sciences*, *158*, 104–110. doi:10.1016/j.lfs.2016.06.030

Chandramohan, D., Dicko, A., Zongo, I., Sagara, I., Cairns, M., Kuepfer, I., Diarra, M., Tapily, A., Issiaka, D., Sanogo, K., Mahamar, A., Sompougdou, F., Yerbanga, S., Thera, I., Milligan, P., Tinto, H., Ofori-Anyinam, O., Ouedraogo, J. B., & Greenwood, B. (2020). Seasonal malaria vaccination: protocol of a phase 3 trial of seasonal vaccination with the RTS,S/AS01(E) vaccine, seasonal malaria chemoprevention and the combination of vaccination and chemoprevention. *BMJ Open*, *10*(9), e035433. doi:10.1136/bmjopen-2019-035433

Chiaradia, E., Tancini, B., Emiliani, C., Delo, F., Pellegrino, R. M., Tognoloni, A., Urbanelli, L., & Buratta, S. (2021). Extracellular vesicles under oxidative stress conditions: Biological properties and physiological roles. *Cells*, *10*(7). doi:10.3390/cells10071763

Coppens, I. (2011). Metamorphoses of malaria: the role of autophagy in parasite differentiation. *Essays in Biochemistry, 51*, 127–136. doi:10.1042/bse0510127

Cordero, J. G., García-Escudero, R., Avila, J., Gargini, R., & García-Escudero, V. (2018). Benefit of oleuropein aglycone for Alzheimer's disease by promoting autophagy. *Oxidative Medicine and Cellular Longevity, 2018*, 5010741. doi:10.1155/2018/5010741

Datoo, M. S., Natama, H. M., Somé, A., Bellamy, D., Traoré, O., Rouamba, T., Tahita, M. C., Ido, N. F. A., Yameogo, P., Valia, D., Millogo, A., Ouedraogo, F., Soma, R., Sawadogo, S., Sorgho, F., Derra, K., Rouamba, E., Ramos-Lopez, F., Cairns, M., Provstgaard-Morys, S., Aboagye, J., Lawrie, A., Roberts, R., Valéa, I., Sorgho, H., Williams, N., Glenn, G., Fries, L., Reimer, J., Ewer, K. J., Shaligram, U., Hill, A. V. S., & Tinto, H. (2022). Efficacy and immunogenicity of R21/Matrix-M vaccine against clinical malaria after 2 years' follow-up in children in Burkina Faso: a phase 1/2b randomised controlled trial. *Lancet Infectious Diseases, 22*(12), 1728–1736. doi:10.1016/s1473-3099(22)00442-x

Devi, S., Negi, S., Sharma, P., Tandel, N., & Tyagi, R. K. (2024). Protocol for oleuropein-induced autophagy mediating drug tolerance in P. falciparum. *STAR Protoc, 5*(3), 103141. doi:10.1016/j.xpro.2024.103141

Ding, Q., Hu, W., Wang, R., Yang, Q., Zhu, M., Li, M., Cai, J., Rose, P., Mao, J., & Zhu, Y. Z. (2023). Signaling pathways in rheumatoid arthritis: implications for targeted therapy. *Signal Transduction and Targeted Therapy, 8*(1), 68. doi:10.1038/s41392-023-01331-9

Duffy, P. E., & Patrick Gorres, J. (2020). Malaria vaccines since 2000: progress, priorities, products. *npj Vaccines, 5*(1), 48.

Fong, K. Y., & Wright, D. W. (2013). Hemozoin and antimalarial drug discovery. *Future Medicinal Chemistry, 5*(12), 1437–1450. doi:10.4155/fmc.13.113

Goulas, V., Exarchou, V., Troganis, A. N., Psomiadou, E., Fotsis, T., Briasoulis, E., & Gerothanassis, I. P. (2009). Phytochemicals in olive-leaf extracts and their antiproliferative activity against cancer and endothelial cells. *Molecular Nutrition & Food Research, 53*(5), 600–608. doi:10.1002/mnfr.200800204

Gupta, H., Macete, E., Bulo, H., Salvador, C., Warsame, M., Carvalho, E., Ménard, D., Ringwald, P., Bassat, Q., Enosse, S., & Mayor, A. (2018). Drug-resistant polymorphisms and copy numbers in Plasmodium falciparum, Mozambique, 2015. *Emerging Infectious Diseases, 24*(1), 40–48. doi:10.3201/eid2401.170864

Hailesilase, G. G., Rajeshwar, Y., Hailu, G. S., Sibhat, G. G., & Bitew, H. (2020). *In Vivo* Antimalarial Evaluation of Crude Extract, Solvent Fractions, and TLC-Isolated Compounds from *Olea europaea* Linn subsp. *cuspidata* (Oleaceae). *Evidence-Based Complementary and Alternative Medicine, 2020*, 6731485. doi:10.1155/2020/6731485

Hain, A. U. P., & Bosch, J. (2013). Autophagy in *Plasmodium, a multifunctional pathway? Computational and Structural Biotechnology Journal, 8*(11), e201308002. doi:https://doi.org/10.5936/csbj.201308002

Hale, A. N., Ledbetter, D. J., Gawriluk, T. R., & Rucker, E. B., 3rd. (2013). Autophagy: regulation and role in development. *Autophagy, 9*(7), 951–972. doi:10.4161/auto.24273

Hamdi, H. K., & Castellon, R. (2005). Oleuropein, a non-toxic olive iridoid, is an anti-tumor agent and cytoskeleton disruptor. *Biochemical and Biophysical Research Communications, 334*(3), 769–778. doi:10.1016/j.bbrc.2005.06.161

Hassen, I., Casabianca, H., & Hosni, K. (2015). Biological activities of the natural antioxidant oleuropein: Exceeding the expectation – A mini-review. *Journal of Functional Foods, 18*, 926–940. doi:https://doi.org/10.1016/j.jff.2014.09.001

Hurley, J. H., & Young, L. N. (2017). Mechanisms of autophagy initiation. *Annual Review of Biochemistry, 86*, 225–244. doi:10.1146/annurev-biochem-061516-044820

Ikegbunam, M. N., Nkonganyi, C. N., Thomas, B. N., Esimone, C. O., Velavan, T. P., & Ojurongbe, O. (2019). Analysis of Plasmodium falciparum Pfcrt and Pfmdr1 genes in parasite isolates from asymptomatic individuals in Southeast Nigeria 11 years after withdrawal of chloroquine. *Malaria Journal, 18*(1), 343. doi:10.1186/s12936-019-2977-6

Imwong, M., Suwannasin, K., Kunasol, C., Sutawong, K., Mayxay, M., Rekol, H., Smithuis, F. M., Hlaing, T. M., Tun, K. M., van der Pluijm, R. W., Tripura, R., Miotto, O., Menard, D., Dhorda, M., Day, N. P. J., White, N. J., & Dondorp, A. M. (2017). The spread of artemisinin-resistant Plasmodium falciparum in the Greater Mekong subregion: A molecular epidemiology observational study. *Lancet Infectious Diseases, 17*(5), 491–497. doi:10.1016/s1473-3099(17)30048-8

Joy, S., Thirunavukkarasu, L., Agrawal, P., Singh, A., Sagar, B. K. C., Manjithaya, R., & Surolia, N. (2018). Basal and starvation-induced autophagy mediates parasite survival during intraerythrocytic stages of Plasmodium falciparum. *Cell Death Discovery, 4*, 43. doi:10.1038/s41420-018-0107-9

Kamil, M., Atmaca, H. N., Unal, S., Kina, U. Y., Burak, P., Deveci, G., Yilmaz, I., & Aly, A. S. I. (2022). An alternative autophagy-related mechanism of chloroquine drug resistance in the Malaria parasite. *Antimicrobial Agents and Chemotherapy, 66*(12), e0026922. doi:10.1128/aac.00269-22

Leleu, I., Alloo, J., Cazenave, P. A., Roland, J., & Pied, S. (2022). Autophagy pathways in the genesis of plasmodium-derived microvesicles: A double-edged sword? *Life (Basel)*, *12*(3). doi:10.3390/life12030415

Liu, L., Ahn, K. S., Shanmugam, M. K., Wang, H., Shen, H., Arfuso, F., Chinnathambi, A., Alharbi, S. A., Chang, Y., Sethi, G., & Tang, F. R. (2019). Oleuropein induces apoptosis via abrogating NF-κB activation cascade in estrogen receptor-negative breast cancer cells. *Journal of Cellular Biochemistry*, *120*(3), 4504–4513. doi:10.1002/jcb.27738

Liu, M., Wang, J., Huang, B., Chen, A., & Li, X. (2016). Oleuropein inhibits the proliferation and invasion of glioma cells via suppression of the AKT signaling pathway. *Oncology Reports*, *36*(4), 2009–2016. doi:10.3892/or.2016.4978

Liu, Y., McKeever, L. C., & Malik, N. S. (2017). Assessment of the antimicrobial activity of olive leaf extract against foodborne bacterial pathogens. *Frontiers in Microbiology*, *8*, 113. doi:10.3389/fmicb.2017.00113

Lover, A. A., Baird, J. K., Gosling, R., & Price, R. N. (2018). Malaria elimination: Time to target all species. *American Journal of Tropical Medicine and Hygiene*, *99*(1), 17–23. doi:10.4269/ajtmh.17-0869

Magrone, T., Spagnoletta, A., Salvatore, R., Magrone, M., Dentamaro, F., Russo, M. A., Difonzo, G., Summo, C., Caponio, F., & Jirillo, E. (2018). Olive leaf extracts act as modulators of the human immune response. *Endocrine, Metabolic & Immune Disorders Drug Targets*, *18*(1), 85–93. doi:10.2174/18715303176 66171116110537

Mantel, P. Y., & Marti, M. (2014). The role of extracellular vesicles in Plasmodium and other protozoan parasites. *Cellular Microbiology*, *16*(3), 344–354. doi:10.1111/cmi.12259

Mbengue, A., Bhattacharjee, S., Pandharkar, T., Liu, H., Estiu, G., Stahelin, R. V., Rizk, S. S., Njimoh, D. L., Ryan, Y., Chotivanich, K., Nguon, C., Ghorbal, M., Lopez-Rubio, J. J., Pfrender, M., Emrich, S., Mohandas, N., Dondorp, A. M., Wiest, O., & Haldar, K. (2015). A molecular mechanism of artemisinin resistance in Plasmodium falciparum malaria. *Nature*, *520*(7549), 683–687. doi:10.1038/nature14412

Meibalan, E., & Marti, M. (2017). Biology of malaria transmission. *Cold Spring Harbor Perspectives in Medicine*, *7*(3). doi:10.1101/cshperspect.a025452

Miles, E. A., Zoubouli, P., & Calder, P. C. (2005). Differential anti-inflammatory effects of phenolic compounds from extra virgin olive oil identified in human whole blood cultures. *Nutrition*, *21*(3), 389–394. doi:10.1016/j.nut.2004.06.031

Milner, D. A., Jr. (2018). Malaria pathogenesis. *Cold Spring Harbor Perspectives in Medicine*, *8*(1). doi:10.1101/cshperspect.a025569

Misganaw, D., Engidawork, E., & Nedi, T. (2019). Evaluation of the anti-malarial activity of crude extract and solvent fractions of the leaves of Olea europaea (Oleaceae) in mice. *BMC Complementary and Alternative Medicine*, *19*(1), 171. doi:10.1186/s12906-019-2567-8

Neafsey, D. E., Juraska, M., Bedford, T., Benkeser, D., Valim, C., Griggs, A., Lievens, M., Abdulla, S., Adjei, S., Agbenyega, T., Agnandji, S. T., Aide, P., Anderson, S., Ansong, D., Aponte, J. J., Asante, K. P., Bejon, P., Birkett, A. J., Bruls, M., Connolly, K. M., D'Alessandro, U., Dobaño, C., Gesase, S., Greenwood, B., Grimsby, J., Tinto, H., Hamel, M. J., Hoffman, I., Kamthunzi, P., Kariuki, S., Kremsner, P. G., Leach, A., Lell, B., Lennon, N. J., Lusingu, J., Marsh, K., Martinson, F., Molel, J. T., Moss, E. L., Njuguna, P., Ockenhouse, C. F., Ogutu, B. R., Otieno, W., Otieno, L., Otieno, K., Owusu-Agyei, S., Park, D. J., Pellé, K., Robbins, D., Russ, C., Ryan, E. M., Sacarlal, J., Sogoloff, B., Sorgho, H., Tanner, M., Theander, T., Valea, I., Volkman, S. K., Yu, Q., Lapierre, D., Birren, B. W., Gilbert, P. B., & Wirth, D. F. (2015). Genetic diversity and protective efficacy of the RTS,S/AS01 malaria vaccine. *New England Journal of Medicine*, *373*(21), 2025–2037. doi:10.1056/NEJMoa1505819

Nediani, C., Ruzzolini, J., Romani, A., & Calorini, L. (2019). Oleuropein, a bioactive compound from Olea europaea L., as a potential preventive and therapeutic agent in non-communicable diseases. *Antioxidants (Basel)*, *8*(12). doi:10.3390/antiox8120578

Negi, S., Tandel, N., Garg, N. K., Sharma, P., Kumar, R., Sharma, P., Kumar, R., Saini, S., Sharma, A., & Tyagi, R. K. (2024). Co-Delivery of Aceclofenac and Methotrexate Nanoparticles Presents an Effective Treatment for Rheumatoid Arthritis. *International Journal of Nanomedicine*, *19*, 2149–2177. doi:10.2147/ijn.S439359

Nlinwe, O. N., Kusi, K. A., Adu, B., & Sedegah, M. (2018). T-cell responses against Malaria: Effect of parasite antigen diversity and relevance for vaccine development. *Vaccine*, *36*(17), 2237–2242.

Nsanzabana, C. (2019). Resistance to Artemisinin Combination Therapies (ACTs): Do Not Forget the Partner Drug! *Tropical Medicine and Infectious Disease*, *4*(1). doi:10.3390/tropicalmed4010026

Omar, S. H. (2010a). Cardioprotective and neuroprotective roles of oleuropein in olive. *Saudi Pharmaceutical Journal*, *18*(3), 111–121. doi:10.1016/j.jsps.2010.05.005

Omar, S. H. (2010b). Oleuropein in olive and its pharmacological effects. *Scientia Pharmaceutica*, *78*(2), 133–154.

Ouattara, A., Kone, A., Adams, M., Fofana, B., Maiga, A. W., Hampton, S., Coulibaly, D., Thera, M. A., Diallo, N., Dara, A., Sagara, I., Gil, J. P., Bjorkman, A., Takala-Harrison, S., Doumbo, O. K., Plowe, C. V., & Djimde, A. A. (2015). Polymorphisms in the K13-propeller gene in artemisinin-susceptible Plasmodium falciparum parasites from Bougoula-Hameau and Bandiagara, Mali. *American Journal of Tropical Medicine and Hygiene*, *92*(6), 1202–1206. doi:10.4269/ajtmh.14-0605

Packard, R. M. (2014). The origins of antimalarial-drug resistance. *New England Journal of Medicine*, *371*(5), 397–399. doi:10.1056/NEJMp1403340

Parra-Perez, A. M., Pérez-Jiménez, A., Gris-Cárdenas, I., Bonel-Pérez, G. C., Carrasco-Díaz, L. M., Mokhtari, K., García-Salguero, L., Lupiáñez, J. A., & Rufino-Palomares, E. E. (2022). Involvement of the PI3K/AKT intracellular signaling pathway in the anticancer activity of hydroxytyrosol, a polyphenol from Olea europaea, in hematological cells and implication of HSP60 levels in its anti-inflammatory activity. *International Journal of Molecular Sciences*, *23*(13). doi:10.3390/ijms23137053

Parzych, K. R., & Klionsky, D. J. (2014). An overview of autophagy: Morphology, mechanism, and regulation. *Antioxid Redox Signal*, *20*(3), 460–473. doi:10.1089/ars.2013.5371

Phillips, M. A., Burrows, J. N., Manyando, C., van Huijsduijnen, R. H., Van Voorhis, W. C., & Wells, T. N. C. (2017). Malaria. *Nature Reviews Disease Primers*, *3*, 17050. doi:10.1038/nrdp.2017.50

Piroddi, M., Albini, A., Fabiani, R., Giovannelli, L., Luceri, C., Natella, F., Rosignoli, P., Rossi, T., Taticchi, A., Servili, M., & Galli, F. (2017). Nutrigenomics of extra-virgin olive oil: A review. *Biofactors*, *43*(1), 17–41. doi:10.1002/biof.1318

Porcu, C., Sideri, S., Martini, M., Cocomazzi, A., Galli, A., Tarantino, G., & Balsano, C. (2018). Oleuropein induces AMPK-dependent autophagy in NAFLD Mice, regardless of the gender. *International Journal of Molecular Sciences*, *19*(12). doi:10.3390/ijms19123948

Rasmussen, C., Alonso, P., & Ringwald, P. (2022). Current and emerging strategies to combat antimalarial resistance. *Expert Review of Anti-Infective Therapy*, *20*(3), 353–372. doi:10.1080/14787210.2021.1962291

Ray, A., Mathur, M., Choubey, D., Karmodiya, K., & Surolia, N. (2022). Autophagy underlies the proteostasis mechanisms of Artemisinin resistance in P. falciparum Malaria. *mBio*, *13*(3), e0063022. doi:10.1128/mbio.00630-22

Reyburn, H. (2010). New WHO guidelines for the treatment of malaria. *BMJ*, *340*, c2637. doi:10.1136/bmj.c2637

Rigacci, S., Miceli, C., Nediani, C., Berti, A., Cascella, R., Pantano, D., Nardiello, P., Luccarini, I., Casamenti, F., & Stefani, M. (2015). Oleuropein aglycone induces autophagy via the AMPK/mTOR signalling pathway: a mechanistic insight. *Oncotarget*, *6*(34), 35344–35357.

Rishmawi, S., Haddad, F., Dokmak, G., & Karaman, R. (2022). A comprehensive review on the anti-cancer effects of Oleuropein. *Life (Basel)*, *12*(8). doi:10.3390/life12081140

Roux, A. T., Maharaj, L., Oyegoke, O., Akoniyon, O. P., Adeleke, M. A., Maharaj, R., & Okpeku, M. (2021). Chloroquine and sulfadoxine-pyrimethamine resistance in sub-Saharan Africa-A review. *Front Genet*, *12*, 668574. doi:10.3389/fgene.2021.668574

RTS, S. C. T. P. (2014). Efficacy and safety of the RTS,S/AS01 malaria vaccine during 18 months after vaccination: a phase 3 randomized, controlled trial in children and young infants at 11 African sites. *PLoS Medicine*, *11*(7), e1001685. doi:10.1371/journal.pmed.1001685

RTS, S. C. T. P. (2015). Efficacy and safety of RTS,S/AS01 malaria vaccine with or without a booster dose in infants and children in Africa: final results of a phase 3, individually randomised, controlled trial. *Lancet*, *386*(9988), 31–45. doi:10.1016/s0140-6736(15)60721-8

Ryu, S. J., Choi, H. S., Yoon, K. Y., Lee, O. H., Kim, K. J., & Lee, B. Y. (2015). Oleuropein suppresses LPS-induced inflammatory responses in RAW 264.7 cell and zebrafish. *Journal of Agricultural and Food Chemistry*, *63*(7), 2098–2105. doi:10.1021/jf505894b

Sain, A., Kandasamy, T., & Naskar, D. (2022). In silico approach to target PI3K/Akt/mTOR axis by selected Olea europaea phenols in PIK3CA mutant colorectal cancer. *Journal of Biomolecular Structure and Dynamics*, *40*(21), 10962–10977. doi:10.1080/07391102.2021.1953603

Santos-Reis, A., & Nina, J. (2022). A difficult diagnosis of plasmodium ovale Malaria. *Acta Medica Portuguesa*, *35*(6), 484–487. doi:10.20344/amp.15814

Sato, S. (2021). Plasmodium-a brief introduction to the parasites causing human malaria and their basic biology. *Journal of Physiological Anthropology*, *40*(1), 1. doi:10.1186/s40101-020-00251-9

Scotece, M., Conde, J., Abella, V., Lopez, V., Pino, J., Lago, F., Smith, A. B., 3rd, Gómez-Reino, J. J., & Gualillo, O. (2015). New drugs from ancient natural foods. Oleocanthal, the natural occurring spicy compound of olive oil: a brief history. *Drug Discov Today*, *20*(4), 406–410. doi:10.1016/j.drudis.2014.10.017

Sharma, P., Tandel, N., Kumar, R., Negi, S., Sharma, P., Devi, S., Saxena, K., Chaudhary, N. R., Saini, S., Kumar, R., Chandel, B. S., Sijwali, P. S., & Tyagi, R. K. (2024). Oleuropein activates autophagy to circumvent anti-plasmodial defense. *iScience*, *27*(4), 109463. doi:10.1016/j.isci.2024.109463

Slater, L. B. (2009). *War and disease: biomedical research on malaria in the twentieth century*. Rutgers University Press.

Somé, A. F., Sorgho, H., Zongo, I., Bazié, T., Nikiéma, F., Sawadogo, A., Zongo, M., Compaoré, Y. D., & Ouédraogo, J. B. (2016). Polymorphisms in K13, pfcrt, pfmdr1, pfdhfr, and pfdhps in parasites isolated from symptomatic malaria patients in Burkina Faso. *Parasite*, *23*, 60. doi:10.1051/parasite/2016069

Stark, A. K., Sriskantharajah, S., Hessel, E. M., & Okkenhaug, K. (2015). PI3K inhibitors in inflammation, autoimmunity and cancer. *Current Opinion in Pharmacology*, *23*, 82–91. doi:10.1016/j.coph.2015.05.017

Su, X. Z., Lane, K. D., Xia, L., Sá, J. M., & Wellems, T. E. (2019). Plasmodium genomics and genetics: New insights into malaria pathogenesis, drug resistance, epidemiology, and evolution. *Clinical Microbiology Reviews*, *32*(4). doi:10.1128/cmr.00019-19

Sun, W., Frost, B., & Liu, J. (2017). Oleuropein, unexpected benefits! *Oncotarget*, *8*(11), 17409. doi:10.18632/oncotarget.15538

Suresh, N., & Haldar, K. (2018). Mechanisms of artemisinin resistance in Plasmodium falciparum malaria. *Current Opinion in Pharmacology*, *42*, 46–54. doi:10.1016/j.coph.2018.06.003

Tandel, N., & Dalai, S. K. (2021). T cell-based vaccines: Hope for malaria elimination. *Current Topics and Emerging Issues in Malaria Elimination*, 285.

Tse, E. G., Korsik, M., & Todd, M. H. (2019). The past, present and future of anti-malarial medicines. *Malaria Journal*, *18*(1), 93. doi:10.1186/s12936-019-2724-z

Tungmunnithum, D., Thongboonyou, A., Pholboon, A., & Yangsabai, A. (2018). Flavonoids and other phenolic compounds from medicinal plants for pharmaceutical and medical aspects: An overview. *Medicines (Basel)*, *5*(3). doi:10.3390/medicines5030093

Tyagi, R. K., Gleeson, P. J., Arnold, L., Tahar, R., Prieur, E., Decosterd, L., Pérignon, J. L., Olliaro, P., & Druilhe, P. (2018). High-level artemisinin-resistance with quinine co-resistance emerges in P. falciparum malaria under in vivo artesunate pressure. *BMC Medicine*, *16*(1), 181. doi:10.1186/s12916-018-1156-x

van der Pluijm, R. W., Amaratunga, C., Dhorda, M., & Dondorp, A. M. (2021). Triple artemisinin-based combination therapies for Malaria – A new paradigm? *Trends in Parasitology*, *37*(1), 15–24. doi:10.1016/j.pt.2020.09.011

Vekemans, J., Schellenberg, D., Benns, S., O'Brien, K., & Alonso, P. (2021). Meeting report: WHO consultation on malaria vaccine development, Geneva, 15–16 July 2019. *Vaccine*, *39*(22), 2907–2916.

Venugopal, K., Hentzschel, F., Valkiūnas, G., & Marti, M. (2020). Plasmodium asexual growth and sexual development in the haematopoietic niche of the host. *Nature Reviews: Microbiology*, *18*(3), 177–189. doi:10.1038/s41579-019-0306-2

Vezza, T., Algieri, F., Rodríguez-Nogales, A., Garrido-Mesa, J., Utrilla, M. P., Talhaoui, N., Gómez-Caravaca, A. M., Segura-Carretero, A., Rodríguez-Cabezas, M. E., Monteleone, G., & Gálvez, J. (2017). Immunomodulatory properties of Olea europaea leaf extract in intestinal inflammation. *Molecular Nutrition & Food Research*, *61*(10). doi:10.1002/mnfr.201601066

Wainstein, J., Ganz, T., Boaz, M., Bar Dayan, Y., Dolev, E., Kerem, Z., & Madar, Z. (2012). Olive leaf extract as a hypoglycemic agent in both human diabetic subjects and in rats. *Journal of Medicinal Food*, *15*(7), 605–610. doi:10.1089/jmf.2011.0243

Walczak, M., Ganesan, S. M., Niles, J. C., & Yeh, E. (2018). ATG8 is essential specifically for an autophagy-independent function in apicoplast biogenesis in blood-stage malaria parasites. *mBio*, *9*(1). doi:10.1128/mBio.02021-17

WHO. (2021). *Global technical strategy for malaria 2016–2030, 2021 update*. Geneva: World Health Organization, 1–40.

WHO. (2022). *World malaria report 2022*. World Health Organization.

Wichers, H. J., Soler-Rivas, C., & Espı, J. C. (2000). Review Oleuropein and related compounds. *Journal of the Science of Food and Agriculture*, *80*, 1013–1023.

Wicht, K. J., Mok, S., & Fidock, D. A. (2020). Molecular mechanisms of drug resistance in Plasmodium falciparum Malaria. *Annual Review of Microbiology*, *74*, 431–454. doi:10.1146/annurev-micro-020518-115546

Yahfoufi, N., Alsadi, N., Jambi, M., & Matar, C. (2018). The immunomodulatory and anti-inflammatory role of polyphenols. *Nutrients*, *10*(11). doi:10.3390/nu10111618

Yan, C. M., Chai, E. Q., Cai, H. Y., Miao, G. Y., & Ma, W. (2015). Oleuropein induces apoptosis via activation of caspases and suppression of phosphatidylinositol 3-kinase/protein kinase B pathway in HepG2 human hepatoma cell line. *Molecular Medicine Reports*, *11*(6), 4617–4624. doi:10.3892/mmr.2015.3266

Zhang, Y. J., Gan, R. Y., Li, S., Zhou, Y., Li, A. N., Xu, D. P., & Li, H. B. (2015). Antioxidant phytochemicals for the prevention and treatment of chronic diseases. *Molecules*, *20*(12), 21138–21156. doi:10.3390/molecules201219753

Section VII

Vector Dynamics Under Climate Change

14 Vector Dynamics Under the Impact of Climate Change

Ramesh C. Dhiman, Poonam Singh, and S.S. Areeb Hussain
ICMR-National Institute of Malaria Research, Dwarka, New Delhi, India

B.K. Tyagi
University Institute of Biotechnology, Chandigarh University, Mohali, India

14.1 INTRODUCTION

Climate change or long-term variations in climatic conditions, come about as a result of natural processes and/or, as observed in the past century, anthropization or human interventions (Tyagi, 2023) (Figure 14.1). Climate change differs from climate variability in terms of the time scales at which they occur, i.e., climate variability is observed in the range of months to decades, climate change occurs over several decades or centuries. Over the course of its history, the Earth's climate has witnessed several large-scale changes in climate that are attributed to natural phenomena such as massive volcanic eruptions, orbital variations, and changes in solar radiation output. Some of these natural forces have caused the global climate to change in long-lasting and profound ways. Over the past half-century or more, the world has witnessed climatic changes at an unprecedented rate, not seen since the last ice age about 11,700 years ago (Gaffney & Steffen, 2017). The observed climate change in the last century is largely a result of the Earth's greenhouse effect, *i.e.*, the trapping of solar radiation in the Earth's atmosphere by various greenhouse gases such as carbon dioxide, methane, nitrous oxide, and water vapors. Natural processes such as volcanic eruptions, animal respiration, and natural decay release these greenhouse gases, which help the Earth to maintain its heat and thereby sustain life. However, human activity such as burning of fossil fuels, deforestation and concentrated animal farming has released unnaturally large amounts of these greenhouse gases into the Earth's atmosphere, resulting in the alarming rise in global temperatures observed today. This increase in the overall temperature of the Earth's atmosphere is called global warming and can have widespread implications for the climate. Global temperatures have already risen by about 1°C as compared to pre-industrial levels (IPCC, 2022), though regional warming much higher than the global average has been observed in many places (e.g., warming in arid, tropical, temperate regions etc.).

Global warming has directly affected the weather patterns all across the Earth, with an immediate consequence like the occurrence of extreme weather conditions such as heat waves, severe droughts, rainstorms, highly destructive hurricanes, forest fires, and large-scale flooding. The Earth is now about 1.1°C (2°F) warmer than it was in the 1800s, and, based on current projections, global temperatures will rise by 2.7°C (4.8°F) by the end of the 21st century (IPCC, 2022). These widespread changes in climate have a direct impact on the biodiversity of the planet, and the effect is more pronounced in ectothermic organisms that are more sensitive to climate. This includes the various insect vectors of infectious diseases, and changes in habitat suitability as a result of climate change can significantly alter the distribution of the diseases they carry (Schmeller et al., 2020). Relationship of the mosquito, *Wyeomyia smithii* with the insectivorous pitcher plant *Sarracenia purpurea* in the

DOI: 10.1201/9781003326991-21

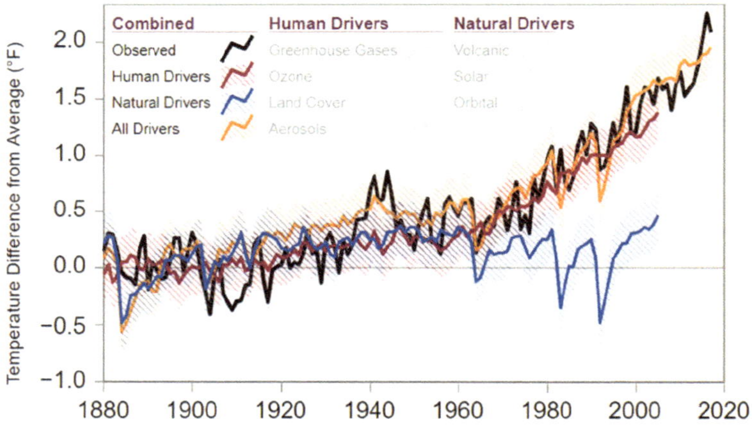

FIGURE 14.1 Human and natural factors both influence the Earth's climate, but the long-term trend observed over the past century can only be explained by the effect of human activities on climate.

***Source*: U.S. Global Change Research Program, Fourth National Climate Assessment, Chapter 2: Our Changing Climate, 2017.**

North America is the world's first animal model to demonstrate the impact of climate change on animals (Bradshaw et al., 2000). Similarly, the desert-turned-urban malaria vector, *Anopheles stephensi*, has during the past decade exhibited an unprecedented advancement in its geographical distribution by invading half a dozen countries in northern Africa, right from east to west, mostly within the Sahara Desert region (Djibouti 2012, Ethiopia 2016, Sudan 2016, Somalia 2019, Nigeria 2020, and Yemen 2021), on one hand, and to some island territories and/or countries such as Lakshadweep and Rameswaram (India) and Sri Lanka. In October 2019, the WHO Global Malaria Program (GMP) released a vector alert that identified the invasion of *Anopheles stephensi* as a major potential threat to malaria control and elimination in the invaded northern Africa and Sri Lanka (WHO, 2023).

Climate change and human health are directly as well as indirectly related to each other. In general, environmental factors like air, temperature, water, and healthy food are climate dependent. As per WHO (https://www.who.int/health-topics/climate-change#tab=tab_1), between the years 2030–2050, approx. 250,000 deaths per year are expected to occur due to malaria, malnutrition, diarrhea, and heat stress only.

14.2 RELATIONSHIP BETWEEN CLIMATIC FACTORS AND INSECT VECTORS OF PUBLIC HEALTH IMPORTANCE

All the vector-borne diseases (VBDs) are climate sensitive, as climatic factors like temperature, precipitation (rainfall), and relative humidity (RH) affect the development and growth of insect vectors as well as pathogens in their body. Worldwide, the estimated number of insect species is approx. 5.5 million and around 7% has been contributed by India. Climate change is a broad term while insects live in favorable conditions of a micro-niche. There is a wide range of distribution of insect species from arid, semi-arid, humid, and hot to cold climates. Further, there is a significant diversity in the type of habitat and micro-environment of the same. Therefore, the interaction between climate and insects has been studied in different paradigms and is still being studied with respect to changes in the distribution of vectors, climate-driven changes in the habitats, and impact on the control measure strategies etc. It is always important to understand the geography of a particular area as it plays a determining role in the insect's ecology. The diversity within the class Insecta, as well as in climate, raises the question of how insects respond to or cope with the changing climate. Long-term monitoring and published literature state that there is a noticeable decline in insect populations due to climate change, along with other factors like habitat destruction, species invasion, drought, and changing precipitation and temperature condition patterns (Wagner, 2020).

There are several categories of insect vectors such as, for example, mosquitoes, sand flies, ticks, and mites, of which studies have been undertaken mostly on mosquitoes in response to climate vicissitudes. The impact of temperature particularly on mosquitoes is demonstrable on anophelines' life history, wherein the difference in temperature tolerance was found among immature stages of *An. arabiensis*, *An. funestus* and *An. quadriannulatus* (Agyekum et al., 2021). The survival and longevity of urban malaria vector, i.e., *An. stephensi* was found to decrease at higher temperatures. The temperature sensitivity among *Anopeheles* mosquitoes was found to be different between the species, emphasizing the need for species-specific studies on different vectors of malaria. The effect of higher temperatures on egg laying in mosquitoes shows a lower number of eggs in the second gonotrophic cycle. Further, the biting rates almost double with increasing temperatures from 21 to 32°C (Shapiro et al., 2017).

14.3 CLIMATIC STRESS ON MOSQUITO PHYSIOLOGY

Climate has been identified as one of the significant factors influencing the physiology of the vectors right from the beginning of emergence. A slight change in the components of climate like temperature, humidity, and rainfall resulted in different courses of disease transmission dynamics. Apart from these three parameters, climate is also dependent on sea level rise, wind speed, sunlight, human activities like migration, agriculture, deforestation, urbanization etc. There are studies describing the direct role of deforestation, urbanization, and other human activities like water storage practices in time-to-time development and survival of the mosquito population. Two examples, one each of the emergence of *Plasmodium falciparum* malaria in the Thar Desert of Western India and that of dengue (DHF) in Kerala state in Southern India from the mid-1990s onwards, offer the best example of how climate change due to human intervention can trigger the emergence of new mosquito-borne disease dynamics, particularly due to new vectors and/or vectorial capacity (Tyagi et al., 1995, 2006).

14.3.1 Temperature as the Most Important Factor Impacting Vector Biology and Pathogen Transmission Potential

Temperature is the only parameter among the whole set of factors inducing climate change that affects mosquitoes the most. In India, a mosquito like *Aedes aegypti*, which until mid-1990s was restricted to the plains and in the urban agglomerations, is now found thriving in the xeric climate of the Thar Desert, while *Ae. albopictus* is projected to expand to the upper and trans-Himalayan regions of the north, triggering the occurrence of dengue epidemics in the newly found lands or *terra incognito* for the vectors (Hussain & Dhiman, 2022). Biological nature of mosquito physiology is dependent on temperature, starting from egg laying, egg hatching, and development of larvae to adult. Temperature has also been held responsible for behavior changes in mosquitoes; their biting rate, blood-feeding, and digestion (Crans et al., 1996), excretion, and even development of pathogens within mosquitoes are temperature-dependent processes (Cator et al., 2020). The impact of temperature on mosquito physiology results in a slow development rate under low-temperature conditions, with longer life. However, at higher temperatures, the rate of development is faster. During unfavorable temperature conditions like overwintering, physiological development of mosquitoes stops, and they go into diapause (Kiarie-Makara et al., 2015; Reisen et al., 2010). The process of desiccation in mosquitoes is reported to be associated with increased temperatures, which is linked to higher mortality in eggs, larvae, and adults as well (Sota & Mogi, 1992a, 1992b). The impact of environmental temperature has also been observed on the immune system of mosquitoes. Major processes of immune responses like phagocytosis, melanization, and transcription of immune genes are affected by temperature by altering the complex network of pathways during the interactions (Murdock et al., 2012).

Exposure of mosquito vectors to the temperature range beyond the optimum conditions, bring biological changes in the physiology to withstand and manage varying thermal stress with the help of heat shock proteins (HSPs). In mosquitoes, a variety of HSPs are expressed to tolerate high- and low-temperature exposures. So far, changes in the expression of particular HSPs have been found to be associated with particular developmental stages like HSPs 31, 33, 38, 43, 44, 51, 57, 62, 69, 71, 113, and 121kD expression at 28°C exposure in the malaria vector; similarly, HSP 62 and 71 during adult development, HSPs31, 43, and 44 during larval and pupae development. Apart from these, expression of HSPs 70 and 90 was found to be high during the process of dehydration and hot blood meal digestion.

Temperature exposure not only affects the biological process in mosquitoes but also physiological changes which lead to the development of thermo-tolerance. If mosquitoes are exposed to high temperatures, followed by re-exposure to increased temperature, they can develop the adaptive thermo-tolerance. The development of thermal tolerance in mosquitoes, up to some extent, is jeopardizing the efforts of control programs in the country. Further, temperature-induced physiological response of mosquitoes toward different insecticides is different, for example higher temperature exposure increases the susceptibility toward permethrin and deltamethrin while low temperature decreases it. The combined impact of all these factors raises the alert toward the importance of temperature on the physiology of the mosquito and their population.

Experimental evidence suggests that temperatures of up to 32°C and 55–75% RH are the most conducive conditions for vector survival (Mayne, 1930; Singh et al., 2022). As per literature, a minimum of 8 days' survival of the mosquitoes is quintessential for parasite development within the vectors. Further, experimental results show that *An. stephensi* could survive up to 12 days at 32°C while *Ae. aegypti* could survive up to 8 days at 36°C, indicating the sturdiness of the vector species. Furthermore, at 42°C survival of *An. stephensi* and *Ae. aegypti* has been observed for only 30 minutes and 3 hours, respectively. Studies analyzing the interaction between relative humidity and vector survival state that less than 60% RH reduces the lifespan of the mosquito.

With respect to Culex mosquito species, it has been observed that all stages of vector development avoid more than 95% RH conditions. The vector remains unaffected with RH variation between 30% and 85% (Thomson, 1938). Similarly, the impact of RH on the flying behavior of *Ae. aegypti* is affected below 30% RH (Rowley & Graham, 1968).

14.3.2 Climate Change and Emergence of *Plasmodium falciparum* Malaria in the Thar Desert, Northwestern India

For several millions of years, the Thar Desert, where the rare and unstable malaria was solely transmissible by *Anopheles stephensi*, has been a scene of harsh, hot, inhospitable xeric ecosystem with very low rainfall and unbearably high ambient temperature. However, a mega-irrigation program comprising the trio: the Gang Canal (operative 1928, with its 1251 km distribution), Bhakra-Sirhind Canal (operative 1955, with 1219 km long distribution system) and Indra Gandhi Canal (operative 1958, with an irrigation tributaries' network of more than 8,000 km) in the first quarter of the 20th century aimed to transform the wastelands of Thar Desert in Rajasthan into agriculturally productive zones along with improvement of afforestation and environment, development and protection of livestock/animal health, human rehabilitation and settlement, and economic growth of the desert people (Anonymous, 1994; Tyagi, 2002, 2004). Over the years these anthropocentric interventions brought about a great deal of changes in the physiography of the Thar Desert; its soil composition, landscape, potential for water retention and climate. As a result, some areas, especially those in the vicinity of the canalized irrigation, became greener and water-rich underground, but also perennial breeding sites for vectors and experiencing annual outbreaks of malaria, particularly that caused by *P. falciparum*. The Thar Desert, which was represented by only one mosquito species, *An. stephensi*, until late 19th century, is now thrived by a dozen species including most dominant Indian malaria vectors like *An. culicifacies, An. stephensi* and *An. fluviatialis*, with *An. culicifacies* being dominant in canal-irrigated villages and *An. stephensi* in the typical desert villages without any outer source of water (Tyagi, 2005). It is noteworthy that during the 1994 nationwide epidemic, the Thar Desert reported an unprecedented escalation in both malaria morbidity and mortality due to *P. falciparum* (Tyagi et al., 1995) (Figure 14.2).

Flowing huge malaria epidemics in the Thar Desert, Bouma and Van der Kaay (1994, 1995, 1996), considering excessive rainfall in the malaria epidemic area, propounded his El Nino Southern Oscillation (ENSO)-based theory where erratically massive downpours affected pathogen transmission potential of the vector mosquitoes. Their hypothesis was repudiated by Tyagi et al. (1995) who instead gave his "multi-vector" theory, of course in association with extensive irrigation and flooding rainfall, being the major cause behind the occurrence of major epidemics in the Thar Desert (Table 14.1); there being a transformation in Anopheles species from the late 19th century, when only *Anopheles stephensi* existed, to about a dozen species (*An. culicifacies, An. stephensi, An. subpictus, An. vagus, An. annularis, An. nigerrimus, An. splendidus, An. d'thali, An. fluviatilis, An. barbirostris*) by the end of the 20th century. As a result of canalized irrigation for nearly a century and preponderance of dominant vector species, a clear-cut chronological relationship amongAnopheline fauna, malaria prevalence and extent of canal irrigation can be drawn in the Thar Desert (Table 14.1).

14.3.3 Climate Change and Emergence of Dengue in Kerala State, South India

The south Indian Kerala state, a *terra incognito* for dengue, chikungunya, and zika virus infections until the mid-1990s, has since the 1950s has undergone an enormous change in respect of both its physiography and climate (Figure 14.3). With a total area of 38,863 sq. km (urban 3365 sq. km and rural 35,498 sq. km), it is one of the smaller states in the country, ironically supporting a very high population density of 819 persons per sq. km. which is nearly three times the national average.

Shaped as a high vertical slope with an average breadth of 50 km, Kerala is endowed with diversified climatic richness, with a pinch of hydrological concern as 41 of its 44 rivers originating in the Western Ghat empty into the Arabian Sea in less than 48 hours after a rain. The state's lowland region, which accounts for 10% of the total area, runs along the coastline and embodies beaches, swamps, and lagoons, besides backwaters, paddy fields, and coconut plantations. Kerala's midland (42% of the total landmass) is primarily made up of valleys with undulating small hills and

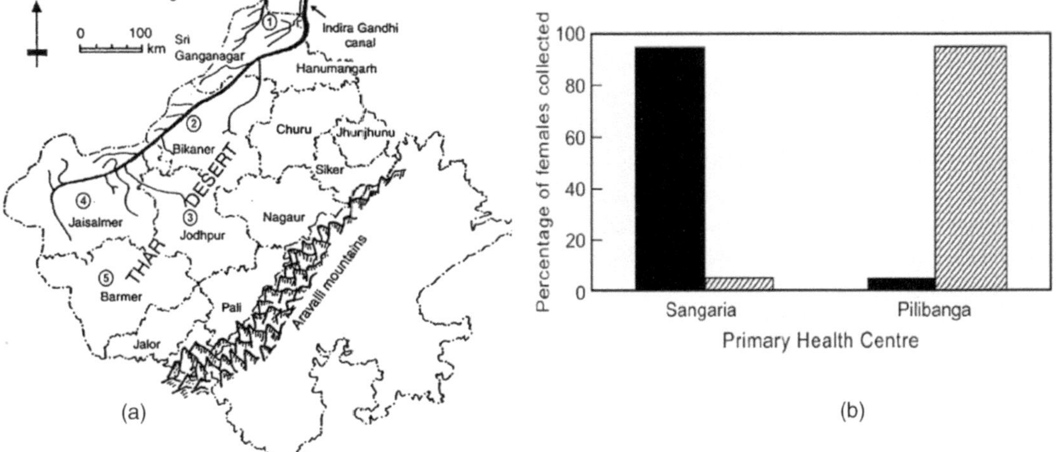

FIGURE 14.2 Thar Desert (a) and the trio of Gang-Bhakra Sirhind and Indira Gandhi Canal systems, and (b) dominant prevalence of *An. stephensi* in desert villages (2–4 in circle; lined column) while *An. culicifacies* in the no-longer canal-irrigated villages (1 in circle; black column) in the Thar Desert.

TABLE 14.1
Chronological Relationship Among Anopheline Fauna, Malaria Prevalence, and Extent of Canal Irrigation in the Thar Desert

	Number of Vector Species[a]		Malaria Cases[b]		*Pf*	Irrigation (in ha 10^6)
Period	**Vectorspecies**	**Non-Vectorspecies**	***Pv***	***Pf***	**(%)**	
pre-1930	*An. stephensi*	-	-	-		-
1930–34	*An. stephensi*	-	-	-		0.24
1935–39	*An. stephensi*	4	-	-		0.24
1940–44	*An. stephensi*	4	-	-		0.24
1945–49	*An. stephensi*	5	-	-		0.24
1950–54	*An. stephensi*	5	-	-		0.24
1955–59	*An. stephensi*	5	-	-		0.32
1960–64	*An. stephensi*	5	-	-		0.37
1965–69	*An. stephensi*	5	-	-		0.71
1970–74	*An. stephensi, An. culicifacies*	5	-	-		0.84
1975–79	*An. stephensi, An. culicifacies*	5	639,754	21,649	(4.9)	1.22
1980–84	*An. stephensi, An. culicifacies*	5	98,242	12,422	(12.6)	1.23
1985–89	*An. stephensi, An. culicifacies*	5	110,746	16,533	(14.9)	1.33
1990–94	*An. stephensi, An. culicifacies, An. subpictus*	6	368,208	106,963	(29.0)	2.17
1995–2000	*An. stephensi, An. culicifacies, An. subpictus*	9	2,45,960	30,077	(10.9)	2.17

[a] *An. stephensi*
[b] In all the Thar Desert districts organized surveillance began only in early seventies.
Pv = Plasmodium vivax; Pf = Plasmodium falciparum

meandering passages. A variety of seasonal, annual and perennial crops are grown in its mainly lateritic soil. Of all these crops, cocoa and pineapple plantations are of tremendous significance as dengue vectors have been found to breed there. The highland region (48% of the total state's area) with steep hills is rife with forests and small streams. Plantation crops like rubber, tea, coffee, and cardamom are grown in this region. It is noteworthy that rubber plantation was introduced in Kerala in the 1950s on the cost of lands used for paddy plantation. The average annual rainfall (3.000 mm) is quite high compared to other Indian states – almost three times higher than that in Karnataka and twice that in Tamil Nadu. The state basically enjoys four types of climate, viz., winter, summer, southwest monsoon and northeast monsoon. The winter season sets in during the month of December and continues till the end of February. During this season there is comparatively less rainfall, especially in the northern parts of the state. Winter is followed by the summer season. It starts in February and continues till May. Temperature is very high during this period. Occasional showers are a characteristic of this season. Kottayam district, tucked in the Western Ghat, ranges receives the

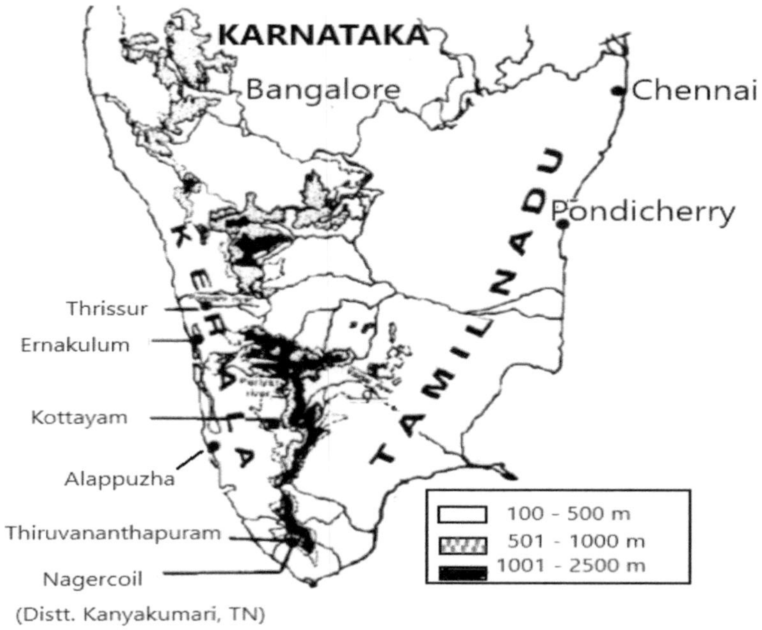

FIGURE 14.3 Kerala state in peninsular India with sites of dengue investigation following a massive epidemic in 2003 and 2004.

Source: **Tyagi et al., 2006.**

highest rainfall during this season. The southwest monsoon, accounting for 60–65% of the total rainfall, begins either at the end of May or the beginning of June, and fades out by September. The northeast monsoon, commencing in October, contributes to 25–30% of the total rainfall. The coastal areas record a maximum temperature of 32°C. The interiors record a maximum of 37°C during summer. In the coastal area it is hot and humid during April–May while cool during December–January. The mountainous and sylvan environs of the Western Ghat ranges on the south-western edge of peninsular India facing the Arabian Sea coast in Kerala of which it covers nearly 90% of the total landmass. The Western Ghat ranges receive a variable amount of rainfall from both the southwest (May–July) and the northeast monsoon (September–November), and is characterized by the Shola forest punctuated with human-grown extensive plantations of rubber (*Hevea brasiliensis*), cocoa (*Theobroma cacao*), coconut (*Cocoa nucifera*) and pineapple (*Ananas comosus*).

Anthropization, particularly human densification, bore a discernible vicissitude in Kerala's forest cover, agricultural practices, traditional water harvesting, demography, urbanization, and human mobility. It would be hard to believe that Kerala experienced a drought in 2003, with the average annual rainfall dipping to an all-time low of 2000 mm! Although Kerala, which enjoys high rainfall, it still gets no more than 50–60 lit. of water for domestic use per person daily. Water allowance per capita depleted miserably in the past few years. The Kerala state has an estimated 77.35 billion cubic meters (BCM) of fresh water, but nearly 40% of the resources are lost as runoff, meaning only 42 BCM of water is available for the whole populace though the state requires a minimum of 49.70 BCM annually for various indispensably vital activities such as irrigation, domestic use, industries and others miscellaneous use. This shortage has led people to resort to exploiting groundwater so severely that the water table in several districts has gone alarmingly low. The water scarcity also induced people to store water in a mélange of containers like pitchers, cement tanks, etc. as long as possible. A cadence in both ambient temperature and RH is considered highly supportive to a wide dissemination of the dengue vector in time and space. There is evidence to prove a decline in rainfall, on one hand, and a marginal yet perceptible rise in ambient temperature on the other. Since 1980 Kerala has suffered seven years of deficient rainfall and another five years below normal rainfall. An analysis of the southwest monsoon since 1998 shows that except Thiruvananthapuram in 2001, all districts have been receiving highly deficient rainfall in most years. Kerala had 2270 mm of rainfall in 2003, still about 33% less than the average. It was among the few states in India declared drought-hit during 2003. It is noteworthy that since 1999 the annual rainfall has never touched the average value of 3107 mm; on an average, annual rainfall has dropped by 15–20%. The southwest monsoon has been particularly poor and in 2002 and 2003 it was deficient by as much as 33.3% and 26.4%, respectively. The year 2003 was further impoverished due to the failure of the northeast monsoon, with no area getting any rain during January–March. A close observation of rainfall data in various districts since 1900 shows that the number of rainy days in Kerala has dropped drastically. A discernible climatic vicissitude in Kerala is possibly linked with the atmospheric El Nino phenomenon which characterizes monsoon-related vagaries. Forests, which help in orographic or convectional rainfall (i.e., rain in hill during evening), play an integral role in holding soil together and retaining rainwater, but large-scale deforestation has been occurring in the state since 1970s, mostly with a view to planting cash crops like rubber plantation. Kerala had an estimated forest cover of 44.45% in 1900 which, according to the recent satellite images, had reduced to 14.7% in 1983 and to a pathetic 9% at present. Kerala originally had a total cropped area spanning over 2 million hectares, with paddy as the major crop covering about 35% of this area, followed by coconut (21%). By 1975 the paddy growing area reduced to 30% of the total cropped area, with only 0.881 million ha growing paddy. In 2003 the area under paddy cultivation nose-dived to a paltry 0.31 million ha. However, the coconut plantation sustained a good growth rate of 3% per annum through 1990s. It is noteworthy that, along with coconut, rubber plantation has *de novo* supported breeding of dengue vector, *Ae. albopictus*. At present while rubber plantations occupy the largest land area in Kerala, paddy fields have reduced and reclaimed largely for construction of human habitation.

Kerala state witnessed the emergence of dengue cases in 1996/ 97, which was followed by a more severe dengue outbreak implicating 67 cases (nearly 5-fold increase) and a toll of 13 human lives (3-fold increase) in 1998, again in the same district. The years of 1999 and 2000 were virtually free from any dengue cases. However, soon in 2001, epidemic dengue resurged mainly in Kottayam, Idukki, and Ernakulam, reporting 70 cases, followed by 219 cases in 2002 with some deaths. But the year 2003 experienced the severest epidemic to date, yielding as many as 3546 confirmed cases (253-fold increase) and a toll of 68 human lives, spread for the first time all over Kerala's 14 districts. The occurrence of DHF and dengue shock syndrome (DSS) was also clinically and serologically documented, along with all four serotypes (DENV1–4) incubation, through the sole vector, *Aedes albopictus*. Kerala is today one of the few states in the country which is endemic in all of its districts.

Aedes albopictus (69.7%) predominance in the Kerala state was so sudden that it could easily be correlated with the catastrophic changes in the climate. It bred predominantly in coconut shells/ plastic cups (79.7%) deployed in collecting latex from rubber trees. Kerala, *Ae. albopictus* exhibited a rather new biological dimension unknown hitherto, particularly with reference to its potential to transmit dengue viruses, without aid of the primary vector, *Ae. aegypti*. Because of its quaint biological attributes, the *Ae. albopictus* differed almost completely from its main ally, *Aedes aegypti*, in its control strategies that were so far entirely based on the vector biology of *Ae. aegypti*. Significantly, soon thereafter many countries in Southeast Asia began reporting similar observations, substantiating strongly the Kerala discovery about both the disease transmission and its vector, *Ae. albopictus*.

14.4 IMPACT OF CLIMATE CHANGE ON MOSQUITO DISTRIBUTION

The non-linear and interactive effects of temperature and other climatic factors (such as precipitation and RH) on the basic physiological and ecological traits of mosquitoes, are expected to either increase, decrease, or more commonly shift the geographical distribution of various mosquito vectors. Studies have forecasted that in some regions with high incidence of mosquito-borne diseases, future warming could exceed the thermal optima and thresholds of mosquitoes, resulting in a decrease in the overall transmission risk of the diseases they carry (Gething et al., 2010; Mordecai et al., 2020; Ryan et al., 2015). At the same time, warming in colder regions could bring temperatures closer to the thermal optima of mosquitoes, leading to an increase in the overall risk of mosquito-borne disease transmission (Hongoh et al., 2012; Hussain & Dhiman, 2022; Samy et al., 2016).

Mosquito species and sub-species differ in their ability to tolerate various temperature extremes. This variation in thermal tolerance is either a result of genetic differences, or due to behavioral differences arising from adaptation to different environments. For example, the European strain of the Asian tiger mosquito *Aedes albopictus* has adapted to temperate climates due to the ability of its eggs to overwinter (evade desiccation in cold temperature conditions) (Hawley et al., 1989). Similarly, the endophilic behavior of *Ae. Aegypti* and *An. stephensi* mosquitoes has enabled them to survive in arid environments by resting indoors and thereby escaping the heat (Martin et al., 2021). Owing to these differences, climate change is believed to have varying effects on different vectors, which could in turn have different implications for the diseases they transmit. Some striking examples are discussed below.

14.4.1 Malaria Vectors, *Anopheles stephensi* and *An. culicifacies*

Malaria is one of the most prevalent infections worldwide with an estimated 241 million cases and 627,000 deaths in 2020 (WHO, 2021). It is transmitted by female Anopheline mosquitoes that carry the causative agent/malarial protozoa, i.e., Plasmodium parasites (mostly *Plasmodium falciparum* and *Plasmodium vivax*). The Anopheline genus consists of about 537 species (Harbach, 2013), of which approximately 70 are able to carry and transmit the malaria parasite (Warrell & Gilles, 2017). Two important Anopheline vectors of malaria are *An. stephensi*, an urban malaria vector, and *An. culicifacies*as a rural vector of malaria in India. *Anopheles stephensi* is an important vector of malaria in India, Pakistan, and Iran (Sinka et al., 2011). It has also been found prevalent in other countries of the Indian sub-continent such as Bangladesh (Bashar et al., 2013), the Arabian Peninsula like Bahrain (Amin, 1989), Oman (Baomar & Mohamed, 2000), and Saudi Arabia (Alahmed, 2012) and the Middle-East, e.g., Iraq (Hantosh et al., 2012). More recently it has invaded new locations such as Sri Lanka (Gayan Dharmasiri et al., 2017) and the Horn of Africa (Balkew et al., 2020; Takken & Lindsay, 2019), where it has caused several outbreaks in urban areas(Seyfarth et al., 2019). *An. stephensi* thrives in arid and semi-arid climates. The optimum temperature for the development of this species is 32°C, with upper and lower thresholds of 15°C and 36°C, respectively (Paaijmans et al., 2013). *Anopheles stephensi*is not able to survive even for a single day beyond 42°C under laboratory conditions (Singh et al., 2022), but in the xeric region in the deserts such as the Thar Desert, the species can stay alive even at 48°C (Tyagi, 2023). The temperature suitability range for transmission of *P. falciparum* through *An. stephensi* has been estimated to be between 15.3 and 37.2°C, while for *P. vivax* it is estimated to be between 15.7 to 32.5°C (Villena et al., 2022).

Higher activity at dusk and the ability to breed in urban environments, where temperatures are higher due to the urban heat island effect, suggest that *An. stephensi* is able to tolerate higher temperatures than other Anopheline vectors, and is more buffered to climatic effects (Sinka et al., 2020). That is why it is able to thrive in arid and semi-arid areas such as the Thar Desert in India and the Arabian Peninsula (Alahmed, 2012; Edalat et al., 2015; Tyagi & Chaudhary, 1997). This tolerance for high temperatures is likely attributable to behavioral adaptations, such as the preference to rest and feed indoors (endophily), the ability to breed in small man-made containers and water tanks, as well as the tendency to bite during twilight hours which enable it to escape the heat (Sinka et al., 2011). Due also to behavioral adaptations, *An. stephensi* has emerged as a highly efficient urban vector for malaria (Surendran et al., 2019). Studies modeling the climatic suitability for malaria transmission report a wider thermal range for *An. stephensi* based on models (Miazgowicz et al., 2019; Villena et al., 2022).

Some of the regions where the vector has invaded are arid deserts that are not typically suitable for vector development, such as Sudan (Abubakr et al., 2022). However, extreme weather events likely brought about by climate change, such as rainstorms and severe flooding, have enabled the vector to grab a foothold in the region and thereby expand the global distribution of *An. stephensi* (Abubakr et al., 2022), indicating the strong invasive potential of the vector species under climate change. In addition to this, the movement of people and livestock has further aided the expansion of the range of this vector westward across the Arabian Peninsula and into Africa (Sinka et al., 2020). Although some studies have assessed the impact of recent changes in climate on the distribution of *An. stephensi*, no projections of the future distribution of this vector under different scenarios of climate change have so far been made. However, indications based on current observations show that the adaptation to urban arid environments may further support the expansion of this vector in Africa under climate change.

An. culicifacies (Giles, 1901) species complex derives its name from the Latin words *culici* (culex) and *facies* (appearance) due to its "culex-like appearance." It was first described in 1901 in Hoshingabad and is one of the most widely distributed malaria vectors in the Indian sub-continent. It is also found in Southeast Asian countries such as Cambodia (Van Bortel et al., 2002), Laos (Kobayashi et al., 1997), Thailand (Kongmee et al., 2012), Myanmar (Zhong et al., 2022) and Sri Lanka (Ramasamy et al., 1994). The vector also has a limited presence in the Arabian Peninsula (Hoosh-Deghati et al., 2017) and Africa (Fletcher et al., 1992).

The species complex is composed of five reproductively distinct sibling species designated as A, B, C, D, and E. Of these, sibling species A, C, and D are considered important vectors of malaria despite showing mostly zoophilic behavior, sibling species E is significantly more anthrophilic and is an efficient vector and sibling species B is generally considered a non-vector due to its very high zoophilic nature (Tripathy et al., 2010). *An. culicifacies*is mainly endophilic in nature, preferring to rest indoors (Rao, 1984), though exophilic behavior has also been observed in some studies (Mojahedi et al., 2020). Feeding behavior of *An. culicifacies* is highly variable and mostly dependent on local ecology and climatic factors (Rao, 1984). During colder months, they prefer to bite during the early hours of the night, whereas, in the summer months, they bite at much later hours (Reisen & Aslamkhan, 1978).

The upper and lower thresholds for the development of *Plasmodium* parasites in *An. culicifacies* mosquitoes have been found to be 15°C and 34°C, respectively. However, the mosquito is able to survive for more than an hour at temperatures of 12.7°C and 41°C (Pal, 1943), and is generally considered to be more physiologically fit than *An. stephensi* (Reisen & Mahmood, 1980).

Recent observations have found an increase in the prevalence of *An. culicifacies* in colder regions of India (Uttara Khand), possibly as a result of increasing temperatures under climate change (Dhiman et al., 2019). However, only a couple of studies at the local level have predicted future distributions of the vector under different scenarios of climate change. Studies in Iran show that the distribution of *An. culicifacies* will increase under low (Representative Concentration Pathway, RCP 2.6) and moderate (RCP 4.5) warming, but would decrease under extreme (RCP 8.5) warming conditions (Hanafi-Bojd et al., 2019; Pakdad et al., 2017). Similarly, modeling of vector density in Odisha, India using the VECTRI reveals that the prevalence of *An. culicifacies* will decrease in lowland areas, but increase in highlands (Parihar et al., 2022).

14.4.2 Dengue/Yellow Fever/Chikungunya/Zika Virus Vectors, *Aedes aegypti* and *Ae.albopictus*

In recent years, arboviral infections such as Dengue and Yellow Fever have greatly expanded their global presence (Brady et al., 2012; Jentes et al., 2011) compounded by the emergence and spread of several new viruses such as chikungunya and Zika (Shragai et al., 2017). The global expansion and/or resurgence of these viruses is attributed to two important arthropod vectors – *Aedes*

(*Stegomyia*) *aegypti* and *Aedes* (*Stegomyia*) *albopictus*. Historically, these vectors have been restricted to distinct regions. However, in recent years they have significantly expanded their global distribution due to the increase in global transportation and urbanization, but also partly due to the effects of climate change (Brown et al., 2014).

Based on the patterns of genetic differentiation, *Ae. aegypti* is speculated to have originated from sub-Saharan Africa (Brown et al., 2011) and has established itself across much of the tropical and sub-tropical parts of the world relatively recently (Brown et al., 2014; Kraemeret al., 2015a). *Ae. aegypti* breeds in a wide range of small natural and artificial containers typically found in human dwellings. They also prefer to rest indoors and feed almost exclusively on humans. These characteristics make the species one of the most efficient and well-adapted urban vector among all known insect vectors (Soper, 1967). Due to this, while the vector is mostly restricted to the tropic and subtropics, it is not always bound by outdoor temperatures, and is also able to thrive in arid environments (Martin et al., 2021). The lower threshold of development of *Ae. aegypti* has been found to be around 10°C (Grech et al., 2015; Tun-Lin et al., 2000), and no eggs survive or larvae develop beyond 40°C (Marinho et al., 2016; Sukiato et al., 2019). The temperature thresholds for adult longevity have been found to range between 16°C and 34°C, respectively (Christophers, 1960), with an optimum temperature of 32°C (Bar-Zeev, 1958).

Presently, it has widespread distribution in Southern and Central America, Southeast Asia as well as limited parts of Europe (Spain and Greece) (Kraemer et al., 2015a). Under the effects of climate change, it is predicted that the species will increase its prevalence in regions that are historically most suitable, i.e., in Southeast Asia, South America, and west-central Africa, albeit with greater inter-annual variability (Iwamura et al., 2020). A marginal increase in suitability for *Ae. aegypti* has also been predicted in the western and northern United States and southern China, particularly under the RCP 8.5 scenario of climate change (Iwamura et al., 2020). In Europe, suitability for *Ae. aegypti* is predicted to increase in isolated patches in the Iberian Peninsula, Italy, and Greece (Iwamura et al., 2020). Climate change is also expected to expand the range of the species to higher altitudes as a result of warming (Lippi et al., 2019). In India, the vector is projected to further increase its range into the Thar Desert region of Rajasthan, with limited increase in distributions in higher altitude Himalayan regions, as a result of future climate change (Hussain and Dhiman, 2022).

Aedes albopictus is a day-biting mosquito that has significant phenotypic plasticity, enabling it to adapt to a large range of environments in both tropical and temperate regions (Tippelt et al., 2020). It is adapted to peri-domestic environments as it feeds on both humans and animals and breeds in a variety of natural and artificial containers. However, it is not as well-adapted to urban environments, as *Ae. aegypti* still thrives in residential environments with enough vegetation and water sources.

Aedes albopictus is native to Southeast Asia and the western Pacific but has recently expanded its global range significantly, spreading to several parts of Africa, the Americas, and even to Europe, where it has been historically absent (Kraemer et al., 2015b). Similar to its African cousin, *Ae. Albopictus* is a day-biting mosquito that breeds in small natural or man-made containers. However, unlike *Ae. aegypti*, it is largely exophilic as well as an opportunistic feeder, biting both humans and animals as and when available (Delatte et al., 2010). Despite this, strong anthropophilic behavior has been observed in some studies, similar to that of *Ae. aegypti* (Delatte et al., 2010; Ponlawat and Harrington, 2005). These characteristics make the vector well adapted to peri-domestic environments with enough vegetation and water sources, though it is not as well adapted to urban environments as *Ae. aegypti*. Nevertheless, the species exhibits a high degree of plasticity and is more ecologically flexible frequently occurring in suburban, rural and agricultural habitats as well.

The minimum temperature threshold for the development of immature stages of *Ae. albopictus* is 10.4°C (Delatte et al., 2009), and eggs belonging to the temperate strain of *Ae. albopictus* undergoing diapause can tolerate temperatures as low as −10°C (Thomas et al., 2012). Temperatures above

35°C have been found to be detrimental to larval development (Monteiro et al., 2007). Nevertheless, *Ae. albopictus* larvae are able to survive at temperatures up to 39 °C, and significant mortality is observed only at temperatures higher than 40°C (Sivan et al., 2019). As compared to its cousin (*Ae. aegypti*), *Ae. albopictus* have significantly better survival rates at and below temperatures of 10°C (Chang et al., 2007), and overall their population growth is more stable regardless of temperature (Farjana et al., 2012). This tolerance for low temperatures in addition to the ability of *Ae. albopictus* to diapause and thereby overwinter, has enabled the vector to invade several colder temperate regions in Northern America and Europe (Benedict et al., 2007). Moreover, a clear distinction in tolerance of low temperatures and ability to diapause is observed between the tropical and temperate strains of the vector, highlighting its strong potential to adapt to the changing environment in a short period (Thomas et al., 2012; Tippelt et al., 2020). In fact, the rapid expansion of the global range of the vector from its native distribution area to all the continents except Antarctica, has occurred only in the span of the last 3–4 decades (Paupy et al., 2009).

While the distribution of *Ae. albopictus* coincides with that of *Ae. aegypti* in some regions of Asia and West Africa, marked differences in range are observed in Europe, East Africa, United States and Australia, with *Ae. albopictus* showing broader distribution potential across temperate regions of Europe and Northern America (Kamal et al., 2018; Kraemer et al., 2015a). Future climate change is expected to result in the northward expansion of the vector in both North America and Europe, at a much larger scale than observed for *Ae aegypti*. In North America, the vector is predicted to expand further into northern United States, also invading several parts of southern Canada, whereas it is predicted to occupy most of Europe by the 2050s and 2070s (Benedict et al., 2007; Kamal et al., 2018; Ryan et al., 2018). True to these predictions, only in the past decade, *Ae. albopictus* has implanted itself and caused dengue/ chikungunya outbreaks in several European countries such as France, Spain, Italy, Greece, Croatia, Bosnia, Albania and Montenegro (Paupy et al., 2009). However, future warming in tropical regions with year-round suitability for the vector presently is expected to exceed the upper thermal limit of *Ae. albopictus* resulting in range contraction in central Africa, northern Australia, Amazon basin, and southern Asia (Ryan et al., 2018), particularly under the moderate and extreme scenarios of climate change (RCP4.5 and RCP 8.5). These predictions are in-line with the observed differences in temperature tolerance of the two species. In some regions, the local spread of *Ae. albopictus* is accompanied by a decline in the population of *Ae. aegypti*. Studies suggest that this is probably linked to inter-specific competition (Daugherty et al., 2000; O'Meara et al., 1995) and or non-reciprocal cross species insemination between the two species (Bargielowski et al., 2013).

14.4.3 Lymphatic Filariasis Vector, *Culex quinquefasciatus*, and Other Allied Culicine Species

The *Culex pipiens* species complex is the most widely distributed mosquito species in the world (Samy et al., 2016). *Culex pipiens* is a polytypic species, that consists of several other sub-species that are morphologically indistinguishable and may also exist as hybrid forms, in regions where they co-exist (Vinogradova & Elena, 2000). While most of the species of this complex are only locally distributed in different regions, two species – *Culex pipiens pipiens* (Linnaeus, 1758) (also called the northern house mosquito) and *Culex pipiens quinquefasciatus* (Say, 1823) (also called the southern house mosquito), show widespread distribution across much of the temperate and tropical/sub-tropical regions of the world, respectively (Farajollahi et al., 2011). These mosquito species are important vectors of the flaviviruses West Nile Virus (WNV) and St. Louis encephalitis as well as filarial and avian malaria parasites (Farajollahi et al., 2011).

Culex p. pipiens is believed to have originated in Africa and colonized the temperate regions of Europe and African highlands following the last glaciation, whereas *Culex p. quinquefasciatus* is postulated to be of tropical south Asian ancestry (Vinogradova &Elena, 2000). Unlike most other species, mosquitoes of the *Culex* genus breed in water sources that are polluted and have high organic content, such as cesspools, ditches and septic tanks.

The upper and lower temperature thresholds of species of the *Culex pipiens* complex have been estimated to be 16°C and 36°C, respectively (Ciota et al., 2014), however, studies have shown that *Cx. p. quiquefasciatus* may have greater tolerance for high temperatures (Oda et al., 1999, 2002). In contrast, *Cx. p. pipiens* are better adapted to colder temperatures largely due to their ability to enter diapause in order to survive freezing winters, *Cx. p. quiquefasciatus* lack this ability. This difference in thermal tolerance of the two species is reflected in their geographic distribution, wherein *Cx. p. pipiens* is widely distributed in temperate regions upwards of 39°N, and *Cx. p. quiquefasciatus* has widespread distribution in tropical areas below 36°N. The region between these latitudes is a hybridization zone, where both vectors may be found inter-breeding, with varying populations of hybrids.

The present distribution of *Culex p. quinquefasciatus* is restricted to southern North America, much of South America, sub-Saharan Africa, South Asia and most of Australia and New Zealand (Samy et al., 2016). It is a major vector in Brazil, tropical Africa and Southeast Asia. Under climate change, future warming is expected to enable a northward expansion of the distribution range of the vector, in the Americas (Fontes et al., 2012), Central Asia and Africa (Kamal et al., 2018). However, extreme temperature rise under the RCP 8.5 scenario could potentially reduce its distribution in the arid and warm tropical regions, where ambient temperatures may rise above the survival thresholds of the vector.

14.5 PROJECTED SCENARIO OF IMPACT OF CLIMATE CHANGE ON MAJOR VECTOR-BORNE DISEASES

Future projections of climate and its consequences are based on the RCP, which is a greenhouse gas concentration trajectory used by the Intergovernmental Panel on Climate Change (IPCC) for climate modeling and research.

The primary purpose of RCPs is to provide a time dependent guideline about the presence of estimated concentrations of atmospheric greenhouse gas (GHG). In simple words, increase in GHG concentrations leads to changes in the climatic factors. The projections of VBDs scenarios are important in the context of climate change as vectors are climate sensitive. Projecting the future scenarios of VBDs is not an easy task therefore, researchers have developed computer-simulated models to assess the change in climate. Model-based projections help in understanding the impact of climate change on transmission dynamics, demography, and re-emergence of vectors. Currently, projections have provided the estimation of the direct effects of climate change on individual species of mosquitoes like Anopheles, Aedes, and the pace of possible changes in their entire ecosystem along with urbanization and deforestation.

Researchers have projected that under climate change, the colder regions will become suitable for arthropod vectors. It has also been expected that under projected scenarios, vector density, survival rate, and biting rate will increase with increasing temperature, as along with the pathogens' development rate in the vector. On the other hand, climate change may not support transmission of the pathogen by the vector, if temperature decreases up to that level where survival of the vector and development of the pathogen is so slow that the vector dies before becoming infective (Rocklöv & Dubrow, 2020). The projected distribution of some alarming mosquito-borne diseases is discussed below.

14.5.1 Kyasanur Forest Disease

Apart from mosquito-borne diseases, other infectious disease like tick-borne Kyasanur Forest Disease (KFD) is also of public health concern in India owing to its vector's expansion to adjoining states from the epicenter of Shivamogga district of Karnataka (Figure 14.4).

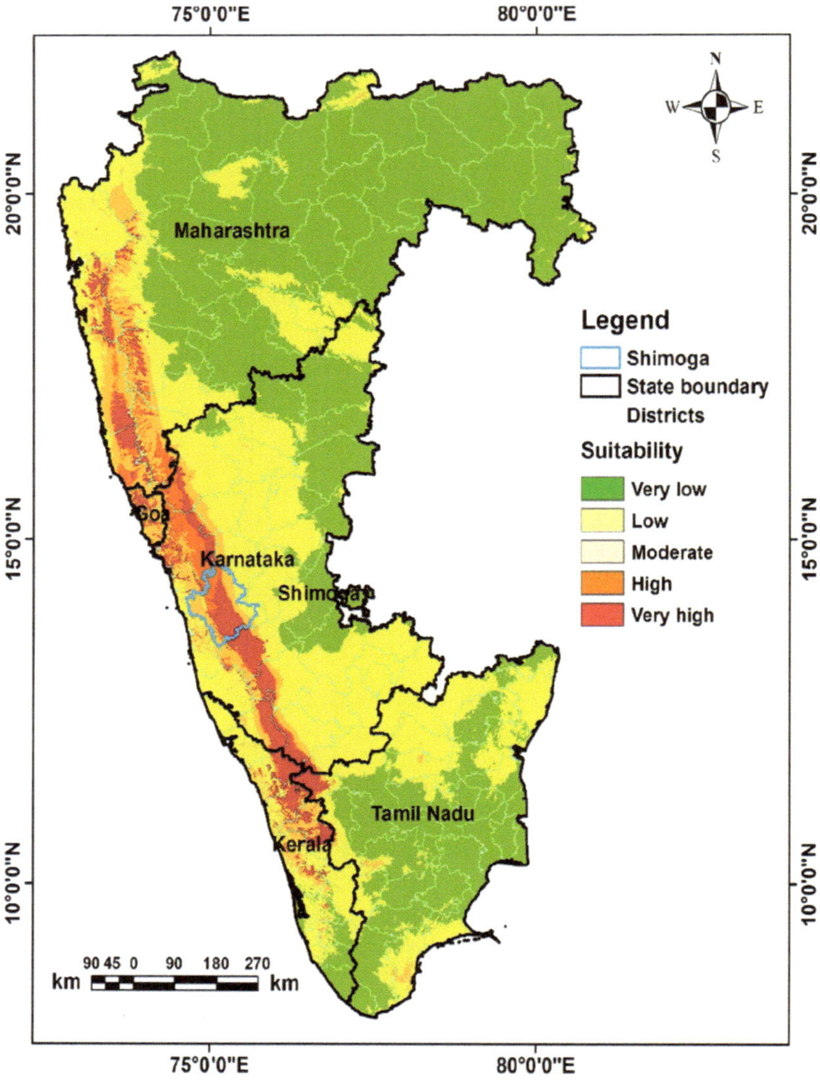

FIGURE 14.4 Predicted distribution of ***Haemaphysalis spinigera***, the vector of KFD in southern India.

***Source*: Pramanik et al., BMC Infect. Dis 2021.**

Pramanik et al. (2021) have predicted the climatic suitability for *Haemaphysalisspinigera* tick by using MaxEnt model. Results suggest that the proportion of bioclimatic suitability areas, falls into five categories, i.e., "very high suitability," "high suitability," "moderate suitability," "low suitability," and "very low suitability." Based on climatic conditions Karnataka state is the highest potential region, followed by Maharashtra, Kerala, Goa, and Tamil Nadu. Further, within the Karnataka state, Shivamogga district is a very high suitable area, followed by Chamrajnagar and Chikmagalur. Similarly, in other states like Kerala, Kozhikode and Wayanad districts, in Maharashtra, Raigad and Ratnagiri districts, in Tamil Nadu, Nilgiris district and North Goa district in Goa state, require continuous vigilance. The district-wise suitability map shows linear spatial clustering along the western Ghats as a very high-risk zone of the potential distribution of H. spinigera. In those districts, extensive survey, continuous vigilance is required, especially from November to March, when most KFD cases occur.

14.5.2 Malaria

At the global scale, the projection of climate change on malaria transmission has been studied while in respect of India information is little. Recently, the projection of malaria by the year 2030s were attempted in India, as an initiative to evaluate the climate change driven dependency of malaria transmission in the country. It may also be considered as a helpful step in achieving the government's National Framework for Malaria Elimination by the year 2030 program.

The projected scenario of malaria by the year 2030 has been attempted to map at district level using regional climate models. The comparison between CORDEX South Asia for Baseline (1976–2005) and RCP 4.5 scenario for future projection by the 2030s (2021–2040) suggest there is a significant rise in spatial transmission windows (TW) months, and seasonal duration as well (Figure 14.5). A new foci of malaria transmission in western Himalayan and rise in more suitable conditions in parts of some southern and eastern states. By the year 2030s, a reduction in the duration of transmission months in Odisha and Andhra Pradesh are also expected. The projected scenarios revealed that climate change will not only increase malaria transmission potential in non-malarious areas but also make it difficult to control as it leads to the changes in parasite and vector growth cycle. Apart from the direct impact of climate change, the related effects of outcomes like social, economic, and sometimes political issues of climate change will also challenge the malaria burden. In other words, climate change is bound to alter the spatial and temporal distribution of malaria in India and globally.

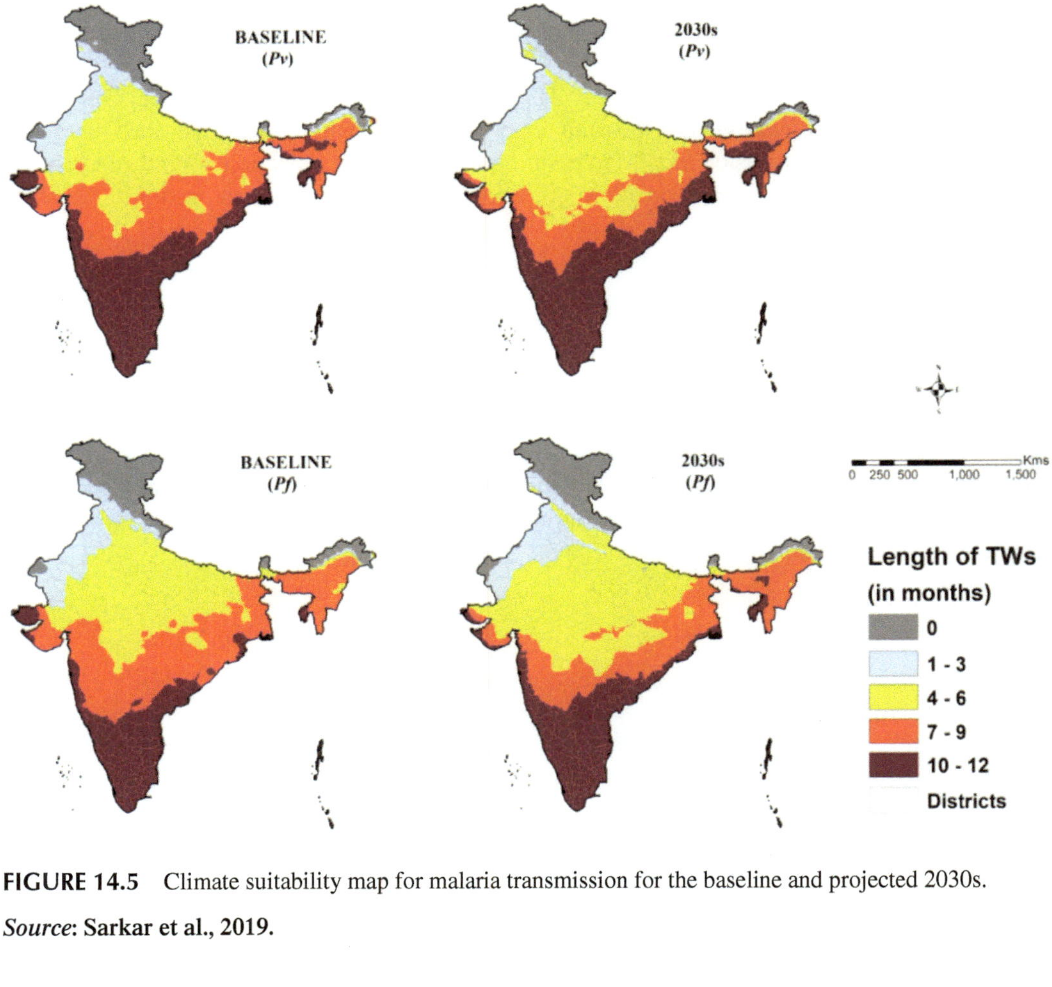

FIGURE 14.5 Climate suitability map for malaria transmission for the baseline and projected 2030s.

Source: Sarkar et al., 2019.

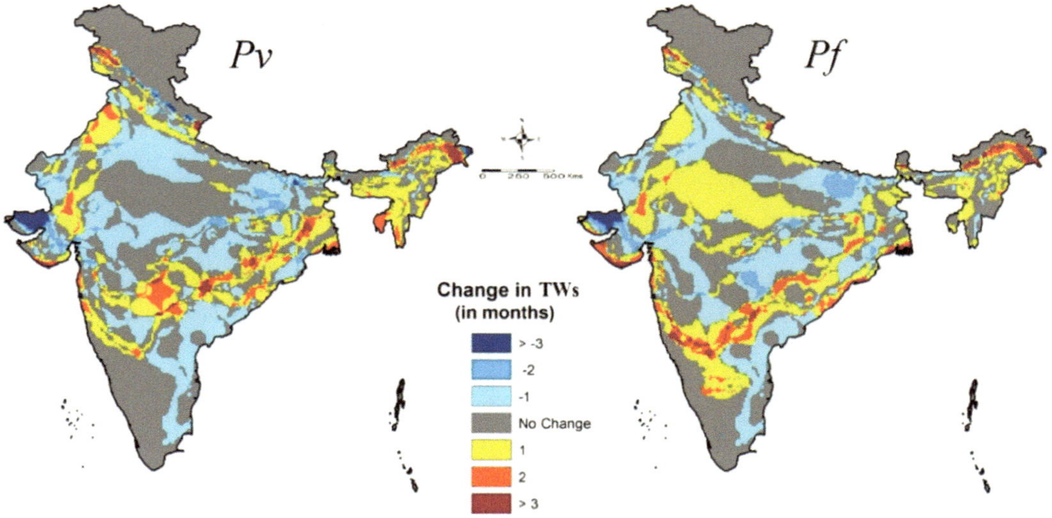

FIGURE 14.6 Projected changes in the number of months in the Transmission Window (TW) for malaria (***Pv*** and ***Pf***) transmission by the 2030s as compared to baseline years.

Source: **Sarkar et al., 2019.**

In India, new foci of transmission of *P. vivax* are visible in parts of Jammu and Kashmir, Uttarakhand, and Himachal Pradesh (Figure 14.6). An increase in the number of months of TWs in north-eastern states and a reduction in parts of Orissa and Gujarat states are also seen. On the other hand, new foci of *P. falciparum* transmission are projected in Jammu & Kashmir and Uttarakhand with an increase in transmission months in north-eastern states and reduction in parts of Orissa and Gujarat states (Figure 14.6). An increase in the TWs of *P. falciparum* by one month in the central part of India is also discernible.

14.5.3 Dengue

Using the criteria of 13–35°C for a window of transmission, the projected scenario of dengue by the year 2030s shows that most parts of India except parts of northern states like Uttarakhand, Himachal Pradesh, Jammu and Kashmir, Arunachal Pradesh, parts of western Rajasthan, and Uttar Pradesh are likely to have 10–12 months of transmission. However, the projected scenario will depend on water availability and storage practices as well as the lifestyle of communities. The projected scenario of Japanese encephalitis is also likely to extend the transmission months in the northern part of the country. However, detailed studies are warranted in this regard.

14.6 CONCLUSION

Climate change impacts on human health and in particular insect vectors and VBDs are a reality. Owing to poikilothermic, the developmental cycles and development of pathogens in mosquitoes, sand flies, ticks, and mites are bound to be affected by climate change. The current evidence for altered distribution of disease vectors during the winter season in northern parts of India and extended distribution of diseases like malaria, dengue, KFD, Leishmaniasis, and scrub typhus, etc. are available. The projected distribution of *Aedes aegypti* and *Aedes albopictus*, the vectors of dengue, chikungunya, and zika virus, in view of climate change, indicate spatial extension. By the year 2030s, malaria and dengue have been projected to extend to the Himalayan regions, which are hitherto free from these diseases. Early preparedness of vector and disease control in projected vulnerable areas is warranted to negate the adverse impacts of climate change in spatial spread.

REFERENCES

Abubakr, M., Sami, H., Mahdi, I., Altahir, O., Abdelbagi, H., Mohamed, N. S., & Ahmed, A. (2022). The phylodynamic and spread of the invasive Asian malaria vectors, *Anopheles stephensi*, in Sudan. *Biology, 11*(3), 409. https://doi.org/10.3390/BIOLOGY11030409/S1

Agyekum, T. P., Botwe, P. K., Arko-Mensah, J., Issah, I., Acquah, A. A., Hogarh, J. N., Dwomoh, D., Robins, T. G., & Fobil, J. N. (2021). A systematic review of the effects of temperature on Anopheles mosquito development and survival: Implications for malaria control in a future warmer climate. *International Journal of Environmental Research and Public Health, 18*(14),7255. https://doi.org/10.3390/ijerph18147255

Alahmed, A.M. (2012). Mosquito fauna (Diptera: Culicidae) of the Eastern Region of Saudi Arabia and their seasonal abundance. *Journal of King Saud University - Science, 24*(1), 55–62. https://doi.org/10.1016/J.JKSUS.2010.12.001

Amin, O.M. (1989). The status of malaria in Bahrain, Arabian Gulf. *Journal of the University of Kuwait (Science), 16*(1), 135–141.

Anonymous (1994). *Resource Atlas of Rajasthan*. Department of Science & Technology, Government of Rajasthan, Jaipur, 231 pp.

Balkew, M., Mumba, P., Dengela, D., Yohannes, G., Getachew, D., Yared, S., Chibsa, S., Murphy, M., George, K., Lopez, K., Janies, D., Choi, S. H., Spear, J., Irish, S. R., & Carter, T. E. (2020). Geographical distribution of Anopheles stephensi in eastern Ethiopia. *Parasites and Vectors, 13*(1). https://doi.org/10.1186/s13071-020-3904-y

Baomar, A. T., & Mohamed, A. G. (2000). Malaria outbreak in a malaria-free region in Oman 1998: unknown impact of civil war in Africa. *Public Health, 114*(6), 480–483. https://doi.org/10.1038/SJ.PH.1900699

Bar-Zeev, M. (1958). The effect of temperature on the growth rate and survival of the immature stages of Aëdes aegypti (L.). *Bulletin of Entomological Research, 49*(1), 157–163. https://doi.org/10.1017/S0007485300053499

Bargielowski, I. E., Lounibos, L. P., & Carrasquilla, M. C. (2013). Evolution of resistance to satyrization through reproductive character displacement in populations of invasive dengue vectors. *Proceedings of the National Academy of Sciences of the United States of America, 110*(8), 2888–2892. https://doi.org/10.1073/pnas.1219599110

Bashar, K., Mahmud, K. H., Reza, S., Goda, P., Rain, F. F., Jesmin, M., & Asaduzzaman. (2013). Revised checklist and distribution maps of *Anopheles* (Insecta: Diptera: Culicidae: Anophelinae) mosquitoes of Bangladesh. *Check List, 9*(2), 211–224. https://doi.org/10.15560/9.2.211

Benedict, M. Q., Levine, R. S., Hawley, W. A., & Lounibos, L. P. (2007). Spread of the tiger: Global risk of invasion by the mosquito Aedes albopictus. *Vector-Borne and Zoonotic Diseases, 7*(1), 76–85. https://doi.org/10.1089/vbz.2006.0562

Bouma, M. J., & Van der Kaay, H. J. (1994) Epidemic malaria in India and the El Nino Southern Oscillation. *Lancet, 334*, 1638–1639.

Bouma, M. J., & Van der Kaay, H. J. (1995) Epidemic malaria in India's Thar Desert. *The Lancet, 373*, 132–133.

Bouma, M. J., & van der Kaay, H. J. (1996) The El Nino Southern Oscillation and the historic malaria epidemics in the Indian sub-Continent and Sri Lanka: an early warning system for future epidemics. *Tropical Medicine & International Health, 1*, 86–96.

Bradshaw, W. E., Fujiyama, S., & Holzapfel, C. M. (2000). Adaptation to the thermal climate of North America by the pitcher-plant mosquito, Wyeomyia smithii. *Ecology, 81*(5), 1262–1272. https://doi.org/10.1890/0012-9658(2000)081[1262:atttco]2.0.co;2

Brady, O. J., Gething, P. W., Bhatt, S., Messina, J. P., Brownstein, J. S., Hoen, A. G., Moyes, C. L., Farlow, A. W., Scott, T. W., & Hay, S. I. (2012). Refining the global spatial limits of dengue virus transmission by evidence-based consensus. *PLOS Neglected Tropical Diseases, 6*(8), e1760. https://doi.org/10.1371/JOURNAL.PNTD.0001760

Brown, J. E., Evans, B. R., Zheng, W., Obas, V., Barrera-Martinez, L., Egizi, A., Zhao, H., Caccone, A., & Powell, J. R. (2014). Human impacts have shaped historical and recent evolution in Aedes aegypti, the dengue and yellow fever mosquito. *Evolution; International Journal of Organic Evolution, 68*(2), 514. https://doi.org/10.1111/EVO.12281

Brown, J. E., Mcbride, C. S., Johnson, P., Ritchie, S., Paupy, C., Bossin, H., Lutomiah, J., Fernandez-Salas, I., Ponlawat, A., Cornel, A. J., Black IV, W. C., Gorrochotegui-Escalante, N., Urdaneta-Marquez, L., Sylla, M., Slotman, M., Murray, K. O., Walker, C., & Powell, J. R. (2011). Worldwide patterns of genetic differentiation imply multiple 'domestications' of Aedes aegypti, a major vector of human diseases. *Proceedings of the Royal Society B: Biological Sciences, 278*(1717), 2446–2454. https://doi.org/10.1098/RSPB.2010.2469

Cator, L. J., Johnson, L. R., Mordecai, E. A., El Moustaid, F., Smallwood, T. R. C., LaDeau, S. L., Johansson, M. A., Hudson, P. J., Boots, M., Thomas, M. B., Power, A. G., & Pawar, S. (2020). The role of vector trait variation in vector-borne disease dynamics. *Frontiers in Ecology and Evolution, 8*. https://doi.org/10.3389/FEVO.2020.00189/FULL

Chang, L. H., Hsu, E. L., Teng, H. J., & Ho, C. M. (2007). Differential survival of Aedes aegypti and Aedes albopictus (Diptera: Culicidae) Larvae exposed to low temperatures in Taiwan. *Journal of Medical Entomology, 44*(2), 205–210. https://doi.org/10.1093/JMEDENT/44.2.205

Christophers, S. R. (1960). *Aedes aegypti (L) the yellow fever mosquito: Its Life, History, Bionomics and Structure*. Cambridge University Press.

Ciota, A. T., Matacchiero, A. C., Kilpatrick, A. M., & Kramer, L. D. (2014). The effect of temperature on life history traits of *Culex*mosquitoes. *Journal of Medical Entomology, 51*(1), 55–62. https://doi.org/10.1603/ME13003

Crans, W. J., Sprenger, D. A., & Mahmood, F. (1996). The blood-feeding habits of Aedes sollicitans (Walker) in relation to eastern equine encephalitis virus in coastal areas of New Jersey II. Results of experiments with caged mosquitoes and the effects of temperature and physiological age on host selection. *Journal of Vector Ecology, 21*(1), 1–5.

Daugherty, M. P., Alto, B. W., & Juliano, S. A. (2000). Invertebrate Carcasses as a Resource for Competing Aedes albopictus and Aedes aegypti (Diptera: Culicidae). *Journal of Medical Entomology, 37*(3), 364–372. https://doi.org/10.1093/JMEDENT/37.3.364

Delatte, H., Desvars, A., Bouétard, A., Bord, S., Gimonneau, G., Vourc'h, G., & Fontenille, D. (2010). Blood-Feeding Behavior of Aedes albopictus, a Vector of Chikungunya on La Réunion. *Vector-Borne and Zoonotic Diseases, 10*(3), 249–258. https://doi.org/10.1089/vbz.2009.0026

Delatte, H., Gimonneau, G., Triboire, A., & Fontenille, D. (2009). Influence of temperature on immature development, survival, longevity, fecundity, and gonotrophic cycles of Aedes albopictus, vector of chikungunya and dengue in the Indian Ocean. *Journal of Medical Entomology, 46*(1), 33–41. https://doi.org/10.1603/033.046.0105

Dhiman, R. C., Singh, P., Yadav, Y., Saraswat, S., Kumar, G., Singh, R., Ojha, V., Joshi, B., & Singh, P. P. (2019). Preparedness for malaria elimination in the wake of climate change in the State of Uttarakhand (India). *Journal of Vector Borne Diseases, 56*(1), 46–52. https://doi.org/10.4103/0972-9062.257774

Edalat, H., Hassan Moosa-Kazemi, S., Abolghasemi, E., Khairandish, S., & Hassan Moosa-Kazemi, S. (2015). Vectorial capacity and Age determination of Anopheles Stephens Liston (Diptera: Culicidae), during the malaria transmission in Southern Iran. *Journal of Entomology and Zoology Studies JEZS, 256*(31), 256–263.

Farajollahi, A., Fonseca, D. M., Kramer, L. D., & Marm Kilpatrick, A. (2011). "Bird biting" mosquitoes and human disease: A review of the role of Culex pipiens complex mosquitoes in epidemiology. *Infection, Genetics and Evolution, 11*(7), 1577–1585. https://doi.org/10.1016/J.MEEGID.2011.08.013

Farjana, T., Tuno, N., & Higa, Y. (2012). Effects of temperature and diet on development and interspecies competition in Aedes aegypti and Aedes albopictus. *Medical and Veterinary Entomology, 26*(2), 210–217. https://doi.org/10.1111/j.1365-2915.2011.00971.x

Fletcher, M., Teklehaimanot, A., & Yemane, G. (1992). Control of mosquito larvae in the port city of Assab by an indigenous larvivorous fish, Aphanius dispar. *Acta Tropica, 52*(2–3), 155–166. https://doi.org/10.1016/0001-706X(92)90032-S

Fontes, G., Leite, A. B., Vasconcelos De Lima, A. R., Freitas, H., Ehrenberg, J. P., & Da Rocha, E. M. M. (2012). Lymphatic filariasis in Brazil: epidemiological situation and outlook for elimination. *Parasites & Vectors, 5*(1). https://doi.org/10.1186/1756-3305-5-272

Gaffney, O., & Steffen, W. (2017). The Anthropocene equation. *The Anthropocene Review, 4*(1), 53–61. https://doi.org/10.1177/2053019616688022

Gayan Dharmasiri, A. G., Perera, A. Y., Harishchandra, J., Herath, H., Aravindan, K., Jayasooriya, H. T. R., Ranawaka, G. R., & Hewavitharane, M. (2017). First record of Anopheles stephensi in Sri Lanka: a potential challenge for prevention of malaria reintroduction. *Malaria Journal, 16*(1), 326. https://doi.org/10.1186/s12936-017-1977-7

Gething, P. W., Smith, D. L., Patil, A. P., Tatem, A. J., Robert, W., & Hay, S. I. (2010). Climate change and the global malaria recession. *Nature, 465*(7296), 342–345. https://doi.org/10.1038/nature09098.Climate

Grech, M. G., Sartor, P. D., Almirón, W. R., & Ludueña-Almeida, F. F. (2015). Effect of temperature on life history traits during immature development of Aedes aegypti and Culex quinquefasciatus (Diptera: Culicidae) from Córdoba city, Argentina. *Acta Tropica, 146*, 1–6. https://doi.org/10.1016/J.ACTATROPICA.2015.02.010

Hanafi-Bojd, A. A., Vatandoost, H., & Yaghoobi-Ershadi, M. R. (2019). Climate Change and the Risk of Malaria Transmission in Iran. *Journal of Medical Entomology*, *X*, 1–15. https://doi.org/10.1093/jme/tjz131

Hantosh, H. A., Hassan, H. M., Ahma, B., & Al-Fatlawy, A. (2012). Mosquito species geographical distribution in Iraq 2009. *Journal of Vector Borne Diseases*, *49*(1), 33–35.

Harbach, R. E. (2013). The phylogeny and classification of Anopheles. In *Anopheles mosquitoes - New insights into malaria vectors*. https://doi.org/10.5772/54695

Hawley, W. A., Pumpuni, C. B., Brady, R. H., & Craig, G. B. (1989). Overwintering Survival of Aedes albopictus (Diptera: Culicidae) Eggs in Indiana. *Journal of Medical Entomology*, *26*(2), 122–129. https://doi.org/10.1093/JMEDENT/26.2.122

Hongoh, V., Berrang-Ford, L., Scott, M. E., & Lindsay, L. R. (2012). Expanding geographical distribution of the mosquito, Culex pipiens, in Canada under climate change. *Applied Geography*, *33*(1), 53–62. https://doi.org/10.1016/j.apgeog.2011.05.015

Hoosh-Deghati, H., Dinparast-Djadid, N., Moin-Vaziri, V., Atta, H., Raz, A. A., Seyyed-Tabaei, S. J., Maleki-Ravasan, N., Alipour, H., Zakeri, S., & Azar-Gashb, E. (2017). Composition of Anopheles Species Collected from Selected Malarious Areas of Afghanistan and Iran. *Journal of Arthropod-Borne Diseases*, *11*(3), 354–362.

Hussain, S. S. A., & Dhiman, R. C. (2022). Distribution expansion of dengue vectors and climate change in India. *GeoHealth*, *6*(6), e2021GH000477. https://doi.org/10.1029/2021GH000477

IPCC. (2022). *Global Warming of 1.5°C*. Cambridge University Press. https://doi.org/10.1017/9781009157940

Iwamura, T., Guzman-Holst, A., & Murray, K. A. (2020). Accelerating invasion potential of disease vector Aedes aegypti under climate change. *Nature Communications*, *11*(1), 1–10. https://doi.org/10.1038/s41467-020-16010-4

Jentes, E. S., Poumerol, G., Gershman, M. D., Hill, D. R., Lemarchand, J., Lewis, R. F., Staples, J. E., Tomori, O., Wilder-Smith, A., & Monath, T. P. (2011). The revised global yellow fever risk map and recommendations for vaccination, 2010: consensus of the Informal WHO Working Group on Geographic Risk for Yellow Fever. *The Lancet Infectious Diseases*, *11*(8), 622–632. https://doi.org/10.1016/S1473-3099(11)70147-5

Kamal, M., Kenawy, M. A., Rady, M. H., & Khaled, A. S. (2018). Mapping the global potential distributions of two arboviral vectors Aedes aegypti and Ae. Albopictus under changing climate. *PLoS ONE*, *13*(12). https://doi.org/10.1371/journal.pone.0210122

Kiarie-Makara, M., Ngumbi, P. M., Kiarie-Makara, M. W., Ngumbi, P. M., & Lee, D.-K. (2015). Effects of temperature on the growth and development of Culex pipiens complex mosquitoes (Diptera: Culicidae). *Researchgate.Net*, *10*(6), 1–10. https://doi.org/10.9790/3008-10620110

Kobayashi, J., Nambanya, S., Miyagi, I., Vanachone, B., Manivong, K., Koubouchan, T., Amano, H., Nozaki, H., Inthakone, S., & Sato, Y. (1997). Collection of anopheline mosquitos in three villages endemic for malaria in Khammouane, Lao PDR. *The Southeast Asian Journal of Tropical Medicine and Public Health*, *28*(3), 20–20.

Kongmee, M., Achee, N. L., Lerdthusnee, K., Bangs, M. J., Chowpongpang, S., Prabaripai, A., & Chareonviriyaphap, T. (2012). Seasonal abundance and distribution of Anopheles larvae in a riparian malaria endemic area of western Thailand. *The Southeast Asian Journal of Tropical Medicine and Public Health*, *43*(3), 601–613.

Kraemer, M. U. G., Sinka, M. E., Duda, K. A., Mylne, A. Q. N., Shearer, F. M., Barker, C. M., Moore, C. G., Carvalho, R. G., Coelho, G. E., Van Bortel, W., Hendrickx, G., Schaffner, F., Elyazar, I. R., Teng, H. J., Brady, O. J., Messina, J. P., Pigott, D. M., Scott, T. W., Smith, D. L., … Hay, S. I. (2015a). The global distribution of the arbovirus vectors *Aedes aegypti* and *Ae. albopictus*. *ELife*, *4*, e08347. https://doi.org/10.7554/eLife.08347

Kraemer, M. U. G., Sinka, M. E., Duda, K. A., Mylne, A., Shearer, F. M., Brady, O. J., Messina, J. P., Barker, C. M., Moore, C. G., Carvalho, R. G., Coelho, G. E., Van Bortel, W., Hendrickx, G., Schaffner, F., Wint, G. R. W., Elyazar, I. R. F., Teng, H.-J., & Hay, S. I. (2015b). The global compendium of *Aedes aegypti* and *Ae. albopictus* occurrence. *Scientific Data*, *2*(1), 150035. https://doi.org/10.1038/sdata.2015.35

Lippi, C. A., Stewart-Ibarra, A. M., Franklin Bajaña Loor, M. E., Dueñas Zambrano, J. E., Espinoza Lopez, N. A., Blackburn, J. K., & Ryan, S. J. (2019). Geographic shifts in Aedes aegypti habitat suitability in Ecuador using larval surveillance data and ecological niche modeling: Implications of climate change for public health vector control. *PLoS Neglected Tropical Diseases*, *13*(4). https://doi.org/10.1371/journal.pntd.0007322

Marinho, R. A., Beserra, E. B., Bezerra-Gusmão, M. A., Porto, V. de S., Olinda, R. A., & dos Santos, C. A. C. (2016). Effects of temperature on the life cycle, expansion, and dispersion of *Aedes aegypti* (Diptera: Culicidae) in three cities in Paraiba, Brazil. *Journal of Vector Ecology*, *41*(1), 1–10. https://doi.org/10.1111/jvec.12187

Martin, J. L., Lippi, C. A., Stewart-Ibarra, A. M., Ayala, E. B., Mordecai, E. A., Sippy, R., Heras, F. H., Blackburn, J. K., & Ryan, S. J. (2021). Household and climate factors influence Aedes aegypti presence in the arid city of Huaquillas, Ecuador. *PLoS Neglected Tropical Diseases*, *15*(11). https://doi.org/10.1371/JOURNAL.PNTD.0009931

Mayne, B. (1930). A study of the Influence of relative humidity on the life and infectibility of the mosquito. *Indian Journal of Medical Research*, *17*(4), 1119 pp.

Miazgowicz, K., Mordecai, E., Ryan, S., & Hall, R. (2019). Mosquito species and age influence thermal performance of traits relevant to malaria transmission 2. *Biorxiv.Org*. https://doi.org/10.1101/769604

Mojahedi, A. R., Safari, R., Yarian, M., Pakari, A., Raeisi, A., Edalat, H., Beniardelan, M., Poudat, A., Zaim, M., & Basseri, H. R. (2020). Biting and resting behaviour of malaria vectors in Bandar-Abbas County, Islamic Republic of Iran. *Eastern Mediterranean Health Journal*, *26*(10), 1218–1226. https://doi.org/10.26719/EMHJ.19.104

Monteiro, L. C. C., De Souza, J. R. B., & De Albuquerque, C. M. R. (2007). Eclosion rate, development and survivorship of Aedes albopictus (Skuse)(Diptera: Culicidae) under different water temperatures. *Neotropical Entomology*, *36*(6), 966–971. https://doi.org/10.1590/S1519-566X2007000600021

Mordecai, E. A., Ryan, S. J., Caldwell, J. M., Shah, M. M., & LaBeaud, A. D. (2020). Climate change could shift disease burden from malaria to arboviruses in Africa. *The Lancet Planetary Health*, *4*(9), e416–e423. https://doi.org/10.1016/S2542-5196(20)30178-9

Murdock, C. C., Paaijmans, K. P., Bells, A. S., King, J. G., Hillyer, J. F., Read, A. F., & Thomas, M. B. (2012). Complex effects of temperature on mosquito immune function. *Proceedings of the Royal Society B: Biological Sciences*, *279*(1741), 3357–3366. https://doi.org/10.1098/RSPB.2012.0638

O'Meara, G. F., Evans, L. F., Gettman, A. D., & Cuda, J. P. (1995). Spread of Aedes albopictus and Decline of Ae. aegypti (Diptera: Culicidae) in Florida. *Journal of Medical Entomology*, *32*(4), 554–562. https://doi.org/10.1093/JMEDENT/32.4.554

Oda, T., Eshita, Y., Uchida, K., Mine, M., Kurokawa, K., Ogawa, Y., Kato, K., & Tahara, H. (2002). Reproductive activity and survival of *Culex pipiens pallens* and *Culex quinquefasciatus* (Diptera: Culicidae) in Japan at high temperature. *Journal of Medical Entomology*, *39*(1), 185–190. https://doi.org/10.1603/0022-2585-39.1.185

Oda, T., Uchida, K., Mori, A., Mine, M., Eshita, Y., Kurokawa, K., Kato, K., & Tahara, H. (1999). Effects of high temperature on the emergence and survival of adult Culex pipiens molestus and Culex quinquefasciatus in Japan. *Journal of the American Mosquito Control Association*, *15*(2), 153–156.

Paaijmans, K. P., Heinig, R. L., Seliga, R. A., Blanford, J. I., Blanford, S., Murdock, C. C., & Thomas, M. B. (2013). Temperature variation makes ectotherms more sensitive to climate change. *Global Change Biology*, *19*(8), 2373–2380. https://doi.org/10.1111/GCB.12240

Pakdad, K., Hanafi-Bojd, A. A., Vatandoost, H., Sedaghat, M. M., Raeisi, A., Moghaddam, A. S., & Foroushani, A. R. (2017). Predicting the potential distribution of main malaria vectors Anopheles stephensi, An. culicifaciess.l. and An. fluviatilis s.l. in Iran based on maximum entropy model. *Acta Tropica*, *169*, 93–99. https://doi.org/10.1016/J.ACTATROPICA.2017.02.004

Pal, R. (1943). On the bionomics of Anopheles culicifacies Giles. Part I. Longevity under controlled conditions of temperature and humidity. *Journal of the Malaria Institute of India*, *5*(1), 77–85.

Parihar, R. S., Bal, P. K., Saini, A., Mishra, S. K., & Thapliyal, A. (2022). Potential future malaria transmission in Odisha due to climate change. *Scientific Reports 2022 12:1*, *12*(1), 1–13. https://doi.org/10.1038/s41598-022-13166-5

Paupy, C., Delatte, H., Bagny, L., Corbel, V., & Fontenille, D. (2009). Aedes albopictus, an arbovirus vector: from the darkness to the light. *Microbes and Infection*, *11*(14–15), 1177–1185. https://doi.org/10.1016/j.micinf.2009.05.005

Ponlawat, A., & Harrington, L. C. (2005). Blood feeding patterns of Aedes aegypti and Aedes albopictus in Thailand. *Journal of Medical Entomology*, *42*(5), 844–849. https://doi.org/10.1093/jmedent/42.5.844

Pramanik, M., Singh, P., & Dhiman, R.C. (2021). Identification of bio-climatic determinants and potential risk areas for Kyasanur forest disease in Southern India using MaxEnt modelling approach. *BMC Infectious Diseases*, 21, 1226. https://doi.org/10.1186/s12879-021-06908-9

Ramasamy, M. S., Kulasekara, R., Srikrishnaraj, K. A., & Ramasamy, R. (1994). Population dynamics of anthropophilic mosquitoes during the northeast monsoon season in the malaria epidemic zone of Sri Lanka. *Medical and Veterinary Entomology*, *8*(3), 265–274. https://doi.org/10.1111/J.1365-2915.1994.TB00508.X

Rao, T. R. (1984). *The Anophelines of India*. Malaria Research Centre, ICMR.

Reisen, W. K., & Aslamkhan, M. (1978). Biting rhythms of some Pakistan mosquitoes (Diptera: Culicidae). *Bulletin of Entomological Research*, *68*(2), 313–330. https://doi.org/10.1017/S0007485300007392

Reisen, W. K., & Mahmood, F. (1980). Horizontal life table characteristics of the malaria vectors Anopheles Culicifacies and Anopheles Stephensi (Diptera: Culicidae). *Journal of Medical Entomology*, *17*(3), 211–217. https://doi.org/10.1093/JMEDENT/17.3.211

Reisen, W. K., Thiemann, T., Barker, C. M., Lu, H., Carroll, B., Fang, Y., & Lothrop, H. D. (2010). Effects of warm winter temperature on the abundance and Gonotrophic activity of Culex (Diptera: Culicidae) in California. *Journal of Medical Entomology*, *47*(2), 230–237. https://doi.org/10.1603/ME09207

Rocklöv, J., & Dubrow, R. (2020). Climate change: an enduring challenge for vector-borne disease prevention and control. *Nature Immunology*, *21*(5), 479–483. https://doi.org/10.1038/S41590-020-0648-Y

Rowley, W. A., & Graham, C. L. (1968). The effect of temperature and relative humidity on the flight performance of female Aedes aegypti. *Journal of Insect Physiology*, *14*(9), 1251–1257. https://doi.org/10.1016/0022-1910(68)90018-8

Ryan, S. J., Carlson, C. J., Mordecai, E. A., & Johnson, L. R. (2018). Global expansion and redistribution of Aedes-borne virus transmission risk with climate change. *PLoS Neglected Tropical Diseases*, *13*(3). https://doi.org/10.1371/journal.pntd.0007213

Ryan, S. J., Mcnally, A., Johnson, L. R., Mordecai, E. A., Ben-Horin, T., Paaijmans, K., & Lafferty, K. D. (2015). Mapping physiological suitability limits for Malaria in Africa under climate change. *Vector Borne and Zoonotic Diseases (Larchmont, N.Y.)*, *15*(12), 718–725. https://doi.org/10.1089/VBZ.2015.1822

Samy, A. M., Elaagip, A. H., Kenawy, M. A., Ayres, C. F. J., Peterson, A. T., & Soliman, D. E. (2016). Climate change influences on the global potential distribution of the mosquito Culex quinquefasciatus, vector of West Nile virus and lymphatic filariasis. *PLoS ONE*, *11*(10), e0163863. https://doi.org/10.1371/journal.pone.0163863

Sarkar, S., Gangare, V., Singh, P., & Dhiman, R.C. (2019). Shift in potential malaria transmission areas in India, using the Fuzzy-Based Climate Suitability Malaria Transmission (FCSMT) model under changing climatic conditions. *International Journal of Environmental Research and Public Health*, *16*(3474), 1–16. https://doi.org/10.3390/ijerph16183474

Schmeller, D. S., Courchamp, F., & Killeen, G. (2020). Biodiversity loss, emerging pathogens and human health risks. *Biodiversity and Conservation*, *29*(11–12), 3095–3102. https://doi.org/10.1007/S10531-020-02021-6/FIGURES/1

Seyfarth, M., Khaireh, B. A., Abdi, A. A., Bouh, S. M., & Faulde, M. K. (2019). Five years following first detection of Anopheles stephensi (Diptera: Culicidae) in Djibouti, Horn of Africa: populations established – malaria emerging. *Parasitology Research*, *118*(3), 725–732. https://doi.org/10.1007/s00436-019-06213-0

Shapiro, L. L. M., Whitehead, S. A., & Thomas, M. B. (2017). Quantifying the effects of temperature on mosquito and parasite traits that determine the transmission potential of human malaria. *PLoS Biology*, *15*(10). https://doi.org/10.1371/JOURNAL.PBIO.2003489

Shragai, T., Tesla, B., Murdock, C., & Harrington, L. C. (2017). Zika and chikungunya: mosquito-borne viruses in a changing world. In *Annals of the New York Academy of Sciences* (Vol. 1399, Issue 1, pp. 61–77). https://doi.org/10.1111/nyas.13306

Singh, P., Pande, V., & Dhiman, R. C. (2022). Revisiting the impact of temperature on survival of Anopheles stephensi and Aedes aegypti and implications on extrinsic incubation period. *Journal of Communicable Diseases*, *54*(1), 60–66. https://doi.org/10.24321/0019.5138.202251

Sinka, M. E., Bangs, M. J., Manguin, S., Chareonviriyaphap, T., Patil, A. P., Temperley, W. H., Gething, P. W., Elyazar, I. R. F., Kabaria, C. W., Harbach, R. E., & Hay, S. I. (2011). The dominant Anopheles vectors of human malaria in the Asia-Pacific region: occurrence data, distribution maps and bionomic précis. *Parasites & Vectors*, *4*(1), 89. https://doi.org/10.1186/1756-3305-4-89

Sinka, M. E., Pironon, S., Massey, N. C., Longbottom, J., Hemingway, J., Moyes, C. L., & Willis, K. J. (2020). A new malaria vector in Africa: Predicting the expansion range of Anopheles stephensi and identifying the urban populations at risk. *Proceedings of the National Academy of Sciences of the United States of America*, *117*(40), 24900–24908. https://doi.org/10.1073/pnas.2003976117

Sivan, A., Shriram, A. N., Vanamail, P., & Sugunan, A. P. (2019). Impact of temperature variant on survival of Aedes albopictus Skuse (Diptera: Culicidae): Implications on thermotolerance and acclimation. *Neotropical Entomology*, *48*(4), 561–571. https://doi.org/10.1007/S13744-019-00680-X

Soper, F. L. (1967). Dynamics of Aedes aegypti distribution and density. Seasonal fluctuations in the Americas. *Bulletin of the World Health Organization*, *36*(4), 536–538. https://www.ncbi.nlm.nih.gov/pmc/articles/PMC2476418/

Sota, T., & Mogi, M. (1992a). Survival time and resistance to desiccation of diapause and non-diapause eggs of temperate Aedes (Stegomyia) mosquitoes. *Entomologia Experimentalis et Applicata*, *63*(2), 155–161. https://doi.org/10.1111/J.1570-7458.1992.TB01570.X

Sota, T., & Mogi, M. (1992b). Interspecific variation in desiccation survival time of Aedes (Stegomyia) mosquito eggs is correlated with habitat and egg size. *Oecologia*, *90*(3), 353–358. https://doi.org/10.1007/BF00317691

Sukiato, F., Wasserman, R. J., Foo, S. C., Wilson, R. F., & Cuthbert, R. N. (2019). The effects of temperature and shading on mortality and development rates of Aedes aegypti (Diptera: Culicidae). *Journal of Vector Ecology*, *44*(2), 264–270. https://doi.org/10.1111/JVEC.12358

Surendran, S. N., Sivabalakrishnan, K., Sivasingham, A., Jayadas, T. T. P., Karvannan, K., Santhirasegaram, S., Gajapathy, K., Senthilnanthanan, M., Karunaratne, S. P., & Ramasamy, R. (2019). Anthropogenic Factors Driving Recent Range Expansion of the Malaria Vector Anopheles stephensi. *Frontiers in Public Health*, *7*(MAR), 53. https://doi.org/10.3389/fpubh.2019.00053

Takken, W., & Lindsay, S. (2019). Increased threat of urban malaria from anopheles stephensi mosquitoes, Africa. *Emerging Infectious Diseases*, *25*(7), 1431–1433.

Thomas, S. M., Obermayr, U., Fischer, D., Kreyling, J., & Beierkuhnlein, C. (2012). Low-temperature threshold for egg survival of a post-diapause and non-diapause European aedine strain, Aedes albopictus (Diptera: Culicidae). *Parasites & Vectors*, *5*(1). https://doi.org/10.1186/1756-3305-5-100

Thomson, R. C. M. (1938). The reactions of mosquitoes to temperature and humidity. *Bulletin of Entomological Research*, *29*(2), 125–140. https://doi.org/10.1017/S0007485300026158

Tippelt, L., Werner, D., & Kampen, H. (2020). Low temperature tolerance of three *Aedes albopictus* strains (Diptera: Culicidae) under constant and fluctuating temperature scenarios. *Parasites & Vectors*, *13*(1), 587. https://doi.org/10.1186/s13071-020-04386-7

Tripathy, A., Samanta, L., Das, S., Parida, S. K., Marai, N., Hazra, R. K., Kar, S. K., & Mahapatra, N. (2010). Distribution of sibling species of Anopheles culicifacies s.l. and Anopheles fluviatilis s.l. and their vectorial capacity in eight different malaria endemic districts of Orissa, India. *Memorias Do Instituto Oswaldo Cruz*, *105*(8), 981–987. https://doi.org/10.1590/S0074-02762010000800006

Tun-Lin, W., Burkot, T. R., & Kay, B. H. (2000). Effects of temperature and larval diet on development rates and survival of the dengue vector Aedes aegypti in north Queensland, Australia. *Medical and Veterinary Entomology*, *14*(1), 31–37. https://doi.org/10.1046/j.1365-2915.2000.00207.x

Tyagi, B. K. (2002) Malaria in the Thar Desert: Facts, figures and future. Agrobios (India), 162 pp.

Tyagi, B. K. (2004). Review of the emergence of Plasmodium falciparum-dominated malaria in irrigated areas of the Thar Desert, India. *Acta Tropica*, 89, 227–239. https://doi.org/10.1016/j.actatropica.2003.09.016

Tyagi, B. K. (2005). *Thar Desert's changing climate and emergence/exacerbation of vector-borne diseases: Past, present and future* (*In*: Changing faunal ecology in the Thar Desert. *Eds*. B.K. Tyagi and Q.H. Baqri) Scientific Publishers (India), Jodhpur, pp. 323–367.

Tyagi, B. K. (2023). *Desert malaria: An emerging malaria paradigm and its global impact on disease elimination*. Springer Nature Singapore.

Tyagi, B. K., & Chaudhary, R. (1997). Outbreak of falciparum malaria in the Thar Desert (India), with particular emphasis on physiographic changes brought about by extensive canalization and their impact on vector density and dissemination. *Journal of Arid Environments*, *36*(3), 541–555. https://doi.org/10.1006/JARE.1996.0188

Tyagi, B. K., Chaudhary, R. C., & Yadav, S. P. (1995). Epidemic malaria in Thar Desert, India. *The Lancet*, *346*, 634–635.

Tyagi, B. K., Hiriyan, J., Samuel, P., Tewari, S. C., & Paramasivan, R. (2006). Dengue in Kerala: A critical review. *ICMR Bulletin 36*(4–5),13–22.

Van Bortel, W., Sochanta, T., Harbach, R. E., Socheat, D., Roelants, P., Backeljau, T., & Coosemans, M. (2002). Presence of Anopheles culicifacies B in Cambodia established by the PCR-RFLP assay developed for the identification of Anopheles minimus species A and C and four related species. *Medical and Veterinary Entomology*, *16*(3), 329–334. https://doi.org/10.1046/j.1365-2915.2002.00376.x

Villena, O. C., Ryan, S. J., Murdock, C. C., & Johnson, L. R. (2022). Temperature impacts the environmental suitability for malaria transmission by Anopheles gambiae and Anopheles stephensi. *Ecology*, *103*(8), e3685. https://doi.org/10.1002/ECY.3685

Vinogradova, E. B., & Elena, B. (2000). *Culex pipiens pipiens mosquitoes: taxonomy, distribution, ecology, physiology, genetic, applied importance and control*.250. https://www.nhbs.com/culex-pipiens-pipiens-mosquitoes-taxonomy-distribution-ecology-physiology-genetics-applied-importance-and-control-book

Wagner, D. L. (2020). Insect declines in the anthropocene. *Annual Review of Entomology*, *65*, 457–480. https://doi.org/10.1146/ANNUREV-ENTO-011019-025151

Warrell, D. A., & Gilles, H. M. (2017). *The Anopheles vector*. 59–84. https://doi.org/10.1201/9780203756621-4
WHO (2021). *World malaria report 2021*. https://apps.who.int/iris/bitstream/handle/10665/350147/9789240040496-eng.pdf?sequence=1
WHO (2023). Vector alert: *Anopheles stephensi* invasion and spread in Africa and Sri Lanka. Information note. World Health Organization, Geneva, 1–4 pp.
Zhong, D., Aung, P. L., Mya, M. M., Wang, X., Qin, Q., Soe, M. T., Zhou, G., Kyaw, M. P., Sattabongkot, J., Cui, L., & Yan, G. (2022). Community structure and insecticide resistance of malaria vectors in northern-central Myanmar. *Parasites & Vectors*, *15*(1). https://doi.org/10.1186/S13071-022-05262-2

Section VIII

Resistance Development in Vectors

15 Insecticide Resistance

Dhiraj Saha
Department of Zoology, University of North Bengal, Siliguri, Darjeeling, India

Minu Bharati
S.B.S. Government College, Dakshin Dinajpur, India

Priyanka Rai, Abhisekh Subba, Manas Pratim Modak, Abhirup Saha, and Subhajit Das
Department of Zoology, University of North Bengal, Siliguri, Darjeeling, India

E. Pushpalatha
University of Calicut, India

Indra Vythilingam
University Malaya, Kuala Lumpur, Malaysia

Nazni bt. Hj. Wasi Ahmad
Institute of Medical Research, Kuala Lumpur, Malaysia

R.S. Sharma
Nalanda Apartments, Sec 21-C, Faridabad, Haryana,

B.K. Tyagi
University Institute of Biotechnology, Chandigarh University, Mohali, India

15.1 INTRODUCTION

More than half a billion of the world's population are at risk of contacting mosquito-borne diseases. Due to the lack of specific medicines and suitable vaccines, as well as the development of drug resistance in pathogens, the only method to minimize the risk of acquiring mosquito-borne diseases remains the prevention of the disease mainly through control of the vector mosquito populations (Phyo et al., 2012; Hassett and Roepe, 2019). In doing so, extensive use of insecticides, targeting the different life stages of the mosquito, was promoted as they are highly effective and cheap to produce. However, this has caused widespread development of insecticide resistance in mosquitoes throughout the world.

With excessive and unscientific use of insecticides against vector populations, the first report on resistance to DDT was reported in *Aedes* mosquitoes in the early 1950s (Liu, 2015). Since then, there have been several reports on resistance to chemical insecticides in mosquito vectors from different corners of the world. The resistance status reported by different research studies is an

DOI: 10.1201/9781003326991-23

important issue and a major topic of concern because as the resistant population grows, the mosquito vectors might become uncontrollable and without proper medical facilities, vector-borne diseases may become difficult to handle, causing many fatalities and economic loss of mankind.

15.1.1 Burden of Mosquito-Borne Diseases in Northern West Bengal

West Bengal is one of the prime states of India which is highly affected by vector-borne diseases. The past four years of data show higher cases of dengue, malaria, and AES (acute encephalitis syndrome) in West Bengal (Table 15.1). This could be attributed to favorable climatic and seasonal requirements for the propagation of their vectors and due to the growing cases of insecticide resistance in these vectors. Out of the 12 lymphatic filariasis endemic districts of West Bengal, two districts, Cooch-Behar and Malda, are in the northern part of West Bengal (NCVBDC, 2022). The State Vector-Borne Diseases Control and Seasonal Influenza Plan 2018 provides some information about northern districts of West Bengal that are prone to mosquito-borne diseases in Table 15.2.

15.1.1.1 Important Mosquito Vectors

We know the northern districts of West Bengal are a hub for vector-borne diseases. As it records seasonal outbreaks of malaria, dengue, and chikungunya, this area houses several different potential mosquito vector species (Table 15.3).

TABLE 15.1
Four Years Data Showing the Number of Cases of Different Vector-Borne Diseases from West Bengal

	Dengue		Malaria		AES		JE		Chikungunya	
Year	Cases	Deaths	Cases	Deaths	Cases	Deaths	Cases	Deaths	Cases	Deaths
2022	239*	0 *	4068 #	0 #	196 #	6 #	2 #	0 #	NA#	NA #
2021	8264	7	28987	3	1240	28	47	6	20	0
2020	5166	0	14049	7	867	26	41	4	82	0
2019	NR	NR	25928	6	1618	89	82	11	NR	0

Source: NCVBDC, 2022.

NR = not recorded; NA = not available; AES = acute encephalitis syndrome; JE = Japanese encephalitis

TABLE 15.2
District-wise mosquito-borne disease profile of the northern part of West Bengal

Name of the Districts	Malaria	Dengue	AES	JE	Lymphatic Filariasis
Alipurduar	✓	–		NA	–
CoochBehar	✓	✓	✓	NA	✓
Dakhin Dinajpur	–	✓	✓	NA	✓
Darjeeling	–	-	✓	NA	–
Jalpaiguri	–	✓	✓	NA	–
Kalimpong	–	–	–	NA	–
Malda	✓	–	–	NA	–
Uttar Dinajpur	–	✓	✓	NA	–

Source: State Vector-Borne Diseases Control and Seasonal Influenza Plan, 2018.

NA = Not Available

TABLE 15.3
Some of the Major Mosquito Vectors from West Bengal

Mosquito-Borne Diseases	Major Mosquito Vectors
Dengue	*Aedes aegypti, Ae. albopictus*
Malaria	*Anopheles culicifacies, An. minimus, An. varuna, An. maculatus, An. fluviatilis, An. vagus, An. hyrcanus, An. barbirostris, An. jeyporiensis*
Lymphatic filariasis	*Culex quinquefasciatus*
Japanese encephalitis and acute encephalitis syndrome	*Culex vishnui* subgroup (*Cx. tritaeniorhynchus, Cx vishnui* and *Cx. pseudovishnui*)
Chikungunya	*Aedes aegypti, Ae. albopictus*

Bharati and Saha 2018a, 2018b; Rudra and Mukhopadhyay, 2010; Malakar et al., 1995; Mariappan et al., 2014; Gurav et al., 2016.

15.1.1.2 Scenario of Insecticides Used in the Public Health and Agriculture Sector

Usually, temephos at 50% EC (emulsion concentrate) is used to control the mosquito larvae in their aquatic habitat. For adulticidal Indoor Residual spray of DDT, malathion and synthetic pyrethroid are commonly used. During the peak transmission season, two rounds of DDT or pyrethroids and three rounds of malathion are sprayed to effectively control the adult population (State Vector-Borne Diseases Control and Seasonal Influenza Plan, 2018).

15.1.2 Drawbacks in Vector Control Through Synthetic Insecticides

Unscientific and overuse of synthetic insecticides has led to the development of insecticide resistance in mosquitoes in the northern districts of West Bengal. The wild population of dengue vectors, *Aedes aegypti* and *Aedes albopictus*, and lymphatic filariasis vector, *Culex quinquefasciatus*, has already developed resistance against synthetic pyrethroids, organophosphates, and DDT (Bharati and Saha, 2018a; Rai et al., 2019). Metabolic detoxification and target-site mutation have been the main reasons behind this (Bharati and Saha, 2018b; Modak and Saha, 2022).

15.2 DEVELOPMENT OF BIOCHEMICAL MECHANISMS OF INSECTICIDE RESISTANCE

Insecticide resistance is mediated through several mechanisms, such as cuticular resistance, behavioral resistance, metabolic resistance, and target-site resistance. These mechanisms may individually or jointly affect the susceptibility status of these vectors. Here we separately discuss each of these mechanisms.

15.2.1 Cuticular Resistance

Insect cuticle acts as a first line of barrier to any foreign pathogen and xenobiotics. Any part of the insect body that comes in direct contact with the insecticides brings about some cuticular changes. It occurs either through the thickening of the cuticle or through changes in the cuticular composition. These changes are brought about by the overexpression of cuticle structural genes or enzymes that promote modification in insect cuticles (Balabanidou et al., 2018). It eventually leads to reduced penetration of insecticides and develops insecticide resistance.

15.2.1.1 Role of Cuticular Hydrocarbon

One way of increasing the cuticular thickness is through the deposition of hydrocarbons on the insect epicuticle. N-alkenes, unsaturated hydrocarbons, and methyl-branched hydrocarbons are some of the cuticular hydrocarbons that are *de novo* synthesized by subepidermal oenocytes of insects (Grigoraki et al., 2020). Some enzymes belonging to the *CYP4G* enzyme family are involved in this process through oxidative decarboxylation of aldehyde (Qiu et al., 2012). In the case of *Anopheles gambiae*, *CYP4G16* and *CYP4G17*, help in synthesizing chitin from C18 aldehyde (Kefi et al., 2019). Interestingly, overexpression of these enzymes was detected in deltamethrin-resistant strains with a thicker epicuticle than in susceptible strains. Besides, experiments with radiolabeled insecticide showed reduced penetration of deltamethrin through such cuticle (Strycharz et al., 2013; Balabanidou et al., 2016). In permethrin-resistant *Anopheles funestes* from South Africa and the neighboring region of South Mozambique, positive correlation was established between cuticular thickness and the knock-down time. Furthermore, female resistant population of *An. funestes* possessed thicker cuticle as compared to the resistant male. Also, the authors commented that the primary mechanism of resistance was through monooxygenase activity, but cuticular thickening slowed the process of insecticide penetration which further enhanced resistance in mosquitoes (Wood et al., 2010). In another study from Cameroon on multiple resistance in *Culex quinquefasciatus* against DDT, permethrin, deltamethrin, bendiocarb, and malathion, a higher amount of cuticular hydrocarbon was observed as compared to the susceptible population. However, it was also reported that other mechanisms were involved in the development of insecticide resistance (Talipouo et al., 2021). On the other hand, qRT-PCR analysis of deltamethrin-resistant *Anopheles gambiae* MRS strain (*kdr* and *Ace1* mutations free strain) showed overexpression (*CPAP3-E*, *CPLCX1*) and continuous transcription of cuticular genes (*CPLCG3*, *CPR124*, *CPR129*, *CPR127*). Transmission electron microscopy further confirmed thicker MRS cuticle than susceptible Kisumu strain (Yahouédo et al., 2017). In case of *Aedes albopictus* and *Aedes aegypti*, population from Cameroon showed some association between higher cuticular hydrocarbon content and DDT and pyrethroid resistance (Yao et al., 2006). In India, scanning electron microscopy of deltamethrin and imidacloprid selected *Aedes aegypti* larvae population displayed thicker cuticle (Samal and Kumar 2021).

15.2.1.2 Role of Cuticular Proteins

Cuticular proteins also play a major role in the development of cuticular resistance in mosquitoes. Several cuticular proteins are over produced on the influence of synthetic insecticides and modifies the cuticular composition. One study suggests that overexpression of *CPLCG5* is responsible for deltamethrin-resistance in *Culex pipiens pallens*, which in turn is regulated by the expression of a transcription factor FTZ-F1 (Fushi tarazu-Factor 1) (Fang et al., 2015; Xu et al., 2020). Overexpression of cuticular proteins encoding genes, like *cpCPR63* and *cpCPR47* in pyrethroid-resistant *Culex pipiens pallens* (Sun et al., 2017) and *CRP30* in deltamethrin-resistant *Anopheles gambiae* (Bonizzoni et al., 2012) was reported from China and West Kenya, respectively. *CRP30* produces soft cuticles mainly found in the active stages of mosquitoes and is located near joints of limbs and wings (Willis, 2010; Vincent and Wegst, 2004). RT-PCR and microarray techniques revealed the role of cuticular proteins coding genes like *CD578215.1*, *CD578215.1*, *Afun010482*, *CD577507.1*, and *Afun013390* in permethrin resistant and *AGAP009480* and *AGAP003382RA* in DDT resistant *Anopheles funestus* population from Nigeria (Atoyebi et al., 2020). Proteome analysis of different life stages of cypermethrin-resistant *Culex pipiens pallens* revealed an enriched amount of 59 encodings and 46 putative cuticular proteins (Zhang et al., 2022).

15.2.1.3 Role of Non-cuticular Proteins in Insecticide Resistance

The modern method of RNAi technology has confirmed the role of the Laccase 2 encoding gene in the sclerotization and pigmentation of cuticles in *Aedes albopictus* eggs (Wu et al., 2013). The *CpLac2* gene, responsible for Laccase 2 protein isolated from *Culex pipiens pallens*, was found to overexpress in fenvalerate (synthetic pyrethroid)-resistant population as compared to a fenvalerate-susceptible population (Pan et al., 2009). Upregulation of genes that code trehalase and chitin synthase, which are involved in the chitin biosynthetic pathway, was observed in deltamethrin-resistant *Culex pipiens pallens*. Knock-down of these genes showed higher mortality even in a deltamethrin-resistant population (Guo et al., 2019).

15.2.2 Behavioral Resistance

Mosquitoes perceive chemical cues through their olfactory receptors distributed in their antennae, palps, and tarsi, which aids them to discriminate between food and toxins. When a mosquito comes into contact with insecticide it usually tries to avoid the area of exposure, broadly termed as behavioral resistance/avoidance. In simple terms, behavioral resistance can be understood as a method of avoiding toxic compounds, either repellent or irritant, through modification in their innate nature. Female mosquitoes basically are attracted to certain chemicals released by their host called kairomones, which stimulate host-seeking and consumption (blood feed). But when mosquitoes are exposed to various ranges of insecticides via human intervention, they sense a threat to their survivability and try to circumvent or avoid the insecticides through changes in their natural behavior. According to Carrasco et al. (2019) mosquitoes modify their behavior in three aspects, such as spatial, trophic, and temporal shifts in their feeding and host-seeking habits.

The use of insecticide-treated nets (ITNs) and indoor residual sprays (IRS) is the most common and highly recommended method worldwide for relegating malaria (World Malaria Report 2005; State Vector-Borne Diseases Control and Seasonal Influenza Plan, 2018). Despite their effectiveness in subsiding the number of malarial incidences worldwide, extensive use of ITNs and IRS for a substantial period of time has led to the development of behavioral plasticity in indoor visiting hematophagous mosquitoes. This kind of behavioral shift ensues very short insecticide exposure time and lower penetration, thus, leading to insecticide resistance in due course. Significantly less work has been conducted on this aspect of insecticide resistance. Behavioral resistance, such as rapid escape response and spatial shifts from endophily to exophily, has been observed from Ethiopia in *Anopheles arabiensis* (Gatton et al., 2013) and from Thailand in *An. minimus* (Potikasikorn et al., 2005). A similar study from Burkina Faso showed that within two and a half years of exposure of deltamethrin-treated LLINs there was a change in the vector composition with a reduced number of *An. gambiae* spp. and the majority of the mosquitoes were found seeking host outdoors in spite of sustained seasonal and spatial parameters (Sanou et al., 2021).

Other modes of insecticide resistance also impact behavioral response toward a particular insecticide. This kind of behavioral plasticity is found in *Anopheles gambiae* from Africa. Here in a semi-field setup, susceptible Kisumu strain mosquitoes normally exhibited exophily in the presence of ITNs but the resistant strain while the resistant population showed no behavioral avoidance against deltamethrin (Machani et al., 2022).

Behavioral response is also affected by the resistance level of mosquitoes (Deletre et al., 2019; He et al., 2019). In a semi-field experiment conducted in Africa, a deltamethrin-susceptible Kisumu strain of *Anopheles gambiae* mosquitoes displayed exophily in the presence of deltamethrin-treated bednets. In the urban Malabo, Equatorial Guinea, IRS of organophosphate failed to produce the desired level of contact irritant in *Culex* spp. that were phenotypical resistant against multiple insecticides (organophosphate, organochlorine, carbamate and pyrethroids). As a result, a higher biting rate was observed in this population (Fuseini et al., 2019).

Another study points out an association between knock-down resistance in Anopheline mosquitoes, and its behavior with respect to blood meal, host-seeking, resting, and exiting houses. When biting time and host-seeking behavior were compared between homozygous susceptible, homozygous resistant, and heterozygous-resistant strains of *Anopheles gambiae*, it was found that their feeding time changed from biting early morning to late night and shifted from endophily to exophily mode. More importantly, the pyrethroid-resistant strain was found to be more anthropophilic than the susceptible population (Githinji et al., 2020). Deletre et al. (2019) showed that in *Anopheles gambiae*, pyrethroid and organophosphate-related target-site mutation (KdrKis and AceKis strains, respectively) influenced and modified their behavioral response toward DEET and natural irritants and repellents. Kawada et al. (2014) have also commented on the influence of target-site mutation in voltage gate sodium channel on the behavioral response in the *Anopheles gambiae* complex toward pyrethroids.

15.2.2.1 Mechanism of Odorant Reception

When a chemical compound binds with one of the olfactory receptors, it activates some voltage-gated ionotropic channels. The extracellular $Ca+^2$ enters into the olfactory receptor cells (neurons), starts depolarization, and produces a signal that is carried out to the antennal lobe for processing (Reisenman et al., 2016; Sato et al., 2008). This will finally bring about some changes in the motor response of the mosquito which directs whether to move away from or toward the stimulus.

Olfactory receptors are of three types: odorant receptors, gustatory receptors, and ionotropic receptors, which bind with different kinds of chemical stimuli and produce appropriate responses (Konopka et al., 2021). Additionally, odorant receptor requires an obligate olfactory receptor co-activator called ORCO which helps in discriminating a very wide range of odors. Mutation in this co-activator causes insensitivity of odorant receptors of mosquitoes toward some repellents (Valbon et al., 2022).

Aedes aegypti has been reported to have developed heritable insensitivity toward N′N′ Diethyl meta tolumide (DEET), a neuroexcitatory compound, that acts via the octopaminergic pathway by activating the octopamine receptor. The electrocardiogram of neuron A also showed that both DEET and pyrethroid are involved in activating short, sharp-tipped (sst) sensilla present in the antenna of hematophagous insects. Changes in these sensilla bring about heritable insensitivity in *Aedes aegypti* (Valbon et al., 2022). Similar heritable insensitivity has also been reported in *Aedes aegypti* against the fast-acting synthetic pyrethroid, transfluthrin (Wagman et al., 2015).

Apart from physiological and behavioral changes, cognitive identification of toxic compounds plays an important role in behavioral resistance. In a study from Malaysia, it was found that when insecticide-conditioned (not resistant) mosquitoes were re-exposed to the same insecticide then they avoided the exposed area better than the unconditioned mosquitoes. This eventually increased their survivability. They suggested that dopamine receptor-1 has some role in the learning behavior of mosquitoes (Sougoufara et al., 2022).

Very recently, the Defence Research and Development Establishment in Gwalior, India has developed a slow-release insecticide paint (SRIP) formulation of chlorpyriphos, deltamethrin, and pyriproxyfen, which elicited behavioral avoidance response in *Aedes aegypti, Culex quinquefasciatus*, and *Anopheles stephensi* (Dhiman et al., 2021). But shorter exposure time and high escape rates result in lower penetration of insecticidal compounds which further leads to insecticide resistance via other faster modes such as metabolic detoxification and target-site mutation.

15.2.2.2 Classification of Chemicals Cues

Insecticides come in different forms producing different consequences. Dethier et al. (1960) classified chemicals according to their mode of action. He proposed five types of chemicals

1. Arrestant: are described as a chemical(s) that cause insects to aggregate when in contact
2. Attractant: make an oriented movement toward a source whereas
3. Locomotor stimulant: chemicals that either start or speed up a movement in response to a stimulus
4. Repellent: make an oriented movement away from a source
5. Deterrent: inhibit certain behaviors such as feeding, mating, oviposition, etc.

In yet another article by Miller et al. (2009), the locomotion of insects was broadly divided into two types, namely, kinesis (random movement with respect to a stimulatory source) and taxis (less random or biased movement with respect to a stimulatory source). They both agreed upon locomotor stimulants, where insects elicit random kinetic movement in the presence of a chemical stimulant and differ from their normal movement, which they referred to as a locomotor initiator (Miller et al., 2009). Well, some classified these chemicals on the basis of consequences that they bring about, such as contact irritant, excite-repellency, and spatial repellent (Table 15.4) (Roberts et al., 1997; Grieco et al., 2007).

15.2.3 Metabolic Resistance

Three major metabolic detoxifying enzymes – esterases, mixed function oxidases (MFO), and glutathione-S-transferases (GSTs) and their elevated activity in mosquitoes – are reported as one of the important mechanisms of resistance development in the mosquito vectors against various classes of insecticides (Hemingway et al., 2004). A similar elevated activity of the three major detoxifying enzymes was associated with resistance in *Culex sp.* and *Anopheles sp.* against DDT, permethrin, and carbosulfan in Africa (Corbel et al., 2007).

15.2.3.1 Esterases

Increased activity of both α and β esterases is often correlated with not only a single insecticide class but with various insecticides belonging to different classes. Moreover, more than one group of enzymes is involved in resistance development against a single class of insecticides (Kudom et al., 2015). Higher activity of CCEs is usually reported to be associated with resistance either to synthetic pyrethroids or organophosphates, or both (Hemingway and Ranson, 2000; Sarkar et al., 2009). Apart from synthetic pyrethroids and organophosphate insecticides, increased activity of α and β esterases correlates with resistance to DDT (an organochlorine) in the mosquito vectors. Qualitative and quantitative mechanisms of esterases, though different in mechanism, are found to occur together in *Cx. tarsalis* (Hemingway et al., 2004).

15.2.3.2 Mixed Function Oxidases

CYP_{450}-dependent monooxygenases activate organophosphates to its toxic oxon form. However, these heme-containing family of metabolic enzymes are important as they catalyze the detoxification of several xenobiotics including synthetic insecticides (Hemingway et al., 2004). As such, CYP_{450} monooxygenases are also associated with the development of insecticide resistance in mosquito vectors. Moreover, the monooxygenase family of metabolic enzymes is particularly linked with resistance to pyrethroids in mosquito vectors owing to various reports on the positive correlation between the overproduction of CYP_{450} monooxygenases and pyrethroid resistance (Delannay et al., 2018; Fagbohun et al., 2019).

15.2.3.3 Glutathione S-transferases (GSTs)

Mosquito vectors showing resistance to DDT, carbosulfan, and permethrin also showed an increased level of GSTs activity along with an elevated esterase activity (Corbel et al., 2007). The higher production of GSTs in the mosquito and the enzyme's elevated activity are mainly associated with resistance toward DDT and pyrethroids in *Culex* sp. (Kudom, 2015). In Assam, India, elevated activity of GSTs in a field-collected population of *Cx. quinquefasciatus* correlated with DDT resistance in the same mosquito population (Sarkar et al., 2009). Similarly, in Zambia, increased GSTs activity was reported as the mechanism behind resistance against an organochlorine – DDT and two pyrethroids (deltamethrin and permethrin) in *Cx. quinquefasciatus.*

15.2.4 Target-Site Resistance

Apart from increased production and elevated activity of major detoxifying enzymes, mosquitoes combat the action of insecticides through a modification in the insecticides' target site

TABLE 15.4
Some Terms Related to Behavioral Responses of Mosquitoes

Terms	Definition
Exophily	Prefers resting outdoors
Endophily	Prefers resting indoors
Exophagy	Prefers outdoor biting/feeding
Endophagy	Prefers indoor biting/feeding
Zoophily	Preferring animal blood feed
Anthropophily	Preferring human blood feed
Excito-repellency	Process of excitation and repellency; mosquitoes exit the house to avoid contacting airborne insecticide
Spatial repellency	Repellent activity of volatile compounds mixed in the air when mosquitoes are in flight
Contact irritant	The irritant activity of compounds when mosquitoes come in direct physical contact. Here, the mosquito enters the house but exits early
Deterrent	Mosquitoes are prevented from entering human dwellings

Source: Grieco et al., 2007; Zhou et al., 2020.

inside the mosquito's body. Insecticides have a particular target site, binding to which it exerts its adverse effect on the insect's physiology. DDT and pyrethroids target the voltage-gated sodium channel, resulting in prolonged opening of the channel and ultimately the death of the mosquito vector. The other two classes – organophosphates and carbamates – target the acetylcholinesterase (AChE) enzyme, resulting in continuous transmission of nerve impulses in the insect's body and death thereafter.

15.2.4.1 Voltage-Gated Sodium Channel (VGSC)

The intensive use of DDT and pyrethroids in combating mosquito vectors and agricultural pests in the past has led to the stabilization of one of the major mechanisms of resistance – knock-down resistance (*kdr*). *Kdr* results from point mutation(s) in the sodium channel making the channel insensitive or with reduced sensitivity to DDT and pyrethroids. In mosquito vectors 61 such point mutations are reported in the *vgsc*, though not all confer resistance against DDT and pyrethroids (Scott, 2019). Point mutation in 1014 codon of *vgsc* resulting in the substitution of leucine by serine, histidine, cysteine, or tryptophan (L1014F/H/C/W) has led to resistance against DDT and pyrethroids in mosquito vectors. Substitution of leucine by phenylalanine in 1014 codon is widely studied in *Cx. quinquefasciatus* and is found to be associated with DDT and pyrethroid resistance (Rai et al., 2019). Mutation from leucine to either serine or histidine involved in insecticide resistance development in *Cx. pipiens* and *An. gambiae* were also reported in the past (Ranson et al., 2000). In a study taking different strains of *Cx. quinquefasciatus* 13 possible combinations of mutation was reported with the coexistence of nine synonymous and non-synonymous mutations (Liu, 2015).

15.2.4.2 Acetylcholinesterase (AChE) Receptor

AChE receptor is the target site of carbamates and organophosphates. Insensitive receptor results in resistance to these insecticide classes in the mosquito vectors being one of the most common mechanisms of resistance against the same (Hemingway et al., 2004). Furthermore, insensitive AChE receptor in insects is reported to usually confer higher resistance against carbamate insecticides as compared to organophosphates (Russell et al., 2004). However, point mutation in 119 codons of *ace1* gene from glycine to serine (G119S) is associated with severe resistance to both carbamate and organophosphate insecticides in various mosquito species – *An. gambiae, Cx. pipiens, Cx. vishnui,* and *Cx. quinquefasciatus* (Liu, 2015).

15.3 INSECTICIDE RESISTANCE IN MAJOR MOSQUITO VECTORS

15.3.1 Resistance of *Culex* Species

Resistance development against chemical insecticides has been well recorded in *Culex* mosquitoes. This chapter however specifically focuses on resistance in *Cx. quinquefasciatus* a major vector of lymphatic filariasis in Asian countries. *Cx. quinquefasciatus* population control usually involves the use of synthetic insecticides like those of *Aedes* and *Anopheles* mosquitoes. Due to the lack of proper knowledge, insecticides against *Aedes* mosquitoes are usually applied in the drains – the breeding habitat of *Culex sp.*

Organophosphate insecticides like temephos, malathion, fenitrothion, fenthion, chlorpyrifos, etc. are widely used in vector management programs and the agricultural sector. The overapplication of organophosphates for the control of medical vectors and agricultural pests has resulted in the development of organophosphate resistance in *Cx. quinquefasciatus* not only in India but around the globe (Delisi et al., 2017,). In India, resistance to organophosphate insecticides in *Cx. quinquefasciatus* are reported from several states – Andhra Pradesh (Mukhopadhyay et al., 2006), Bengaluru (Shetty et al., 2012), Chhattisgarh, Gujarat (Raghavendra et al., 2011), Haryana (Thomas et al., 2013), Punjab (Suman et al., 2010), Rajasthan (Bansal and Singh, 2007), Uttar Pradesh (Kumar et al., 2011), and West Bengal (Rai et al., 2019). On the contrary, a susceptible larval population showing 100% mortality to temephos and fenitrothion was reported from Andhra Pradesh (Mukhopadhyay et al., 2006).

Pyrethroids are the main component of insecticide-treated bednets (ITNs) and other domestic mosquitocidal products. As such, resistance development against this insecticide is common among mosquito vectors. Accumulation of pyrethroids from the agricultural run-off into drains further aggravates the problem of pyrethroids resistance in *Culex* mosquitoes as these vectors usually inhabit drains and sewers.

Pyrethroid resistance in mosquito vectors in India poses a serious threat in combating different vector-borne disease. Pyrethroid-resistant *Cx. quinquefasciatus* populations are reported from Assam (Sarkar et al., 2009), Bengaluru (Shetty et al., 2012), Chhattisgarh, Gujarat (Raghavendra et al., 2011), Maharashtra (Karlekar et al., 2013), Mysore (Fakoorziba et al., 2009), Uttar Pradesh (Kumar et al., 2011) and West Bengal (Rai et al., 2019). Pyrethroid resistance in *Cx. quinquefasciatus* is also widespread and frequently studied around the globe (Jones et al., 2012; Kudom et al., 2015; Richards et al., 2017; Yuan et al., 2019).

DDT and dieldrin are the most commonly used organochlorine class of insecticide in the control of mosquito vectors. The first DDT-resistant *Cx. quinquefasciatus* population of India was reported from Delhi in 1952 (Gopalakrishnan and Veer, 2018). Since then, resistance to DDT in the mosquito vector has been reported from different parts of the country – Andhra Pradesh (Mukhopadhyay et al., 2006), Assam (Sarkar et al., 2009), Chhattisgarh, Gujarat (Raghavendra et al., 2011), Maharashtra (Karlekar et al., 2013) and Uttar Pradesh (Kumar et al., 2011) and West Bengal (Rai et al., 2019). Severe resistance to DDT was reported from Andhra Pradesh, Chhattisgarh and West Bengal with zero mortality.

A similar resistance pattern against DDT and dieldrin is prevalent in several tropical and subtropical countries – Burkina Faso (Skovmand and Sanogo, 2018), La Reunion (Tantely et al., 2010), Benin (Corbel et al., 2007), Sri Lanka (Wondji et al., 2008), Zanzibar (Jones et al., 2012), and Ghana (Kudom et al., 2013).

Carbamate insecticides are a derivative of carbamic acid and are most commonly used for control of household pests. Resistance to carbamates is popular in the mosquito vectors though less frequently reported than the other insecticide classes. Carbamate resistance in *Cx. quinquefasciatus* are reported from four Indian states – Chhattisgarh, Gujarat (Raghavendra et al., 2011), Karnataka, and West Bengal (Rai et al., 2019). Carbamate resistance in *Culex* vectors has also been reported from around the globe – Burkina Faso (Skovmand and Sanogo, 2018), Benin (Yadouleton et al., 2015), Malaysia (Chen et al., 2013), and Tanzania (Jones et al., 2012).

15.3.2 Resistance of *Aedes* spp.

Resistance against DDTs, the first insecticide to be used, has been noted to be widespread throughout different countries of the world in both *Ae. aegypti* and *Ae. albopictus* (Vontas et al., 2012: Bharati and Saha, 2021). Such resistance has mainly been shown to be conferred by either metabolic detoxification through xenobiotic detoxifying enzymes or mutations in voltage-gated sodium channels (VGSC), or a combination of both. In some reports, overexpression of CCEs and/or GSTs has been shown to provide some degree of protection against DDT selection pressure in *Ae. aegypti* (Aponte et al., 2013, 2019). Some other studies have reported the elevated activity of α-CCEs, β-CCEs, CYP450s, and GSTs either solely or in combination have been found to confer reduced susceptibility against DDT in *Ae. aegypti* (Aguirre-Obando et al., 2016; Nugapola et al., 2021; Mukhtar and Ibrahim, 2022). Another report has shown that in *Ae. aegypti*, elevated activities of GSTs and CYP450s might have a synergistic role in conferring resistance against DDT (Fagbohun et al., 2020; Leong et al., 2019). Advanced studies have shown the possible involvement of two CYP genes, *CYP6BB2* and *CYP6P12*, with DDT resistance (Marcombe et al., 2019). In *Aedes albopictus*, activity levels of GSTs have been found to be elevated in DDT-resistant strains (Nugapola et al., 2021; Marcombe et al., 2014). Another such report has indicated the combined role of detoxifying enzyme groups, i.e., GSTs and β-CCEs, to provide resistance against DDT (Das and Dutta, 2014).

Some of the commonly used insecticides belonging to the organophosphate group of insecticides include temephos, malathion, chlorpyrifos, dichlorvos, etc. In most of the countries, temephos is the choice of insecticide to control mosquito larvae. In *Aedes* mosquitoes, resistance against temephos has now become a common phenomenon throughout the world (Vontas et al., 2012; Bharati and Saha, 2021). Role of α- CCEs and β-CCEs have been confirmed in providing resistance against temephos in *Ae. aegypti* (Muthusamy and Shivakumar, 2015), along with GSTs (Bellinato et al., 2016) and CYP450s (Putra et al., 2016). In *Ae. albopictus*, enhanced quantitative levels of CCEs (Bharati & Saha, 2017) along with GSTs (Ngoagouni et al., 2016) were shown to confer temephos resistance. Two CCE genes have been pinpointed responsible for conferring resistance against temephos in *Ae. aegypti*, namely *CCEae3A*, *CCEae6A* (Poupardin et al., 2014). Same candidate genes have also been shown to have the same effect on *Ae. albopictus* mosquitoes by gene amplification (Grigoraki et al., 2015, 2016). Similarly, some genes of the CYP450 family, namely *CYP6M11*, *CYP6F3*, *CYP6N12*, and *CYP9M9* have also been shown to be involved in providing resistance against temephos in *Ae. aegypti* (McAllister et al., 2012; Yougang et al., 2017). Some studies have reported the upregulation of both CYP450s and CCEs in temephos resistant strain of *Ae. aegypti*. Furthermore, another major detoxifying enzyme family, *i.e.* GSTs have also been found to be involved behind such resistance in *Ae. aegypti* as polymorphism in *GSTe2* has been found to be associated with resistance against temephos (Helvecio et al., 2020). Studies with Quantitative trait loci have also supported the role of CCEs in providing resistance against temephos (Paiva et al., 2016). Cross resistance between temephos and malathion is a topic of controversy with some studies supporting it, with others rejecting it (Putra et al., 2016; Goindin et al., 2017). Differences have been noted between the resistance pattern against temephos, malathion, fenitrothion, or other organophosphates, owing to the difference in their current structure (Hemingway and Ranson, 2000; Goindin et al., 2017). However, resistance against malathion have been attributed to CCEs alone (Widiastuti and Ikawati, 2016), along with GSTs (Choovattanapakorn et al., 2017) or along with CYP450s (Putra et al., 2016). It has also been reported that synergistic action among different detoxification enzymes is responsible for resistance against other organophosphate insecticides (Muthusamy et al., 2014; Seixas et al., 2017). The same gene associated with temephos resistance has also been found to be associated with resistance against malathion (Marcombe et al., 2019).

The involvement of all the major detoxifying enzymes has been noted behind resistance against synthetic pyrethroids. Particularly one enzyme family, i.e., *CYP450s*, has been found to be predominantly associated with such resistance either alone (Mukhtar & Ibrahim, 2022; Amelia-Yap et al.,

2019; Bharati and Saha, 2017; Ishak et al., 2015) or with GSTs or CCEs (Marcombe et al., 2012). Against, deltamethrin, the sole activity of CCEs (Bisset et al., 2013), combined effect of i) GSTs and CYP450s (Alvarez et al., 2013) or ii) GSTs and CCEs (Aponte et al., 2013; Contreras-Perera et al., 2020) or not attributed to a single enzyme but a combination of all the major enzyme groups (Rocha et al., 2015). An overexpression of GSTe2 has been shown to provide resistance against deltamethrin in *Ae. aegypti* (Goindin et al., 2017). Against Lambdacyhalothrin, resistance was shown to be developed due to the activity of CCEs (Muthusamy et al., 2014); for cypermethrin, combined action of both CCEs and GSTs together were found to play role in development of resistance (Li et al., 2015). Against Permethrin, different studies have shed light upon different enzyme groups in providing resistance to *Ae. aegypti* with some studies imparting on the involvement of α-CCEs (Alvarez Gonzalez et al., 2016), combined action of both CYP450s and GSTs (Fagbohun et al., 2020) whereas others rejecting the role of detoxifying enzymes in such resistance (Choovattanapakorn et al., 2017; Muthusamy and Shivakumar, 2015). Advanced studies have confirmed that some CYP450 and GSTs enzymes, namely *CYP9J32*, *CYP9J20*, *CYP9J28*, *CYP9M9* and *GSTE7* have the potentiality to metabolize/detoxify deltamethrin (Stevenson et al., 2012; Marcombe et al., 2012). Recently, *CYP6BB2* and *CYP6P12* have been shown to be associated with pyrethroid resistance in *Ae. aegypti* (Marcombe et al., 2019). In case of *Ae. albopictus*, there are mixed reports concerning the resistance status against synthetic pyrethroids, with the severe resistance pattern noted against permethrin (Karunaratne et al., 2013; Arslan et al., 2016; Ishak et al., 2015) and moderately against deltamethrin (Ngoagouni et al., 2016). Such resistance against permethrin and deltamethrin was found to be mediated via enhanced detoxification by enzyme groups mainly CYP450s (Bharati et al., 2019; Ngoagouni et al., 2016; Ishak et al., 2015).

Among carbamate insecticides, mainly Bendiocarb and Propoxur have been used against *Aedes* mosquitoes throughout the world. Generally, CCEs are believed to confer resistance against carbamates (Hemingway and Karunaratne, 1998). Such correlation between elevated activity of CCEs and resistance against propoxur has been noted in various studies in *Ae. aegypti* (Bharati and Saha, 2018b) and in both *Ae. aegypti* and *Ae. albopictus* (Ali et al., 2020) However, elevated activity of CYP450s has been reported to be responsible for a part of the observed resistance against Bendiocarb in both the species of *Aedes* (Ishak et al., 2015).

The altered structure of insecticide target sites is now a common phenomenon in mosquitoes. At present, 16 different point mutations are known to be prevalent in VGSC in *Ae. aegypti* throughout the world (Kandel et al., 2019). In recent works, *kdr* mutations, V1016I and F1534C have been found to provide resistance against DDT, deltamethrin and permethrin (Aponte et al., 2013). In a Colombian population of *Ae. aegypti kdr* V1016L was found to be present, providing resistance against a wide range of synthetic pyrethroid insecticide, Organochlorines and Organophosphate insecticides (Maestre-Serrano et al., 2014). In an *Ae. aegypti* population, resistant against DDT, lambdacyhalothrin, permethrin, bendiocarb, malathion, and dieldrin, two major *kdr* mutations F1534C and V1016G were found to exist (Ishak et al. 2015). Similarly high frequency of *kdr* F1534C has been noted to confer resistance against DDT and Deltamethrin in *Ae. aegypti* (Kushwah et al., 2015; Fagbohun et al., 2022). The *kdr* mutation V1016I has been noted to be predominant in a temephos-resistant population of *Ae. aegypti* thus indicating the correlation between the two (Aguirre-Obando et al., 2015, 2016). Resistance against deltamethrin has been found to be inferred by the high allele frequency of V1016L and F1534L (Goindin et al., 2017). Highly altered AChE has been found to provide resistance against temephos and malathion in *Ae. aegypti* collected from French West Indies (Goindin et al., 2017). Fixation of F1534C allele in *Ae. aegypti* has been shown to provide resistance against a wide range of insecticide (Badolo et al., 2018). Another study has reported that resistance against synthetic pyrethroids in *Ae. aegypti* can be inferred by the *kdr* mutations V1016G and S989P (Demok et al., 2019). In a population of *Ae. aegypti* from Cameroon both F1534C and V1016I was found to provide resistance against a wide range of insecticides, with the same population reporting the presence of V410L for the first time in the country, where these insecticides are not generally used to control mosquito population thus pinpointing the presence of such mutation in areas even where these insecticides are not used (Montogomery et al., 2022). In *Ae. aegypti*, single *kdr* mutation has been linked to resistance against different insecticides, *e.g.*, mutation in F1534C has been reported to confer resistance against deltamethrin (Toe et al., 2022), DDT (Mukhtar & Ibrahim, 2022), permethrin (Granada et al., 2021). Presence of dual mutations of F1534C and V1016G have been noted to confer greater resistance against synthetic pyrethroid insecticides (Sombie et al., 2019; Abu Bakar et al., 2021). It has also been reported that resistance against lambdacyhalothrin is related to the presence of more than one mutations in the VGSC known as super *kdr* phenomenon in other insects (Freeman et al., 2021; Granada et al., 2021). The reports of *kdr* mutations in *Ae. albopictus* are few and very recent. As in case of *Ae. aegypti*, the mutation F1534C has been implicated in providing resistance against DDT and synthetic pyrethroids in *Ae. albopictus* (Rath et al., 2018; Chen et al., 2016). Similarly, F1534L was found to confer resistance against DDT (Marcombe et al., 2014). Mutations in the acetylcholinesterase leading to insensitive AChE has been reported to confer resistance against Propoxur in *Ae. albopictus* (Ngoagouni et al., 2016).

15.3.3 Resistance of *Anopheles* spp.

Several species of Anopheline mosquitoes are important malarial vectors in different regions of the world. The uncontrolled use of various synthetic insecticides for controlling the mosquito population in order to combat malaria in the past has led to resistance development in *Anopheles sp.* against the insecticides. The problem of growing insecticide resistance in *Anopheles sp.* is not new in India and continues to pose a challenge in the malarial control programs. *An. culicifacies* was reported to show resistance against DDT, malathion, and deltamethrin along with higher KDT (knock-down time) value in Chhattisgarh (Sharma et al., 2007). The authors highlighted the spraying of pyrethroids in the state for controlling mosquitoes to be the reason behind the observed resistance and high KDT value. Similar trends of resistance to DDT, malathion, and deltamethrin from different districts of Madhya Pradesh (Mishra et al., 2012) and Odisha (Sahu et al., 2019) in the same vector were also reported. *An. culicifacies* were found to be DDT- and malathion-resistant in Maharashtra, Gujarat, and Uttar Pradesh. However, this species of *Anopheles* showed susceptibility to malathion in Haryana and tolerance in Jharkhand (Bhatt et al., 2012). From Northeast India, Dhiman et al. (2014) reported severe resistance to DDT with a high KDT value in *An. annularis*. However, *An. annularis* populations were found to be susceptible to deltamethrin with higher mortality rates and lower KDT values in the same study. Data on resistance provided in this chapter dates to a few years back. Therefore, the recent trend of insecticide resistance in *Anopheles sp.* in India can be clearly highlighted only if resistance mapping at the regional level is carried out throughout the country on regular basis.

As insecticide resistance in mosquito vectors is not limited to India, resistant populations of *An. gambiae* are reported from various countries. In Cote d Ivoire *An. gambiae* was found to be resistant to pyrethroids, organochlorines, and carbamate insecticides with a high frequency of L1014F *kdr* mutation and a lower ace-1^R mutation frequency along with an increased activity of esterase, mixed function oxidases, and GSTs (Camara et al., 2018). Similar involvement of metabolic enzymes, higher *kdr* frequencies, and ace-1^R mutation in resistance development to insecticides in *An. gambiae* are reported from Benin (Corbel et al., 2007), Burkina Faso (Namountougou et al., 2012), Mali and Mozambique. Apart from *An. gambia*e, resistance against DDT, deltamethrin, lambdacyhalothrin, fenitrothion, and bendiocarb was also reported in *An. coluzzii* and *An. arabiensis* with higher *kdr* frequency and comparably lower ace-1^R mutation frequency in Mali (Cisse et al., 2015). In Mozambique, resistant populations of *An. gambiae* and *An. arabiensis* to all four classes of chemical insecticides were found to have elevated activity of metabolic enzymes (Casimiro et al., 2014). Likewise, elevated activity of metabolic enzymes was observed in *An. stephensi* populations resistant to pyrethroids, organophosphates, and carbamates in Afghanistan with the occurrence of L1014S and L1014F *kdr* frequencies (Safi et al., 2017).

15.4 OVERCOMING ISSUES OF RESISTANCE DEVELOPMENT IN DISEASE VECTOR MOSQUITOES

In order to combat the challenges of insecticide resistance in disease vector mosquitoes the primary step should be to create insecticide resistance map from different part of affected areas and to take note of different kinds of insecticides being used in these areas, both in agricultural field and public health sectors. Then, with this knowledge introduction of classes of insecticides with different mode of action should be used. Rotation of these insecticides may also curb the risk of increasing the diagnostic dose of a single insecticide. The use of detoxifying enzyme inhibitors along with insecticides also decreases the insecticidal load on the environment. The application of more than one insecticide in a sustainable formulation also results in better management of mosquito-borne diseases (Dhiman et al., 2021). The use of phytochemicals has also provided promising results in mosquito control strategies (Isman 2020; Kamaraj et al., 2011). They have some advantages over synthetic insecticides in that they are biodegradable, some are non-toxic to non-target animals, and due to their diverse and redundant nature, they provide an unending source of chemical arsenal against mosquitoes.

15.5 CONCLUSION

The vector control program is an unending war between human and disease vectors where humans have prevailed in a few battles and mosquitoes in the most due to their rapid adaptability. The use of synthetic insecticide is like a necessary evil; the more we use, the more is the risk of developing resistance, and the less we use threatens us with contacting mosquito-borne diseases. Overexploitation of synthetic insecticides has led to several challenges and complications, such as insecticide resistance development in vectors, environmental pollution, and several unwanted diseases. However, intellectual planning and application of insecticide can aid in reducing both mosquito-borne disease rates, as well as relegate insecticide resistance.

BIBLIOGRAPHY

Abu Bakar, A., Ahmad Mokhtar, A., Mat Jusoh, T.N., Shueb, R.H. (2021). Evaluation of the DDT and Pyrethroid Resistance Status of *Aedes aegypti* (Linnaeus) (Diptera: Culicidae) in Kota Bharu, Kelantan. *Transactions on Science and Technology*, 8(3), 128–136.

Aguirre-Obando, O.A., Bona, A.C., Duque, L.J.E., Navarro-Silva, M.A. (2015). Insecticide resistance and genetic variability in natural populations of *Aedes (Stegomyia) aegypti* (Diptera: Culicidae) from Colombia. *Zoologia* (Curitiba), 32, 14–22.

Aguirre-Obando, O.A., Pietrobon, A.J., Bona, A.C., Navarro-Silva, M.A. (2016). Contrasting patterns of insecticide resistance and knockdown resistance (kdr) in *Aedes aegypti* populations from Jacarezinho (Brazil) after a Dengue Outbreak. *Revista Brasileira de Entomologia*, 60, 94–100.

Ali, W.N., Ahmad, R., Nor, Z.M., Hassan, A.F. (2020). Spatial distribution, enzymatic activity, and insecticide resistance status of *Aedes aegypti* and *Aedes albopictus* from dengue hotspot areas in Kuala Lumpur and Selangor, Malaysia. Malaysia. *Serangga*, 25, 65–92.

Alvarez, L.C., Ponce, G., Oviedo, M., Lopez, B., Flores, A.E. (2013). Resistance to malathion and deltamethrin in Aedes aegypti (Diptera: Culicidae) from western Venezuela. *Journal of Medical Entomology*, 50(5), 1031–1039.

Alvarez Gonzalez, L., Ponce Garcia, G., Flores Suarez, A. (2016). Mechanisms of permethrin resistance in two populations of *Aedes aegypti* in western Venezuela. *Boletín de Malariología y Salud Ambiental*, 56(1), 43–52.

Amelia-Yap, Z.H., Sofian-Azirun, M., Chen, C.D., Suana, I.W., Lau, K.W., Elia-Amira, N.M., Haziqah-Rashid, A., Tan, T.K., Lim, Y.A., Low, V.L. (2019). Pyrethroids use: threats on metabolic-mediated resistance mechanisms in the primary dengue vector *Aedes aegypti* (Diptera: Culicidae). *Journal of Medical Entomology*, 56(3), 811–816.

Aponte, A., Penilla, R.P., Rodríguez, A.D., Ocampo, C.B. (2019). Mechanisms of pyrethroid resistance in *Aedes (Stegomyia) aegypti* from Colombia. *Acta Tropica*, 191, 146–154.

Aponte, H.A., Penilla, R.P., Dzul-Manzanilla, F., Che-Mendoza, A., López, A.D., Solis, F., Manrique-Saide, P., Ranson, H., Lenhart, A., McCall, P.J., Rodríguez, A.D. (2013). The pyrethroid resistance status and mechanisms in *Aedes aegypti* from the Guerrero state, Mexico. *Pesticide Biochemistry and Physiology*, 107(2), 226–34.

Arslan, A., Rathor, H.R., Mukhtar, M.U., Mushtaq, S., Bhatti, A., Asif, M., Arshad, I., Ahmad, J.F. (2016). Spatial distribution and insecticide susceptibility status of *Aedes aegypti* and *Aedes albopictus* in dengue affected urban areas of Rawalpindi, Pakistan. *Journal of Vector Borne Diseases*, 53(2), 136–143.

Atoyebi, S.M., Tchigossou, G.M., Akoton, R., Riveron, J.M., Irving, H., Weedall, G., Tossou, E., Djegbe, I., Oyewole, I.O., Bakare, A.A., Wondji, C.S. (2020). Investigating the molecular basis of multiple insecticide resistance in a major malaria vector Anopheles funestus (sensu stricto) from Akaka-Remo, Ogun State, Nigeria. *Parasites & Vectors*, 13, 1–14.

Badolo, A., Sombie, A., Pignatelli, P., Wangrawa, D., Yameogo, F., Sanon, A., Sanon, A., Kanuka, H., McCall, P., Weetman, D. (2018). Variable resistance to insecticide in *Aedes aegypti* is explained by combined kdr mutations and metabolic gene overexpression in Burkina Faso. *American Journal of Tropical Medicine and Hygiene*, 99(4), 266–267.

Balabanidou, V., Grigoraki, L., Vontas, J. (2018). Insect cuticle: a critical determinant of insecticide resistance. *Current Opinion in Insect Science*, 27, 68–74.

Balabanidou, V., Kampouraki, A., MacLean, M., Blomquist, G.J., Tittiger, C., Juárez, M.P., Mijailovsky, S.J., Chalepakis, G., Anthousi, A., Lynd, A., Antoine, S. (2016). Cytochrome P450 associated with insecticide resistance catalyzes cuticular hydrocarbon production in *Anopheles gambiae*. *Proceedings of the National Academy of Sciences*, 113(33), 9268–9273.

Bansal, S.K., Singh, K.V. (2007). Relative susceptibility of some common mosquito vector larvae to synthetic insecticidal compounds in north-western Rajasthan. *Journal of Environmental Biology*, 28(4), 829–832.

Bellinato, D.F., Viana-Medeiros, P.F., Araújo, S.C., Martins, A.J., Lima, J.B., Valle, D. (2016). Resistance status to the insecticides temephos, deltamethrin, and diflubenzuron in Brazilian *Aedes aegypti* populations. *BioMed Research International*, 2016, 8603263.

Bharati, M., Rai, P., Saha, D. (2019). Insecticide resistance in *Aedes albopictus* Skuse from sub-Himalayan districts of West Bengal, India. *Acta Tropica*, 192, 104–11.

Bharati, M., Saha, D. (2017). Insecticide susceptibility status and major detoxifying enzymes' activity in *Aedes albopictus* (Skuse), vector of dengue and chikungunya in Northern part of West Bengal, India. *Acta Tropica*, 170, 112–9.

Bharati, M., Saha, D. (2018a). Multiple insecticide resistance mechanisms in primary dengue vector, *Aedes aegypti* (Linn.) from dengue endemic districts of sub-Himalayan West Bengal, India. *PLoS One*, 13(9), e0203207.

Bharati, M., Saha, D. (2018b). Assessment of insecticide resistance in primary dengue vector, *Aedes aegypti* (Linn.) from Northern Districts of West Bengal, India. *Acta Tropica*, 187, 78–86.

Bharati, M., Saha, D. (2021). Insecticide resistance status and biochemical mechanisms involved in *Aedes* mosquitoes: A scoping review. *Asian Pacific Journal of Tropical Medicine*, 14(2), 52.

Bhatt, R.M., Sharma, S.N., Barik, T.K., Raghavendra, K. (2012). Status of insecticide resistance in malaria vector, *Anopheles culicifacies* in Chhattisgarh state, India. *Journal of Vector Borne Diseases*, 49(1), 36.

Bisset, J.A., Marín, R., Rodríguez, M.M., Severson, D.W., Ricardo, Y., French, L., Díaz, M., Pérez, O. (2013). Insecticide resistance in two *Aedes aegypti* (Diptera: Culicidae) strains from Costa Rica. *Journal of Medical Entomology*, 50(2), 352–61.

Bonizzoni, M., Afrane, Y., Dunn, W.A., Atieli, F.K., Zhou, G., Zhong, D., Li, J., Githeko, A. Yan, G. (2012). Comparative transcriptome analyses of deltamethrin-resistant and-susceptible *Anopheles gambiae* mosquitoes from Kenya by RNA-Seq. *PLoS One*, 7(9), e44607.

Brady, O.J., Gething, P.W., Bhatt, S., Messina, J.P., Brownstein, J.S., Hoen, A.G., Moyes, C.L., Farlow, A.W., Scott, T.W., Hay, S.I. (2012). Refining the global spatial limits of dengue virus transmission by evidence-based consensus. *PLoS Neglected Tropical Diseases*, 6(8), e1760.

Camara, S., Koffi, A.A., Ahoua Alou, L.P., Koffi, K., Kabran, J.P., Koné, A., Koffi, M.F., N'Guessan, R., Pennetier, C. (2018). Mapping insecticide resistance in *Anopheles gambiae* (sl) from Côte d'Ivoire. *Parasites & Vectors*, 11(1), 1–11.

Carrasco, D., Lefèvre, T., Moiroux, N., Pennetier, C., Chandre, F., Cohuet, A. (2019). Behavioural adaptations of mosquito vectors to insecticide control. *Current Opinion in Insect Science*, 34, 48–54.

Casimiro, S., Coleman, M., Hemingway, J., Sharp, B. (2014). Insecticide resistance in *Anopheles arabiensis* and *Anopheles gambiae* from Mozambique. *Journal of Medical Entomology*, 43(2), 276–282.

Chen, C.D., Low, V.L., Lau, K.W., Lee, H.L., Nazni, W.A., Heo, C.C., Azidah, A.A., Sofian-Azirun, M. (2013). First report on adulticide susceptibility status of *Aedes albopictus*, *Culex quinquefasciatus*, and *Culex vishnui* from a pig farm in Tanjung Sepat, Selangor, Malaysia. *Journal of the American Mosquito Control Association*, 29(3), 243–250.

Chen, H., Li, K., Wang, X., Yang, X., Lin, Y., Cai, F., Zhong, W., Lin, C., Lin, Z., Ma, Y. (2016). First identification of kdr allele F1534S in VGSC gene and its association with resistance to pyrethroid insecticides in *Aedes albopictus* populations from Haikou City, Hainan Island, China. *Infectious Diseases of Poverty*, 5(03), 40–47.

Choovattanapakorn, N., Yanola, J., Lumjuan, N., Saingamsook, J., Somboon, P. (2017). Characterization of metabolic detoxifying enzymes in an insecticide resistant strain of *Aedes aegypti* harboring homozygous S989P and V1016G kdr mutations. *Medical Entomology and Zoology*, 68(1), 19–26.

Cisse, M., Keita, C., Dicko, A., Dengela, D., Coleman, J., Lucas, B., Mihigo, J., Sadou, A., Belemvire, A., George, K., Fornadel, C. (2015). Characterizing the insecticide resistance of *Anopheles gambiae* in Mali. *Malaria Journal*, 14(1), 1–10.

Contreras-Perera, Y., Ponce-Garcia, G., Villanueva-Segura, K., Lopez-Monroy, B., Rodríguez-Sanchez, I.P., Lenhart, A., Manrique-Saide, P., Flores, A.E. (2020). Impact of deltamethrin selection on kdr mutations and insecticide detoxifying enzymes in *Aedes aegypti* from Mexico. *Parasites & Vectors*, 13(1), 1–22.

Corbel, V., N'Guessan, R., Brengues, C., Chandre, F., Djogbenou, L., Martin, T., Akogbeto, M., Hougard, J.M., Rowland, M. (2007). Multiple insecticide resistance mechanisms in *Anopheles gambiae* and *Culex quinquefasciatus* from Benin, West Africa. *Acta Tropica*, 101(3), 207–216.

Das, M., Dutta, P. (2014). Status of insecticide resistance and detoxifying enzyme activity of *Aedes albopictus* population in Sonitpur district of Assam, India. *International Journal of Mosquito Research*, 1(4), 35–41.

Delannay, C., Goindin, D., Kellaou, K., Ramdini, C., Gustave, J., Vega-Rúa, A. (2018). Multiple insecticide resistance in *Culex quinquefasciatus* populations from Guadeloupe (French West Indies) and associated mechanisms. *PLoS One*, 13(6), e0199615.

Deletre, E., Martin, T., Duménil, C., Chandre, F. (2019). Insecticide resistance modifies mosquito response to DEET and natural repellents. *Parasites & Vectors*, 12, 1–10.

DeLisi, N., Ottea, J., Healy, K. (2017). Susceptibility of *Culex quinquefasciatus* (Diptera: Culicidae) in southern Louisiana to larval insecticides. *Journal of Economic Entomology*, 110(6), 2562–2567.

Demok, S., Endersby-Harshman, N., Vinit, R., Timinao, L., Robinson, L.J., Susapu, M., Makita, L., Laman, M., Hoffmann, A., Karl, S. (2019). Insecticide resistance status of *Aedes aegypti* and *Aedes albopictus* mosquitoes in Papua New Guinea. *Parasites and Vectors*, 12(1), 333.

Dethier, V.G., Browne, B.L., Smith, C.N. (1960). The designation of chemicals in terms of the responses they elicit from insects. *Journal of Economic Entomology*, 53(1), 134–136.

Dhiman, S., Rabha, B., Goswami, D., Das, N.G., Baruah, I., Bhola, R.K., Veer, V. (2014). Insecticide resistance and human blood meal preference of *Anopheles annularis* in Asom-Meghalaya border area, northeast India. *Journal of Vector Borne Diseases*, 51(2),133.

Dhiman, S., Yadav, K., Acharya, B.N., Ahirwar, R.K., Sukumaran, D. (2021). Behavioural response of mosquito vectors *Aedes aegypti*, *Anopheles stephensi* and *Culex quinquefasciatus* to synthetic pyrethroid and organophosphorus-based slow-release insecticidal paint. *Parasites & Vectors*, 14(1), 1–12.

Fagbohun, I.K., Idowu, E.T., Olakiigbe, A.K., Oyeniyi, A.T., Otubanjo, O.A., Awolola, T.S. (2020). Metabolic resistance mechanism in *Aedes aegypti* from Lagos State, Nigeria. *The Journal of Basic and Applied Zoology*, 81(1), 1–7.

Fagbohun, I.K., Oyeniyi, T.A., Idowu, E.T., Nwanya, O., Okonkwo, F., Adesalu, K.O., Jimoh, R.T., Oladosu, Y., Otubanjo, O.A., Adeogun, A.O. (2022). Detection and co-occurrence of kdr (F1534C and S989P) mutations in multiple insecticides resistant Aedes aegypti (Diptera: Culicidae) in Nigeria. *Journal of Medical Entomology*, 59(5), 1741–1748.

Fagbohun, I.K., Oyeniyi, T.A., Idowu, T.E., Otubanjo, O.A., Awolola, S.T. (2019). Cytochrome P450 monooxygenase and resistance phenotype in DDT and deltamethrin-resistant *Anopheles gambiae* (Diptera: Culicidae) and *Culex quinquefasciatus* in Kosofe, Lagos, Nigeria. *Journal of Medical Entomology*, 56(3), 817–821.

Fakoorziba, M.R., Eghbal, F., Vijayan, V.A. (2009). Synergist efficacy of piperonylbutoxide with deltamethrin as pyrethroid insecticide on *Culex tritaeniorhynchus* (Diptera: Culicidae) and other mosquito species. *Environmental Toxicology: An International Journal*, 24(1), 19–24.

Fang, F., Wang, W., Zhang, D., Lv, Y., Zhou, D., Ma, L., Shen, B., Sun, Y., Zhu, C. (2015). The cuticle proteins: A putative role for deltamethrin resistance in *Culex pipiens pallens*. *Parasitology Research*, 114(12), 4421–4429.

Freeman, J.C., San Miguel, K., Scott, J.G. (2021). All resistance alleles are not equal: the high fitness cost of super-kdr in the absence of insecticide. *Pest Management Science*, 77(8), 3693–7.

Fuseini, G., Nguema, R.N., Phiri, W.P., Donfack, O.T., Cortes, C., Von Fricken, M.E., Meyers, J.I., Kleinschmidt, I., Garcia, G.A., Maas, C., Schwabe, C. (2019). Increased biting rate of insecticide-resistant *Culex* mosquitoes and community adherence to IRS for malaria control in urban Malabo, Bioko Island, Equatorial Guinea. *Journal of Medical Entomology*, 56(4), 1071–1077.

Gatton, M.L., Chitnis, N., Churcher, T., Donnelly, M.J., Ghani, A.C., Godfray, H.C.J., Gould, F., Hastings, I., Marshall, J., Ranson, H., Rowland, M. (2013). The importance of mosquito behavioural adaptations to malaria control in Africa. *Evolution*, 67(4), 1218–1230.

Githinji, E.K., Irungu, L.W., Ndegwa, P.N., Machani, M.G., Amito, R.O., Kemei, B.J., Murima, P.N., Ombui, G.M., Wanjoya, A.K., Mbogo, C.M., Mathenge, E.M. (2020). Impact of Insecticide Resistance on *P. falciparum* Vectors' Biting, Feeding, and Resting Behaviour in Selected Clusters in Teso North and South Subcounties in Busia County, Western Kenya. *Journal of Parasitology Research*, 2020, 9423682.

Goindin, D., Delannay, C., Gelasse, A., Ramdini, C., Gaude, T., Faucon, F., David, J.P., Gustave, J., Vega-Rua, A., Fouque, F. (2017). Levels of insecticide resistance to deltamethrin, malathion, and temephos, and associated mechanisms in *Aedes aegypti* mosquitoes from the Guadeloupe and Saint Martin islands (French West Indies). *Infectious Diseases of Poverty*, 6(1), 1–5.

Gopalakrishnan, R., Veer, V. (2018). Bioecology, Insecticide Susceptibility and Management of *Culex quinquefasciatus* Say, 1823: A Major Vector of Lymphatic Filariasis in India. In: *Lymphatic Filariasis: epidemiology, treatment and prevention - the Indian perspective*. Ed Tyagi BK. Springer, Singapore; 199–210.

Granada, Y., Mejía-Jaramillo, A.M., Zuluaga, S., Triana-Chávez, O. (2021). Molecular surveillance of resistance to pyrethroids insecticides in Colombian *Aedes aegypti* populations. *PLoS Neglected Tropical Diseases*, 15(12), e0010001.

Grieco, J.P., Achee, N.L., Chareonviriyaphap, T., Suwonkerd, W., Chauhan, K., Sardelis, M.R., Roberts, D.R. (2007). A new classification system for the actions of IRS chemicals traditionally used for malaria control. *PLoS One*, 2(8), e716.

Grigoraki, L., Balabanidou, V., Meristoudis, C., Miridakis, A., Ranson, H., Swevers, L., Vontas, J. (2016). Functional and immunohistochemical characterization of CCEae3a, a carboxylesterase associated with temephos resistance in the major arbovirus vectors *Aedes aegypti* and *Ae. albopictus*. *Insect Biochemistry and Molecular Biology*, 74, 61–67.

Grigoraki, L., Grau-Bové, X., Carrington Yates, H., Lycett, G.J., Ranson, H. (2020). Isolation and transcriptomic analysis of Anopheles gambiae oenocytes enables the delineation of hydrocarbon biosynthesis. *Elife*, 9, e58019.

Grigoraki, L., Lagnel, J., Kioulos, I., Kampouraki, A., Morou, E., Labbe, P., Weill, M., Vontas, J. (2015). Transcriptome profiling and genetic study reveal amplified carboxylesterase genes implicated in temephos resistance, in the Asian tiger mosquito *Aedes albopictus*. *PLoS Neglected Tropical Diseases*, 9(5), e0003771.

Guo, J., Xu, Y., Yang, X., Sun, X., Sun, Y., Zhou, D., Ma, L., Shen, B., Zhu, C. (2019). TRE1 and CHS1 contribute to deltamethrin resistance in *Culex pipiens pallens*. *Archives of Insect Biochemistry and Physiology*, 100(4), e21538.

Gurav, Y.K., Bondre, V.P., Tandale, B.V., Damle, R.G., Mallick, S., Ghosh, U.S., Nag, S.S. (2016). A large outbreak of Japanese encephalitis predominantly among adults in northern region of West Bengal, India. *Journal of Medical Virology*, 88(11), 2004–2011.

Guzman, M.G., Harris, E. (2015). Dengue. *The Lancet*, 385(9966), 453–65.

Hassett, M.R., Roepe, P.D. (2019). Origin and spread of evolving artemisinin-resistant *Plasmodium falciparum* malarial parasites in Southeast Asia. *The American Journal of Tropical Medicine and Hygiene*, 101(6), 1204.

He, Z., Zhang, J., Shi, Z., Liu, J., Zhang, J., Yan, Z., Chen, B. (2019). Modification of contact avoidance behaviour associated with pyrethroid resistance in Anopheles sinensis (Diptera: Culicidae). *Malaria Journal*, 18, 1–11.

Helvecio, E., Romão, T.P., de Carvalho-Leandro, D., de Oliveira, I.F., Cavalcanti, A.E., Reimer, L., de Paiva Cavalcanti, M., de Oliveira, A.P., Paiva, P.M., Napoleão, T.H., Wallau, G.L. (2020). Polymorphisms in GSTE2 is associated with temephos resistance in *Aedes aegypti*. *Pesticide Biochemistry and Physiology*, 165, 104464.

Hemingway, J., Hawkes, N.J., McCarroll, L., Ranson, H. (2004). The molecular basis of insecticide resistance in mosquitoes. *Insect Biochemistry and Molecular Biology*, 34(7), 653–665.

Hemingway, J., Karunaratne, S.H. (1998). Mosquito carboxylesterases: A review of the molecular biology and biochemistry of a major insecticide resistance mechanism. *Medical and Veterinary Entomology*, 12(1), 1–2.

Hemingway, J., Ranson, H. (2000). Insecticide resistance in insect vectors of human disease. *Annual Review of Entomology*, 45(1), 371–91.

Ishak, I.H., Jaal, Z., Ranson, H., Wondji, C.S. (2015). Contrasting patterns of insecticide resistance and knockdown resistance (kdr) in the dengue vectors *Aedes aegypti* and *Aedes albopictus* from Malaysia. *Parasites & Vectors*, 8(1), 1–3.

Isman, M.B. (2020). Botanical insecticides in the twenty-first century – fulfilling their promise? *Annual Review of Entomology*, 65(1), 233–249.

Jones, C.M., Machin, C., Mohammed, K., Majambere, S., Ali, A.S., Khatib, B.O., Mcha, J., Ranson, H., Kelly-Hope, L.A. (2012). Insecticide resistance in *Culex quinquefasciatus* from Zanzibar: implications for vector control programmes. *Parasites & Vectors*, 5(1), 78–87.

Kamaraj, C., Bagavan, A., Elango, G., Zahir, A.A., Rajakumar, G., Marimuthu, S., Santhoshkumar, T., Rahuman, A.A. (2011). Larvicidal activity of medicinal plant extracts against Anopheles subpictus & Culex tritaeniorhynchus. *Indian Journal of Medical Research*, 134(1), 101–106.

Kandel, Y., Vulcan, J., Rodriguez, S.D., Moore, E., Chung, H.N., Mitra, S., Cordova, J.J., Martinez, K.J., Moon, A.S., Kulkarni, A., Ettestad, P. (2019). Widespread insecticide resistance in *Aedes aegypti* L. from New Mexico, USA. *PLoS One*, 14(2), e0212693.

Karlekar, S.R., Deshpande, M.M., Andrew, R.J. (2013). Present susceptibility status of *Culex quinquefasciatus* Say to three insecticides in Nagpur District of India. *Indian Journal of Scientific Research and Technology*, 1, 12–14.

Karunaratne, S.H., Weeraratne, T.C., Perera, M.D., Surendran, S.N. (2013). Insecticide resistance and, efficacy of space spraying and larviciding in the control of dengue vectors *Aedes aegypti* and *Aedes albopictus* in Sri Lanka. *Pesticide Biochemistry and Physiology*, 107(1), 98–105.

Kasai, S., Komagata, O., Itokawa, K., Shono, T., Ng, L.C., Kobayashi, M., Tomita, T. (2014). Mechanisms of pyrethroid resistance in the dengue mosquito vector, *Aedes aegypti*: target site insensitivity, penetration, and metabolism. *PLoS Neglected Tropical Diseases*, 8(6), e2948.

Kawada, H., Ohashi, K., Dida, G.O., Sonye, G., Njenga, S.M., Mwandawiro, C., Minakawa, N. (2014). Insecticidal and repellent activities of pyrethroids to the three major pyrethroid-resistant malaria vectors in western Kenya. *Parasites & Vectors*, 7(1), 1–9.

Kefi, M., Balabanidou, V., Douris, V., Lycett, G., Feyereisen, R., Vontas, J. (2019). Two functionally distinct CYP4G genes of *Anopheles gambiae* contribute to cuticular hydrocarbon biosynthesis. Insect *Biochemistry and Molecular Biology*, 110, 52–59.

Konopka, J.K., Task, D., Afify, A., Raji, J., Deibel, K., Maguire, S., Lawrence, R., Potter, C.J. (2021). Olfaction in Anopheles mosquitoes. *Chemical Senses*, *46*, bjab021.

Kudom, A. (2015). *Insecticide resistance status of Culex species in urban areas in Ghana and efficacy of long lasting nets against them* (Doctoral dissertation, lmu).

Kudom, A.A., Mensah, B.A., Froeschl, G., Rinder, H., Boakye, D. (2015). DDT and pyrethroid resistance status and laboratory evaluation of bio-efficacy of long lasting insecticide treated nets against *Culex quinquefasciatus* and *Culex decens* in Ghana. *Acta Tropica*, 150, 122–130.

Kudom, A.A., Mensah, B.A., Nunoo, J. (2013). Assessment of anti-mosquito measures in households and resistance status of *Culex* species in urban areas in southern Ghana: Implications for the sustainability of ITN use. *Asian Pacific Journal of Tropical Medicine*, 6(11), 859–864.

Kumar, K., Sharma, A.K., Kumar, S., Patel, S., Sarkar, M., Chauhan, L.S. (2011). Multiple insecticide resistance/susceptibility status of *Culex quinquefasciatus*, principal vector of bancroftian filariasis from filaria endemic areas of northern India. *Asian Pacific Journal of Tropical Medicine*, 4(6), 426–429.

Kushwah, R.B., Dykes, C.L., Kapoor, N., Adak, T., Singh, O.P. (2015). Pyrethroid-resistance and presence of two knockdown resistance (kdr) mutations, F1534C and a novel mutation T1520I, in Indian *Aedes aegypti*. *PLoS Neglected Tropical Diseases*, 9(1), e3332.

Leong, C.S., Vythilingam, I., Liew, J.W., Wong, M.L., Wan-Yusoff, W.S., Lau, Y.L. (2019). Enzymatic and molecular characterization of insecticide resistance mechanisms in field populations of *Aedes aegypti* from Selangor, Malaysia. *Parasites & Vectors*, 12(1), 1–7.

Li, C.X., Kaufman, P.E., Xue, R.D., Zhao, M.H., Wang, G., Yan, T., Guo, X.X., Zhang, Y.M., Dong, Y.D., Xing, D., Zhang, H.D. (2015). Relationship between insecticide resistance and kdr mutations in the dengue vector *Aedes aegypti* in Southern China. *Parasites & Vectors*, 8(1), 1–9.

Liu, N. (2015). Insecticide resistance in mosquitoes: impact, mechanisms, and research directions. *Annual Review of Entomology*, 60, 537–559.

Machani, M.G., Ochomo, E., Amimo, F., Mukabana, W.R., Githeko, A.K., Yan, G., Afrane, Y.A. (2022). Behavioral responses of pyrethroid resistant and susceptible *Anopheles gambiae* mosquitoes to insecticide treated bed net. *PloS One*, 17(4), e0266420.

Maestre-Serrano, R., Gomez-Camargo, D., Ponce-Garcia, G., Flores, A.E. (2014). Susceptibility to insecticides and resistance mechanisms in Aedes aegypti from the Colombian Caribbean Region. *Pesticide Biochemistry and Physiology*, 116, 63–73.

Malakar, P., Das, S., Saha, G.K., Dasgupta, B., Hati, A.K. (1995). Anophelines of Siliguri-Naxalbari block, Darjeeling, West Bengal. *Indian Journal of Malariology*, 32(4), 133–139.

Marcombe, S., Farajollahi, A., Healy, S.P., Clark, G.G., Fonseca, D.M. (2014). Insecticide resistance status of United States populations of Aedes albopictus and mechanisms involved. *PloS One*, 9(7), e101992.

Marcombe, S., Fustec, B., Cattel, J., Chonephetsarath, S., Thammavong, P., Phommavanh, N., David, J.P., Corbel, V., Sutherland, I.W., Hertz, J.C., Brey, P.T. (2019). Distribution of insecticide resistance and mechanisms involved in the arbovirus vector *Aedes aegypti* in Laos and implication for vector control. *PLoS Neglected Tropical Diseases*, 13(12), e0007852.

Marcombe, S., Mathieu, R.B., Pocquet, N., Riaz, M.A., Poupardin, R., Sélior, S., Darriet, F., Reynaud, S., Yebakima, A., Corbel, V., David, J.P. (2012). Insecticide resistance in the dengue vector *Aedes aegypti* from Martinique: distribution, mechanisms and relations with environmental factors. *PloS One*, 7(2), e30989.

Mariappan, T., Samuel, P.P., Thenmozhi, V., Paramasivan, R., Sharma, P.K., Biswas, A.K., Tyagi, B.K. (2014). Entomological investigations into an epidemic of Japanese encephalitis (JE) in northern districts of West Bengal, India (2011–2012). *Indian Journal of Medical Research*, 139(5), 754–761.

McAllister, J.C., Godsey, M.S., Scott, M.L. (2012). Pyrethroid resistance in *Aedes aegypti* and *Aedes albopictus* from Port-au-Prince, Haiti. *Journal of Vector Ecology*, 37(2), 325–32.

Miller, J.R., Siegert, P.Y., Amimo, F.A., Walker, E.D. (2009). Designation of chemicals in terms of the locomotor responses they elicit from insects: an update of Dethier et al. (1960). *Journal of Economic Entomology*, 102(6), 2056–2060.

Mishra, A.K., Chand, S.K., Barik, T.K., Dua, V.K., Raghavendra, K. (2012). Insecticide resistance status in *Anopheles culicifacies* in Madhya Pradesh, central India. *Journal of Vector Borne Diseases*, 49(1), 39.

Modak, M.P., Saha, D. (2022). First report of F1534C kdr mutation in deltamethrin resistant *Aedes albopictus* from northern part of West Bengal, India. *Scientific Reports*, 12, 13653.

Montogomery, M., Harwood, J.F., Yougang, A.P., Wilson-Bahun, T.A., Tedjou, A.N., Keumeni, C.R., Kilpatrick, A.M., Wondji, C.S., Kamgang, B. (2022). Spatial distribution of insecticide resistant populations of *Aedes aegypti* and *Ae. albopictus* and first detection of V410L mutation in Ae. aegypti from Cameroon. *Infectious Diseases of Poverty*, 11(1), 1–3.

Mukhopadhyay, A.K., Patnaik, S.K., Babu, P.S. (2006). Susceptibility status of some culicine mosquitoes to insecticides in Rajahmundry town of Andhra Pradesh, India. *Journal of Vector Borne Diseases*, 43(1), 39–41.

Mukhtar, M.M., Ibrahim, S.S. (2022). Temporal Evaluation of Insecticide Resistance in Populations of the Major Arboviral Vector *Aedes aegypti* from Northern Nigeria. *Insects*, 13(2), 187.

Murray, N.E., Quam, M.B., Wilder-Smith, A. (2013). Epidemiology of dengue: past, present and future prospects. *Clinical Epidemiology*, 5, 299.

Muthusamy, R., Ramkumar, G., Karthi, S., Shivakumar, M.S. (2014). Biochemical mechanisms of insecticide resistance in field population of Dengue vector *Aedes aegypti* (Diptera: Culicidae). *International Journal of Mosquito Research*, 1(2), 1–4.

Muthusamy, R., Shivakumar, M.S. (2015). Susceptibility status of *Aedes aegypti* (L.)(Diptera: Culicidae) to temephos from three districts of Tamil Nadu, India. *Journal of Vector Borne Diseases*, 52(2), 159.

Namountougou, M., Simard, F., Baldet, T., Diabaté, A., Ouédraogo, J.B., Martin, T., Dabiré, R.K. (2012). Multiple insecticide resistance in *Anopheles gambiae* sl populations from Burkina Faso, West Africa. *PLoS One*, 7(11), e48412.

NCVBDC, (2022). Dengue, Chikungunya, Malaria, Acute Encephalities Syndrome, Japanese Encephalities. Retrieved from https://nvbdcp.gov.in/index4.php?lang=1&level=0&linkid=407&lid=3683 on 22.08.2022

Ngoagouni, C., Kamgang, B., Brengues, C., Yahouedo, G., Paupy, C., Nakouné, E., Kazanji, M., Chandre, F. (2016). Susceptibility profile and metabolic mechanisms involved in *Aedes aegypti* and *Aedes albopictus* resistant to DDT and deltamethrin in the Central African Republic. *Parasites & Vectors*, 9(1), 1–3.

Nugapola, N.W., De Silva, W.A., Weeraratne, T.C., Karunaratne, S.H. (2021). kdr type mutations and enhanced GST based insecticide resistance in dengue vector mosquitoes *Aedes aegypti* and *Aedes albopictus*. *International Journal of Tropical Insect Science*, 41(1), 409–17.

Paiva, M.H., Lovin, D.D., Mori, A., Melo-Santos, M.A., Severson, D.W., Ayres, C.F. (2016). Identification of a major Quantitative Trait Locus determining resistance to the organophosphate temephos in the dengue vector mosquito *Aedes aegypti*. *Genomics*, 107(1), 40–8.

Pan, C., Zhou, Y. Mo, J. (2009). The clone of laccase gene and its potential function in cuticular penetration resistance of Culex pipiens pallens to fenvalerate. *Pesticide Biochemistry and Physiology*, 93(3), 105–111.

Phyo, A.P., Nkhoma, S., Stepniewska, K., Ashley, E.A., Nair, S., McGready, R., Ler Moo, C., Al-Saai, S., Dondorp, A.M., Lwin, K.M., Singhasivanon, P., Nosten, F. (2012). Emergence of artemisinin-resistant malaria on the western border of Thailand: a longitudinal study. *The Lancet*, 379(9830), 1960–1966.

Potikasikorn, J., Chareonviriyaphap, T., Bangs, M.J., Prabaripai, A. (2005). Behavioral responses to DDT and pyrethroids between *Anopheles minimus* species A and C malaria vectors in Thailand. Kasetsart Univ Bangkok (Thailand). *American Journal of Tropical Medicine and Hygiene*, 73(2), 343–349.

Poupardin, R., Srisukontarat, W., Yunta, C., Ranson, H. (2014). Identification of carboxylesterase genes implicated in temephos resistance in the dengue vector *Aedes aegypti*. *PLoS Neglected Tropical Diseases*, 8(3), e2743.

Putra, R.E., Ahmad, I., Prasetyo, D.B., Susanti, S., Rahayu, R., Hariani, N. (2016). Detection of insecticide resistance in the larvae of some *Aedes aegypti* (Diptera: Culicidae) strains from Java, Indonesia to temephos, malathion and permethrin. *International Journal of Mosquito Research*, 3, 23–8.

Qiu, Y., Tittiger, C., Wicker-Thomas, C., Le Goff, G., Young, S., Wajnberg, E., Fricaux, T., Taquet, N., Blomquist, G.J. Feyereisen, R. (2012). An insect-specific P450 oxidative decarbonylase for cuticular hydrocarbon biosynthesis. *Proceedings of the National Academy of Sciences*, 109(37), 14858–14863.

Raghavendra, K., Barik, T.K., Bhatt, R.M., Srivastava, H.C., Sreehari, U., Dash, A.P. (2011). Evaluation of the pyrrole insecticide chlorfenapyr for the control of *Culex quinquefasciatus* Say. *Acta Tropica*, 118(1), 50–55.

Rai, P., Bharati, M. and Saha, D. (2019b). Assessment of insecticide resistance in Culex quinquefasciatus Say with first report on the presence of L1014F mutation from northern districts of West Bengal, India. *International Journal of Tropical Insect Science*, 39, 301–309.

Rai, P., Bharati, M., Subba, A., Saha, D. (2019). Insecticide resistance mapping in the vector of lymphatic filariasis, *Culex quinquefasciatus* Say from northern region of West Bengal, India. *PLoS One*, 14(5), e0217706.

Ranson, H., Jensen, B., Vulule, J.M., Wang, X., Hemingway, J., Collins, F.H. (2000). Identification of a point mutation in the voltage-gated sodium channel gene of Kenyan *Anopheles gambiae* associated with resistance to DDT and pyrethroids. *Insect Molecular Biology*, 9(5), 491–497.

Rath, A., Mohanty, I., Hazra, R.K. (2018). Insecticide susceptibility status of invasive *Aedes albopictus* across dengue endemic districts of Odisha, India. *Pest Management Science*, 74(6), 1431–1440.

Reisenman, C.E., Lei, H., Guerenstein, P.G. (2016). Neuroethology of olfactory-guided behavior and its potential application in the control of harmful insects. *Frontiers in Physiology*, 7, 271.

Richards, S.L., Balanay, J.A., Fields, M., Vandock, K. (2017). Baseline insecticide susceptibility screening against six active ingredients for *Culex* and *Aedes* (Diptera: Culicidae) mosquitoes in the United States. *Journal of Medical Entomology*, 54(3), 682–695.

Roberts, D.R., Chareonviriyaphap, T., Harlan, H.H., Hshieh, P. (1997). Methods of testing and analyzing excito-repellency responses of malaria vectors to insecticides. *Journal of the American Mosquito Control Association*, 13(1), 13–17.

Rocha, H.D., Paiva, M.H., Silva, N.M., de Araújo, A.P., da Moura, A.J., Gómez, L.F., Ayres, C.F., de Melo Santos, M.A. (2015). Susceptibility profile of *Aedes aegypti* from Santiago Island, Cabo Verde, to insecticides. *Acta Tropica*, 152, 66–73.

Rudra, S.K., Mukhopadhyay, A. (2010). Mosquito species composition of the Dooars of West Bengal, India. *Proceedings of the Zoological Society*, 63(1), 21–25.

Russell, R.J., Claudianos, C., Campbell, P.M., Horne, I., Sutherland, T.D., Oakeshott, J.G. (2004). Two major classes of target site insensitivity mutations confer resistance to organophosphate and carbamate insecticides. *Pesticide Biochemistry and Physiology*, 79(3), 84–93.

Safi, N.H., Ahmadi, A.A., Nahzat, S., Ziapour, S.P., Nikookar, S.H., Fazeli-Dinan, M., Enayati, A., Hemingway, J. (2017). Evidence of metabolic mechanisms playing a role in multiple insecticides resistance in *Anopheles stephensi* populations from Afghanistan. *Malaria Journal*, 16(1), 1–10.

Sahu, S.S., Thankachy, S., Dash, S., Swaminathan, S., Kasinathan, G., Purushothaman, J. (2019). Multiple insecticide resistance in *Anopheles culicifacies* sl (Diptera: Culicidae) in east-central India. *Pathogens and Global Health*, 113(8), 352–358.

Samal, R.R., Kumar, S. (2021). Cuticular thickening associated with insecticide resistance in dengue vector, *Aedes aegypti* L. *International Journal of Tropical Insect Science*, 41(1), 809–20.

Sanou, A., Nelli, L., Guelbéogo, W.M., Cissé, F., Tapsoba, M., Ouédraogo, P., Sagnon, N.F., Ranson, H., Matthiopoulos, J., Ferguson, H.M. (2021). Insecticide resistance and behavioural adaptation as a response to long-lasting insecticidal net deployment in malaria vectors in the Cascades region of Burkina Faso. *Scientific Reports*, 11(1), 17569.

Sarkar, M., Bhattacharyya, I.K., Borkotoki, A., Goswami, D., Rabha, B., Baruah, I., Srivastava, R.B. (2009). Insecticide resistance and detoxifying enzyme activity in the principal bancroftian filariasis vector, *Culex quinquefasciatus*, in northeastern India. *Medical and Veterinary Entomology*, 23(2), 122–131.

Sato, K., Pellegrino, M., Nakagawa, T., Nakagawa, T., Vosshall, L.B., Touhara, K. (2008). Insect olfactory receptors are heteromeric ligand-gated ion channels. *Nature*, 452(7190), 1002–1006.

Scott, J.G. (2019). Life and death at the voltage-sensitive sodium channel: Evolution in response to insecticide use. *Annual Review of Entomology*, 64, 243–257.

Seixas, G., Grigoraki, L., Weetman, D., Vicente, J.L., Silva, A.C., Pinto, J., Vontas, J., Sousa, C.A. (2017). Insecticide resistance is mediated by multiple mechanisms in recently introduced *Aedes aegypti* from Madeira Island (Portugal). *PLoS Neglected Tropical Diseases*, 11(7), e0005799.

Sharma, S.N., Shukla, R.P., Mittal, P.K., Adak, T., Subbarao, S.K. (2007). Insecticide resistance in malaria vector *Anopheles culicifacies* in some tribal districts of Chhattisgarh, India. *Current Science*, 10, 1280–1282.

Shetty, V., Sanil, D., Shetty, N.J. (2012). Insecticide susceptibility status in three medically important species of mosquitoes, *Anopheles stephensi*, *Aedes aegypti* and *Culex quinquefasciatus*, from Bruhat Bengaluru Mahanagara Palike, Karnataka, India. *Pest Management Science*, 69(2), 257–267.

Skovmand, O., Sanogo, E. (2018). Resistance of *Culex quinquefasciatus* to selected chemical and biological pesticides. *Medical Research Archives*, 6(4), 1–9.

Sombie, A., Saiki, E., Yaméogo, F., Sakurai, T., Shirozu, T., Fukumoto, S., Sanon, A., Weetman, D., McCall, P.J., Kanuka, H., Badolo, A. (2019). High frequencies of F1534C and V1016I kdr mutations and association with pyrethroid resistance in Aedes aegypti from Somgandé (Ouagadougou), Burkina Faso. *Tropical Medicine and Health*, 47(1), 1–8.

Sougoufara, S., Yorkston-Dives, H., Aklee, N.M., Rus, A.C., Zairi, J., Tripet, F. (2022). Standardised bioassays reveal that mosquitoes learn to avoid compounds used in chemical vector control after a single sub-lethal exposure. *Scientific Reports*, 12(1), 1–12.

State Vector-Borne Diseases Control and Seasonal Influenza Plan, (2018). Retrieved from https://www.wbhealth.gov.in/upload_files/ticker/State_Vector_Borne_2018.pdf on 17.08.2022

Stevenson, B.J., Pignatelli, P., Nikou, D., Paine, M.J. (2012). Pinpointing P450s associated with pyrethroid metabolism in the dengue vector, *Aedes aegypti*: developing new tools to combat insecticide resistance. *PLoS Neglected Tropical Diseases*, 6(3), e1595.

Strycharz, J.P., Lao, A., Li, H., Qiu, X., Lee, S.H., Sun, W., Yoon, K.S., Doherty, J.J., Pittendrigh, B.R., Clark, J.M. (2013). Resistance in the highly DDT-resistant 91-R strain of Drosophila melanogaster involves decreased penetration, increased metabolism, and direct excretion. *Pesticide Biochemistry and Physiology*, 107(2), 207–217.

Suman, D.S., Tikar, S.N., Parashar, B.D., Prakash, S. (2010). Development of insecticide resistance in *Culex quinquefasciatus* mosquito (Diptera: Culicidae) from different locations in India. *Journal of Pesticide Science*, 35(1), 27–32.

Sun, X., Guo, J., Ye, W., Guo, Q., Huang, Y., Ma, L., Zhou, D., Shen, B., Sun, Y., Zhu, C. (2017). Cuticle genes CpCPR63 and CpCPR47 may confer resistance to deltamethrin in *Culex pipiens pallens*. *Parasitology Research*, 116(8), 2175–2179.

Talipouo, A., Mavridis, K., Nchoutpouen, E., Djiappi-Tchamen, B., Fotakis, E.A., Kopya, E., Bamou, R., Kekeunou, S., Awono-Ambene, P., Balabanidou, V., Balaska, S. (2021). High insecticide resistance mediated by different mechanisms in *Culex quinquefasciatus* populations from the city of Yaoundé, Cameroon. *Scientific Reports*, 11(1), 1–11.

Tantely, M.L., Tortosa, P., Alout, H., Berticat, C., Berthomieu, A., Rutee, A., Dehecq, J.S., Makoundou, P., Labbé, P., Pasteur, N., Weill, M. (2010). Insecticide resistance in *Culex pipiens quinquefasciatus* and *Aedes albopictus* mosquitoes from La Reunion Island. *Insect Biochemistry and Molecular Biology*, 40(4), 317–324.

Thomas, T.G., Prakash, V., Singh, S., Mandal, A.K., Chauhan, L.S. (2013). Insecticide susceptibility status of *Culex quinquefasciatus* Say, the vector of bancroftian filariasis against temephos in Delhi and National Capital Region. *Japanese Journal of Infectious Diseases*, 66(3), 238–240.

Toe, H.K., Zongo, S., Guelbeogo, M.W., Kamgang, B., Viana, M., Tapsoba, M., Sanou, A., Traoré, A., McCall, P.J., Sagnon, N.F. (2022). Multiple insecticide resistance and first evidence of V410L kdr mutation in Aedes (Stegomyia) aegypti (Linnaeus) from Burkina Faso. *Medical and Veterinary Entomology*, 36(3), 309–319.

Valbon, W., Andreazza, F., Oliveira, E.E., Liu, F., Feng, B., Hall, M., Klimavicz, J., Coats, J.R., Dong, K. (2022). Bioallethrin activates specific olfactory sensory neurons and elicits spatial repellency in Aedes aegypti. *Pest Management Science*, 78(2), 438–445.

Vincent, J.F., Wegst, U.G., (2004). Design and mechanical properties of insect cuticle. *Arthropod Structure & Development*, 33(3), 187–199.

Vontas, J., Kioulos, E., Pavlidi, N., Morou, E., Della Torre, A., Ranson, H. (2012). Insecticide resistance in the major dengue vectors *Aedes albopictus* and *Aedes aegypti. Pesticide Biochemistry and Physiology*, 104(2), 126–31.

Wagman, J. M., Achee, N. L., & Grieco, J. P. (2015). Insensitivity to the spatial repellent action of transfluthrin in Aedes aegypti: A heritable trait associated with decreased insecticide susceptibility. *PLoS Neglected Tropical Diseases*, *9*(4), e0003726.

Widiastuti, D., Ikawati, B. (2016). *Resistensi malathion dan aktivitas enzim esterase pada populasi nyamuk Aedes aegypti di Kabupaten Pekalongan.* Balaba: Jurnal Litbang Pengendalian Penyakit Bersumber Binatang Banjarnegara, 61–70.

Willis, J.H. (2010). Structural cuticular proteins from arthropods: annotation, nomenclature, and sequence characteristics in the genomics era. *Insect Biochemistry and Molecular Biology*, 40(3), 189–204.

Wondji, C.S., Priyanka De Silva, W.A., Hemingway, J., Ranson, H., Parakrama Karunaratne, S.H. (2008). Characterization of knockdown resistance in DDT-and pyrethroid-resistant *Culex quinquefasciatus* populations from Sri Lanka. *Tropical Medicine & International Health*, 13(4), 548–555.

Wood, O.R., Hanrahan, S., Coetzee, M., Koekemoer, L.L., Brooke, B.D. (2010). Cuticle thickening associated with pyrethroid resistance in the major malaria vector *Anopheles funestus. Parasites & Vectors*, 3(1), 1–7.

World Malaria Report (2005). *World Health Organization and UNICEF.* https://www.who.int/publications/i/item/9241593199

Wu, X., Zhan, X., Gan, M., Zhang, D., Zhang, M., Zheng, X., Wu, Y., Li, Z., He, A. (2013). Laccase2 is required for sclerotization and pigmentation of *Aedes albopictus* eggshell. *Parasitology Research*, 112(5), 1929–1934.

Xu, Y., Yang, X., Sun, X., Li, X., Liu, Z., Yin, Q., Ma, L., Zhou, D., Sun, Y., She, B., Zhu, C. (2020). Transcription factor FTZ-F1 regulates mosquito cuticular protein CPLCG5 conferring resistance to pyrethroids in *Culex pipiens pallens. Parasites & Vectors*, 13(1), 1–13.

Yadouleton, A., Badirou, K., Agbanrin, R., Jöst, H., Attolou, R., Srinivasan, R., Padonou, G., Akogbéto, M. (2015). Insecticide resistance status in *Culex quinquefasciatus* in Benin. *Parasites & Vectors*, 8(17), 1–6.

Yahouédo, G.A., Chandre, F., Rossignol, M., Ginibre, C., Balabanidou, V., Mendez, N.G.A., Pigeon, O., Vontas, J., Cornelie, S. (2017). Contributions of cuticle permeability and enzyme detoxification to pyrethroid resistance in the major malaria vector *Anopheles gambiae. Scientific Reports*, 7(1),1–10.

Yang, X., Quam, M.B., Zhang, T., Sang, S. (2021). Global burden for dengue and the evolving pattern in the past 30 years. *Journal of Travel Medicine*, 28(8), 146.

Yao, X.H., Min, H., Lü, Z.H., Yuan, H.P. (2006). Influence of acetamiprid on soil enzymatic activities and respiration. *European Journal of Soil Biology*, 42(2), 120–126.

Yougang, A.P., Kamgang, B., Tchoupo, M., Riveron, J.M., Wondji, C. (2017). Temporal distribution and insecticide resistance profile of two major arbovirus vectors *Aedes aegypti* and *Aedes albopictus* in Yaoundé, the capital city of Cameroon. *Parasites & Vectors*, 10(1), 1–9.

Yuan, L., Yang, X., Yu, X., Wu, Y., Jiang, D. (2019). Resistance to insecticides and synergistic and antagonistic effects of essential oils on dimefluthrin toxicity in a field population of *Culex quinquefasciatus* Say. *Ecotoxicology and Environmental Safety*, 169, 928–936.

Zhang, C., Guo, X., Li, T., Cheng, P., Gong, M. (2022). New Insights into Cypermethrin Insecticide Resistance Mechanisms of *Culex pipiens pallens* by Proteome Analysis. *Pest Management Science*, 78(11), 4579–4588.

Zhou, G., Yu, L., Wang, X., Zhong, D., Lee, M.C., Kibret, S., Yan, G. (2020). Behavioral response of insecticide-resistant mosquitoes against spatial repellent: A modified self-propelled particle model simulation. *PloS One*, 15(12), e0244447.

16 Target Site Mediated Insecticide Resistance in Major Mosquito Vectors

Dhiraj Saha, Manas Pratim Modak, Subhajit Das, Abhirup Saha, Nilu Limboo, Prerana Bhujel, Prapti Das, Priyanka Rai, and Debayan Raha
Department of Zoology, University of North Bengal, Siliguri, Darjeeling, India

S.L. Hoti
ICMR-Regional Medical Research Centre, Belagavi, India

Vasuki Venkatesan
ICMR-Vector Control Research Centre, Puducherry, India

B.K. Tyagi
University Institute of Biotechnology, Chandigarh University, Mohali, India

16.1 INTRODUCTION

Mosquitoes are important carriers of many diseases such as malaria, lymphatic filariasis, yellow fever, dengue, chikungunya, zika, and various kinds of encephalitis (WHO, 2017, 2020, 2022a, 2022b), mostly in tropical and sub-tropical countries where they pose a potential threat to global public health and economy (Achee et al., 2019; Franklinos et al., 2019; Filho et al., 2019; Dahmana and Mediannikov, 2020).

India has a diverse climatic condition which makes it a suitable habitat for a wide variety of vectors and pathogens of medical and veterinary importance. Transmission of these vector- borne diseases (VBDs) are closely linked to regional temperature, rainfall, and humidity and climate change in the recent times also gives a major push for the globalization of VBDs (Caminade et al., 2019; Rocklöv and Dubrow, 2020). Moreover, rapid urbanization, high rates of population growth, insufficient water supply, inefficient sewer and waste management systems in India spread up this transmission more rapidly (Singh, 2020; Anoopkumar and Aneesh, 2022). The high rate of disease transmission, emergence of new dengue serotypes, drug-resistant pathogens, and insecticide-resistant vectors pose a severe threat to the current vector management strategy (Mustafa et al., 2015; Anoopkumar and Aneesh, 2022). In India, vector control largely depends upon the use of chemical insecticides and larval habitat management (NCVBDC, 2022). Among these chemical insecticides, synthetic pyrethroids are the major arsenal used for the control of these vectors as it has low mammalian toxicity and are less persistent in the environment (Norris et al., 2018; Matsuo, 2019). As the use of these chemical insecticides increased sharply, the development of insecticide resistance started to occur. Mosquito populations have acquired insecticide resistance as a result of

 DOI: 10.1201/9781003326991-24

the indiscriminate application of insecticides directly targeted at them, as well as indirect exposure to insecticides sprayed on agricultural fields (Bharati and Saha, 2018; Bharati and Saha, 2021). The amount of resistance in insect vector populations depends on the amount and frequency of insecticide application and also the inherent characteristics of the insect species. Tsetse flies, for instance, were controlled by wide-scale spraying of DDT for several years; however, DDT resistance has never developed in this species (Burroughs et al., 2003). In distinction, mosquitoes have all the characteristics suited to rapid resistance development, including short life cycles with abundant progeny (Abate et al., 2020; Hemingway and Ranson, 2000; Karunaratne et al., 2018). Recent studies have indicated that multiple resistance mechanisms are involved in mosquito resistance development. One of the major ways of gaining such resistance is target site insensitivity, in which the target site for the particular insecticide may alter in such a way that insecticidal agents are no longer able to bind with it (IRAC, 2022). Largely used DDT and synthetic pyrethroids are bind with PyR1 and PyR2 sites on the voltage-gated sodium channel of neuronal axon; depolarize its nerve membranes and producing a *knock down effect* (Zhorov and Dong, 2017; Chen et al., 2018). Insecticides belong to group organophosphate and carbamate target acetylcholinesterase (AChE), a crucial enzyme in the neurological system that hydrolyzes cholinergic neurotransmitters and stops nerve impulses (Colovic et al., 2013). Cyclodiene insecticides including dieldrin and fipronil targets insect gamma-aminobutyric acid (GABA) receptor subunit encoded by the *Rdl* gene which plays a central role in neuronal signaling and other physiological processes (Yang et al., 2017). Other insecticides like neonicotinoids and spinosyns target the nicotinic acetylcholine receptor (nAChE) and exert their insecticidal activity (Yu, 2014).

Insecticide resistance is an issue that is becoming more and more of a problem worldwide, so monitoring and management plans must be appropriately prepared to combat this issue. In order to delay the development of insecticide resistance, insecticide management tools such as rotations, mosaics, and mixtures have been proposed and mathematical models have been introduced to estimate their optimal use (Dusfour et al., 2019; Madgwick and Kanitz 2022). Field tests of resistance management strategies are now more feasible due to the recent advent of numerous biochemical and molecular approaches for determining the frequency of resistance allele in vector populations. However, this technique suffers greatly from the practical challenges of assessing changes in resistance gene frequencies in large sample of insects. Additionally, genomic diversity, differential upregulation and post-transcriptional editing of resistance conforming genes, different binding specificities, and the presence of different subspecies with varying vectorial capacities makes the situation more critical in present scenario. However, with the ability to manipulate the insect genome and the availability of advanced molecular tools, new strategies in recent future may resolve this insecticide resistance saddle of mosquito vectors.

16.2 CONVENTIONAL INSECTICIDES USED IN INDIA AND THEIR MODE OF ACTION

Insecticides are chemical or biological substances that are intended to harm, kill, repel, or control a particular type of insect. In India, there are many different formulations of insecticides with several different modes of action (Table 16.1). In some facets of the vector control component of disease elimination in India, recent advancements in biotechnology have even integrated bacterial genes that code for insecticidal proteins (Federici et al., 2003; NCVBDC, 2017). Some insecticides affect the neurological system (Figure 16.1), while others might damage the exoskeletons of the insects, make them flee, or exert other forms of control.

The effectiveness of each insecticide product might be increased or decreased by using more than one in the same area. Broad-spectrum insecticides are effective against all insects, even the good ones whereas other insecticides target specific insects. Using a targeted insecticide minimizes the risk to beneficial or non-target insects. Some insecticides work immediately to kill insects while others may need some time to take effect. Insect growth regulators like pyriproxyfen and

TABLE 16.1
NCVBDC Recommended Insecticides for the Control of Mosquito Vectors in India and Their Mode of Action

Insecticide	Group	Mode of Action
DDT	Organochloride	Sodium channel modulators
Deltamethrin Cyfluthrin Alpha-cypermethrin Lambda-cyhalothrin Bifenthrin Cyphenothrin	Synthetic Pyrethroids	(Alteration of voltage gated sodium channel (*vgsc*) due to point mutation causes prolonged depolarization at neuronal axon in insect species causes state of intoxication and partial paralysis which usually precedes death)
Malathion Temephos	Organophosphate	Inhibition of acetylcholinesterase (AChE) enzyme, which is responsible for the degradation of the excitatory neurotransmitter, acetylcholine, thereby terminating transmission of nerve impulses at cholinergic synapses, resulting hyper-excitation and eventual death
Bacillus thuringiensis var israelensis	Biocontrol agent	Microbial disruptors of insect midgut membranes
Pyriproxyfen	Insect Growth regulator	Juvenile hormone mimics (Pyriproxyfen mimics a natural hormone in insects and disrupts their growth. It is a type of insect growth regulator that affects mostly young insects and eggs)
Diflubenzuron		It inhibits the production of chitin, which is used by an insect to build exoskeleton

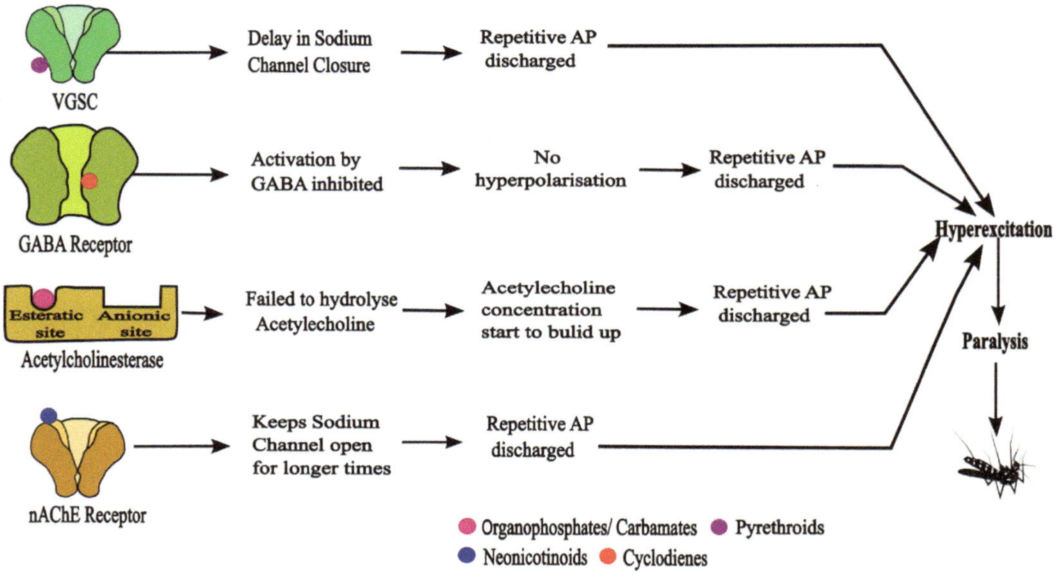

FIGURE 16.1 The mode of action of conventionally used insecticides. In general, these insecticides induce repeating action potential, resulting in hyperexcitation and death (AP: Action Potential).

methoprene prevent insects from properly molting (growing) or laying eggs. According to Insecticide Resistance Action Committee (IRAC), more than thirty types of insecticides are globally used based on their mode of action. These insecticides can be broadly categorized based on the insect physiology they are targeting, such as nerve and muscle, growth and development, respiration, and the midgut.

16.3 DEVELOPMENT OF INSECTICIDE RESISTANCE AND TARGET SITE

16.3.1 Insensitivity

Insecticide resistance is a long standing and expanding problem for vector control measures. It is a heritable change in the vector population resulted in repeated failure of insecticide exposure when treated with recommended dosages. Development of insecticide resistance is an evolutionary process depending directly on genetic, physiological, behavioral, and ecological factors and indirectly on the volume and frequency of applications of insecticides (Sarwar and Salman, 2015; Zhu et al., 2016). A study on an *Ae. albopictus* population from Malaysia further concluded that insecticide resistance is autosomally inherited and incompletely dominant (Chan and Zairi, 2013). In mosquito vectors, insecticide resistance is a pre-adaptive phenomenon in a way that, prior to exposure to an insecticide, there already exist rare individuals who carry one or more resistance alleles (such as polymorphisms in the resistance allele sequence or increased expression of the resistance allele) that allow them to survive exposure to that insecticide. The offspring of the resistant individuals will have an increased survival rate, and eventually resistant individuals will be the predominant group in the population (Liu et al., 2007). Studies from recent times suggested that insecticide resistance in the Indian sub-continent is largely due to the target site insensitivity together with an effect of increased metabolic detoxification (Rai et al., 2019).

In insects, target site insensitivity results from the structural modification or mutation (point mutation) of genes encoding target proteins that interact with insecticides and is currently the most conferred mode of insecticide resistance (Casida and Durkin, 2013) (Figure 16.2). Four different target site alterations have been reported in insect species; however, in mosquito vectors, only three sites were reported to date. Target site insensitivity in mosquito vectors is primarily acquired by a non-silent point mutation in the structural genes (Karunaratne et al., 2018). A single nucleotide change in the DNA sequence results in alteration to amino acid sequences of the functional protein, thereby causing reduced insensitivity for insecticides. In both *Ae. aegypti* and *Ae. albopictus*, changes from TTC to TGC in IIIS6 *vgsc* gene results in phenylalanine to cystine alteration, commonly known as F1534C *kdr* mutation, because of its position at 1534 codon on exon no 31 (Kushwah et al., 2015; Modak and Saha, 2022). As few alterations can diminish pesticide sensitivity without disrupting the physiological functions of the target location, the number of possible amino acid replacements is limited. As a result, identical resistance-related mutations are frequently discovered in significantly divergent taxa (Karunaratne et al., 2018). The most effective plague vector, the oriental rat flea (*Xenopsylla cheopis*), as well as various mosquito populations, including *Cx. quinquefasciatus* and *An. culicifacies*, is found to have the classic leucine to phenylalanine mutation (L1014F) in the *vgsc* gene. This indicates an independent origin of the same mutation in two different species that are geographically isolated (Liu et al., 2019). Mutations in target sites alter its binding affinities to different insecticides depending on the molecular structure of the insecticide (O'Reilly et al., 2006).

16.3.2 Voltage-Gated Sodium Channel

Voltage-gated sodium channel is located on the neuronal axon of insect species and binding of insecticide with this channel causes a prolonged depolarization during nerve impulse propagation and

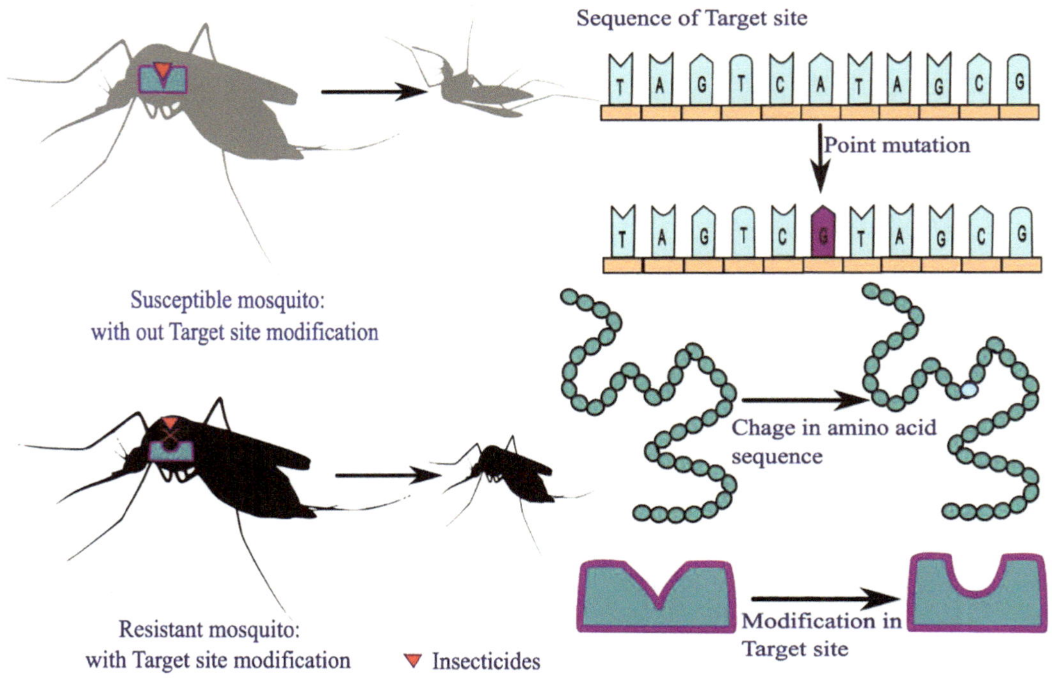

FIGURE 16.2 Development of target site resistance through point mutation in the target site gene, which resulted in a change in amino acid sequence and, eventually, a modification in the insecticide binding site.

thus results a knock-down effect. Mutation in this *vgsc* gene (*kdr* mutation) causes alteration in this channel in a way that insecticides are no longer capable of binding with it. Knock-down resistance (*kdr*), the most widely discussed resistance mechanism in mosquito vectors, is first described in house flies. *Kdr* has been documented globally in a large number of arthropods of relevant agricultural and health interest. In particular, DDT and pyrethroids trigger the *kdr* phenotype. Pyrethroids are synthetic analogs of pyrethrin, used worldwide as broad-spectrum chemical insecticides with a neurotoxic activity. The *kdr* is due to point mutations in the voltage-gated sodium channel encoded by the *vgsc* gene. The channel is involved in the beginning and propagation of action potentials in excitable cells, such as nerve cells. DDT and pyrethroids alter the gating kinetics of *vgsc* by binding the activated form of the channel and blocking it in the open status. Thus, they disrupt the physiological transmission of the nerve impulse (Narahashi, 1996).

The sodium channel protein contains four homologous repeats, each with six transmembrane segments (S1–S6) with a circular radial arrangement in which a central ion pore is formed. The most conserved segment is S4, present in each repeated domain, which contains a unique motif of a positively charged amino acid residue followed by two non-polar residues that repeat four to eight times in each helix. The S4 structures are suggested to participate in the voltage-sensing mechanism (von Stein et al., 2013). The loop connecting domains III and IV is the region suggested to participate in the fast inactivation. A hydrophobic sequence (IFM) in the intracellular segment connecting domains III and IV is required for fast inactivation and serves as an inactivation particle of the sodium channel (Silver et al., 2010). Several different *vgsc* mutations have been reported from Indian mosquito populations. Most of them are located in the transmembrane segments IIS5, IIS6, and IIIS6. Even linker regions connecting S4-S5 of domain-II have been associated with substitutions (Figure 16.3) (Vontas et al., 2012; Oliveira et al., 2013).

The first *kdr* mutation was discovered in *Musca domestica*, a substitution of Lysine with Phenylalanine (L1014F), which was thought to be associated with pyrethroid resistance (Williamson et al., 1996). This L1014F *kdr* mutation in mosquitoes was first detected in pyrethroid-resistant

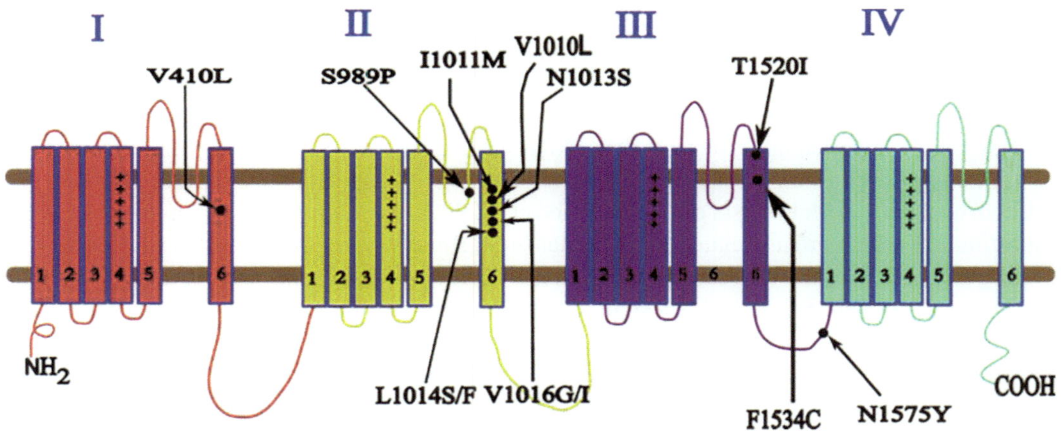

FIGURE 16.3 Schematic representation of voltage-gated sodium channel in mosquito species with major *kdr* mutations. Roman numerals (I–IV) denote each domain of VGSC; Arabic numbers (1–6) refer to segments of respective domains. The "+" sign depicts the voltage-sensing segment of each domain.

An. gambiae lineages (selected in permethrin for 15 generations) (Martinez-Torres et al., 1998), after that it was reported in other *Anopheles* mosquitoes. Another substitution on the same position by serine (L1014S) was reported in *An. gambiae* from Kenya (Ranson et al., 2000). L1014F and L1014S are commonly referred to as "*kdr-west*" and "*kdr-east*," respectively (Silva et al., 2014), because the former was reported from West Africa and the latter from East Africa. In addition to these substitutions, another substitution, L1014W (leucine to tryptophan), at the same location to *An. sinensis* was reported from China along with another substitution N1013S (asparagine to serine), located immediately before the conventional L1014 (Tan et al., 2014). In 2010, a novel substitution from Valine to Leucine (V1010L) was reported in *An. culicifacies* population from India (Singh et al., 2010). Among all *kdr* mutations, L1014F/S are widely distributed, reported from three continents. These classical substitutions are also reported to contribute to pyrethroid and DDT resistance in different *Culex* mosquitoes. L1014F/S was mostly reported in *Cx. quinquefasciatus* populations from Sri Lanka, Thailand, Uganda, USA, French West Indies, and Colombia (Wondji et al., 2008; Yanola et al., 2015; Martins et al., 2017; Lucas et al., 2020; Kothera et al., 2019; Delannay et al., 2018; Maestre-Serrano et al., 2020) and in *Cx. pipiens pallens*, these mutations were reported from China and Belgium (Liu et al., 2019; Wang et al., 2022). Earlier it was believed that these substitutions are solely responsible for pyrethroid and DDT resistance in *Culex* mosquitoes, but studies in recent times further added two more substitutions (L932F and I936V) along with L1014F/S was reported in *Cx. quinquefasciatus* from Brazil and thought to be associated with pyrethroid resistance (Sugiura et al., 2021). In case of *Aedes* more than 15 haplotype reported to have contributed in resistance; among these, F1534C in domain III is widely distributed. Another widely reported substitution is V1016G in domain-II, this position reported to have another substitution V1016I. V1016G in *Ae. aegypti* are previously thought to be restricted to Asia but recently it had been reported in Panama (Murcia et al., 2019). V1016G is alone can contribute to pyrethroid resistance but in combination with another mutation S989P it becomes highly potent. S989P and V1016G mostly co-occur but sometimes they are reported along with F1534C (Chen et al., 2020). When these three mutations co-occur, mosquitoes show extreme case of resistance and the channel sensitivity to permethrin reduces by 1100-fold (Hirata et al., 2014). A novel mutation T1520I is also reported in *Ae. aegypti*; this mutation generally co-occurs with F1534L, reported from India and Myanmar (Kushwah et al., 2020; Naw et al., 2022). I1011 site in domain-II in *Ae. aegypti* reported to have two resistance haplotypes, one is I1011M and other one is I1011V. I1011M was reported with another substitution in same domain G923V from Brazil, Guyana and French overseas territories (Brengues et al., 2003; Dusfour et al., 2015). The same mutation was also reported to occur alone from Latin America and Brazil (Saavedra-Rodriguez et al., 2007; Garcia et al., 2018; De Araújo et al., 2019). I1011V was reported from Latin America, Thailand and Vietnam (Bingham et al., 2011). Two other new substitutions reported in *Ae. aegypti* are A1007G and V410L. A1007G in IIS6 of *vgsc* reported in *Ae. aegypti* from Malaysia (Zuharah and Sufian, 2021; Akhir et al., 2022), V410L, only substitution in domain I, reported in *Ae. aegypti* population from Brazil, Burkina Faso (Haddi et al., 2017; Melo Costa et al., 2020; Toé et al., 2022). *Ae. albopictus* is reported to have four substitutions which are associated with resistance. In *Ae. albopictus*, the position of 1534 at IIIS6 *vgsc* gene is very sensitive, because of the substitution of phenylalanine with lysine, cysteine and serine (Marcombe et al., 2022; Chen et al., 2016;). Another substitution T1520I reported frequently in *Ae. aegypti*, was recently been reported in *Ae. albopictus* from India (Modak and Saha, 2022). Another classical substitution V1016G was also reported in *Ae. albopictus* from China, Vietnam and Italy (Kasai et al., 2019; Zhou et al., 2019; Pichler et al., 2021). Functional characterization of these mutations in *Xenopus* oocytes revealed that most *kdr* mutations alone or in combination are involved in pyrethroid resistance (Chen et al., 2020).

16.3.3 Acetylcholinesterase (AChE)

Acetylcholinesterase (EC 3.1.1.7) is a cholinergic enzyme primarily found at postsynaptic neuromuscular junctions, especially in muscles and nerves. It immediately hydrolyzes the neurotransmitter acetylcholine to end synaptic transmission in the cholinergic nervous system of most animals, including arthropods (Colovic et al., 2013). Insecticides like organophosphate (OP) and carbamate (CB) target AChE as it has a critical function in nerve impulse transmission. However, mosquitoes are rapidly evolved insecticide resistance against these groups of insecticides by introducing numerous point mutations on *ace* gene, coding AChE enzyme (Tmimi et al., 2018). The first insect *ace* gene (now called *ace-2*) was isolated from *Drosophila melanogaster* (Hall and Spierer, 1986). In other insects (*Aphis gossypii*, *Nephotettix cincticeps*, *Helicoverpa armigera* and probably others), including some dipteran species (*Cx. tritaeniorynchus*, *Cx.pipiens*), the gene homologous to the *Drosophila ace-2* is not involved in resistance (Malcolm et al., 1998; Charpentier et al., 2000; Tomita et al., 2000; Mori et al., 2001; Ren et al., 2002). Another *ace* gene (*ace-1*) was shown to be responsible for AChE insensitivity in case of mosquito vectors. These two *ace* genes are the result of an ancient duplication probably older than the emergence of arthropods (Weill et al., 2004).

Although several mutations related to the development of insecticide resistance in other insect species have been reported but in mosquito species, only three have been linked to insensitive AChE mediated resistance: G119S, F290V, F331W (according to *Torpedo californica* AChE nomenclature) (Alout and Weill 2008; Liebman et al., 2015). A single amino acid substitution from a Glycine to Serine at the position 119 of *ace-1* gene, causing G119S mutation in the close proximity of catalytic serine residue (S200), results in steric hindrance leading to reduced accessibility of insecticide (Figure 16.4(a)). G119S mutation in the *ace-1* gene causes about 60% reduction of catalytic activity of acetylcholinesterase compared to the wild-type enzyme. G119S substitution has been reported in several mosquito species, including *An. gambiae, An. albimanus, Cx. vishnui, Cx. pipiens*, and *Cx. quinquefasciatus* (Wang et al., 2022; Alout et al., 2007; Talipouo et al., 2021; Dabiré et al., 2015; Muthusamy and Shivakumar, 2015a). Together with G119S, another mutation – namely F290V in *ace-1* gene – further contributed to resistance development in natural populations of *Cx. pipens* from Cyprus (Alout et al., 2009). Substitution of this phenylalanine to valine (F290V) also reduces insecticides interaction efficiency with catalytic serine (S200), probably due to the orientation of valine side chain. It involves a substitution at the acyl binding pocket and in substrate specificity, likely as a result of hindering the gorge entrance. The natural population of Cx. *tritaeniorhynchus* from India revealed the presence of F331W mutation in *ace-2*, which reduces the accessibility of insecticides due to the bigger size of tryptophan (Figure 16.4(a)) (Ordentlich et al., 1993, Misra and Gore, 2015). Another mutation, F455W (homologous to the Phe331 of AChE in

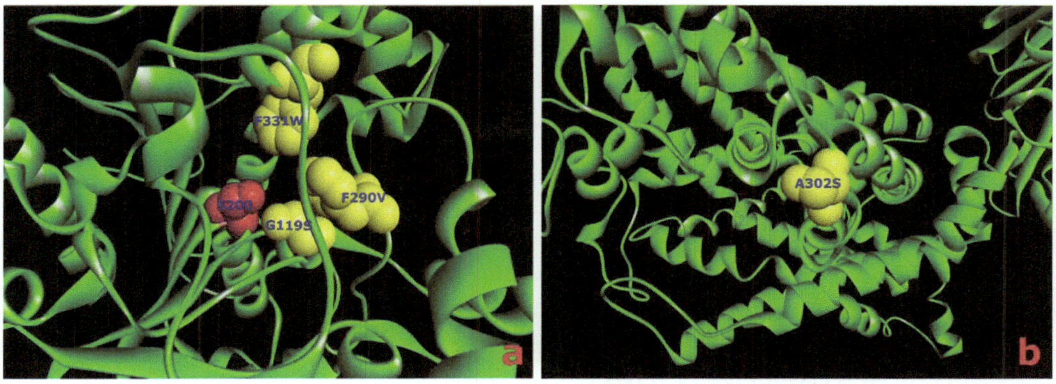

FIGURE 16.4 (a) A 3D viewpoint of the catalytic gorge of acetylcholinesterase (PDB ID: 1ACL). The backbone of the enzyme structure is rendered as a ribbon. The red sphere denotes catalytic serine (S200) and the yellow spheres denote three mutation sites G119S, F290V, F331W. (b) A 3D viewpoint of the GABA receptor (PDB ID: 7WGT). The backbone of the receptor structure is rendered as a ribbon. Yellow spheres denote the mutation site A302S.

T. californica) in the *ace-2* gene, is also found to be responsible for insecticide insensitivity of AChE in *Cx. tritaeniorhynchus* (Nabeshima et al., 2004). In addition, this mutation was also found in *Myzus persicae*, *Tetranychus urtica*, and *Bemisia tabaci* and thus concluding the significance of this mutation in resistant insects generally.

16.3.4 Gamma-Aminobutyric Acid (GABA)

GABA is the principal inhibitory neurotransmitter in mosquito vectors and serves as the agonist for opening the pentameric transmembrane chlorine channel. Cyclodiene insecticides, such as dieldrin, phenyl pyrazoles, and fipronil target the GABA receptor, receptor for the neurotransmitter γ-aminobutyric acid. Each GABA receptor consists of five subunits (two α-subunits, two β-subunits, and one γ-subunit) to form a central ion pore. Each subunit comprises of four transmembrane domains (TM1–TM4) and a N-terminal extracellular domain that serves as a neurotransmitter (GABA) binding site (Yu, 2014). A point mutation resulting in a substitution of Alanine to Serine (A296S) is a key mutation (*rdl*) in the GABA receptor for dieldrin resistance in mosquito vectors like *An. rampae* population from Laos (Marcombe et al., 2020). The same type of substitution of Alanine to Serine at position 302 in the GABA receptor has been reported in *Ae. aegypti*, *Ae. albopictus*, and several *Anopheles* species around the globe (Du et al., 2005, Ffrench-Constant et al., 2000; Low et al., 2015). A previous study by Thompson et al., 1993 in *D. melanogaster* showed that the A302S mutation is connected to resistance development and it may be because alanine 302 is a conserved residue in the M2 transmembrane domain, which lines the chloride ion channel (Figure 16.4(b)). This same A302S mutation in GABA receptor of *Ae. albopictus* and *Cx. pipiens* were also reported from France and Malaysia (Tantely et al., 2010; Amelia-Yap et al., 2020). India banned dieldrin in 2003 and thus reflects the absence of such mutation in India.

16.3.5 Nicotinic Acetylcholine Receptor (nAChR)

Insecticides like neonicotinoids and spinosyns exert their insecticidal activity by binding with the nicotinic acetylcholine receptor (nAChR). As nAChRs also belong to a superfamily of ligand-gated ion channels known as Cys-loop receptors like GABA receptor, they are composed of typically five subunits: two α and three non-α subunits (β, γ, δ) that together form an ion pore. Each subunit consists of four transmembrane domains (TM1-TM4) with an N-terminal extracellular domain for acetylcholine binding site (Yu, 2014). The number of subunits in nAChR varies in insect species as in *D. melanogaster* have ten subunits while the *Neoprotopalmeria capitata* contains 11 subunits (Ureña et al., 2019). Various agricultural pests with nAChR mutations have been reported around the globe till date. The R81T mutation in the nicotinic acetycholine receptor of *Aphis gossypii*, the cotton aphid is associated with neonicotinoid resistance (Hirata et al., 2017). Mutation like Y151S has been found to be associated with resistance to imidacloprid in *Nilaparvata lugens* and substitution of Glycine to Glutamic acid (G275E) in *Tuta absoluta* (Liu et al., 2005). Though a spinosad-resistant *Ae. albopictus* population has been reported from Pakistan (Khan et al., 2011), till date no mutation associated with insecticide resistance in nAChR has been reported in mosquito species around the globe.

16.4 DISTRIBUTION OF TARGET SITE MEDIATED RESISTANCE IN MOSQUITOES FROM INDIA

Rising of dengue etc. in different districts of India is likely a clear indication that the vector resistance monitoring program should be reviewed. Mapping of target site resistance mutations in major mosquito vectors in India is a need of the hour. In the global context, four different target site mutation, i.e., *kdr* mutation, *ace* gene mutation, *Rdl* mutation, and mutation in nAChR, are widely distributed among insect species, but in India, mutation in *vgsc* gene and *ace* gene only confers insecticide resistance in mosquito vectors. *Kdr* mutation is the most extensively studied in India and is reported in different species of *Anopheles*, *Aedes*, and *Culex* mosquitoes. No report of a mutation in the GABA and nAChE receptors is available in India so far.

16.4.1 Distribution of *kdr* Mutation in India

India is the source of many reports on *kdr* mutations in mosquito vectors (Table 16.2). *Aedes* mosquitoes, particularly *Ae. aegypti*, have been found to have the largest numbers of *kdr* mutations. Mutations like F1534C, F1534L, V1016G, and T1520I as well as certain non-synonymous

TABLE 16.2
Major Mosquito Vectors of India with Reported *kdr* Mutations

Mosquito Species		Geographical Location	Insecticide Against Which Resistance Has Developed	*Kdr* Mutation	References
Aedes	*aegypti*	West Bengal	DDT, Deltamethrin	V1016**G** F1534**C** T1520**I**	Saha et al., 2019
		Tamil Nadu	Permethrin	F1534C	Muthusamy and Shivakumar, 2015b
		Delhi		F1534C T1520I	Kushwah et al., 2015
		Karnataka	DDT, Permethrin, Deltamethrin	S989P V1016G F1534C F1534L	Kushwah et al., 2020
	albopictus	West Bengal	DDT, Deltamethrin, Permethrin, Lambda- cyhalothrin	F1534C	Modak and Saha, 2022
		West Bengal	Deltamethrin, Lambdacyhalo-thrin, Permethrin, DDT	L1014F	Rai et al., 2019
		Assam	DDT Deltamethrin	L1014F	Sarkar et al. 2009

(*Continued*)

TABLE 16.2 (Continued)

Mosquito Species	Geographical Location	Insecticide Against Which Resistance Has Developed	*Kdr* Mutation	References
Anopheles culicifacies	Gujarat	Pyrethroid DDT	L1014F	Singh et al., 2009
	Odisha	_	L1014F L1014S V1010L	Singh et al., 2010
	Gujarat, Odisha, Chhattisgarh, Haryana, Rajasthan	Deltamethrin DDT	L1014F L1014S	Dykes et al., 2015
	Odisha,	DDT	–	Hoti et al., 2006
stephensi	Rajasthan	_	L1014F L1014S	Singh et al., 2011
	Rajasthan, Haryana, Chhattisgarh, Tamil Nadu	DDT	L1014F L1014S	Dykes et al., 2016
subpictus	Chhattisgarh	_	L1014F	Singh et al., 2015

mutations were discovered in *Ae. aegypti* populations from different regions of India (Kushwah et al., 2015; Muthusamy and Shivakumar, 2015b; Saha et al., 2019; Kushwah et al., 2020). The F1534C *kdr* mutation has recently been discovered in populations of *Ae. albopictus* from sub-Himalayan region of India (Modak and Saha, 2022). A conventional L1014F *kdr* mutation was also reported in three distinct *Anopheles* populations (Hoti et al., 2006; Singh et al., 2010, 2011; 2015; Dykes et al., 2015, 2016;) and *Cx. quinquefasciatus* population from India (Sarkar et al. 2009; Rai et al., 2019). (Figure 16.5, left).

16.4.2 Distribution of Mutation in *ace* Gene in India

As compared to *kdr* mutation, very few reports of *ace* mutation were reported from India. Mosquitoes belonging to the genus *Aedes* and *Culex* are only found to carry such mutations. A single *Ae. aegypti*

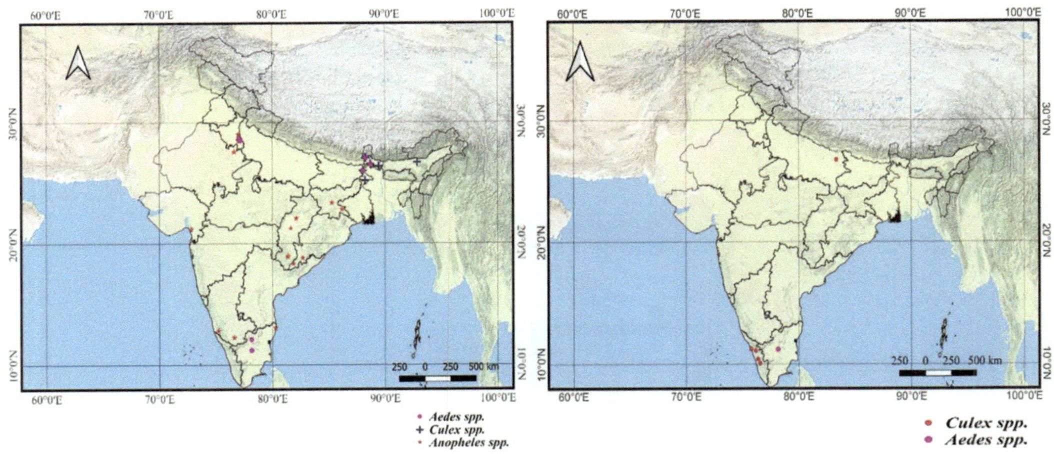

FIGURE 16.5 Distribution of *kdr* mutation (left) and mutation in *ace gene* (right) in mosquitoes from India.

population from the Namakkal district of Tamilnadu was found to carry G119S mutation (Muthusamy and Shivakumar, 2015a). Two different *Culex* populations namely *Cx. tritaeniorhynchus* from Uttar Pradesh and *Cx. quinquefasciatus* from both Uttar Pradesh and Kerala were reported to carry G119S mutation (Misra and Gore, 2015; Viswan and Pushpalatha, 2019). *Culex tritaeniorhynchus* from Uttarpradesh was also reported to carry F331W mutation in *ace-2* (Misra and Gore, 2015) (Figure 16.5(right)). No *Anopheles* spp. from India was found to carry such mutations to date.

16.5 MOLECULAR TOOLS TO ANALYZE TARGET SITE INSENSITIVITY IN MAJOR MOSQUITO VECTORS

Detection of resistance patterns is very important to design an effective vector control strategy. Larval and adult bioassay following the protocol of the World Health Organization is the sole way to determine the susceptibility status of mosquito vectors. Studies from different parts of India reveal that both phenotypically resistant and susceptible mosquitoes bear the same genetic configuration. To overcome this issue molecular mechanism behind this resistance development was studied thoroughly. Synergist assay is a key method to determine the comparative role of target site insensitivity and metabolic detoxification in resistance development. Biochemical assays, like different enzymatic assays, can reveal the role of major detoxifying enzymes in major mosquito vectors. For the detection of target site mutation, Allele-specific PCR (AS-PCR) and RFLP PCR-based approaches are widely accepted throughout the globe. In AS-PCR, a mixture of primers was used in the same reaction whereas in RFLP PCR, a restriction digestion was applied to the amplified gene product in order to determine the point mutation (Figure 16.6). In recent times, modern techniques like qPCR and rt-PCR are employed for this purpose, as it has less chance of giving false positive results. One of the most effective ways for detecting these mutations is the use of sequence-based approaches like Sanger sequencing and next-generation sequencing, where the target site was amplified, sequenced, and then finally aligned with other homologous sequences in order to

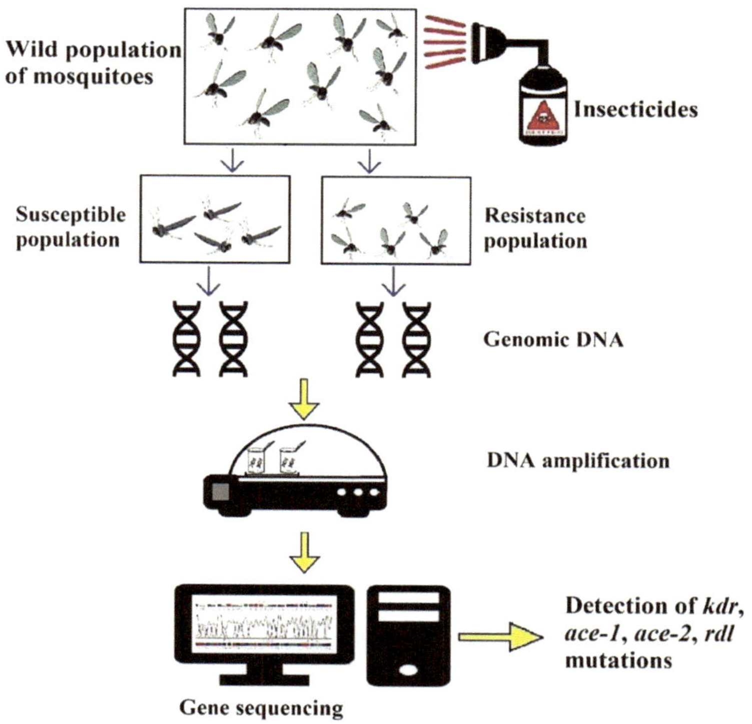

FIGURE 16.6 Molecular methods to identify target site mutations.

determine the point mutation. This approach is highly useful for detecting novel mutations in the target site. The introduction of modern techniques like amplification refractory mutation system (ARMS), and primer-introduced restriction analysis PCR (PIRA-PCR) may also contribute to the mutation detection in the insecticide target site.

16.6 CURRENT APPROACHES FOR INSECTICIDE RESISTANCE MANAGEMENT

Growing insecticide resistance in major mosquito vectors in India is a major concern these days. To overcome this problem a country-wide insecticide resistance screening is needed. Since most of the chemical insecticides are no longer effective in vector control management, new strategies may be incorporated, like mosaic, rotation, combination, or mixtures. In the mosaic approach, spatial alternating of two or more insecticides with various modes of action is applied. In this technique, different insecticides are used in adjacent areas within the same time period. Depending on the flight range of the mosquitoes, the mosaic method can be applied in various-sized mosquito breeding grounds even in a broad range like villages or cities of different districts. In rotation, temporal alternation of different types of insecticides takes place in a single habitat. Rotations must be performed at a reasonably high frequency because developing resistance is supposed to take a while to establish in a mosquito population, and the number of resistant individuals should decrease if an alternative insecticide is applied. Another strategy is the use of insecticides in combination or mixture in which multiple insecticides with a diverse mode of action are employed simultaneously. This strategy will ensure that mosquitoes will come in contact with different classes of insecticides. Mixtures are not currently available for control of *Anopheles* but can be proved to be useful in insecticide resistance management. In addition to all of the aforementioned approaches, a distinct set of rules that permit the assessment of the effectiveness of commonly used insecticides under specific local environmental conditions must be developed. Besides regular resistance monitoring, more investigation is required for the development of more efficient chemical substances with less harmful environmental affect. The agricultural industry should work with vector control authorities in an intersectoral manner to avoid employing insecticide classes that are frequently used for public health. Lastly, human awareness is the most crucial component of vector control. A healthy society can only be created by educating people about mosquito habitat, seasonal abundance, and control measures through an awareness campaign organized by the local government.

16.7 CONCLUSION

The ongoing rise and reemergence of mosquito-borne diseases has made controlling mosquito vectors extremely crucial. However, vector control programs are seriously endangered by the emergence of pesticide resistance in mosquitoes. In order to build an effective vector control strategy, insecticide resistance management should be taken into account. The continuous vector control campaign will be successful if mosquito vector susceptibility against conventionally used insecticides is regularly monitored in disease-endemic areas, and insecticide resistance is mapped in a region-specific way. The World Health Organization has initiated different missions to fight against mosquito-borne diseases, like the "Global Malaria Program (GMP)" for controlling and eliminating Malaria. The Indian government has also established a "National Framework for Malaria Elimination in India (2016–2030)" with the goal of eliminating malaria (zero endemic cases) in the country. In 1997, the Indian government signed the World Health Assembly Resolution for the Global Eradication of Lymphatic Filariasis. In National Health Policy 2002, 2015 was set as the goal year for the elimination of disease from India through annual Mass Drug Administration and home-based care of cases of lymphoedema. As these diseases are yet to be eliminated, it is being advocated that, in addition to the ongoing interventions, vector control must also be incorporated into integrated disease (lymphatic filariasis) management. This will automatically summon compulsory monitoring of the insecticide resistance development in mosquito vectors, with appropriate and judicious use of insecticides to manage the development of insecticide resistance issues in vector control.

BIBLIOGRAPHY

Abate, A., Getu, E., Wale, M., Hadis, M., Mekonen, W. (2020). Impact of bendiocarb 80% WP indoor residual spraying on insecticide resistance status of *Anopheles arabiensis. Ethiopian Journal of Science and Technology*, 13(3), 215–228.

Achee, N.L., Grieco, J.P., Vatandoost, H., Seixas, G., Pinto, J., Ching-Ng, L., Martins, A.J., Juntarajumnong, W., Corbel, V., Gouagna, C., David, J.P. (2019). Alternative strategies for mosquito-borne arbovirus control. *PloS Neglected Tropical Diseases*, 13(1), e0006822.

Akhir, M.A.M., Wajidi, M.F., Lavoué, S., Azzam, G., Jaafar, I.S., Awang Besar, N.A.U., Ishak, I.H. (2022). Knockdown resistance (kdr) gene of *Aedes aegypti* in Malaysia with the discovery of a novel regional specific point mutation A1007G. *Parasites and Vectors*, 15(1), 1–15.

Alout, H., Berthomieu, A., Hadjivassilis, A., Weill, M., (2007). A new amino-acid substitution in acetylcholinesterase 1 confers insecticide resistance to Culex pipiens mosquitoes from Cyprus. *Insect Biochemistry and Molecular Biology*, 37(1), 41–47.

Alout, H., Labbé, P., Berthomieu, A., Pasteur, N., Weill, M. (2009). Multiple duplications of the rare ace-1 mutation F290V in *Culex pipiens* natural populations. *Insect Biochemistry and Molecular Biology*, 39(12), 884–891.

Alout, H., Weill, M. (2008). Amino-acid substitutions in acetylcholinesterase 1 involved in insecticide resistance in mosquitoes. *Chemico-Biological Interactions*, 175(1–3), 138–141.

Amelia-Yap, Z.H., Tan, T.K., Prakash, B.K., Chen, C.D., Sofian-Azirun, M., Low, V.L. (2020). First report of Rdl mutant alleles in *Culex quinquefasciatus* (Diptera: Culicidae) in Malaysia. *Journal of Vector Borne Diseases*, 57(4), 359.

Anoopkumar, A.N., Aneesh, E.M. (2022). A critical assessment of mosquito control and the influence of climate change on mosquito-borne disease epidemics. *Environment, Development and Sustainability*, 24(6), 8900–8929.

Bharati, M., Saha, D. (2018). Multiple insecticide resistance mechanisms in primary dengue vector, *Aedes aegypti* (Linn.) from dengue endemic districts of sub-Himalayan West Bengal, India. *Plos One*, 13(9), e0203207.

Bharati, M., Saha, D. (2021). Insecticide resistance status and biochemical mechanisms involved in *Aedes* mosquitoes: A scoping review. *Asian Pacific Journal of Tropical Medicine*, 14(2), 52.

Bingham, G., Strode, C., Tran, L., Khoa, P.T., Jamet, H.P. (2011). Can piperonyl butoxide enhance the efficacy of pyrethroids against pyrethroid-resistant *Aedes aegypti*?s *Tropical Medicine & International Health*, 16(4), 492–500.

Brengues, C., Hawkes, N.J., Chandre, F., McCarroll, L., Duchon, S., Guillet, P., Manguin, S., Morgan, J.C.,Hemingway, J. (2003). Pyrethroid and DDT cross-resistance in *Aedes aegypti* is correlated with novel mutations in the voltage-gated sodium channel gene. *Medical and Veterinary Entomology*, 17(1), 87–94.

Burroughs, T., Najafi, M., Lemon, S.M.,Knobler, S.L. (2003). *The resistance phenomenon in microbes and infectious disease vectors: implications for human health and strategies for containment: workshop summary*. National Academies Press. Washington, DC.

Caminade, C., McIntyre, K.M., Jones, A.E. (2019). Impact of recent and future climate change on vector-borne diseases. *Annals of the New York Academy of Sciences*, 1436(1), 157–173.

Casida, J.E., Durkin, K.A. (2013). Neuroactive insecticides: targets, selectivity, resistance, and secondary effects. *Annual Review of Entomology*, 58, 99–117.

Chan, H.H., Zairi, J. (2013). Permethrin resistance in *Aedes albopictus* (Diptera: Culicidae) and associated fitness costs. *Journal of Medical Entomology*, 50(2), 362–370.

Charpentier, A., Menozzi, P., Marcel, V., Villatte, F., Fournier, D. (2000). A method to estimate acetylcholinesterase-active sites and turnover in insects. *Analytical Biochemistry*, 285(1), 76–81.

Chen, H., Li, K., Wang, X., Yang, X., Lin, Y., Cai, F., Zhong, W., Lin, C., Lin, Z., Ma, Y. (2016). First identification of kdr allele F1534S in VGSC gene and its association with resistance to pyrethroid insecticides in *Aedes albopictus* populations from Haikou City, Hainan Island, China. *Infectious Diseases of Poverty*, 5(3), 40–47.

Chen, M., Du, Y., Nomura, Y., Zhorov, B.S., Dong, K. (2020). Chronology of sodium channel mutations associated with pyrethroid resistance in *Aedes aegypti. Archives of Insect Biochemistry and Physiology*, 104(2), e21686.

Chen, M., Du, Y., Zhu, G., Takamatsu, G., Ihara, M., Matsuda, K., Zhorov, B.S., Dong, K. (2018). Action of six pyrethrins purified from the botanical insecticide pyrethrum on cockroach sodium channels expressed in Xenopus oocytes. *Pesticide Biochemistry and Physiology*, 151, 82–89.

Colovic, M.B., Krstic, D.Z., Lazarevic-Pasti, T.D., Bondzic, A.M., Vasic, V.M. (2013). Acetylcholinesterase inhibitors: pharmacology and toxicology. *Current Neuropharmacology*, 11(3), 315–335.

Dabiré, R.K., Namountougou, M., Diabaté, A., Soma, D.D., Bado, J., Toé, H.K., Bass, C., Combary, P. (2015). Correction: Distribution and frequency of kdr mutations within *Anopheles gambiae* populations and first report of the ace. 1 G119S mutation in *Anopheles arabiensis* from Burkina Faso (West Africa). *Plos One*, 10(11), e0141645.

Dahmana, H., Mediannikov, O. (2020). Mosquito-borne diseases emergence/resurgence and how to effectively control it biologically. *Pathogens*, 9(4), 310.

De Araújo, A.P., Paiva, M.H.S., Cabral, A.M., Cavalcanti, A.E.H.D., Pessoa, L.F.F., Diniz, D.F.A., Helvecio, E., da Silva, E.V.G., da Silva, N.M., Anastácio, D.B., Pontes, C. (2019). Screening *Aedes aegypti* (Diptera: Culicidae) populations from Pernambuco, Brazil for resistance to temephos, diflubenzuron, and cypermethrin and characterization of potential resistance mechanisms. *Journal of Insect Science*, 19(3), 16.

Delannay, C., Goindin, D., Kellaou, K., Ramdini, C., Gustave, J., Vega-Rúa, A. (2018). Multiple insecticide resistance in *Culex quinquefasciatus* populations from Guadeloupe (French West Indies) and associated mechanisms. *PloS One*, 13(6), e0199615.

Du, W., Awolola, T.S., Howell, P., Koekemoer, L.L., Brooke, B.D., Benedict, M.Q., Coetzee, M., Zheng, L. (2005). Independent mutations in the Rdl locus confer dieldrin resistance to *Anopheles gambiae* and *An. arabiensis*. *Insect Molecular Biology*, 14(2), 179–183.

Dusfour, I., Vontas, J., David, J.P., Weetman, D., Fonseca, D.M., Corbel, V., Raghavendra, K., Coulibaly, M.B., Martins, A.J., Kasai, S., Chandre, F. (2019). Management of insecticide resistance in the major Aedes vectors of arboviruses: Advances and challenges. *PloS Neglected Tropical Diseases*, 13(10), e0007615.

Dusfour, I., Zorrilla, P., Guidez, A., Issaly, J., Girod, R., Guillaumot, L., Robello, C., Strode, C. (2015). Deltamethrin resistance mechanisms in *Aedes aegypti* populations from three French overseas territories worldwide. *PloS Neglected Tropical Diseases*, 9(11), e0004226.

Dykes, C.L., Das, M.K., Eapen, A., Batra, C.P., Ghosh, S.K., Vijayan, V.A., Mishra, S., Singh, O.P. (2016). Knockdown resistance (*kdr*) mutations in Indian *Anopheles stephensi* (Diptera: Culicidae) populations. *Journal of Medical Entomology*, 53(2), 315–320.

Dykes, C.L., Kushwah, R.B.S., Das, M.K., Sharma, S.N., Bhatt, R.M., Veer, V., Agrawal, O.P., Adak, T., Singh, O.P. (2015). Knockdown resistance (*kdr*) mutations in Indian *Anopheles culicifacies* populations. *Parasites and Vectors*, 8(1), 1–7.

Federici, B.A., Park, H.W., Bideshi, D.K., Wirth, M.C., Johnson, J.J. (2003). Recombinant bacteria for mosquito control. *Journal of Experimental Biology*, 206(21), 3877–3885.

Ffrench-Constant, R.H., Anthony, N., Aronstein, K., Rocheleau, T., Stilwell, G. (2000). Cyclodiene insecticide resistance: from molecular to population genetics. *Annual Review of Entomology*, 45(1), 449–466.

Ffrench-Constant, R.H., Williamson, M.S., Davies, T.E., Bass, C. (2016). Ion channels as insecticide targets. *Journal of Neurogenetics*, 30(3–4), 163–177.

Filho, W.L., Scheday, S., Boenecke, J., Gogoi, A., Maharaj, A., Korovou, S. (2019). Climate change, health and mosquito-borne diseases: Trends and implications to the pacific region. *International Journal of Environmental Research and Public Health*, 16(24), 5114.

Franklinos, L.H., Jones, K.E., Redding, D.W., Abubakar, I. (2019). The effect of global change on mosquito-borne disease. *The Lancet Infectious Diseases*, 19(9), e302–e312.

Garcia, G.D.A., David, M.R., Martins, A.D.J., Maciel-de-Freitas, R., Linss, J.G.B., Araújo, S.C., Lima, J.B.P., Valle, D. (2018). The impact of insecticide applications on the dynamics of resistance: The case of four *Aedes aegypti* populations from different Brazilian regions. *PloS Neglected Tropical Diseases*, 12(2), e0006227.

Haddi, K., Tomé, H.V., Du, Y., Valbon, W.R., Nomura, Y., Martins, G.F., Dong, K., Oliveira, E.E. (2017). Detection of a new pyrethroid resistance mutation (V410L) in the sodium channel of *Aedes aegypti*: a potential challenge for mosquito control. *Scientific Reports*, 7(1), 1–9.

Hall, L.C., Spierer, P. (1986). The Ace locus of Drosophila melanogaster: structural gene for acetylcholinesterase with an unusual 5′ leader. *The EMBO Journal*, 5(11), 2949–2954.

Hemingway, J., Ranson, H. (2000). Insecticide resistance in insect vectors of human disease. *Annual Review of Entomology*, 45(1), 371–391.

Hemingway, J. (2000). The molecular basis of two contrasting metabolic mechanisms of insecticide resistance. *Insect Biochemistry and Molecular Biology*, 30(11), 1009–1015.

Hirata, K., Jouraku, A., Kuwazaki, S., Shimomura, H., Iwasa, T. (2017). Studies on *Aphis gossypii* cytochrome P450s CYP6CY22 and CYP6CY13 using an in vitro system. *Journal of Pesticide Science*, 42(3), D17–006.

Hirata, K., Komagata, O., Itokawa, K., Yamamoto, A., Tomita, T., Kasai, S. (2014). A single crossing-over event in voltage-sensitive Na+ channel genes may cause critical failure of dengue mosquito control by insecticides. *PloS Neglected Tropical Diseases*, 8(8), e3085.

Hoti, S.L., Vasuki, V., Jambulingam, P., Sahu, S.S. (2006). *Kdr* allele-based PCR assay for detection of resistance to DDT in *Anopheles culicifacies* sensulato Giles population from Malkangiri District, Orissa, India. *Current Science*, 91(5), 658–661.

IRAC (Insecticide Resistance Action Committee). (2022). The IRAC Mode of Action Classification Online. https://irac-online.org/mode-of-action/classification-online/

Karunaratne, S.H.P.P., De Silva, W.A.P.P., Weeraratne, T.C., Surendran, S.N. (2018). Insecticide resistance in mosquitoes: development, mechanisms and monitoring. *Ceylon Journal of Science*, 47(4), 299–309.

Karunaratne, S.H.P.P., Hawkes, N.J., Perera, M.D.B., Ranson, H., Hemingway, J. (2007). Mutated sodium channel genes and elevated monooxygenases are found in pyrethroid resistant populations of Sri Lankan malaria vectors. *Pesticide Biochemistry and Physiology*, 88(1), 108–113.

Kasai, S., Caputo, B., Tsunoda, T., Cuong, T.C., Maekawa, Y., Lam-Phua, S.G., Pichler, V., Itokawa, K., Murota, K., Komagata, O., Yoshida, C. (2019). First detection of a Vssc allele V1016G conferring a high level of insecticide resistance in *Aedes albopictus* collected from Europe (Italy) and Asia (Vietnam), 2016: a new emerging threat to controlling arboviral diseases. *Eurosurveillance*, 24(5), 1700847.

Khan, H.A.A., Akram, W., Shehzad, K., Shaalan, E.A. (2011). First report of field evolved resistance to agrochemicals in dengue mosquito, *Aedes albopictus* (Diptera: Culicidae), from Pakistan. *Parasites and Vectors*, *4*(1), 1–11.

Kothera, L., Phan, J., Ghallab, E., Delorey, M., Clark, R., Savage, H.M. (2019). Using targeted next-generation sequencing to characterize genetic differences associated with insecticide resistance in *Culex quinquefasciatus* populations from the southern US. *PloS One*, *14*(7), e0218397.

Kushwah, R.B.S., Dykes, C.L., Kapoor, N., Adak, T., Singh, O.P. (2015). Pyrethroid-resistance and presence of two knockdown resistance (*kdr*) mutations, F1534C and a novel mutation T1520I, in Indian *Aedes aegypti*. *PloS Neglected Tropical Diseases*, 9(1), e3332.

Kushwah, R.B.S., Kaur, T., Dykes, C.L., Ravi Kumar, H., Kapoor, N., Singh, O.P. (2020). A new knockdown resistance (*kdr*) mutation, F1534L, in the voltage-gated sodium channel of *Aedes aegypti*, co-occurring with F1534C, S989P and V1016G. *Parasites and Vectors*, 13(1), 1–12.

Li, T., Liu, N. (2014). Inheritance of permethrin resistance in *Culex quinquefasciatus*. *Journal of Medical Entomology*, 47(6), 1127–1134.

Liebman, K.A., Pinto, J., Valle, J., Palomino, M., Vizcaino, L., Brogdon, W., Lenhart, A., (2015). Novel mutations on the ace-1 gene of the malaria vector *Anopheles albimanus* provide evidence for balancing selection in an area of high insecticide resistance in Peru. *Malaria Journal*, 14(1), 1–10.

Liu, H., Xie, L., Cheng, P., Xu, J., Huang, X., Wang, H., Song, X., Liu, L., Wang, H., Kou, J., Yan, G. (2019). Trends in insecticide resistance in *Culex pipiens* pallens over 20 years in Shandong, China. *Parasites and Vectors*, 12(1), 1–9.

Liu, N. (2015). Insecticide resistance in mosquitoes: impact, mechanisms, and research directions. *Annual Review of Entomology*, 60(1), 537–559.

Liu, N., Liu, H., Zhu, F., Zhang, L. (2007). Differential expression of genes in pyrethroid resistant and susceptible mosquitoes, *Culex quinquefasciatus* (S.). *Gene*, 394(1–2), 61–68.

Liu, Z., Williamson, M.S., Lansdell, S.J., Denholm, I., Han, Z., Millar, N.S. (2005). A nicotinic acetylcholine receptor mutation conferring target-site resistance to imidacloprid*in Nilaparvata lugens* (brown planthopper). *Proceedings of the National Academy of Sciences*, 102(24), 8420–8425.

Low, V.L., Vinnie-Siow, W.Y., Lim, Y.A.L., Tan, T.K., Leong, C.S., Chen, C.D., Azidah, A.A., Sofian-Azirun, M. (2015). Research Note First molecular genotyping of A302S mutation in the gamma aminobutyric acid (GABA) receptor in *Aedes albopictus* from Malaysia. *Tropical Biomedicine*, 32(3), 554–556.

Lucas, K.J., Bales, R.B., McCoy, K., Weldon, C. (2020). Oxidase, esterase, and *KDR*-associated pyrethroid resistance in *Culex quinquefasciatus* field collections of Collier County, Florida. *Journal of the American Mosquito Control Association*, 36(1), 22–32.

Madgwick, P.G., Kanitz, R. (2022). Modelling new insecticide-treated bed nets for malaria-vector control: how to strategically manage resistance? *Malaria Journal*, 21(1), 1–20.

Maestre-Serrano, R., Lara-Cobos, J., Gomez-Camargo, D., Ponce-Garcia, G., Pareja-Loaiza, P., Flores, A.E. (2020). Susceptibility to Pyrethroids and the First Report of L1014F *kdr* Mutation in *Culex quinquefasciatus* (Diptera: Culicidae) in Colombia. *Journal of Medical Entomology*, 57(6), 1830–1834.

Malcolm, C.A., Bourguet, D., Ascolillo, A., Rooker, S.J., Garvey, C.F., Hall, L.M.C., Pasteur, N., Raymond, M. (1998). A sex-linked Ace gene, not linked to insensitive acetylcholinesterase-mediated insecticide resistance in *Culex pipiens*. *Insect Molecular Biology*, 7(2), 107–120.

Marcombe, S., Shimell, K., Savage, R., Howlett, E., Luangamath, P., Nilaxay, S., Vungkyly, V., Baby, A., King, M., Clarke, J., Jeffries, C. (2022). Detection of pyrethroid resistance mutations and intron variants in the voltage-gated sodium channel of *Aedes (Stegomyia) aegypti* and *Aedes (Stegomyia) albopictus* mosquitoes from Lao People's Democratic Republic. *Medical and Veterinary Entomology*, 36(1), 1–11.

Marcombe, S., Thammavong, P., Luangamath, P., Chonephetsarath, S., Phommavanh, N., Lakeomany, K., Nilaxay, S., Rahmani, Z., Saverton, P.J., Abdullateef, O.H., Forward, J. (2020). Malaria and dengue mosquito vectors from Lao PDR show a lack of the rdl mutant allele responsible for Cyclodiene insecticide resistance. *Journal of Medical Entomology*, 57(3), 815–823.

Martinez-Torres, D., Chandre, F., Williamson, M.S., Darriet, F., Berge, J.B., Devonshire, A. L., Guillet, P., Pasteur, N., Pauron, D. (1998). Molecular characterization of pyrethroid knockdown resistance (kdr) in the major malaria vector *Anopheles gambiae* s.s. *Insect Molecular Biology*, 7(2), 179–184.

Martins, W.F.S., Subramaniam, K., Steen, K., Mawejje, H., Liloglou, T., Donnelly, M.J., Wilding, C.S. (2017). Detection and quantitation of copy number variation in the voltage-gated sodium channel gene of the mosquito *Culex quinquefasciatus*. *Scientific Reports*, *7*(1), 1–10.

Matsuo, N. (2019). Discovery and development of pyrethroid insecticides. *Proceedings of the Japan Academy, Series B*, 95(7), 378–400.

McCart, C., Buckling, A. (2005). DDT resistance in flies carries no cost. *Current Biology*, 15(15), R587–R589.

Melo Costa, M., Campos, K.B., Brito, L.P., Roux, E., Melo Rodovalho, C., Bellinato, D.F., Lima, J.B.P., Martins, A. J. (2020). Kdr genotyping in *Aedes aegypti* from Brazil on a nation-wide scale from 2017 to 2018. *Scientific Reports*, 10(1), 1–12.

Misra, B.R., Gore, M. (2015). Malathion resistance status and mutations in acetylcholinesterase gene (Ace) in Japanese encephalitis and filariasis vectors from endemic area in India. *Journal of Medical Entomology*, 52(3), 442–446.

Modak, M.P., Saha, D. (2022). First report of F1534C kdr mutation in deltamethrin resistant *Aedes albopictus* from northern part of West Bengal, India. *Scientific Reports*, 12(1), 13653.

Mori, A., Tomita, T., Hidoh, O., Kono, Y., Severson, D.W. (2001). Comparative linkage map development and identification of an autosomal locus for insensitive acetylcholinesterase-mediated insecticide resistance in *Culex tritaeniorhynchus*. *Insect Molecular Biology*, 10(3), 197–203.

Murcia, O., Henríquez, B., Castro, A., Koo, S., Young, J., Márquez, R., Pérez, D., Cáceres, L., Valderrama, A. (2019). Presence of the point mutations Val1016Gly in the voltage-gated sodium channel detected in a single mosquito from Panama. *Parasites and Vectors*, 12(1), 1–7.

Mustafa, M.S., Rasotgi, V., Jain, S., Gupta, V. (2015). Discovery of fifth serotype of dengue virus (DENV-5): A new public health dilemma in dengue control. *Medical Journal Armed Forces India*, 71(1), 67–70.

Muthusamy, R., Shivakumar, M.S. (2015a). Susceptibility status of *Aedes aegypti* (L.)(Diptera: Culicidae) to temephos from three districts of Tamil Nadu, India. *Journal of Vector Borne Diseases*, 52(2), 159.

Muthusamy, R., Shivakumar, M.S. (2015b). Involvement of metabolic resistance and F1534C *kdr* mutation in the pyrethroid resistance mechanisms of *Aedes aegypti* in India. *Acta Tropica*, 148, 137–141.

Nabeshima, T., Mori, A., Kozaki, T., Iwata, Y., Hidoh, O., Harada, S., Kasai, S., Severson, D. W., Kono, Y., Tomita, T. (2004). An amino acid substitution attributable to insecticide-insensitivity of acetylcholinesterase in a Japanese encephalitis vector mosquito, *Culex tritaeniorhynchus*. *Biochemical and Biophysical Research Communications*, 313(3), 794–801.

Narahashi, T. (1996). Neuronal ion channels as the target sites of insecticides. *Pharmacology & Toxicology*, 79(1), 1–14.

Naw, H., Võ, T.C., Lê, H.G., Kang, J.M., Mya, Y.Y., Myint, M.K., Kim, T.S., Shin, H.J., Na, B. K. (2022). Knockdown resistance mutations in the voltage-gated sodium channel of *Aedes aegypti* (Diptera: Culicidae) in Myanmar. *Insects*, 13(4), 322.

Norris, E., Johnson, J., Gross, A., Bartholomay, L., Coats, J. (2018). Plant essential oils enhance diverse pyrethroids against multiple strains of mosquitoes and inhibit detoxification enzyme processes. *Insects*, 9(4), 132.

National Center for Vector Borne Diseases Control (2017). Compendium on entomological surveillance & vector control in India. https://ncvbdc.mohfw.gov.in/WriteReadData/l892s/Compendium-Entomological-Surveillance&Vector-Control-India.pdf

National Center for Vector Borne Diseases Control (2022). Indoor Residual Spray. https://ncvbdc.mohfw.gov.in/WriteReadData/l892s/Indoor%20Residual%20Spray%20-%205.doc

Oliveira, E.E., Du, Y., Nomura, Y., Dong, K. (2013). A residue in the transmembrane segment 6 of domain I in insect and mammalian sodium channels regulate differential sensitivities to pyrethroid insecticides. *Neuro Toxicology*, 38, 42–50.

Ordentlich, A., Barak, D., Kronman, C., Flashner, Y., Leitner, M., Segall, Y., Ariel, N., Cohen, S., Velan, B., Shafferman, A. (1993). Dissection of the human acetylcholinesterase active center determinants of substrate specificity. Identification of residues constituting the anionic site, the hydrophobic site, and the acyl pocket. *Journal of Biological Chemistry*, 268(23), 17083–17095.

O'Reilly, A.O., Khambay, B.P., Williamson, M.S., Field, L.M., Wallace, B.A., Davies, T. E. (2006). Modelling insecticide-binding sites in the voltage-gated sodium channel. *Biochemical Journal*, 396(2), 255–263.

Pichler, V., Mancini, E., Micocci, M., Calzetta, M., Arnoldi, D., Rizzoli, A., Lencioni, V., Paoli, F., Bellini, R., Veronesi, R., Martini, S. (2021). A novel allele specific polymerase chain reaction (AS-PCR) assay to detect the V1016G knockdown resistance mutation confirms its widespread presence in *Aedes albopictus* populations from Italy. *Insects*, 12(1), 79.

Rai, P., Bharati, M., Saha, D. (2019). Assessment of insecticide resistance in *Culex quinquefasciatus* Say with first report on the presence of L1014F mutation from northern districts of West Bengal, India. *International Journal of Tropical Insect Science*, 39(4), 301–309.

Ranson, H., Jensen, B., Vulule, J.M., Wang, X., Hemingway, J., Collins, F.H. (2000). Identification of a point mutation in the voltage-gated sodium channel gene of Kenyan *Anopheles gambiae* associated with resistance to DDT and pyrethroids. *Insect Molecular Biology*, 9(5), 491–497.

Ren, X., Han, Z., Wang, Y. (2002). Mechanisms of monocrotophos resistance in cotton bollworm, Helicoverpa armigera (Hübner). *Archives of Insect Biochemistry and Physiology: Published in Collaboration with the Entomological Society of America*, 51(3), 103–110.

Rocklöv, J., Dubrow, R. (2020). Climate change: an enduring challenge for vector-borne disease prevention and control. *Nature Immunology*, 21(5), 479–483.

Saavedra-Rodriguez, K., Urdaneta-Marquez, L., Rajatileka, S., Moulton, M., Flores, A.E., Fernandez-Salas, I., Bisset, J., Rodriguez, M., McCall, P.J., Donnelly, M.J., Ranson, H. (2007). A mutation in the voltage-gated sodium channel gene associated with pyrethroid resistance in Latin American *Aedes aegypti. Insect Molecular Biology*, 16(6), 785–798.

Saha, P., Chatterjee, M., Ballav, S., Chowdhury, A., Basu, N., Maji, A.K. (2019). Prevalence of *kdr* mutations and insecticide susceptibility among natural population of *Aedes aegypti* in West Bengal. *PloS One*, 14(4), e0215541.

Sarkar, M., Borkotoki, A., Baruah, I., Bhattacharyya, I.K., Srivastava, R.B. (2009). Molecular analysis of knock down resistance (*kdr*) mutation and distribution of *kdr* genotypes in a wild population of *Culex quinquefasciatus* from India. *Tropical Medicine and International Health*, 14(9), 1097–1104.

Sarwar, M., Salman, M. (2015). Insecticides resistance in insect pests or vectors and development of novel strategies to combat its evolution. *International Journal of Bioinformatics and Biomedical Engineering*, 1(3), 344–351.

Silva, A.P.B., Santos, J.M.M., Martins, A.J. (2014). Mutations in the voltage-gated sodium channel gene of anophelines and their association with resistance to pyrethroids–a review. *Parasites and Vectors*, *7*(1), 1–14.

Silver, K.S., Song, W., Nomura, Y., Salgado, V.L., Dong, K. (2010). Mechanism of action of sodium channel blocker insecticides (SCBIs) on insect sodium channels. *Pesticide Biochemistry and Physiology*, 97(2), 87–92.

Singh, O.P., Bali, P., Hemingway, J., Subbarao, S.K., Dash, A.P.,Adak, T. (2009). PCR-based methods for the detection of L1014 *kdr* mutation in *Anopheles culicifacies* sensulato. *Malaria Journal*, 8(1), 1–8.

Singh, O.P., Dykes, C.L., Das, M.K., Pradhan, S., Bhatt, R.M., Agrawal, O.P., Adak, T. (2010). Presence of two alternative *kdr*-like mutations, L1014F and L1014S, and a novel mutation, V1010L, in the voltage gated Na+ channel of *Anopheles culicifacies* from Orissa, India. *Malaria Journal*, *9*(1), 1–6.

Singh, O.P., Dykes, C.L., Lather, M., Agrawal, O.P., Adak, T. (2011). Knockdown resistance (*kdr*)-like mutations in the voltage-gated sodium channel of a malaria vector *Anopheles stephensi* and PCR assays for their detection. *Malaria Journal*, 10(1), 1–7.

Singh, O.P., Dykes, C.L., Sharma, G., Das, M.K. (2015). L1014F-*kdr* mutation in Indian *Anopheles subpictus* (Diptera: Culicidae) arising from two alternative transversions in the voltage-gated sodium channel and a single PIRA-PCR for their detection. *Journal of Medical Entomology*, 52(1), 24–27.

Singh, S. (2020). Solid waste management in urban india: imperatives for improvement. *ORF Occasional Paper*, 283.

Sugiura, M., Kimoto, F., Itokawa, K., Kasai, S. (2021). Novel CONCOMITANT mutations L932F and I936V in the voltage-gated sodium channel and its association with pyrethroid resistance in *Culex quinquefasciatus* (Diptera: Culicidae). *Journal of Medical Entomology*, 58(2), 798–806.

Talipouo, A., Mavridis, K., Nchoutpouen, E., Djiappi-Tchamen, B., Fotakis, E.A., Kopya, E., Bamou, R., Kekeunou, S., Awono-Ambene, P., Balabanidou, V., Balaska, S. (2021). High insecticide resistance mediated by different mechanisms in *Culex quinquefasciatus* populations from the city of Yaoundé, Cameroon. *Scientific Reports*, 11(1), 1–11.

Tan, W.L., Li, C.X., Wang, Z.M., Liu, M.D., Dong, Y.D., Feng, X.Y., Wu, Z.M., Guo, X. X., Xing, D., Zhang, Y.M., Wang, Z.C. (2014). First detection of multiple knockdown resistance (kdr)-like mutations in voltage-gated sodium channel using three new genotyping methods in *Anopheles sinensis* from Guangxi Province, China. *Journal of Medical Entomology*, 49(5), 1012–1020.

Tantely, M.L., Tortosa, P., Alout, H., Berticat, C., Berthomieu, A., Rutee, A., Dehecq, J.S., Makoundou, P., Labbé, P., Pasteur, N., Weill, M. (2010). Insecticide resistance in *Culex pipiens quinquefasciatus* and *Aedes albopictus* mosquitoes from La Reunion Island. *Insect Biochemistry and Molecular Biology*, 40(4), 317–324.

Thompson, M., Shotkoski, F., Ffrench-Constant, R. (1993). Cloning and sequencing of the cyclodiene insecticide resistance gene from the yellow fever mosquito *Aedes aegypti*: conservation of the gene and resistance associated mutation with *Drosophila*. *FEBS Letters*, 325(3), 187–190.

Tmimi, F.Z., Faraj, C., Bkhache, M., Mounaji, K., Failloux, A.B. (2018). Insecticide resistance and target site mutations (G119S ace-1 and L1014F *kdr*) of *Culex pipiens* in Morocco. *Parasites and Vectors*, 11(1), 1–9.

Toé, H.K., Zongo, S., Guelbeogo, M.W., Kamgang, B., Viana, M., Tapsoba, M., Sanou, A., Traoré, A., McCall, P.J., Sagnon, N.F. (2022). Multiple insecticide resistance and first evidence of V410L kdr mutation in *Aedes (Stegomyia) aegypti* (Linnaeus) from Burkina Faso. *Medical and Veterinary Entomology*, 36(3), 309–319.

Tomita, T., Hidoh, O., Kono, Y. (2000). Absence of protein polymorphism attributable to insecticide-insensitivity of acetylcholinesterase in the green rice leafhopper, *Nephotettix cincticeps*. *Insect Biochemistry and Molecular Biology*, 30(4), 325–333.

Ureña, E., Guillem-Amat, A., Couso-Ferrer, F., Beroiz, B., Perera, N., López-Errasquín, E., Castañera, P., Ortego, F., Hernández-Crespo, P. (2019). Multiple mutations in the nicotinic acetylcholine receptor Ccα6 gene associated with resistance to spinosad in medfly. *Scientific Reports*, 9(1), 1–12.

Viswan, A., Pushpalatha, E. (2019). Biochemical and molecular analysis of acetylcholine esterase to rule out Organophosphorus detoxification level in field populations of *Culex quiquefasciatus* say. *Journal of Communicable Diseases*, 51(2), 22–27.

Von Stein, R.T., Silver, K.S., Soderlund, D.M. (2013). Indoxacarb, metaflumizone, and other sodium channel inhibitor insecticides: mechanism and site of action on mammalian voltage-gated sodium channels. *Pesticide Biochemistry and Physiology*, 106(3), 101–112.

Vontas, J., Kioulos, E., Pavlidi, N., Morou, E., Della Torre, A., Ranson, H. (2012). Insecticide resistance in the major dengue vectors *Aedes albopictus* and *Aedes aegypti*. *Pesticide Biochemistry and Physiology*, 104(2), 126–131.

Wang, L., Soto, A., Remue, L., Rosales Rosas, A.L., De Coninck, L., Verwimp, S., Bouckaert, J., Vanwinkel, M., Matthijnssens, J., Delang, L. (2022). First Report of Mutations Associated With Pyrethroid (L1014F) and Organophosphate (G119S) Resistance in Belgian *Culex* (Diptera: Culicidae) Mosquitoes. *Journal of Medical Entomology*, 59(6), 2072–2079.

Weill, M., Malcolm, C., Chandre, F., Mogensen, K., Berthomieu, A., Marquine, M., Raymond, M. (2004). The unique mutation in *ace-1* giving high insecticide resistance is easily detectable in mosquito vectors. *Insect Molecular Biology*, 13(1), 1–7.

Williamson, M.S., Martinez-Torres, D., Hick, C.A., Devonshire, A.L. (1996). Identification of mutations in the housefly para-type sodium channel gene associated with knockdown resistance (*kdr*) to pyrethroid insecticides. *Molecular and General Genetics MGG*, 252(1), 51–60.

Wondji, C.S., Priyanka De Silva, W.A.P., Hemingway, J., Ranson, H., Parakrama Karunaratne, S.H.P. (2008). Characterization of knockdown resistance in DDT-and pyrethroid-resistant *Culex quinquefasciatus* populations from Sri Lanka. *Tropical Medicine and International Health*, 13(4), 548–555.

World Health Organization (2020). Vector-borne Diseases. https://www.who.int/news-room/fact-sheets/detail/vector-borne-diseases

World Health Organization. (2022a). Dengue and severe dengue. https://www.who.int/news-room/fact-sheets/detail/dengue-and-severe-dengue

World Health Organization. (2022b). Malaria Fact-sheets. News-room/fact-sheets/detail/malaria.

World Health Organization. (2017). Global Vector Control Response 2017-2030. https://apps.who.int/iris/bitstream/handle/10665/259205/9789241512978-eng.pdf

Yang, C., Huang, Z., Li, M., Feng, X., Qiu, X. (2017). RDL mutations predict multiple insecticide resistance in *Anopheles sinensis* in Guangxi, China. *Malaria Journal*, 16(1), 1–13.

Yanola, J., Chamnanya, S., Lumjuan, N., Somboon, P. (2015). Insecticides resistance in the *Culex quinquefasciatus* populations from northern Thailand and possible resistance mechanisms. *Acta Tropica*, 149, 232–238.

Yu, S. (2014). *The Toxicology and Biochemistry of Insecticides, Second Edition*. Hoboken: CRC Press.
Zhorov, B.S., Dong, K. (2017). Elucidation of pyrethroid and DDT receptor sites in the voltage-gated sodium channel. *Neurotoxicology*, 60, 171–177.
Zhou, X., Yang, C., Liu, N., Li, M., Tong, Y., Zeng, X., Qiu, X. (2019). Knockdown resistance (*kdr*) mutations within seventeen field populations of *Aedes albopictus* from Beijing China: first report of a novel V1016G mutation and evolutionary origins of *kdr* haplotypes. *Parasites and Vectors*, 12(1), 1–16.
Zhu, F., Lavine, L., O'Neal, S., Lavine, M., Foss, C., Walsh, D. (2016). Insecticide resistance and management strategies in urban ecosystems. *Insects*, 7(1), 2.
Zuharah, W.F., Sufian, M. (2021). The discovery of a novel knockdown resistance (kdr) mutation A1007G on *Aedes aegypti* (Diptera: Culicidae) from Malaysia. *Scientific Reports*, 11(1), 1–9.

17 Effect of Insecticide Resistance on Biology and Reproductive Fitness of Mosquito Vectors

Dhiraj Saha, Nilu Limboo, Prerana Bhujel, Prapti Das, Abhisekh Subba, and Debayan Raha
Department of Zoology, University of North Bengal, Siliguri, Darjeeling, India

Minu Bharati
S.B.S Government College, Dakshin Dinajpur, India

B.K. Tyagi
Chandigarh University, Mohali, India

17.1 INTRODUCTION

Resistance against insecticide is a pre-adaptive trait resulting from random genetic mutation. This trait favors the survivability and reproducibility of mosquitoes under insecticide resistance, which is a major hindrance in vector control and management programs. Mechanisms behind such resistance development is being thoroughly studied worldwide (Hemingway & Ranson, 2000, Liu, 2015, Martinez-Torres et al., 1998, Weill et al., 2004, Du et al., 2005), but there have been relatively few studies on the fitness of vectors (Martins et al., 2012, Platt et al., 2015), despite the fact that such data are essential for understanding how populations develop resistance to insecticides. Resistance in mosquito vectors has been mainly associated with metabolic resistance and target site modification along with cuticular thickening and behavioral modification (Figure 17.1).

Metabolic resistance results from an increase in detoxifying enzyme activity, i.e., overexpression of detoxifying genes such as Cytochrome P450 monooxygenases, Esterases, and Glutathione-S-Transferases (GSTs), which inactivate insecticides by changing it into less toxic form and smaller molecules which can easily be eliminated from the body (Hemingway and Ranson, 2000, Liu, 2015). While target site modification results from mutations in genes encoding protein binding sites on which the insecticides would act (*kdr*, *Ace*, and *Rdl*) (Martinez-Torres et al., 1998, Weill et al., 2004, Du et al., 2005). However, to acquire this ability to resist insecticide, mosquito vectors have to compromise several fitness traits as most of their energy is invested in survival and reproduction. Hatchability, immature stage survivability, and development duration, pupa and adult emergence rate, adult longevity, sex ratio, fecundity, fertility, and male competitiveness are some of the life history traits studied in resistant mosquitoes. However, these studies on fitness traits and insecticide resistance are mainly done under laboratory conditions or semi-field conditions as it is difficult to monitor all the fitness parameters in field populations. As fitness is dependent on the environment and under

DOI: 10.1201/9781003326991-25

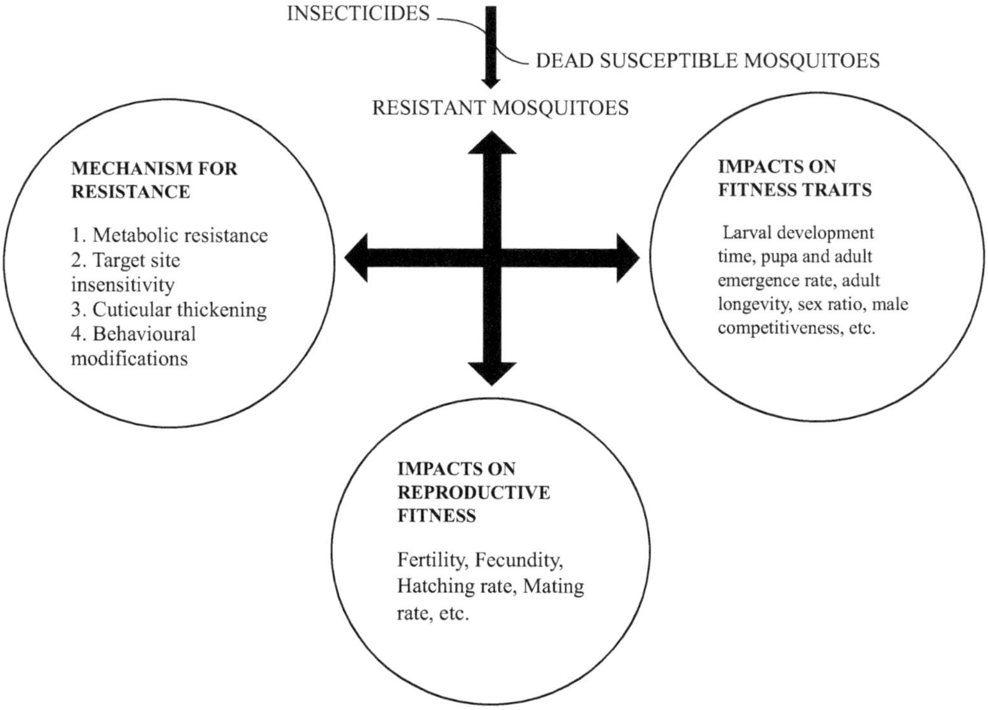

FIGURE 17.1 Flow diagram showing the impact of insecticide resistance on mosquitoes.

the laboratory conditions the number of biotic and abiotic stressors are more limited than the field conditions, so the choice of conditions is to be considered carefully (Hardstone et al., 2009). Fitness studies in the laboratory need to have a well-representative collection from the field. A laboratory lineage of susceptible population should be reared in parallel, as a control of experimental conditions. The knowledge of alterations in physiology and reproductive biology is generally achieved by comparing life trait parameters between susceptible and resistant individuals (Belinato and Martins, 2016). Understanding the association between resistant genotypes and their fitness traits is important to the development of more effective resistance management programs to manage vector-borne diseases. Therefore, in this chapter we shall discuss insecticide resistance and associated fitness traits, as well as the effect of insecticide selection pressure on the detoxifying enzyme genes.

17.2 EFFECT OF INSECTICIDE SELECTION ON THE DEVELOPMENTAL AND REPRODUCTIVE FITNESS OF MOSQUITO

The most crucial criterion for describing the selection pressure brought on by the use of insecticides is the biological fitness of the target vector mosquitoes since it frequently influences the success of vector control strategies (Campanhola et al., 1991). Many resistance management tactics, such as rotation of pesticides assume that resistance results in a fitness cost on mosquitoes, such that when the insecticidal pressure is withdrawn, insecticide susceptibility is reverted in a short period of time, however little is known about the fitness of disease-transmitting vectors (Platt et al., 2015).

It is observed that even populations of the same species can vary in fitness costs over time and geography. The fitness cost can be evaluated by assessing the specific life history components or by tracking the resistance levels or resistance allele frequencies through time.

17.2.1 Life History Parameters

Studying the effect of insecticide on the fitness component of the mosquito provides us with baseline information of the vector's population dynamics after it develops resistance. Some of the crucial life trait parameters used to assess the fitness of the target vectors are as follows:

1. Developmental parameters
   (i) *Immature stage developmental time*: One of the important criteria used to assess the fitness components of mosquito populations is the time it takes for the immature stages to develop. There are six immature stages of mosquitoes which include an egg, four larval instars, and a pupal stage. As the growth and survival of mosquito larvae are influenced by a variety of environmental cues, including temperature, nutrition, and larval density (Clements, 1992), a good and clear knowledge of the immature duration of the vector is essential to understand the trend of population fluctuation of immature mosquitoes as well as adults.

   Most studies (Rahim et al., 2017; Martins et al., 2012) report the disadvantage in the fitness of resistant populations when continuous insecticidal exposure occurs whereas some studies show no variations in the time of development from larva to pupa (Plernsub et al., 2013). The occurrence of such variation in development could be due to interference of insecticide on the production of growth regulatory hormones or alteration in the regulatory mechanism involved (Venkatesan et al., 2014).

   (ii) *Hatchability*: Egg hatchability information of any mosquito species is essential for estimating the survival of the population, which is a key vector characteristic influencing disease transmission. Apart from this, it also provides information about the viability of the eggs and their tolerance to insecticide pressure.

   (iii) *Longevity*: Mostly environmental and/or climatic conditions determine the mosquito vector's adult lifespan. Changes in the mosquito's physiological state and changes in environmental factors like temperature, humidity, and the availability of food might affect the mosquito's capacity to survive (Waldock et al., 2013). Given the increased likelihood of infective bites to hosts, vector longevity is one of the most important parameters in disease transmission (McCarroll and Hemingway, 2002; Aly et al., 2009). The chances of disease transmission and its spreading rise gradually with the age of a mosquito because the longer it lives, the greater number of blood meals it takes (Tchouakui et al., 2018). Insecticide resistance is generally thought to increase the longevity of resistant vectors, thereby increasing the infectiousness of parasites and threatening vector control.

   (iv) *Sex ratio*: The role of male and female mosquitoes are different in the disease-transmitting process, thus the knowledge of the sex should be taken into account for a better understanding of their contribution to maintaining the population. Most of the time, the females are considered the main culprit behind disease transmission as they require blood meal for egg production, and thus, in the process of taking blood from the host, they transmit disease. However, the hormone 20-hydroxyecdysone (20E), which primarily drives both the induction of egg laying and further mating, is male-induced (Gabrieli et al., 2014).

2. Reproductive Parameters
   (i) *Blood-Feeding*: Blood-feeding arthropods including mosquitoes use their olfactory receptors for host recognition and location. Given that disease transmission and the blood-feeding behavior of mosquitoes are closely related, it is important to investigate the role of insecticides in this behavior.

   (ii) Fecundity: Fecundity is the predominant fitness feature since it specifies the efficiency of conversion of the blood meal to eggs and also influences the number of offspring from a single female mosquito that is likely to survive to become adults. Since neurohormones regulate reproduction, it is believed that neurohormonal imbalance from insecticide poisonings could interfere with the normal operation of reproduction (Xiao et al., 2015). Thus, fecundity is one of the important criteria to be taken into account if the biological fitness of the vector is to be known.

(iii) *Mating competitiveness*: Mating competition in males is regarded as the crucial reproductive parameter, as the only fit and healthy males get the chance to mate and spreading their genes in the population. As a result of insecticide resistance, only those males which possess the resistant alleles or other resistance mechanism are the one that are benefited and are likely to mate. Apart from male, female mosquitoes also undergo behavioral changes after mating, including induction of egg laying and refractoriness to further insemination (Klowden and Russell, 2004).

(iv) Gonotrophic cycle: Gonotrophic cycle is directly linked with the survival of vectors. Females with long life typically experience more gonotrophic cycles, which results in more offspring. These modifications may lead to more blood meals being consumed and an increased risk of disease transmission.

17.2.2 Reports on the Fitness Costs in Response to Insecticide Selection

Fitness costs due to insecticide resistance can be studied by two methods (Bourguet et al., 2004):

(i) Comparing the fitness traits between resistant and susceptible mosquito populations.
(ii) Studying the changes in the frequency of resistant alleles in mosquito populations for several generations which are maintained without any insecticide exposure.

Such studies from different parts of the world have reported different fitness costs already mentioned above in many laboratory-reared resistant and susceptible mosquito populations and naturally occurring resistant mosquito populations. These fitness costs were correlated with different resistance mechanisms like target site insensitivity, metabolic resistance, cuticular resistance, etc. In France, organophosphate-resistant *Culex pipiens* populations, with altered acetylcholinesterase allele (*ace-1*) and overproduction of esterase enzymes, were reported to have shorter wing length and longer developmental time (Bourguet et al., 2004). Laboratory-reared strains of *Cx. pipiens* with the same genetic background (*ace-1* allele and esterase overproduction) were also found to have reduced energy reserves compared to the susceptible population (Rivero et al., 2011). In another study malathion (an organophosphate)-resistant *Cx. pipiens* populations from Egypt showed reduced levels of carbohydrate and protein content and high phenoloxidase activity (Elhawary et al., 2022).

Pyrethroid- and organophosphate-resistant field populations of *Aedes aegypti* from Brazil showed reduced acceptance of blood meal, lower blood ingestion, reduced number of eggs, and a smaller number of inseminated female mosquitoes. These fitness costs were also correlated with the presence of *kdr* alleles in the population (Belinato et al., 2012). A laboratory-reared pyrethroid-resistant strain of *Ae. aegypti*, named Rock-kdr (homozygous for two *kdr* mutations, Val1016Ile and Phe1534Cys), was reported to have longer larval developmental time, reduced female fertility, and lower number of laid eggs (Brito et al., 2013). In 2015, Diniz et al. (2015) reported that the larval stage of the temephos-resistant *Ae. aegypti* in Brazil was extended by an extra 5–7 days. In Venezuela, the selection of field populations of *Ae. aegypti* with deltamethrin for 16 generations resulted in the overexpression of cytochrome P_{450} monooxygenases as well as decreased hatchability, lower rates of pupation, decreased fertility, and lower survivability rate (Alvarez-Gonzalez et al., 2017). In another study, a laboratory-reared strain of *Ae. aegypti* resistant to both organophosphate and pyrethroids showed lower lifespan of female mosquitoes and reduced hatchability of the eggs. This strain also showed higher activities of esterase and glutathione S-transferase enzymes as well as possessed *kdr* mutations (Val1016Ile and Phe1534Cys) (dos Santos et al., 2020). In Malaysia, *Ae. albopictus* population, showing moderate resistance to permethrin and deltamethrin mainly due to enhanced monooxygenase activity, was selected with permethrin for seven generations and the selection pressure led to longer developmental time, reduced rate of adult emergence and lower survivability of adult mosquitoes (Chan and Zairi, 2013). Another study with a deltamethrin-resistant strain of *Ae. albopictus* showed that the resistant females had shorter lifespans and higher body weights than the susceptible ones and the overall survival time of the resistant adults was significantly lower (Deng et al., 2021).

A study in Kenya showed that *Anopheles gambiae* populations, resistant to organophosphate (homozygous for *ace-1* mutation) and pyrethroid and DDT (homozygous for *kdr* mutation Leu1014Phe) respectively, had lower fecundity than the susceptible population (Alout et al., 2016). Deltamethrin resistant field *Anopheles labranchiae* populations from Tunisia showed lower hatchability, longer larval developmental time and higher larval mortality (Tabbabi and Daaboub, 2018). In the study of fitness cost in resistant *An. coluzzii* population from Cameroon, it was reported that the blood-feeding rate in the resistant population was quite low when compared to their control population (Nkahe et al., 2020). A study by Osoro et al. (2021) reported that the resistant colony of *An. gambiae*, selected against 0.05% deltamethrin for six generations, showed longer developmental duration from first instar to fourth instar, first instar to pupa formation as well as egg hatching to adult emergence than that of susceptible colony while there were no significant changes in pupation rate and adult emergence rate between the two.

In India, very few studies have been done regarding fitness costs in resistant mosquitoes. In 2009, a study in Delhi showed that deltamethrin selection of field *Ae. aegypti* population resulted in a significant reduction in the duration of the gonotrophic cycle, lower fecundity, and reduced hatchability rates (Kumar et al., 2009). An experiment was conducted in Bangalore to observe the effect of sublethal exposure to propoxur and temephos in *An. stephensi's* reproductive fitness where the hatchability in the vector was reduced (Sanil and Shetty, 2012). Reduction in fecundity was also demonstrated in *Ae. aegypti* treated with alpha-cypermethrin in New Delhi (Gupta et al., 2021).

17.3 EFFECT OF INSECTICIDE SELECTION ON THE DETOXIFYING ENZYMES OF MOSQUITO VECTORS

Insects are exposed to various toxins and other compounds at different life stages and therefore possess an array of detoxification enzymes that protect them from these xenobiotics by metabolizing them into relatively less toxic products that can be easily excreted from the body. Thus, detoxifying enzymes can play a major role in the development of insecticide resistance by an increased rate of insecticide metabolism before the insecticides can reach their target sites (Nkya et al., 2013). This is called metabolic resistance and is quite prominent in major mosquito vectors like *Anopheles*, *Culex*, and *Aedes* (Black IV et al., 2021). Metabolic resistance can occur in two ways (Vontas et al., 2020):

(i) Over production of enzymes due to increased expression of enzyme coding genes.
(ii) Enhanced catalytic activity of the enzymes due to mutations in the corresponding genes.

These two processes can act together, i.e., the detoxification enzymes can be overexpressed as well as structurally altered at the same time (Vontas et al., 2020). Such changes in the enzyme coding genes can either be constitutive or can be induced by the selection pressure due to continuous exposure to different classes of insecticides (Black IV et al., 2021).

Three classes of detoxification enzymes are associated with insecticide metabolism in mosquitoes – cytochrome P_{450} monooxygenases (CYPs), GSTs, and non-specific esterases (α/β) (Hemingway et al., 2004). Studies showed that CYPs are more closely associated with the metabolism of pyrethroid and carbamate, GSTs are correlated mostly with the metabolism of DDT and esterases are responsible for organophosphate metabolism (David et al., 2013). Selection of different mosquito species using these insecticide classes also showed that continuous exposure for only one generation or several generations can lead to overexpression of the above-mentioned enzymes (Vontas et al., 2020).

17.3.1 OVEREXPRESSION OF CYTOCHROME P_{450} GENES IN RESPONSE TO INSECTICIDE SELECTION

Insect cytochrome P_{450} monooxygenases are well known for metabolizing xenobiotics such as insecticides, and these toxic compounds can also induce the expression of CYP genes, leading to their overexpression and therefore overproduction of the corresponding enzymes resulting in the increased metabolism of insecticides. Thus, induction of CYP genes by insecticides and their consequent overexpression can lead to the development of insecticide resistance (Feyereisen, 2012). Moreover, they can also act as activators for the organophosphates (Amezian et al., 2021). The role of P_{450} enzymes in pyrethroid resistance in mosquitoes was revealed through synergistic studies. In these studies, mosquito vectors, which are already resistant to synthetic pyrethroids (SPs), showed increased susceptibility to these insecticides, when pre-exposed to piperonylbutoxide (PBO) – an inhibitor of P_{450} enzymes (Liu et al., 2006). Besides, elevated levels of P_{450} enzymes in different populations of *Anopheles*, *Culex*, and *Aedes* mosquitoes, resistant to conventionally used type I and type II synthetic pyrethroids, also provided evidence for the involvement of these enzymes in the development of resistance to SPs (Nkya et al., 2013; David et al., 2013).

There are several families of cytochrome P_{450} monooxygenases but specifically two families, CYP6 and CYP9, are mostly found to be overexpressed in mosquito populations under the continuous selection pressure of SPs (Vontas et al., 2020). Moreover, different CYP genes were identified to be associated with pyrethroid resistance in different mosquito species as well as in different populations of the same mosquito species (Table 17.1). For example, *Anopheles gambiae* populations from African countries, resistant to pyrethroids, showed elevated expression levels of *CYP6P3*, *CYP6M2*, *CYP6Z1*, and *CYP6Z2* (Kwiatkowska et al., 2013; Adolfi et al., 2019; Gueye et al., 2020). On the other hand, pyrethroid resistant wild populations of *An. funestus* from African countries were found to over-express *CYP6M7*, *CYP6P4*, *CYP6P9a* and *CYP6P9b* genes (Wondji et al., 2009; Riveron et al., 2014). In *An. minimus* populations from Thailand, *CYP6AA3* and *CYP6P7* were found to be involved in the metabolism of permethrin, cypermethrin, and deltamethrin (Duangkaew et al., 2011). Recently in India, a deltamethrin-resistant population of *An. culicifacies* has been found showing overexpression of *CYP6Z1* (Kareemi et al., 2022). Overexpression of four CYP genes – *CYP6AA7*, *CYP9J34*, *CYP9J40*, and *CYP9M10* – was found to be associated with permethrin resistance in a laboratory-reared pyrethroid-resistant strain of *Culex quinquefasciatus* (Liu et al., 2011a). Elevated expression of *CYP9M10* was also observed in permethrin-resistant, laboratory-reared *Cx. quinquefasciatus* populations from Japan, South Africa, and the USA. Use of RNAi technique on the laboratory-reared permethrin resistant *Cx. quinquefasciatus* population revealed the association of *CYP6BZ2* and *CYP9J35* in addition to the previously identified genes (Yang et al., 2021). Numerous studies in *Aedes* mosquitoes showed that CYP6 and CYP9 families are also involved with pyrethroid resistance. In Malaysia, overexpression of *CYP6P12* and *CYP6N3* was associated with permethrin and deltamethrin resistance in wild populations of *Ae. albopictus* (Ishak et al., 2016). In China, another CYP gene, *CYP6A8*, was found to be associated with deltamethrin resistance in natural populations of *Ae. albopictus* (Xu et al., 2018). *CYP6M6*, and *CYP6Z6* were overexpressed in permethrin-resistant *Ae. aegypti* from French West Indies (Marcombe et al., 2009). *CYP6BB2, CYP9J9, CYP9J10, CYP9J26, CYP9J27, and CYP9J28* genes showed differential expression in pyrethroid resistant *Ae. aegypti* from Cuba and Cayman Islands (Bariami et al., 2012). In *Ae. aegypti*, four CYP genes in *CYP9J* subfamily – *CYP9J24, CYP9J26, CYP9J28* and *CYP9J32* were identified as pyrethroid metabolizers (Stevenson et al., 2012). *CYP9J24* and *CYP9J26* were found to be overexpressed in pyrethroid-resistant *Ae. aegypti* in Singapore (Vontas et al., 2020), whereas *CYP9J32* was found to be overexpressed in pyrethroid resistant populations of *Ae. aegypti* in Vietnam and Mexico. *CYP9J24*, *CYP9J26*, and *CYP9J32* were associated with pyrethroid resistance in *Ae. aegypti* populations from Thailand (Stevenson et al., 2012). Overexpression of *CYP9J26*, *CYP9J27*, *CYP6CB1*, and *CYP9M4* was observed in the pyrethroid-resistant wild *Aedes aegypti* populations from Malaysia (Ishak et al., 2017). In a laboratory-reared permethrin-resistant strain of *Ae. aegypti*, overexpression of another four CYP genes – *CYP6BB2, CYP6Z8*, *CYP9M5*, and *CYP9M6* – was found to be associated with resistance (Smith et al., 2018). Another study in Vietnam showed overexpression of *CYP9E2*, *CYP6A8*, *CYP6D3* genes in pyrethroid-resistant *Ae. aegypti* (Lien et al., 2019).

Very few studies have been done to identify CYP genes associated with the selection pressure of carbamate insecticides. In *An. gambiae*, *CYP6P3*, *CYP6Z3*, *CYP6Z1*, *CYP12F2*, *CYP6M3*, and *CYP6P4* genes were found to be overexpressed in bendiocarb-resistant wild populations of

TABLE 17.1
CYP Genes Associated with Pyrethroid Resistance in Major Mosquito Vectors

Mosquito Species	Insecticide Class	Insecticide Name	CYP Genes Associated	Expression Level	Country	Reference
Anopheles gambiae	Type I Synthetic pyrethroid	Permethrin	*CYP6P3*, *CYP6Z2*	High	Burkina Faso	Kwiatkowska et al. (2013)
	Type II Synthetic pyrethroid	Deltamethrin, Lambda-cyhalothrin				
	Type I Synthetic pyrethroid	Permethrin	*CYP6M2*, *CYP6P3*	High	–	Adolfi et al. (2019)
	Type II Synthetic pyrethroid	Deltamethrin				
	Type I Synthetic pyrethroid	Permethrin	*CYP6Z1*, *CYP6Z2*	High	Senegal	Gueye et al. (2020)
	Type II Synthetic pyrethroid	Alpha-cypermethrin, Deltamethrin, Lambda-cyhalothrin				
An. funestus	Type I Synthetic pyrethroid	Permethrin	*CYP6P4*, *CYP6P9*	High	South Africa	Wondji et al. (2009)
	Type I Synthetic pyrethroid	Permethrin	*CYP6M7*, *CYP6P9a*, *CYP6P9b*	High	Zambia	Riveron et al. (2014)
	Type II Synthetic pyrethroid	Deltamethrin				
An. minimus	Type I Synthetic pyrethroid	Permethrin	*CYP6AA3*, *CYP6P7*	High	Thailand	Duangkaew et al. (2011)
	Type II Synthetic pyrethroid	Cypermethrin, Deltamethrin				
An. culicifacies	Type II Synthetic pyrethroid	Deltamethrin	*CYP6Z1*	High	India	Kareemi et al. (2022)
Culex quinquefasciatus	Type I Synthetic pyrethroid	Permethrin	*CYP6AA7*, *CYP9J34*, *CYP9J40*, *CYP9M10*	High	–	Liu et al. (2011a)
			CYP6BZ2, *CYP9J35*	High	–	Yang et al. (2021)
Aedes aegypti	Type I Synthetic pyrethroid	Permethrin	*CYP6M6*, *CYP6Z6*	High	French West Indies	Marcombe et al. (2009)
	Type I Synthetic pyrethroid	Permethrin	*CYP6BB2*, *CYP9J9*, *CYP9J10*, *CYP9J26*, *CYP9J27*, *CYP9J28*	High	Cuba	Bariami et al. (2012)
	Type II Synthetic pyrethroid	Deltamethrin			Cayman Islands	
	Type I Synthetic pyrethroid	Permethrin	*CYP9J24*, *CYP9J26*, *CYP9J32*	High	Thailand	Stevenson et al. (2012)
	Type II Synthetic pyrethroid	Deltamethrin				
	Type I Synthetic pyrethroid	Permethrin	*CYP9J32**	High	Vietnem	
	Type II Synthetic pyrethroid	Deltamethrin	*CYP9J32*	High	Mexico	

(*Continued*)

TABLE 17.1 (Continued)

Mosquito Species	Insecticide Class	Insecticide Name	CYP Genes Associated	Expression Level	Country	Reference
	Type I Synthetic pyrethroid	Permethrin	*CYP9J26, CYP9J27, CYP6CB1, CYP9M4*	High	Malaysia	Ishak et al. (2017)
	Type I Synthetic pyrethroid	Permethrin	*CYP6BB2, CYP6Z8, CYP9M5 CYP9M6*	High	–	Smith et al. (2018)
	Type I Synthetic pyrethroid Type II Synthetic pyrethroid	Permethrin Deltamethrin	*CYP9E2, CYP6A8, CYP6D3*	High	Vietnam	Lien et al. (2019)
	Type II Synthetic pyrethroid	Deltamethrin	*CYP9J24 CYP9J26*	High	Singapore	Vontas et al. (2020)
Ae. albopictus	Type I Synthetic pyrethroid Type II Synthetic pyrethroid	Permethrin Deltamethrin	*CYP6N3, CYP6P12*	High	Malaysia	Ishak et al. (2016)
	Type II Synthetic pyrethroid	Deltamethrin	*CYP6A8*	High	China	Xu et al. (2018)

'–' indicates the study was done in laboratory-reared, pyrethroid resistant mosquito population.

Cameroon (Antonio-Nkondjio et al., 2016; Ngangue-Siewe et al., 2022). *CYP6Z1* was also reported to be overexpressed in bendiocarb-resistant *An. funestus* from Malawi (Ibrahim et al., 2016).

Some studies reported differential expression of CYP genes in organophosphate-resistant mosquitoes. In Colombia, *CYP6N12, CYP6F3* and *CYP6M11*were identified as candidate genes associated with temephos resistance in *Ae. aegypti* (Grisales et al., 2013). Another study in India also suggested the strong association between temephos resistance and overexpression of *CYP6M2* and *CYP6Z2* in *An. stephensi* (Vivekanandhan et al., 2021).

17.3.2 Overexpression of Glutathione S-Transferase Genes in Response to Insecticide Selection

Glutathione-S-transferase is another important family of detoxification enzymes in insects which can be divided into six classes, namely, delta, epsilon, omega, sigma, theta, and zeta. Among these, delta and epsilon classes are crucial in xenobiotic detoxification and therefore play an important role in insecticide resistance development (Hemingway et al., 2004). Overexpression of epsilon class GST genes *GSTe2* was associated with DDT and organophosphate resistance in *An. gambiae* (Adolfi et al., 2019) and association of *GSTD1, GSTe2* and *GSTe7* in DDT resistant *Ae. aegypti* (Liu, 2015). Detailed molecular studies have identified four epsilon class GST genes – *GSTe2, GSTe3, GSTe4,* and *GSTe7* – to be associated with pyrethroid resistance in *Ae. aegypti* (Strode et al., 2008; Lumjuan et al., 2011) and three GST genes – *GSTs1, GSTt3,* and *GSTt4* were reported to be overexpressed in permethrin-resistant wild *Ae. albopictus* populations in Malaysia (Ishak et al., 2016).

17.3.3 Overexpression of Non-specific Esterase Genes in Response to Insecticide Selection

Non-specific esterases (α/β) or carboxylesterases (CCEs) break the ester bonds and therefore are supposed to metabolize ester compounds like organophosphates and pyrethroids (Bhatt et al., 2020). Studies have identified a number of α- and β- esterase genes in organophosphate or pyrethroid-resistant mosquito populations. In *Cx. pipiens*, overproduction of *Estα2, Estα8, Estβ1, Estβ2*, and *Estβ8* was reported from China (Qiao et al., 1999). Another study in China reported that the selection pressure of pyrethroids and organophosphates can induce multiple esterase alleles *like* $Ester^{B1}$, $Ester^{2}$, $Ester^{8}$, $Ester^{9}$, $Ester^{B10}$, and $Ester^{11}$ in *Cx. pipiens* (Liu et al., 2011b). The esterase B1 gene was reported to be amplified about 250-fold in temephos-resistant *Cx. quinquefasciatus* from California (Mouchès et al., 1986). Co-amplification of *Estα2* and *Estβ2* genes was common in organophosphate-resistant *Cx. quinquefasciatus* (Paton et al., 2000). *Estβ1* and *Estβ2* were strongly correlated with chlorpyriphos resistance in *Cx. tritaeniorhynchus* from Sri Lanka (Karunaratne et al., 1998). In Florida, *Temsha est-1* gene was found to be differentially induced by organophosphate exposure in *Cx. nigripalpus* (Shin and Smartt, 2016). Two esterase genes, α-esterase and esterase β1, were overexpressed in temephos-resistant larvae of *An. stephensi* from India (Vivekanandhan et al., 2021). Elevated expression of carboxylesterase genes: *CCEae1C, CCEae3A, CCEae5C,* and *CCEae6C* was observed in laboratory-reared temephos-resistant *Ae. aegypti* (Paiva et al., 2016). Overexpression of *CCEae3a* and *CCEae6a* was also reported in temephos-resistant *Ae. albopictus* from Greece (Grigoraki et al., 2015).

In some studies, esterase genes were also implicated in pyrethroid resistance. Two α-esterase genes – AAEL013543 and AAEL015264-RA were found to be overexpressed in permethrin-resistant wild *Ae. albopictus* populations of Malaysia (Ishak et al., 2016). CCE genes, *AsAe10 and AsAce1* were strongly correlated with pyrethroid resistance in *An. sinensis* from China (Wu et al., 2018).

17.4 CONCLUSION

The selection pressure of continuous insecticide exposure results in the development of insecticide resistance in mosquitoes, which can cause a reduction of biological fitness when the selection pressure is not there. Studying the developmental and reproductive changes in the life history of insecticide-resistant mosquitoes, aids in a better understanding of the alterations in the behavior of resistant vectors and population dynamics, which can enhance the effectiveness of vector control programs. However, such studies are mostly conducted with laboratory-reared resistant mosquito strains under laboratory conditions and therefore the results could be different from those in natural conditions. This problem can be solved by directly studying the fitness costs in field populations of resistant mosquitoes (Bourguet et al., 2004). Another important aspect is to correlate the level of resistance with the fitness cost whether the cost is more severe in the mosquito populations with increasing level of resistance. Therefore, testing of the resistance status should be done while studying fitness costs. It is evident that mosquito populations relying on different resistance mechanisms show various fitness costs but whether mosquito populations relying on the same resistance mechanisms carry the same fitness costs is not clearly understood. Moreover, some mosquito populations rely on multiple resistance mechanisms. The pleiotropic effect of all those resistance alleles on fitness needs to be studied for clear understanding (Freeman et al., 2021). These knowledge gaps should be given more importance and discussed thoroughly in future studies, as such information can help delay the onset of resistance in mosquito populations.

Identification of detoxification enzyme genes in resistant mosquitoes is very crucial to combat the development of insecticide resistance relying on metabolic detoxification. As differential selection pressure of insecticides can lead to the selection of different enzyme coding genes in different mosquito populations, more population-specific studies need to be done to identify particular enzyme genes responsible for the development of resistance (Black IV et al., 2021). These particular genes can be targeted to restore the susceptibility status of the corresponding mosquito populations. Several techniques have emerged as potent candidates for this purpose such as the use of enzyme blockers or RNAi (inhibitory RNA) method (Vontas et al., 2020). Although in most resistant mosquito populations, change in the gene copy number is responsible for the overexpression of the enzyme coding genes, alteration of the cis- or transregulation of gene expression can also be a very crucial factor for the changes in transcription rate of these genes. Detailed studies regarding the regulation of gene expression should gain more importance in the future. A few studies in *Cx. quinquefasciatus* have reported altered levels of some transcription factors in resistant strains showing overexpression of monooxygenases (Amezian et al., 2021). Such studies should be done in other mosquito species also to clearly understand the development of metabolic resistance in response to selection pressure of insecticides.

REFERENCES

Adolfi, A., Poulton, B., Anthousi, A., Macilwee, S., Ranson, H., and Lycett, G.J. (2019). Functional genetic validation of key genes conferring insecticide resistance in the major African malaria vector, *Anopheles gambiae*. *Proceedings of the National Academy of Sciences*, 116(51), 25764–25772.

Alout, H., Dabiré, R.K., Djogbénou, L.S., Abate, L., Corbel, V., Chandre, F. and Cohuet, A. (2016). Interactive cost of Plasmodium infection and insecticide resistance in the malaria vector *Anopheles gambiae*. *Scientific Reports*, 6(1), 1–11.

Alvarez-Gonzalez, L.C., Briceño, A., Ponce-Garcia, G., Villanueva-Segura, O.K., Davila-Barboza, J.A., Lopez-Monroy, B., Gutierrez-Rodriguez, S.M., Contreras-Perera, Y., Rodriguez-Sanchez, I.P. and Flores, A.E. (2017). Assessing the effect of selection with deltamethrin on biological parameters and detoxifying enzymes in *Aedes aegypti* (L.). *Pest Management Science*, 73(11), 2287–2293.

Aly, A.S., Vaughan, A.M. and Kappe, S.H. (2009). Malaria parasite development in the mosquito and infection of the mammalian host. *Annual Review of Microbiology*, 63, 195.

Amezian, D., Nauen, R. and Le Goff, G. (2021). Transcriptional regulation of xenobiotic detoxification genes in insects-An overview. *Pesticide Biochemistry and Physiology*, 174, 104822.

Antonio-Nkondjio, C., Poupardin, R., Tene, B.F., Kopya, E., Costantini, C., Awono-Ambene, P. and Wondji, C.S. (2016). Investigation of mechanisms of bendiocarb resistance in *Anopheles gambiae* populations from the city of Yaoundé, Cameroon. *Malaria journal*, 15(1), 1–11.

Bariami, V., Jones, C.M., Poupardin, R., Vontas, J. and Ranson, H. (2012). Gene amplification, ABC transporters and cytochrome P450s: unraveling the molecular basis of pyrethroid resistance in the dengue vector, *Aedes aegypti*. *PLoS Neglected Tropical Diseases*, 6(6), e1692.

Belinato, T.A., Martins, A.J. and Valle, D. (2012). Fitness evaluation of two Brazilian *Aedes aegypti* field populations with distinct levels of resistance to the organophosphate temephos. *Memorias do Instituto Oswaldo Cruz*, 107, 916–922.

Belinato, T.A. and Martins, A.J. (2016). Insecticide resistance and fitness cost. *Insecticides resistance*. London, UK: InTech.

Bhatt, P., Bhatt, K., Huang, Y., Lin, Z. and Chen, S. (2020). Esterase is a powerful tool for the biodegradation of pyrethroid insecticides. *Chemosphere*, 244, 125507.

Black IV, W.C., Snell, T.K., Saavedra-Rodriguez, K., Kading, R.C. and Campbell, C.L. (2021). From global to local—new insights into features of pyrethroid detoxification in vector mosquitoes. *Insects*, 12(4), 276.

Bourguet, D., Guillemaud, T., Chevillon, C. and Raymond, M. (2004). Fitness costs of insecticide resistance in natural breeding sites of the mosquito *Culex pipiens*. *Evolution*, 58(1), 128–135.

Brito, L.P., Linss, J.G., Lima-Camara, T.N., Belinato, T.A., Peixoto, A.A., Lima, J.B.P., Valle, D. and Martins, A.J. (2013). Assessing the effects of *Aedes aegypti* kdr mutations on pyrethroid resistance and its fitness cost. *PloS one*, 8(4), e60878.

Campanhola, C., McCutchen, B.F., Baehrecke, E.H. and Plapp Jr, F.W. (1991). Biological constraints associated with resistance to pyrethroids in the tobacco bud worm (Lepidoptera: Noctuidae). *Journal of Economic Entomology*, 84(5), 1404–1411.

Chan, H.H. and Zairi, J. (2013). Permethrin resistance in *Aedes albopictus* (Diptera: Culicidae) and associated fitness costs. *Journal of Medical Entomology*, 50(2), 362–370.

Clements, A.N. (1992). *The biology of mosquitoes. Volume 1: development, nutrition and reproduction*. UK: Chapman and Hall.

David, J.P., Ismail, H.M., Chandor-Proust, A. and Paine, M.J.I. (2013). Role of cytochrome P450s in insecticide resistance: impact on the control of mosquito-borne diseases and use of insecticides on Earth. *Philosophical Transactions of the Royal Society B: Biological Sciences*, 368(1612), 20120429.

Deng, J., Guo, Y., Su, X., Liu, S., Yang, W., Wu, Y., Wu, K., Yan, G. and Chen, X. G. (2021). Impact of deltamethrin-resistance in *Aedes albopictus* on its fitness cost and vector competence. *PLoS Neglected Tropical Diseases*, 15(4), e0009391.

Diniz, D.F.A., Melo-Santos, M.A.V.D., Santos, E.M.D.M., Beserra, E.B., Helvecio, E., de Carvalho-Leandro, D., dos Santos, B.S., de Menezes Lima, V.L. and Ayres, C.F.J. (2015). Fitness cost in field and laboratory *Aedes aegypti* populations associated with resistance to the insecticide temephos. *Parasites & Vectors*, 8(1), 1–15.

Dos Santos, C.R., de Melo Rodovalho, C., Jablonka, W., Martins, A.J., Lima, J.B.P., dos Santos Dias, L., da Silva Neto, M.A.C. and Atella, G.C. (2020). Insecticide resistance, fitness and susceptibility to Zika infection of an interbred *Aedes aegypti* population from Rio de Janeiro, Brazil. *Parasites & Vectors*, 13(1), 1–14.

Du, W., Awolola, T.S., Howell, P., Koekemoer, L.L., Brooke, B.D., Benedict, M.Q., Coetzee, M. and Zheng, L. (2005). Independent mutations in the Rdl locus confer dieldrin resistance to *Anopheles gambiae* and *An. arabiensis*. *Insect Molecular Biology*, 14(2), 179–183.

Duangkaew, P., Kaewpa, D. and Rongnoparut, P. (2011). Protective efficacy of *Anopheles minimus* CYP6P7 and CYP6AA3 against cytotoxicity of pyrethroid insecticides in *Spodoptera frugiperda* (Sf9) insect cells. *Tropical Biomedicine*, 28(2), 293–301.

Elhawary, N.A., Soliman, M.A., Seif, A.I. and Meshrif, W.S. (2022). Fitness cost of malathion resistance in Egyptian *Culex pipiens* populations. *Medical and Veterinary Entomology*, 36(3), 338–346.

Feyereisen, R. (2012). Insect CYP genes and P450 enzymes. *Insect molecular Biology and Biochemistry*. USA: Elsevier.

Freeman, J.C., Smith, L.B., Silva, J.J., Fan, Y., Sun, H. and Scott, J.G. (2021). Fitness studies of insecticide resistant strains: Lessons learned and future directions. *Pest Management Science*, 77(9), 3847–3856.

Gabrieli, P., Kakani, E.G., Mitchell, S.N., Mameli, E., Want, E.J., Mariezcurrena Anton, A., Serrao, A., Baldini, F. and Catteruccia, F. (2014). Sexual transfer of the steroid hormone 20E induces the postmating switch in *Anopheles gambiae*. *Proceedings of the National Academy of Sciences*, 111(46), 16353–16358.

Grigoraki, L., Lagnel, J., Kioulos, I., Kampouraki, A., Morou, E., Labbe, P., Weill, M. and Vontas, J. (2015). Transcriptome profiling and genetic study reveal amplified carboxylesterase genes implicated in temephos resistance, in the Asian tiger mosquito *Aedes albopictus*. *PLoS Neglected Tropical Diseases*, 9(5), e0003771.

Grisales, N., Poupardin, R., Gomez, S., Fonseca-Gonzalez, I., Ranson, H. and Lenhart, A. (2013). Temephos resistance in *Aedes aegypti* in Colombia compromises dengue vector control. *PLoS Neglected Tropical Diseases*, 7(9), e2438.

Gueye, O.K., Tchouakui, M., Dia, A.K., Faye, M.B., Ahmed, A.A., Wondji, M.J., Nguiffo, D.N.J., Mugenzi, L. M., Tripet, F., Konaté, L. and Diabate, A. (2020). Insecticide resistance profiling of *Anopheles Coluzzii* and *Anopheles Gambiae* populations in the Southern Senegal: Role of target sites and metabolic resistance mechanisms. *Genes*, 11(12), 1403.

Gupta, A., Samal, R.R. and Kumar, S. (2021). Physiological and reproductive fitness cost in *Aedes aegypti* on exposure to toxic xenobiotics in New Delhi, India. *Journal of Applied and Natural Science*, 13(1), 71–78.

Hardstone, M.C., Lazzaro, B.P. and Scott, J.G. (2009). The effect of three environmental conditions on the fitness of cytochrome P450 monooxygenase-mediated permethrin resistance in *Culex pipiens quinquefasciatus*. *BMC Evolutionary Biology*, 9(1), 1–13.

Hemingway, J., Hawkes, N.J., McCarroll, L. and Ranson, H. (2004). The molecular basis of insecticide resistance in mosquitoes. *Insect Biochemistry and Molecular Biology*, 34(7), 653–665.

Hemingway, J. and Ranson, H. (2000). Insecticide resistance in insect vectors of human disease. *Annual Review of Entomology*, 45(1), 371–391.

Ibrahim, S.S., Ndula, M., Riveron, J.M., Irving, H. and Wondji, C.S. (2016). The P450 CYP6Z1 confers carbamate/pyrethroid cross-resistance in a major African malaria vector beside a novel carbamate-insensitive N485I acetylcholinesterase-1 mutation. *Molecular Ecology*, 25(14), 3436–3452.

Ishak, I.H., Kamgang, B., Ibrahim, S.S., Riveron, J.M., Irving, H. and Wondji, C.S. (2017). Pyrethroid resistance in Malaysian populations of dengue vector *Aedes aegypti* is mediated by CYP9 family of cytochrome P450 genes. *PLoS Neglected Tropical Diseases*, 11(1), e0005302.

Ishak, I.H., Riveron, J.M., Ibrahim, S.S., Stott, R., Longbottom, J., Irving, H. and Wondji, C.S. (2016). The Cytochrome P450 gene CYP6P12 confers pyrethroid resistance in kdr-free Malaysian populations of the dengue vector *Aedes albopictus*. *Scientific Reports*, 6(1), 1–13.

Kareemi, T.I., Mishra, A.K., Chand, S.K., Nirankar, J.K., Vishwakarma, A.K., Tiwari, A. and Bharti, P.K. (2022). Analysis of the insecticide resistance mechanism in *Anopheles culicifacies sensu lato* from a malaria-endemic state in India. *Transactions of the Royal Society of Tropical Medicine and Hygiene*, 116(3), 252–260.

Karunaratne, S.H.P.P., Vaughan, A., Paton, M.G. and Hemingway, J. (1998). Amplification of a serine esterase gene is involved in insecticide resistance in Sri Lankan *Culex tritaeniorhynchus*. *Insect Molecular Biology*, 7(4), 307–315.

Klowden, M.J. and Russell, R.C. (2004). Mating affects egg maturation in *Anopheles gambiae* Giles (Diptera: Culicidae). *Journal of Vector Ecology*, 29, 135–139.

Kumar, S., Thomas, A., Samuel, T., Sahgal, A., Verma, A. and Pillai, M.K.K. (2009). Diminished reproductive fitness associated with the deltamethrin resistance in an Indian strain of dengue vector mosquito, *Aedes aegypti* L. *Tropical Biomedicine*, 26(2), 55–64.

Kwiatkowska, R.M., Platt, N., Poupardin, R., Irving, H., Dabire, R.K., Mitchell, S., Jones, C.M., Diabaté, A., Ranson, H. and Wondji, C.S. (2013). Dissecting the mechanisms responsible for the multiple insecticide resistance phenotype in *Anopheles gambiae* ss, M form, from Vallee du Kou, Burkina Faso. *Gene*, 519(1), 98–106.

Lien, N.T.K., Ngoc, N.T.H., Lan, N.N., Hien, N.T., Van Tung, N., Ngan, N.T.T., Hoang, N.H. and Binh, N.T.H. (2019). Transcriptome sequencing and analysis of changes associated with insecticide resistance in the dengue mosquito (*Aedes aegypti*) in Vietnam. *The American Journal of Tropical Medicine and Hygiene*, 100(5), 1240.

Liu, N. (2015). Insecticide resistance in mosquitoes: impact, mechanisms, and research directions. *Annual Review of Entomology*, 60(1), 537–559.

Liu, N., Li, T., Reid, W.R., Yang, T. and Zhang, L. (2011a). Multiple cytochrome P450 genes: their constitutive overexpression and permethrin induction in insecticide resistant mosquitoes, *Culex quinquefasciatus*. *Plos one*, 6(8), e23403.

Liu, N., Xu, Q., Zhu, F. and Zhang, L.E.E. (2006). Pyrethroid resistance in mosquitoes. *Insect Science*, 13(3), 159–166.

Liu, Y., Zhang, H., Qiao, C., Lu, X. and Cui, F. (2011b). Correlation between carboxylesterase alleles and insecticide resistance in *Culex pipiens* complex from China. *Parasites and Vectors*, 4(1), 1–7.

Lumjuan, N., Rajatileka, S., Changsom, D., Wicheer, J., Leelapat, P., Prapanthadara, L.A., Somboon, P., Lycett, G. and Ranson, H. (2011). The role of the *Aedes aegypti* Epsilon glutathione transferases in conferring resistance to DDT and pyrethroid insecticides. *Insect Biochemistry and Molecular Biology*, 41(3), 203–209.

Marcombe, S., Poupardin, R., Darriet, F., Reynaud, S., Bonnet, J., Strode, C., Brengues, C., Yébakima, A., Ranson, H., Corbel, V. and David, J.P. (2009). Exploring the molecular basis of insecticide resistance in the dengue vector *Aedes aegypti*: A case study in Martinique Island (French West Indies). *BMC Genomics*, 10, 1–14.

Martinez-Torres, D., Chandre, F., Williamson, M.S., Darriet, F., Berge, J.B., Devonshire, A.L., Guillet, P., Pasteur, N. and Pauron, D. (1998). Molecular characterization of pyrethroid knockdown resistance (kdr) in the major malaria vector *Anopheles gambiae* ss. *Insect Molecular Biology*, 7(2), 179–184.

Martins, A.J., Ribeiro, C.D.E.M., Bellinato, D.F., Peixoto, A.A., Valle, D. and Lima, J.B.P. (2012). Effect of insecticide resistance on development, longevity and reproduction of field or laboratory selected *Aedes aegypti* populations. *PloS one*, 7(3), e31889.

McCarroll, L. and Hemingway J. (2002). Can insecticide resistance status affect parasite transmission in mosquitoes? *Insect Biochemistry and Molecular Biology* 32, 1345–1351.

Mouchès, C., Pasteur, N., Bergé, J.B., Hyrien, O., Raymond, M., Robert de Saint Vincent, B., De Silvestri, M. and Georghiou, G.P. (1986). Amplification of an esterase gene is responsible for insecticide resistance in a California *Culex* mosquito. *Science*, 233(4765), 778–780.

Ngangue-Siewe, I.N., Ndjeunia-Mbiakop, P., Kala-Chouakeu, N.A., Bamou, R., Talipouo, A., Djamouko-Djonkam, L., Vontas, J., Mavridis, K., Tombi, J., Tchuinkam, T. and Mbida-Mbida, J.A. (2022). Bendiocarb and Malathion resistance in two major malaria vector populations in Cameroon is associated with high frequency of the G119S mutation (Ace-1) and overexpression of detoxification genes. *Pathogens*, 11(8), 824.

Nkahe, D.L., Kopya, E., Djiappi-Tchamen, B., Toussile, W., Sonhafouo-Chiana, N., Kekeunou, S., Mimpfoundi, R., Awono-Ambene, P., Wondji, C.S. and Antonio-Nkondjio, C. (2020). Fitness cost of insecticide resistance on the life-traits of a *Anopheles coluzzii* population from the city of Yaoundé, Cameroon. *Wellcome Open Research*, 5.

Nkya, T.E., Akhouayri, I., Kisinza, W., and David, J.P. (2013). Impact of environment on mosquito response to pyrethroid insecticides: Facts, evidences and prospects. *Insect Biochemistry and Molecular Biology*, 43(4), 407–416.

Osoro, J.K., Machani, M.G., Ochomo, E., Wanjala, C., Omukunda, E., Munga, S., Githeko, A.K., Yan, G. and Afrane, Y.A. (2021). Insecticide resistance exerts significant fitness costs in immature stages of *Anopheles gambiae* in western Kenya. *Malaria Journal*, 20(1), 1–7.

Paiva, M.H., Lovin, D.D., Mori, A., Melo-Santos, M.A., Severson, D.W. and Ayres, C.F. (2016). Identification of a major Quantitative Trait Locus determining resistance to the organophosphate temephos in the dengue vector mosquito *Aedes aegypti*. *Genomics*, 107(1), 40–48.

Paton, M.G., Karunaratne, S.P., Giakoumaki, E., Roberts, N. and Hemingway, J. (2000). Quantitative analysis of gene amplification in insecticide-resistant *Culex* mosquitoes. *Biochemical Journal*, 346(1), 17–24.

Platt, N., Kwiatkowska, R.M., Irving, H., Diabaté, A., Dabire, R. and Wondji, C.S. (2015). Target-site resistance mutations (kdr and RDL), but not metabolic resistance, negatively impact male mating competiveness in the malaria vector *Anopheles gambiae*. *Heredity*, 115(3), 243–252.

Plernsub, S., Stenhouse, S.A., Tippawangkosol, P., Lumjuan, N., Yanola, J., Choochote, W. and Somboon, P. (2013). Relative developmental and reproductive fitness associated with F1534C homozygous knockdown resistant gene in *Aedes aegypti* from Thailand. *Tropical Biomedicine*, 30(4), 621–630.

Qiao, C.L., Sun, Z.Q. and Liu, J.E. (1999). New esterase enzymes involved in organophosphate resistance in *Culex pipiens* (Diptera: Culicidae) from Guang Zhou, China. *Journal of Medical Entomology*, 36(6), 666–670.

Rahim, J., Ahmad, A.H. and Maimusa, A.H. (2017). Effects of temephos resistance on life history traits of *Aedes albopictus* (Skuse) (Diptera: Culicidae), a vector of arboviruses. *Revista Brasileira de Entomologia*, 61, 312–317.

Rivero, A., Magaud, A., Nicot, A. and Vézilier, J. (2011). Energetic cost of insecticide resistance in *Culex pipiens* mosquitoes. *Journal of Medical Entomology*, 48(3), 694–700.

Riveron, J.M., Ibrahim, S.S., Chanda, E., Mzilahowa, T., Cuamba, N., Irving, H., Barnes, K.G., Ndula, M. and Wondji, C.S. (2014). The highly polymorphic CYP6M7 cytochrome P450 gene partners with the directionally selected CYP6P9a and CYP6P9b genes to expand the pyrethroid resistance front in the malaria vector *Anopheles funestus* in Africa. *BMC Genomics*, 15, 1–19.

Sanil, D. and Shetty, N.J. (2012). The effect of sublethal exposure to temephos and propoxur on reproductive fitness and its influence on circadian rhythms of pupation and adult emergence in *Anopheles stephensi* Liston—a malaria vector. *Parasitology Research*, 111(1), 423–432.

Shin, D. and Smartt, C.T. (2016). Assessment of esterase gene expression as a risk marker for insecticide resistance in Florida *Culex nigripalpus* (Diptera: Culicidae). *Journal of Vector Ecology*, 41(1), 63–71.
Smith, L.B., Tyagi, R., Kasai, S. and Scott, J.G. (2018). CYP-mediated permethrin resistance in *Aedes aegypti* and evidence for trans-regulation. *PLoS Neglected Tropical Diseases*, 12(11), e0006933.
Stevenson, B.J., Pignatelli, P., Nikou, D. and Paine, M.J. (2012). Pinpointing P450s associated with pyrethroid metabolism in the dengue vector, *Aedes aegypti*: Developing new tools to combat insecticide resistance. *PLoS Neglected Tropical Diseases*, 6(3), e1595.
Strode, C., Wondji, C.S., David, J.P., Hawkes, N.J., Lumjuan, N., Nelson, D.R., Drane, D.R., Karunaratne, S.P., Hemingway, J., Black IV, W.C. and Ranson, H. (2008). Genomic analysis of detoxification genes in the mosquito *Aedes aegypti*. *Insect Biochemistry and Molecular Biology*, 38(1), 113–123.
Tabbabi, A. and Daaboub, J. (2018). Fitness cost in field *Anopheles labranchiae* populations associated with resistance to the insecticide deltamethrin. *Revista Brasileira de Entomologia*, 62, 107–111.
Tchouakui, M., Riveron, J.M., Djonabaye, D., Tchapga, W., Irving, H., Takam, P.S., Njiokou, F. and Wondji, C.S. (2018). Fitness costs of the Glutathione S-transferase Epsilon 2 (L119F-GSTe2) mediated metabolic resistance to insecticides in the major African Malaria Vector *Anopheles funestus*. *Genes*, 9(12), 1–15.
Venkatesan, R., Ravindran, J., Eapen, A. and William, J. (2014). Insecticidal and growth regulating activity of crude leaf extracts of *Cassia occidentalis* L. (Caesalpinaceae) against the urban malaria vector, *Anopheles stephensi* Liston (Diptera: Culicidae). *Asian Pacific Journal of Tropical Disease*, 4, 578–582.
Vivekanandhan, P., Thendralmanikandan, A., Kweka, E.J. and Mahande, A.M. (2021). Resistance to temephos in *Anopheles stephensi* larvae is associated with increased cytochrome P450 and α-esterase genes overexpression. *International Journal of Tropical Insect Science*, 41(4), 2543–2548.
Vontas, J., Katsavou, E. and Mavridis, K. (2020). Cytochrome P450-based metabolic insecticide resistance in *Anopheles* and *Aedes* mosquito vectors: Muddying the waters. *Pesticide Biochemistry and Physiology*, 170, 104666.
Waldock, J., Chandra, N.L., Lelieveld, J., Proestos, Y., Michael, E., Christophides, G. and Parham, P.E. (2013). The role of environmental variables on *Aedes albopictus* biology and chikungunya epidemiology. *Pathogens and Global Health*, 107(5), pp.224–241.
Weill, M., Malcolm, C., Chandre, F., Mogensen, K., Berthomieu, A., Marquine, M. and Raymond, M. (2004). The unique mutation in ace-1 giving high insecticide resistance is easily detectable in mosquito vectors. *Insect Molecular Biology*, 13(1), pp.1–7.
Wondji, C.S., Irving, H., Morgan, J., Lobo, N.F., Collins, F.H., Hunt, R.H., Coetzee, M., Hemingway, J. and Ranson, H. (2009). Two duplicated P450 genes are associated with pyrethroid resistance in *Anopheles funestus*, a major malaria vector. *Genome Research*, 19(3), 452–459.
Wu, X.M., Xu, B.Y., Si, F.L., Li, J., Yan, Z.T., Yan, Z.W., He, X. and Chen, B. (2018). Identification of carboxylesterase genes associated with pyrethroid resistance in the malaria vector *Anopheles sinensis* (Diptera: Culicidae). *Pest Management Science*, 74(1), 159–169.
Xiao, D., Yang, T., Desneux, N., Han, P. and Gao, X. (2015). Assessment of sublethal and transgenerational effects of pirimicarb on the wheat aphids *Rhopalosiphum padi* and Sitobion avenae. *PLoS One*, 10(6), e0128936.
Xu, J., Su, X., Bonizzoni, M., Zhong, D., Li, Y., Zhou, G., Nguyen, H., Tong, S., Yan, G. and Chen, X.G. (2018). Comparative transcriptome analysis and RNA interference reveal CYP6A8 and SNPs related to pyrethroid resistance in *Aedes albopictus*. *PLoS Neglected Tropical Diseases*, 12(11), e0006828.
Yang, T., Li, T., Feng, X., Li, M., Liu, S. and Liu, N. (2021). Multiple cytochrome P450 genes: Conferring high levels of permethrin resistance in mosquitoes, *Culex quinquefasciatus*. *Scientific Reports*, 11(1), 1–10.

Index

Pages in *italics* refer to figures and pages in **bold** refer to tables.

J

K

L

M

N

O

P

X

Y

Z